Henrik Steinmann
Dermaptera
Catadermaptera II

DAS TIERREICH

The Animal Kingdom

Eine Zusammenstellung und Kennzeichnung der rezenten Tierformen

A Compilation and Characterization of the Recent Animal Groups

Πάντα ῥεῖ Sine systemate chaos

Herausgeber

Heinz Wermuth (Wirbeltiere)

Edwin Möhn · Maximilian Fischer (Wirbellose)

Teilband **105** Part

Walter de Gruyter · Berlin · New York 1989

Henrik Steinmann

Dermaptera

Catadermaptera II

Walter de Gruyter · Berlin · New York 1989

Author

Dr. Henrik Steinmann
Hungarian Natural History Museum
Zoological Department
Baross u. 13
H-1088 Budapest
Hungary

Editors

Dr. Heinz Wermuth
Falkenweg 1
D-7149 Freiberg

Dr. Maximilian Fischer
Naturhistorisches Museum Wien
2. Zoologische Abt. (Insekten)
Burgring 7
Postfach 417
A-1014 Wien

Managing Editor

Johanna Wermuth
Falkenweg 1
D-7149 Freiberg

Citation

Das Tierreich, Teilband 105
Henrik Steinmann
Dermaptera
Catadermaptera II
Verlag Walter de Gruyter, Berlin · New York 1989

Deutsche Bibliothek Cataloguing in Publication Data

Das Tierreich : eine Zusammenstellung und Kennzeichnung der rezenten Tierformen = The animal kingdom / Hrsg. Heinz Wermuth ... – Berlin ; New York : de Gruyter.
Teilw. hrsg. von Robert Mertens ; Willi Hennig

NE: Wermuth, Heinz [Hrsg.]; Mertens, Robert [Hrsg.]; PT

Teilbd. 105. Dermaptera : Catadermaptera. – 2. Henrik Steinmann. – 1989
ISBN 3-11-010611-6 (Berlin, New York) Gb.
ISBN 0-89925-536-1 (New York) Gb.
NE: Steinmann, Henrik [Mitverf.]

∞ printed on acid free paper

ISBN 3110106116 Walter de Gruyter · Berlin · New York
ISBN 0899255361 Walter de Gruyter, Inc., New York
ISSN 0040-7305

Typesetting and Printing: Tutte Druckerei GmbH, Salzweg-Passau. Binding: Dieter Mikolai, Berlin.

To my beloved wife
Mrs. *Vilma Steinmann*
whose love, sympathy and aid
in countless ways
has made this work possible

Systematic Index

Section 2: **Mesodermaptera** sectio nova

Superfamilia typica: Carcinophoroidea STEINMANN 1975.
1911 Labiduriales (partim) ZACHER, Zool. Jahrb., Jena, **30**: 333.
1911 Labiduridae (partim) BURR, Genera Insectorum, Bruxelles, **122**: 10.
1954 Labiduroidea BRUES, MELANDER & CARPENTER (partim), Classification of Insects, Cambridge, Mass.: 115.
1965 Labioidea (partim) POPHAM, Entomologist, London, **98**: 126.
1975 Carcinophoroidea (partim) STEINMANN, Acta zool. Acad. Sci. hung., Budapest, **21**: 203.
Literature: SAKAI (as Carcinophoridae and Labiduridae) 1971.

Description: In the majority these are wingless insects with a more or less flattened or cylindrical body. The forceps are thicker, frequently asymmetrical for male, and symmetrical in female. The pygidium curves downwards between the forceps, the telson is reduced. The group cannot be separated from the Eudermaptera on the basis of external morphological characteristics, since they are both of the forficuloid-type, but only after a thorough examination. Their separation is justified only by the different structure, or development, of the male genitalia. They have two functional genital lobes (one of which may be reduced). They have been classified in various ways, e.g. the Paradermaptera had been regarded as a family of this group, or as a subfamily of the Labiduridae.

MELANDER et al., in their system published in 1954, separated in essence the same species-groups under the name Labiduroidea, but they also assigned the Esphalmeninae with family rank, which, however, as they are of the blattoid-type, are members on the subfamily level of the Pygidicranidae.

1 superfamilia.

Superfamily 2: **Carcinophoroidea** STEINMANN

1902 Labiduridae VERHOEFF (partim), Zool. Anz., Leipzig, No. **665**: 189.
1909 Labiduridae BURR (partim), Dtsch. ent. Zeitschr. Berlin, **1909**: 324.
1911 Labiduriales ZACHER (partim), Zool. Jahrb., Jena, **30**: 333.
1975 Carcinophoroidea STEINMANN, Acta zool. Acad. Sci. hung., Budapest, **21**: 199. – Familia typica: Carcinophoridae POPHAM 1965.
Literature: SAKAI (as Carcinophoridae and Labiduridae) 1971.

Description: Wingless, sometimes winged species. Body dorso-ventrally depressed; surface covered with shorter pubescence, body accordingly lustrless. The taxonomy is largely based on the male genitalia, since often there are few

satisfactory external characteristics, and those which occur often belong to the male only. Male genitalia with two genital lobes on paramere, one lobe directed anteriorad, the other posteriorad. The abdomen is fusiform and cylindrical, narrowed both anteriorly and posteriorly, sometimes slightly depressed dorsally, with the lateral tubercles on the third and fourth abdominal tergites usually small, occasionally absent, and rarely well developed. Male forceps more or less trigonal basally, cylindrical apically, and curved, symmetrical or asymmetrical. Female forceps approximately symmetrical, tapering, contiguous, slender.
2 families.

Identification key to the families

1 Tarsal joint 2 normal, not elongated under ventral surface of joint 3, scapiform (shaped like the antennal scapus). Male genital lobe with or without virgae, its base never with a basal vesicle, but a basal sclerotized plate, pouch or sac may occasionally occur .... Family 2: **Carcinophoridae**, p. **2**

1′ Tarsal joint 2 distally slightly elongated and discernibly, originating on ventral surface of joint 3 (metatarsus), hence tarsus not scapiform. Male genitalia with two genital lobes, with virgae invariably present, the base always containing a smaller or larger but easily recognizable basal vesicle .... Family 3: **Labiduridae**, p. **398**

Family 2: **Carcinophoridae** POPHAM

1902 Anisolabidae VERHOEFF (partim), Sber. Ges. naturf. Fr. Berlin, **1902**: 9.
1909 Psalinae BURR, Dtsch. ent. Zeitschr. Berlin, **1909**: 325. – Genus typicum: *Psalis* AUDINET-SERVILLE 1831.
1911 Labiduridae ZACHER (partim), Zool. Jahrb., Jena, **30**: 333.
1954 Psalididae BRUES, MELANDER & CARPENTER, Classification of Insects, Cambridge, Mass.: 116.
1965 Carcinophoridae POPHAM, Entomologist, London, **98**: 135. – Genus typicum: *Carcinophora* SCUDDER 1876.

Literature: BRINDLE 1978; POPHAM 1965; SAKAI 1971; STEINMANN 1975, 1977, 1984.

Description: Mainly wingless species, even tegmina undeveloped, often present approximately as lateral flaps. Dorsum nude, but also some winged species occur. The common characteristics of the species constituting the family is the presence of two genital lobes in the males, except for the species of Idolopsalinae in which one male genital lobe is reduced. The groups diverged phylogenetically in its evolution from the other species of the family as did the Karschiellinae in the Pygidicranidae (the reduction of the genital lobe shows a certain transition to the Labiidae species with a single genital lobe). Metatarsus normal. Head usually transverse, sometimes strongly so, and more or less convex dorsally; postfrontal and coronal sutures well developed, or absent. Each antenna has the first joint much shorter than in the Brachylabiinae, and the second joint is transverse or quadrate; the third joint is usually about twice as long as wide, and often more or less cylindrical, but it tends to vary, even in the antennae of a single specimen; the fourth joint is usually about as wide as long, and the fifth somewhat longer. The antennal joints towards the base of the antennae tend to be relatively uniform in their proportions, although the distal joints show greater variety in the various species. The number of antennal joints is much greater than in the Brachylabiinae, possibly up to 30, but it is rare that the antennae are complete in dried specimens. Abdomen cylindrical, more or less expanded to last tergite or abdominal tergites 5–6. At the sides the tergite may have a lateral longitudinal keel, which is usually oblique and rugoso-striate. There may also be a dorsolateral ridge, which may form an oblique crest, tending towards the midline posteriorly, or it may be represented by a tubercle. Pygidium in both sexes is usually flat but may be protruding. Forceps in both sexes are small, but in some the curvature of the male forceps is very strong and the forceps may be strongly asymmetrical. In most males the forceps are only weakly asymmetrical and in all females the forceps are more or less symmetrical, and nearly contiguous, although one branch may be shorter than the other.

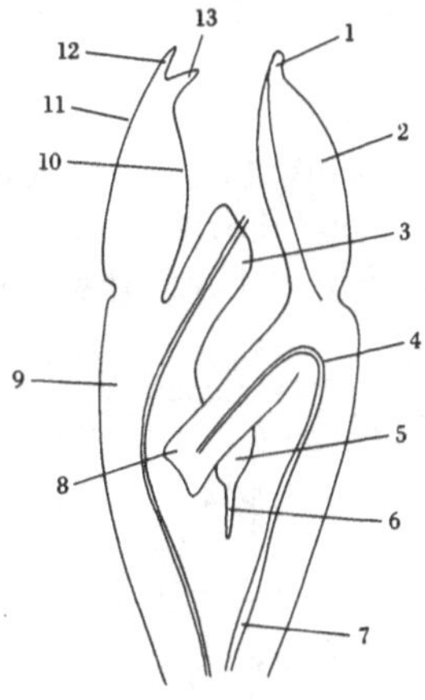

Fig. 1 Male genital armature of Carcinophorids-type. Original.
1 apex of external paramere
2 external paramere or metaparamere
3 left genital or distal lobe
4 virga or ejaculatory duct
5 median incision or rima genitalis
6 fissura medialis
7 base of virga
8 right genital or distal lobe
9 central or unpaired paramere
10 outer margin of external paramere
11 inner margin of external paramere
12 inner tooth of external paramere
13 outer or fore tooth of external paramere

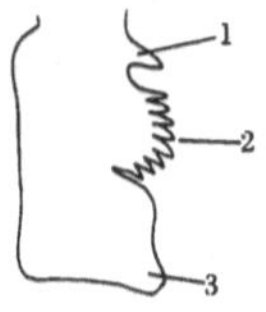

Fig. 2 Male genital lobe in apical portion of *Euborellia*-type. Original.
1 denticulus or denticle of genital lobe
2 reticular comb or denticulation pads of genital lobe
3 apex of genital lobe

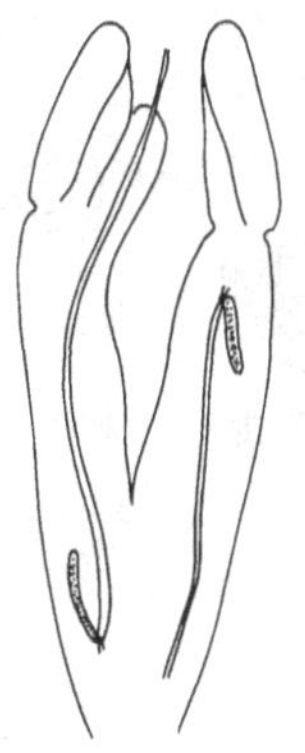

Fig. 3 Male genital armature of *Idolopsalis*-type. Original.

The male genitalia of the Carcinophoridae consist of unpaired central parameres with two functional genital lobes or only one functional and one reduced genital lobe. Genital lobes with or without virgae. As in other families of the order, the female genitalia have not been systematically studied, and it is the male genitalia which have been used in taxonomy.

The paired "Carcinophora-type" genital armature has two genital lobes which in dorsal view may be called "left" and "right". The left lobe projects forwards (Fig. 1: 3) and even in resting position it is next to the left external paramere (Fig. 1). On the other hand the right lobe (Fig. 1: 8) is directed backwards, pointing towards the base of the paramere. In the genital lobe, in the majority of the species, a virga (Fig. 1: 4) or an ejaculatory duct is present, at the base of which there is no basal vesicle. Base of virga (Fig. 1: 7) is simple. On the anterior margin of the unpaired paramere (Fig. 1: 9) a shallow or deep (Fig. 1: 5) rima genitalis or anterior incision is observable, dividing the paramere into two distinct parts. Thus, the paramere may be narrow or broad V-shaped, or Y-shaped. Below the medial incision a narrow fissure, the so-called fissura medialis (Fig. 1: 6) is present. On the anterior-lateral corners of the unpaired paramere which is highly variable both in shape and size, paired metaparameres or external parameres (Fig. 1: 2) are present. The inner margin of the external paramere (Fig. 1: 11) and the outer margin (Fig. 1: 10) bear very important features, such as the inner (Fig. 1: 12) or the fore tooth (Fig. 1: 13), very reliable in determination.

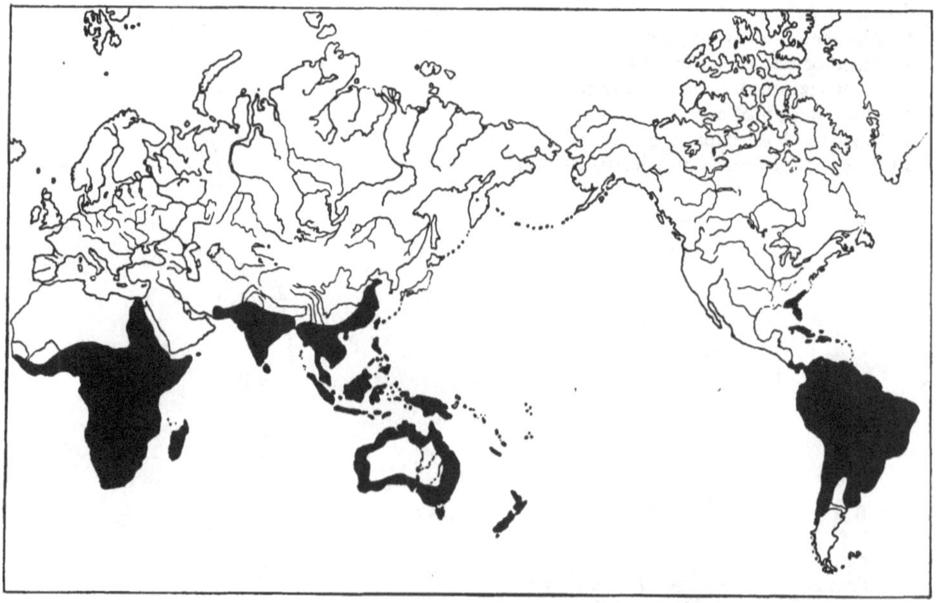

Fig. 4 Approximate recorded distribution of the family Carcinophoridae (without cosmopolitan species). Original.

The apex of the external paramere (Fig. 1: 1), also known as acumen or cuspis, varies from species to species, and is usually obtuse or acute. In the *Euborellia* species of Carcinophorinae the genital lobe is characteristic (Fig. 2), especially its apex (Fig. 2: 1), since in many species a so-called reticular comb (Fig. 2: 2) may be present, preceded generally by a denticulus lobi genitalis (Fig. 2: 3). Male genitalia of species of the Idolopsalinae STEINMANN are characteristic, one male genital lobe more or less reduced (Fig. 3), virga present or absent.

Some authors treated the taxon as subfamily Psalinae in the Labiduridae, until POPHAM gave it family rank in 1965, on the basis of the genital characteristics outlined above.

Distribution: nearly all over the World (Fig. 4).

9 subfamilies.

Identification key to the subfamilies

1 Body extremely flattened, dorsal and ventral surfaces nearly parallel. Tegmina and wings present, latter just reaching base of articulation of hind leg. Ultimate tergite wide, medially slightly impressed. Male forceps strongly curved, female's elongated. Male genital armature elongate, narrow; external parameres rather wide .... subfamily 1: **Platylabiinae**, p. 6

1′ Body only slightly flattened, rather cylindrical **2**

2 Prosternum evenly attenuating posteriorad; plate in a ventral view rounded cordiform. Male forceps curved like a finger, female's more slightly arcuate and with a discernible tooth on inner margins. Both male genital lobes well developed. Pygidium varying per species .... subfamily 2: **Gonolabiinae**, p. 8

2′ Prosternum after a slight constriction usually widening posteriorad, plate not cordiform **3**

3 Posterior margin of mesosternum in both sexes rounded arcuately in a semicircle **4**

3′ Posterior margin of mesosternum in both sexes straight, transversely truncate **5**

4 Head without eyes, blind species; body long and narrow, legs short. Antennae 13-jointed. External paramere of male genitalia vestigial, strongly reduced, or even absent. The single species inhabiting the Galapagos Islands .... subfamily 3: **Anophthalmolabiinae**, p. 13

4′ Head with well-developed eyes. External paramere of male genitalia well developed, longer or shorter, but never absent. Median incision of anterior margin of male paramere very deep, or medium deep, and median section of external paramere occasionally characteristically widened. Species winged or wingless, or only with tegmina .... subfamily 4: **Carcinophorinae**, p. 15

5 First antennal joint (scape) long, longer than distance between antennal bases **6**

5′ First antennal joint short, shorter than the distance between antennal bases or sometimes as long as the frontal section between scapes **7**

6 Lateral margins of mesonotum with a ridge-like edge. Ultimate tergite visibly excised. Paramere of male genitalia wide or narrower; external parameres acute and small, or larger. Genital lobes wide and short, or elongate .... subfamily 5: **Brachylabiinae**, p. 280

6′ Lateral margins of mesonotum without longitudinal ridges. Paramere of male genitalia attenuating like a spoon, its external parameres shorter or longer. Paramere occasionally also wider, its external parameres small, or elongate .... subfamily 6: **Isolabiinae**, p. 311

7 Male genitalia united along all the midline, virga within genital lobe conspicuously sclerotized. Eyes very small; body strongly pubescent; forceps with branches cylindrical throughout and symmetrical. Very small species, body length 5 mm .... subfamily 7: **Antisolabiinae**, p. 339

7′ Genital lobes of male genitalia united only at base, divergent distally, virga slender and not strongly sclerotized, often very inconspicuous **8**

8 Both male genital lobes well developed, functional. Abdomen strongly widened, oval. Male

paramere relatively wide, genital lobes nearly equal in size, and occasionally hightly characteristic in contrast to those of the other species in the family .... subfamily 8: **Parisolabiinae**, p. 371

8′ One male genital lobe more or less reduced, but male genital apparatus of Carcinophorid-type. Forceps of male asymmetrical, or symmetrical, but comparatively small.. subfamily 9: **Idolopsalinae**, p. 386

Subfamily 1: **Platylabiinae** BURR

1876 Labidophora SCUDDER, Proc. Boston, Soc. nat. Hist., **18**: 297.

1910 Palicinae BURR, Fauna Brit. India, Dermaptera, London, Bombay: 67. – Genus typicum: *Palex* BURR 1910.

1911 Platylabiinae BURR, Genera Insectorum, Bruxelles, **122**: 43. – Genus typicum: *Platylabia* DOHRN 1867.

1954 Platylabiidae BRUES, MELANDER & CARPENTER, Classification of Insects, Cambridge, Mass.: 115.

1954 Palicidae = Platylabiidae BRUES, MELANDER & CARPENTER, Classification of Insects, 115.

Literature: BURR 1912, 1915; POPHAM & BRINDLE 1966; SAKAI 1970; STEINMANN 1975, 1977; ZACHER 1911.

Description: Body very strongly depressed. Antennae 19 or 20-jointed; first joint moderately small, a little shorter than the distance between antennal bases. Tarsi slender, third segment longer than first and second together. Tegmina and wings present. Abdomen elongated, flattened; ultimate tergite broad, strongly depressed medially. Forceps sickle-shaped. Male genitalia characteristic, comparatively narrow; external parameres specific; lateral margins almost parallel, apex with a small prolongation.

Distribution: Oriental Region.

1 genus.

Genus 1: **Platylabia** DOHRN

1867 *Platylabia* DOHRN, Stettiner ent. Ztg., **28**: 347. – Species typica: *Platylabia major* DOHRN 1867.

1876 *Labidophora* SCUDDER (new name for *Platylabia* DOHRN, because of its similarity to *Platylabus* WESMAEL 1845). Proc. Boston, Soc. nat. Hist., **18**: 297. – Species typica: *Platylabia major* DOHRN 1867.

1910 *Palex* BURR, Fauna Brit., India, Dermaptera, London, Bombay: 68. – Species typica: *Platylabia sparattoides* BORMANS 1900.

Literature: BURR 1902, 1904, 1911, 1915; HANDLIRSCH 1925; SAKAI 1970; STEINMANN 1977; TOWNES 1945; ZACHER 1915.

Description: Body strongly flattened. Head smooth and depressed medially. Antennae with 19 or 20 joints; first joint long, subconical; second small, transverse, third joint long and cylindrical. Pronotum subquadrate, and convex anteriorly. Tegmina perfectly developed, but rather short, rounded at the apex, feebly developed at the axillary angles, thus exposing a small scutellum; ecarinate laterally. Mesosternum about as long as broad, more or less rounded, posterior margin truncate. Legs not very long; femora broad and compressed; tibiae short and slender; first tarsal segment short. Male forceps strongly curved.

Distribution: Oriental Region.

Single species.

1. Platylabia major Dohrn

1867 *Platylabia major* Dohrn, Stettiner ent. Ztg., **28**: 347. – Terra typica: Burma (Type male: Mus. civ. Stor. nat., Genova).

1876 *Labidophora major* Scudder, Proc. Boston, Soc. nat. Hist., **18**: 297. New name for *Platylabia major* Dohrn because of its similarity to *Platylabus* Wesmael 1845 – Hymenoptera.

1900 *Platylabia sparattoides* Bormans, Ann. Mus. civ. Stor. nat. Genova, (2) **20**: 459. – Terra typica: Burma (Type male: Mus. civ. Stor. nat., Genova).

1910 *Platylabia major* Burr, Fauna Brit. India, Dermaptera, London, Bombay: 125. Burma, Ceylon.

1911 *Platylabia major* Burr, Genera Insectorum, Bruxelles, **122**: 43; pl. 4, fig. 6 (male), 6a (tarsus). Burma, Sumatra, Java, India.

Literature: Burr 1902, 1908, 1911; Hebard 1927; Kirby 1904; Popham 1966; Sakai 1970; Steinmann 1977, 1984.

Description: Male colour dark brown. Head, pronotum, and abdomen depressed, flattened. Head smooth, frons a little tumid, occiput depressed medially. Postfrontal and coronal sutures obsolete. Eyes comparatively small, shorter than the length of head behind eyes. First antennal joint normal, shorter than the distance between antennal bases. Pronotum more or less narrower than the head, and convex anteriorly; lateral sides parallel; anterior angles rounded; posterior margin truncate; median longitudinal furrow distinct. Tegmina very flat, and well developed, rounded at apex, feebly developed at the axillary angles; ecarinate. Wings short, but usually visible. Prosternum of thorax parallel-sided, scarcely constricted; meso- and metasternum about as broad as long, more or less rounded, and truncate posteriorly. Legs not very long, femora broad and compressed; tibiae short and slender; tarsi slender, first segment short, the third longer than first and second combined. Abdomen depressed, without lateral tubercles, abdominal tergites a little expanded to tergites 5–6. Ultimate tergite ample, simple, subquadrate, smooth. Pygidium short, broad, flattened, vertical, fused with dorsal segment and without marked suture. Forceps (Fig. 5) remote at base, nearly straight at first, then bowed; inner margin with large tooth medially. Penultimate sternite quadrate, ample. Genitalia (Fig. 6) characteristic; paramere

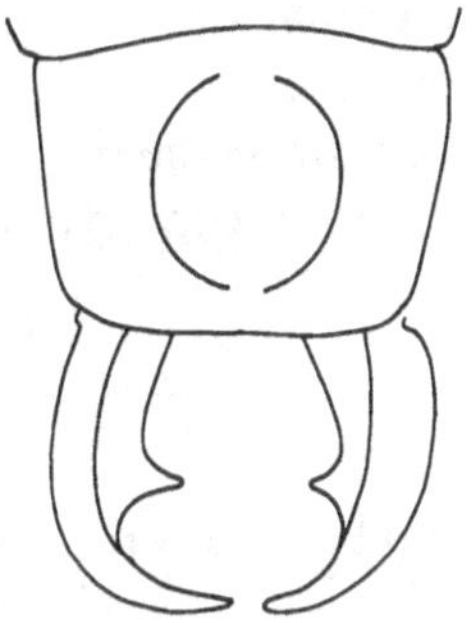

Fig. 5 Male ultimate tergite with forceps of *Platylabia major* Dohrn. After Burr.

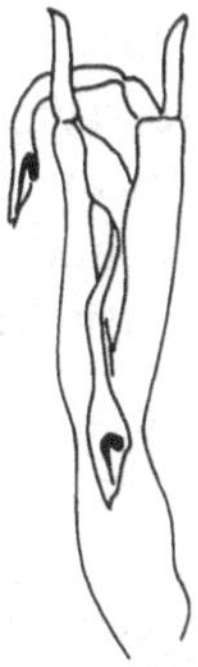

Fig. 6 Male genital armature of *Platylabia major* Dohrn. After Srivastava.

elongate, median longitudinal incision between external parameres small. Genital lobes undeveloped, without virgae. External parameres with parallel inner and outer margins laterally. – **Female** similar to male, but forceps simple, contiguous, tapering. – **Length** of body with forceps: male: 11:12; female: 11–13,5 mm.

Distribution: India, Sri Lanka, Burma, Thailand, Vietnam, Java, Sumatra, and Celebes.

Subfamily 2: **Gonolabiinae** POPHAM & BRINDLE

1966 Gonolabinae POPHAM & BRINDLE, Entomologist, London **99**: 277. – Genus typicum: *Gonolabina* VERHOEFF 1902.
1975 Gonolabiinae – STEINMANN, Acta zool. Acad. Sci. hung., Budapest, **21**: 204.
Literature: SAKAI 1970; STEINMANN 1977, 1984.

Description: Apterous. Prosternum cordiform. Ultimate tergite of abdomen flattened and fused with the pygidium to form an anal process between the forceps. Forceps of male strongly curved, asymmetrical, elliptical in cross-section; female forceps remote at base, broad proximally, tapering distally, short, with or without inner tooth. Male genitalia very characteristic, immature-like, simple, as in fig. 8.

Distribution: Neotropical Region.

1 genus.

Genus 2: **Gonolabina** VERHOEFF

1902 *Gonolabina* VERHOEFF, Sber. Ges. naturf. Fr. Berlin, **1902**: 8 – Species typica: *Gonolabina kuhlgatzi* VERHOEFF 1902.
Literature: BRINDLE 1965, 1967; BURR 1908, 1911; KIRBY 1904; SAKAI 1970; STEINMANN 1977; ZACHER 1911.

Description: Wingless, body broad and depressed. Rudimentary tegmina are usually present on the lateral margins of the mesonotum. Abdomen narrower at the base, widening about the third segment, most of the abdomen parallel-sided. The last two abdominal segments are narrower, and this feature distinguishes *Gonolabina* from *Esphalmenus* (Pygidicranidae). Colour blackish or dark brown, with the thoracic segments, legs, and head lighter in colour. Male forceps strongly asymmetrical; left branch being sharply curved, while the right branch is only gently curved and may be almost straight. Female forceps symmetrical.

Distribution: South America.

3 species.

Identification key to the species

1 Posterior margin of male ultimate tergite with two tubercles, the distance between these being about the same as the width of a branch of the forceps at base (Fig. 7). Penultimate sternite of male as broad as long with margin evenly rounded; male genitalia with large sclerite in each genital lobe. Female pygidium long, more or less triangular, with apex pointed; female forceps without inner tooth .... 1. **Gonolabina spectabilis**, p. 9

1′ Posterior margin of male ultimate tergite with three tubercles (Fig. 10), or with two tubercles more widely spaced (Fig. 13). Penultimate sternite of male about half as long as broad; male genitalia with two narrow unequal sclerites in each genital lobe. Female pygidium rounded; female forceps with inner tooth (Fig. 15) **2**

2 Pygidium of male with three tubercles (Fig. 10). Penultimate sternite of male with margin evenly rounded; genitalia of male with smaller sclerite in each genital lobe with a curved base (Fig. 11). Female pygidium transverse, not globular; female forceps with inner tooth nearer base .... 2. **Gonolabina trinodosa**, p. 10

2′ Pygidium of male with two tubercles, the distance between these being greater than width of forceps at base (Fig. 13); penultimate sternite of male with distal emargination. Genitalia of male with smaller sclerite not possessing a curved base (Fig. 14). Female forceps (Fig. 15) with inner tooth almost at middle .... 3. **Gonolabina bidonosa**, p. 11

1. Gonolabina spectabilis (Philippi)

1863 *Forficula spectabilis* Philippi, Z. Naturwiss., Leipzig, **21**: 218. – Terra typica: Chile (Type male: Santiago Mus.).

1902 *Gonolabina kuhlgatzi* Verhoeff, Sber. Ges. naturf. Fr. Berlin, **1902**: 8. – Terra typica: Chile (Type male: Mus. Naturkunde, Berlin).

1911 *Gonolabina kuhlgatzi* – Burr, Genera Insectorum, Bruxelles, **122**: 26, pl. 2, fig. 11 (ultimate tergite and forceps), 11 a (sternum of thorax).

1911 *Anisolabis spectabilis* – Burr, Genera Insectorum, Bruxelles, **122**: 30.

1933 *Gonolabina kuhlgatzi* – Rehn, Trans. amer. ent. Soc., Philadelphia, **59**: 162 (proposed synonymy with *spectabilis* Philippi).

1967 *Gonolabina spectabilis* – Brindle, Entomologist's month. Mag., London, **103**: 8, fig. 1 (ultimate tergite and forceps of male). 7 (genital sclerite).

Literature: Brindle 1965, 1967; Burr 1910; Kirby 1904; Popham & Brindle 1966; Reichard 1968; Sakai 1970.

Description: Male colour dark brown; pronotum yellowish-brown, abdomen shiny. Head rather longer than broad; postfrontal and coronal sutures distinct. Antennae ll-jointed; first joint dilated distally, and about half as long as distance between antennal bases; second joint quadrate, and equal to fourth in length.

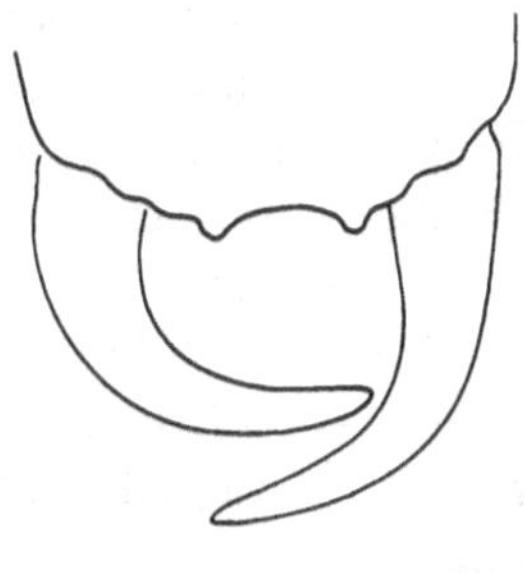

Fig. 7 Posterior margin of male ultimate tergite and forceps of *Gonolabina spectabilis* (Philippi). After Brindle.

Fig. 8 Genital sclerite of male genital lobe of *Gonolabina spectabilis* (Philippi). After Brindle.

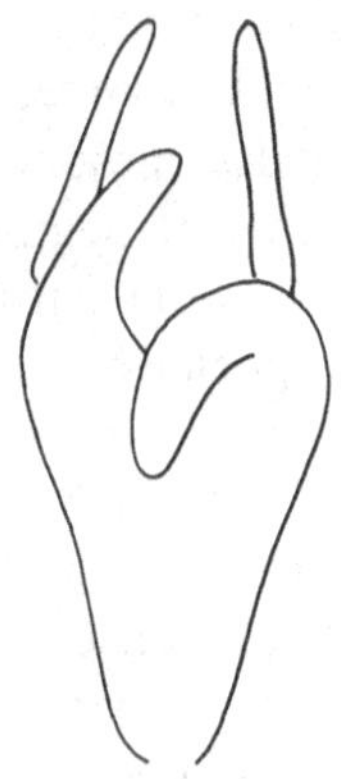

Fig. 9 Male genital armature of *Gonolabina spectabilis* (Philippi). After Brindle.

Eyes small, triangular. Pronotum transverse, anterior angles rounded, posterior angles sharp; all margins straight; median longitudinal furrow impressed; sparsely punctate medially. Mesonotum with rudimentary tegmina on lateral margins. Wings entirely absent. Legs short. Abdomen strongly punctate, expanded to segments 7–8. Ultimate tergite transverse, rugose, with median area strongly declivous, posterior margin, so-called pygidium, with two tubercles (Fig. 7). Forceps remote at base, sparsely covered with deep punctation; left branch sharply curved mesad about mid-point; right branch slightly curved. Penultimate sternite broad, evenly rounded, about as long as broad at base. Genitalia (Fig. 9) characteristic, of *Gonolabina*-type, immature-like, genital lobe with a large double chitinous sclerite (Fig. 8). – **Female** similar to male, but ultimate tergite narrower, rugose, with median longitudinal impressed sulcus, slightly declivious on posterior border; forceps tapering distally, short, without inner tooth. – **Length** of body with forceps: male: 26–28 mm; female: 20–23 mm.

Distribution: Chile: Valparaiso, Santiago, San Antonio.

2. Gonolabina trinodosa Brindle

1967 *Gonolabina trinodosa* Brindle, Entomologist's month. Mag., London, **103**: 9; fig. 2 (ultimate tergite and forceps), 4 (posterior margin of ultimate tergite with forceps of female), 8 (genital sclerite of male genitalia). – Terra typica: Peru (Type male: Collect. F. Blancas, Univ. Mus. Lima, Peru).

Literature: Sakai 1970, Steinmann 1984.

Description: Male colour reddish-brown, antennae yellowish-brown, abdomen and forceps dark brown. Head broad, slightly wider than pronotum. Frons tumid, vertex with transverse depression. Postfrontal and coronal sutures distinct. Antennae 18-jointed, first one long, longer than distance between antennal bases; second quadrate. Eyes moderately small, shorter than the length of head behind eyes. Pronotum transverse, median longitudinal furrow distinct, all angles rounded, and all margins straight. Thoracic nota and legs as in *spectabilis*: mesonotum with rudimentary tegmina, wings absent. Abdomen broadened to segment 5, then parallel-sided to last tergite which is narrower. Ultimate tergite transverse, rugose, with slight median longitudinal impressed sulcus; posterior margin with three tubercles (Fig. 10). Forceps asymmetrical, left branch more sharply curved than in spectabilis, and right branch almost straight. Penultimate sternite broad and rounded posteriorly. Genitalia (Fig. 12) immature-like, of *Gonolabina*-type, genital lobe with apical denticulate area, with two sclerites, one long and narrow, the other short and broad (Fig. 11). – **Female** similar to male, but abdominal tergites less strongly punctate. Ultimate tergite narrow, not declivious as in male; pygidium rounded distally; forceps tapering, with ventral tooth on each inner margin near base. – **Length** of body with forceps: male: 23 mm; female: 22 mm.

Distribution: Peru.

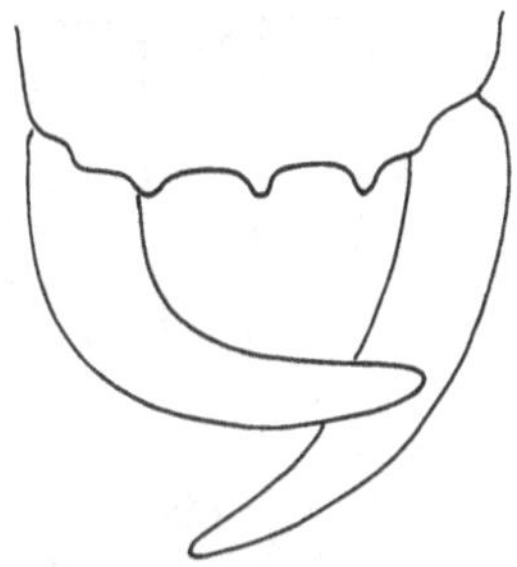

Fig. 10 Posterior margin of holotype ultimate tergite and forceps of *Gonolabina trinodosa* BRINDLE. After BRINDLE.

Fig. 11 Genital sclerite of holotype genital lobe of *Gonolabina trinodosa* BRINDLE. After BRINDLE.

Fig. 12 Genital armature of holotype of *Gonolabina trinodosa* BRINDLE. After BRINDLE.

3. Gonolabina binodosa BRINDLE

1967 *Gonolabina binodosa* BRINDLE, Entmologist's month. Mag. London, **103**: 9; fig. 3 (posterior margin of male ultimate tergite and forceps), 5 (posterior margin of female ultimate tergite and forceps), 9 and 10 (genital sclerite). – Terra typica: Peru (Type male: Collect. F. BLANCAS, Peru).

Literature: SAKAI 1970.

Description: Apparently a variable species. BRINDLE described two series: a larger and broader one (forma *binodosa*), body length 16–20 mm, body width 4 mm, and a smaller and more narrow one (forma *llanguena*), body length 14–16 mm, and body width 3 mm. Forma *binodosa* possess rudimentary tegmina on the lateral margins of the mesonotum, but no trace of these are to be seen on the smaller species.

Distribution: South America (Peru).

3a. Gonolabina binodosa bidonosa BRINDLE

1967 *Gonolabina binodosa bidonosa* BRINDLE, Entomologist's month. Mag., London, **103**: 10.

Description: Original description by BRINDLE: "**Male**: variable in colour, ranging from almost uniformly dark brown to specimens in which the thoracic segments are lighter brown than the abdomen. Very similar to *trinodosa* in external structure, from which it differs by having only two tubercles on the pygidium. Antennae yellowish-brown, 25-jointed in type. Pronotum rather less transverse than in *trinodosa*. Mesonotum usually with rudimentary elytra on the lateral margins. Ultimate tergite rugose, blackish, declivous, with two widely spaced tubercles on pygidium. Forceps reddish-brown, shorter than in *trinodosa* but somewhat variable, most specimens having forceps similar to those of

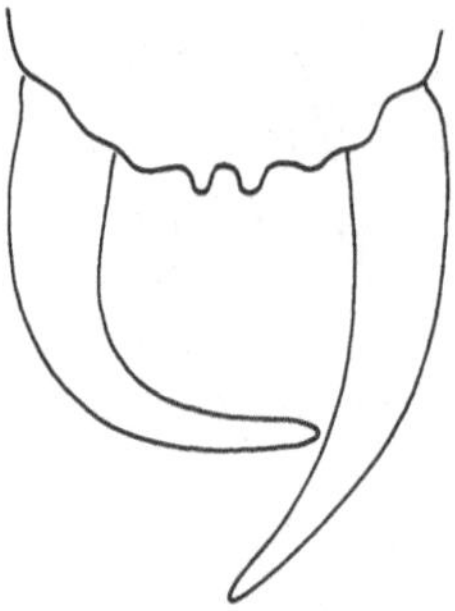

Fig. 13 Posterior margin of holotype ultimate tergite and forceps of *Gonolabina binodosa* BRINDLE. After BRINDLE.

Fig. 14 Genital sclerite of holotype of *Gonolabina binodosa* BRINDLE. After BRINDLE.

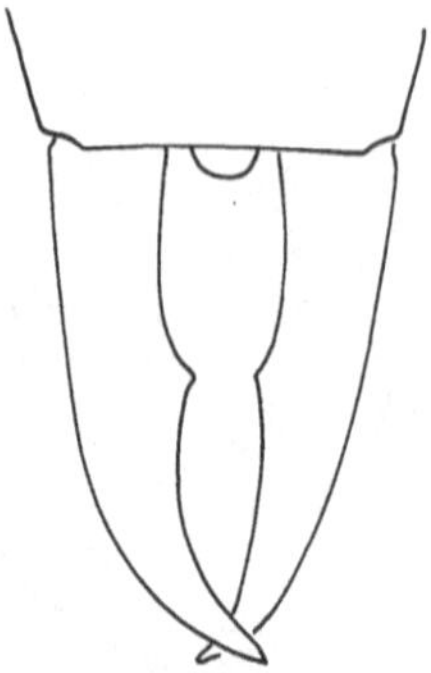

Fig. 15 Posterior margin of female ultimate tergite and forceps of *Gonolabina binodosa* BRINDLE. After BRINDLE.

trinodosa (Fig. 13), in which the right branch is almost straight. Genitalia of usual type (Fig. 12), distal lobe with two sclerites arising from the apical denticulate area, the sclerites unequal, one long and narrow, the second short and broad (Fig. 14). – **Female** as male but ultimate tergite narrowed posteriorly, and not declivous; pygidium globose, forceps almost straight, short, broad at base, tapering distally, each branch with prominent ventral tooth on inner margin, placed almost at mid-point." – **Length** of body with forceps: male: 19–22,5 mm, female: 19–21 mm.

Distribution: Peru: Zaráte.

3b. Gonolabina binodosa llanguena BRINDLE

1967 *Gonolabina binodosa llanguena* BRINDLE, Entomologist's month. Mag., London, **103**: 10.

Description: Original description by BRINDLE: "**Male**: narrower than in form *binodosa*, and without any trace of rudimentary elytra on the mesonotum, otherwise similar in external structure. Head brown, thoracic nota yellowish-brown with darker transverse and longitudinal bands and spots, variable in pattern. Legs brownish-yellow, and abdomen blackish-brown. The structure of the ultimate tergite, pygidium and forceps appears to be identical to those of the form *bidonosa* except that both branches of the forceps are usually curved (Fig. 13) and short. Genitalia as in form *binodosa*, with the smaller sclerite narrow in dorsal view (Fig. 14). – **Female** as male, but ultimate tergite narrowed, not declivous, pygidium globose. Forceps almost straight, broad proximally, narrowed distally, with ventral tooth on each inner margin near the mid-point (Fig. 15)." – **Lenght** of body with forceps: male: 16–18 mm, female: 16–17 mm.

Distribution: Peru: Hda Llanguen.

Subfamily 3: **Anophthalmolabiinae** STEINMANN

1975 Anophthalmolabiinae STEINMANN, Acta zool. Acad. Sci. hung., Budapest, **21**: 205. – Genus typicum: *Anophthalmolabis* BRINDLE 1968.

Description: Small depressed blind earwigs. Prosternum after a slight constriction usually widening posteriorad, thereby plate not cordiform. Body long and narrow, legs short, eyes absent. Posterior margin of mesosternum rounded arcuately in a semicircle. External paramere of male genitalia vestigial, strongly reduced, or even absent in *leleupi* BRINDLE; virgae furcate (Fig. 16).

Distribution: Galapagos Islands and Argentina.

Single genus.

Genus 3: **Anophthalmolabis** BRINDLE

1968 *Anophthalmolabis* BRINDLE, Miss. zool. belge, Galapagos-Ecuador, **1**: 172. – Species typica: *Anophthalmolabis leleupi* BRINDLE 1968.

Literature: SAKAI 1970; STEINMANN 1977, 1978.

Description: Slender, elongated species; body ant-like, depressed dorso-ventrally, yellowish or reddish-brown in colour. Head rather cordiform in shape, without any trace of eyes or ocelli. Pronotum longer than broad, mesonotum transverse, with or withouth traces of lateral folds; metanotum transverse, simple, posterior margin straight or concave. Prosternum narrow, long, posterior margin truncate; mesosternum broad, posterior margin rounded; metasternum broad with posterior margin truncate. Legs with femora strongly dilated. Forceps simple in both sexes, broad at base, more or less contiguous, narrowing distally. Male genitalia very short and with almost aborted external parameres.

Distribution: Galapagos Islands and Argentina.

2 species.

Identification key to the species (after BRINDLE)

1 Smaller, body length at most 7 mm, antennae 13-jointed; metanotum with posterior margin straight. Abdominal segments without lateral ridges; ultimate tergite more transverse and without a median longitudinal sulcus. Forceps broad, relatively shorter, and completely contiguous .... 1. **Anophthalmolabis leleupi**, p. 13

1′ Larger, body length at least 9 mm, antennae 16-jointed; metanotum with posterior margin concave. Abdominal segments 5–9 (males) and 5–8 (females) with lateral ridges; ultimate tergite more quadrate, and with a prominent median longitudinal sulcus. Forceps more slender, relatively longer, and not completely contiguos ... 2. **Anophthalmolabis caeca**, p. 14

1. **Anophthalmolabis leleupi** BRINDLE

1968 *Anophthalmolabis leleupi* BRINDLE, Miss. zool. belge, Galapagos-Ecuador, **1**: 172; fig. 1 (male), 2 (genitalia of holotype). – Terra typica: Galapagos Islands: Santa Cruz (Type male: Inst. r. Sci. nat. belg., Bruxelles).

Literature: SAKAI 1982; STEINMANN 1975, 1977, 1978.

Description: Male colour yellowish or reddish-brown. Head longer than broad, tumid, rather cordiform; postfrontal and coronal sutures obsolete; posterior margin of head rounded. Eyes entirely absent. Antennae 13-jointed; first antennal joint or scapus relatively long, about as long as the distance between antennal bases; second joint quadrate, third joint rather longer than broad, fourth joint quadrate but wider than third; the distal joints about twice as long as broad, narrower basally. Pronotum longer than broad, a little widened posteriorly; median longitudinal furrow absent, posterior margin almost straight. Tegmina and wings absent. Mesonotum transverse, with indications of lateral folds on each side; metanotum longer than mesonotum, posterior margin straight. Legs with femora strongly dilated, especially the anterior pair. Abdomen elongated, almost parallel-sided laterally, segments without lateral ridges. Ultimate tergite a little transverse, posterior margin angular, with a median emargination. Forceps (Fig. 16) short, broad, symmetrical. Penultimate sternite simple, posterior margin straight. Genitalia (Fig. 17) characteristic; paramere broad, oval, without median longitudinal incision between external parameres; one genital lobe directed posteriorly, and one anteriorly; virga partially sclerotized and with an associated hookshaped sclerite, distal part of virga narrower and membranous. External parameres reduced to membranous plate, with sclerotized margins confined to base, externally. – **Female** similar to male. – **Length** of body with forceps: in both sexes: 6,5 mm.

Distribution: Galapagos Islands: Santa Cruz.

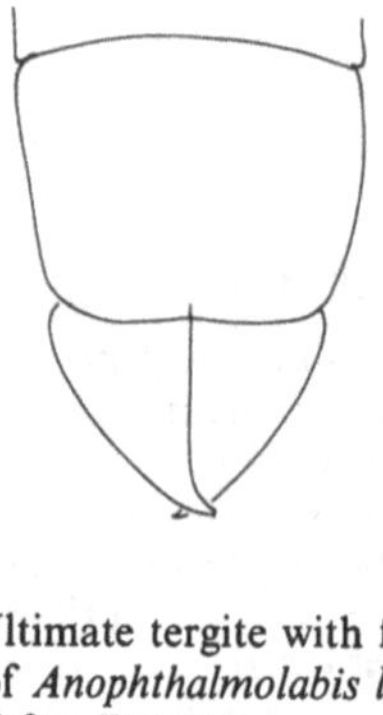

Fig. 16 Ultimate tergite with forceps of holotype of *Anophthalmolabis leleupi* BRINDLE. After BRINDLE.

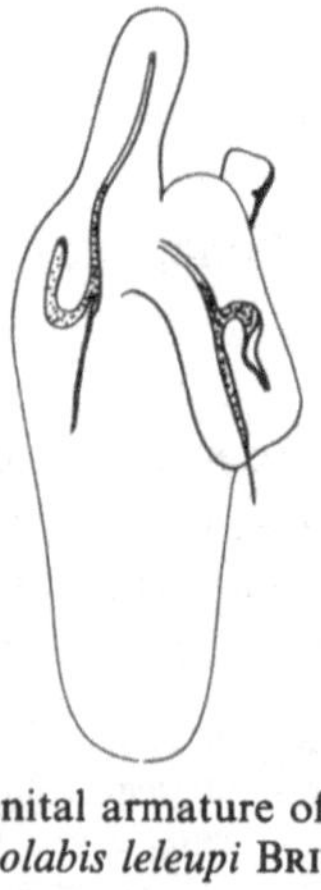

Fig. 17 Genital armature of holotype of *Anophthalmolabis leleupi* BRINDLE. After BRINDLE.

2. Anophthalmolabis caeca (BORELLI) comb. nov.

1902 *Anisolabis caeca* BORELLI, Boll. Mus. Zool. Anat. comp. Univ. Torino, **17**: 2; fig. 1 (female). – Terra typica: Argentina: Prov. Santa Fé (Type female: SILVESTRI's Collect. Univ. Mus., Milano).

Literature: BRINDLE 1968; SAKAI 1970; STEINMANN 1977, 1978.

Description: Female general colour yellowish-brown. Body elongate, finely punctate, a little shining. Head oval, longer than broad; postfrontal and coronal sutures finely distinct, not well marked. Eyes absent. Antennae 16-jointed, first antennal joint comparatively long, about as long as the distance between antennal bases; second transverse, the distal joints about twice as long as broad. Pronotum oval, all margins rounded; median longitudinal furrow present. Tegmina and wings entirely absent. Mesonotum with posterior margin straight, metanotum with posterior margin concave. Legs moderately short, femora dilated. Abdomen a little expanded to middle; abdominal tergites 5–8 with lateral longitudinal ridges. Ultimate tergite more quadrate; median longitudinal sulcus present. Forceps more slender, trigonal basally, cylindrical apically; relatively longer, and not completely contiguous. Penultimate sternite simple. – **Male** similar to female, but abdominal tergites 5–9 with lateral longitudinal ridges; forceps a little shorter, contiguous. Genitalia unknown. – **Length** of body with forceps: in both sexes: 9 mm.

Distribution: Argentina: Province of Santa Fé.

Subfamily 4: **Carcinophorinae** HINCKS

1902 Anisolabidae VERHOEFF, Sber. Ges. naturf. Fr. Berlin, **1902**: 9. – Genus typicum: *Anisolabis* FIEBER 1853.
1909 Psalinae BURR, Dtsch. ent. Zeitschr., Berlin, **1909**: 325. – Genus typicum: *Psalis* AUDINET-SERVILLE 1831.
1911 Anisolabinae ZACHER, Zool. Jahrb., Jena, **30**: 370.
1915 Psalidae BURR, Journ. r. microsc. Soc., London, **1915**: 422.
1954 Psalididae BRUES, MELANDER & CARPENTER, Classification of Insects, Cambridge, Mass.: 116.
1954 Carcinophorinae HINCKS, Ark. Zool. Uppsala, **394**: 4. – Genus typicum: *Carcinophora* SCUDDER 1876.
1959 Anisolabinae – BEIER in: BRONN's Klassen und Ordnungen des Tierreichs, 4B, **3**: 562.

Literature: BORELLI 1932; BORMANS 1900; BRINDLE 1978; BURR 1908, 1910; GÜNTHER 1930; HEBARD 1923; HINCKS 1947; REICHARDT 1968; SAKAI 1970; STEINMANN 1975.

The Carcinophorinae are, beyond any doubt, the most problematic taxon of Dermaptera, especially if we consider their present generic arrangement. The superspecific system may be constructed only on the base of a solid taxonomic knowledge of the species. BURR & ZACHER first prepared a generic system and subfamiliar level, rendering a generic outline of the family not modified since the sixties. The external morphological features were usefully complemented by the thorough investigation of the genital apparatus. In 1911 ZACHER, subsequently, in 1915–1916, BURR published very comprehensive works in which they employed extensively the specific differences perceived in the genital paramere and external parameres of the males. In the 1960, POPHAM & BRINDLE, in a series of contribution: "Genera and species of the Dermaptera" drew together in several genera in the various families including Carcinophorinae. In other words, they synonymized genera erected by BURR and ZACHER, and the majority of the species were placed under two genera only: *Anisolabis* FIEBER 1853, and *Euborellia* BURR 1910.

My extensive, comparative studies of the higher taxa have shown that the old classification is better than the modern one. The evaluation of common characteristics at the order or superfamily level has revealed that BURR's system was basically right. Accordingly, I have accepted ZACHER and BURR's works as suitable for separating taxa higher than species, all the more so, since several experts still use the old genera in their publications.

Description: The head is usually transverse, sometimes strongly so, and is more or less convex dorsally. Postfrontal sutures are usually distinct, coronal suture present but may be weak or not visible. Antennae multisegmentated, antennal joints 4, 5 and 6 short. Eyes present, moderately prominent. Prosternum subparallel. Posterior margin of mesosternum rounded arcuately in a semicircle. Metasternum broad, truncate posteriorly or produced to a prominent rounded lobe. Tegmina and wings fully developed, rudimentary or absent. Legs normal, short. Abdomen feebly dilated about the middle or at the apex. The sides of the posterior abdominal tergites of the male may be produced laterally, forming blunt rugoso-striate lateral longitudinal ridges; in many species these tergites have a narrower well marked and more or less smooth lateral longitudinal ridge which is not necessarily associated with a rugosostriate cuticle. Ultimate tergite with a median longitudinal sulcus. The pygidium is usually flat but may be protruding. Penultimate sternite broad, various, usually simple and rounded. Male genital armature which have been used in taxonomy. Paramere with two genital lobes, left lobe directed distally, short, the right lobe directed basally. If the lobes are erected, however, the short left lobe becomes elongated. Genital lobes with or without virgae. The virga is seldom prominent, and is usually visible as a slender tube. Median incision between external parameres well developed, deep. External parameres various in size.

Distribution: Nearly in all through out the world.

11 genera.

Identification key to the genera

1 External parameres of male genitalia extremely long, at least 10 times longer than wide; apex pointed .... Genus 4: **Epilandex**, p. 17

1′ External parameres of male genitalia significatly shorter, length at most 6–7, 3–4 times, or twice as long as wide, or quadrate, as long as wide **2**

2 External parameres at antero-lateral angles of the parameral plate about 6–7 times longer than their width **3**

2′ External parameres at anterior-lateral angles of central parameral plate less than 5 times longer than wide **6**

3 Paired genital lobes of male genital armature very long, much longer than the length of central parameral plate; virgae within genital lobe very long **4**

3′ Paired genital lobes of male genital armature comparatively short, and without virgae ... **5**

4 Very large species, body length with forceps more than 25 mm. Tegmina and wings entirely

absent. Forceps of male more or less straight. Indo-Austral group .... Genus 5: **Titanolabis**, p. 25
4′ Smaller species, body length with forceps less than 20 mm. Tegmina and wings fully developed. Forceps of male strongly curved. Neotropical group . Genus 6: **Capralabis**, p. 27
5 Genital lobes of male genitalia without sclerotized plates; genital lobes comparatively small, narrow, undeveloped, significatly shorter than the external paramere .... Genus 7: **Zacheria**, p. 30
5′ Genital lobes of male genitalia with sclerotized plates; genital lobes fully developed, large and broad .... Genus 8: **Thekalabis**, p. 32
6 External parameres of male genitalia 3–4 times longer than broad **7**
6′ External parameres of male genitalia twice as long as wide or quadrate **11**
7 Apices of external parameres of male genital armature bifid. Ultimate tergite with characteristic lateral carinae. Forceps of male of *Anisolabis*-type .. Genus 9: **Indolabis**, p. 33
7′ Apices of external parameres of male genital armature not bifid, pointed or obtuse **8**
8 Apex of external paramere of male genitalia curved, characteristic; genital lobes with slender virgae **9**
8′ Apex of external paramere of male genitalia more or less straight, not curved strongly outwards **10**
9 External paramere more or less parallel-sided medially, not expanded; genital lobes without sclerotized plate apically .... Genus 10: **Flexiolabis**, p. 34
9′ External paramere strongly expanded medially; genital lobes with characteristic sclerotized plates apically .... Genus 11: **Paraflexiolabis**, p. 37
10 Male genitalia elongated, narrow, external parameres originating on frontal margin, small, about a third as long as the paramere itself, or less elongated, broader, stouter; external parameres originating on frontal margin larger, at least twice longer than those of latter taxon. Virga of genital lobe cordiform, usually thick, well discernible. Tegmina developed, wings present or absent .... Genus 12: **Carcinophora**, p. 38
10′ Male genitalia robust, genital lobe with unchitinized sclerotized plate, sometimes dentate; genital lobes usually characteristic, external parameres gently obtuse .... Genus 13: **Anisolabis**, p. 60
11 External paramere about twice longer than wide **12**
11′ External paramere quadrate, more or less as long as wide ..... Genus 22: **Euborellia**, p. 231
12 Vertex with conspicuous, deep cavea between the frons and occiput .... Genus 14: **Foramenolabis**, p. 152
12′ Vertex without cavea **13**
13 External paramere with characteristic inner lobe basally .... Genus 15: **Aborolabis**, p. 155
13′ External paramere without conspicuous inner lobe at basal margin **14**
14 External parameres excised apically (Fig. 271) .... Genus 16: **Ornatolabis**, p. 166
14′ External paramere not excised apically **15**
15 External paramere egg-shaped (Fig. 274) .... Genus 17: **Anisolabella**, p. 168
15′ External paramere not egg-shaped **16**
16 Virga within genital lobe with specific slerotized plate basally (Fig. 274) .... Genus 18: **Heterolabis**, p. 169
16′ Virga without specific sclerotized plate basally **17**
17 Virga within genital lobe thickened (Fig. 277) .... Genus 19: **Epilabis**, p. 170
17′ Virga within genital lobe not thickened **18**
18 Abdomen parallel-sided laterally in dorsal view, not expanded strongly to last tergite..... Genus 20: **Metalabis**, p. 180
18′ Abdomen strongly expanded to last tergite in dorsal view ..... Genus 21: **Gonolabis**, p. 186

Genus 4: **Epilandex** HEBARD

1915 *Landex* BURR, J. r. microsc. Soc., London, **1915**: 445. – Species typica: *Labidura femoralis* DOHRN 1863.

1927 *Epilandex* HEBARD, Proc. Acad. nat. Sci. Philadelphia, **79**: 26. – Species typica: *Landex burri* BORELLI 1921.
1954 *Landicinae* BOESEMAN, Zool. Verh., Leiden, **21**: 46.
Literature: BRINDLE 1970; SAKAI 1970; STEINMANN 1979; TOWNES 1945.

Notes by HEBARD: "This *Epilandex* name is proposed for the genus described under *Landex* by BURR and later more fully characterized by BORELLI in the discussion before the description of his *burri*. In his genus *Landex* BURR originally included without question only one species: "*Psalis*" *femoralis* DOHRN, and *femoralis* is, therefore, inevitably the genotype. BORELLI has found that BURR confused two species belonging to very distinct genera, but unfortunately did not consider the generotype situation, which obliges the use of *Landex* for DOHRN's *femoralis*. As a result his *burri* can not be referred to that genus and we propose *Epilandex* for its reception."

Description: Rather small species, reddish or yellowish brown in colour, and shiny; more or less glabrous. Tegmina always distinct, wings present or absent. Abdomen moderately elongate, ultimate tergite broad; male forceps asymmetrical, trigonal basally, curved and cylindrical apically; those of female symmetrical. Male genitalia characteristic; paramere with distinct median incision between external parameres; the latter very long, nearly 10 times longer than their basal width; characteristically attenuating, apically acicular and slightly exclinate. Genital lobes with strikingly long virgae, projecting far from genital lobes even at rest, their basal section wide, terminating in a concave, strongly sclerotized curvature.

Distribution: Oriental and Indo-austral species.

6 species.

Identification key to the species

1 Indo-Australian endemic species, known from the Solomon Islands. External parameres of male genitalia about ten times longer than its basal width (Fig. 19); virga long, widening and dividing at the base. Carinae of penultimate sternite of male weak, not produced into a process. Male forceps (Fig. 18) with specific left branch, the latter with conspicuous inner teeth medially .... 1. **Epilandex solomonensis**, p. 19

1′ Oriental species 2

2 External parameres of male genitalia originating on antero-lateral margin of parameral central plate thin and long, more than ten times longer than their basal width 3

2′ External parameres of male genitalia originating on antero-lateral margin of paramere long and narrow, at most ten times as long as their basal width. Penultimate sternite of male with a distinct carina on caudal half, produced into a process 4

3 Male forceps (Fig. 20) very long, slender; inner margin concave based, with a distinct hump at middle, gently curved beyond, smooth; .... 2. **Epilandex handschini**, p. 20

3′ Male forceps (Fig. 22) short, robust; inner margin entire, left branch abruptly narrowed near apex and with or without a sub-apical tooth. Male genitalia as in fig. 23.......... 3. **Epilandex peterseni**, p. 21

4 No sclerotized plate, appendage or lobe near basal section of virga in genital lobe (Fig. 25). Forceps of male (Fig. 24) with a distinct cleft based on inner margin.......... 4. **Epilandex undulata**, p. 22

4′ A sclerotized plate or appendage near basal section of virga in genital lobe present 5

5 A longer (median) section of external paramere parallel-sided, attenuating only insignificatly, narrowing only short of apex; apex therefore rapidly acute (Fig. 26). Virga short and thick, without a characteristic basal widening. Sclerotized plate wide .... 5. **Epilandex burri**, p. 23

5′ External paramere evenly and gradually attenuating from wide base to aciculate apex (Fig. 28). Virga thin and long, basally conspicuously widening, terminating in a characteristic, concave, strongly sclerotized curvature. A narrow, distinct, sclerotized plate present .... 6. **Epilandex bazyluki**, p. 23

1. Epilandex solomonensis Brindle

1970 *Epilandex solomonensis* Brindle, Pacific Insects, Honolulu, (3) **12**: 654; fig. 9 (male), 10 (male genitalia), 11 (male forceps of form minor), 12 (female forceps), 13 (penultimate sternite of male). – Terra typica: Solomon Islands (Type male: Brit. Mus. nat. Hist., London).
Literature: Steinmann 1979, 1984.

Description: Male dark reddish brown to blackish brown; lateral margins of pronotum yellow, basal third of distal antennal joints, apices of femora and tibiae, and tarsi, yellow. Head broad, frons tumid, vertex with small depression in area around coronary suture. Eyes comparatively small, shorter than the length of head behind eyes, not protruding. First antennal joint long, but a little shorter than the distance between antennal bases, second transverse. Cuticle of head almost entirely glabrous. Pronotum more or less as broad as long, slightly widened posteriorly; lateral margins straight, posterior margin convex, all angles rounded. Tegmina well developed, a little longer than pronotum, wings absent or concealed, cuticle of tegmina coriaceous, glabrous. Abdomen depressed, broad, finely punctate and pubescent, hairs short and yellow; abdominal tergites 7–8 each with well-defined longitudinal ridges laterally, and tergite 5 with a small projection on lateral margin on each side. Ultimate tergite transverse, forming a depressed area on each side, the depression being bordered on median side by a longitudinal ridge which extends for distal 2/3 of tergite. Forceps (Fig. 18)

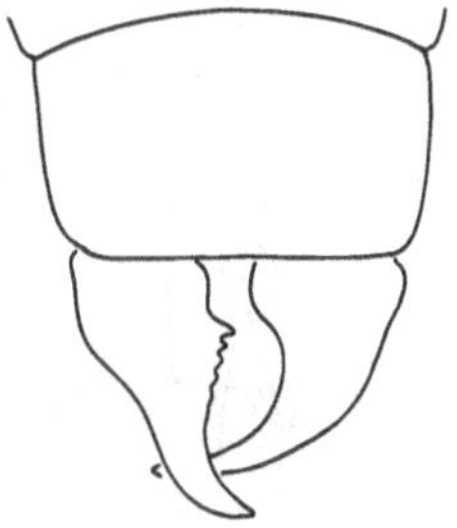

Fig. 18 Ultimate tergite with forceps of holotype of *Epilandex solomonensis* Brindle. After Brindle.

Fig. 19. Genital armature of holotype of *Epilandex solomonensis* Brindle. After Brindle.

asymmetrical; triangular basally, and cylindrical apically; left branch with specific dentation at inner margin medially. Penultimate sternite broad, posterior margin convex. Genitalia (Fig. 19) characteristic; paramere narrow, median longitudinal incision of anterior margin deep; genital lobes well developed, virga very long, but basal section of virga without sclerotized plates. External parameres about 10 times longer than broad, apices acute. – **Female** similar to male, but only 7 tergites visible, no lateral longitudinal ridges on any tergites, nor longitudinal ridges on ultimate tergite; forceps simple, straight, broad at base, evenly narrowed distally, apex incurved, inner margin dentated. – **Length** of body with forceps: male: 12–12,5 mm, female: 9,5–10,5 mm.

Distribution: Solomon Islands: Guadalcanal.

2. Epilandex handschini HINCKS

1954 *Epilandex handschini* HINCKS, Verh. naturf. Ges. Basel, **65**: 12; fig. 3 (male genitalia). – Terra typica: West Sumba (Type male: Naturhist. Mus., Basel).
Literature: POPHAM & BRINDLE 1966; SAKAI 1970; STEINMANN 1979, 1984.

Description: Male colour uniformly brown, including antennae, legs, mouthparts, sides of pronotom and small inner apical spot on wingscales somewhat lighter. Head more or less as broad as long, frons moderately convex, postfrontal and coronal sutures absolete. Eyes comparatively small, shorter than the length of head behind eyes. Antennae 16-jointed; first long, slender, but a little shorter

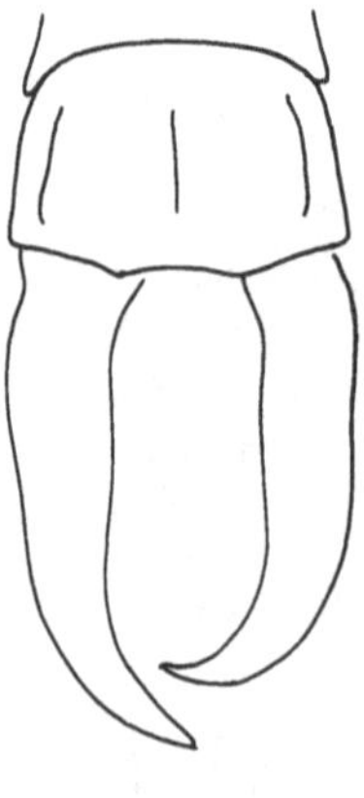

Fig. 20 Male ultimate tergite with forceps of *Epilandex handschini* HINCKS. Original.

Fig. 21 Genital armature of holotype of *Epilandex handschini* HINCKS. After HINCKS.

than the distance between antennal bases. Pronotum broader than long, lateral margins divergent caudad; posterior margin broadly rounded. Tegmina present, ample, twice as long as pronotum, truncate distad, very shallowly and sparsely punctate. Wings well developed, slightly longer than pronotum. Abdomen having 6th to 9th tergites very acute and carinate laterad. Ultimate tergite with lateral carinae; posterior margin slightly concave mesad, concavity terminating laterad in a tubercle which coincides with a projection on the upper median ridge of forceps. Forceps (Fig. 20) trigonal basally, asymmetrical. Penultimate sternite with caudal margin rounded, bearing mesad a very short, feeble rudiment of the longitudinal median process present in *E. burri*. Genitalia (Fig. 21) specific; paramere very narrow, genital lobes with very long virgae. External parameres very slender, long, as in fig. 21. – **Female** unknown. – **Length** of body with forceps: 13 mm.

Distribution: West Sumba.

3. Epilandex peterseni RAMAMURTHI

1967 *Epilandex peterseni* RAMAMURTHI, Ent. Meddr., København, **35**: 234; fig. 6 (ultimate tergite and forceps of male), 7 (ultimate tergite and forceps of female), 8 (male genitalia). – Terra typica: Philippines: Tawi Tawi (Type male: Zool. Mus., København).

Literature: STEINMANN 1979, 1984.

Description: Male colour black and brownish, paler: head black, rest of the body brownish, legs paler. Head transverse, smooth, tumid; postfrontal and coronal sutures indistinct. Eyes comparatively small, shorter than the length of head behind eyes. Antennae 17-jointed; first long, but a little shorter than the distance between antennal bases; clavate, joints 12–17 whitish and more narrowly elongated basally than the rest. Pronotum a little longer than broad, narrower than head anteriorly, slightly expanded caudad; sides truncate. Tegmina less than one and a half time longer than pronotum; wings well developed. Abdomen expanded to last tergite; tergites 6–9 acute and carinate laterad. Ultimate tergite transverse, with well marked lateral longitudinal

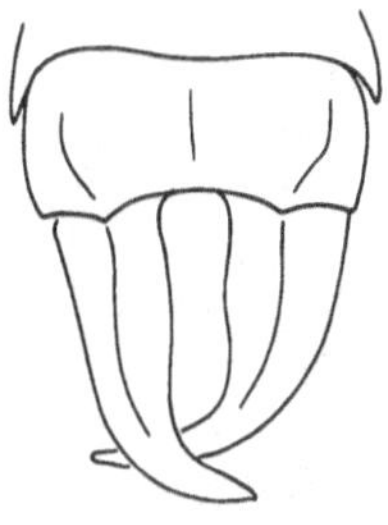

Fig. 22 Ultimate tergite and forceps of holotype of *Epilandex peterseni* RAMAMURTHI. After RAMAMURTHI.

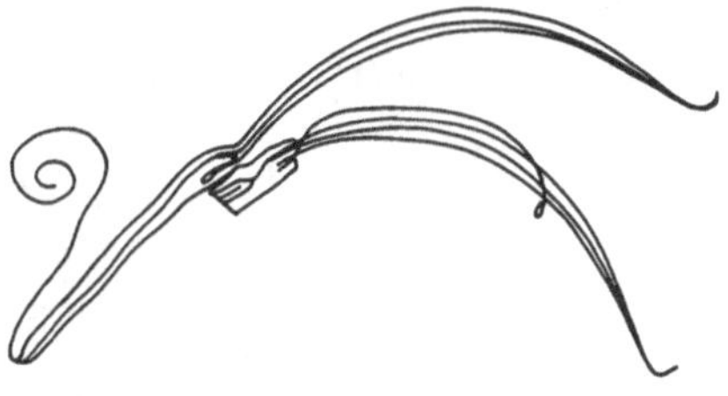

Fig. 23 Genital armature of holotype of *Epilandex peterseni* RAMAMURTHI. After RAMAMURTHI.

carinae; plate with a distinct median longitudinal sulcus and shallow caudal depression. Forceps (Fig. 22) asymmetrical, trigonal basally, cylindrical and strongly curved apically; left branch abruptly narrowed at apex, right branch gently arcuate. Penultimate sternite with broadly rounded posterior margin. Genitalia (Fig. 23) typical to genus; paramere very narrow, genital lobes with very long virgae; external parameres very similar to *handschini*. – **Female** similar to male, but forceps symmetrical, trigonal, very broad at base. – **Length** of body with forceps: male: 7,5–9 mm, female: 8–10 mm.

Distribution: Philippines: Tawi Tawi, Palawan, Bismarck Islands, New Britain, New Ireland.

4. Epilandex undulata RAMAMURTHI

1964 *Epilandex undulata* RAMAMURTHI, Ann. Mag. nat. Hist., London, (13) **6**: 672; fig. 1 (ultimate tergite and forceps of male), 2 (caudal margin of penultimate sternite of male), 3 (male genitalia). – Terra typica: India: Coimbatore (Type male: RAMAMURTHI's Collect.).
Literature: SAKAI 1970; STEINMANN 1979.

Description: Male general colour reddish-brown. Head as wide as long; frons distinctly tumid; postfrontal and coronal sutures absolete. Eyes prominent, longer than genae. Antennae 13-jointed (+?); first joint comparatively long, but a little shorter than the distance between antennal bases. Pronotum broader than long; prozona tumid, median longitudinal furrow weak; posterior margin broadly rounded, sides truncate. Brachypterous. Legs normal. Abdomen moderately depressed, gradually widened caudad. Ultimate tergite ample, transverse; median longitudinal sulcus distinct; lateral margins carinate. Forceps (Fig. 24) characteristic, with a deep incision near the base at inner margins. Penultimate sternite broadly rounded. Genitalia (Fig. 25) specific; paramere

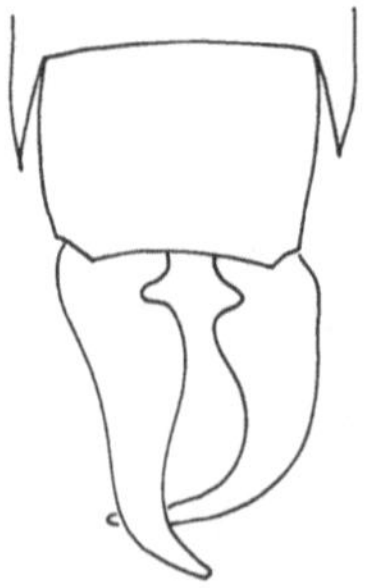

Fig. 24 Ultimate tergite and forceps of holotype of *Epilandex undulata* RAMAMURTHI. After RAMAMURTHI.

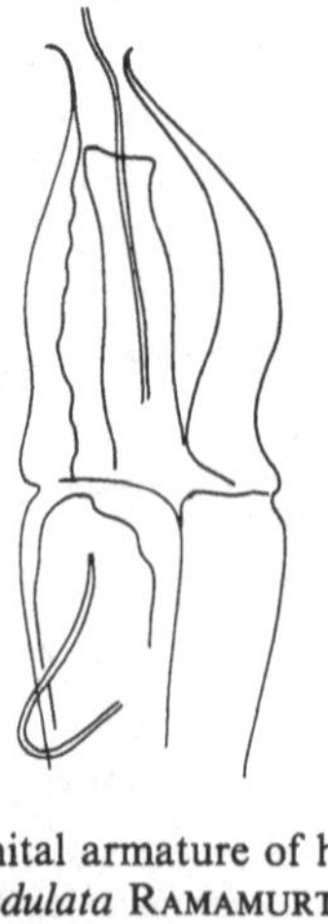

Fig. 25 Genital armature of holotype of *Epilandex undulata* RAMAMURTHI. After RAMAMURTHI.

broad, median longitudinal incision not visible; genital lobes well developed, virga long, basal section without sclerotized plate. External parameres as in fig. 25. – **Female** unknown. – **Length** of body with forceps: 11 mm.

Distribution: India.

5. Epilandex burri (BORELLI)

1921 *Landex burri* BORELLI, Bull. Mus. nation. Hist. nat., Paris, **1921**: 81. – Terra typica: Siam (Type male: Acad. nat. Sci., Philadelphia).
1927 *Epilandex burri* – HEBARD, Proc. Acad. nat. Sci., Philadelphia, **79**: 27.
Literature: BRINDLE 1971; KAPOOR 1967; SAKAI 1970; STEINMANN 1979.

Description: Male colour brown or reddish-brown. Head transverse; frons distinctly tumid. Postfrontal sutures and coronal suture present. Eyes prominent, a little longer than genae. First antennal joint long, but a little shorter than the distance between antennal bases. Pronotum broader than long, prozona tumid, median longitudinal furrow weak; posterior margin rounded. Tegmina well developed, wings concealed, not visible, but the tegmina are nearly one and one-half times the length of the pronotum. Legs normal. Abdomen moderatelly depressed, widened to last tergite. Ultimate tergite ample, quadrate. Pygidium short and quadrangular with slightly spinous angles. Forceps typical, asymmetrical; trigonal basally, cylindrical apically. Penultimate sternite broad, rounded caudad. Genitalia (Fig. 26) relatively broad; paramere well developed, its basal section with sclerotized plate, genital lobes with short virgae, median section of external paramere long, parallel-sided, narrowing only short of apex. – **Female** similar to male, but forceps symmetrical, tapering. – **Length** of body with forceps: male: 11–12 mm, female: 12–13 mm.

Distribution: Thailand and Sri Lanka, Sumatra, Java (var. *brachyptera* BORELLI 1921: Sri Lanka).

Fig. 26 Male genital armature of *Epilandex burri* (BORELLI). After BRINDLE.

6. Epilandex bazyluki STEINMANN

1979 *Epilandex bazyluki* STEINMANN, Fol. ent. hung., Budapest, **32**: 155; fig. 15 (male pronotum, tegmina and wings), 16 (male ultimate tergite and forceps), 17 (genitalia of holotype). – Terra typica: Ceylon (Type male: Zool. Inst., Warszawa).

Description: Male colour pale yellowish brown. Head and pronotum as well as tegmina, brown. Head broad, frons tumid, occiput depressed medially. Postfrontal and coronal sutures obsolete. Antennae broken in holotype, but first joint long, more or less as long as the distance between antennal bases. Pronotum a little longer than broad, slightly expanded to posterior margin: lateral margins straight; posterior margin arcuately rounded. Tegmina comparatively long, nearly twice as long as pronotum. Wings fully developed, pale brown, projecting far from under tegmina. Abdomen a little shiny, normal, ultimate tergite wide, posterior margin apparently sinuous owing to the two smaller projections of forceps at median line. Forceps (Fig. 27) asymmetrical, basally only moderately widened, basal section of inner margin with an irregular small tooth or concavity. Penultimate sternite typical. Genitalia (Fig. 28, gen. prep. No. 366, det. Dr. H. STEINMANN) considerably elongate; paramere broad, median longitudinal incision between external parameres insignificant. Genital lobes with strikingly long virgae projecting far from genital lobes even at rest, their basal section wide, terminating in a concave, strongly sclerotized curvature. Base of virgae with sclerotized narrow lobes, their rounded apices slightly dentate. External parameres about ten times longer than their basal width. – **Female** unknown. – **Length** of body with forceps: 14 mm.

Distribution: Sri Lanka.

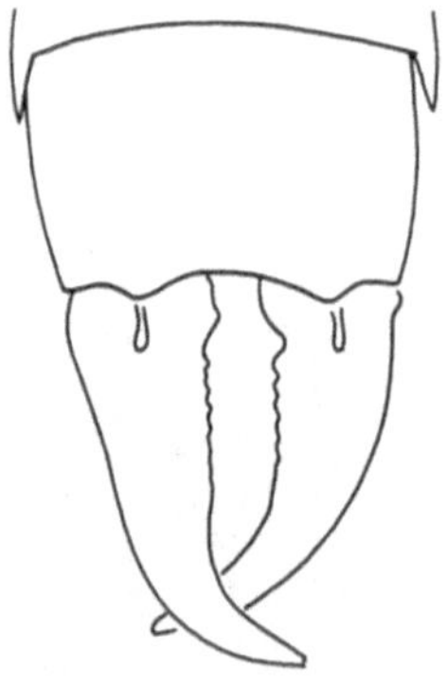

Fig. 27 Ultimate tergite and forceps of holotype of *Epilandex bazyluki* STEINMANN. After STEINMANN.

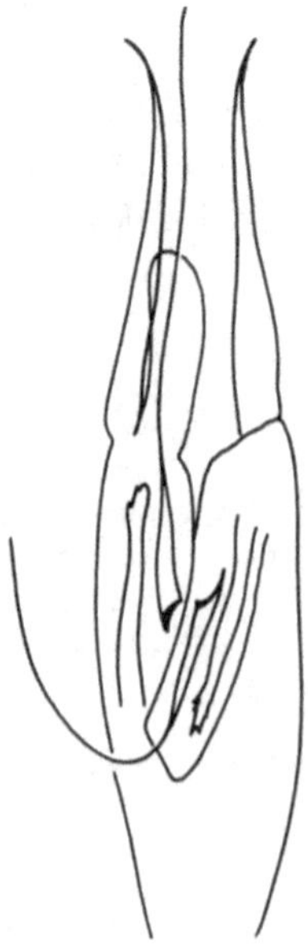

Fig. 28 Genital armature of holotype of *Epilandex bazyluki* STEINMANN. After STEINMANN.

Genus 5: **Titanolabis** Burr

1910 *Titanolabis* Burr, Trans. ent. Soc., London, **1910**: 168. – Species typica: *Forcinella colossea* Dohrn 1864.

Literature: Burr 1911, 1915; Günther 1930; Popham & Brindle 1966; Sakai 1970; Steinmann, 1975, 1977, 1978; Townes 1945; Zacher 1911.

Description: Body very large; wingless species. Metasternum produced posteriorly into a rounded lobe; agrees otherwise with *Anisolabis* Fieber. Colour dark, reddish or dark yellowish brown. Abdomen very long, a little expanded to abdominal tergite 6. Male genitalia very large, anterior margin of paramere deeply incised but parts close to each other, central plate narrow, V-shaped. Lobus genitalis extremely long and thin.

Distribution: Indo-Australia.

2 species.

Identification key to the species

1 Pronotum quadrate or a little longer than broad. Male genitalia (Fig. 30) with long genital lobes, as long as paramere; external parameres comparatively large .... 1. **Titanolabis colossea**, p. 25

1′ Pronotum less quadrate or a little broader than long. Male genitalia (Fig. 32) with very long genital lobes, which are twice as long as the central paramere; external parameres comparatively small .... 2. **Titanolabis gigas**, p. 26

1. Titanolabis colossea (Dohrn)

1864 *Forcinella colossea* Dohrn, Stettiner ent. Ztg., **25**: 286. – Terra typica: Australia: Queensland (Type male: unknown locality, paratype male: Inst. zool., Warszawa).

1894 *Anisolabis colossea* – Bormans, Ann. Mus. civ. Stor. nat., Genova, (2) **14**: 379.

1900 *Anisolabis colossea* – Bormans, Das Tierreich, Berlin, **11**: 47; fig. 19 (male) Neu-Kaledonien, Neu-Hebriden, Fidschi-Inseln, Birma.

1902 *Anisolabis colossea* var. *minor* Burr, Természetrajzi Füzetek, Budapest, **25**: 479. – Terra typica: Australia: New South Wales (a pair of usually small size, the length of the male, including the forceps, is only 18 mm, and of the female, 13,5 mm).

1910 *Titanolabis colossea* – Burr, Trans. ent. Soc., London, **1910**: 168.

1911 *Titanolabis colossea* – Burr, Genera Insectorum, Bruxelles, **122**: 27; pl. 2., fig. 16 (pro-, meso- and metasternum). Australia, ? Burma.

1915 *Titanolabis colossea* – Burr, Journ. r. microsc. Soc. London, **1915**: 532; pl. 10, fig. 2 (male genitalia).

1979 *Titanolabis colossea* – Steinmann, Fol. ent. hung., Budapest, **32**: 155; fig. 13 (lectotype, abdominal end with forceps of male), 14 (male genitalia of lectotype; gen. prep. No. 347, det. Dr. H. Steinmann).

Literature: Boeseman 1954; Hebard 1933; Sakai 1970; Steinmann 1977; Tillyard 1926; Zacher 1911.

Description: Male large, dark yellowish or reddish-brown. Head rounded, broad, broader than pronotum; frons and vertex tumid; postfrontal and coronal sutures distinct; eyes relatively short, shorter than the first antennal joint. First antennal joint broad, comparatively small, shorter than the distance between antennal bases. Pronotum slightly longer than broad; lateral margins paralell, straight; posterior margin truncate, all angles rounded. Median longitudinal furrow present, long. Tegmina and wings entirely absent. Meso- and metanotum normal, narrow. Abdomen expanded to 5th and 6th tergites. Metasternum long and narrow, lobe well produced between posterior coxae, and rounded. Ultimate tergite as long as wide; posterior margin straight. Forceps (Fig. 29) more or less symmetrical, trigonal basally, cylindrical apically. Genitalia (Fig. 30) large; paramere long and relatively narrow, median incision of anterior margin deep and wide or narrow, plate therefore V-shaped. Genital lobes elongate, but shorter than central parameral plate; virga present. External paramere comparatively large, about two and a half to three times longer than wide; basally not widened, but slightly narrowed, apically rounded. – **Female** very similar to male, forceps with branches subcontiguous at the base, stout and robust; trigonal, broad, gradually tapering, with the inner margin crenulate. – **Length** of body with forceps: male and female: size variable from 20 to 50 mm.

Distribution: Australia: New South Wales, Queensland; New Caledonia, New Hebrides, Fiji-Islands.

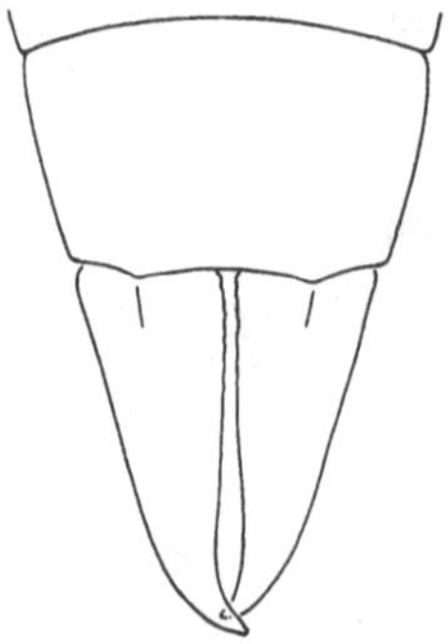

Fig. 29 Male ultimate tergite with forceps of *Titanolabis colossea* (DOHRN). Original.

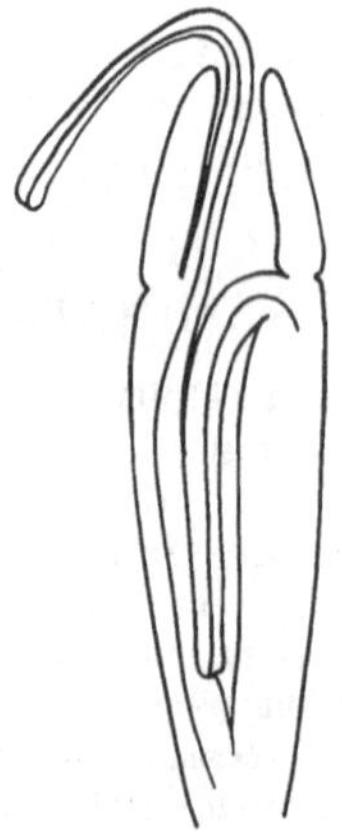

Fig. 30 Male genital armature of *Titanolabis colossea* (DOHRN). Original.

2. Titanolabis gigas sp. nov.

Terra typica: Australia, New South Wales, Mt. Corigudges, 20.11.1967, legit: HANGAY, gen. prep. No. 577, det. Dr. H. STEINMANN. (Type male: Természettudományi Múz., Budapest; 1 female in the same museum).

Description: Male general colour dark reddish-brown; legs brown, forceps black. Head rounded, large; frons tumid, postfrontal and coronal sutures distinct. Eyes very small, shorter than the first antennal joint, or the length of head behind eyes. Antennae broken, first joint normal, shorter than the distance between antennal bases; second transverse, third joint twice as long as the fourth joint. Pronotum a little transverse; lateral margins straight, but a little expanded to posterior angles; caudal margin truncate; median longitudinal furrow distinct. Tegmina and wings entirely absent. Meso- and metanotum typical. Legs comparatively short. Abdomen elongated, a little expanded to median segments. Ultimate tergite well developed, simple; forceps (Fig. 31) symmetrical, trigonal basally, cylindrical apically. Genitalia (Fig. 32) characteristic; paramere with deep median incision between external parameres; plate V-shaped, genital lobes fully developed, very long, about twice as long as the length of central parameral plate; virga very long. External parameres comparatively small, as in fig. 32. – **Female** similar to male, but forceps a little longer, tapering, straight. – **Length** of body with forceps: male: 29 mm, female: 35 mm.

Distribution: Australia.

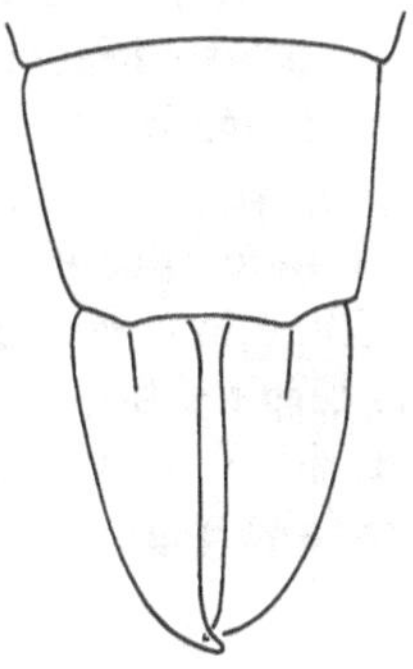

Fig. 31 Ultimate tergite and forceps of holotype of *Titanolabis gigas* sp.n. Original.

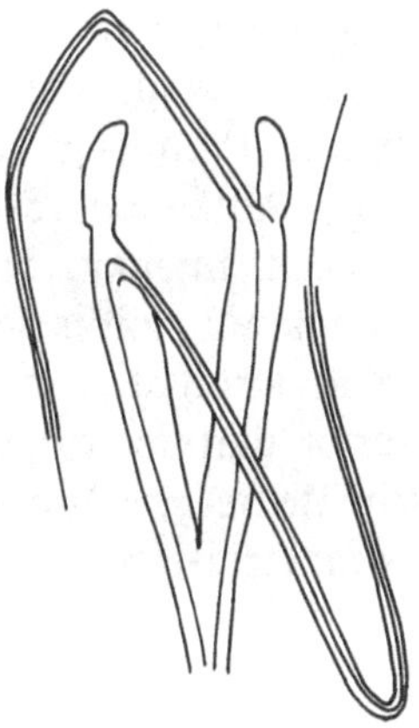

Fig. 32 Genital armature of holotype of *Titanolabis gigas* sp.n. Original.

Genus 6: **Capralabis** BRINDLE

1981 *Capralabis* BRINDLE, Entomologist's month. Mag., London, **117**: 77. – Species typica: *Capralabis ashmolei* BRINDLE 1981.

Description: Body colour dark, more or less glabrous and rather shiny. Head a little elongated, eyes small or normal. Pronotum quadrate or a little longer than broad. Tegmina short and wings absent or concealed, or tegmina and wings fully developed. Male forceps asymmetrical. Male genitalia characteristic, external parameres very long, both genital lobes directed posteriorly at rest.

Distribution: Central and South America.

3 species.

Identification key to the species

1 Wings absent or concealed. Eyes comparatively large, but shorter than length of head behind eyes. Tegmina smooth. Postfrontal and coronal sutures almost absent .... 1. **Capralabis ashmolei**, p. 28

1′ Wings visible **2**

2 Tegmina wrinkled, wings broadly pale basally; male penultimate sternite with apex rounded; dorso-lateral ridge on ultimate tergite absent or almost so. Male penultimate sternite transverse, rounded .... 2. **Capralabis steinmanni**, p. 29

2′ Tegmina smooth or almost so, wings dark; male penultimate sternite triangular; dorso-lateral ridge on ultimate tergite strongly present; smaller species ... 3. **Capralabis srivastavai**, p. 29

1. Capralabis ashmolei Brindle

1981 *Capralabis ashmolei* Brindle, Entomologist's month. Mag., London, **117**: 77; fig. 1 (holotype), 2 (penultimate sternite of male), 4 (holotype genital armature). – Terra typica: Ecuador (Type male: Manchester Mus.).

Description: Male general colour dark brown; body shiny. Head slightly longer than broad; frons tumid, postfrontal and coronal sutures finely; eyes moderately large, but shorter than length of head behind eyes, and a little longer than first antennal joint. Antennae 15-jointed (+?); first joint long, but shorter than distance between antennal bases; second quadrate, third joint a little long, cylindrical, remaining joints conical. Pronotum with rounded posterior margin; lateral margins straight, but expanded posteriorly. Tegmina well developed, shiny, posterior margin truncate. Abdomen slender, tergites 8–9 with lateral spines. Ultimate tergite broad, lateral longitudinal ridges distinct. Forceps (Fig. 33) asymmetrical, strongly trigonal basally, curved and cylindrical apically.

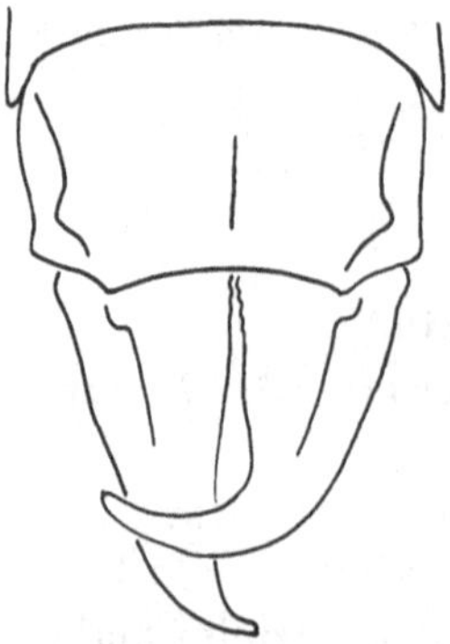

Fig. 33 Holotype ultimate tergite and forceps of *Capralabis ashmolei* Brindle. After Brindle.

Fig. 33a Holotype genital armature of *Capralabis ashmolei* Brindle. After Brindle.

Penultimate sternite with posterior margin broadly rounded. Genitalia (Fig. 33a) characteristic, strongly of *Epilandex*-type, external parameres very long; genital lobes directed posteriorly; virga very long. – **Female** similar to male, but posterior abdominal tergites not produced, nor with lateral longitudinal ridges; forceps symmetrical. – **Length** of body with forceps: male: 13,5 mm, female 13 mm.

Distribution: Ecuador.

2. Capralabis steinmanni Brindle

1981 *Capralabis steinmanni* Brindle, Entmologist's month. Mag., London, **117**: 80. – Terra typica: Panama (Type male: Természettudományi Múz., Budapest; holotype without penis).

Description: Male bicolorous, body dark brown, head black, wings yellow, and legs lighter brown. Head a little elongated, postfrontal and coronal sutures distinct. Eyes large, a little longer than first antennal joint. Antennae broken, 14-jointed (+?) in holotype; first antennal joint long, but shorter than distance between antennal bases; second transverse, third long, longer than fourth; joints 11–12 whitish. Pronotum more or less quadrate; lateral margins expanded posteriorly; posterior margin rounded; median longitudinal furrow distinct. Tegmina and wings fully developed. Abdomen of *Euborellia*-type, tergites 6–9 with lateral spines. Ultimate tergite broad. Forceps (Fig. 34) strongly a-symmetrical; trigonal basally, cylindrical and curved apically. Penultimate sternite with tubercle at posterior margin medially. Distal lobes fully developed, similar to *ashmolei* Brindle, very long; external parameres absent from genitalia. – **Female** similar to male, but lateral longitudinal ridges and spines absent at abdominal tergites 6–9; forceps symmetrical. – **Length** of body with forceps: male: 14,5 mm, female: 12–13,5 mm.

Distribution: Panama to Venezuela.

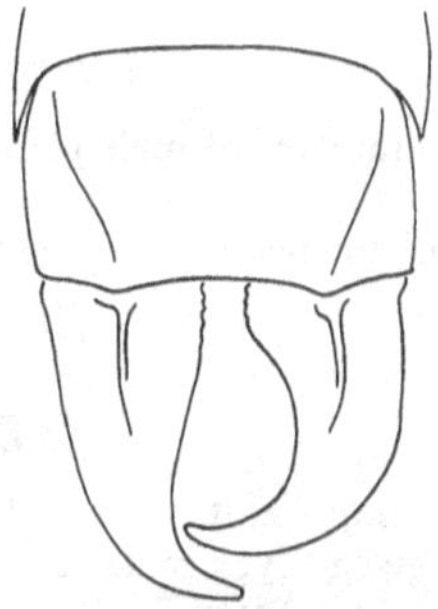

Fig. 34 Holotype ultimate tergite and forceps of *Capralabis steinmanni* Brindle. After Brindle.

3. Capralabis srivastavai Brindle

1981 *Capralabis srivastavai* Brindle, Entomologist's months. Mag., London, **117**: 79. – Terra typica: Panama (Type male: Nation. Collect. zool. Survey India, Calcutta).

Description: Male colour dark brown; distal antennal joints paler. Head slightly longer than broad; frons tumid; postfrontal and coronal sutures indistinct. Eyes moderately large, but shorter than length of head behind eyes. First antennal joint long, but shorter than distance between antennal bases. Pronotum with lateral margins straight, and posterior margin rounded. Tegmina and wings fully developed, tegmina almost impunctate. Abdomen slender, closely and finely punctate; lateral margins of abdominal tergites 7–9 produced and with weak longitudinal ridges. Ultimate tergite with dorso-lateral keel. Forceps less asymmetrical than in *ashmolei* BRINDLE. Penultimate sternite rounded. Genitalia with very long narrow external parameres, but complete genitalia have not yet been obtained. – **Female** similar to male, but abdominal tergites without produced ridges. Forceps symmetrical. – **Length** of body with forceps: male: 7–8 mm, female: 7,5–8,5 mm.

Distribution: Panama.

Genus 7: **Zacheria** STEINMANN

1911 *Eulabis* ZACHER (nec BURR 1915), Zool. Jahrb., Jena, **30**: 377. – Species typica: *Eulabis polita* ZACHER 1911.

1966 *Eulabis* – POPHAM & BRINDLE, Entomologist, London, **99**: 272 [= *Carcinophora* SCUDDER 1876].

1975 *Zacheria* STEINMANN, Acta zool. Acad. Sci. hung., Budapest, **21**: 205. Nomen novum pro *Eulabis* ZACHER 1911; nomen preoccupied by ESCHSCHOLTZ 1929 in Coleoptera.

Literature: BURR 1915; SAKAI 1970.

Description: External parameres of male genital armature with or without small outer tooth at apex; parameres at least twice, often from four to five times, as long as broad and apically pointed. Genital lobe of male paramere small, significantly shorter than external parameres (Fig. 35).

Distribution: Australia.

2 species.

Identification key to the species

1 External parameres of male genitalia with small outer tooth apically (Fig. 35) .... 1. **Zacharia polita**, p. 30

1′ External parameres of male genitalia without small outer tooth apically (Fig. 36) .... 2. **Zacheria dentata**, p. 31

1. **Zacheria polita** (ZACHER)

1911 *Eulabis polita* ZACHER, Zool. Jahrb., Jena, **30**: 376; fig. Z (male genitalia). – Terra typica: no locality given (Type male: unknown locality).

1915 *Eulabis polita* – BURR, J. r. microsc. Soc., London, **1915**: 537; pl. XI, fig. 3 (male genitalia, but in caption of figures: *Eulabis kirbyi* BURR).

1966 *Carcinophora polita* – POPHAM & BRINDLE, Entomologist, London, **99**: 275.

1975 *Zacheria polita* – STEINMANN, Acta zool. Acad. Sci. hung., Budapest, **21**: 205.

Description: Not described by author. **Male** genitalia (Fig. 35) characteristic; central parameral plate relatively short, broad; genital lobe small, without virgae; median longitudinal incision between external parameres very deep. External parameres moderately long, longer than paramere, strongly widened basally; apex of external parameres with single outer tooth. – **Female** unknown. – **Length** of body with forceps: unknown.

Distribution: unknown.

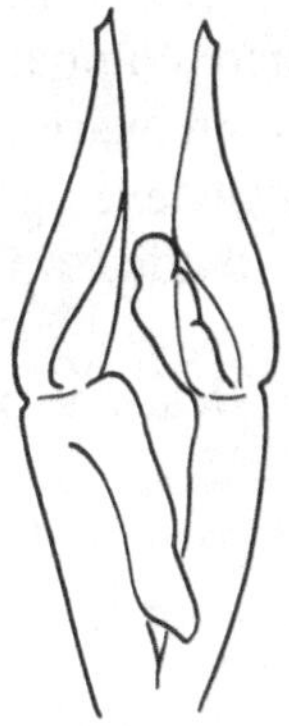

Fig. 35 Genital armature of holotype of *Zacheria polita* (ZACHER). After ZACHER.

2. Zacheria dentata (BURR)

1908 *Gonolabis woodwardi* var. *dentata* BURR, South West Australian Dermaptera: 75; pl. 1, fig. 1 (male). – Terra typica: West Australia (Type male: unknown locality).
1911 *Eulabis dentata* – ZACHER, Zool. Jahrb. Jena, **30**: 376, fig. Y[1] (male genitalia).
1975 *Zacharia dentata* – STEINMANN, Acta. zool. Acad. Sci. hung., Budapest **21**: 205.
Literature: BURR 1915, 1911.

Description: Male colour reddish-brown; antennae and legs lighter brown or yellowish-brown. Head a little elongate, frons tumid, occiput a little depressed medially. Postfrontal and coronal sutures distinct. Eyes normal, a little shorter

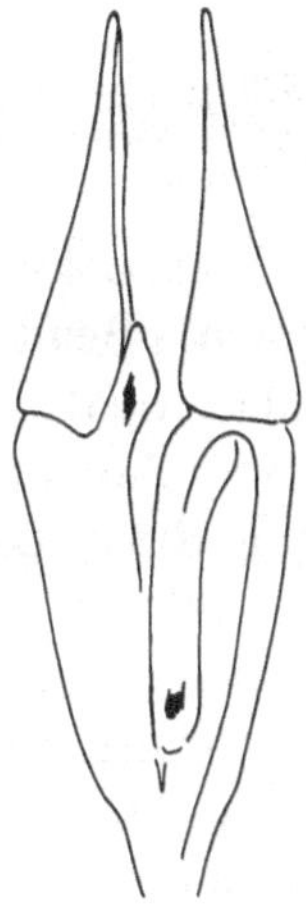

Fig. 36 Male genital armature of *Zacheria dentata* (BURR). After ZACHER.

than the length of head behind eyes. Antennae broken, first joint typical, slightly shorter than the distance between antennal bases; second joint transverse, third long, longer than fourth. Pronotum more or less quadrate; lateral margins straight, posterior margin truncate; median longitudinal furrow distinct. Tegmina and wings absent. Abdomen normal, a little expanded to median segments. Ultimate tergite broad, forceps asymmetrical. Genitalia (Fig. 36) specific; paramere comparatively short, about as long as the length of external parameres; median incision of anterior margin very deep, central plate V-shaped. Genital lobes well developed, virga present. External parameres very long, broad basally, and extended apically. – **Female** unknown. – **Length** of body with forceps: 13 mm.

Distribution: West Australia.

Genus 8: **Thekalabis** KAPOOR

1967 *Thekalabis* KAPOOR, J. nat. Hist., London, **1**: 355. – Species typica: *Thekalabis genitalis* KAPOOR 1967.
Literature: STEINMANN 1975, 1977.

Description: Apterous insects; body slightly depressed, very similar to Pygidicranids. Head tumid; 4th and 5th antennal joints smaller than 3rd. Pronotum longer than broad; mesonotum very slightly sinuate posteriorly; metanotum deeply concave posteriorly. Abdomen widened beyond middle, a few segments produced distad. Ultimate tergite transverse with longitudinal linear depression and oblique carinae on the lateral sides. Forceps asymmetrical, stout. Male genitalia with asymmetrical external parameres, and genital lobes without virgae.

Distribution: India.

Single species.

1. **Thekalabis genitalis** KAPOOR

1967 *Thekalabis genitalis* KAPOOR, J. nat. Hist., London, **1**: 355, fig. 1 (male), 2 (penultimate sternite of male), 3 (male genitalia). – Terra typica: India (Type male: Dept. Zool., Agra College).
Literature: STEINMANN 1977, 1984.

Description: Male colour yellowish brown; legs, antennal joints and sides of pronotum brown. Head well developed, large, broader than pronotum; postfrontal and coronal sutures deep. Eyes large, but shorter than the length of head behind eyes. Antennae broken in holotype, but first joint club-shaped, about as long as the distance between antennal bases; second joint quadrate, third comparatively short, but longer than fourth. Frons tumid, occiput depressed medially, the head similar to *Cranopygia*. Pronotum moderately small; lateral margins parallel; posterior margin a little rounded; median longitudinal furrow present, well marked. Mesonotum about as wide as pronotum, tegmina absent. Wings not developed. Legs very long and slender, posterior leg about as long as abdomen. Abdomen expanded to last tergite, sides of tergites 5–9 produced into processes distad. Ultimate tergite characteristic, not of Carcinophorids-type, surface strongly and irregularly sculptured; posterior margin subsinuate, exterior angles well produced, with convex prolongations over the roots of the forceps and a pair of small tubercles near the outer angles of the forceps. Pygidium present. Forceps (Fig. 37) subcontiguous, with a pair of small teeth near the base, right branch bent distad and somewhat touching the apical portion of the left one. Penultimate sternite triangular. Genitalia (Fig. 38) specific; paramere with external parameres very similar to the larve of the Pygidicranidae. Parameres with deep incision at anterior margin; genital lobes without virgae. External parameres asymmetrical, immature-shaped, widest near the base, and then gradually narrowed towards the apex. – **Female** unknown. – **Length** of body with forceps: 14 mm.

Distribution: South India.

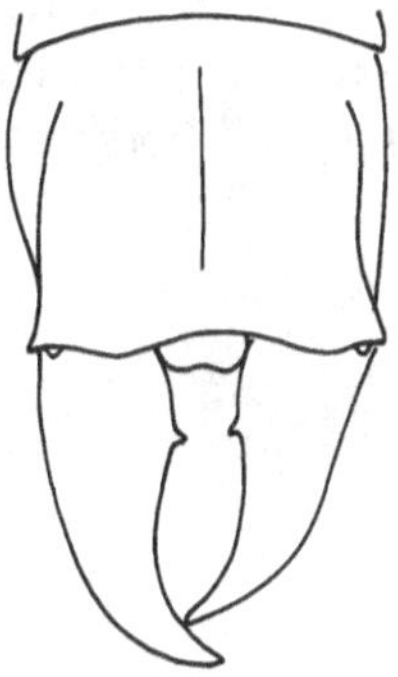

Fig. 37 Ultimate tergite and forceps of holotype of *Thekalabis genitalis* KAPOOR. After KAPOOR.

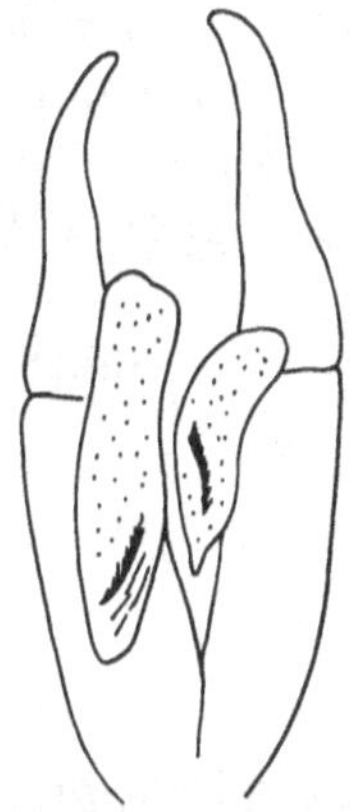

Fig. 38 Genital armature of holotype of *Thekalabis genitalis* KAPOOR. After KAPOOR.

Genus 9: **Indolabis gen. nov.**

Species typica: *Indolabis papua* **sp. nov.**

Description: Body of medium size, similar to *Euborellia*. Head rounded, pronotum quadrate, tegmina and wings entirely absent. Abdomen elongate, tergites progressively widened to the seventh, then narrowed, sides of tergite 6–9 produced into processes distad. Ultimate tergite with strongly lateral carinae. Forceps of *Anisolabis*-type. Male genitalia very characteristic; paramere elongated, genital lobes very slender, long, external parameres bifid apically.

Distribution: New Guinea.

Single species.

1. Indolabis papua sp. nov.

Terra typica: New Guinea, Wau, Mt. Wilhelm, 1968, legit: Dr. J. BALOGH; gen. prep. No. 209, det. Dr. H. STEINMANN. Type male: Természettudományi Múz., Budapest.

Description: **Male** colour dark reddish-brown; antennae brown, legs light brown. Head a little longer than broad; frons tumid, occiput depressed; postfrontal and coronal suture distinct. Antennae broken in holotype, first joint sub-conical, shorter than the distance between antennal bases; second joint quadrate, third a little longer than fourth. Eyes comparatively small, slightly shorter than the length of head behind eyes. Pronotum more or less as long as broad; lateral margins straight, a little broader in posterior angles as in anterior ones; posterior margin truncate; median longitudinal furrow long and distinct. Tegmina and wings entirely absent. Meso- and metanotum typical. Abdomen slender, tergites 7–9 with lateral carinae. Ultimate tergite broad, with lateral keels dorsally. Forceps (Fig. 39) asymmetrical, strongly trigonal basally, cylindrical and curved apically. Penultimate sternite broad, excised at posterior margin medially. Genitalia (Fig. 40) characteristic; central parameral plate elongate, narrow, median longitudinal incision betwen external parameres very deep; genital lobes very long, with thin virgae. External paramere broaded medially, bifid at apex. – **Female** unknown. – **Length** of body with forceps: 14,5 mm.

Distribution: New Guinea.

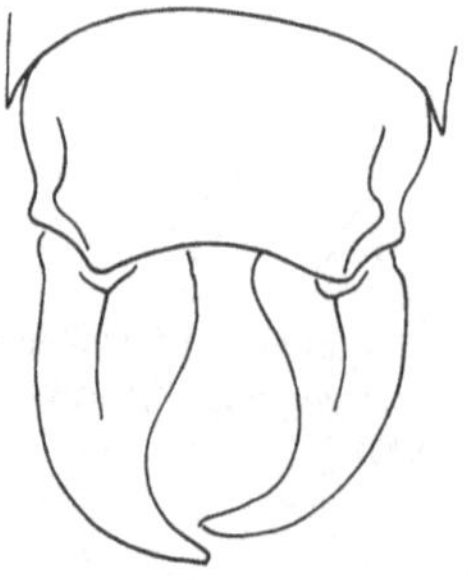

Fig. 39 Ultimate tergite and forceps of holotype of *Indolabis papua* sp.n. Original.

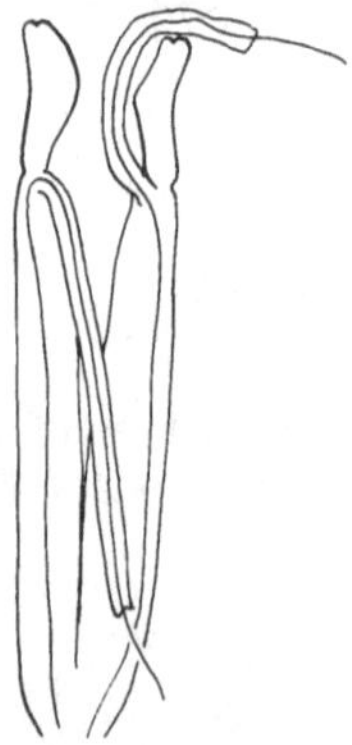

Fig. 40 Genital armature of holotype of *Indolabis papua* sp.n. Original.

Genus 10: **Flexiolabis gen. nov.**

Species typica: *Anisolabis uncinata* BRINDLE 1978.

Description: Black, blackish-brown, rather shiny species; head broad, rounded; pronotum transverse, tegmina and wings entirely absent. Abdomen broad, slightly expanded to 6th tergite. Male forceps asymmetrical, of

Anisolabis-type. Male genitalia with very characteristic external parameres, the latter falciform, strongly curved outwards apically. Penultimate sternite with a projection or process medially at posterior margin (Fig. 42).

Distribution: Africa.

2 species.

Identification key to the species

1 Abdominal tergites 8–9 rugoso-striate at sides, but without lateral longitudinal ridges. External parameres of male genitalia a little narrow basally and apex ending in a re-curved process-like hook (Fig. 43) .... 1. **Flexiolabis uncinata**, p. 35

1′ Abdominal tergites 6–8 rugoso-striate at sides, with lateral longitudinal ridges. External parameres of male genitalia conspicuous, inner and outer margins parallel, distal part directed outwards and forming a hook-shaped or falciform process (Fig. 45) .... 2. **Flexiolabis hamata**, p. 36

1. Flexiolabis uncinata (Brindle) comb. nov.

1978 *Anisolabis uncinata* Brindle, Dermaptera of Africa, Tervuren, **2**: 88; fig. 110 (holotype male penultimate sternite in ventral view), 111 (holotype male penultimate sternite in lateral view), 112 (male forceps), 236 (external paramere of male genitalia). – Terra typica: Zaïre (Type male: Mus. r. Afr. centr., Tervuren).

Description: Male colour blackish-brown, antennae blackish, first three joints yellowish-brown, and some distal joints whitish; legs yellow, and pronotum yellow laterally. Head transverse, rounded; frons tumid, postfrontal and coronal sutures weak. Eyes comparatively small, shorter than the length of head behind eyes. Antennae broken, but first joint shorter than the distance between antennal bases. Pronotum broad, transverse, slightly broader posteriorly, lateral margins slightly sinuate; median longitudinal furrow distinct. Tegmina and wings absent. Mesonotum long, metanotum short. Abdomen slightly widened medially and

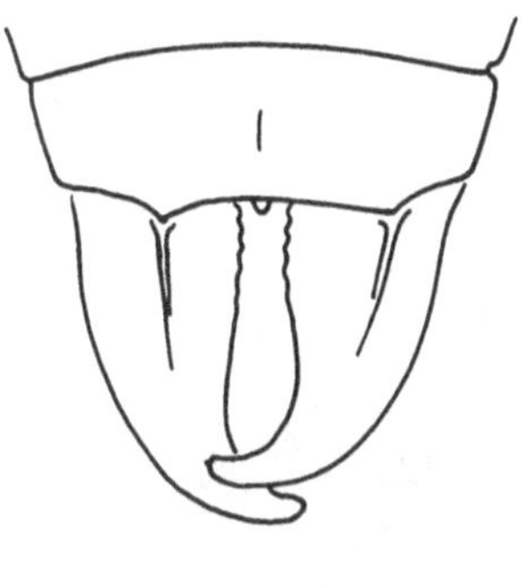

Fig. 41 Ultimate tergite and forceps of holotype of *Flexiolabis uncinata* (Brindle). After Brindle.

Fig. 42 Posterior margin of male penultimate sternite of *Flexiolabis uncinata* (Brindle). After Brindle.

Fig. 43 External paramere of holotype genital armature of *Flexiolabis uncinata* (Brindle). After Brindle.

depressed, tergites 8–9 rugoso-striate at sides but without lateral longitudinal ridges. Ultimate tergite broad, median longitudinal sulcus present but short; weakly and sparsely punctate, depressed medially. Forceps (Fig. 41) short, trigonal basally, cylindrical apically. Penultimate sternite (Fig. 42) somewhat triangular posteriorly, posterior margin straight to apex, with a projection or process medially. Genitalia (Fig. 43) with narrow parameral plate; median incision of intermediate length between external parameres; genital lobes typical, simple; external paramere specific, broad, inner and outer margins not parallel, and apex ending in a re-curved hook. – **Female** unknown. – **Length** of body with forceps: 10 mm.

Distribution: Zaïre.

2. Flexiolabis hamata (Brindle) comb. nov.

1978 *Anisolabis hamata* Brindle, Dermaptera of Africa, Tervuren, **2**: 132; fig. 201 (male), 202 (male penultimate sternite), 203 (genitalia of holotype), 204 (female forceps), 272 (external paramere of male genitalia). – Terra typica: Ethiopia (Type male: Mus. r. Afr. centr., Tervuren).

Description: Male colour black, antennae dark brown, pronotum narrowly yellow at sides, and legs yellowish-brown, femora darkened. Head transverse, rounded in dorsal view; frons tumid, vertex depressed. Postfrontal sutures present, coronal suture deep; cuticle of head mainly smooth but with large shallow isolated punctation. Eyes comparatively small, slightly shorter than the length of head behind eyes. Antennae broken in holotype; first joint moderately short, shorter than the distance between antennal bases. Pronotum transverse, lateral margins more or less parallel; posterior margin a little rounded; median longitudinal furrow present; cuticle of pronotum coriaceous, with isolated punctation. Tegmina and wings absent. Meso- and metanotum normal. Abdomen relatively short, depressed and strongly widened medially; tergites 6–8

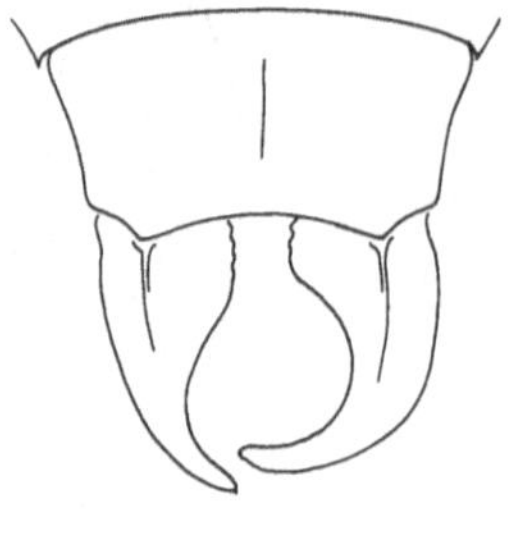

Fig. 44 Ultimate tergite and forceps of holotype of *Flexiolabis hamata* (Brindle). After Brindle.

Fig. 45 Genital armature of holotype of *Flexiolabis hamata* (Brindle). After Brindle.

with well-defined lateral longitudinal ridges. Ultimate tergite very broad, median longitudinal sulcus distinct, the line deep and included in a median depression. Forceps (Fig. 44) asymmetrical, trigonal basally, and cylindrical apically, with well marked dorsal ridges. Penultimate sternite very similar to *uncinata*, but broadly rounded, apex convex. Genitalia (Fig. 45) characteristic; paramere narrow, median longitudinal incision between external parameres of intermediate size; genital lobes comparatively short, virga present. External parameres conspicuous, distal part directed outwards and forming a hook-shaped or falciform process; external parameres with parallel inner and outer margins. – **Female** similar to male, but sides of abdominal tergites not rugoso-striate; forceps straight, slender, tapering. – **Length** of body with forceps: male: 14–15,5 mm, female: 14–16 mm.

Distribution: Ethiopia.

Genus 11: **Paraflexiolabis gen. nov.**

Species typica: *Paraflexiolabis ornata* **sp. nov.**

Description: Dark brownish, reddish-black species; head broad, rounded, shiny, tegmina and wings entirely absent, abdomen similar to *Euborellia*. Male forceps characteristic, symmetrical, a little expanded basally, and with strong dorsal ridge in basal section. Male genitalia conspicuous, paramere V-shaped, and external parameres with curved apices directed outwards.

Distribution: Hawaiian Islands.

Single species.

1. Paraflexiolabis ornata sp. nov.

Terra typica: Hawaii, Hilo Mout, 5–6, VII. 1971, legit: Dr. J. BALOGH, gen. prep. No. 166 (holotype). 167, and 199 (paratypes), det. Dr. H. STEINMANN, males, and one female. Type male: Természettudományi Múz., Budapest.

Description: Male colour dark brownish, reddish-black; head black, thorax and abdomen dark reddish-brown, forceps dark red; legs yellow with femora bicoloured, latter yellow but with dark brown band medially. Head rounded, but a little longer than broad; frons tumid, occiput depressed medially; postfrontal and coronal sutures distinct. Antennae 16-jointed; first joint moderately long, longer than eyes, but shorter than the distance between antennal bases; second transverse, third joint long, longer than fourth. Pronotum slightly longer than broad; lateral margins more or less parallel, or a little concave; median longitudinal furrow present; posterior margin truncate. Tegmina and wings absent. Meso- and metanotum of *Euborellia*-type. Abdomen a little expanded to median segments. Ultimate tergite characteristic, with specific dorsal lobe or process medially at posterior margin. Forceps (Fig. 46) symmetrical; trigonal basally, with dorsal ridge at basal section; inner margin crenulate. Penultimate sternite simple, rounded at posterior margin. Genitalia (Fig. 47) very characteristic; paramere strongly V-shaped, median longitudinal incision between external parameres very deep and wide; genital lobes comparatively large, elongate, with sclerotized plates apically. External paramere conspicuous, slightly expanded medially, and curved at apex; the latter process directed outwards. – **Female** similar to male, but forceps slender, tapering. – **Length** of body with forceps: male: 12–17 mm, female: 16,5 mm.

Distribution: Hawaii.

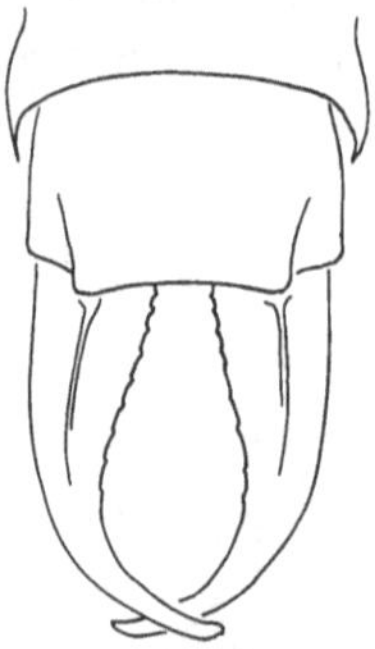

Fig. 46 Ultimate tergite and forceps of holotype of *Paraflexiolabis ornata* sp.n. Original.

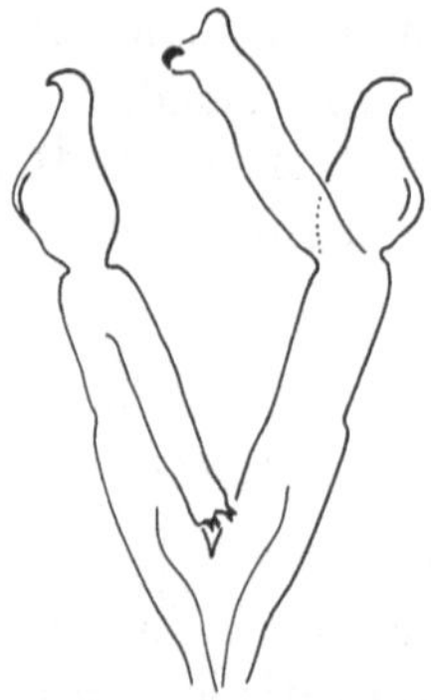

Fig. 47 Genital armature of holotype of *Paraflexiolabis ornata* sp.n. Original.

Genus 12: **Carcinophora** SCUDDER

1831 *Psalis* AUDINET-SERVILLE, Ann. Sci. nat., Paris, **22**: 34. – No type-species cited; preoccupied by *Psalis* HÜBNER 1823 in Lepidoptera.

1876 *Carcinophora* SCUDDER, Proc. Boston Soc. nat. Hist., **18**: 291. – Species typica: *Chelidura robusta* SCUDDER 1869.

1911 *Psalis* – BURR, Genera Insectorum, Bruxelles, **122**: 31. – Species typica: *Psalis americana* PALISOT de BEAUVOIS 1817.

1915 *Mandex* BURR, J. r. microsc. Soc., London, **1915**: 533. – Species typica: *Anisolabis peruvianus* BORMANS 1880.

1915 *Spandex* BURR, J. r. microsc. Soc., London, **1915**: 537. – Species typica: *Psalis pulchra* REHN 1903.

1915 *Eulabis* BURR (nec ZACHER 1911), J. r. microsc. Soc., London, **1915**: 537. – Species typica: *Gonolabis michaelseni* BURR 1915 (nec BURR 1908 = *Anisolabis occidentalis* KIRBY 1896).

1915 *Apolabis* BURR (partim), J. r. microsc. Soc., London, **1915**: 539.

1954 *Notolabis* HINCKS, Acta Univ. Lund Avdeling, (n. F.) **250** (4): 5. – Species typica: *Anisolabis occidentalis* KIRBY 1896.

1966 *Metalabis* – POPHAM & BRINDLE, Entomologist, London, **99**: 272 [= *Carcinophora* SCUDDER].

1966 *Epilabis* – POPHAM & BRINDLE, Entomologist, London, **99**: 272 [= *Carcinophora* SCUDDER].

Literature: BORMANS 1900; BRINDLE 1967, 1971; BURR 1908, 1910, 1911; KIRBY 1904; POPHAM & BRINDLE 1966; REICHARDT 1968; SAKAI 1970; STEINMANN 1975, 1977, 1978, 1979; TOWNES 1945.

Description: Rather small to very large species. Tegmina always present and usually visible, but sometimes rudimentary, and represented by lateral flaps, or normally developed; dark brown to blackish, sometimes partly yellow or red;

tegmina with yellow spots ot patches or entirely dark, wings present or absent. Pronotum relatively small; abdomen not greatly widened or depressed. Male forceps symmetrical or asymmetrical, but gently trigonal basally and cylindrical apically. Male genitalia various, a little similar to those of *Anisolabis*, but virga usually more prominent and more or less indurated; external parameres generally acuminate.

Distribution: Neotropic, Oriental and Indo-Australian faunal regions.

22 species.

Identification key to the species

1 Neotropical species-group **2**
1′ Oriental or Indo-Australian species **19**
2 Tegmina rudimentary, represented by lateral flaps, well separated from each other **3**
2′ Tegmina at least touching in the mid-line **7**
3 Legs entirely yellow or with femora partly darker **4**
3′ Legs yellow, darker at joints **5**
4 Pronotum quadrate, all margins straight and lateral margins parallel; legs entirely yellow. Male forceps as in fig. 48 .... 1. **Carcinophora spitzi**, p. 40
4′ Pronotum a little longer than broad, lateral margins parallel; legs yellow, but femora partly darker .... 2. **Carcinophora boesemani**, p. 41
5 Male forceps (Fig. 51) asymmetrical, strongly expanded basally. Male genital lobes with very long virgae (Fig. 52) .... 3. **Carcinophora festae**, p. 42
5′ Male forceps symmetrical, inner margin with very distinct tooth medially **6**
6 Male ultimate tergite with broad depression at posterior margin dorsally (Fig. 53); inner tooth of male forceps as in fig. 53. Male genitalia specific (Fig. 54) .... 4. **Carcinophora peruviana**, p. 43
6′ Male ultimate tergite normal, without large and broad depression dorsally; forceps straight, as in fig. 55 .... 5. **Carcinophora apolinari**, p. 44
7 Tegmina present, but short, wings absent or concealed **8**
7′ Tegmina and wings well or fully developed **11**
8 Tegmina with a median yellow spot. Male unknown. Female forceps straight, symmetrical, as in fig. 56 .... 6. **Carcinophora festiva**, p. 45
8′ Tegmina unicolourous **9**
9 Small species, body about 7–7,5 mm. Male unknown. Female forceps as in fig. 57 .... 7. **Carcinophora waddyi**, p. 45
9′ Larger species, 10 mm or more **10**
10 Tegmina longer than pronotum laterally, but shorter than pronotum along sutures. Male forceps specific, as in fig. 58 .... 8. **Carcinophora venezuelica**, p. 46
10′ Tegmina smooth, shorter than pronotum laterally. Male forceps strongly trigonal basally, dorsal ridge priminent (Fig. 60) .... 9. **Carcinophora burri**, p. 47
11 Tegmina not entirely unicolourous, at least median part in each tegmina, lighter in colour **12**
11′ Tegmina unicolourous, dark brown or blackish **15**
12 Head and lateral margins of pronotum yellow. Male forceps straight, sligthly trigonal basally, with dorsal ridge (Fig. 61). Male genitalia of *Carcinophora*-type, with acuminate apices, and genital lobes with long virgae (Fig. 62) .... 10. **Carcinophora percheroni**, p. 48
12′ Head and pronotum dark brown or black **13**
13 Tegmina with large, broad yellow spot, which is larger than pronotum. Male forceps (Fig. 63) asymmetric, strongly trigonal basally. Male genitalia of *Anisolabis*-type (Fig. 64), external parameres with obtuse apex .... 11. **Carcinophora americana**, p. 49
13′ Tegmina with small yellowish spot, which is shorter than the length of pronotum. Male genitalia of *Carcinophora*-type, external parameres with acute apices **14**

14 Abdomen a little expanded to 5th and 6th abdominal tergites. Male forceps as in fig. 65. Male genitalia (Fig. 66) with comparatively shorter external parameres........................ .... 12. **Carcinophora rosenbergi**, p. 51
14′ Abdomen a little expanded to 7th and 8th abdominal tergites. Male forceps very similar to *rosenbergi*, but a little longer (Fig. 67). Male genitalia (Fig. 68) with moderately long external parameres .... 13. **Carcinophora haenschi**, p. 51
15 Very small species, body length, including forceps, only 4,5 mm. Male forceps (Fig. 69) similar to *rosenbergi*, but inner margin a little flattened basally .... 14. **Carcinophora minima**, p. 52
15′ Larger species, at least 10 mm in length **16**
16 Wings obscurely yellow, either at base or medially **17**
16′ Wings entirely dark **18**
17 Wings obscurely yellow at base. Male forceps asymmetrical, as in fig. 70................ .... 15. **Carcinophora scudderi**, p. 53
17′ Wings obscurely yellow medially. Male unknown .... 16. **Carcinophora nigra**, p. 54
18 Male forceps (Fig. 71) broad, slightly flattened, asymmetrical. Virga within genital lobe with specific sclerotized plate basally (Fig. 72) .... 17. **Carcinophora gagatina**, p. 55
18′ Male forceps (Fig. 73) flattened at basal section, but slender at apical half. Male genitalia unknown to me .... 18. **Carcinophora brasiliensis**, p. 56
19 Indo-Australian species **20**
19′ Oriental species **21**
20 Inner tooth of male forceps acute (Fig. 74); ultimate tergite with process-like emargination at posterior margin. Male genitalia (Fig. 75) of *Carcinophora*-type......................... .... 19. **Carcinophora occidentalis**, p. 57
20′ Inner tooth of male forceps bifid (Fig. 76); ultimate tergite without process-like emargination at posterior margin. Male genitalia not of *Carcinophora*-type, as in fig. 77............... .... 20. **Carcinophora venusta**, p. 58
21 Formosan species. Male forceps comparatively small, asymmetrical, as in fig. 78. Tegmina well developed .... 21. **Carcinophora kawakamii**, p. 59
21′ Indian species. Male forceps moderately elongate, symmetrical, as in fig. 79. Tegmina short. Male genitalia of *Carcinophora*-type .... 22. **Carcinophora castetsi**, p. 59

1. Carcinophora spitzi (Menozzi)

1932 *Anisolabis spitzi* Menozzi, Rev. Ent., São Paulo, **2**: 151; fig. 1 (dorsal view and male genitalia). – Terra typica: Brazil (Type male: location unknown).
1954 *Anisolabis surinamensis* Boeseman, Zool. Verh., Leiden, **21**: 41; fig. 11 a (antenna of female), 11 b (pro-, meso- and metanotum of male), 11 c (male penultimate sternite). 11 d (male ultimate tergite and forceps), 11 e (apical end of male abdomen in lateral view). – Terra typica: Surinam (Type male: Inst. taxonom. Zool., Amsterdam). – Holotype not male (Surinam, Lucie River, VII–VIII, 1926, Surinam Expedition), but an immature female.
1966 *Carcinophora spitzi* – Popham & Brindle, Entomologist, London, **99**: 275.
1981 *Carcinophora spitzi* Steinmann, Acta zool. Acad. Sci. hung., Budapest, **27**: 190; fig. 11 (ultimate tergite of male), 12 (male genitalia). Brazil, Surinam.
Literature: Reichardt 1968; Sakai 1970.

Description: Male general colour yellowish brown or castaneous, vertex and ultimate tergite reddish brown. Head tumid, occiput a little depressed medially. Postfrontal sutures and coronal suture distinct. Eyes large, but one eye a little shorter in diameter than distance between antennal bases. Antennae 17-jointed, first joint relatively large, long yet shorter than the distance between antennal bases, and about as long as coronal suture. Second joint transverse, third longer, but shorter than first. Pronotum quadrate, all margins straight and lateral margins parallel; median longitudinal furrow distinct. Tegmen and wings entirely absent or rudimentary, represented by lateral flaps. Abdomen punctate, sides subangular, not very sharp. Pygidium situated between and in front of proximal origin of internal ridges of forceps. Ultimate tergite broad, with lateral keels; forceps (Fig. 48) symmetrical, trigonal basally, curved and cylindrical apically. Penultimate sternite rounded at posterior margin. Genitalia (Fig. 49) of *Carcinophora*-type, parameres V-shaped, median incision between external parameres deep and wide; genital lobes with slender virgae. External parameres trigonal, and triangular, broad basally, acute apically. – **Female** similar to male, but forceps slenderer, tapering. – **Length** of body with forceps: male: 13–14,5 mm, female: 15 mm.

Distribution: Brazil, Surinam.

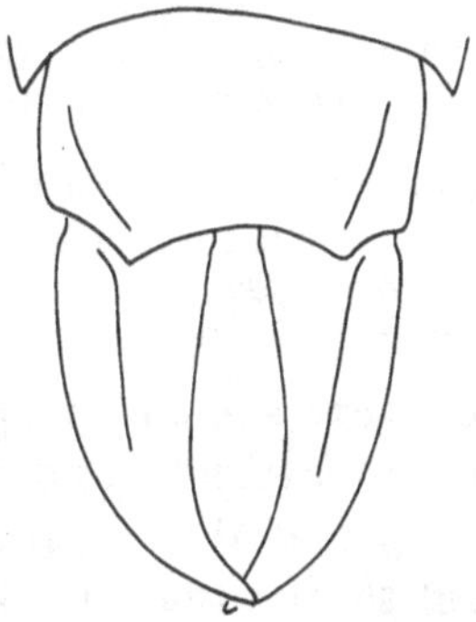

Fig. 48 Ultimate tergite and forceps of *Carcinophora spitzi* (MENOZZI). After STEINMANN.

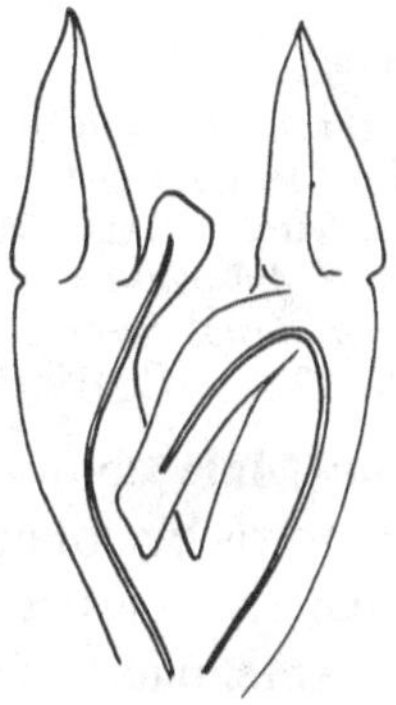

Fig. 49 Male genital armature of *Carcinophora spitzi* (MENOZZI). After STEINMANN.

2. Carcinophora boesemani STEINMANN

1981 *Carcinophora boesemani* STEINMANN, Acta zool. Acad. Sci. hung., Budapest **27**: 191; fig. 13 (forceps of holotype). – Terra typica: Surinam (Type female: Inst. taxonom. Zool., Amsterdam). – Holotype designated by M. BOESEMAN as allotype of *Anisolabis surinamensis* BOESEMAN 1954.

Description: Female colour dark brown or reddish brown; antennae brown, basal joints and, more or less, second and proximal part of third joint lighter; legs testaceous yellow. Head broad, tumid; frons and vertex a little punctate; postfrontal sutures and coronal suture distinct. Eyes large, but a little shorter than length of head behind eyes. Antennae 15-jointed (broken in holotype): first joint about as long as eye, second transverse, third joint long, more or less twice as long as fourth. Pronotum a little longer than broad, lateral margins parallel; posterior margin a little rounded; median longitudinal furrow fine. Tegmina presented by lateral flaps, similar to *spitzi*. Meso- and metanotum of *Anisolabis*-type. Abdomen a little expanded to middle, punctate. Ultimate tergite broad, as in fig. 50. Forceps symmetrical, trigonal basally, dorsal ridge very prominent and long. – **Male** unknown. – **Length** of body with forceps: 14 mm.

Distribution: Surinam.

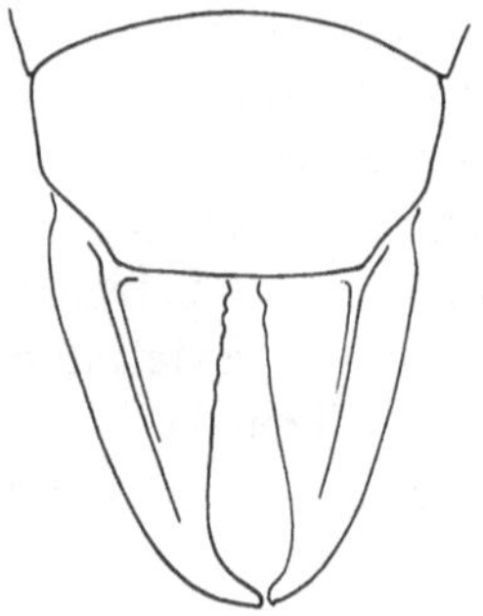

Fig. 50 Female ultimate tergite and forceps of *Carcinophora boesemani* STEINMANN. After STEINMANN.

3. Carcinophora festae (BORELLI)

1904 *Anisolabis festae* BORELLI, Boll. Mus. Zool. Anat. comp. Univ. Torino, **19** No. 475: 3; fig. a (male), b (female ultimate tergite and forceps). – Terra typica: Ecuador (Type male, without penis: Mus. Ist. Zool. sist. Univ., Torino).

1910 *Anisolabis festae* – BURR, Trans. ent. Soc., London, **1910**: 175. – In key, with incorrect indication of *Anisolabis incerta* BORMANS being a melanic form of this species.

1966 *Carcinophora festae* – POPHAM & BRINDLE, Entomologist, London, **99**: 275.

Literature: BURR 1914, 1915; HEBARD 1924; REICHARDT 1968; SAKAI 1970; STEINMANN 1978, 1984.

Description: Male colour reddish brown, body very similar to *apolinari* (HEBARD). Head visibly longer than broad; rounded. Frons tumid, occiput a little depressed medially. Eyes comparatively small, shorter than the length of head behind eyes. Antennae broken; first joint long, but shorter than the distance between antennal bases; second transverse. Pronotum quadrate; median longitudinal furrow distinct; lateral margins straight, parallel, posterior margin rounded. Tegmina rudimentary, represented by lateral flaps. Wings entirely absent. Mesonotum transverse, metanotum with caudal margin broadly concave. Abdomen expanded to middle tergites, expansion depressed dorso-laterally. Ultimate tergite transverse, smooth. Forceps (Fig. 51) trigonal pro-

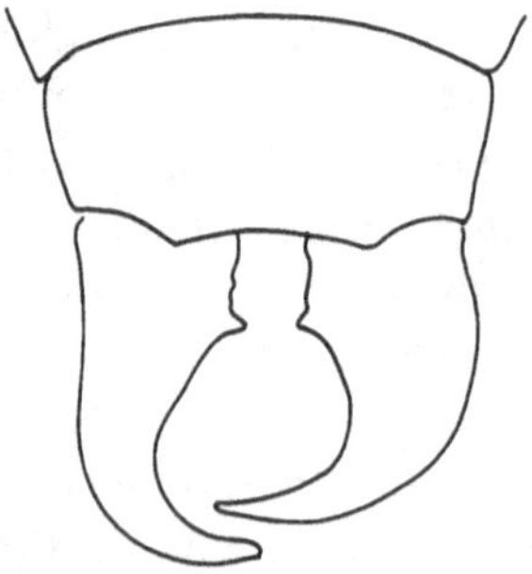

Fig. 51 Male ultimate tergite with forceps of *Carcinophora festae* (BORELLI). After STEINMANN.

Fig. 52 Male genital armature of *Carcinophora festae* (BORELLI). Original.

ximad, strongly curved at apex; asymmetrical, inner margin with very distinct large tooth at basal section. Penultimate sternite simple, rounded posteriorad. Genitalia (Fig. 52) of *Carcinophora*-type, paramere broad, median incision of anterior margin deep; genital lobes fully developed, virga distinct. External paramere well developed, broad, lateral margins parallel medially. – **Female** very similar to male, but forceps slenderer, symmetrical; trigonal basally, cylindrical apically, dorsal ridge prominent. – **Length** of body with forceps: male: 22–25 mm, female: 21–23 mm.

Distribution: Ecuador.

4. Carcinophora peruviana (BORMANS)

1880 *Anisolabis peruvianum* BORMANS, Acta Soc. esp. Hist. nat., Madrid, **9**: 505. – Terra typica: Peru (Type male: unknown locality).
1909 *Borellia peruviana* – BURR, Dtsch. ent. Zeitschr., Berlin, **1909**: 320; pl. 4 (male).
1911 *Euborellia peruviana* – BURR, Genera Insectorum, Bruxelles, **122**: 30; pl. 3, fig. 3 (male).
1915 *Mandex peruviana* – BURR, J. r. microsc. Soc., London, **1915**: 533; pl. 10, fig. 4 (male genitalia).
1966 *Carcinophora peruviana* – POPHAM & BRINDLE, Entomologist., London., **99**: 275.
1968 *Carcinophora peruviana* – REICHARDT, Pap. avulsos Zool. São Paulo, **22**: 42, holotype deposited in the Zool. Inst., Warszawa).

Literature: BORMANS 1900; BURR 1910; REICHARDT 1968; SAKAI 1970; STEINMANN 1975, 1977, 1979.

Description: Male general colour black; antennae and legs dark brown. Head broad, visibly broader than pronotum; postfrontal and coronal sutures distinct. Eyes prominent, large, a little longer than the length of head behind eyes. Frons tumid, vertex smooth. First antennal joint long, more or less as long as the

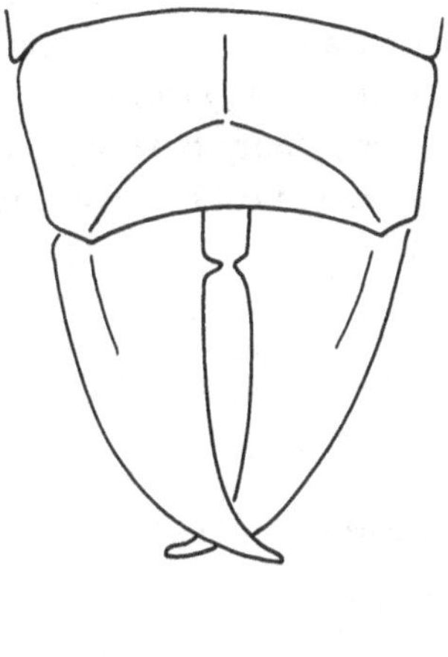

Fig. 53 Male ultimate tergite with forceps of *Carcinophora peruviana* (BORMANS). After BURR.

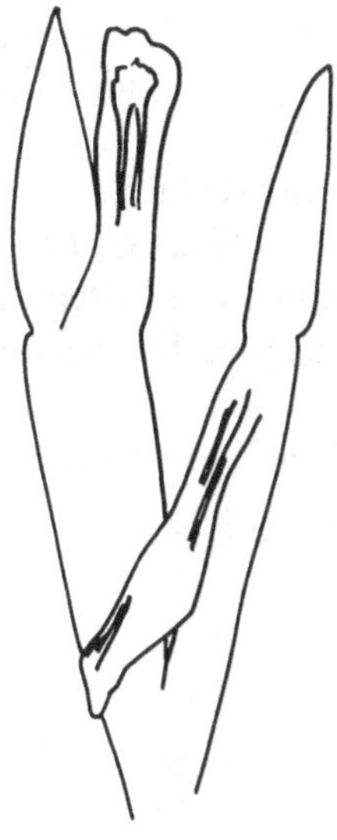

Fig. 54 Male genitalia of *Carcinophora peruviana* (BORMANS). After BURR.

distance between antennal bases. Pronotum quadrate, lateral margins straight, and parallel; median longitudinal furrow distinct; posterior margin a little rounded. Tegmina represented by lateral flaps; mesonotum transverse. Abdomen expanded to middle, shiny. Ultimate tergite with characteristic depression medially at posterior margin; transverse. Forceps (Fig. 53) symmetrical, trigonal basally, cylindrical and curved apically; inner margin with one prominent tooth at basal section. Penultimate sternite simple. Genitalia (Fig. 54) specific; paramere with deep median incision at anterior margin, central plate V-shaped; genital lobe well developed, large, stouter including some sclerotized plate. External parameres of *Carcinophora*-type, pointed at apex. – **Female** very similar to male, but forceps without tooth at inner margin basally. – **Length** of body with forceps: male: 13–14 mm, female: 12–15 mm.

Distribution: Peru.

5. Carcinophora apolinari (HEBARD)

1910 *Psalis apolinari* HEBARD, Trans. amer. ent. Soc., Philadelphia, **45**: 90; pl. 16, fig. 1 (female). – Terra typica: Colombia (Type female: HEBARD's Collect., Type Nr. 441, Nat. Mus., Philadelphia).
1921 *Mandex apolinari* – HEBARD, Trans. amer. ent. Soc., Philadelphia, **47**: 108.
1966 *Carcinophora apolinari* – POPHAM & BRINDLE, Entomologist, London, **99**: 275.
Literature: HEBARD 1933; REICHARDT 1968; SAKAI 1970; STEINMANN 1977, 1984.

Description: **Female** robust. Head a little longer than broad; postfrontal and coronal sutures distinct. Eyes small, shorter than the length of head behind eyes. Antennae broken, first joint very elongate, about as long as the distance between antennal bases; second joint small, quadrate; third elongate, about three times as long as width. Pronotum subquadrate; lateral margins weakly cingulate and feebly diverging caudad; median longitudinal furrow present, but fine. Tegmina represented by lateral flaps; wings absent. Abdomen smooth, broadening to 5th segment. Ultimate tergite very broad, smooth, with a weak median longitudinal sulcus. Forceps (Fig. 55) flattened; inner margin with a heavy tooth just beyond end of proximal third. Penultimate sternite with distal margin nearly rectangulate with apex broadly rounded. – **Male** unknown. – **Length** of body with forceps: 16 mm.

Distribution: Colombia.

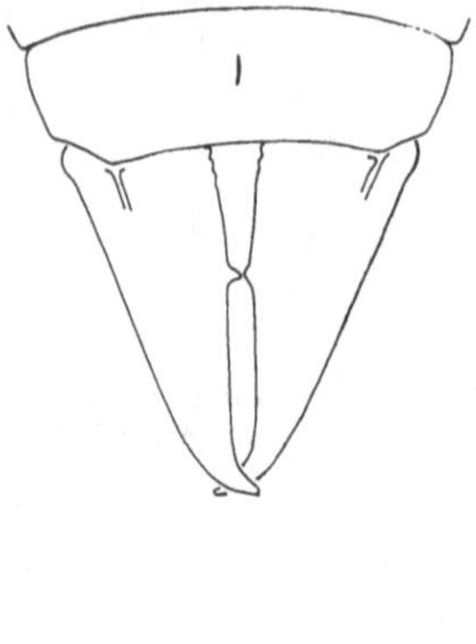

Fig. 55 Female ultimate tergite and forceps of *Carcinophora apolinari* (HEBARD). After HEBARD.

6. Carcinophora festiva (BURR)

1910 *Psalis festiva* BURR, Trans. ent. Soc., London, **1910**: 182. – Terra typica: Brazil (Type female: Naturhist. Mus., Wien).
1915 *Spandex festiva* – BURR, J. r. microsc. Soc., London, **1915**: 524.
1966 *Carcinophora festiva* – POPHAM & BRINDLE, Entomologist, London, **99**: 275.
Literature: BRINDLE 1967; HEBARD 1917; REICHARDT 1968; SAKAI 1970; STEINMANN 1979, 1984.

Description: Female colour dark brown for body, legs light brown or yellowish-brown. Head tumid, frons smooth, occiput a little depressed medially; postfrontal and coronal sutures a little distinct. Eyes relatively long, longer than the length of head behind eyes. First antennal joint long, about as long as the distance between antennal bases. Pronotum quadrate; lateral margins straight, more or less parallel; median longitudinal furrow present. Tegmina rudimentary, represented by lateral flaps; wings absent. Abdomen more or less parallel-sided, a little expanded to last tergite. Ultimate tergite broad, simple. Forceps (Fig. 56) straight, symmetrical; a little flattened basally. Penultimate sternite rounded posteriorad. – **Male** unknown. – **Length** of body with forceps: 22 mm.
Distribution: Ecuador, Venezuela, Colombia, Brazil, Peru.

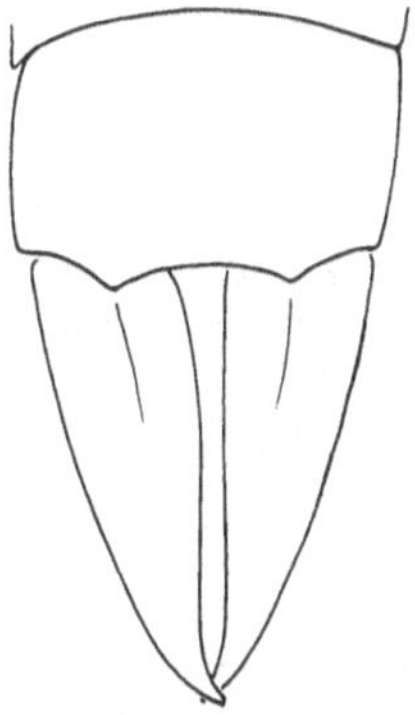

Fig. 56 Female ultimate tergite and forceps of *Carcinophora festiva* (BURR). After BURR.

7. Carcinophora waddyi BURR

1904 *Carcinophora waddyi* BURR, Trans. ent. Soc. London, **1904**: 290. – Terra typica: Martinique (Type female: Mus. nation. Hist. nat., Paris).
1910 *Psalis waddyi* – BURR, Trans. ent. Soc., London, **1910**: 182.
1966 *Carcinophora waddyi* – POPHAM & BRINDLE, Entomologist, London, **99**: 275.
Literature: BRINDLE 1967; REICHARDT 1968; SAKAI 1970; STEINMANN 1977, 1984.

Description: Female colour dark brown. Small species. Head black, antennae dark brown, but antennal joints 1–3, pale; lateral margins of pronotum yellow; legs yellow. Head broad, frons tumid, occiput a little depressed medially. Postfrontal and coronal sutures distinct. Eyes small, much less than the length of head behind eyes. Antennae 13-jointed, first joint long, about as long as the distance between antennal bases. Pronotum quadrate, lateral margins parallel-sided; posterior margin truncate; median longitudinal furrow present. Tegmina rather short, wings absent or concealed. Abdomen slender, but a little expanded to median segments; ultimate tergite broad, forceps (Fig. 57) with branches contiguous. – **Male** unknown. – **Length** of body with forceps: 8,5–9 mm.

Distribution: Martinique.

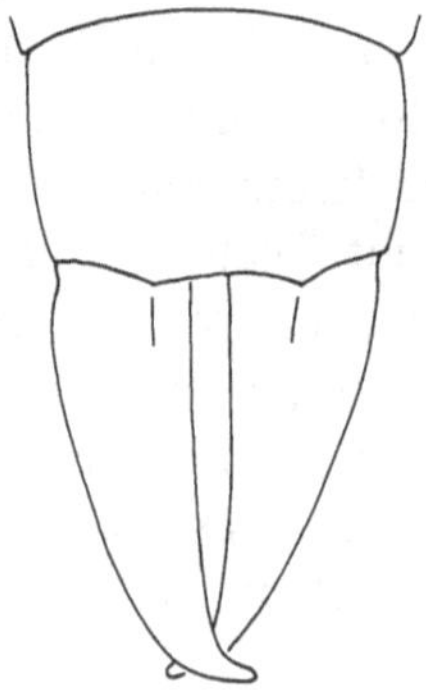

Fig. 57 Female ultimate tergite with forceps of *Carcinophora waddyi* BURR. Original.

8. Carcinophora venezuelica BRINDLE

1968 *Carcinophora venezuelica* BRINDLE, Acta biol. venez., Caracas, **6** (1): 86; fig. 15 (genitalia of holotype), 16 (male forceps). – Terra typica: Venezuela (Type male: Inst. Zool. tropical, Univ. Venezuela, Caracas).

Literature: STEINMANN 1977, 1984.

Description: Male shiny black, legs yellowish-brown, femora banded with black, antennae dark brown, palpi yellow. Head tumid, about as long as broad; postfrontal and coronal sutures distinct. Eyes small, shorter than the length of head behind eyes. Antennae 16-jointed; first joint more or less as long as the distance between antennal bases; second quadrate, third almost three times as long as wide. Pronotum slightly transverse; lateral margins straight and parallel; posterior margin convex; surface of pronotum coriaceous. Tegmina well developed, longer than pronotum laterally, but shorter than pronotum along sutures. Abdomen widening to segment 3, thence parallel-sided. Ultimate tergite large, broad, posterior margin truncate, median longitudinal sulcus present. Forceps (Fig. 58) short, with well-developed dorsal ridge; asymmetrical, inner margins a little flattened basally. Penultimate sternite normal. Genitalia (Fig. 59) of *Carcinophora*-type; paramere broad, median longitudinal incision between external parameres deep; genital lobes fully developed, virga of *Forcipula*-type. External paramere acute. – **Female** unknown. – **Length** of body with forceps: 15 mm.

Distribution: Venezuela.

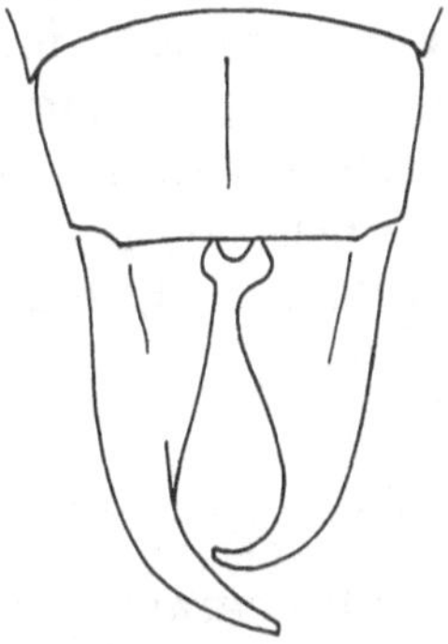

Fig. 58 Ultimate tergite and forceps of holotype of *Carcinophora venezuelica* BRINDLE. After BRINDLE.

Fig. 59 Genital armature of holotype of *Carcinophora venezuelica* BRINDLE. After BRINDLE.

9. Carcinophora burri (BORELLI)

1905 *Psalis burri* BORELLI, Boll. Mus. Zool. Anat. comp. Univ. Torino, **20**, No. 516: 2, fig. (female). – Terra typica: Paraguay (Type female: Mus. Ist. Zool. sist., Torino).
1910 *Psalis burri* – BURR, Trans. ent. Soc., London, **1910**: 182 (in key).
1917 *Euborellia burri* – HEBARD, Trans. amer. ent. Soc., Philadelphia, **43**: 233.
1919 *Psalis compacta* HEBARD, Trans. amer. ent. Soc., Philadelphia, **45**: 92; pl. XVI, fig. 2 (male), 3 (female ultimate tergite and forceps). – Terra typica: Colombia (Type male: HEBARD's Collect., Typ No. 442, Nat. Mus., Philadelphia).
1966 *Carcinophora burri* – POPHAM & BRINDLE, Entomologist, London, **99**: 275.
1970 *Euborellia burri* – SAKAI, Dermapt. Cat. Prael., Tokyo, **1**: 833.
1981 *Capralabis burri* – BRINDLE, Entomologist's month. Mag., London, **117**: 79.
Literature: BRINDLE 1967; BURR 1911; REICHARDT 1968; STEINMANN 1977.

Description: Male colour dark brown; antennae brown, legs yellowish brown. Head tumid; postfrontal and coronal sutures represented by faint lines; occiput smooth. Eyes comparatively small, much shorter than genae. Antennae 14-jointed; first joint moderately long, about as long as the distance between antennal bases; second transverse, third long, longer than fourth. Pronotum

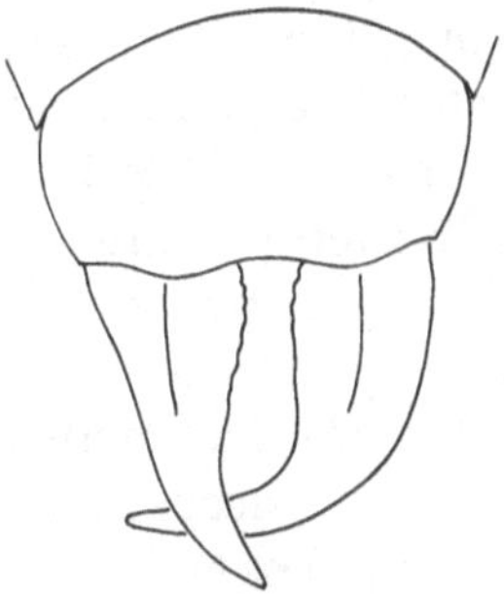

Fig. 60 Male ultimate tergite and forceps of *Carcinophora burri* (BORELLI). Original.

subquadrate; lateral margins parallel, straight; posterior margin rounded. Tegmina present, but short, shorter than the pronotum; smooth; wings absent. Abdomen broad, depressed, expanded to 5th abdominal tergite. Ultimate tergite broad; forceps (Fig. 60) asymmetrical; trigonal and flattened basally, curved and cylindrical apically. Genitalia unknown to me. – **Female** similar to male, but forceps straight, contiguous, tapering. – **Length** of body with forceps: male: 18–21 mm, female: 18–22 mm.

Distribution: Paraguay, Colombia.

10. Carcinophora percheroni (GUÉRIN & PERCHERON)

1838 *Forficula percheroni* GUÉRIN-MÉNEVILLE & PERCHERON, Genera Insectorum, Bruxelles, **6**: 4. – Terra typica: French Guiana (Type female: unknown locality).
1838 *Forficula flexuosa* BURMEISTER, Handb. Ent., Jena, **2**: 753. – Terra typica: French Guiana (Type: unknown locality).
1838 *Forficula elegans* KLUG in BURMEISTER, Handb. Ent., Jena, **2**: – Terra typica: West Indies: St. Johanna (Type: unknown locality).
1862 *Spongophora bipunctata* SCUDDER, J. Boston, Soc. nat. Hist., **1862**: 415. – Terra typica: USA: Massachusetts (Type male: unknown locality).
1903 *Psalis pulchra* REHN, Proc. Acad. nat. Sci., Philadelphia, **1903**: 303. – Terra typica: Costa Rica (Type male: U.S. nation. Mus., Washington).
1906 *Labia pictipennis* BRUNER, J. N. York ent. Soc., **14**: 138. – Terra typica: Trinidad (Type male: BRUNER's Collect., Univ. Mus. Washington).
1907 *Psalis pulchra* – CAUDELL, J. N. York ent. Soc., **15**: 167. – Trinidad.
1910 *Psalis percheron* – BURR, Trans. ent. Soc., London, **1910**: 181. South and Central America.
1915 *Spandex pulchra* – BURR, J. r. micr. Soc., London, **1915**: 537, pl. XI, fig. 5 (male genitalia).
1917 *Spandex percheron* – HEBARD, Proc. Acad. nat. Sci., Philadelphia, **1917**: 232. Trinidad (type of *Labia pictipennis* BRUNER 1906).
1966 *Carcinophora percheron* – POPHAM & BRINDLE, Entomologist, London, **99**: 275.
1971 *Carcinophora percheroni* – BRINDLE, Stud. Fauna Curacao, Carib. Isl., **38**: 21. – Brazil, French Guiana, Guyana, Panama, Costa Rica, Trinidad, Grenada, Dominica, Guadeloupe.

Literature: BRINDLE 1967, 1971; BURR 1910, 1911; HEBARD 1931; KIRBY 1904; REICHARDT 1968; SAKAI 1970; STEINMANN 1977, 1980, 1984.

Description: Male colourful; shiny black, head usually reddish, occasionally darker; antennae brown to dark brown, basal joints yellow, pronotum with yellow lateral margins, or sometimes partly or entirely reddish-yellow; tegmina black with an anterior yellow spot, and wings black, broadly yellow at base. Legs yellow. Head tumid, rounded; frons tumid. Eyes small, shorter than first antennal joint. First antennal joint yellow, long, about as long as the distance between antennal bases; second joint yellow, short. Pronotum relatively small, as broad as long or a little longer the broad, lateral margins straight, widest anteriorly; posterior margin rounded. Tegmina and wings fully developed. Abdominal tergites punctate. Ultimate tergite broad, forceps (Fig. 61) strongly trigonal basally, dorsal ridge distinct; apical section curved and cylindrical. Genitalia (Fig. 62) characteristic, paramere broad, median longitudinal incision of anterior margin deep; genital lobes well developed, virga within genital lobe elongate. External parameres similar to *americana*, but more acuminate distally.

– **Female** very similar to male, but branches of forceps straighter and more contiguous. – **Length** of body with forceps: male: 15–20,5 mm, female: 16–21 mm.

Distribution: Brazil, French Guiana, Surinam, West Indies: Trinidad, Dominica, Grenada, Guadeloupe; southern part of Central America: Panama, Costa Rica, Nicaragua; USA: Massachusetts.

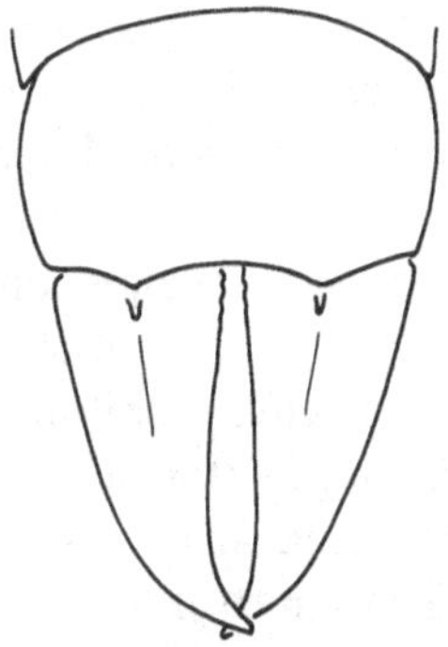

Fig. 61 Male ultimate tergite with forceps of *Carcinophora percheroni* (GUÉRIN & PERCHERON). Original.

Fig. 62 Male genital armature of *Carcinophora percheroni* (GUÉRIN & PERCHERON). After BURR.

11. Carcinophora americana (PALISOT DE BEAUVOIS)

1817 *Forficula americana* PALISOT DE BEAUVOIS, Insectes rec. Afrique Amérique, Orth.: 165; pl. 14, fig. 1 (dorsal view of male). – Terra typica: Hispaniola (Type male: unknown locality).
1838 *Forficula procera* BURMEISTER, Handb. Ent., Jena, **2**: 753. – Terra typica: West Indies (Type male: unknown locality).
1856 *Forficula distincta* GUÉRIN-MENEVILLE, in RAMON DE LA SAGRA, Historia Insecta de Cuba, **7**: 136. – Terra typica: Cuba (Type female: unknown locality).
1839 *Forficesila americana* AUDINET-SERVILLE, Hist. nat. Ins., Orth.,: 22. Dominica
1855 *Forficesila elegans* STÅL (nec KLUG), Öfvers K. svenska Vetensk. Akad. Förh., Stockholm, **12**: 348 – Terra typica: Colombia (Type female: Naturhist. Riksmus., Stockholm).
1863 *Labidura americana* – DOHRN, Stettiner ent. Ztg., **24**: 319.
1869 *Chelidura robusta* SCUDDER, Proc. Boston, Soc. nat. Hist., **12**: 344. – Terra typica: Ecuador (Type male: unknown locality).
1883 *Psalis colombiana* BORMANS, Ann. Soc. ent. Belge, Bruxelles, **27**: 61; pl. 2, fig. 2 (male). – Terra typica: Colombia (Type male: Természettudományi Múz., Budapest).
1888 *Psalis americana* – BOLIVAR, Mém. Soc. zool. France, Paris, **1**: 4. – Cuba.
1903 *Apterygida buschi* REHN (partim), Trans. amer. ent. Soc., Philadelphia, **29**: 129.
1904 *Psalis fusca* BORELLI, Boll. Mus. Zool. Anat. comp. Univ., Torino, **19**, No. 475: 1. – Terra typica: Ecuador (Type female: Mus. Ist. Zool. sist., Torino).
1915 *Psalis americana* – BURR, J. r. microsc. Soc., London, **1915**: 533; pl. X, fig. 5 (male genitalia).

1966 *Carcinophora americana* – POPHAM & BRINDLE, Entomologist, London, **99**: 275.
1971 *Carcinophora americana* – BRINDLE, Stud. Fauna Curacao Caribbean Islands, den Haag, **38**: 19; fig. 7 (male genitalia), 8 (male).

Literature: BRINDLE 1971; BURR 1911; HEBARD 1917; KIRBY 1904; REICHARDT 1968; SAKAI 1970; STEINMANN 1977, 1979, 1984.

Description: Male large and robust; usually dark brown and blackish; antennae brown, but basal joints yellow, both tegmina with a large broad transverse yellow, sometimes orange, band, or with this patch obscured; wings dark, legs yellow or with femora darkened. Head broad and tumid; postfrontal and coronal sutures distinct. Eyes comparatively small, a little shorter than the length of head behind eyes. First antennal joint long, more or less as long as the distance between antennal bases. Pronotum relatively small, as broad as long; lateral margins straight, posterior margin convex, rounded. Tegmina well developed, longer than pronotum, wings fully developed or rather short. Abdomen broad and depressed. Ultimate tergite broad, median longitudinal sulcus distinct. Forceps (Fig. 63) asymmetrical, relatively long, branches slender; strongly trigonal basally, curved and cylindrical apically. Genitalia (Fig. 64) characteristic; paramere well developed, large, incision of anterior margin wide, and deep; fissura medialis present. Genital lobes with indurated virgae. External parameres of *Carcinophora*-type. – **Female** very similar to male, but forceps symmetrical, contiguous; trigonal basally, cylindrical apically. – **Length** of body with forceps: male: 20–35 mm, female 22–38 mm.

Distribution: West Indies: Cuba, Jamaica, Dominica, Haiti, Puerto Rico, Hispaniola, and recorded from Guatemala, Nicaragua, Costa Rica, Panama, Ecuador, Venezuela, Colombia and Peru.

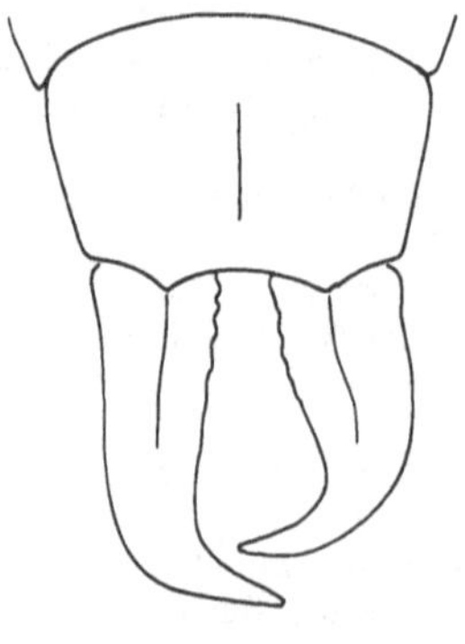

Fig. 63 Male ultimate tergite with forceps of *Carcinophora americana* (PALISOT DE BEAUVOIS). Original.

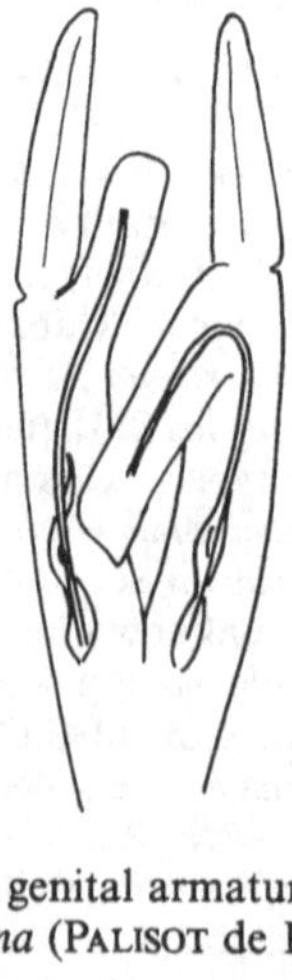

Fig. 64 Male genital armature of *Carcinophora americana* (PALISOT de BEAUVOIS). Original.

12. Carcinophora rosenbergi (Burr)

1899 *Psalis rosenbergi* Burr, Ann. Mag. nat. Hist., London, (7) **4**: 253. – Terra typica: Ecuador (Type male: Brit. Mus. nat. Hist., London).
1915 *Spandex rosenbergi* – Burr, J. r. microsc. Soc., London, **1915**: 524.
1966 *Carcinophora rosenbergi* – Popham & Brindle, Entomologist, London, **99**: 275.
Literature: Burr 1910, 1911; Hebard 1921; Kirby 1904; Reichardt 1968; Sakai 1970; Steinmann 1977, 1979, 1981, 1984.

Description: Male dark brown, except tegmina with large spots, wings with large bands, and legs with brown bands, yellow. Head tumid, shiny, with frontal or postfrontal and coronal sutures. Eyes relatively large, about as long as first antennal joint. Antennae with long, conical joints; first joint long, but a little shorter than the distance between antennal bases; second transverse, third normal, a little longer than 4th. Pronotum comparatively small; lateral margins concave, posterior margin rounded; median longitudinal furrow distinct anteriorly. Tegmina and wings fully developed; wings yellow, but inner posterior angles dark brown. Legs normal. Abdomen nearly parallel, a little expanded to 5th and 6th abdominal tergites. Ultimate tergite broad, median longitudinal sulcus present. Forceps asymmetrical (Fig. 65); trigonal basally, cylindrical and curved apically. Penultimate sternite bifurcated at posterior margin medially. Genitalia (Fig. 66, gen. prep. No. 576, det. Dr. H. Steinmann) of *Carcinophora*-type; paramere well developed, median incision of anterior margin very deep; genital lobes fully developed, virgae elongated; external parameres with pointed apices. – **Female** similar to male, but forceps contiguous, tapering. – **Length** of body with forceps: male: 11–15 mm, female: 10,5–16 mm.

Distribution: Ecuador, Colombia.

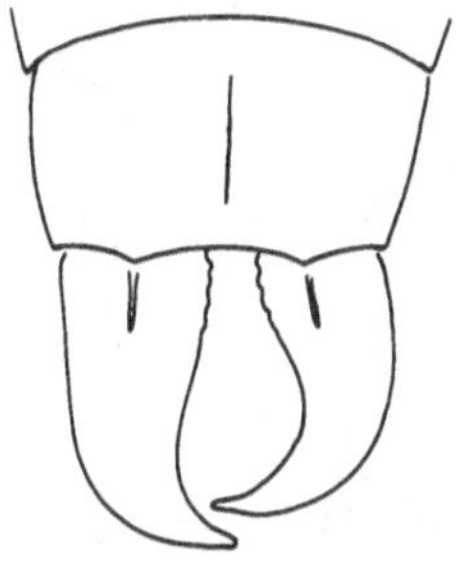

Fig. 65 Male ultimate tergite with forceps of *Carcinophora rosenbergi* (Burr). Original.

Fig. 66 Male genital armature of *Carcinophora rosenbergi* (Burr). Original.

13. Carcinophora haenschi (BURR)

1912 *Psalis haenschi* BURR, Sber. Ges. naturf. Fr., Berlin, **1912**: 317; fig. 3 (penultimate sternite of male). – Terra typica: Ecuador (Type male: Mus. Naturkunde, Berlin).
1915 *Spandex haenschi* – BURR, J. r. microsc. Soc., London **1915**: 537; pl. XI, fig. 6 (male genitalia).
1966 *Carcinophora haenschi* – POPHAM & BRINDLE, Entomologist, London, **99**: 276.
1977 *Spandex haenschi* – STEINMANN, Ann. hist.-nat. Mus. nation. hung., Budapest, **69**: 92.
Literature: REICHARDT 1968; SAKAI 1970; STEINMANN 1984.

Description: Male general colour dark brown; antennae, legs yellowish or yellowish-brown. Head tumid, broad, subpyriform in outline when seen from dorsum, its greatest length subequal to width across eyes. Postfrontal and coronal sutures distinct. Eyes moderately prominent. First antennal joint long, but a little shorter than the distance between antennal bases; second joint transverse, third normal, a little longer than fourth. Pronotum subquadrate, greatest length very slightly more than caudal width; lateral margins straight, all angles rounded. Tegmina and wings well developed. Abdomen depressed, expanded to 7th and 8th abdominal tergites. Ultimate tergite broad, large; median longitudinal sulcus present. Forceps robust, asymmetrical; trigonal basally, curved and cylindrical apically, as in fig. 67. Genitalia (Fig. 68) characteristic; paramere broad at anteriorly; incision of anterior margin relatively short, narrow; genital lobes very long, narrow, with long and slender virgae. Apices of external parameres pointed, but outer and inner lateral margins more or less parallel. – **Female** similar to male, but forceps slenderer, symmetrical, tapering, contiguous. – **Length** of body with forceps: male: 11–14,5 mm, female: 12–17 mm.

Distribution: Ecuador.

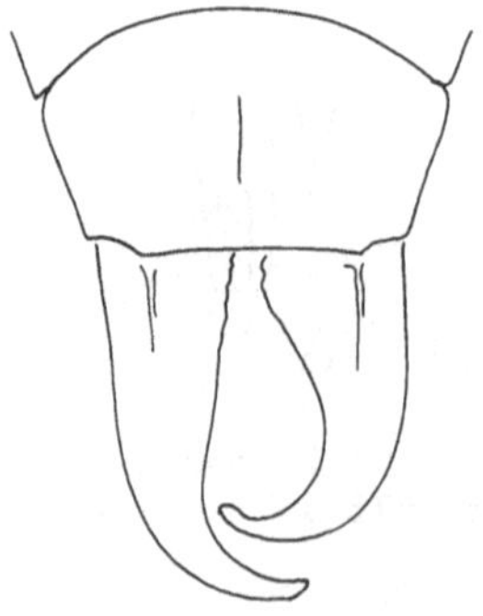

Fig. 67 Male ultimate tergite with forceps of *Carcinophora haenschi* (BURR). Original.

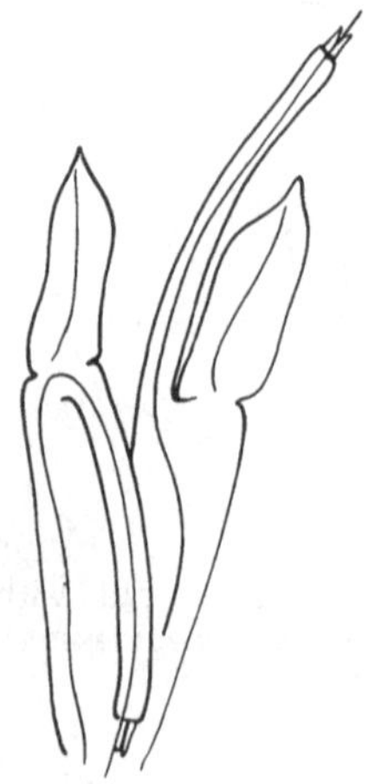

Fig. 68 Male genital armature of *Carcinophora haenschi* (BURR). After BURR.

14. Carcinophora minima (MOREIRA)

1931 *Psalis minima* MOREIRA, Rev. Ent., São Paulo, **1**: 250; fig. 4 (dorsal view of male). – Terra typica: Brazil (Type male: Biol. Defesa agricola, Rio de Janeiro).
1966 *Carcinophora minima* – POPHAM & BRINDLE, Entomologist, London, **99**: 275.
Literature: REICHARDT 1968; SAKAI 1970; STEINMANN 1977, 1984.

Description: Male general colour very dark brown, or brownish-red. Body small. Head tumid, postfrontal and coronal sutures distinct, visible. Eyes relatively small, a little shorter than first antennal joint. Scapus long, about as long as the distance between antennal bases. Pronotum quadrate; lateral margins more or less parallel, all angles rounded; median longitudinal furrow present. Tegmina well developed, unicolourous, dark brown; wings fully developed, without yellowish spots. Abdomen a little depressed, shiny, expanded to middle abdominal segments. Ultimate tergite broad, posterior margin concave. Forceps (Fig. 69) characteristic; strongly curved, more or less symmetrical, not of *Carcinophora*-type; a little depressed basally, inner margins crenulate at basal section. Genitalia unknown to me. – **Female** unknown. – **Length** of body with forceps: 4 mm.

Distribution: Brazil.

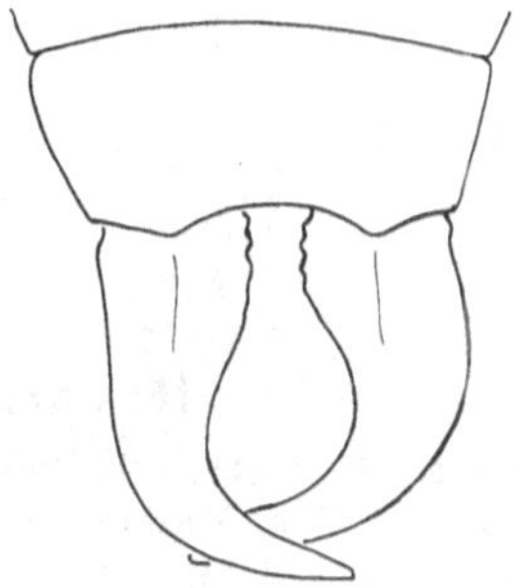

Fig. 69 Male ultimate tergite with forceps of *Carcinophora minima* (MOREIRA). After MOREIRA.

15. Carcinophora scudderi (BORMANS)

1900 *Psalis scudderi* BORMANS, Ann. Mus. civ. Stor. nat. Genova, (2) **20**: 449. – Terra typica: Paraguay (Type female: Mus. civ. Stor. nat., Genova).
1917 *Euborellia scudderi* – HEBARD, Proc. Acad. nat. Sci., Philadelphia, **69**: 233 (description of male from Peru).
1967 *Carcinophora scudderi* – BRINDLE, Ent. Mitt. zool. Inst. Mus., Hamburg, Nr. 56, **3**: 193.
1968 *Euborellia scudderi* – BRINDLE, Stud. Fauna Surinam and Guyanas, den Haag, **10**: 22.
1968 *Carcinophora scudderi* – REICHARDT, Pap. avulsos Zool., São Paulo, **22**: 42. Paraguay; Brazil; Amazonas; French Guiana.
Literature: BURR 1910, 1915; HEBARD 1917; POPHAM & BRINDLE 1966; SAKAI 1970.

Description: Male general colour dark brown, but legs, and anterior part of pronotum yellowish; femora with brownish spots laterally. Head tumid, occiput a little depressed; postfrontal and coronal sutures well marked. Eyes comparatively small, shorter than the length of head behind eyes. Antennae with short small joints. First joint very short, shorter than the distance between antennal bases; second quadrate, third joint small, not of *Carcinophora*-type. Pronotum more or less quadrate, lateral margins nearly parallel; median longitudinal furrow present. Tegmina unicolour, normally developed, wings well developed, obscurely yellow, either at base or medially. – One male specimen from Peru (Chanchamayo) with label: "*Euborellia scudderi* (BORMANS), det. HEBARD, 1927" deposited in the Hungarian Natural History Museum, Budapest without tegmina and without wings. – Abdomen similar to *minima* (MOREIRA), a little depressed, shiny, and expanded to middle. Ultimate tergite large, median longitudinal sulcus present. Forceps (Fig. 70) asymmetrical; strongly trigonal basally, dorsal ridge well marked, and curved apically. Genitalia unknown to me. – **Female** similar to male, but forceps symmetrical, tapering. – **Length** of body with forceps: male 10–12 mm, female: 11–13 mm.

Distribution: Paraguay, Peru, Brazil, French Guiana.

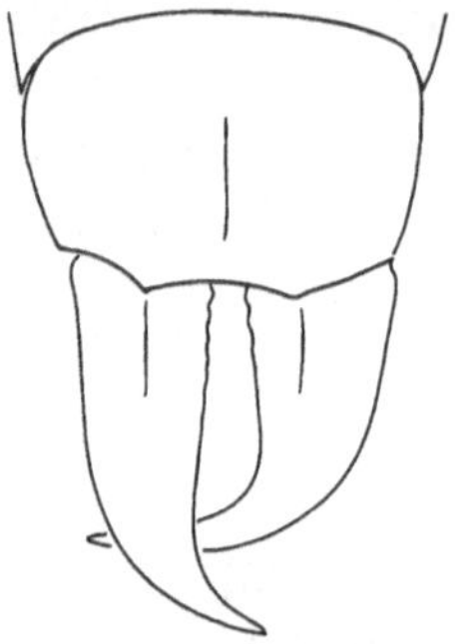

Fig. 70 Male ultimate tergite with forceps of *Carcinophora scudderi* (BORMANS). Original.

16. Carcinophora nigra (CAUDELL)

1907 *Psalis nigra* CAUDELL, J. N. York ent. Soc., **15**: 167. – Terra typica: Trinidad (Type female: U.S. nation. Mus., Washington).

1915 *Spandex nigra* – BURR, J. r. microsc. Soc., London, **1915**: 537.

1966 *Carcinophora nigra* – POPHAM & BRINDLE, Entomologist, London, **99**: 275.

1978 *Carcinophora nigra* – STEINMANN, Dtsch. ent. Zeitschr., Berlin, (n. F.) **25**: 174; fig. 11 (male genitalia - misquotation).

1981 *Capralabis nigra* – BRINDLE, Entomologist's month. Mag., London, **117**: 80.

Literature: BRINDLE 1971; BURR 1910, 1911; REICHARDT 1968; SAKAI 1970.

Description: Female general colour black; antennal joints dark brown, joints 13–14 or 12–14 yellow; wings brown, with central part slightly yellowish; legs light brown, faintly marked longitudinally with darker brown. Head tumid, normal, frons a little depressed; postfrontal and coronal sutures distinct. Eyes about as long as genae. First antennal joint normal, a little shorter than the distance between antennal bases. Pronotum quadrate, lateral margins more or less parallel, posterior margin convex. Median longitudinal furrow finely marked. Tegmina and wings normally developed. Abdomen a little depressed and expanded to median portion, shiny. Ultimate tergite broad, posterior margin truncate. Forceps symmetrical, relatively short, trigonal basally, curved caudad. – **Male** similar to female, but genitalia not examined. – **Length** of body with forceps: 11,5 mm.

Distribution: Trinidad: Montserrat.

17. Carcinophora gagatina (KLUG)

1838 *Forficula gagatina* KLUG in BURMEISTER, Handb. Ent., 2: 753. – Terra typica: Puerto Rico (Type not designated).
1869 *Chalidura robusta* SCUDDER (partim), Proc. Boston Soc., nat. Hist., **12**: 344.
1883 *Psalis colombiana* BORMANS (partim), Ann. Soc. ent. Belge, Bruxelles, **27**: 61, pl. 2, fig. 1 (male).
1903 *Apterygida buscki* REHN, Trans. amer. ent. Soc., Philadelphia, **29**: 129. – Terra typica: Puerto Rico (Type female: U.S. nation. Mus., Washington).
1909 *Psalis gagatina* – BURR, Ann. Mag. nat. Hist., London, (8) **4**: 126.
1911 *Psalis gagatina* – BURR, Genera Insectorum, Bruxelles, **122**: 31, pl. 3, fig. 8 (sternum). – South and Central America, West Indies.
1915 *Psalis gagatina* – BURR, J. r. microsc., Soc., London, **1915**: 533; pl. X, fig. 6 (male genitalia).
1967 *Carcinophora gagatina* – BRINDLE, Ent. Mitt. zool. Inst. Mus., Hamburg, **3**: 191.
1968 *Carcinophora gagathina* BURMEISTER & REICHARDT, Pap. avulsos Zool., São Paulo, **22**: 40.
1970 *Carcinophora gagatina* – SAKAI, Dermapt. Cat. Prael., Tokyo, **2**: 75 [= *Carcinophora americana* (PALISOT DE BEAUVOIS 1817)].
1971 *Carcinophora gagatina* – BRINDLE, Stud. Fauna Curacao and Caribbean Islands, den Haag, **38**: 20 [= *Carcinophora americana* (PALISOT DE BEAUVOIS 1817)].

Literature: BOESEMAN 1954; STEINMANN 1977, 1978, 1979.

Description: Male colour dark brown; legs yellowish brown; first antennal joint reddish brown. Head tumid, shiny. Both tegmina with a large broad transverse yellow or orange band, or with this patch obscured. Head broad and tumid; postfrontal and coronal sutures distinct. Eyes relatively short, visibly shorter than first antennal joint. First antennal joint comparatively long, about as long as or a little longer than the distance between antennal bases; second joint quadrate, third longer than fourth. Pronotum a little longer than wide; lateral margins parallel; posterior margin rounded; median longitudinal furrow present in anterior section. Tegmina normally developed or a little abbreviated, wings well developed or absent. Tegmina generally short, about one and half times as long as wide; posterior margin truncate. Abdomen expanded to 5th and 6th

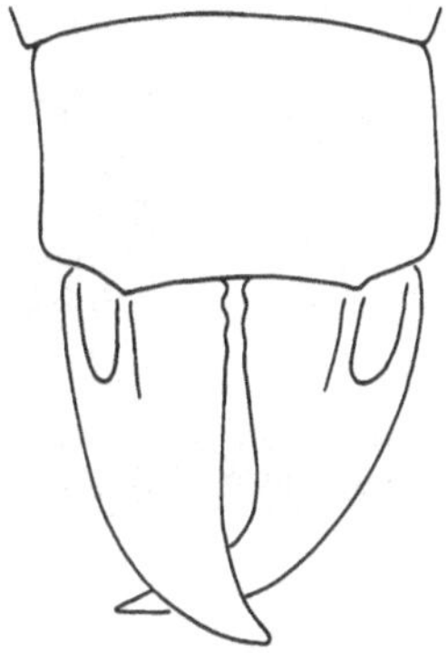

Fig. 71 Male ultimate tergite with forceps of *Carcinophora gagatina* (KLUG). Original.

Fig. 72 Male genital armature of *Carcinophora gagatina* (KLUG). Original.

abdominal tergites. Ultimate tergite broad, robust, surface without median longitudinal sulcus. Forceps (Fig. 71) robust, with very strong dorsal ridge basally; branches a little depressed, but cylindrical apically. Penultimate sternite with rounded posterior margin. Genitalia (Fig. 72) specific, gen. prep. No. 358, det. Dr. H. STEINMANN as in fig. 72; paramere with narrow median fissure; genital lobes well developed, virga broad with sclerotized plate basally. External parameres broad, pointed apically. – **Female** very similar to male, but forceps slender and symmetrical. – **Length** of body with forceps: male: 20–22 mm, female: 18–23 mm.

Distribution: South and Central America, West Indies.

18. Carcinophora brasiliensis (MOREIRA)

1931 *Psalis brasiliensis* MOREIRA, Rev. Ent., São Paulo, **1**: 260, fig. 3 (dorsal view of male). – Terra typica: Brazil (Type male: Inst. biol. de Defesa Agricola, Rio de Janeiro).
1966 *Carcinophora brasiliensis* – POPHAM & BRINDLE, Entomologist, London, **99**: 275.
Literature: REICHARDT 1968; SAKAI 1970; STEINMANN 1977, 1978, 1984.

Description: Male general colour dark brown; first antennal joint reddish, legs yellowish-brown; body shiny. Head tumid, frons and occiput a little depressed; postfrontal and coronal sutures distinct. Eyes comparatively prominent, but small, shorter than the length of head behind eyes. First antennal joint normal, shorter than the distance between antennal bases, but longer than the length of eye. Pronotum quadrate; lateral margins parallel, posterior angles and margin rounded; median longitudinal furrow present. Tegmina and wings well developed. Abdomen a little depressed, expanded to 5th and 6th abdominal tergites. Ultimate tergite broad, median longitudinal sulcus distinct; posterior margin concave. Forceps (Fig. 73) more or less symmetrical, strongly trigonal basally, cylindrical and curved apically. Penultimate sternite normal, posterior margin rounded. Genitalia unknown to me. – **Female** unknown. – **Length** of body with forceps: 12 mm.

Distribution: Brazil.

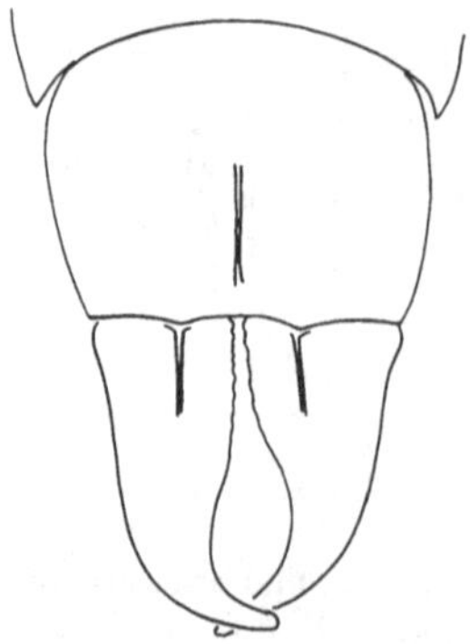

Fig. 73 Male ultimate tergite with forceps of *Carcinophora brasiliensis* (MOREIRA). After MOREIRA.

19. **Carcinophora occidentalis** (Kirby)

1896 *Anisolabis occidentalis* KIRBY, J. Linn. Soc., London, **25**: 525; pl. XX, fig. 5 (female). – Terra typica: Western Australia (Type female: Brit. Mus. nat. Hist., London).
1908 *Gonolabis michaelseni* BURR, Die Fauna Südwest-Australiens 2 (**5**): 71; pl. XI, fig. 5a (female). – Terra typica: Western Australia (Type female: Brit. Mus. nat. Hist., London).
1915 *Eulabis michaelseni* – BURR, J. r. microsc. Soc., London, **1915**: 537; pl. XI, fig. 4 (male genitalia).
1954 *Notolabis occidentalis* – HINCKS, Lunds Univ. Arsskrift, (4) **66**: 5.
1966 *Carcinophora occidentalis* – POPHAM & BRINDLE, Entomologist, London, **99**: 275.
Literature: BORMANS 1900; BURR 1910, 1911; SAKAI 1970; STEINMANN 1977, 1978, 1984.

Description: Female general colour dark reddish-brown; antennae brownish or yellowish; first abdominal segment blackish brown, with a dark red shine. Head tumid, rounded, broad, frons and occiput a little depressed. Postfrontal and coronal sutures distinct. Eyes moderately small, shorter than the length of head behind eyes, and shorter than the length of scapus. Antennae 20-jointed; first joint normal, but a little shorter than the distance between antennal bases; stouter than the flagellum; second joint transverse, third joint twice as long as wide. Pronotum quadrate, lateral margins parallel; posterior margin rounded; median longitudinal furrow present. Tegmina and wings absent in holotype. Mesonotum and metanotum transverse. Abdomen finely punctate, expanded to 6th and 7th abdominal tergites. Ultimate tergite broad, with two dorsal protuberances and two ear-shaped keels. Forceps (Fig. 74) of *Carcinophora*-type; inner margins with basal crenulation, trigonal basally, curved apically. – **Male** similar to female, but forceps not of *Carcinophora*-type, inner margin with one basal large tooth, strongly trigonal basally, and curved apically. Genitalia (Fig. 75 and 76) well developed; paramere V-shaped; anterior margin with deep

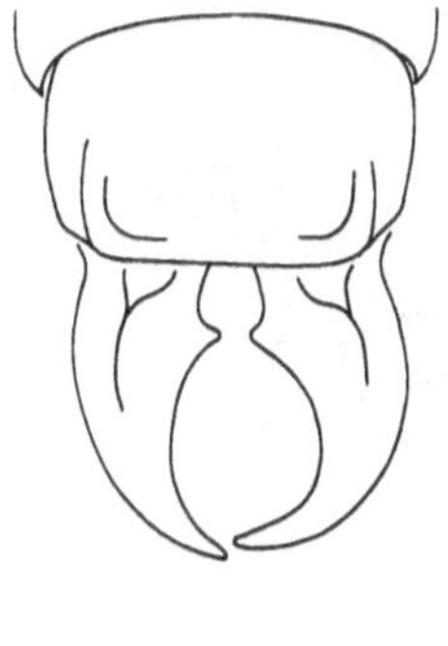

Fig. 74 Male ultimate tergite with forceps of *Carcinophora occidentalis* (KIRBY). Original.

Fig. 75 Male genitalia of *Carcinophora occidentalis* (KIRBY). After BURR.

Fig. 76 Male genital armature of *Carcinophora occidentalis* (KIRBY). After STEINMANN.

median incision between external parameres. Genital lobes relatively short, with slender virgae. External parameres typical, of *Carcinophora*-type, acute apically. – **Length** of body with forceps: female: 23–25 mm, male: 22–24 mm.

Distribution: Western Australia.

20. Carcinophora venusta sp. nov.

Terra typica: Australia, Melbourne, Sheerbrook Forest, 24. VIII, 1968, legit: Dr. J. Balogh, gen. prep. No. 174, det. Dr. H. Steinmann, and three paratypes, ditto, two females, and one immature (Type male: Természettudományi Múz., Budapest).

Description: Male colour reddish-brown; head red, shiny, antennae and legs brown, abdomen and forceps blackish-red. Head rounded; postfrontal and coronal sutures faintly visible; eyes comparatively small, shorter than first antennal joint, or genae. Antennae 19-jointed; first joint long and narrow, but shorter than the distance between antennal bases, second quadrate, third joint twice as long as wide. Pronotum quadrate, all margins straight, lateral margins widened posteriorly; all angles rounded; median longitudinal furrow distinct. Tegmina and wings entirely absent. Meso- and metanotum normal. Abdomen punctate, ultimate tergite striato-punctate longitudinally. Abdominal tergites expanded to middle. Ultimate tergite with lateral keels posteriorly which are smaller than *occidentalis* (Kirby). Forceps (Fig. 77) asymmetrical; inner margin with specific tooth, apex of tooth similar to *occidentalis* (Kirby), but with two small protuberances; branches trigonal basally, strongly curved apically. Penultimate sternite normal, posterior margin rounded. Genitalia (Fig. 78) characteristic; paramere very narrow, and long; median incision of anterior margin very deep; genital lobe narrow, without virgae. External parameres of *Carcinophora*-type, but apex pointed and tapering. – **Female** similar to male, but forceps slenderer, symmetrical, and with small inner tooth basally. – **Length** of body with forceps: male: 15,5–16 mm, female: 14,5–16 mm.

Distribution: Australia: Victoria.

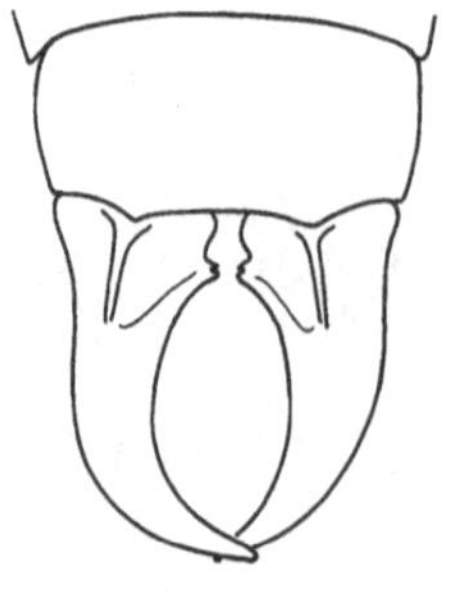

Fig. 77 Ultimate tergite and forceps of holotype of *Carcinophora venusta* sp.n. Original.

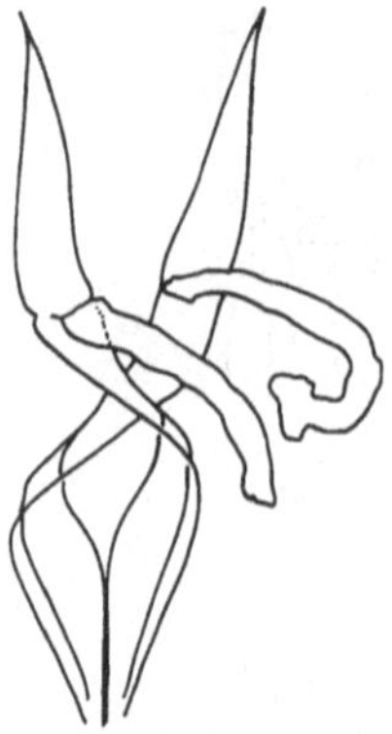

Fig. 78 Genital armature of holotype of *Carcinophora venusta* sp.n. Original.

21. Carcinophora kawakamii (SHIRAKI)

1928 *Psalis kawakamii* SHIRAKI, Insecta Matsumurana, Sapporo, (1) **3**: 5, fig. 2 (pronotum with tegmina and ultimate tergite and forceps of male). – Terra typica: Formosa (Type male: unknown locality).
1970 *Carcinophora kawakamii* – SAKAI, Dermapt. Cat. Prael., Tokyo, **2**: 77.

Description: Male general colour reddish-brown, head with yellowish mouthparts; end of abdomen and forceps blackish or blackish-brown. Head normal, elongate; postfrontal and coronal sutures present; eyes normal, about as long as first antennal joint. Antennae 17-jointed, with light yellowish apical joints; first joint long, but a little shorter than the distance between antennal bases; second joint quadrate, third twice as long as broad. Pronotum quadrate, all angles rounded; lateral margins parallel, posterior margin rounded; median longitudinal furrow finely visible. Tegmina well developed, longer than pronotum; wings very small. Abdomen normal, a little expanded to 6th tergite. Ultimate tergite broad, posterior margin concave. Forceps (Fig. 79) of *Carcinophora*-type, asymmetrical, trigonal basally, curved and cylindrical apically. Penultimate sternite with rounded posterior margin. Genitalia unknown to me. – **Female** unknown. – **Length** of body with forceps: 10 mm.
Distribution: Taiwan.

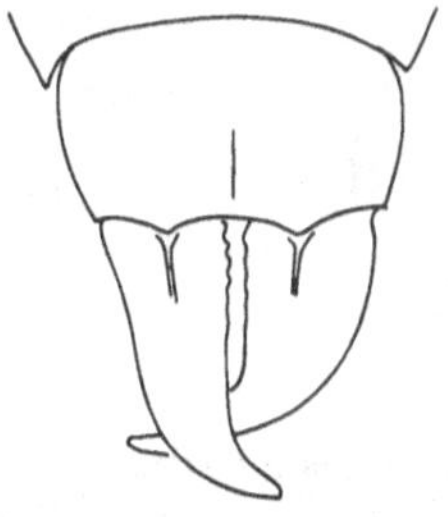

Fig. 79 Male ultimate tergite and forceps of *Carcinophora kawakamii* (SHIRAKI). After SHIRAKI.

22. Carcinophora castetsi BORMANS

1897 *Carcinophora castetsi* BORMANS apud BOLIVAR, Ann. Soc. ent. France, Paris, **66**: 284. – Terra typica: India (Type male: unknown locality).
1910 *Psalis castetsi* – BURR, Fauna Brit. India, Dermaptera, London, Bombay: 77. Madras.
1911 *Euborellia astruci* BURR, J. asiat. Soc. Bengal, Calcutta, **7**: 779. – Terra typica: South India (Type male: Brit. Mus. nat. Hist., London).
1923 *Paralabis castetsi* – HEBARD, Mem. Dept. Agric. India ent. Ser., Pusa, **7**: 204; pl. 19, fig. 4 (male).
1929 *Apolabis mnemrosyne* MENOZZI, Mem. Soc. ent. ital., Genova, **8**: 8, fig. 1 (male). – Terra typica: India (Type male: unknown locality).
1931 *Apolabis castetsi* – BORELLI, Rev. suisse Zool., Genève, (2) **38**: 295; fig. 2 (male genitalia).
1966 *Carcinophora castetsi* – POPHAM & BRINDLE, Entomologist, London, **99**: 275.
Literature: BORMANS 1900; BURR 1910, 1915; KIRBY 1904; SAKAI 1970; STEINMANN 1977, 1978.

Description: Male body shiny, dark reddish-brown; posterior margin of pronotum, legs, and antennae light brown. Head tumid, occiput a little depressed; postfrontal and coronal sutures distinct. Eyes small, shorter than the first antennal joint. First antennal joint long, but shorter than the distance between antennal bases; second joint transverse, third short, but longer than wide. Pronotum quadrate; lateral margins parallel, posterior angles and margin rounded. Tegmina very short, truncate posteriorly, excavate at axillary angle so as to expose a minute scutellum. Abdomen normal, expanded to middle section; ultimate tergite broad; forceps (Fig. 80) asymmetrical, depressed basally, trigonal at base, and cylindrical apically. Penultimate sternite rounded posteriorly. Genitalia (Fig. 81, gen. prep. No. 354, det. Dr. H. STEINMANN) characteristic; of *Carcinophora*-type; paramere well developed, broad, genital lobes without virgae, external parameres specific, broadest basally, and strongly pointed apically. – **Female** similar to male, but forceps (Fig. 82) symmetrical, trigonal basally, broad and a little depressed apically. – **Length** of body with forceps: male: 11–13 mm, female: 13 mm.

Distribution: India: Madras.

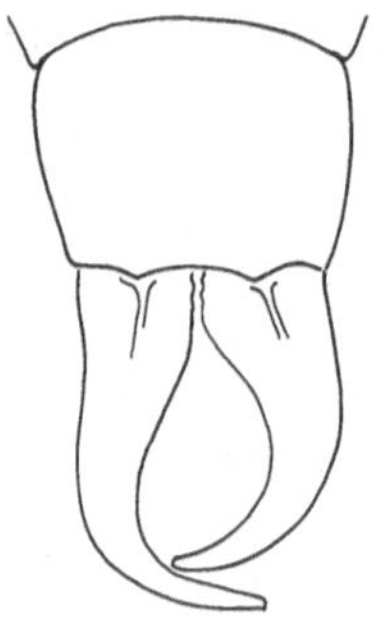

Fig. 80 Male ultimate tergite and forceps of *Carcinophora castetsi* BORMANS. After STEINMANN.

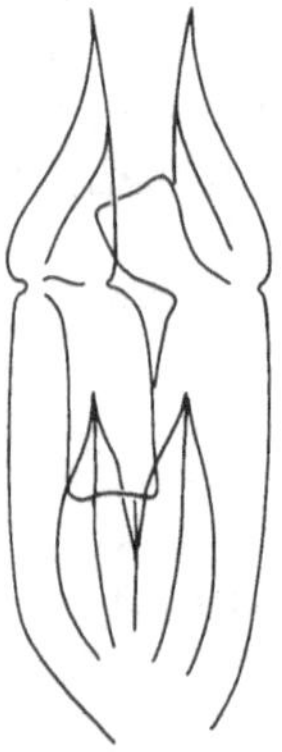

Fig. 81 Male genital armature of *Carcinophora castetsi* BORMANS. Original.

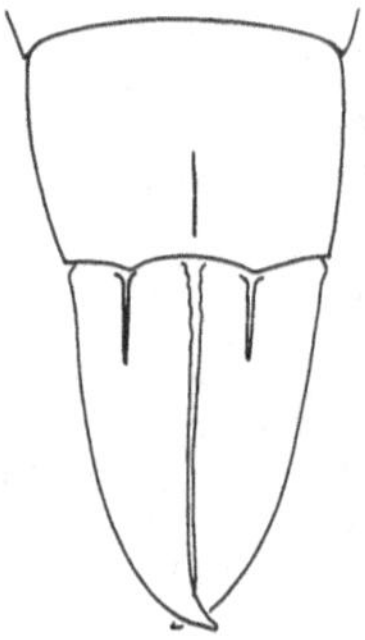

Fig. 82 Female ultimate tergite and forceps of *Carcinophora castetsi* BORMANS. Original.

Genus 13: **Anisolabis** FIEBER

1853 *Anisolabis* FIEBER, Lotos, Praha, **3**: 257. – Species typica: *Forficula maritima* BONELLI in GÉNÉ 1832.

1862 *Forcinella* DOHRN, Stettiner ent. Ztg., **23**: 226. – Species typica: *Forficula maritima* BONELLI in GÉNÉ 1832 (original designation). *Anisolabis* has the same type species.

1879 *Labidurodes* DUBRONY, Ann. Mus. civ. Stor. nat., Genova, **14**: 355. – Species typica: *Labidurodes robustus* DUBRONY 1879 (a pseudonym of BORMANS).

1911 *Logicolabis* ZACHER, Zool. Jahrb., Jena, **30**: 375. – Species typica: *Anisolabis vosseleri* BURR 1907.

1911 *Horridolabis* ZACHER, Zool. Jahrb., Jena, **30**: 384. – Species typica: *Horridolabis paradoxura* ZACHER 1911.

1911 *Homoeolabis* BORELLI, Boll. Mus. Zool. Anat. comp. Univ., Torino, **26**: 1. – Species typica: *Homoeolabis maindroni* BORELLI 1911.

1911 *Gelotolabis* ZACHER, Zool. Jahrb., Jena, **30**: 385. – Species typica: *Gelotolabis burri* ZACHER, 1911.

1914 *Spondox* Burr, Nova Caledonia, Zool., Wellington, **1**: 317. – Species typica: *Spondox sarasini* Burr 1914.
1915 *Apolabis* Burr, J. r. microsc. Soc., London, **1915**: 538. – Species typica: *Forcinella hottentotta* Dohrn 1864.
1915 *Paralabis* Burr, J. r. microsc. Soc., London, **1915**: 540. – Species typica: *Anisolabis oweni* Burr 1915.
1959 *Plancolabis* Bey-Bienko, Ent. Obozr., Moskva, **38**: 608. – Species tyica: *Plancolabis mira* Bey-Bienko 1959.
1966 *Parapsalis* – Popham & Brindle, Entomologist, London, **99**: 272 [= *Anisolabis* Fieber 1853].
1978 *Anisolabella* – Brindle, Dermaptera of Africa, Tervuren, **2**: 71 [= *Anisolabis* Fieber 1853].

Literature: Borelli 1904, 1907, 1932; Bormans 1900; Brindle 1978; Brues, Melander & Carpenter 1954; Burr 1897, 1908, 1910, 1911, 1915; Caudell 1913; Grassi 1907; Handlirsch 1925; Kirby 1904; Popham & Brindle 1966; Rehn 1924; Sakai 1970; Sharp 1901; Steinmann 1975, 1977, 1979; Tillyard 1926; Townes 1945; Verhoeff 1902, 1904; Zacher 1911.

Description: Body convex, rather elongate, generally stout. Small to large or very large insects, typically dark coloured, smooth, any pubescence short, sparse, and inconspicuous. Head large, broad or longer than broad, more or less convex dorsally, although the vertex may be somewhat depressed; postfrontal and coronal sutures are usually distinct but may be weak or not visible. Antennae with from 15 to 25 joints. First antennal joint or scapus comparatively long, second joint or pedicellus transverse or quadrate. Eyes generally prominent, vary in size, small (very much smaller than the length of the head behind eyes), fairly large (the length of the eyes about half the length of the head behind the eyes or somewhat larger) or large (nearly as long as the length of the head behind the eyes or longer). Pronotum longer than wide, rectangular or shorter than wide. Tegmina and wings either entirely absent, represented by lateral flaps, or well-developed. Legs rather short, wide. Sternal plates of thorax rather narrow, longer than wide; metasternum somewhat produced between posterior coxae. Mesosternum rounded posteriorly. Abdomen a little depressed, rather broad, generally expanded to 5th and 6th abdominal tergites. Ultimate tergite of male ample, subrectangular, but wider than long, less wide than the preceding segments; in the female similar to male, but somewhat narrower; median longitudinal sulcus present or absent. Penultimate sternite of male variously shaped and simple in females. Branches of forceps in the male of two types: contiguous or subcontiguous, stout, nearly straight, tapering, or remote, straigt at first and then strongly and abruptly bent inwards, the right branch often more strongly than the left. Female forceps contiguous, straight, stout, tapering. Male genitalia are relatively simple, both penes or genital lobes united at the base only, the virga often inconspicuous, and the genital lobes sometimes have small sclerites or denticulations, but these are usually weakly developed. Generally the left genital lobe directed distally, and the right lobe directed basally. Genital lobes with or without virgae. Median incision of anterior margin between external parameres well-developed, median fissure present or absent. The external parameres of the central parameral plate vary in shape and are the final check for the determination of a species. External parameres various in size, about 3–4 times longer than wide.

Distribution: nearly whole the world.

82 species.

Identification key to the species

1 External parameres of male genitalia long and narrow, with a characteristic narrowing before actual apex at outer margin (Fig. 85). Abdominal tergites mainly smooth, but with isolated and very sparse shallow large punctate; male forceps strongly curved and asymmetrical (Fig. 83). Cosmopolitan .... 1. **Anisolabis maritima**, p. 69
1′ Not *maritima*, not cosmopolitan **2**
2 African species (partially after BRINDLE) **3**
2′ Not distributed in the Ethiopian Region **45**
3 Penultimate sternite with a ventral process or crest (Figs. 87, 90, 97) **4**
3′ Penultimate sternite without process at posterior margin **8**
4 Process of penultimate sternite does not reach posterior margin (Fig. 87). Abdomen strongly widened medially, somewhat of *Gonolabis*-type; abdominal tergites very finely punctate, sides of tergites 6–9 rugoso-striate and with lateral longitudinal ridges. External parameres of male genitalia rounded at outer margins (Fig. 88) .... 2. **Anisolabis cristata**, p. 70
4′ Process of penultimate sternite reaching to, or extending beyond, posterior margin **5**
5 Process of posterior margin of penultimate sternite broad, at least in part, and extending slightly beyond posterior margin of sternite **6**
5′ Process of posterior margin of penultimate sternite narrower, and extending well beyond posterior margin of sternite **7**
6 Process spatulate (Figs 90, 91); abdomen more strongly widened, tergites almost smooth; sides of posterior abdominal tergites impunctate. External parameres of male genitalia (Fig. 92) long, acuminate distally, somewhat of *Carcinophora*-type, but inner membranes ending well beyond apices .... 3. **Anisolabis spatula**, p. 72
6′ Process semi-circular in section, tunnel-like (Fig. 94); abdomen weakly widened, tergites finely punctate, sides of posterior abdominal tergites strongly punctate and somewhat rugoso-striate. External parameres of male genitalia long, narrowed distally (Fig. 95), apices rounded, inner membranes extending nearly to apices, median membranes less widened .... 4. **Anisolabis cunicula**, p. 73
7 Process porrect, produced ventrally, ventral margin more or less straight, apex not sharply pointed (Fig. 97). External parameres short, broad, almost parallel for most part, outer margins sinuate, apices blunt (Fig. 98) .... 5. **Anisolabis porrectella**, p. 74
7′ Process not porrect, curving dorsally, ventral margin sinuate or not, apex sharply pointed. Posterior margin of penultimate sternite curved, process wider and sinuate on dorsal margin (Fig. 100). External parameres (Fig. 101) longer than broad, narrowed distally, apices bluntly rounded .... 6. **Anisolabis rostrata**, p. 75
8 Penultimate sternite triangular, pointed at apex or with apex slightly rounded **9**
8′ Penultimate sternite not so shaped **12**
9 Smaller species, body length 9–12 mm. Pronotum smooth; femora darkened on apical half. Abdominal punctation regular, tergites 9–10 almost impunctate. Penultimate sternite pointed apically (Fig. 103). External parameres of male genitalia more narrowed distally (Fig. 104) .... 7. **Anisolabis acutiventris**, p. 76
9′ Larger species, body length 12–16 mm. Legs yellow, femora not darkened on apical half **10**
10 Abdominal punctation irregular, punctures of varying sizes and varying distances apart. Pronotum rugose, meso- and metanota rather rugose anteriorly and rugoso-striate posteriorly. Penultimate sternite very similar to *acutiventris*, as in fig. 106. External parameres longer, weakly narrowed distally (Fig. 107) .... 8. **Anisolabis oweni**, p. 77
10′ Abdominal punctation regular, punctures similar in size and a similar distance apart **11**

11 Pronotum less transverse and less widened posteriorly, cuticle almost smooth;, meso- and metanota partially punctate but with areas of smooth cuticle. Apex of penultimate sternite rounded (Fig. 109). External parameres of male genitalia (Fig. 110) relatively broad .... 9. **Anisolabis rugosa**, p. 78

11′ Pronotum more transverse and more widened posteriorly, cuticle rugose; meso- and metanota weakly rugose and punctate laterally and anteriorly. Male unknown .... 10. **Anisolabis pluto**, p. 80

12 Penultimate sternite excised at apex (Figs. 113, 117) **13**

12′ Penultimate sternite not excised at apex **14**

13 General colour blackish; femora yellow; abdominal punctation irregular. Excision of penultimate sternite right-angled or nearly so (Fig. 113). External parameres of male genitalia shorter, apices well rounded (Fig. 114). .... 11. **Anisolabis incisa**, p. 80

13′ General colour reddish-brown; femora darkened medially; abdominal punctation regular. Excision of penultimate sternite obtuse-angled, i.e. more than 90°. External parameres of male genitalia longer and apices as in fig. 118 .... 12. **Anisolabis incisiodes**, p. 82

14 Penultimate sternite with posterior margin concave **15**

14′ Penultimate sternite with posterior margin truncate or rounded **18**

15 Sides of posterior abdominal tergites without lateral longitudinal ridges **16**

15′ Sides of posterior abdominal tergites with lateral longitudinal ridges **17**

16 Sides of posterior abdominal tergites punctate or only slightly rugoso-striate. Eyes small; head reddish, legs yellow. Dorso-lateral ridge on last tergite prominent. External parameres of male genitalia broad medially (Fig. 120), narrowed distally, apices rounded .... 13. **Anisolabis quadricollis**, p. 83

16′ Sides of posterior abdominal tergites strongly rugoso-striate. Abdomen strongly widened medially, of *Gonolabis*-type; sides of abdominal tergites 6–9 rugoso-striate. Male forceps (Fig. 121) strongly curved and asymmetrical. External parameres of male genitalia (Fig. 122) rounded or with a small outer angle .... 14. **Anisolabis felix**, p. 84

17 Pronotum more or less quadrate; abdomen regularly punctate; sides of abdominal tergites punctate, not rugoso-striate, with a well defined lateral longitudinal ridge on tergites 6–8, and a weak ridge on tergite 9. External parameres of male genitalia (Fig. 125) oval, apices bluntly rounded .... 15. **Anisolabis bintumanensis**, p. 87

17′ Pronotum strongly transverse; abdomen irregularly punctate; sides of abdominal tergites 7–9 rugoso-striate, 7–8 with a lateral longitudinal ridge, 9 with an incipient ridge. External parameres of genitalia of *Carcinophora*-type, narrowed distally, apices long and pointed (Fig. 127) .... 16. **Anisolabis hottentotta**, p. 88

18 Penultimate sternite with apex truncate **19**

18′ Penultimate sternite with apex rounded **32**

19 Sides of posterior abdominal tergites without lateral longitudinal ridges **20**

19′ Sides of posterior abdominal tergites with lateral longitudinal ridges **28**

20 Larger species, body length 24–33 mm **21**

20′ Smaller species, body length 20 mm or less **22**

21 Blackish, with thoracic sclerites wholly or partially yellowish or reddish-brown, head sometimes partly reddish. Male forceps (Fig. 218) strongly curved and asymmetrical, laminated basally at inner margins. External parameres of genitalia short and broad .... 17. **Anisolabis rufescens**, p. 89

21′ Entirely black or blackish, head and thoracic sclerites not paler in colour; shiny. Pronotum quadrate. Male forceps (Fig. 131) with more or less straight branches. Male genitalia unknown. .... 18. **Anisolabis atra**, p. 90

22 Head and thoracic sclerites strongly punctate **23**

22′ Head and thoracic sclerites impunctate or almost so **24**

23 Less strongly punctate, punctation on thoracic sclerites close but rarely confluent. Male forceps (Fig. 132) strongly asymmetrical. External parameres of male genitalia angular on external margin (Fig. 133); genital lobes comparatively small ... 19. **Anisolabis infelix**, p. 91

23′ More strongly punctate, punctation on thoracic sclerites deep and close, that on pronotum prominently rugoso-striate medially and punctures on other tergites often confluent. Male forceps (Fig. 134) more or less symmetrical. External parameres of male genitalia rounded on external margin (Fig. 136) .... 20. **Anisolabis straeleni,** p. 92

24 Eyes large; abdomen finely punctate. Penultimate sternite triangular, with apex truncate (Fig. 138). External paramere of genitalia long, both inner and median membranes large (Fig. 139) .... 21. **Anisolabis pagana**, p. 93

24′ Eyes small **25**

25 Sides of posterior abdominal tergites strongly and regularly punctate, not rugoso-striate; dorso-lateral ridge on ultimate tergite prominent. External parameres of male genitalia (Fig. 120) broad medially, narrowed distally, apices rounded .... 13. **Anisolabis quadricollis**, p. 83

25′ Sides of posterior abdominal tergites rugoso-striate; dorso-lateral ridge on ultimate tergite small or absent **26**

26 Sides of abdominal tergites 7–9 weakly rugoso-striate. Male forceps with well-separated branches and almost symmetrical (Fig. 140). Apex of penultimate sternite wide. External parameres of male genitalia broad basally (Fig. 141), narrowed distally, apices rounded... .... 22. **Anisolabis duplicata**, p. 94

26′ Sides of abdominal tergites 6–9 rugoso-striate; branches of forceps not widely separated and less symmetrical. Apex of penultimate sternite narrower **27**

27 Blackish species, head partly reddish; legs reddish, sometimes with apices of femora darkened. Sides of tergites 6–9 more strongly rugoso-striate. Pygidium flat and forceps (Fig. 142) relatively longer, asymmetrical, not of *Anisolabis*-type. External parameres of genitalia long, inner and outer margins parallel, apices narrowed (Fig. 143) .... 23. **Anisolabis nimbaensis**, p. 96

27′ Blackish species, including head; legs yellow, femora banded with black. Sides of abdominal tergites 6–7, only slightly rugoso-striate. Pygidium protruding and forceps relatively short (Fig. 144). External parameres of male genitalia comparatively short and broad (Fig. 145). .... 24. **Anisolabis punctulata**, p. 97

28 Lateral longitudinal ridges on abdominal tergites 6–9, ridges blunt and rugoso-striate. Body dark brown to blackish, head reddish. Male forceps (Fig. 146) with slender branches, not of *Anisolabis*-type, not strongly curved. External parameres of genitalia long, inner and outer margins parallel (Fig. 147) .... 25. **Anisolabis jeanneli**, p. 98

28′ Lateral longitudinal ridges on abdominal tergites 6–7, 6–8, 6–9, or 7–9. Forceps less slender **29**

29 Lateral longitudinal ridges on posterior abdominal tergites blunt and rugoso-striate **30**

29′ Lateral longitudinal ridges on posterior abdominal tergites narrow, well defined, and more or less smooth **31**

30 Darker in colour, antennae yellow. Pronotum more strongly widened posteriorly. Sides of abdominal tergites 5–9 rugoso-striate, 6–8 with blunt lateral longitudinal ridge. Male forceps (Fig. 148) asymmetrical, strongly trigonal basally. External parameres of genitalia almost rectangular in shape, apices pointed (Fig. 149) .... 26. **Anisolabis burri**, p. 99

30′ Paler in colour, antennae dark brown. Pronotum less widened posteriorly. Sides of abdominal tergites 6–9 rugoso-striate, 6–8 with blunt lateral longitudinal ridge. Male forceps as in fig. 151. External parameres of genitalia long and narrow (Fig. 152) .... 27. **Anisolabis scotti**, p. 100

31 Lateral longitudinal ridges on abdominal tergites 6–7. Eyes small. Body blackish, head reddish or partly reddish. Male forceps not strongly curved, relatively longer (Fig. 153). External parameres of genitalia widened medially, narrow basally, apices pointed (Fig. 154) .... 28. **Anisolabis laeta**, p. 101

31′ Lateral longitudinal ridges on abdominal tergites 7–9. Eyes moderately large. Body blackish-brown, abdomen very dark red. Male forceps more curved (Fig. 156), relatively short. External parameres of genitalia broad at base, narrowed distally, apices rounded (Fig. 157) .... 29. **Anisolabis tumida**, p. 102

32 Ultimate tergite with a dorso-lateral ridge (Figs. 158, 160) **33**

32′ Ultimate tergite without a dorso-lateral ridge **34**

33 Pronotum transverse. Abdominal punctation stronger; abdominal tergites 6–9 rugoso-striate at sides, and with lateral longitudinal ridges. Penultimate sternite broadly rounded but

somewhat sinuate at sides. External parameres of genitalia (Fig. 159) pointed apically .... 30. **Anisolabis royi**, p. 103

33′ Pronotum less transverse. Abdominal punctation weaker; abdominal tergites 5–9 rugoso-striate at sides, and with lateral longitudinal ridges. Penultimate sternite broadly rounded, not sinuate at sides. External parameres of genitalia (Fig. 161) rounded apically .... 31. **Anisolabis turgida**, p. 104

34 Sides of posterior abdominal tergites without lateral longitudinal ridges **35**

34′ Sides of posterior abdominal tergites with lateral longitudinal ridges **41**

35 Sides of posterior abdominal tergites punctate, not rugoso-striate **36**

35′ Sides of posterior abdominal tergites rugoso-striate **38**

36 Pronotum strongly transverse. Abdominal tergites almost impunctate and shiny. Eyes comparatively large. Body dark reddish-brown to black, head reddish or partly reddish. Branches of male forceps (Fig. 162) unusually thick. External parameres of male genitalia long and narrow, apex of each with a triangular process (Fig. 163) 32. **Anisolabis elongata**, p. 105

36′ Pronotum less strongly transverse. Abdominal tergites obviously punctate, punctures relatively large but shallow **37**

37 Body length with forceps less than 18 mm. Ultimate tergite and forceps shorter (Fig. 164), broad. Male genitalia with broader external parameres, genital lobes as in fig. 165........ .... 33. **Anisolabis isomorpha**, p. 106

37′ Body length with forceps more than 18 mm. Ultimate tergite and forceps slenderer (Fig. 166). Male genitalia elongate, with very long external parameres, genital lobes as in fig. 167 .... 34. **Anisolabis brindlei**, p. 107

38 Sides of posterior abdominal tergites 6–9 rugoso-striate, tergite 6 weakly sp. Ultimate tergite without lateral longitudinal ridges. Abdomen strongly punctate. Body blackish in colour, legs yellow, femora vaguely darkened. Male forceps similar to *gestri* BORELLI, as in fig. 168. External parameres of male genitalia long, rather narrow, apices obtuse (Fig. 169) .... 35. **Anisolabis neavei**, p. 108

38′ Sides of posterior abdominal tergites 7–9 rugoso-striate **39**

39 Shiny black species with head reddish or mainly reddish; abdominal punctation weak. Male forceps short, very similar to *neavei* BRINDLE, as in fig. 168 ...... 36. **Anisolabis gestri**, p. 109

39′ Blackish species, head not reddish or partly reddish. Male genitalia with very deep median incision between external parameres **40**

40 Outer and inner margins of external parameres of male genitalia parallel at medially (Fig. 171). Pronotum transverse, weakly widened posteriorly. Penultimate sternite long, triangular, posterior margin convex .... 37. **Anisolabis vosseleri**, p. 110

40′ Outer and inner margins of external parameres of male genitalia arched medially (Fig. 172). Pronotum as long as wide. Penultimate sternite with rounded posterior margin .... 38. **Anisolabis caesarea**, p. 111

41 Eyes comparatively large. Lateral longitudinal ridges on abdominal tergites 6–9; body dark reddish-brown, head paler in colour. Abdominal tergites strongly punctate at sides but only slightly rugoso-striate on tergites 7–8. Male forceps as in fig. 173. External parameres of genitalia long (or short by BRINDLE 1978; Figs. 174, 175) ..... 39. **Anisolabis quarens**, p. 112

41′ Eyes comparatively small **42**

42 Sides of abdominal tergites 8–9 almost rugoso-striate; lateral longitudinal ridges on tergites 8–9; abdominal punctation fairly strong. Body blackish in colour. Male ultimate tergite and forceps as in fig. 176. External parameres of male genitalia broad basally, acuminate distally (Fig. 177) .... 40. **Anisolabis lamottei**, p. 113

42′ Sides of abdominal tergites 5–9 or 6–9 rugoso-striate **43**

43 Lateral longitudinal ridges on abdominal tergites 6–7. Black species. Pronotum strongly transverse and slightly widened posteriorly. Abdominal punctation fine basally, more strongly punctate distally. Forceps (Fig. 178) rather slender, inner margins with tubercle-like projections medially, branches strongly curved and asymmetrical. External parameres of male genitalia (Fig. 179) long and narrow .... 41. **Anisolabis rougemonti**, p. 114

43′ Lateral longitudinal ridges on abdominal tergites 5–6 .... **44**

44 Ultimate tergite of male very transverse; forceps (Fig. 180) with rather slender branches,

strongly curved and asymmetrical, inner margins with small spines medially. External parameres of male genitalia similar to *simiensis* BRINDLE, as in fig. 181 .... 42. **Anisolabis kristenseni**, p. 115

44′ Ultimate tergite of male more or less quadrate; forceps (Fig. 182) with rather slender branches, strongly curved and asymmetrical, inner margins with some small projections medially. External parameres of male similar to *kristenseni* BURR, as in fig. 183 .... 43. **Anisolabis simiensis**, p. 116

45 Malgassian species. Pronotum quadrate, tegmina distinct, ample, truncate apically; wings visible. Female forceps slender, tapering, contiguous (Fig. 184). Male unknown .... 44. **Anisolabis nitida**, p. 117

45′ Not Malgassian species **46**

46 Indo-Australian species **47**

46′ Not Indo-Australian species **66**

47 Australian group **48**

47′ New Guinean, Melanesian, Micronesian or Polynesian groups **52**

48 Pronotum transverse. Lateral longitudinal ridges on abdominal tergites 3–9. Meso- and metanota, and abdominal tergites, very strongly punctate dorsally. Male forceps relatively well developed, straight basally, and strongly curved at apical portion (Fig. 185). Male genitalia (Fig. 186) characteristic, genital lobes with very long virgae; genital muscle loop visible within central parameral plate .... 45. **Anisolabis australis**, p. 117

48′ Pronotum quadrate or longer than wide **49**

49 Pronotum longer than wide. First antennal joint as long as the distance between antennal bases. Tegmina present, represented by lateral flaps. Forceps broad basally, trigonal (Fig. 187). Male genitalia specific, as in fig. 188 .... 46. **Anisolabis nigrofusca**, p. 118

49′ Pronotum quadrate. Tegmina entirely absent **50**

50 Head bicoloured, yellow with a large brown spot at vertex; postfrontal sutures and coronal suture very distinct, deep. Inner margins of female forceps with a small, but well-developed tooth, basally (Fig. 189) .... 47. **Anisolabis flavocapitata**, p. 119

50′ Head unicoloured, black, reddish-brown, or brown. Postfrontal and coronal sutures present but not deep, or absent **51**

51 First antennal joint or scape long, longer than the distance between antennal bases. Pro-, and mesonota bicoloured, anterior part yellow, the posterior one brown. Male forceps (Fig. 190) asymmetrical, trigonal basally, and curved apically, similar to *westralica* BURR. Male genitalia (Fig. 191) with very long genital lobes, virgae also very long .. 48. **Anisolabis littorea**, p. 120

51′ First antennal joint or scape short, shorter than the distance between antennal bases. Pro- and mesonota unicoloured, brown, brownish-red or black. Male forceps (Fig. 192) similar to *littorea* WHITE, as in fig. 192. Male genitalia (Fig. 193) with short genital lobes, and genital lobes without virgae .... 49. **Anisolabis westralica**, p. 121

52 New Guinean group **53**

52′ Melanesian, Micronesian or Polynesian groups **59**

53 Tegmina and wings present, well developed **54**

53′ Tegmina and wings entirely absent. Genital lobes of male central paramere without virgae (Figs. 198, 200, 202, 204, 206) **55**

54 Antennae multisegmented, more than 20 joints. Male forceps (Fig. 194) characteristic, not of *Anisolabis*-type; asymmetrical, with 2–4 teeth at inner margins. Male genitalia large, genital lobes with virgae (Fig. 195) .... 50. **Anisolabis robusta**, p. 122

54′ Antennae less than 20 joints. Male forceps well developed, inner margins crenulate, as in fig. 196. Male genitalia unknown to me .... 51. **Anisolabis egregius**, p. 123

55 Larger species, length of body with forceps more than 15 mm. Male forceps (Fig. 197) asymmetrical, relatively long, trigonal basally, cylindrical apically. Male genitalia of *Anisolabis*-type (Fig. 198), median incision of anterior margin of central paramere very deep and broad, paramere V-shaped .... 52. **Anisolabis perissa**, p. 124

55′ Smaller species, length of body with forceps less than 15 mm **56**

56 Penultimate sternite with a deep semi-circular emargination at posterior margin. Male forceps asymmetrical, trigonal basally, cylindrical apically, as in fig. 199. Male genitalia large, characteristic; genital lobes long, more or less longer or a little shorter than the median

incision of anterior margin of central parameral plate (Fig. 200) ....
.... 53. **Anisolabis horvathi**, p. 125

56′ Penultimate sternite without semi-circular emargination posteriorly **57**

57 Ultimate tergite with large lateral ridge or keel at dorsal surface (Fig. 201). Male forceps asymmetrical, of *Anisolabis*-type. Male genitalia (Fig. 202) well developed, central parameral plate oval, median incision of anterior margin very deep, longer than the genital lobe ....
.... 54. **Anisolabis modesta**, p. 126

57′ Ultimate tergite without longitudinal lateral keel or ridge dorsally **58**

58 Femora long, its apical end as long as abdominal tergites 3–4. Ultimate tergite broad, forceps relatively long, almost asymmetrical; outer margins straight at middle (Fig. 203). External parameres of male genitalia with longitudinal chitinous spine, apices not broadening (Fig. 204) .... 55. **Anisolabis baloghi**, p. 127

58′ Femora comparatively short, its apical end as long as abdominal tergites 1–2. Ultimate tergite transverse, forceps almost asymmetrical, as in fig. 205. External parameres of male genitalia without longitudinal chitinous spine, apices visibly broadening (Fig. 206) ....
.... 56. **Anisolabis pilosa**, p. 128

59 Melanesian group **60**

59′ Micronesian or Polynesian groups **63**

60 Male forceps with large tooth at inner margin basally (Fig. 207); trigonal basally, cylindrical apically, both branches narrowed apically, apex incurved, not of *Anisolabis*-type. Male genitalia with very long external parameres, central parameral plate broad, genital lobes well developed with slender virgae (Fig. 208) .... 57. **Anisolabis bifida**, p. 129

60′ Male forceps without large tooth at inner margin basally. External parameres of male genitalia shorter than the *bifida* Brindle **61**

61 Male forceps symmetrical, with strongly dorsal ridge as in fig. 209. Male genitalia of *Anisolabis*-type, paramere narrow, median incision of anterior margin very deep; genital lobes with slender virgae (Fig. 210) .... 58. **Anisolabis verhoeffi**, p. 130

61′ Male forceps asymmetrical, right branch shorter than left (Figs. 211, 213) **62**

62 General colour blackish. Forceps asymmetrical, with much more pronounced teeth on inner margins of branches (Fig. 211). Male genitalia characteristic (Fig. 212), parameral plate very broad, median incision of anterior margin deep, but very similar to *sarasini* Burr.......
.... 59. **Anisolabis canaca**, p. 131

62′ General colour reddish-brown to dark brown; antennae palpi and legs reddish-yellow. Forceps asymmetrical, trigonal at base, cylindrical apically; inner margin weakly dentated, branches weakly curved (Fig. 213). Male genitalia specific (Fig. 214), central parameral plate V-shaped, similar to *canaca* Brindle .... 60. **Anisolabis sarasini**, p. 132

63 Micronesian group **64**

63′ Polynesian group **65**

64 Colour dark brown to reddish-brown. Postfrontal and coronal sutures distinct. Head more or less tumid, longer than broad. Pronotum quadrate. Male forceps (Fig. 215) asymmetrical, trigonal basally, cylindrical apically .... 61. **Anisolabis tegminata**, p. 133

64′ Colour blackish, body shiny. Postfrontal and coronal sutures indistinct. Head a little transverse. Pronotum transverse, widened posteriorly. Male forceps asymmetrical, as in fig. 216. Male genitalia (Fig. 217) of *Anisolabis*-type, large, with deep median incision at anterior margin of paramere .... 62. **Anisolabis minutissima**, p. 133

65 Pronotum slightly longer than wide, gently widened posteriorly. Male forceps more or less symmetrical (Fig. 218), inner margins straight, crenulate. Male genitalia not of *Anisolabis*-type, as in fig. 219 .... 63. **Anisolabis pacifica**, p. 134

65′ Pronotum quadrate, lateral margins more or less parallel. Male forceps asymmetrical, inner margins a little lamellate basally with some projections (Fig. 220). Male genitalia unknown to me .... 64. **Anisolabis subarmata**, p. 136

66 Palaearctical, and North-east Asian species **67**

66′ Oriental group **68**

67 Kamtschatka species. Male forceps (Fig. 221) asymmetrical, of *Anisolabis*-type. Original description by Eschscholtz very short .... 65. **Anisolabis pectoralis**, p. 136

67′ Ryukyu species. Male forceps elongated, slender, as in fig. 222. Genitalia characteristic, somewhat of *Gonolabis*-type; genital lobes with hyaline denticulation pads (Fig. 223)... 66. **Anisolabis ryukyuensis**, p. 137
68 Pronotum longer than wide **69**
68′ Pronotum quadrate or wider than long **77**
69 Tegmina present, wings visible or absent **70**
69′ Tegmina entirely absent or present but represented by lateral flaps **74**
70 Penutltimate sternite with posterior margin emarginate. Central parameral plate of male genitalia very broad, widened V-shaped (Fig. 224, after BEY-BIENKO: genitalia of holotype) or narrow (Fig. 225, after STEINMANN, genitalia of paratype) 67. **Anisolabis montshadskii**, p. 138
70′ Penultimate sternite with posterior margin rounded or truncate **71**
71 Ultimate tergite with specific, short, longitudinal sculpture dorsally (Fig. 226). Male forceps broad and short, characteristic; inner margins crenulate. Male genitalia (Fig. 227) with V-shaped central parameral plate, genital lobes armed with a pair of chitinous denticulate pads apically .... 68. **Anisolabis aborensis**, p. 139
71′ Ultimate tergite without short, longitudinal sculpture dorsally, its surface smooth or a little rugoso-striate. Genital lobes of male genitalia armed with minute chitinous teeth **72**
72 Tegmina with posterior margins rounded. Ultimate tergite with characteristic depression dorsally (Fig. 228). Genital lobes of male genitalia with specific chitinous serration, and large chitinous plate (Fig. 229) .... 69. **Anisolabis antennata**, p. 140
72′ Tegmina with posterior margins truncate. Ultimate tergite normal, without semicircle depression at posterior margin. Genital lobes of male genitalia without chitinous serration **73**
73 Forceps relatively long, longer than ultimate tergite; the latter with posterior margin concave (Fig. 230). External parameres relatively short, genital lobe and paramere similar to *dohrni* (KIRBY), as in fig. 231 .... 70. **Anisolabis lefroyi**, p. 141
73′ Forceps relative short, more or less equal in length to ultimate tergite; the latter with straight posterior margin (Fig. 232). External parameres relatively long, genital armature very similar to *lefroyi* (BURR), as in fig. 233 .... 71. **Anisolabis dohrni**, p. 142
74 Tegmina represented by lateral flaps. Postfrontal and coronal sutures indistinct; pronotum longer than wide. Ultimate tergite ample, transverse, forceps (Fig. 234) symmetrical, not of *Anisolabis*-type. Genitalia (Fig. 235) characteristic, without median fissure .... 72. **Anisolabis orientalis**, p. 144
74′ Tegmina and wings entirely absent **75**
75 Posterior margin of 9th abdominal sternite, so-called penultimate sternite, with a characteristic process: posterior margin medially with a longer, parallel-sided, bank-like plate formation pointig backwards (Fig. 236). Genital armature as in fig. 237... 73. **Anisolabis mira**, p. 145
75′ Posterior margin of 9th abdominal sternite normal, without any plate formation **76**
76 Antennae 21-jointed. Postfrontal and coronal sutures indistinct. Pronotum subrectangular, slightly longer than wide, lateral margins slightly broader posteriorly than anteriorly. Forceps subcontiguous, stout. Genitalia unknown .... 74. **Anisolabis dubronyi**, p. 145
76′ Antennae 19-jointed. Postfrontal and coronal sutures distinct. Pronotum a little longer than wide, lateral margins parallel. Forceps asymmetrical, branches robust, very strongly curved apically .... 75. **Anisolabis vitalisi**, p. 146
77 Pronotum transverse, wider than long. Pronotum wider than long, lateral margins straight, concave, all angles rounded. Forceps asymmetrical, of *Anisolabis*-type, trigonal basally, cylindrical apically. Male genitalia unknown .... 76. **Anisolabis penetrans**, p. 146
77′ Pronotum quadrate **78**
78 Tegmina present, narrow at base, exposing a broad, short, transverse area of the mesonotum. Forceps subcontiguous, weakly asymmetrical, inner margins with a faint tooth at about middle or nearer to the base. Genitalia simple, paramere oval, median incision of anterior margin deep and wide, genital lobe without virgae, as in fig. 238 77. **Anisolabis greeni**, p. 147
78′ Tegmina entirely absent **79**
79 General colour brown, antennae light brown, legs yellowish-brown. Head with distinct postfrontal and coronal sutures. Ultimate tergite transverse, forceps asymmetrical, of *Anisolabis*-type, trigonal basally, cylindrical apically .... 78. **Anisolabis recurva**, p. 148

79′ Colour black or blackish-red **80**

80 Head red; antennae 15-jointed, blackish. Body black, shiny. Pronotum quadrate, female forceps (Fig. 239) stout, depressed, subcontiguous, straight, tapering. Male unknown .... 79. **Anisolabis gaudens**, p. 148

80′ Head black **81**

81 Genital lobes of male genital armature with virgae (Figs. 340, 341, 342). Body very similar to *Euborellia stali* (Dohrn). Forceps trigonal basally, curved and cylindrical apically. Lateral margins of pronotum parallel .... 80. **Anisolabis maindroni**, p. 149

81′ Genital lobes of male genital armature without virgae, genitalia immature-like **82**

82 Pronotum nearly as long as wide, widened posteriorly. Forceps very broad basally, and laminate, as in fig. 243. Genitalia very similar to *kudagae* Burr, as in fig. 244 .... 81. **Anisolabis rubella**, p. 150

82′ Pronotum strongly transverse. Forceps typical, as in fig. 245. Male genitalia similar to *rubella* Brindle, as in fig. 246 .... 82. **Anisolabis kudagae**, p. 151

1. Anisolabis maritima (Bonelli)

1832 *Forficula maritima* Bonelli in Géné, Ann. Sci. nat. Regno Lombardo-Veneto, Padova, **2**: 224. – Terra typica: Genua (Type male: location unknown).

1839 *Forficesilla maritima* – Buffon, Ins. Orth., 27.

1842 *Forficula* (*Forficesilla*) *maritima* – de Haan, Verhandl. nederlandse Overz. Bezitt., den Haag; 240.

1853 *Labidura maritima* – Fischer, Orth. Eur., 68; pl. 6, fig. 4 (male).

1853 *Anisolabis maritima* – Fieber, Lotos, Praha, **3**: 257.

1862 *Forcinella maritima* – Dohrn, Stettiner ent. Ztg., **22**: 226.

1863 *Labidura advena* Meinert, Naturhist. Tidskr., København, 13, **5**: 279. – Terra typica: Jamaica (Type male: location unknown). **syn. nov.**

1864 *Brachylabis maritima* – Dohrn, Stettiner ent. Ztg., **25**: 293.

1891 *Anisolabis advena* Kirby, J. Linn. Soc. Zool., London, **23**: 517.

1905 *Labidurodes singularis* Shiraki, Trans. Sapporo nat. Hist. Soc., **1**: 8. – Terra typica: Formosa (Type male: Shiraki's Collect. Mus. Formosa?) – **syn. nov.**

1905 *Anisolabis piceus* Shiraki, Trans. Sapporo nat. Hist. Soc., **1**: 9. – Terra typica: Formosa (Type male: Shiraki's Collect., Mus. Formosa?). – **syn. nov.**

1911 *Anisolabis maritima* – Burr, Genera Insectorum, Bruxelles, **122**: 29; pl. 3, fig. 6 (sternal plates of thorax); Cosmopolitan.

1913 *Anisolabis addita* Burr, Ent. Mitt., Berlin, **2**: 66. – Terra typica: Formosa (Type male: Dtsch. ent. Inst., Eberswalde). – **syn. nov.**

1946 *Anisolabis quadrata* Liu, China Border Res. Soc., Shanghai, (B) **16**: 19, fig. 1 (male). – Terra typica: China (Type male: unknown locality). – **syn. nov.**

1968 *Anisolabis maritima* – Reichardt, Pap. avulsos Zool. São Paulo, **22**: 46. (*Forficula*; type-locality "France", South America, Central America, West Indies and United States).

1975 *Anisolabis maritima* – Steinmann, Acta zool. Acad. Sci. hung., Budapest, **21**: 205, fig. 27 (male genitalia).

Literature: Borelli 1915; Bormans 1894, 1900; Brindle 1978; Burr 1910, 1915; Caudell 1908; Chopard 1922, 1951; Comstock 1933; Finot 1890; Günther 1930; Hebard 1917; Hincks 1947, 1949; Kirby 1904; Matsumura 1935; Menozzi 1927; Morse 1921; Popham & Brindle 1966; Rehn 1906, 1924; Sakai 1970; Sharp 1901; Steinmann 1977, 1978, 1980; Walker 1919.

Description: Male colour black or dark brown. Antennae and mouthparts yellow or yellowish-brown; legs uniformly yellow. Head tumid, postfrontal and coronal sutures visible, distinct. Eyes relatively short, shorter than first antennal joint. Antennae 19-jointed; first joint more or less as long as the distance between antennal bases; second joint quadrate, third joint twice as long as fourth. Thoracal sternal plates with rounded posterior margins. Pronotum a little longer than wide, slightly broadening posteriorly; median longitudinal furrow present. Tegmina and wings absent. Abdomen expanded to middle and rather strongly depressed, tergites smoothly coriaceous, finely punctate at bases which are mainly hidden from view, most of tergites with scattered, very isolated punctures; tergites 6–8 or 6–9 rugoso-striate at sides but without lateral longitudinal ridges. Ultimate tergite sparsely punctate, punctures arranged in longitudinal irregular rows separated by smooth cuticle; median longitudinal sulcus weak. Pygidium flat. Forceps (Fig. 83) asymmetrical, widened at base, margin at base rugose on inner side, left branch straighter than right branch which also has a more sharply pointed process on inner margin. Penultimate sternite broadly triangular, posterior margin truncate. Genitalia (Fig. 84; gen. prep. No. 376, det. DR. H. STEINMANN) very simple, but specific; central parameral plate oval, broad; median incision of anterior margin between external parameres very deep and wide, median fissure present. Genital lobes well developed, virgae slender, elongated, the latter without sclerotized plates. External parameres (Fig. 85), of so-called *Anisolabis*-type: three or four times as long as wide, obtuse apically, with a characteristic narrowing before actual apex. – **Female** similar to male, but forceps symmetrical, and somewhat narrower, subcontiguous, trigonal, unarmed and tapering. Abdominal punctation even weaker, tergites 6–9 not rugoso-striate at sides. Pygidium slightly protruding. – **Length** of body with forceps: male: 15–26,5 mm, female: 18–27 mm.

Distribution: Cosmopolitan.

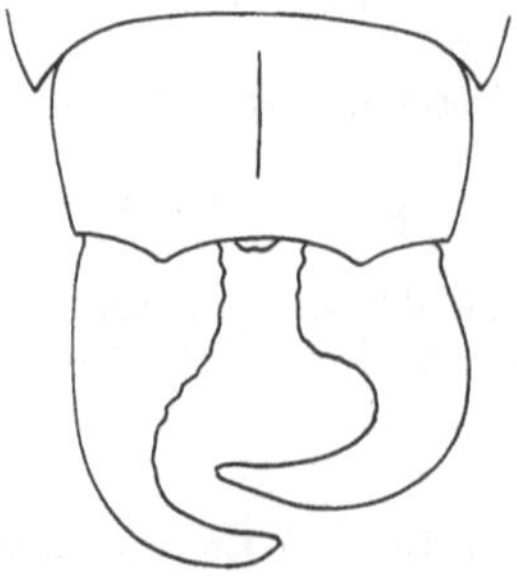

Fig. 83 Male ultimate tergite and forceps of *Anisolabis maritima* (BONELLI). Original.

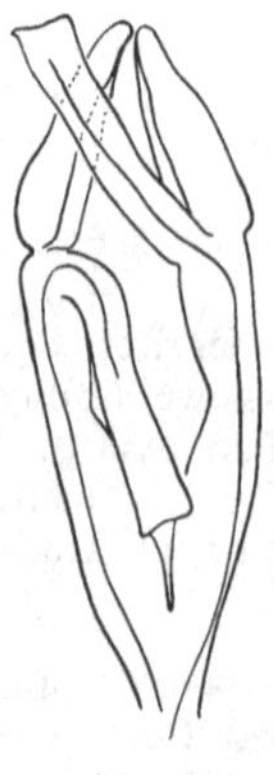

Fig. 84 Male genital armature of *Anisolabis maritima* (BONELLI). Original.

Fig. 85 External paramere of male genitalia of *Anisolabis maritima* (BONELLI). Original.

2. Anisolabis cristata (HINCKS)

1954 *Apolabis cristata* HINCKS, Mém. Inst. fr. Afr. noire, Paris, Dakar, **40**: 104, fig. 2 (male), 3 (male genitalia), 4 (penultimate sternite, dorsal view), and 5 (lateral view). – Terra typica: Guinea (Mt. Nimba) (Type male: Mus. nat. Hist. nat., Paris).

1966 *Carcinophora cristata* – POPHAM & BRINDLE, Entomologist, London, **99**: 275.
1978 *Anisolabis cristata* – BRINDLE, Dermaptera of Africa, Tervuren, **2**: 109; fig. 95 (male), 96 and 97 (penultimate sternite ventrally and laterally), 98 (male genitalia), 241 (external paramere), Guinea.
Literature: SAKAI 1970; STEINMANN 1984.

Description: Male general colour dark brown; legs, mouthparts, proximal antennal joints, and sides of pronotum, brown. Head tumid, postfrontal and coronal sutures distinct; head in dorsal view rounded, eyes comparatively small, but not prominent, much shorter than first antennal joint. Antennae 19-jointed, first joint long, about as long as the distance between antennal bases, second joint very short, transverse, third longer than fourth. Pronotum longer than wide, median longitudinal furrow present, lateral margins straight, but a little expanded to posterior angles. Tegmina and wings absent; mesonotum about as wide as pronotum, with scattered punctures. Abdominal tergites of *Gonolabis*-type, progressively widened to 7th, then narrowed, Ultimate tergite transverse, median longitudinal sulcus distinct, depressed dorsally, laterad with strong, crest-like carina. Pygidium small, triangular. Forceps asymmetrical (Fig. 86), powerful, unarmed, right bent distad in a right angle and crossing left. Penultimate sternite (Fig. 87) specific, strongly rugose, distal margin bisinuate, thus presenting a rounded median lobe. Genitalia (Fig. 88) somewhat V-shaped; paramere moderately narrow, median incision of anterior margin deep, genital lobes well developed, with slender, long virgae. External parameres not of *Anisolabis*-type, somewhat longer than wide, broadest near base, strongly curved on outer side; inner margin nearly straight. – **Female** similar to male in colour and general characters, but punctation less strong, ultimate tergite without lateral carinae, and forceps simple, subtrigonal, tapering, symmetrical. – **Length** of body with forceps: male: 15–20 mm, female: 14–18 mm.

Distribution: Guinea.

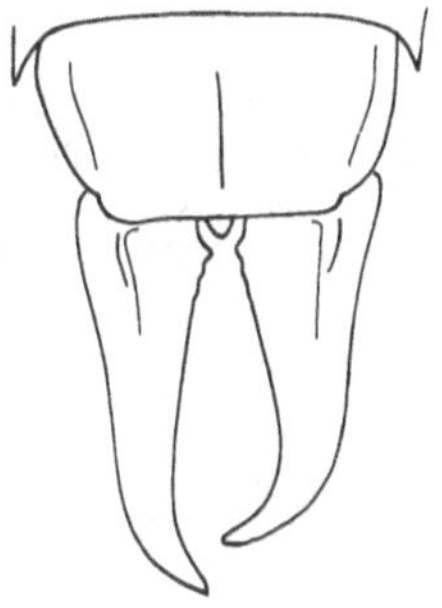

Fig. 86 Male ultimate tergite with forceps of *Anisolabis cristata* (HINCKS). After BRINDLE.

Fig. 87 Posterior margin of penultimate sternite of *Anisolabis cristata* (HINCKS). After HINCKS.

Fig. 88 Male genital armature of holotype of *Anisolabis cristata* (HINCKS). After HINCKS.

3. **Anisolabis spatula** Brindle

1978 *Anisolabis spatula* Brindle, Dermaptera of Africa, Tervuren, **2**: 106, fig. 103 (penultimate sternite in ventral view), 104 (penultimate sternite in lateral view), 105 (male forceps), 240 (external paramere of male genitalia). – Terra typica: Cameroon (Type male: Mus. nation. Hist. nat., Paris).

Description: Male colour blackish, head reddish, antennae brown, basal joints yellow; legs reddish-yellow. Head rather strongly transverse; postfrontal sutures weak, coronal suture fine. Eyes rather large, about as long as the first antennal joint. Antenna normal, first joint long, about as long as the distance between antennal bases. Pronotum transverse, shiny, almost parallel-sided, median longitudinal furrow distinct and depressed anteriorly; cuticle coriaceous, smooth. Tegmina and wings absent. Mesonotum with a few isolated punctures, metanotum coriaceous. Abdomen gradually widened to tergites 8–9, depressed; posterior tergites impunctate at sides, and without lateral longitudinal ridges. Ultimate tergite almost smooth, median longitudinal sulcus distinct in centre, lateral longitudinal ridge well marked and rugoso-striate; dorso-lateral carina absent. Forceps asymmetrical, strongly trigonal basally, curved and cylindrical apically; inner margin almost smooth (Fig. 89) and crenulate. Penultimate sternite finely and closely punctate at base, almost smooth distally, but very weak striations, posterior margin rounded (Fig. 90); about one third of the way from base a median longitudinal process arise, increasing in height posteriorly, broad from a ventral view. And narrower in centre, the distal part being spatulate and

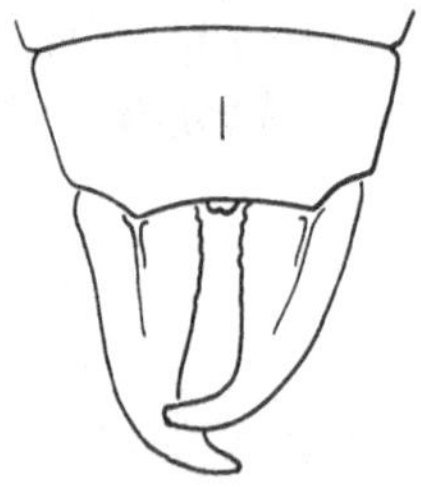

Fig. 89 Male ultimate tergite and forceps of *Anisolabis spatula* Brindle. After Brindle.

Fig. 90 Posterior margin of male penultimate sternite of *Anisolabis spatula* Brindle. After Brindle.

Fig. 91 Posterior margin of male penultimate sternite of *Anisolabis spatula* Brindle, in lateral view. After Brindle.

Fig. 92 External paramere of male genitalia of *Anisolabis spatula* Brindle. After Brindle.

slightly convace on ventral surface; lateral view as in fig. 91. Genitalia (Fig. 92) with very specific external parameres of *Carcinophora*-type, acute apically. Paramere broad, median longitudinal incision of anterior margin deep. Genital lobes well developed. – **Female** similar to male, but abdominal tergites impunctate, no lateral longitudinal ridge on last tergite. Forceps symmetrical, tapering, subcontiguous. – **Length** of body with forceps: male: 15–17 mm, female: 15–17,5 mm.

Distribution: Cameroon, Equatorial Guinea and Rio Muni.

4. Anisolabis cunicula Brindle

1978 *Anisolabis cunicula* Brindle, Dermaptera of Africa, Tervuren, **2**: 113, fig. 113 (penultimate sternite in ventral view), 114 (penultimate sternite in lateral view), 115 (male forceps), 239 (external paramere of male genitalia). – Terra typica: Nigeria (Type male: Mus. r. Afr. centr., Tervuren).

Description: Male colour black, rather shining, head dark red, blackish medially; antennae dark brown, basal joints paler, legs dark yellow. Head a little transverse; postfrontal and coronal sutures weak; eyes small, shorter than the length of head behind eyes. Antennae typical, first joint long, but somewhat shorter than the distance between antennal bases. Pronotum slightly transverse, and only slightly widened posteriorly; median longitudinal furrow depressed anteriorly and posteriorly; cuticle almost smooth but with isolated small punctures. Tegmina and wings absent. Mesonotum with fine punctures on anterior half and laterally, metanotum more weakly punctate. Abdomen weakly widened medially and depressed, tergites finely punctate, sides of tergites 7–9 with larger punctures, which are closer together and sometimes coalesce to become rugoso-striate; without lateral longitudinal ridges. Ultimate tergite smooth but with faint punctures tending to form longitudinal rows with smooth cuticle in-between; lateral longitudinal ridge oblique, well marked. Forceps

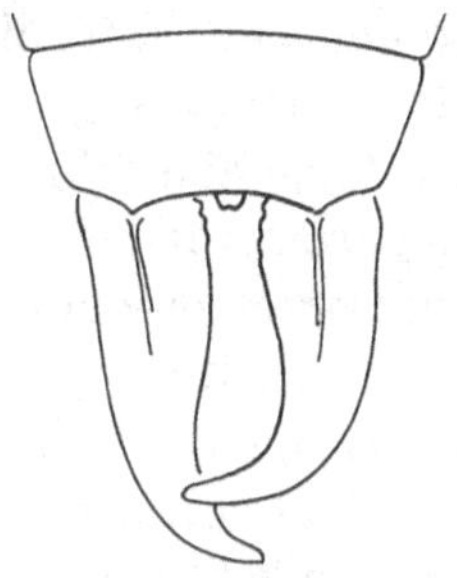

Fig. 93 Male ultimate tergite and forceps of *Anisolabis cunicula* Brindle. After Brindle.

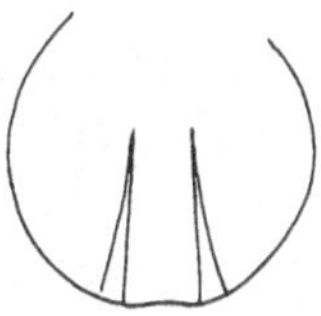

Fig. 94 Posterior margin of male penultimate sternite of *Anisolabis cunicula* Brindle. After Brindle.

Fig. 95 External paramere of holotype genital armature of *Anisolabis cunicula* Brindle. After Brindle.

(Fig. 93) with rather short branches, right branch curved more than left. Penultimate sternite (Fig. 94) characteristic; broad, posterior margin rounded; from a point slightly basal of midpoint the cuticle along the midline is smooth and gradually raised ventrally, forming a ridge, semi-circular in section, which is highest at the posterior end of the plate, the striation widely separated. Genitalia specific; paramere broad, robust, median longitudinal incision of anterior margin weak, deep; genital lobes well developed. External parameres (Fig. 95) long and narrow, except near base, inner membrane wide and white on inner margin, median membrane extending from about midpoint and wider basally. – **Length** of body with forceps: 13,5–16,5 mm. – **Female** unknown.

Distribution: Nigeria, Benin, Togo.

5. Anisolabis porrectella Brindle

1978 *Anisolabis porrectella* Brindle, Dermaptera of Africa, Tervuren, **2**: 86, fig. 99 (holotype in dorsal view), 100 (penultimate sternite in ventral view), 101 (penultimate sternite in lateral view), 238 (external paramere of male genitalia). – Terra typica: Zaire (Type male: Mus. r. Afr. centr., Tervuren).

Description: Male general colour blackish-brown; rather shiny, antennae blackish, basal joints brown and some distal joints white; pronotum yellow laterally, legs yellowish-brown. Head transverse, well rounded; postfrontal and coronal sutures distinct. Eyes comparatively small, shorter than the length of head behind eyes. First antennal joint about as long as the distance between antennal bases. Pronotum rather strongly transverse, widened posteriorly, coriaceous, smooth; median longitudinal furrow narrow and depressed anteriorly. Tegmina and wings absent, meso- and metanotum normal. Abdomen widened medially, somewhat depressed, sides of posterior tergites more strongly punctate but not rugoso-striate, although a few punctures are confluent, and there are no lateral longitudinal ridges. Ultimate tergite with scattered and sparse hairs; medial longitudinal sulcus distinct and deep medially, lateral carinae absent, and without dorso-lateral ridges. Forceps (Fig. 96) asymmetrical, comparatively short, trigonal basally, cylindrical apically. Penultimate sternite (Fig. 97) with a prominent process arising from midpoint and extending posteriorly, widened and becoming higher posteriorly, the apex triangular from a ventral view, and with long yellow hairs around the apex. Genitalia (Fig. 98) specific; paramere broad, median incision between external parameres deep and wide, genital lobes with well-marked virgae. External parameres moderately broad and short, external margin sinuate and ending in a blunt tip, inner membrane fairly wide, median membrane extending to about two-thirds of paramere. – **Female** very similar to male, but punctation finer; branches of forceps slenderer and almost straight; penultimate sternite rounded posteriorly and without a process. – **Length** of body with forceps: male: 9–10,5 mm, female: 10–11 mm.

Distribution: Zaire.

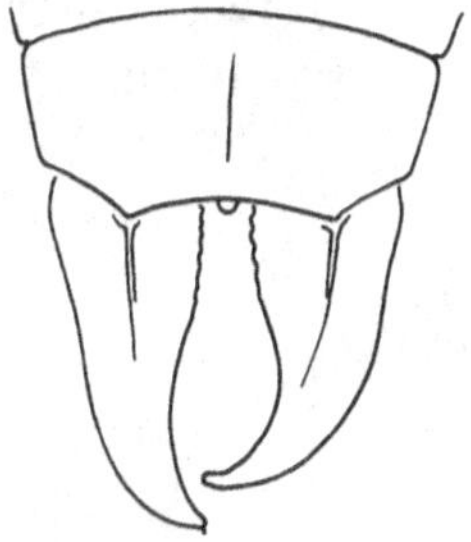

Fig. 96 Male ultimate tergite and forceps of *Anisolabis porrectella* BRINDLE. After BRINDLE.

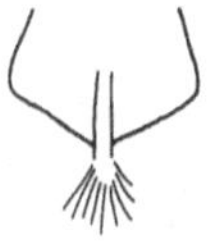

Fig. 97 Posterior margin of male penultimate sternite of *Anisolabis porrectella* BRINDLE. After BRINDLE.

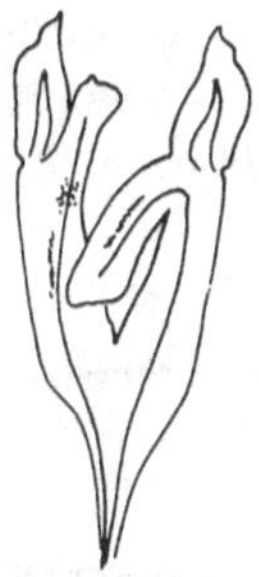

Fig. 98 Genital armature of holotype of *Anisolabis porrectella* BRINDLE. After BRINDLE.

6. Anisolabis rostrata BRINDLE

1978 *Anisolabis rostrata* BRINDLE, Dermaptera of Africa, Tervuren, **2**: 87, fig. 107 (penultimate sternite of male, in ventral view), 108 (penultimate sternite in lateral view), 237 (external paramere of male genitalia). – Terra typica: Zaire (Type male: Mus. r. Afr. centr., Tervuren).

Description: Male colour blackish-brown, rather shiny; antennae brown, basal joints yellowish, and some distal joints white; lateral margins of pronotum and legs yellow. Head transverse, somewhat convex; postfrontal and coronal sutures distinct. Eyes moderately small, shorter than the length of head behind eyes. Antennae normal, first joint long, about as long as the distance between antennal bases. Pronotum strongly transverse, widened posteriorly, expanded to posterior angles; median longitudinal furrow distinct. Tegmina and wings absent. Mesonotum long, metanotum short. Cuticle of head and thoracic tergites smooth, coriaceous, and almost impunctate. Abdomen more or less parallel-sided, not strongly depressed, first abdominal tergite impunctate, rest of tergites punctate; sides of tergites not rugoso-striate, nor with lateral longitudinal ridges. Ultimate tergite with median longitudinal sulcus in centre; lateral longitudinal ridge weak, no dorso-lateral ridge. Forceps (Fig. 99) rather slender, narrowed distally, inner margin slightly roughened, branches almost symmetrical. Penultimate sternite rounded posteriorly, and with a darkened longitudinal process arising from distal third of sternite, extending well beyond posterior margin, the process (Fig. 100) long, narrow, with ventral edge curved dorsally towards tip, dorsal edge sinuate, forming a beak-like projection. Genitalia typical, paramere broad, median incision between external parameres deep and wide; genital lobes with virgae. External parameres (Fig. 101) narrowed distally, ending in a narrow curved apex, inner membrane relatively narrow distally, median membrane widened medially. – **Female** similar to male, but abdominal punctation weaker, penultimate sternite without projection; both branches of forceps almost straight. – **Length** of body forceps: male: 8–9,6 mm, female: 8,5–10 mm.

Distribution: Zaire.

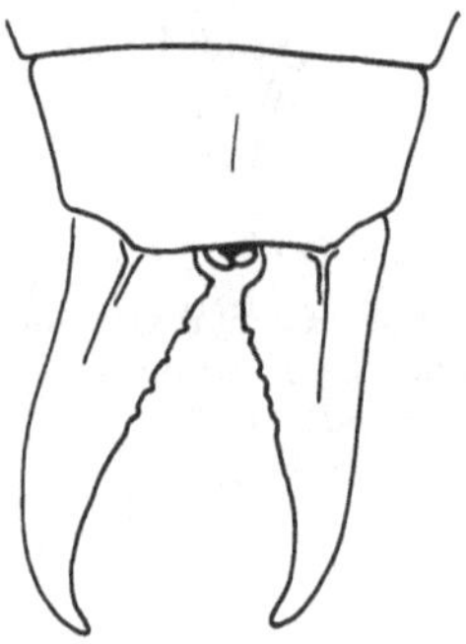

Fig. 99 Male ultimate tergite and forceps of *Anisolabis rostrata* BRINDLE. After BRINDLE.

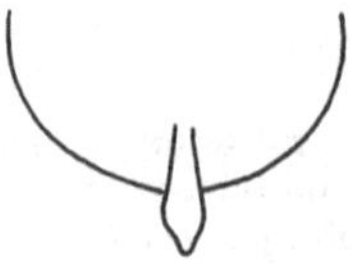

Fig. 100 Posterior margin of male penultimate sternite of *Anisolabis rostrata* BRINDLE. After BRINDLE.

Fig. 101 External paramere of male genitalia of *Anisolabis rostrata* BRINDLE. After BRINDLE.

7. **Anisolabis acutiventris** (HINCKS)

1954 *Apolabis acutiventris* HINCKS, Mém. Inst. fr. Afr. noire, Paris, Dakar, **40**: 107; fig. 7 (male genitalia), 8 (penultimate sternite). – Terra typica: Guinea (Type male: Mus. nation. Hist. nat., Paris).

1966 *Carcinophora acutiventris* – POPHAM & BRINDLE, Entomologist, London, **99**: 275.

1978 *Apolabis acutiventris* – STEINMANN, Dtsch. ent. Zeitschr., Berlin, n. F. **25**: 184.

1978 *Anisolabis acutiventris* – BRINDLE, Dermaptera of Africa, Tervuren, **2**: 88.

Literature: SAKAI 1970; STEINMANN 1975, 1984.

Description: Male general colour dark brown, or black; antennae, especially proximal joints, mouthparts, sides of pronotum, legs brown. Head tumid, occiput depressed, postfrontal and coronal sutures present but not strong; eyes rather large, nearly as long as genae. First antennal joint long, about as long as the distance between antennal bases; cuticle of head somewhat rugose. Pronotum about as long as broad, slightly widened posteriorly; posterior margin almost truncate; median longitudinal furrow moderately strong, almost complete. Tegmina and wings absent. Meso- and metanotum with clearly marked longitudinal median sulcus continuing that of pronotum. Abdomen a little expanded caudad, punctures of tergites regular, and usually separated by about their own diameter, but occasional punctures are confluent; tergites 6–7, each with a weak lateral longitudinal ridge on distal third of each. Ultimate tergite smooth, shiny, with very few punctures; no lateral longitudinal ridge or dorso-lateral ridge. Forceps (Fig. 102) short and broad, asymmetrical; lower surface very flat; inner margin crenulate basally. Penultimate sternite triangular, finely and closely punctate at base; apex of posterior margin produced and pointed (Fig. 103). Genitalia (Fig. 104) oval; paramere simple, median incision between external parameres deep and narrow; genital lobes without virgae; external parameres rather narrow and tapering distally. – **Female** unknown. – **Length** of body with forceps: 11–13 mm.

Distribution: Guinea, Sierra Leone, Ivory Coast.

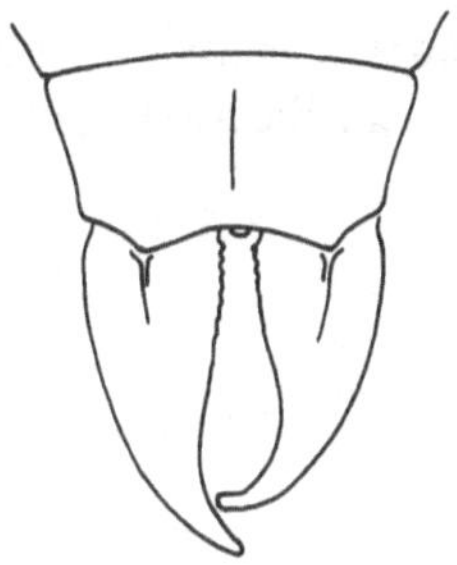

Fig. 102 Male ultimate tergite and forceps of *Anisolabis acutiventris* (HINCKS) Original.

Fig. 103 Posterior margin of male penultimate sternite of *Anisolabis acutiventris* (HINCKS). After HINCKS.

Fig. 104 Genital armature of holotype of *Anisolabis acutiventris* (HINCKS). After HINCKS.

8. Anisolabis oweni BURR

1911 *Anisolabis oweni* BURR, Ann. Mag. nat. Hist., London, (6) **8**: 40. – Terra typica: Liberia (Type male: Brit. Mus. nat. Hist., London).
1915 *Paralabis oweni* – BURR, J. r. microsc. Soc., London, **1915**: 540; fig. 62 (male genitalia).
1954 *Anisolabis oweni* – BOESEMAN, Zool. Verhand., Leiden, **21**: 38.
1954 *Paralabis oweni* – HINCKS, Mém. Inst. fr. Afr. noire, Paris, Dakar, **40**: 108; fig. 9 (male genitalia), 10 (penultimate sternite of male). Guinea.
1961 *Apolabis acutiventris* – HINCKS, Bull. Inst. fr. Afr. noire, Paris, Dakar, **23**: 29. Ivory Coast.
1971 *Anisolabis rugosus* BRINDLE (partim), Mém. Inst. fr. Afr. noire, Paris, Dakar, **86**: 274.
1978 *Anisolabis oweni* – BRINDLE, Dermaptera of Africa, Tervuren, **2**: 89, fig. 126 (penultimate sternite), 127 (male abdominal posterior segments laterally), 128 (male forceps), 243 (external paramere of male genitalia). Liberia, Ivory Coast.

Literature: BURR 1911; SAKAI 1970; STEINMANN 1977, 1984.

Description: Male colour blackish or very dark reddish-brown; antennae dark brown, legs yellowish to yellowish-brown. Head transverse, convex; eyes large, postfrontal and coronal sutures indistinct. Antennae normal, first antennal joint long, about as long as the distance between antennal bases. Pronotum transverse, widened posteriorly, median longitudinal furrow distinct; cuticle rugose and somewhat rugoso-striate posteriorly. Tegmina and wings absent. Meso- and metanotum rather rugose anteriorly and rugoso-striate posteriorly. Abdomen weakly widened medially and not strongly depressed; tergite 8 less strongly punctate and tergite 9 almost impunctate. Tergites 6–8 rugoso-striate at sides and with lateral longitudinal ridges. Ultimate tergite smooth, with sparse and faint punctures; forceps (Fig. 105) asymmetrical, each branches weakly curved, left branch curved less the right. Penultimate sternite (Fig. 106) with a triangular apex. Genitalia (Fig. 107) specific; paramere V-shaped, median incision of anterior margin of parameral plate very deep and wide; genital lobes without virgae, well developed. External paramere normal, of *Anisolabis*-type, obtuse apically. – **Female** similar to male, but abdominal punctation weaker, sides of posterior abdominal tergites not rugoso-striate, nor with lateral longitudinal ridges. – **Length** of body with forceps: male: 16–19 mm, female: 16,5–20,5 mm.

Distribution: Western Africa: Ivory Coast, Guinea, Liberia, Sierra Leone.

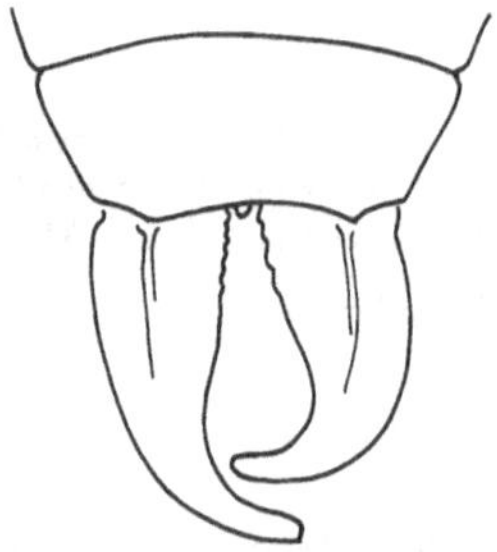

Fig. 105 Male ultimate tergite and forceps of *Anisolabis oweni* BURR. After BRINDLE.

Fig. 106 Penultimate sternite of male of *Anisolabis oweni* BURR. After HINCKS.

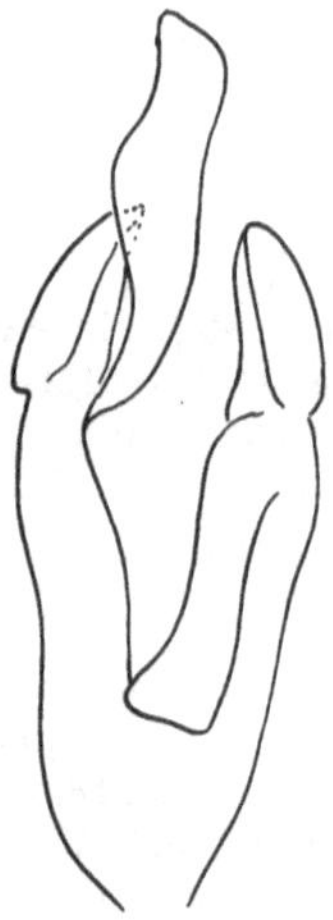

Fig. 107 Male genital armature of *Anisolabis oweni* BURR. After HINCKS.

9. Anisolabis rugosa BRINDLE

1971 *Anisolabis rugosa* BRINDLE, Mém. Inst. fr. Afr. noire, Paris, Dakar, **86**: 274, fig. 16 (head and pronotum of male), 17 (male forceps), 18 (male genitalia), 19 (penultimate sternite of male). – Terra typica: Sierra Leone (Type male: Mus. nation. Hist. nat., Paris).

Literature: BRINDLE 1978, 1982; STEINMANN 1984.

Description: Male shiny black or dark reddish-brown, head and pronotum sometimes dark red; antennae brown, basal joints yellow, forceps dark red. Head transverse; postfrontal and coronal sutures distinct. Eyes relatively long, about as long as first antennal joint; the latter long, more or less as long as the distance between antennal bases. Pronotum slightly transverse, a little widened posteriorly, with a faint median longitudinal impressed line. Cuticle of pronotum coriaceous, almost smooth but with a slightly irregular cuticle posteriorly. Tegmina and wings entirely absent. Meso- and metanotum partially punctate. Abdomen weakly widened and depressed; tergites strongly and closely punctate; basal two segments more weakly punctate; tergites 6–8 not rugose laterally, but each with a short lateral longitudinal ridge on both sides. Ultimate tergite transverse, almost impunctate, with a well defined median longitudinal sulcus. Forceps (Fig. 108) with branches trigonal at base, narrowed distally and cylindrical; inner margin with small crenulations basally; branches not widely separated. Penultimate sternite (Fig. 109) triangular, apex rounded; cuticle finely and closely punctate at base, distal part with very weak and shallow punctures and striations. Genitalia (Fig. 111) characteristic; paramere oval, broad, incision of anterior margin deep and wide; genital lobes well developed, with virgae. External parameres (Fig. 110) broad, of *Anisolabis*-type. – **Female** as male, but tergites 6–8 without lateral longitudinal ridges; ultimate tergite without weakly

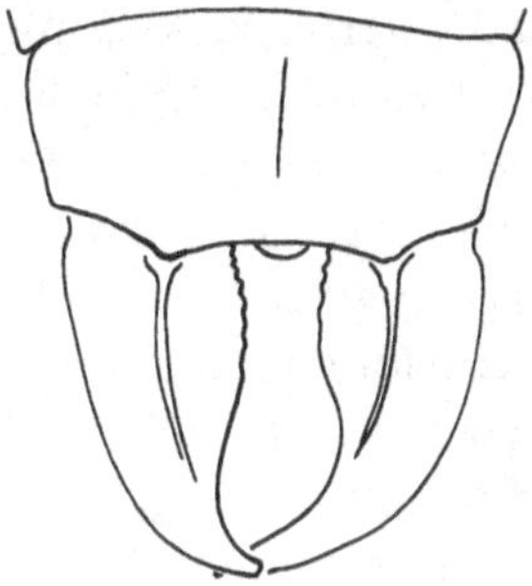

Fig. 108 Male ultimate tergite and forceps of *Anisolabis rugosa* BRINDLE. After BRINDLE.

Fig. 109 Posterior margin of male penultimate sternite of *Anisolabis rugosa* BRINDLE. After BRINDLE.

Fig. 110 External paramere of male genitalia of *Anisolabis rugosa* BRINDLE. After BRINDLE.

Fig. 111 Male genital armature of *Anisolabis rugosa* BRINDLE. After BRINDLE.

developed dorso-lateral ridges; branches of forceps straighter. – **Length** of body with forceps: male: 16,6 mm, female: 14–16 mm.

Distribution: Sierra Leone.

10. Anisolabis pluto Rehn

1905 *Anisolabis pluto* Rehn, Proc. U.S. nation. Mus., Washington, **29**: 506. – Terra typica: Liberia (Type female: U.S. nation. Mus., Washington).

1911 *Anisolabis pluto* – Burr, Genera Insectorum, Bruxelles, **122**: 30 (as synonym of *Anisolabis angulifera* (Dohrn)).

1966 *Anisolabis pluto* – Popham & Brindle, Entomologist, London, **99**: 274.

Literature: Brindle 1978; Burr 1906, 1910, 1915; Sakai 1970; Steinmann 1984.

Description: Female colour dark reddish-brown, head dark red, antennae yellowish-brown, basal joints yellow; legs yellowish, and pronotum narrowly yellow laterally. Head strongly transverse, lateral sides rounded, convex; postfrontal and coronal sutures rather weak; eyes large, about as long as length of head behind eyes. Antennae normal, first antennal joint more or less as long as the distance between antennal bases. Pronotum strongly transverse, widened posteriorly, median longitudinal furrow distinct. Tegmina and wings absent. Meso- and metanotum rugose and punctate laterally and anteriorly. Abdomen weakly widened medially and depressed, distinctly narrowed from tergite 6 to apex; cuticle of abdominal tergites strongly punctate, basal two segments with smaller punctures which are separated by rather more than their own diameter medially; remaining tergites with larger punctures. Sides of tergites punctate but punctation less strong than on dorsum; sides without lateral longitudinal ridges. Ultimate tergite transverse, median longitudinal sulcus short and weak; punctation laterally larger than on dorsum but separated by similar longitudinal bands of smooth cuticle. Penultimate sternite triangular, rounded apically. Forceps short, broad, inner margins very slightly crenulated, apex short and incurved. – **Male** unknown. – **Length** of body with forceps: 18 mm.

Distribution: Liberia.

11. Anisolabis incisa Borelli

1914 *Anisolabis incisa* Borelli, Boll. Lab. Zool. gen. agr., Portici, **8**: 268, fig. 2 (penultimate sternite of male and ultimate tergite with forceps of male). – Terra typica: Guinea (Type male without genitalia: Mus. Ist. Zool. sist. Univ., Torino).

1915 *Euborellia incisa* – Burr, J. r. microsc. Soc., London, **1915**: 542, fig. 63 (male genitalia).

1978 *Gonolabis incisa* – Steinmann, Fol. ent. hung., Budapest, **31**: 192. Guinea, Mauritania, Liberia.

1978 *Anisolabis incisa* – Brindle, Dermaptera of Africa, Tervuren, **2**: 93, fig. 133 (penultimate sternite of male), 134 (male in dorsal view), 248 (external paramere of male genitalia). Guinea, Sierra Leone.

Literature: Boeseman 1954; Sakai 1970; Steinmann 1978, 1984.

Description: Male general colour black, shiny; antennae dark brown, two distal joints (12–13 or 13–14) yellowish; legs yellow. Head transverse;

postfrontal and coronal sutures deep, well marked. Antennae 19-jointed; first joint long, but shorter than the distance between antennal bases. Eyes comparatively large, but shorter than the genae. Pronotum quadrate or more or less quadrate, all angles rounded, and all margins straight, lateral margins expanded posteriorly; median longitudinal furrow distinct. Cuticle of thoracic tergites coarsely coriaceous, somewhat rugose posteriorly, and sometimes rugoso-striate. Tegmina and wings absent. Meso- and metanotum normal, simple. Abdomen of *Gonolabis*-type, expanded to last tergite; abdominal tergites rather depressed, tergites 3–9 strongly and irregularly punctate, punctures separated by about their own diameter, but there is some variation and the punctation tends to be strongest on tergites 4–7. Sides of tergites 5–8 with a narrow smooth lateral longitudinal ridge. Ultimate tergite broad, median longitudinal sulcus well marked; posterior margin convex. Forceps (Fig. 112) with very strong branches, whitch are trigonal basally, cylindrical apically. Penultimate sternite (Fig. 113) broadly triangular, angles rounded, and apices excised, the angle of excision being about 90° Genitalia (Fig. 114 after BRINDLE, and fig. 115 after

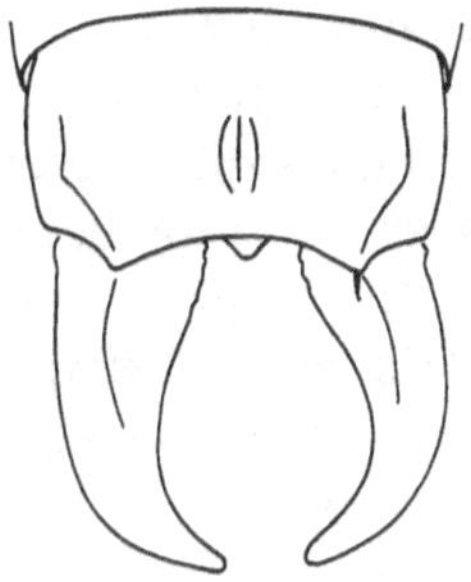

Fig. 112 Male ultimate tergite and forceps of *Anisolabis incisa* BORELLI. Original.

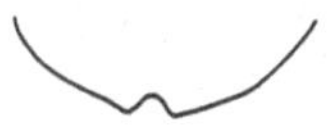

Fig. 113 Posterior margin of penultimate sternite of *Anisolabis incisa* BORELLI. After BORELLI.

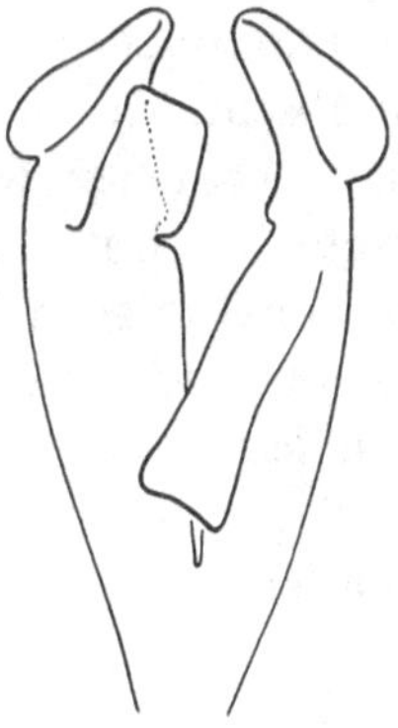

Fig. 114 Male genital armature of *Anisolabis incisa* BORELLI. After BRINDLE.

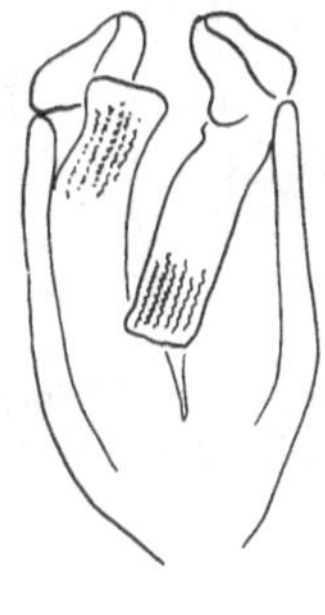

Fig. 115 Male genital armature of *Anisolabis incisa* BORELLI. After BURR.

Burr) broad, robust; paramere with very deep, wide median incision between external parameres. Genital lobes without virgae. External parameres of *Anisolabis*-type, apex rounded. – **Female** very similar to male, but forceps symmetrical, tapering, contiguous; ultimate tergite without dorso-lateral keels. – **Length** of body with forceps: male: 20–21 mm, female: 17–19 mm.

Distribution: Guinea, Liberia, Mauritania, Sierra Leone.

12. Anisolabis incisiodes Brindle

1978 *Anisolabis incisiodes* Brindle, Dermaptera of Africa, Tervuren, **2**: 94; fig. 136 (penultimate sternite of holotype), 137 (forceps of male), 247 (external paramere of male genitalia). – Terra typica: Zaire (Type male: Mus. r. Afr. centr., Tervuren).

Description: Male dark brownish-red to blackish, rather shiny; head sometimes dark red, antennae brown to dark brown, with one or more distal joints (12, 13, or 14), paler; lateral margin of pronotum yellow, legs yellowish, femora darkened medially. Head transverse, convex; postfrontal and coronal sutures indistinct; eyes large, but a little shorter than the length of head behind eyes. Antennae normal, first antennal joint long, more or less as long as the distance between antennal bases. Pronotum broad, transverse, slightly widened posteriorly, median longitudinal furrow distinct and depressed anteriorly; tegmina and wings absent. Mesonotum rugose and sparsely punctate anteriorly, metanotum normal. Abdomen weakly widened medially; sides of tergites 5–8 strongly punctate but not rugoso-striate, and each with an almost straight lateral longitudinal ridge. Ultimate tergite smooth, median longitudinal sulcus deep; lateral longitudinal keels absent. Forceps (Fig. 116) short, weakly curved, the left branch less curved than the right. Penultimate sternite (Fig. 117) broadly triangular, excised at apex, the excision angle obtuse. Genitalia normal, simple, parameral plate broad, median incision between external parameres deep and

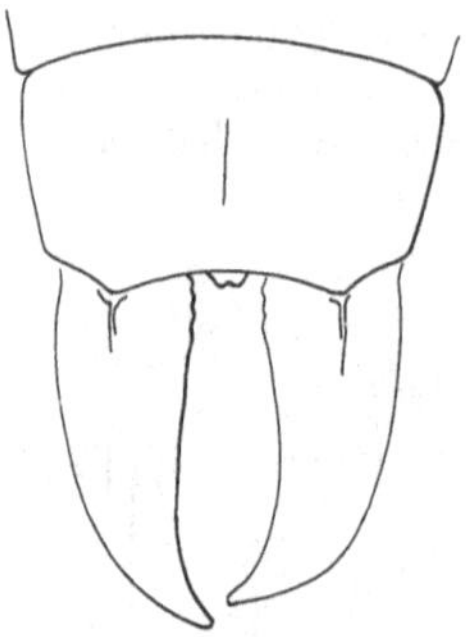

Fig. 116 Male ultimate tergite and forceps of *Anisolabis incisiodes* Brindle. After Brindle.

Fig. 117 Posterior margin of penultimate sternite of *Anisolabis incisiodes* Brindle. After Brindle.

Fig. 118 External paramere of male genitalia of *Anisolabis incisiodes* Brindle. After Brindle.

wide. Genital lobes well-developed, elongate. External parameres (Fig. 118) more triangular than in *incisa* BORELLI, and the apex narrower. – **Female** similar to male, but abdominal punctation weaker, and no lateral longitudinal ridges on posterior abdominal tergites; branches of forceps straight, tapering. – **Length** of body with forceps: male: 12–13 mm, female: 12–14,5 mm.

Distribution: Zaire, Cameroon.

13. Anisolabis quadricollis (HINCKS)

1954 *Apolabis quadricollis* HINCKS, Mém. Inst. fr. Afr. noire, Paris, Dakar, **40**: 106, fig. 6 (genitalia of holotype). – Terra typica: Guinea (Type male: Mus. nation. Hist. nat., Paris).

1966 *Carcinophora quadricollis* – POPHAM & BRINDLE, Entomologist, London, **99**: 275.

1978 *Anisolabis quadricollis* – BRINDLE, Dermaptera of Africa, Tervuren, **2**: 114, fig. 173 (male forceps), 174 (penultimate sternite of male), 175 (female forceps), 262 (external paramere of male genitalia). Guinea, Liberia.

Literature: SAKAI 1970; STEINMANN 1978, 1984.

Description: Male colour brown; head, pronotum sometimes reddish; antennae and legs yellowish-brown. Head moderately tumid; postfrontal and coronal sutures distinct. Eyes not prominent, shorter than first antennal joint. Antennae 20-jointed; first joint long, more or less as long as the distance between antennal bases. Pronotum somewhat variable, more or less quadrate, as long as broad or slightly broader than long or slightly longer than broad; weakly widened posteriorly; cuticle almost smooth but with slight striations near median furrow. Tegmina and wings absent. Mesonotum large, broad, metanotum transverse. Abdomen widened to middle, lateral margins of tergites 7–9 obtusely promi-

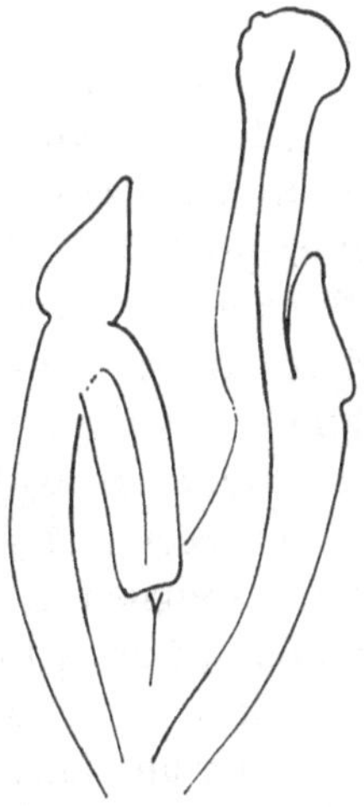

Fig. 119 Genital armature of holotype of *Anisolabis quadricollis* (HINCKS). After HINCKS.

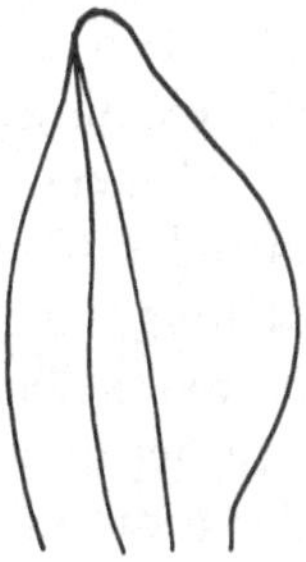

Fig. 120 External paramere of male genitalia of *Anisolabis quadricollis* (HINCKS). After BRINDLE.

nent, moderately strongly to rugosely punctate but without a trace of longitudinal keels. Ultimate tergite transverse, longitudinally impressed mesad. Penultimate sternite transverse, sides oblique; caudal margin very slightly emarginate mesad. Forceps asymmetrical, trigonal basally, curved and cylindrical apically. Genitalia (Fig. 119) oval; paramere broad, median longitudinal incision of anterior margin wide, genital lobes elongate, with slender virgae. External paramere (Fig. 120) wider medially, narrowed distally, apex rounded. – **Female** similar to male, punctation weaker, lateral longitudinal ridges and dorsolateral ridges on last tergite almost absent. Forceps straight, tapering. – **Length** of body with forceps; male: 13–16 mm, female: 10–13 mm.

Distribution: Guinea, Liberia, Sierra Leone, Togo, Ghana.

14. Anisolabis felix Burr

1907 *Anisolabis felix* Burr, in: Sjöstedt, Kilimanjaro Orth.: 17; pl. 4, fig. 4 (male). – Terra typica: Tanganyika (Type male: Brit. Mus. nat. Hist., London).
1914 *Horridolabis felix* – Burr, Ann. Mag. nat. Hist., London, (8) **13**: 581.
1915 *Gelotolabis felix* – Burr, J. r. microsc. Soc., London, **1915**: 541.
1964 *Gelotolabis erythrocephala* Brindle, Ann. Mag. nat. Hist., London, (13) **7**: 753, fig. 7 (forceps of holotype), 14 (external paramere and genital lobe of male genitalia). – Terra typica: Tanganyika (Type: Mus. r. Afr. centr., Tervuren).
1966 *Genolabis felix* – Popham & Brindle, Entomologist, London, **99**: 273.
1968 *Euborellia felix felix* – Brindle, Ark. Zool., Uppsala, **20**: 537.
1978 *Anisolabis felix* – Brindle, Dermaptera of Africa, Tervuren, **2**: 127, fig. 139 (male in dorsal view), 144 (penultimate sternite of male), 256 (external paramere of male genitalia).
Literature: Borelli 1915; Burr 1910, 1911; Sakai 1970; Steinmann 1978.

Description: Male colour blackish to black, head reddish; antennae brown, basal joints paler; legs reddish yellow. Head transverse, frons tumid, occiput a little depressed. Postfrontal and coronal sutures present. Eyes comparatively small, not prominent, smaller than the length of scape. Antennae with small joints. First joint very short, shorter than the distance between antennal bases; second joint quadrate, third longer, a little longer than fourth. Pronotum more or less quadrate, with a short median longitudinal impressed furrow; lateral margins slightly expanded to posterior angles. Tegmina and wings entirely absent. Meso- and metanotum normal. Abdomen expanded to last tergite or median tergites. Sides of tergites 6–9 very strongly and deeply rugoso-striate but without lateral longitudinal ridges. Ultimate tergite very broad, transverse, median longitudinal sulcus present. Forceps (Fig. 121) specific, asymmetrical; both branches rather long, dorsal ridge prominent and roughened; left branch with a triangular tooth at base on inner margin, and curved on last third; right branch wider at base with a larger and more acute inner tooth, branch very sharply curved just beyond midpoint. Penultimate sternite finely punctate basally, rest rugoso-striate; posterior margin concave. Genitalia (Fig. 122) V-shaped, genital lobes well developed, without virgae. – **Female** similar to male, but penultimate sternite with apex evenly rounded; both branches of forceps straighter, inner margins crenulated. – **Length** of body with forceps: in both sexes: 23–28 mm.

Notes by BRINDLE (1978): "The above is the type form of *felix*, but very closely related *Anisolabis* occour on other isolated mountain areas in Tanzania, Uganda, and Kenya. They differ mainly in colouration and slightly in the shape of the male external parameres, but the available material is not sufficient to confirm that these are really distinct species. In view of the isolation of the relevant mountain areas, however, the present view is that they should be regarded as subspecies."

Distribution: Eastern Africa.

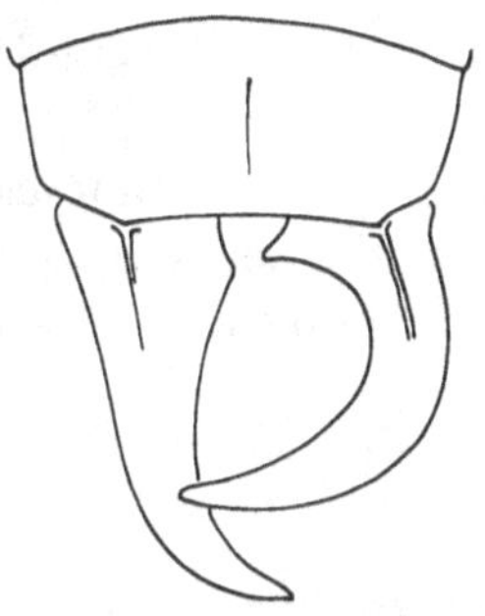

Fig. 121 Male ultimate tergite and forceps of *Anisolabis felix* BURR. Original.

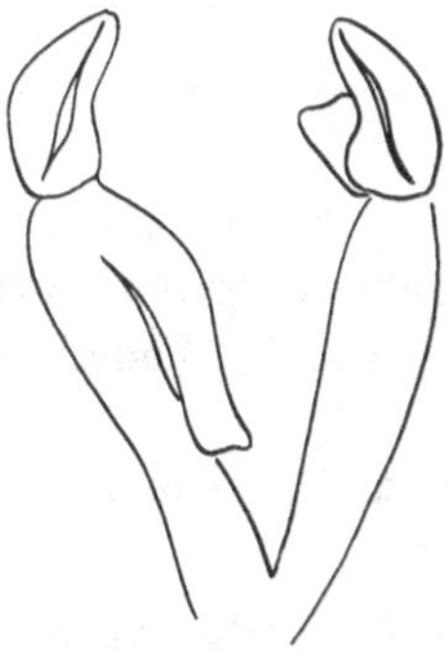

Fig. 122 Male genitalia of *Anisolabis felix* BURR. Original.

Identification key to the subspecies

1 Length of body more than 20 mm **2**

1′ Length of body less than 20 mm **3**

2 External parameres of male genitalia moderately short, rounded externally, inner membrane wide basally, and median membrane narrow. Distributed in Tanzania .... 14a. **Anisolabis felix felix**, p. 85

2′ External parameres of male genitalia are somewhat distorted in the mounts but appear to resemble those of *paradoxura* (ZACHER). Distribution in Uganda .... 14b. **Anisolabis felix elgonensis**, p. 86

3 Distributed in Tanzania. External parameres of male genitalia as *felix* but median membrane wider .... 14c. **Anisolabis felix meruensis**, p. 86

3′ Distributed in Kenya **4**

4 External parameres of male genitalia as *felix* but outer angle more pronounced and median membrane narrower .... 14d. **Anisolabis felix paradoxura**, p. 86

4′ External parameres of male genitalia narrower than in *felix*, but this may be somewhat distorted .... 14e. **Anisolabis felix chyuluensis**, p. 86

14a. Anisolabis felix felix BURR

The same as principal form, see page 84

14b. Anisolabis felix elgonensis BRINDLE

1978 *Anisolabis felix elgonensis* BRINDLE, Dermaptera of Africa, Tervuren, **2**: 128. – Terra typica: Uganda (Type male: Mus. r. Afr. centr., Tervuren).

Original description by BRINDLE: "As *felix felix*, but the head is reddish-brown, the thoracic tergites partly reddish-brown, and the abdomen is more strongly widened and smoother. External parameres of genitalia are somewhat distorted in the mounts but appear to resemble those of *paradoxura*." – **Female** unknown. – **Length** of body with forceps: 27,5–28 mm.

Distribution: Uganda.

14c. Anisolabis felix meruensis BRINDLE

1978 *Anisolabis felix meruensis* BRINDLE, Dermaptera of Africa, Tervuren, **2**: 129. – Terra typica: Tanzania (Type male: Mus. r. Afr. centr., Tervuren).

Original description by BRINDLE: "Dark brown, head and thoracic tergites reddish-brown. External parameres of male genitalia as felix but median membrane wider." – **Female** unknown. – **Length** of body with forceps: 19 + 4 mm.

Distribution: Tanzania.

14d. Anisolabis felix paradoxura (ZACHER)

1911 *Horridolabis paradoxura* ZACHER, Zool. Jahrb., Jena, **30**: 384, fig. L^2 (genitalia of holotype), M^2 (male forceps). – Terra typica: Dtsch. Ost-Afrika (Type male: Mus. Naturkunde, Berlin).
1915 *Gelotolabis felix* – BURR, J. r. microsc. Soc., London, **1915**: 541 (synonymy).
1978 *Horridolabis paradoxura* – STEINMANN, Dtsch. ent. Zeitschr., Berlin, **25**: 185, fig. 38 (male genitalia).
1978 *Anisolabis felix paradoxura* – BRINDLE, Dermaptera of Africa, Tervuren, **2**: 128.

Original description by ZACHER: "Vom Kopf bis Metanotum hell rötlichbraun, Abdomen dunkelbraun, fast schwarz, Beine gelb. Pronotum etwas länger als breit. Sein Seitenrand in der Mitte etwas eingebuchtet, neben der Einbuchtung eine Grube. Brustplatte des Microthorax sehr groß, gleich 1/2 des Prosternum. Zangen sehr asymmetrisch, die rechte fast rechtwinklig geknickt, die linke fast gerade, beide innen mit starkem Basalzahn. Vorderrand der Supraanalplatte etwas konkav. 8. Abdominalsternit hinten sehr flach ausgeschnitten." – **Female** unknown. – **Length** of body with forceps: 18 + 4 mm.

Distribution: Kenya.

14e. Anisolabis felix chyuluensis BRINDLE

1978 *Anisolabis felix chyuluensis* BRINDLE, Dermaptera of Africa, Tervuren, **2**: 128. – Terra typica: Kenya (Type male: Mus. r. Afr. centr., Tervuren).

Original description by BRINDLE: "Blackish-brown to dark reddish-brown, head red, darkened medially; antennae brown, basal segments paler. As *felix* but pronotum more strongly transverse, abdominal punctation and cuticle of penultimate sternite smoother, and abdomen more evenly widened medially; paramere of holotype narrower than in *felix felix*, but this may be somewhat distorted." – **Female** unknown. – **Length** of body with forceps: 18 + 3 mm.

Distribution: Kenya.

15. Anisolabis bintumanensis Brindle

1971 *Anisolabis bintimanensis* Brindle, Mém. Inst. fr. Afr. noire, Paris, Dakar, **86**: 272, fig. 10 (head and pronotum of male), fig. 11 (male forceps), fig. 12 (genitalia of holotype), fig. 13 (male penultimate sternite). – Terra typica: Sierra Leone (Type male: Mus. nation. Hist. nat., Paris).

Literature: Brindle 1978; Steinmann 1978.

Description: Male shiny black or dark reddish-black; antennae dark brown, basal two joints yellow; legs yellow. Head transverse, rounded, about as long as broad; postfrontal and coronal sutures visible, distinct. Eyes comparatively small, shorter than first antennal joint. Cuticle of head coriaceous, almost impunctate. First antennal joint long, more or less as long as the distance between antennal bases. Pronotum less transverse, and slightly widened posteriorly; median longitudinal furrow weak, all margins straight, and all angles rounded. Tegmina and wings absent. Meso- and metanotum transverse. Abdomen weakly punctate, basal segments almost impunctate; tergites 7–9 with lateral longitudinal ridges; sides of posterior tergites punctate, not rugoso-striate. Ultimate tergite smooth, transverse, sparsely and unevenly punctate; median longitudinal sulcus present. Forceps (Fig. 123) trigonal basally, cylindrical distally, not strongly curved, more or less symmetrical. Penultimate sternite broad, posterior margin (Fig. 124) rounded, and with apical part concave. Genitalia (Fig. 125) oval; paramere broad, median longitudinal incision between external parameres wide and deep; genital lobes well developed, with virgae.

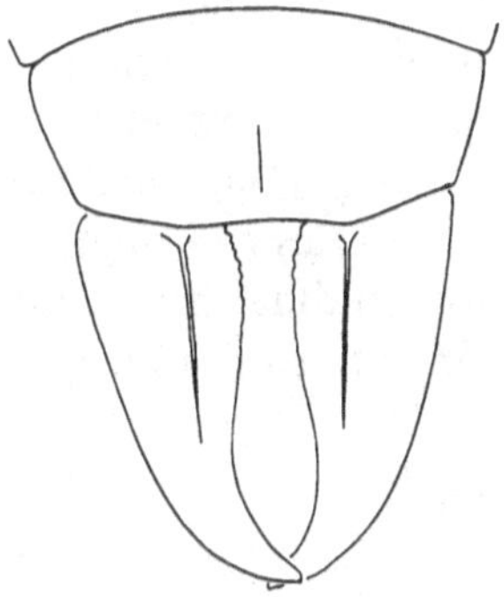

Fig. 123 Male ultimate tergite with forceps of *Anisolabis bintumanensis* Brindle. After Brindle.

Fig. 124 Posterior margin of male penultimate sternite of *Anisolabis bintumanensis* Brindle. After Brindle.

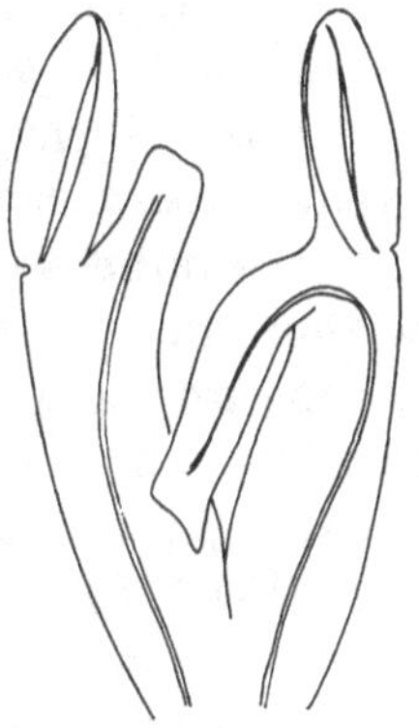

Fig. 125 Genital armature of holotype of *Anisolabis bintumanensis* Brindle. After Brindle.

External parameres evenly narrowed distally; obtuse apically. – **Female** very similar to male, but tergites 7–9 without lateral ridges; ultimate tergite narrower and without dorso-lateral ridges. Forceps similar to male, but branches straighter, contiguous. – **Length** of body with forceps: male: 13–14 mm, female: 11–12 mm.

Distribution: Sierra Leone.

16. Anisolabis hottentotta (DOHRN)

1867 *Forcinella hottentotta* DOHRN, Stettiner ent. Ztg., **28**: 344. – Terra typica: Caffraria (Type male: PAN Zoolog. Inst., Warszawa).

1904 *Anisolabis hottentotta* – KIRBY, A Synonymic Cat. Orthopt., London, **1**: 18.

1907 *Anisolabis hottentotta* – BORELLI, Ann. Mus. civ. Stor. nat., Genova, (3) **3**: 14. Fernando Po.

1911 *Anisolabis hottentotta* – BURR, Genera Insectorum, Bruxelles, **122**: 30; pl. 3, fig. 5 (male in dorsal view). South Africa.

1915 *Apolabis hottentotta* – BURR, J. r. microsc. Soc., London, **1915**: 528; pl. XI, fig. 10 (male genitalia).

1966 *Carcinophora hottentotta* – POPHAM & BRINDLE, Entomologist, London, **99**: 275.

1978 *Anisolabis hottentotta* – BRINDLE, Dermaptera of Africa, Tervuren, **2**: 115; fig. 285 (external paramere of male genitalia), 286 (male forceps), 187 (male penultimate sternite). – Natal.

1979 *Apolabis hottentotta* – STEINMANN, Fol. ent. hung., Budapest, **32**: 155; fig. 11 (holotype, end of abdomen with forceps), 12 (genitalia of holotype; gen. prep. No. 494, det. Dr. H. STEINMANN). Caffraria.

Literature: BURR 1911, 1913; HINCKS 1957; SAKAI 1970; STEINMANN 1977, 1978, 1979, 1984.

Description: Male colour blackish; antennae dark brown, basal joints yellow or reddish-orange, and one or more distal joints whitish (11 or 12–13); sides of pronotum yellow; legs yellow, femora with basal two thirds or so brown to dark brown. Male very similar to *Euborellia annulipes* (LUCAS), but head moderately broad, transverse, tumid, shiny. Postfrontal and coronal sutures distinct. Eyes prominent, rounded, small, shorter than the first antennal joint, long, but shorter than the distance between antennal bases at frons. Cuticle of head almost smooth. Pronotum quadrate or a little transverse; lateral margins concave, widening a little to posterior margin; all angles rounded; median longitudinal furrow weak. Tegmina and wings absent. Mesonotum broad, metanotum transverse. Abdomen widened to middle, depressed; tergite 1 almost impunctate, tergites 2–4 rather sparsely punctate, distal tergites more strongly punctate but punctation, although fairly close, is rather fine but irregular; tergites 7–9 rugoso-striate towards, and on sides, 7–8 each with a lateral longitudinal ridge and tergite 9 with an incipient ridge. Ultimate tergite broad, large; posterior margin nearly straight, disregarding the weak curvature observable at lateral angles. Forceps (Fig. 126) a little asymmetrical, thick, short, hardly twice as long as the rather short ultimate tergite. Penultimate sternite transverse, narrowed distally, posterior margin concave. Genitalia (Fig. 127) medium elongate; median longitudinal incision of anterior margin of central parameral plate deep, extending considerably beyond median line of paramere. Genital lobes well developed, without virgae; one genital lobe recurving beyond middle of paramere, the other short, only slightly longer than wide; both harbouring a small sclerotized plate (in holotype). – **Female** very similar to male, but forceps simple, contiguous, tapering. – **Length** of body with forceps: male: 8,5–9,5 mm, female: 9 mm.

Distribution: Southern Africa.

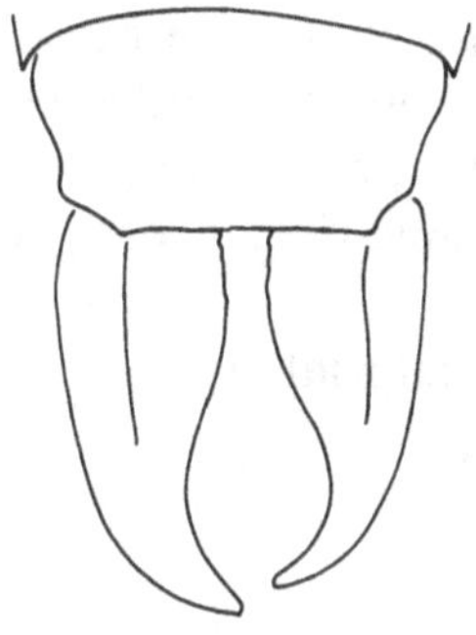

Fig. 126 Male ultimate tergite with forceps of *Anisolabis hottentotta* (DOHRN). After STEINMANN.

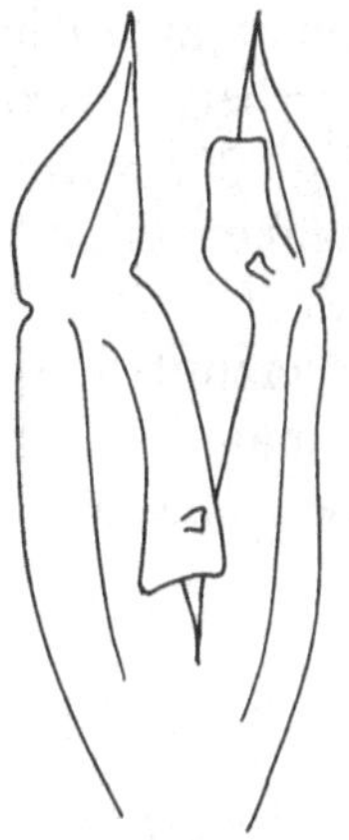

Fig. 127 Genital armature of holotype of *Anisolabis hottentotta* (DOHRN). After STEINMANN.

17. Anisolabis rufescens KIRBY

1891 *Anisolabis rufescens* KIRBY, J. Linn. Soc. Zool., London, **23**: 517; pl. XII, fig. 10 (male). – Terra typica: Cameroun (Type male: Brit. Mus. nat. Hist., London).

1978 *Anisolabis rufescens* – BRINDLE, Dermaptera of Africa, Tervuren, **2**: 136; fig. 182 (male), 183 (male penultimate sternite), 184 (female forceps), 185 (distal antennal joints), 264 (external paramere of male genitalia). Cameroun, Gabon.

Literature: BORELLI 1907; BORMANS 1900; BURR 1911; POPHAM & BRINDLE 1966; REHN 1906; SAKAI 1970; STEINMANN 1977, 1978, 1984.

Description: Male colour reddish-brown, black above frons, and the red part behind marked with several black lines to the occiput; a red spot on each side of the vertex, within each eye; clypeus brown, bordered with yellowis; mandibles and extremity of the labrum red. Antennae reddish-brown, thoracic tergites and legs reddish or yellowish-brown; abdomen black, shiny. Head a little broader than pronotum, transverse, sides almost parallel and postero-lateral angles rounded. Postfrontal and coronal sutures distinct. Antennae 21-jointed, first moderately short, more shorter than the distance between antennal bases. Eyes comparatively large, about as long as the first antennal joint. Pronotum quadrate or a little transverse, parallel-sided, posterior margin rounded; median longitudinal furrow present. Tegmina and wings absent. Meso- and metanotum normal. Abdomen of *Gonolabis*-type, strongly widened posteriorly and depressed; tergites 7–9 rugoso-striate at sides but 7 only weakly so, no lateral longitudinal ridges. Ultimate tergite large, broad; lateral longitudinal ridge present, median longitudinal sulcus deep. Forceps (Fig. 128) more or less symmetrical (!), thick, broadened beyond the base, but without a basal tooth; the inside of the curve denticulated, the right side rather more curved and shorter than the left. Pygidium exposed, bifid at the extremity. Penultimate sternite bluntly triangular, apex truncate. Genitalia broad, central parameral plate wide, median incision between external parameres deep; genital lobes normal. External parameres (Fig. 130) of *Anisolabis*-type. – **Female** similar to male, but abdominal punctation weaker; ultimate tergite with lateral longitudinal ridge less prominent; branches of forceps (Fig. 129) longer, straighter, inner margins tending to have stronger crenulations, branches slightly asymmetrical. – **Length** of body with forceps: male 33–35 mm, female: 33 mm.

Distribution: Cameroon, Rio Muni, Gabon, Equatorial Guinea.

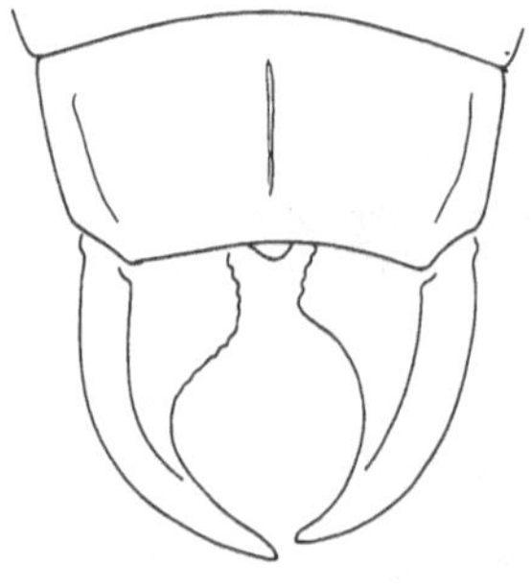

Fig. 128 Male ultimate tergite and forceps of *Anisolabis rufescens* KIRBY. After BRINDLE.

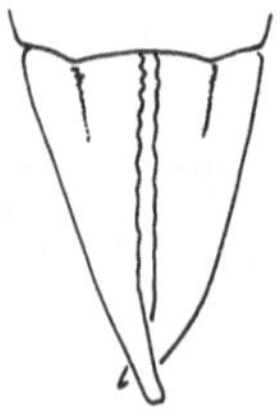

Fig. 129 Female forceps of *Anisolabis rufescens* KIRBY. After BRINDLE.

Fig. 130 External paramere of male genitalia of *Anisolabis rufescens* KIRBY. After BRINDLE.

18. Anisolabis atra BORELLI

1907 *Anisolabis atra* BORELLI, Ann. Mus. civ. Stor. nat., Genova, **43**: 361. – Terra typica: Cameroun (Type male: Mus. civ. Stor. nat., Genova).

1978 *Anisolabis atra* – BRINDLE, Dermaptera of Africa, Tervuren, **2**: 124. Central Africa.

Literature: BURR 1911; POPHAM & BRINDLE 1966; SAKAI 1970; STEINMANN 1977, 1978.

Description: Male colour black, shiny; femora and tibiae reddish-brown, strongly marked with black, tarsi yellowish; forceps dark brown. Head transverse, mainly convex but depressed along postfrontal sutures; coronal suture weak. Eyes comparatively large, about as long as the scape. First antennal joint normal, shorter than the distance between antennal bases. Pronotum quadrate, more or less as wide as head, margins more or less straight; median longitudinal furrow distinct. Tegmina and wings absent. Meso- and metanotum normal, simple. Abdomen more or less parallel-sided, finely punctate. Ultimate tergite quadrangular, median longitudinal sulcus distinct, cuticle with longitudinal irregular bands of punctures, with smooth cuticle between. Forceps (Fig. 131) simple, trigonal basally, cylindrical apically; inner margins with crenulations at base. Penultimate sternite almost triangular, apex truncate. Genitalia unknown. – **Female** unknown. – **Length** of body with forceps: 27–33 mm.

Distribution: Cameroon.

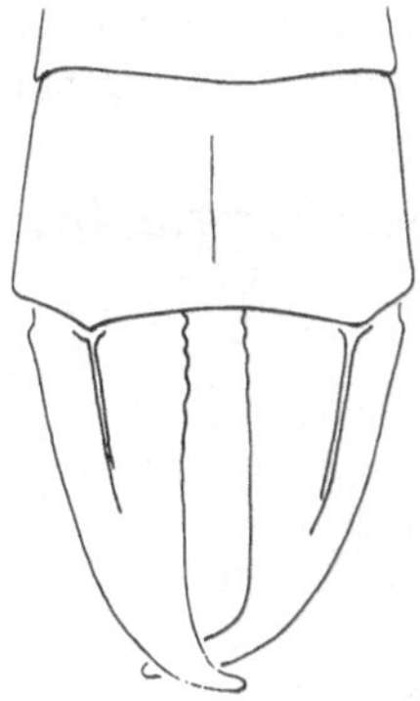

Fig. 131 Male ultimate tergite with forceps of *Anisolabis atra* BORELLI. Original.

19. Anisolabis infelix BURR

1907 *Anisolabis infelix* BURR, Wiss. Ergeb. Kilimanjaro-Meru Exp., **17**: 6. – Terra typica: Tanganyika (Type male: Brit. Mus. nat. Hist., London).
1915 *Gelotolabis infelix* – BURR, J. r. microsc. Soc., London, **1915**: 541; pl. 12, fig. 5 (male genitalia). Central Africa.
1965 *Paralabis infelix* – BRINDLE, Ann. Mag. nat. Hist., London, (13) **7**: 175.
1966 *Gonolabis infelix* – POPHAM & BRINDLE, Entomologist, London, **99**: 273. Africa.
1978 *Anisolabis infelix* – BRINDLE, Dermaptera of Africa, Tervuren, **2**: 94; fig. 192 (penultimate sternite of male), 269 (external paramere of male genitalia). Malawi, Tanzania, Zambia.
Literature: BURR 1910, 1911; SAKAI 1970; STEINMANN 1978, 1984.

Description: Male dark brown to black, rather shiny, legs yellow, apices of femora somewhat darkened. Head broad and tumid; postfrontal and coronal sutures fine or absent. Eyes moderately large, prominent, about as long as the length of scape. Antennae with first joint long but shorter than the distance between antennal bases; second joint quadrate, third long, cylindrical. Pronotum slightly transverse and slightly expanded posteriorly, lateral margins somewhat sinuate; median longitudinal furrow present, but short, and depressed for basal half, raised posteriorly as a narrow line. Tegmina and wings absent. Meso- and metanotum typical. Abdomen widened to last tergite, of *Gonolabis*-type, depressed; tergites 6–9 rugoso-striate at sides but without lateral longitudinal ridges. Ultimate tergite with prominent alternating bands of punctures and smooth cuticle medially; median longitudinal sulcus strongly depressed medially;

lateral longitudinal ridge present and rugoso-striate, dorso-lateral ridge represented by a raised tubercle above the base of each branch of the forceps. Pygidium small, apex excised. Forceps (Fig. 132) strongly asymmetrical; both branches of forceps short, left branch weakly curved, right branch more strongly curved, inner margins of both weakly crenulated. Penultimate sternite short, broad, and rounded; posterior margin truncate. Genitalia (Fig. 133) specific; parameres very long, large, median longitudinal incision between external parameres very deep, central parameral plate V-shaped. Genital lobes short, small, without virgae. External parameres simple, obtuse apically. – **Female** similar to male, but punctation weaker, forceps usually symmetrical, tapering and contiguous. – **Length** of body with forceps: male: 18–22 mm, female: 18–24,5 mm.

Distribution: Zambia, Malawi, North Mozambique, Tanzania.

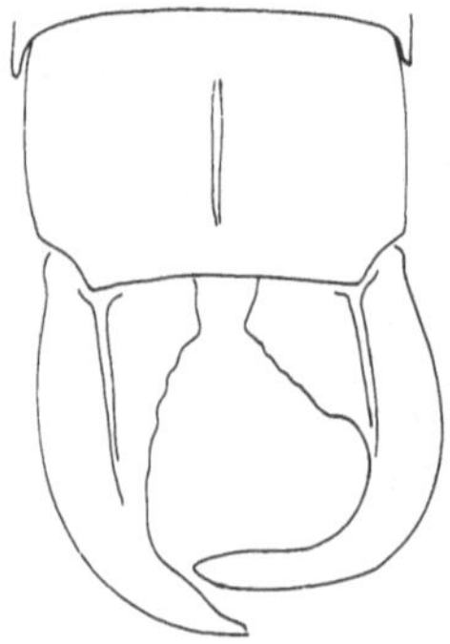

Fig. 132 Male ultimate tergite and forceps of *Anisolabis infelix* BURR. Original.

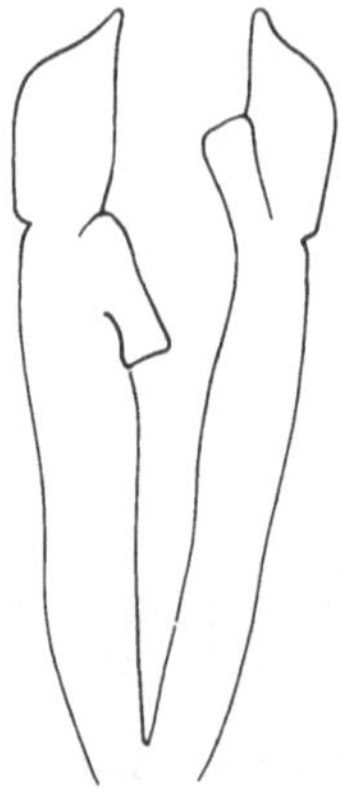

Fig. 133 Male genitalia of *Anisolabis infelix* BURR. After BURR.

20. Anisolabis straeleni (HINCKS)

1955 *Apolabis straelini* HINCKS, Explor. Parc Nat. Upemba, Bruxelles, **35** (1): 7, fig. 1 (male), 2 (male genitalia). – Terra typica: Congo (Type male: Brit. Mus. nat. Hist., London).
1966 *Carcinophora straeleni* – POPHAM & BRINDLE, Entomologist, London, **99**: 275.
1978 *Apolabis straeleni* – STEINMANN, Dtsch. ent. Zeitschr., Berlin, n. F. **25**: 184.
1978 *Anisolabis straeleni* – BRINDLE, Dermaptera of Africa, Tervuren, **2**: 96; fig. 179 (male), 180 (male penultimate sternite), 268 (external paramere of male genitalia). Central Africa.

Description: Male colour black, rather shiny; antennae dark brown to blackish-brown, with one distal joint white (10 or 11); legs yellowish, femora with a dark spot near apex or vaguely darkened. Head transverse, or about as long as broad; moderately convex, Postfrontal sutures obsolete, coronal suture rather short, distinct. Eyes comparatively large, more or less as long as the scape; the latter moderately short, shorter than the distance between antennal bases; antennae 16-jointed (?). Pronotum almost quadrate, slightly broader than long at caudal margin which is slightly wider than cephalic margin; median longitudinal furrow present. Tegmina and wings absent. Meso- and metanotum slightly broader than pronotum and similarly sculptured. Abdomen widened to 8th tergite then slightly contracted to ultimate tergite; lateral sculpture of tergites 6–9 very rugose but without lateral longitudinal carinae or ridges. Ultimate tergite transverse, median longitudinal sulcus moderately distinct; punctation somewhat variable, and with longitudinal smooth bands medially. Forceps (Fig. 134) with broad and short branches; inner margins almost smooth, branches more or less contiguous. Penultimate sternite triangular, sides rounded as well as apex. Genitalia (Fig. 135) characteristic; paramere V-shaped, median longitudinal incision of anterior margin very deep; genital lobes without virgae. External parameres (Fig. 136) broad, outer margin sinuate, apex acicular. – **Female** similar to male, but punctation weaker, forceps symmetrical, straight, except at apex. – **Length** of body with forceps: male: 19–21 mm, female: 20–24 mm.

Distribution: Congo Republic, Zaire.

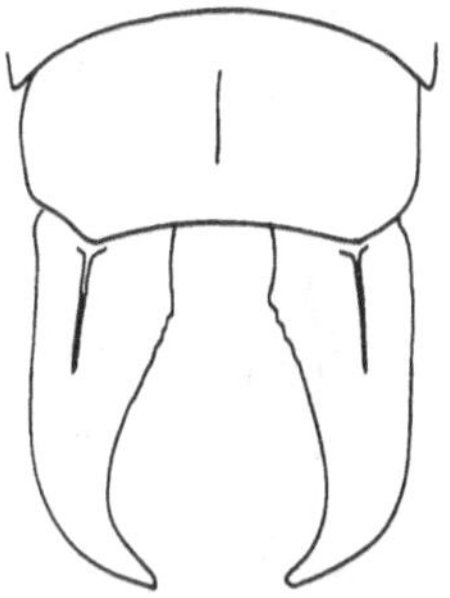

Fig. 134 Male ultimate tergite and forceps of *Anisolabis straeleni* (HINCKS). Original.

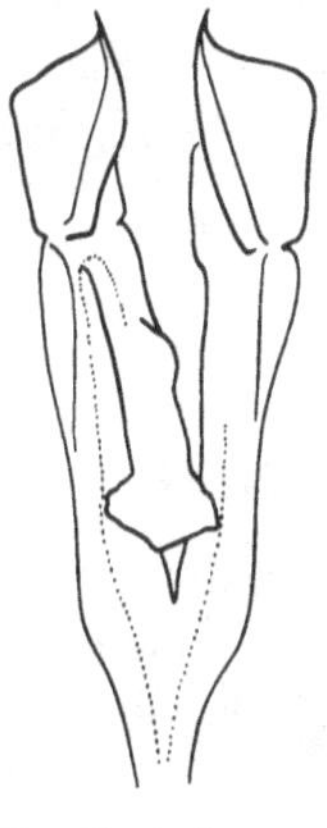

Fig. 135 Genital armature of holotype of *Anisolabis straeleni* (HINCKS). After HINCKS.

Fig. 136 External paramere of male genitalia of *Anisolabis straeleni* (HINCKS). After BRINDLE.

21. Anisolabis pagana BURR

1915 *Anisolabis pagana* BURR, J. r. microsc. Soc., London, **1916**: 535, fig. 61 (apex of abdomen with forceps of male, lateral view). – Terra typica: Cameroon (Type male: Mus. Naturkunde, Berlin).

1978 *Anisolabis pagana* – BRINDLE, Dermaptera of Africa, Tervuren, **2**: 102, fig. 167 (male), 267 (external paramere of male genitalia).

Literature: POPHAM & BRINDLE 1966; SAKAI 1970; STEINMANN 1978, 1984.

Description: Male colour black, shiny; antennae and mouthparts dark brown, basal joint of antennae yellow, legs yellowish, femora with a dark band slightly basal of centre. Head strongly transverse, postfrontal and coronal sutures weak; eyes comparatively large, but a little shorter than the length of scape. First antennal joint very short, shorter than the distance between antennal bases. Pronotum broad, slightly widened posteriorly; lateral margins straight, all angles rounded; median longitudinal furrow distinct; cuticle of pronotum coriaceous. Tegmina and wings absent. Meso- and metanotum normal. Abdomen weakly widened medially, depressed; sides of tergites punctate, and sides of tergite 8 slightly rugoso-striate, but without lateral longitudinal ridge. Ultimate tergite broad, median longitudinal sulcus in a deep depression; posterior margin concave. Forceps (Fig. 137) short, asymmetrical, strongly trigonal basally, apex narrowed and incurved; inner margins almost simple. Penultimate sternite (Fig. 138) broadly triangular, apex truncate, angles rounded. Genitalia broad, median longitudinal incision between external parameres deep; genital lobes normal. External parameres (Fig. 139) of *Anisolabis*-type, long, inner membrane wide, margin sinuate and extending well below the parameres outer edge; median membrane very wide in centre. – **Female** similar to male, but sides of posterior tergites less strongly punctate, and both branches of forceps longer than those of male, almost straight, inner margins almost simple. – **Length** of body with forceps: male 14–17 mm, female: 14–19 mm.

Distribution: Cameroon, Rio Muni, Gabon, Equatorial Guinea, Congo Republic.

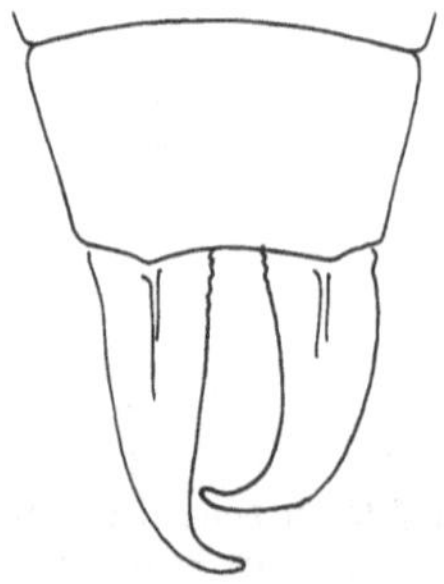

Fig. 137 Male ultimate tergite and forceps of *Anisolabis pagana* BURR. Original.

Fig. 138 Posterior margin of male penultimate sternite of *Anisolabis pagana* BURR. After BRINDLE.

Fig. 139 External paramere of male genitalia of *Anisolabis pagana* BURR. After BRINDLE.

22. Anisolabis duplicata nom. nov.

Anisolabis duplicata **nov. nom.** pro *Gonolabis picea* BORELLI 1907; secundary junior homonymy of *picea* SHIRAKI, 1905.

1907 *Gonolabis picea* BORELLI, Boll. Mus. Zool. Anat. comp. Univ., Torino, **22** No. 572: 1. – Terra typica: Uganda (Type male: Mus. Ist. Zool. sist. Univ., Torino).

1915 *Apolabis picea* – BURR, J. r. microsc. Soc., London, **1915**: 539, pl. XI, fig. 8 (male genitalia).
1966 *Gonolabis picea* – POPHAM & BRINDLE, Entomologist, London, **99**: 273.
1966 *Carcinophora picea* – POPHAM & BRINDLE, Entomologist, London, **99**: 275.
1978 *Anisolabis picea* – BRINDLE, Dermaptera of Africa, Tervuren, **2**: 116, fig. 161 (male), 162 (penultimate sternite), 263 (external paramere of male genitalia).
Literature: BURR 1911; SAKAI 1970; STEINMANN 1977, 1978, 1981.

Description: Male general colour black, antennae dark brown to black, legs yellowish-brown, femora darkened. Head transverse, rounded; postfrontal and coronal sutures not visible; eyes small, comparatively shorter than the scape; antennae 14-jointed, first joint small, shorter than the distance between antennal bases. Pronotum quadrate or a little transverse, widened posteriorly; median longitudinal furrow present; lateral margins straight, posterior angles rounded, cuticle of pronotum smoother than head. Tegmina and wings absent. Meso- and metanotum coriaceous, relatively smooth. Abdomen slender, widened posteriorly, sometimes more, sometimes less gradually widened from base; tergites 7–9 rather rugoso-striate, but without lateral longitudinal ridges. Ultimate tergite broad, median longitudinal sulcus well marked and depressed; lateral longitudinal ridge prominent and rugoso-striate, dorso-lateral ridge represented by a swelling on each side towards the posterior margin. Forceps (Fig. 140) short, strongly trigonal basally, cylindrical apically; symmetrical or nearly so. Penultimate sternite broad, somewhat pentagonal, angles rounded, posterior margin wide and truncate. Genitalia (Fig. 141) specific; central parameral plate V-shaped, median incision of anterior margin very deep; genital lobes well-developed, without virgae. External parameres moderately short, broad,

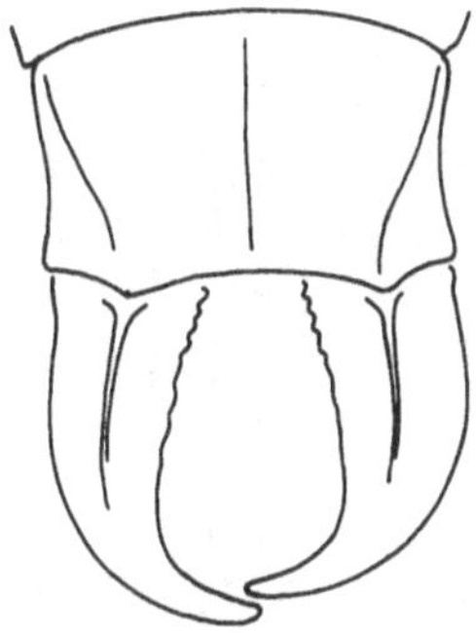

Fig. 140 Male ultimate tergite and forceps of *Anisolabis duplicata* nom.nov. Original.

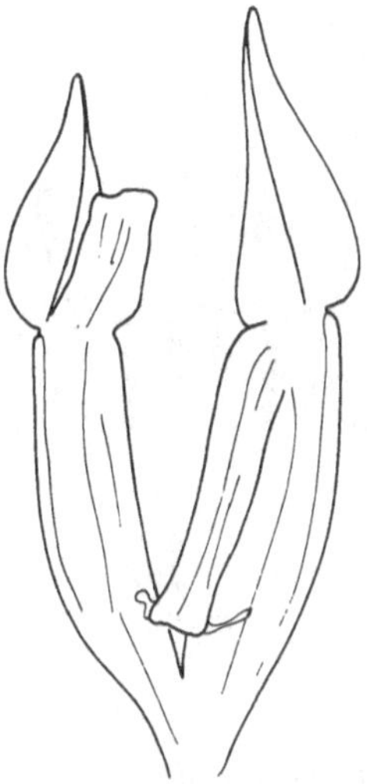

Fig. 141 Male genitalia of *Anisolabis duplicata* nom.nov. Original.

narrowed distally, apex obtuse. – **Female** similar to male, but forceps symmetrical, contiguous, tapering. – **Length** of body with forceps: male: 13,5–16,5; female: 14–16 mm.

Distribution: Uganda, Ethiopia, Cameroon.

23. Anisolabis nimbaensis HINCKS

1954 *Anisolabis nimbaensis* HINCKS, Mém. Inst. fr. Afr. noire, Paris, Dakar, **40**: 103; fig. 1 (male genitalia). – Terra typica: Guinea (Type male: Mus. nation. Hist. nat., Paris).

1978 *Anisolabis nimbaensis* – BRINDLE, Dermaptera of Africa, Tervuren, **2**: 121; fig. 170 (male), 171 (penultimate sternite of male), 266 (external paramere of male genitalia).

Literature: SAKAI 1970; STEINMANN 1978, 1984.

Description: Male colour black or blackish-brown; lateral and posterior parts of head reddish, shiny; antennae dark brown, legs reddish, sometimes yellowish red. Head more or less as broad as long, slightly depressed between antennae; postfrontal and coronal sutures distinct. Eyes normal, small, shorter than the length of head behind eyes. First antennal joint long, but shorter than the distance between antennal bases. Pronotum narrower then head, subquadrate, sides straight, slightly divergent posteriorly. Median longitudinal furrow present, or almost absent. Tegmina and wings entirely absent. Meso- and metanotum with feeble scattered punctation. Abdomen weakly widened and depressed; sides of tergites 6–9 rugoso-striate but without longitudinal lateral ridges. Ultimate tergite transverse, weakly and sparsely punctate; lateral longitudinal sulcus present. Forceps (Fig. 142) trigonal, more or less distant at base, asymmetrical, right branch sharply incurved distad. Penultimate sternite broadly triangular, apex slightly truncate. Genitalia (Fig. 143) narrow; paramere relatively short,

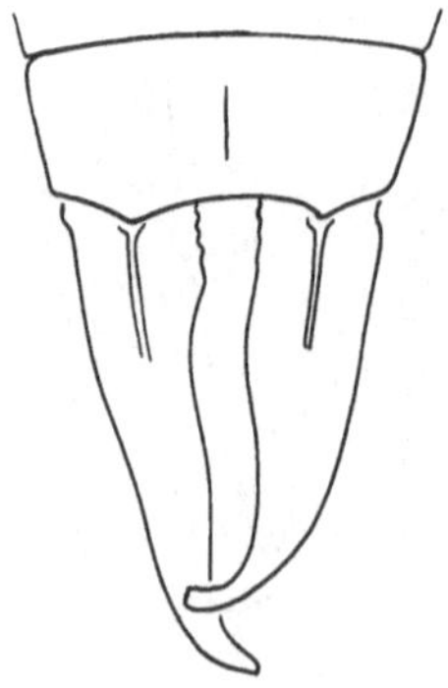

Fig. 142 Male ultimate tergite and forceps of *Anisolabis nimbaensis* HINCKS. After BRINDLE.

Fig. 143 Genital armature of holotype of *Anisolabis nimbaensis* HINCKS. After HINCKS.

incision of anterior margin deep and very narrow; genital lobes fully developed, with virgae. External parameres well-developed, inner and outer margins parallel at medial section; apex more or less of *Carcinophora*-type. – **Female** as male, but forceps simple, trigonal basally, cylindrical apically, symmetrical. – **Length** of body with forceps: male: 18–23 mm, female: 16–24 mm.

Distribution: Guinea, Ivory Coast, Liberia.

24. Anisolabis punctulata Brindle

1978 *Anisolabis punctulata* Brindle, Dermaptera of Africa, Tervuren, **2**: 114, fig. 193 (male), 194 (penultimate sternite of male), 283 (external paramere of male genitalia). – Terra typica: Zaire (Type male: Mus. r. Afr. centr., Tervuren).

Description: Male colour blackish, first antennal joint and legs yellow, femora with a wide median blackish band; pronotum narrowly yellow at sides; forceps red. Head rounded, almost as broad as long; postfrontal and coronal sutures distinct. Eyes comparatively small, slightly shorter than the length of scape. First antennal joint long, but shorter than the distance between antennal bases; cuticle of vertex coriaceous; smooth shiny areas occur around occiput where the postfrontal sutures meet and on posterior margin. Pronotum quadrate, weakly widened posteriorly; median longitudinal furrow present in centre; lateral margins straight, all angles rounded. Tegmina and wings absent. Mesonotum long, coarsely coriaceous, with two small shiny smooth areas on each side of midline; metanotum similar. Abdomen weakly widened medially and depressed; sides of tergites 6–9 more strongly punctate and sides of 6–7 slightly rugoso-striate, but without lateral longitudinal ridges. Ultimate tergite broad; with cuticle smooth and shiny; lateral longitudinal ridge prominent. Forceps (Fig. 144) short, broad, asymmetrical; trigonal basally, cylindrical apically. Penultimate sternite broadly triangular, apex rather sharply truncate. Genitalia normal, parameres broad, median longitudinal incision between external parameres typical; genital lobes well developed. External parameres (Fig. 145)

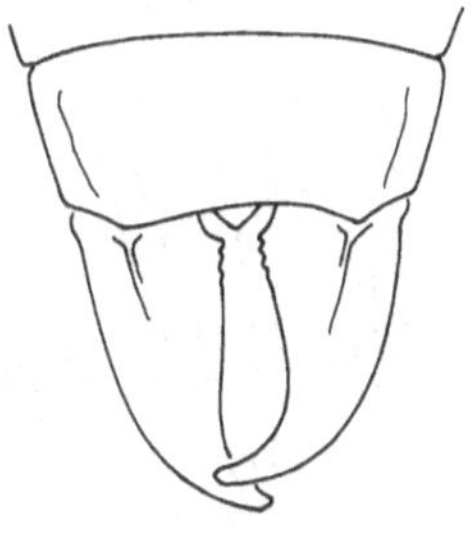

Fig. 144 Male ultimate tergite and forceps of *Anisolabis punctulata* Brindle. After Brindle.

Fig. 145 External paramere of male genitalia of *Anisolabis punctulata* Brindle. After Brindle.

almost quadrangular, median membrane narrow, inner membrane wide. – **Female** unknown. – **Length** of body with forceps: 13 mm.

Distribution: Zaire.

25. Anisolabis jeanneli (MENOZZI)

1938 *Apolabis jeanneli* MENOZZI, Mém. Miss. Sci. de l'Omo, Nairobi, **4**: 135, fig. 1 (male). – Terra typica: Kenya (Type male: Mus. nation. Hist. nat., Paris).
1966 *Carcinophora jeanneli* – POPHAM & BRINDLE, Entomologist, London, **99**: 275.
1978 *Anisolabis jeanneli* – BRINDLE, Dermaptera of Africa, Tervuren, **2**: 120; fig. 164 (male), 163 (male genital armature), 165 (penultimate sternite of male).
Literature: SAKAI 1970; STEINMANN 1978, 1984.

Description: Male colour blackish-brown to almost dark reddish-brown; head red, antennae dark brown, basal joints yellow, legs and lateral margins of pronotum yellow. Head rounded, widened behind eyes; postfrontal and coronal sutures weak; eyes moderately small, a little shorter than the length of first antennal joint; latter shorter than the distance between antennal bases. Pronotum transverse, strongly widened posteriorly; lateral margins straight, slightly sinuate, median longitudinal furrow depressed anteriorly and slightly posteriorly. Cuticle coriaceous, and with sparse punctures. Tegmina and wings entirely absent. Meso- and metanotum typical. Abdomen strongly widened medially and depressed, a little of *Gonolabis*-type; tergites 5–9 rugoso-striate towards and on sides, each with a rugoso-striate blunt lateral longitudinal ridge. Ultimate tergite relatively small, strongly and regularly punctate and rather flattened, median longitudinal sulcus almost absent, and no lateral or dorso-lateral ridges. Forceps (Fig. 146) not of *Anisolabis*-type, each branch slender, dorsal ridge rounded and short, inner margins almost smooth, the right branch

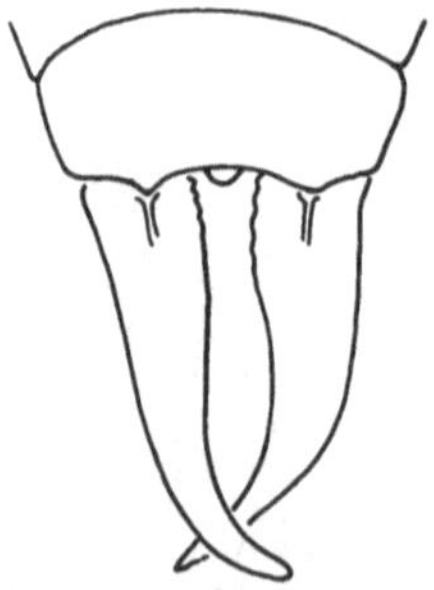

Fig. 146 Male ultimate tergite and forceps of *Anisolabis jeanneli* (MENOZZI). After MENOZZI.

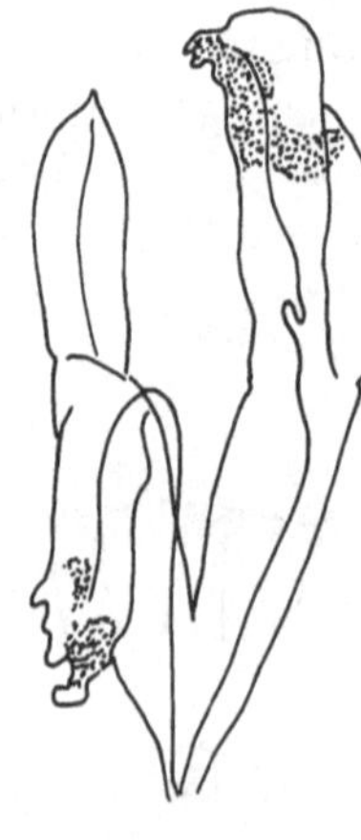

Fig. 147 Male genitalia of *Anisolabis jeanneli* (MENOZZI). After BRINDLE.

curved more than the left. Penultimate sternite triangular, rounded distally, apex almost truncate and straight, but very slightly concave. Genitalia (Fig. 147) characteristic, not of *Anisolabis*-type, central parameral plate elongated, median incision very deep and wide; genital lobes fully developed, with virgae and some chitinous granules. External parameres very long, outer and inner margins more or less parallel, apex acuminate. – **Female** very similar to male, but abdomen less widened and with weaker punctation; penultimate sternite evenly rounded distally, branches of forceps more slender, only weakly curved and closer together at base, right branch more curved than left branch. – **Length** of body with forceps: male: 13–15 mm, female: 14–16 mm.

Distribution: Kenya, Tanzania.

26. Anisolabis burri (ZACHER)

1911 *Gelotolabis burri* ZACHER, Zool. Jahrb., Jena, **30**: 386, fig. P (male genitalia), Q (male forceps). – Terra typica: North-east Africa (Type male: Mus. Naturkunde, Berlin).

1911 *Anisolabis aethiopica* BURR, Ann. Mag. nat. Hist., London, (8) **8**: 44. – Terra typica: Abyssinia (Type male: Brit. Mus. nat. Hist., London).

1915 *Gelotolabis burri* – BURR, J. r. microsc. Soc., London, **1915**: 541, pl. XII, fig. 4 (male genitalia). Synonymy of *aethopica* with *burri*.

1966 *Gonolabis burri* – POPHAM & BRINDLE, Entomologist, London, **99**: 273.

1978 *Gelotolabis burri* – STEINMANN, Dtsch. ent. Zeitschr., Berlin, n. F. **25**: 185, fig. 33 (male genitalia).

1978 *Anisolabis burri* – BRINDLE, Dermaptera of Africa, Tervuren, **2**: 129, fig. 152 (male), 258 (external paramere of male genitalia). Ethiopia.

Literature: BURR 1911; SAKAI 1970; STEINMANN 1978, 1981, 1984.

Description: Male reddish-brown to blackish; antennae yellowish, pronotum with dark brown line anteriorly; femora and tibiae dark brown. Head tumid, strongly transverse, and depressed along epicranial sutures, the postfrontal and coronal sutures distinct. Eyes comparatively small, shorter than first antennal joint, the latter small, shorter than the distance between antennal bases. Pronotum transverse, widened posteriorly, all margins straight, and all angles rounded; cuticle coriaceous and with weak and shallow punctures posteriorly; median longitudinal furrow depressed anteriorly and slightly posteriorly. Tegmina and wings absent. Meso- and metanotum transverse, coriaceous. Abdomen depressed, rather strongly widened to tergites 6–7; sides of tergites 5–9 rugoso-striate, and tergites 6–8 each with a very broad blunt lateral longitudinal ridge. Ultimate tergite very broad, median longitudinal sulcus deep and included in a triangular depression. Forceps (Fig. 148) asymmetrical, inner margins with very prominent tooth at basal section; dorsal keel distinct. Penultimate sternite broadly rounded, slightly truncate apically, closely and finely punctate at base. Genitalia (Fig. 149; gen. prep. No. 703, det. DR. H. STEINMANN) specific, paramere very long, V-shaped, genital lobes well developed but without virgae. External parameres (Fig. 150) short, broad, almost quadrangular, apex pointed. – **Female** similar to male; larger with widened abdomen, and weaker punctation; both branches of forceps straight for most part, distinctly narrowed before apical cylindrical part; inner margins crenulated. – **Length** of body with forceps: male: 24–27 mm, female: 24,5–28,6 mm.

Distribution: Ethiopia, Egypt, Somalia.

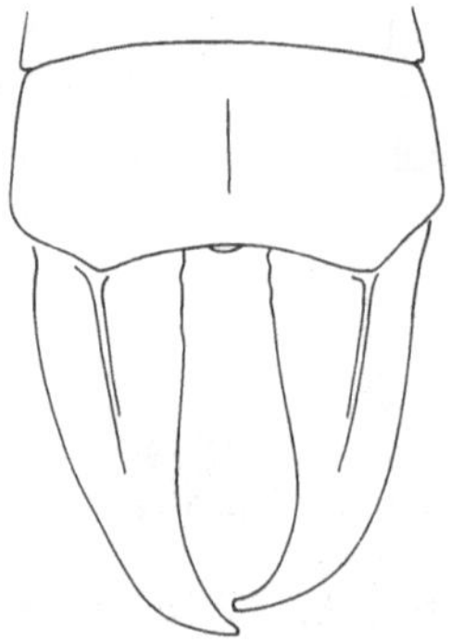

Fig. 148 Male ultimate tergite and forceps of *Anisolabis burri* (ZACHER). After ZACHER.

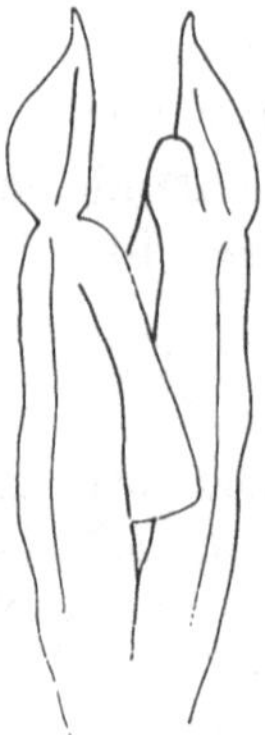

Fig. 149 Male genitalia of *Anisolabis burri* (ZACHER). Original.

Fig. 150 External paramere of male genitalia of *Anisolabis burri* (ZACHER). After BRINDLE.

27. Anisolabis scotti BRINDLE

1978 *Anisolabis scotti* BRINDLE, Dermaptera of Africa, Tervuren, **2**: 131, fig. 195 (holotype), 196 (penultimate sternite of male), 260 (external paramere of male genitalia). – Terra typica: Ethiopia (Type male: Brit. Mus. nat. Hist., London).

Description: Male colour dark reddish-brown; head blackish-brown, antennae dark brown, basal joints yellowish-brown. Head strongly transverse, convex; postfrontal and coronal sutures weak; frons tumid, occiput depressed; eyes moderately small, shorter than the scape; the latter small, shorter than the distance between antennal bases. Pronotum quadrate or a little transverse, widened posteriorly, lateral margins straight, median longitudinal furrow distinct. Tegmina and wings absent. Meso- and metanotum typical. Abdomen strongly widened medially and depressed; tergites 3–9 with fine, close punctation basally and with stronger punctation distally; sides of tergites 6–9 very strongly rugoso-striate, and tergites 6–8 each with a blunt longitudinal lateral ridge. Ultimate tergite with median longitudinal sulcus and strongly depressed towards posterior margin. Forceps (Fig. 151) short, asymmetrical; left branch curved distally, right branch curved sharply beyond midpoint. Penultimate sternite triangular, apex truncate. Genitalia typical, paramere broad, median incision of anterior margin deep; genital lobes well developed. External parameres

(Fig. 152) long, inner membrane wide, median membrane wide in centre, narrowed distally. – **Female** unknown. – **Length** of body with forceps: 15–18 mm.

Distribution: Ethiopia.

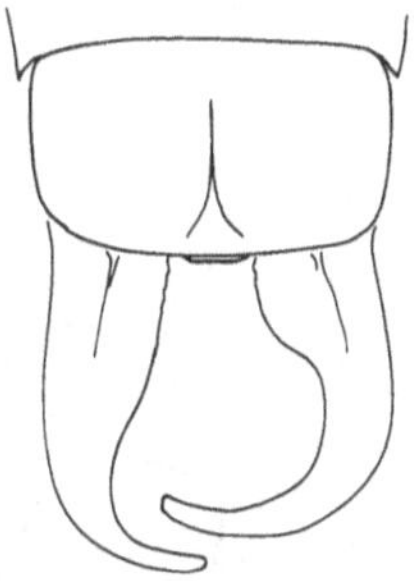

Fig. 151 Male ultimate tergite and forceps of *Anisolabis scotti* BRINDLE. After BRINDLE.

Fig. 152 External paramere of male genitalia of *Anisolabis scotti* BRINDLE. After BRINDLE.

28. Anisolabis laeta (GERSTAECKER)

1869 *Forficula laeta* GERSTAECKER, Arch. Naturgesch., Berlin, **35**: 221. – Terra typica: East Africa (Type?; without abdomen: Staatl. Mus., Greifswald).
1873 *Anisolabis laeta* – DECKEN, Reisen, (2) **3**: 49; pl. 3, fig. 8 (male).
1915 *Apolabis laeta* – BURR, J. r. microsc. Soc., London, **1915**: 539.
1966 *Carcinophora laeta* – POPHAM & BRINDLE, Entomologist, London, **99**: 275.
1978 *Anisolabis laeta* – BRINDLE, Dermaptera of Africa, Tervuren, **2**: 119; fig. 176 (male forceps), 177 (penultimate sternite), 261 (external paramere of male genitalia). Tanzania.

Literature: BORMANS 1900; BURR 1910, 1911; KIRBY 1904; SAKAI 1970; STEINMANN 1978, 1984.

Description: Male colour blackish, rather dull, head reddish or partly reddish; antennae yellowish-brown to brown, pronotum narrowly yellow laterally, legs yellow, femora darkened; forceps red. Head transverse, and very slightly depressed on vertex; frons tumid; postfrontal sutures indistinct, coronal suture present; eyes small, shorter than the length of head behind eyes. First antennal joint long, but shorter than the distance between antennal bases. Pronotum transverse, widened posteriorly; median longitudinal furrow well marked medially; cuticle of pronotum coriaceous. Tegmina and wings absent. Meso- and metanotum simple. Abdomen scarcely widened medially, rather depressed; tergites 6–7 with well-defined lateral longitudinal ridges. Ultimate tergite broad, large, weakly punctate, punctures shallow; median longitudinal sulcus present, but no well defined lateral longitudinal ridge or keel. Forceps (Fig. 153) more or less symmetrical, strongly trigonal basally, and cylindrical apically. Penultimate sternite broadly rounded, apex truncate. Genitalia (Fig. 154) narrow, but large, central parameral plate with very deep median incision between external parameres, genital lobes well developed, without virgae. External parameres (Fig. 155) broad, acuminate and sharply pointed at apex, inner membrane narrow, median membrane wide. – **Female** similar to male, but penultimate sternite rounded at apex, and branches of forceps straighter, inner margins with small denticulations. – **Length** of body with forceps: male: 14,5–17, female: 15–18 mm.

Distribution: Tanzania.

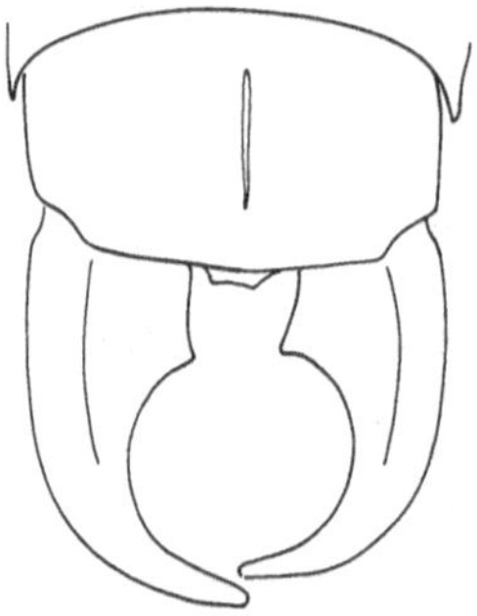

Fig. 153 Male ultimate tergite and forceps of *Anisolabis laeta* (GERSTAECKER). After BRINDLE.

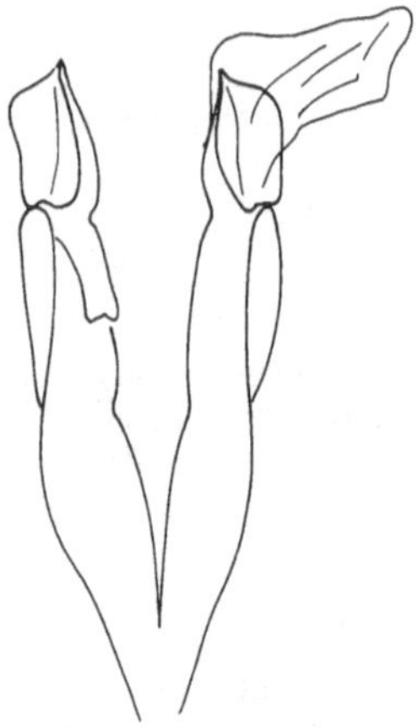

Fig. 154 Male genitalia of *Anisolabis laeta* (GERSTAECKER). Original.

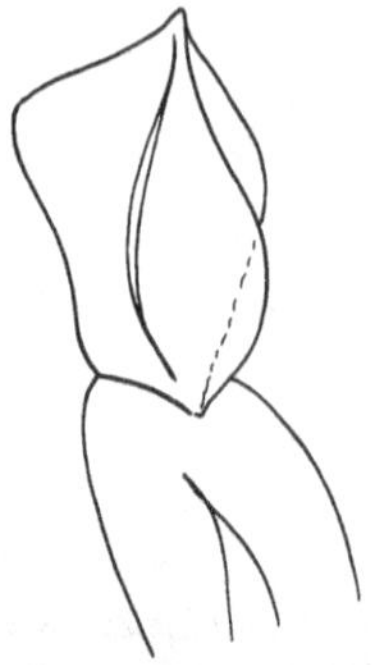

Fig. 155 External paramere of male genitalia of *Anisolabis laeta* (GERSTAECKER). After BRINDLE.

29. Anisolabis tumida BORELLI

1907 *Anisolabis tumida* BORELLI, Ann. Mus. civ. Stor. nat., Genova, **43**: 364. – Terra typica: Fernando Po (Type male: Museo civ. Stor. nat., Genova).

1978 *Anisolabis tumida* – BRINDLE, Dermaptera of Africa, Tervuren, **2**: 99; fig. 190 (male), 191 (penultimate sternite), 282 (external paramere of male genitalia). Congo Republic, Cameroon.

Literature: BURR 1910, 1911; POPHAM & BRINDLE 1966; SAKAI 1970; STEINMANN 1978, 1984.

Description: Male colour blackish-brown, shiny, antennae dark brown, but three basal joints yellow, abdomen red. Head strongly transverse, convex, frons tumid, occiput a little depressed; postfrontal and coronal sutures distinct. Eyes large, more or less as long as the length of head behind eyes. First antennal joint small, much shorter than the distance between antennal bases, apex of 11th joint, and all joint 12 or 11–12 completely white. Pronotum strongly transverse, almost parallel-sided, lateral margins weakly convex; median longitudinal furrow weak but depressed anteriorly. Tegmina and wings absent. Meso- and metanotum normal. Abdomen slender, depressed, regularly punctate, punctures fairly small and separated by rather more than their own diameter; sides of tergites 7–9 with well defined lateral longitudinal ridges. Ultimate tergite very transverse, median longitudinal sulcus present and depressed near posterior margin. Forceps (Fig. 156) short, asymmetrical, strongly curved, inner margin laminate basally. Penultimate sternite with apex broadly truncate. Genitalia typical, paramere broad, median incision between external parameres deep; genital lobes well developed. External parameres (Fig. 157) broad basally and narrowed distally, apex obtuse. – **Female** unknown to me. – **Length** of body with forceps: 15 mm.

Distribution: Equatorial Guinea, Cameroon, Rio Muni, Congo Republic.

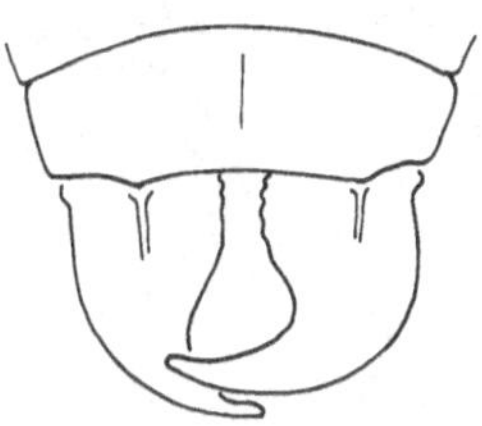

Fig. 156 Male ultimate tergite and forceps of *Anisolabis tumida* BORELLI. After BRINDLE.

Fig. 157 External paramere of male genitalia of *Anisolabis tumida* BORELLI. After BRINDLE.

30. Anisolabis royi BRINDLE

1971 *Anisolabis royi* BRINDLE, Mém. Inst. fr. Afr. noire, Paris, Dakar, **86**: 268; fig. 1 (male), 2 (posterior abdominal segments of male in lateral view), 3 (male genitalia), 4 (penultimate sternite of male). – Terra typica: Sierra Leone (Type male: Mus. nation. Hist. nat., Paris).
Literature: BRINDLE 1978; STEINMANN 1978, 1984.

Description: Male shiny blackish-brown; head reddish, antennae dark brown, but basal joints yellow; legs yellow, forceps dark red. Head transverse, convex, posterior margin slightly concave; postfrontal and coronal sutures visible, well marked. Eyes comparativelly small, shorter than first antennal joint, the latter long, but a little shorter than the distance between antennal bases. Pronotum transverse, slightly widened posteriorly, all margins straight, and all angles rounded; median longitudinal furrow weak; cuticle of pronotum coriaceous. Tegmina and wings absent. Meso- and metanotum less smooth than pronotum and with sparse punctures anteriorly. Abdomen expanded caudad, depressed medially, basal two tergites sparsely punctate, remaining tergites becoming more strongly punctate posteriorly; sides of tergites 6–9 with well-defined lateral longitudinal ridges. Ultimate tergite broad, median longitudinal sulcus present, lateral longitudinal ridge distinct but short. Forceps (Fig. 158) short, asymmetrical, strongly trigonal basally, inner margins smooth or with fine crenulations. Penultimate sternite broadly rounded, somewhat sinuate at sides. Genitalia (Fig. 159) very broad; paramere wide, median incision deep, genital lobes with virgae. External parameres broad at base and pointed apically. – **Female** very similar to male, but abdomen widened to median abdominal tergites and more weakly punctate; ultimate tergite narrower and without dorso-lateral ridges; forceps straighter, inner margins slightly dentated basally. – **Length** of body with forceps: male: 15–17 mm, female: 15,5–18 mm.

Distribution: Sierra Leone.

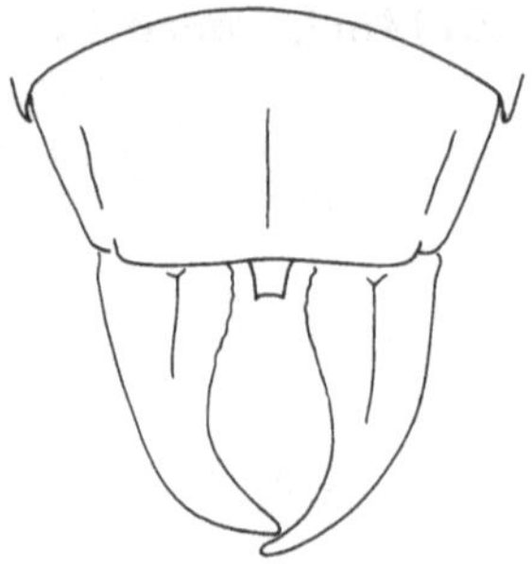

Fig. 158 Male ultimate tergite and forceps of *Anisolabis royi* BRINDLE. After BRINDLE.

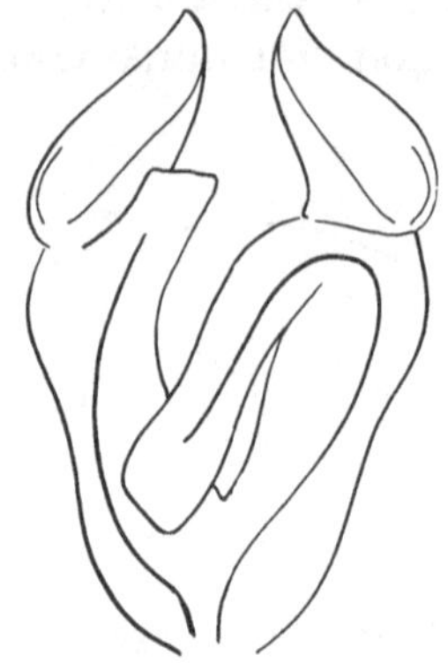

Fig. 159 Genital armature of holotype of *Anisolabis royi* BRINDLE. After BRINDLE.

31. Anisolabis turgida BURR

1911 *Anisolabis turgida* BURR, Ann. Mag. nat. Hist., London, (8) **8**: 41. – Terra typica: Cameroon (Type male: Mus. Naturkunde, Berlin).
1915 *Apolabis turgida* – BURR, J. r. microsc. Soc., London, **1915**: 539; pl. XI, fig. 11 (male genitalia). Western Africa.
1966 *Carcinophora turgida* – POPHAM & BRINDLE, Entomologist, London, **99**: 275.
1978 *Apolabis turgida* – STEINMANN, Dtsch. ent. Zeitschr., Berlin, n. F. **25**: 184.
1978 *Anisolabis turgida* – BRINDLE, Dermaptera of Africa, Tervuren, **2**: 109, fig. 284 (external paramere of male genitalia).

Literature: BURR 1911, 1912; SAKAI 1970; STEINMANN 1977, 1984.

Description: Male colour dark reddish-brown; antennae yellowish-brown, head reddish, legs yellow. Head transverse, a little convex; postfrontal sutures indistinct, coronal suture present. Eyes small, shorter than the length of head behind eyes. First antennal joint long, but a little shorter than the distance between antennal bases. Pronotum quadrate or a little transverse, all angles rounded; lateral margins weakly widened posteriorly; median longitudinal furrow distinct, depressed anteriorly; cuticle of pronotum more or less smooth but with sparse fine punctures. Tegmina and wings absent. Meso- and metanotum transverse, cuticle similar to pronotum. Abdomen weakly widened and depressed, finely punctate, tergites 5–9 rugoso-striate at sides and each with a lateral longitudinal keel. Ultimate tergite broad, large; lateral longitudinal ridge present, and median sulcus distinct; plate irregularly punctate. Forceps (Fig. 160) asymmetrical, dorsal keels very prominent; each branch short, broad at base, arcuate and narrowed distally, inner margins crenulated. Penultimate sternite broadly rounded. Genitalia comparatively short, paramere with very wide median incision between external parameres; genital lobes normal, without virgae. External parameres (Fig. 161) short, broader basally, narrowed distally, apex obtuse. – **Female** unknown. – **Length** of body with forceps: 16–18,5 mm.

Distribution: Cameroon.

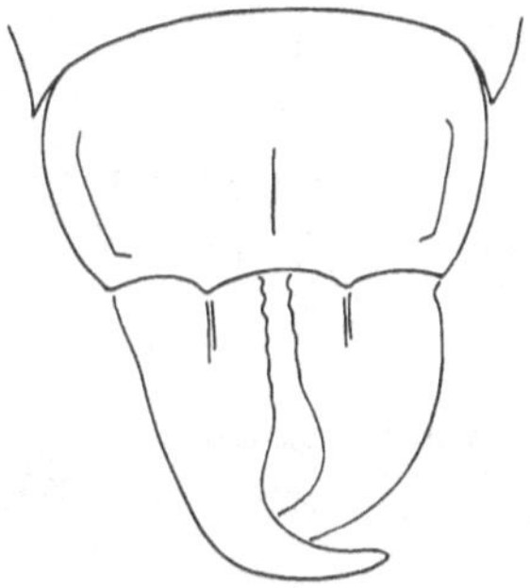

Fig. 160 Male ultimate tergite with forceps of *Anisolabis turgida* BURR. Original.

Fig. 161 External paramere of male genitalia of *Anisolabis turgida* BURR. After BRINDLE.

32. Anisolabis elongata BRINDLE

1978 *Anisolabis elongata* BRINDLE, Dermaptera of Africa, Tervuren, **2**: 103, fig. 205 (holotype), 206 (penultimate sternite of male), 278 (external paramere of male genitalia). – Terra typica: Cameroon (Type male: Calif. Acad. Sci., San Francisco).

Description: Male colour dark reddish-brown to blackish, shiny; antennae dark brown, basal two joints yellow, lateral margins of pronotum yellowish, legs yellow, femora darkened on sides. Head transverse, rounded; postfrontal and coronal sutures distinct. Eyes large, more or less as long as the scape; the latter moderately small, slightly shorter than the distance between antennal bases. Pronotum strongly transverse, lateral margins widened posteriorly, median longitudinal furrow distinct and depressed anteriorly; cuticle of head and pronotum smooth or finely punctate. Tegmina and wings absent. Meso- and metanotum normal, transverse. Abdomen gradually but only weakly widened from base; slender, a little depressed; sides of tergites 8–9 more strongly punctate but not rugoso-striate, nor with lateral longitudinal ridges. Ultimate tergite very broad, with very isolated punctures, median longitudinal sulcus distinct. Forceps (Fig. 162) small, asymmetrical, strongly trigonal basally, cylindrical apically. Penultimate sternite broadly rounded. Genitalia typical, paramere broad, median longitudinal incision of anterior margin deep. Genital lobes normal. External parameres (Fig. 163) characteristic; elongated, narrow, with a small projection near tip, both inner and median membranes almost equal in size. – **Female** similar to male, but sides of tergites 8–9 without lateral longitudinal ridges; forceps longer and slenderer than in male. – **Length** of body with forceps: male: 13,5–15 mm, female: 14–17 mm.

Distribution: Cameroon.

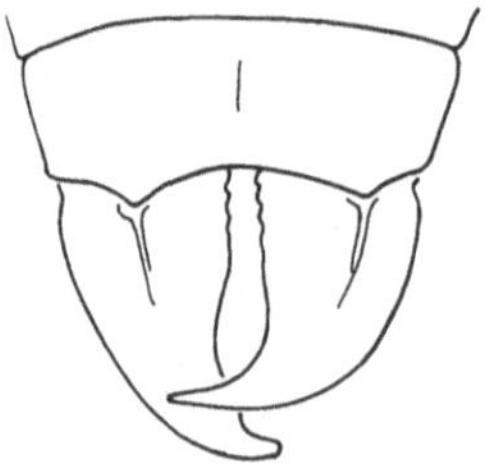

Fig. 162 Holotype ultimate tergite and forceps of *Anisolabis elongata* Brindle. After Brindle.

Fig. 163 External paramere of male genitalia of *Anisolabis elongata* Brindle. After Brindle.

33. Ansiolabis isomorpha Borelli

1907 *Anisolabis isomorpha* Borelli, Ann. Mus. civ. Stor. nat., Genova, **43**: 362. – Terra typica: Portuguese Guinea (Type male: Mus. civ. Stor. nat., Genova).

1915 *Apolabis isomorpha* – Burr, J. r. microsc. Soc., London, **1915**: 528; pl. XI, fig. 12 (male genitalia).

1966 *Carcinophora isomorpha* – Popham & Brindle, Entomologist, London, **99**: 275.

1978 *Apolabis isomorpha* – Steinmann, Dtsch. ent. Zeitschr., Berlin, n. F. **25**: 184.

1978 *Anisolabis isomorpha* – Brindle, Dermaptera of Africa, Tervuren, **2**: 104; fig. 253 (external paramere of male genitalia).

Literature: Borelli 1914; Burr 1911; Sakai 1970; Steinmann 1984.

Description: Male dark brown to reddish brown; head dark red, antennae dark brown, basal joints yellow and joints 15–16 whitish; legs yellow; lateral margins of pronotum yellowish. Head tumid, transverse, more or less convex and flattened along postfrontal sutures; coronal suture distinct. Eyes large, about as long as genae. Antennae 17-jointed, first joint long, but shorter than the distance between antennal bases. Pronotum quadrate or a little transverse, slightly widened posteriorly; median longitudinal furrow distinct. Tegmina and wings absent. Meso- and metanotum transverse. Abdomen typical, scarcely widened medially; sides of tergites 6–9 more strongly punctate, but without longitudinal ridges. Ultimate tergite broad, median longitudinal sulcus present; cuticle almost smooth at the sides and there is no lateral longitudinal ridge. Forceps as in fig. 164, strongly trigonal basally, curved and cylindrical apically. Penultimate sternite with posterior margin rounded. Genitalia (Fig. 165) specific; paramere

V-shaped, genital lobes well developed, with some chitinous granulates. – **Female** similar to male, but forceps slenderer, tapering, contiguous. – **Length** of body with forceps: male: 15,5–18 mm, female: 16–17,5 mm.

Distribution: Gambia, Bissau-Guinea, Liberia, Togo, Cameroon.

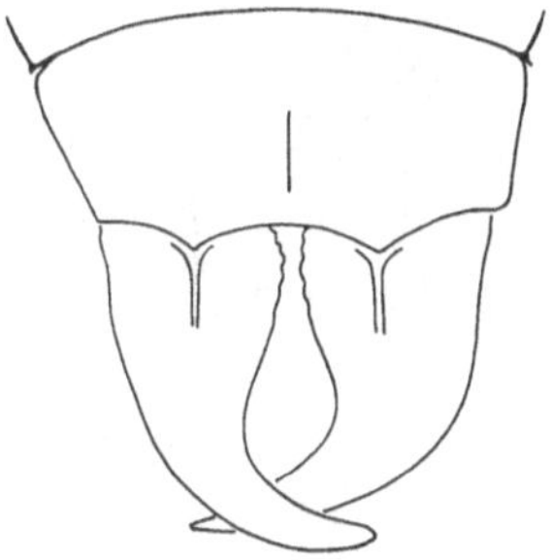

Fig. 164 Male ultimate tergite and forceps of *Anisolabis isomorpha* BORELLI. Original.

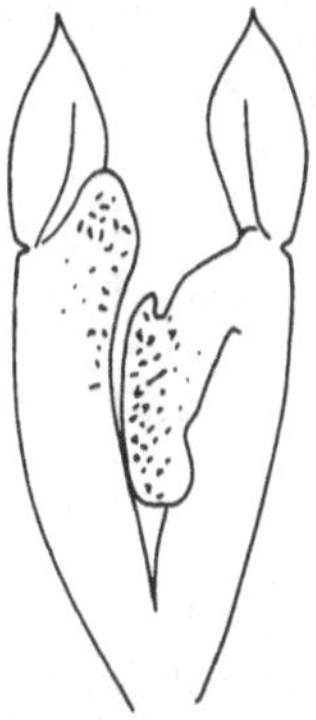

Fig. 165 Male genitalia of *Anisolabis isomorpha* BORELLI. After BURR.

34. Anisolabis brindlei CAPRA

1978 *Anisolabis brindlei* CAPRA, Ann. Mus. civ. Stor. nat., Genova, **5**: 1; fig. 1 (holotype), 2 (ultimate tergite with forceps of male), 3 (genital armature of holotype), 4 (penultimate sternite of male). – Terra typica: Island Perim in Red Sea (Type male: Museo Januense, Genova).

Description: Male colour reddish-brown; antennae light brown, abdomen blackish-red, legs yellowish-brown. Head rounded, transverse; postfrontal and coronal sutures distinct; eyes comparatively small, slightly shorter than the length of head behind eyes. Antennae 21-jointed (?), first joint long, about as long as the distance between antennal bases. Pronotum quadrate or a little transverse; lateral margins straight, slightly widened posteriorly; median longitudinal furrow present. Tegmina and wings absent. Meso- and metanotum normal. Cuticle of head and thoracic nota coriaceous, shiny. Abdomen slender, sides of tergites 5–9 rugoso-striate, with slender and narrow lateral longitudinal ridges. Ultimate tergite broad, median longitudinal sulcus well marked. Forceps (Fig. 166) similar to *maritima* (BONELLI), asymmetrical, strongly trigonal basally, curved and cylindrical apically; inner margins crenulated. Penultimate tergite broadly rounded. Genitalia (Fig. 167) characteristic; central parameral plate narrow, V-shaped, median longitudinal incision of anterior margin very deep, genital lobes well developed. External parameres specific, narrow, elongate. – **Female** unknown. – **Length** of body with forceps: 24–25 mm.

Distribution: Island Perim (Red Sea).

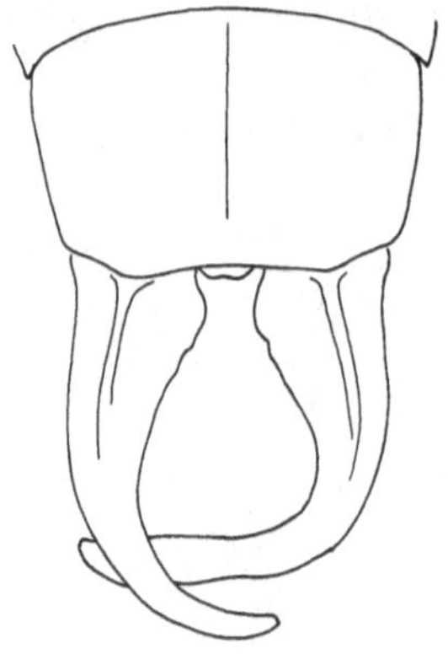

Fig. 166 Holotype ultimate tergite and forceps of *Anisolabis brindlei* CAPRA. After CAPRA.

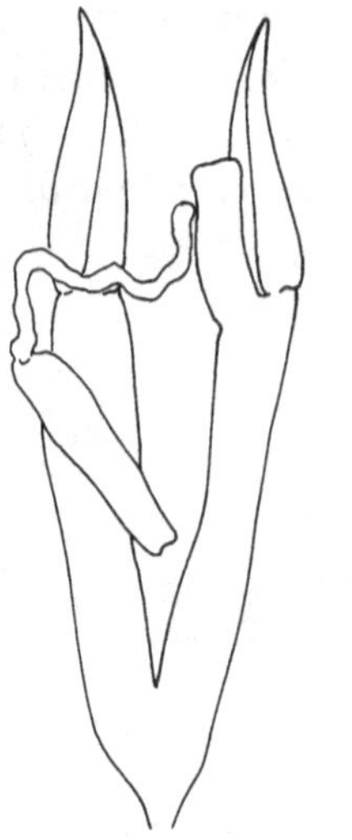

Fig. 167 Genital armature of holotype of *Anisolabis brindlei* CAPRA. After CAPRA.

35. Anisolabis neavei BRINDLE

1978 *Anisolabis neavei* BRINDLE, Dermaptera of Africa, Tervuren, **2**: 97; fig. 212 (male forceps), 213 (penultimate sternite of male), 276 (external paramere of male genitalia). – Terra typica: Tanzania (Type male: Brit. Mus. nat. Hist., London).

Description: Male colour blackish, body shiny; antennae dark yellowish-brown, basal joints scarcely paler; legs yellow. Head transverse, convex; frons tumid, occiput depressed; postfrontal and coronal sutures well marked. Eyes large, but a little shorter than the length of head behind eyes. First antennal joint typical, long, but a little shorter than the distance between antennal bases. Pronotum transverse, almost parallel-sided; median longitudinal furrow distinct. Cuticle of head and pronotum with punctation. Tegmina and wings absent. Meso- and metanotum strongly punctate, more evenly and closer than on pronotum. Abdomen gradually widened medially, strongly punctate; sides of tergites 6–9 rugoso-striate, tergite 6 weakly so, but without lateral longitudinal ridges. Ultimate tergite very broad; median longitudinal sulcus deep; cuticle punctate, punctures tending to form irregular longitudinal bands separated by smooth cuticle. Forceps (Fig. 168) short, asymmetrical, dorsal ridge very strong; inner margin crenulated. Penultimate sternite broadly rounded. Genitalia normal, paramere broad, median incision of anterior margin deep; genital lobes well developed. External parameres (Fig. 169) long, median membrane wide in centre. – **Female** similar to male, but punctation weaker; penultimate sternite rounded; branches of forceps elongated, almost straight except at apices. – **Length** of body with forceps: male: 21–23 mm, female: 22,5–25 mm.

Distribution: Tanzania.

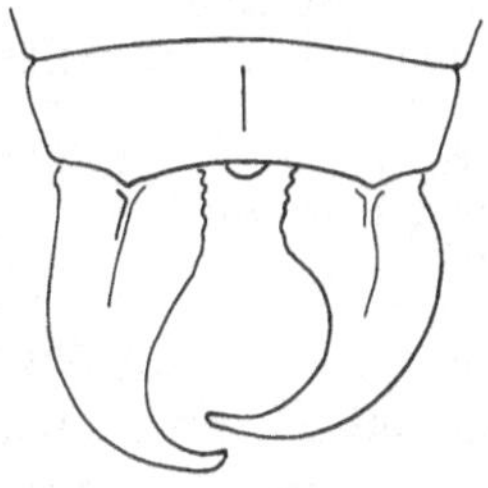

Fig. 168 Ultimate tergite and forceps of *Anisolabis neavei* BRINDLE. After BRINDLE.

Fig. 169 External paramere of male genitalia of *Anisolabis neavei* BRINDLE. After BRINDLE.

36. Anisolabis gestri BORELLI

1907 *Anisolabis gestri* BORELLI, Ann. Mus. civ. Stor. nat., Genova, **43**: 359. – Terra typica: Cameroon (Type male: Mus. civ. Stor. nat., Genova).
1966 *Anisolabis gestroi* – POPHAM & BRINDLE, Entomologist, London **99**: 274.
1970 *Anisolabis gestroi* – SAKAI, Dermapt. Cat. Prael., Tokyo, **I**: 61.
1978 *Anisolabis gestroi* – STEINMANN, Dtsch. ent. Zeitschr., Berlin, n. F. **25**: 183.
1978 *Anisolabis gestri* – BRINDLE, Dermaptera of Africa, Tervuren, **2**: 124; fig. 211 (male forceps).
Literature: BURR 1911, 1912, 1915; STEINMANN 1977.

Description: Male colour dark brown to black; head reddish; antennae black, basal joints red; legs dark red; forceps almost blackish. Head transverse, mainly convex but slightly flattened on vertex; frons tumid; postfrontal and coronal sutures weak; eyes large, but a little shorter than the length of head behind eyes. First antennal joints long, but shorter than the distance between antennal bases. Pronotum quadrate or a little transverse; lateral margins widened posteriorly; all angles rounded; median longitudinal furrow distinct. Tegmina and wings absent. Meso- and metanotum normal; transverse. Abdomen weakly widened medially, tergites with punctation. Ultimate tergite broad, median longitudinal sulcus present. Forceps very similar to *neavei* BRINDLE, as in fig. 168; asymmetrical, inner margins with low but distinct crenulations. Penultimate sternite rounded posteriorly. Genitalia unknown to me. – **Female** similar to male, but ultimate tergite with double bands of rugose punctation arranged longitudinally, separated by bands of smooth cuticle. – **Length** of body with forceps: male: 28–31 mm, female: 26–35 mm.

Distribution: Cameroon.

37. Anisolabis vosseleri Burr

1907 *Anisolabis vosseleri* Burr, Berl. ent. Zeitschr., **52**: 203. – Terra typica: Tanganyika (Type male: Mus. Naturkunde, Berlin).
1911 *Logicolabis vosseleri* – Zacher, Zool. Jahrb., Jena, **30**: 376, fig. X[1] (genital lobe of male genitalia).
1915 *Logicolabis vosseleri* – Burr, J. r. microsc, Soc., London, **1915**: 533, pl. X, fig. 7 (male genitalia).
1976 *Anisolabis vosseleri* – Brindle, Rev. Zool. Bot. afr., Bruxelles, **89**: 686.
1977 *Logicolabis vosseleri* – Steinmann, Ann. hist.-nat. Mus. nation. hung., Budapest, **69**: 91; fig. 11 (male genitalia).
1978 *Anisolabis vosseleri* – Brindle, Dermaptera of Africa, Tervuren, **2**: 122; fig. 217 (male), 218 (penultimate sternite of male), 277 (external paramere of male genitalia).
Literature: Burr 1911, 1912; Sakai 1970; Steinmann 1978, 1981, 1984.

Description: Male colour black, body rather shiny; antennae dark brown, joints 1–3 yellowish; legs yellow, femora of second and third pairs with a dark longitudinal streak on each side, femora of first pair scarcely marked. Head rounded; postfrontal and coronal sutures distinct; eyes comparatively large, about as long as the length of head behind eyes. First antennal joint a little shorter than the distance between antennal bases. Pronotum quadrate; median longitudinal furrow present; lateral margins parallel; cuticle smoothly coriaceous. Tegmen and wings absent. Meso- and metanotum transverse. Abdomen long, more or less parallel-sided; tergites 7–9 rugoso-striate, but without lateral longitudinal ridges. Ultimate tergite broad; lateral longitudinal ridge not prominent but rugoso-striate; plate with irregular punctation with longitudinal bands of smooth cuticle between. Forceps not of *Anisolabis*-type; slender and long (Fig. 170), trigonal basally, cylindrical apically; inner margins with inconspicuous crenulations, Penultimate sternite long, triangular, with convex posterior margin. Genitalia (Fig. 171) characteristic, unknown to me, but fig. 171 after Burr 1915; parameres narrow, V-shaped, median incision of anterior margin very deep; genital lobes fully developed, with slender virgae.

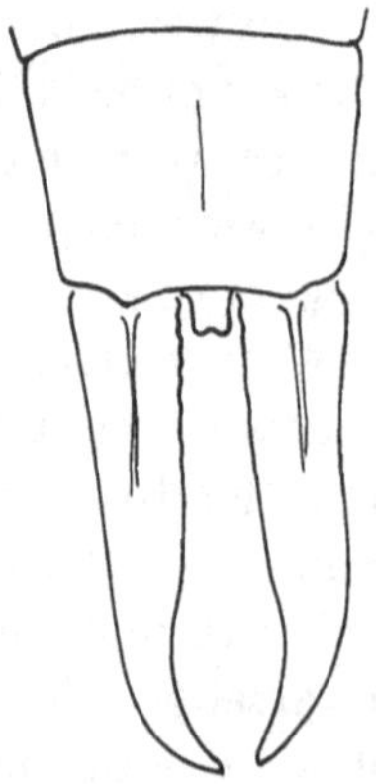

Fig. 170 Male ultimate tergite and forceps of *Anisolabis vosseleri* Burr. After Brindle.

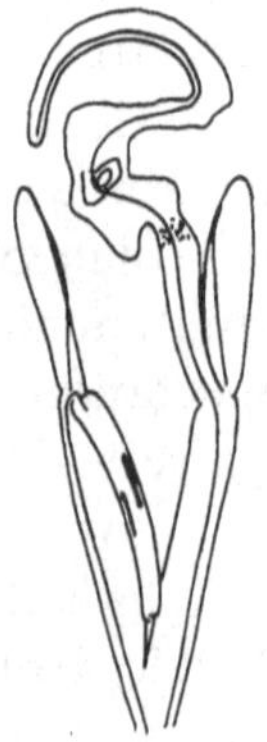

Fig. 171 Male genitalia of *Anisolabis vosseleri* Burr. After Burr.

External parameres long, inner membrane broad at base, narrowed distally, median membrane wide. – **Female** similar to male, but abdominal tergites smoother, tergites 7–9 not rugoso-striate at sides; forceps more or less straight except towards apex, inner margins almost smooth. – **Length** of body with forceps: 22–26 mm, female: male: 23–29 mm.

Distribution: Tanzania.

38. Anisolabis caesarea (ZACHER)

1911 *Logicolabis caesarea* ZACHER, Zool. Jahrb., Jena, **30**: 377. – Terra typica: Deutsch Ost-Afrika (Type male: Mus. Naturkunde, Berlin).
1966 *Carcinophora caesarea* – POPHAM & BRINDLE, Entomologist, London, **99**: 275.
1977 *Logicolabis caesarea* – STEINMANN, Ann. hist.-nat. Mus. nation. hung., Budapest **69**: 91.
1978 *Anisolabis caesarea* – BRINDLE, Dermaptera of Africa, Tervuren, **2**: 123.
Literature: BURR 1911, 1915; SAKAI 1970; STEINMANN 1978, 1984.

Description: Male colour brown, legs yellow, middle of thoracal nota, and first abdominal tergite, yellowish. Head rounded; postfrontal and coronal sutures distinct; eyes moderately large, about as long as the length of head behind eyes. First antennal joint typical, a little shorter than the distance between antennal bases. Pronotum quadrate, weakly widened posteriorly; median longitudinal furrow deep anteriorly. Tegmina and wings absent. Meso- and metanotum normal. Head smooth, thoracal nota with very fine, regular punctures. Abdomen slender, a little widened medially. Ultimate tergite broad; median longitudinal sulcus distinct. Forceps slender, left branch almost straight, right branch weakly curved, apices crossed, inner margins crenulated. Penultimate sternite with rounded posterior margin. Genitalia (Fig. 172) characteristic; not of *Anisolabis*-type; paramere slender, median incision of anterior margin very deep; genital

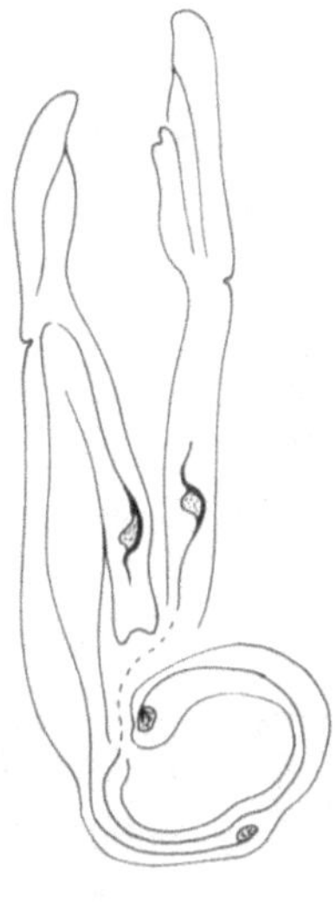

Fig. 172 Male genital armature of *Anisolabis caesarea* (ZACHER). Original.

lobes fully developed, with virgae, the latter with well-marked sclerotized vesicle basally. External parameres elongated, apex obtuse. – **Female** unknown. – **Length** of body with forceps: 22 mm.

Distribution: Tanzania.

39. Anisolabis quarens BURR

1915 *Anisolabis quarens* BURR, J. r. microsc. Soc., London, **1915**: 530, fig. 60 (apex of abdomen of male, lateral and ventral view). – Terra typica: Congo (Type male: Mus. Naturkunde, Berlin).

1924 *Anisolabis quarens* – REHN, Bull. amer. Mus. nat. Hist., New York, **49**: 367 [= *Anisolabis pluto* REHN].

1978 *Anisolabis quarens* – BRINDLE, Dermaptera of Africa, Tervuren, **2**: 105; fig. 228 (male), 229 (penultimate sternite), 230 (external paramere of male genitalia). Nigeria, Cameroon.

Literature: SAKAI 1970; STEINMANN 1977, 1978.

Description: Male reddish-brown, head paler, pronotum almost blackish-brown; antennae brown, basal joints yellow and joints 10–11 whitish; legs yellow. Head rounded; postfrontal sutures indistinct, coronal suture prominent. Eyes large, about as long as the length of head behind eyes. First antennal joint a little shorter than the distance between antennal bases. Pronotum transverse, almost paralell-sided; median longitudinal furrow distinct anteriorly and posteriorly. Tegmina and wings absent. Meso- and metanotum finely punctate, especially laterally. Abdomen weakly widened medially and depressed; sides of tergites 6–9 with lateral longitudinal ridges, the ridges well marked and shortest on tergite 9. Ultimate tergite broad, strongly punctate, median longitudinal sulcus indistinct but depressed near posterior margin. Forceps (Fig. 173) short, broad, inner margin with a flattened flange; trigonal basally and medially,

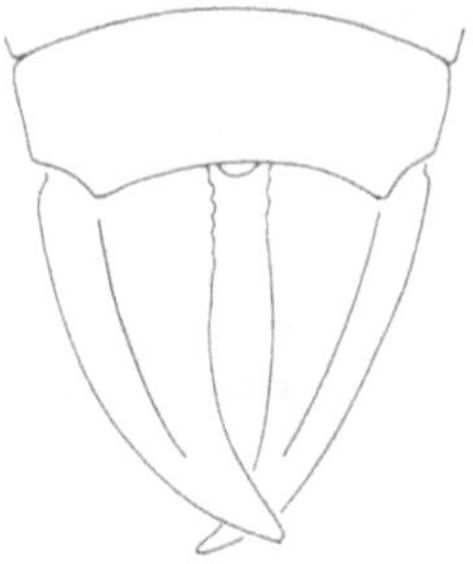

Fig. 173 Male ultimate tergite and forceps of *Anisolabis quarens* BURR. After BRINDLE.

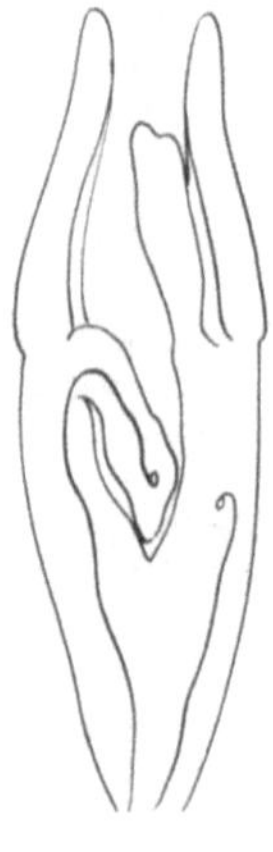

Fig. 174 Male genital armature of *Anisolabis quarens* BURR. After BURR.

Fig. 175 External paramere of male genitalia of *Anisolabis quarens* BURR. After BRINDLE.

cylindrical apically. Penultimate sternite triangular, apex rounded. Genitalia (Fig. 174) narrow, paramere normal, median incision of anterior margin short but wide; genital lobes well developed, with virgae. External parameres very much elongated (after BURR, as in fig. 174) or short and broad (after BRINDLE, as in fig. 175). – **Female** unknown. – **Length** of body with forceps: 12,5 mm.

Distribution: Nigeria, Togo, Equatorial Guinea, Cameroon, Congo Republic.

40. Anisolabis lamottei BRINDLE

1971 *Anisolabis lamottei* BRINDLE, Mém. Inst. fr. Afr. noire, Paris, Dakar, **86**: 270; fig. 5 (male forceps), 6 (female forceps), 7 (male genitalia), 9 (penultimate sternite). – Terra typica: Sierra Leone (Type male: Mus. nation. Hist. nat., Paris).

Literature: BRINDLE 1978; STEINMANN 1977, 1978, 1984.

Description: Male shiny black; antennae dark brown, one or more distal joints paler, legs yellowish-brown. Head almost quadrate or a little transverse, convex, and slightly depressed along postfrontal sutures, the latter and coronal suture distinct. Eyes small, shorter than first antennal joint. Cuticle of head coriaceous, mainly smooth but punctate anteriorly. First antennal joint long, more or less as long as the distance between antennal bases. Pronotum transverse, all margins straight; median longitudinal furrow present, impressed for almost the whole length. Tegmina and wings absent. Meso- and metanotum normal. Cuticle of thoracal nota mainly smooth. Abdomen weakly widened and depressed; sides of tergites 8–9 almost rugoso-striate, and each with a weak lateral longitudinal ridge. Ultimate tergite broad, punctate; lateral longitudinal sulcus deep; longitudinal lateral ridge almost absent. Forceps characteristic (Fig. 176), both branches weakly curved, right branch curved more sharply than left, narrowed

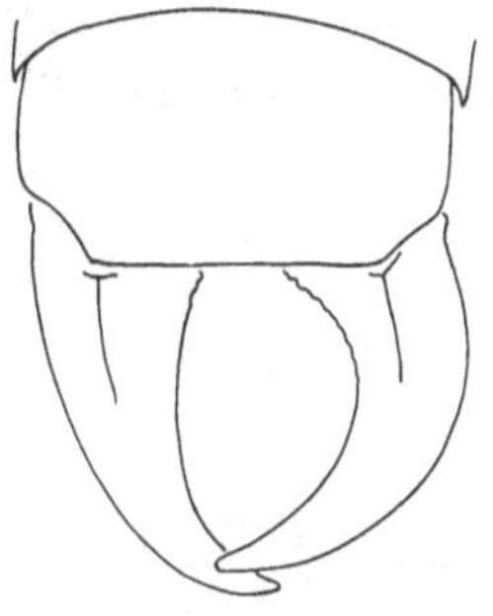

Fig. 176 Holotype ultimate tergite and forceps of *Anisolabis lamottei* BRINDLE. After BRINDLE.

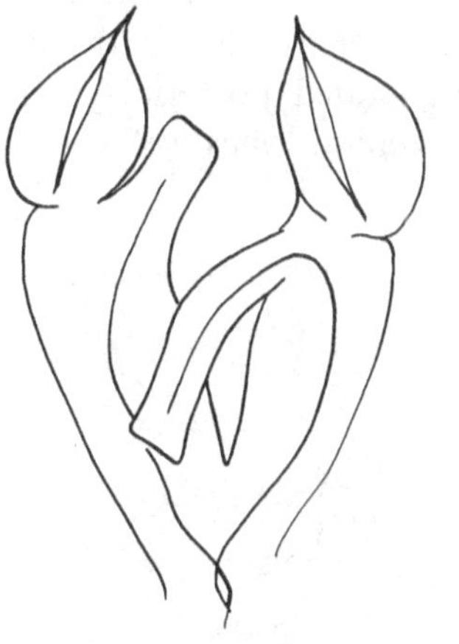

Fig. 177 Genital armature of holotype of *Anisolabis lamottei* BRINDLE. After BRINDLE.

distally, apex pointed; inner margins crenulated. Penultimate sternite broadly rounded. Genitalia (Fig. 177) specific, not of *Anisolabis*-type; central parameral plate broad, median incision of anterior margin deep and wide; genital lobes fully developed, with slender virgae. External parameres a little rounded, broad medially, pointed apically. – **Female** as male, but ultimate tergite narrower, tergites 6–7 not rugose, and tergites 8–9 without lateral longitudinal ridges. – **Length** of body with forceps: male: 16–16,5 mm, female: 15–17 mm.

Distribution: Sierra Leone.

41. Anisolabis rougemonti Brindle

1978 *Anisolabis rougemonti* Brindle, Dermaptera of Africa, Tervuren, **2**: 135; fig. 225 (holotype, male), 226 (penultimate sternite), 270 (external paramere of male genitalia). – Terra typica: Ethiopia (Type male: Manchester Mus.).

Description: Male general colour black, body shiny; antennae dark brown, basal joints paler; lateral margin of pronotum yellow, legs yellowish-brown. Head transverse, rounded; postfrontal and coronal sutures distinct; eyes comparatively small, shorter than the length of head behind eyes. First antennal joint shorter than the distance between antennal bases. Pronotum strongly transverse; lateral margins parallel; plate coriaceous, rather smooth and with a few isolated punctures. Median longitudinal furrow of pronotum distinctly in centre. Tegmina and wings entirely absent. Meso- and metanotum transverse. Abdomen widened and depressed rather strongly; sides of tergites 6–9 rugoso-striate, 6 with a curved rugose ridge, rather blunt, 7 with a similar but shorter and less well-marked ridge, tergites 8–9 without these lateral longitudinal ridges. Ultimate tergite broad, cuticle rugose; median longitudinal sulcus indistinct but in a longitudinal deep depression which extends to posterior margin. Forceps (Fig. 178) short, asymmetrical; strongly trigonal basally, curved and cylindrical apically; inner margins with small projections, unevenly spaced. Penultimate sternite rounded posteriorly. Genitalia typical, paramere with deep median incision; genital lobes well developed. External parameres (Fig. 179) long and

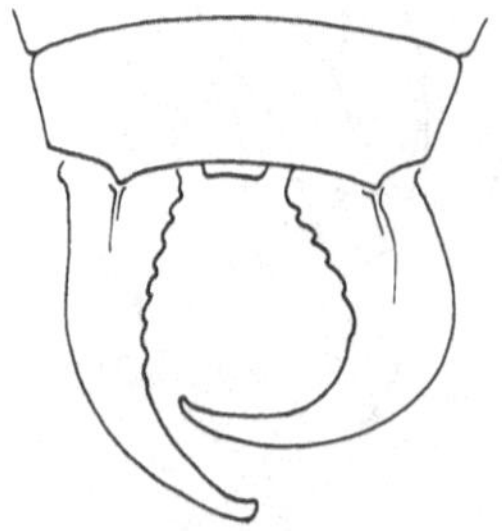

Fig. 178 Holotype ultimate tergite and forceps of *Anisolabis rougemonti* Brindle. After Brindle.

Fig. 179 External paramere of male genitalia of *Anisolabis rougemonti* Brindle. After Brindle.

narrow. – **Female** similar to male, but forceps broader, inner margins with prominent triangular teeth, distal part somewhat abruptly narrowed before apex. – **Length** of body with forceps: male: 17,5–19 mm, female: 18–22 mm.

Distribution: Ethiopia.

42. Anisolabis kristenseni Burr

1911 *Anisolabis kristenseni* Burr, Ann. Mag. nat. Hist., London, (8) **8**: 45. – Terra typica: Ethiopia (Type male: Brit. Mus. nat. Hist., London).

1915 *Gelotolabis kristenseni* – Burr, J. r. microsc. Soc., London, **1915**: 541.

1966 *Gonolabis kristenseni* – Popham & Brindle, Entomologist, London, **99**: 273.

1978 *Anisolabis kristenseni* – Brindle, Dermaptera of Africa, Tervuren, **2**: 133; fig. 208 (male forceps), 209 (penultimate sternite), 275 (external paramere of male genitalia).

Literature: Brindle 1965; Sakai 1970; Steinmann 1977, 1978, 1984.

Description: Male colour black or blackish-brown; head reddish, legs reddish-yellow, antennae yellowish-brown, joints 4–7 and apex of 3rd darker, joints 1 and 2 yellow. Head transverse, rounded; postfrontal and coronal sutures distinct; eyes small, shorter than the length of head behind eyes. First antennal joints shorter than the distance between antennal bases. Pronotum strongly transverse, slightly widened posteriorly; median longitudinal furrow present but short; lateral margins straight. Tegmina and wings absent. Meso- and metanotum transverse, rather smoother than pronotum. Abdomen strongly widened and depressed medially; sides of tergites 5–9 rugoso-striate and tergites 5–6 with a blunt ridge. Ultimate tergite very broad; lateral longitudinal ridges absent but sides of tergite strongly rugoso-striate. Forceps (Fig. 180) asymmetrical; branches widely separated at base, relatively slender, dorsal ridge prominent at base, inner margin with isolated small projections. Penultimate sternite broadly triangular. Genitalia unknown to me. External parameres (Fig. 181) long, inner membrane broad, median membrane small. – **Female** similar to male, sides of tergites 6–7 with blunt ridges; abdomen strongly widened, forceps less strongly curved, inner margins with prominent teeth. – **Length** of body with forceps: male: 24–27 mm, female: 25–28 mm.

Distribution: Ethiopia.

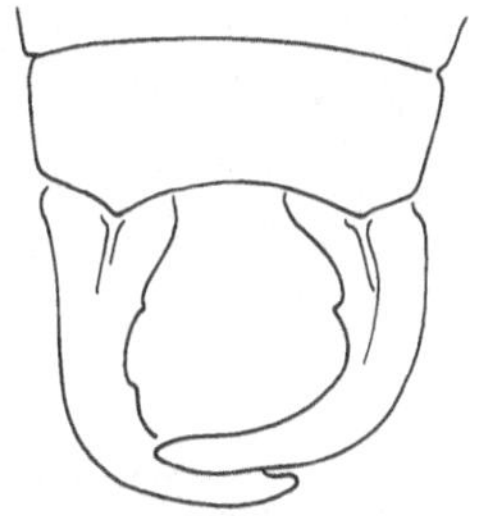

Fig. 180 Male ultimate tergite and forceps of *Anisolabis kristenseni* Burr. After Brindle.

Fig. 181 External paramere of male genitalia of *Anisolabis kristenseni* Burr. After Brindle.

43. Anisolabis simiensis BRINDLE

1978 *Anisolabis simiensis* BRINDLE, Dermaptera of Africa, Tervuren, **2**: 134; fig. 231 (holotype, male), 232 (penultimate sternite of male), 233 (female forceps), 234 (external paramere of male genitalia). – Terra typica: Ethiopia (Type male: Brit. Mus. nat. Hist., London).

Description: Male general colour black; antennae dark brown, legs brown. Head strongly transverse, rounded; postfrontal and coronal sutures distinct. Eyes small, shorter than the length of head behind eyes. First antennal joints slightly shorter than the distance between antennal bases. Pronotum strongly transverse, widened posteriorly; lateral margins straight. Median longitudinal furrow of pronotum present in anterior section. Tegmina and wings absent. Meso- and metanotum transverse; cuticle of head, pronotum and mesonotum coriaceous, metanotum with isolated punctation. Abdomen strongly widened medially; sides of tergites 5–9 rugoso-striate, tergite 7 with an incipient ridge, but 8–9 without one. Ultimate tergite broad, with prominent longitudinal smooth bands; median longitudinal sulcus distinct, lateral longitudinal ridge almost absent. Forceps (Fig. 182) asymmetrical, dorsal ridge prominent and rugose, left branch less curved than right, which is sharply curved about midpoint, left branch with more prominent crenulations on inner margin. Penultimate sternite broadly triangular. Genitalia normal, paramere with deep median incision, genital lobes well developed. External parameres (Fig. 183) broad, inner membrane wider basally, median membrane wide. – **Female** similar to male, but abdomen narrowed distally, sides of tergites 5–7 rugoso-striate and tergite 6 with an incipient lateral longitudinal ridge. Forceps straighter, inner margins with more distinct crenulations. – **Length** of body with forceps: male: 25–27 mm, female: 26–28 mm.

Distribution: Ethiopia.

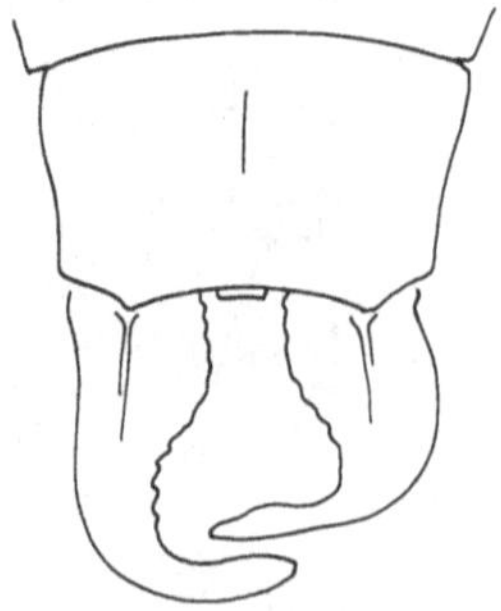

Fig. 182 Holotype ultimate tergite and forceps of *Anisolabis simiensis* BRINDLE. After BRINDLE.

Fig. 183 External paramere of male genitalia of *Anisolabis simiensis* BRINDLE. After BRINDLE.

44. **Anisolabis nitida** (Burr)

1904 *Labidurodes nitidus* Burr, Trans. ent. Soc., London, **1904**: 290. – Terra typica: Madagascar (Type female: Brit. Mus. nat. Hist., London).
1966 *Anisolabis nitidus* – Popham & Brindle, Entomologist, London, **99**: 275.
Literature: Hincks 1953; Sakai 1970.

Original description by Burr: "After, nitidus, laevis; caput convexum; antennae segmentis 14; segmentis 8–11 et 14 nigris, ceteris pallidis, 2 minimo, 3 longo, 4 and 5 brevis, rotundata, ceteris elongatis. Pronotum quadratum, antice quam postice paullo angustius, margine antico subrotundato. Elytra ampla, apice truncata; alae prominentes; pedes breves; femora valde incrassata; tibiae apice pallidae; tarsi pallidi, segmento primo cetera 2 unita longitudine aequanti, 2° minimo, haud lobato. Abdomen cylindricum, segmento tertio tuberculo laterali magno instructo; segmentum ultimum angustum, declivum. Pygidium haud perspicuum. Forcipis (Fig. 184) bracchia paulo remota, recta, triquetra, valida, conica, apicem versus sensim attenuata atque incurva, apice decussata, inermia."

This species is unknown to me. It is supposedly identical with *Marava nitida* (Burr) in the family Labiidae.

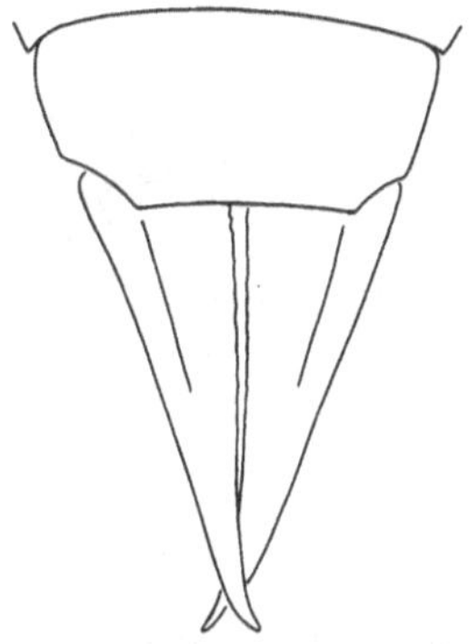

Fig. 184 Female ultimate tergite and forceps of *Anisolabis nitida* (Burr). After Burr.

45. **Anisolabis australis** Tindale

1923 *Anisolabis australis* Tindale, Trans. ent. Soc., London, **47**: 363, fig. 1 (male). – Terra typica: South Australia (Type male: Brit. Mus. nat. Hist., London).
1979 *Anisolabis australis* – Steinmann, Reichenbachia, Dresden, **17**: 65; fig. 14 (male genitalia).
Literature: Popham & Brindle 1966; Sakai 1970; Steinmann 1977, 1978, 1984.

Description: Male colour dark brown; antennae and legs brown. From mesonotum to ultimate tergite densely punctate dorsally. Head more or less wider than pronotum; rounded posteriorly; tumid, postfrontal and coronal sutures distinct. Eyes small, shorter than first antennal joint; the latter long, more or less as long as the distance between antennal bases. Pronotum transverse, lateral margins paralell, but posterior margin rounded; median longitudinal furrow present in centre. Tegmina and wings absent. Meso- and metanotum normal, transverse. Abdomen expanded, very strongly widened to middle; abdominal tergites 3–7 with lateral longitudinal ridges. Ultimate tergite broad, but only a half times broader than 5th abdominal tergite. Forceps (Fig. 185) relatively short, straight basally, strongly curved at apical portion; dorsal ridges prominent, not of *Anisolabis*-type. Penultimate sternite rounded posteriorly. Genitalia (Fig. 186) highly specific; paramere three times longer than external parameres in length; median longitudinal incision of anterior margin comparatively short, parameral plate Y-shaped, genital lobes well developed, with virgae; base of virga very long and broad, muscle loop visible. External parameres of *Anisolabis*-type, widened in basal section, obtuse apically. – **Female** similar to male, but forceps slenderer, contiguous, tapering. – **Length** of body with forceps: male: 16–17 mm, female: 15–16 mm.

Distribution: South Australia.

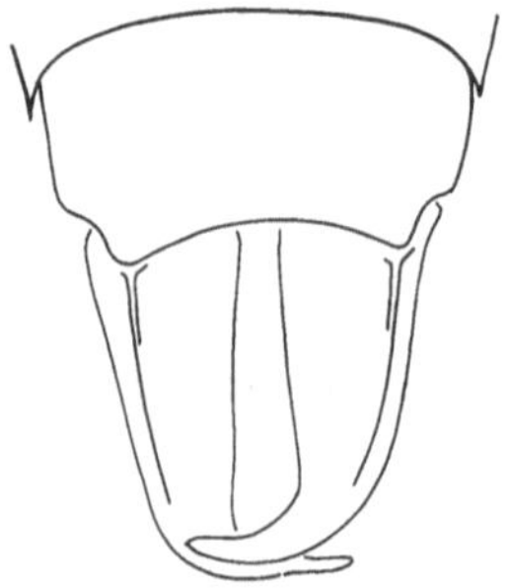

Fig. 185 Male ultimate tergite and forceps of *Anisolabis australis* TINDALE. Original.

Fig. 186 Male genitalia of *Anisolabis australis* TINDALE. Original.

46. Anisolabis nigrofusca STEINMANN

1979 *Anisolabis nigrofusca* STEINMANN, Reichenbachia, Dresden, **17**: 67; fig. 19 (male forceps), 20 (genital armature of holotype; gen. prep. Nr. 581, det. Dr. H. STEINMANN). – Terra typica: New South Wales (Type male: Természettudományi Múz., Budapest).

Description: Male colour dark reddish-black; antennae brown, joints 10–11 whitish, legs yellow. Head large, broader than pronotum; postfrontal sutures and coronal suture faintly distinct. Eyes relatively large, but shorter than first antennal joint. Occiput a little depressed. Antennae with long first joint, the latter more or less as long as the distance between antennal bases. Pronotum visibly longer than broad; lateral margins a little convex; all angles rounded; median longitudinal furrow present in centre. Tegmina present, represented by lateral flaps. Wings absent. Mesonotum normal, metanotum broad. Abdomen slender, but a little expanded to middle; sides of tergites 6–9 rugoso-striate, and punctate dorsally. Ultimate tergite broad, simple. Forceps (Fig. 187) moderately wide and short, not of *Anisolabis*-type; symmetrical, strongly trigonal basally, cylindrical apically; straight in basal section and curved apically. Penultimate sternite broad, rounded posteriorly. Genitalia (Fig. 188) very characteristic; central parameral plate broad, median incision of anterior margin deep, wide apically, narrow basally. Genital lobes very long, without virgae. External parameres of *Anisolabis*-type, obtuse. – **Female** unknown. – **Length** of body with forceps: 18 mm.

Distribution: Australia: New South Wales.

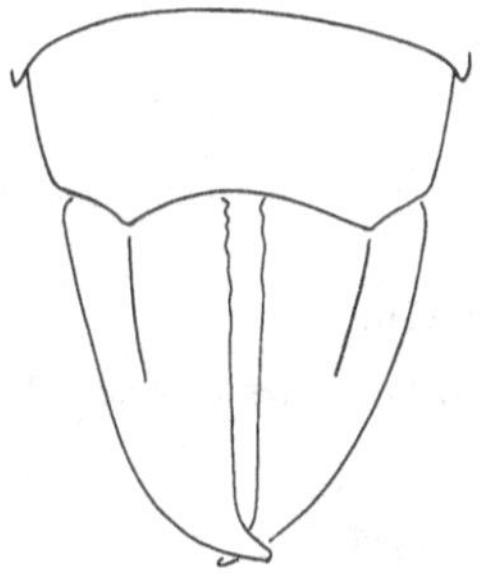

Fig. 187 Holotype ultimate tergite and forceps of *Anisolabis nigrofusca* STEINMANN. After STEINMANN.

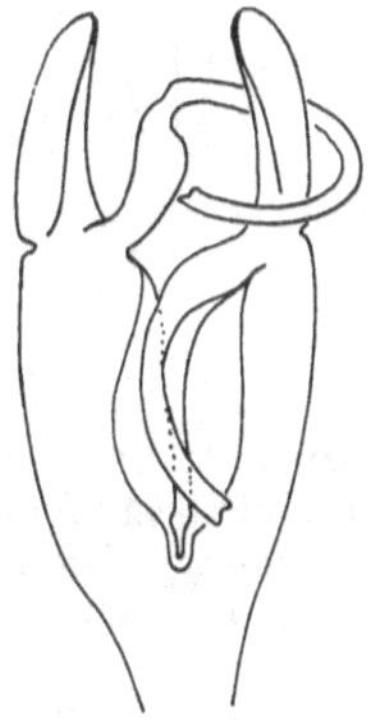

Fig. 188 Genital armature of holotype of *Anisolabis nigrofusca* STEINMANN. After STEINMANN.

47. Anisolabis flavocapitata STEINMANN

1979 *Anisolabis flavocapitata* STEINMANN, Reichenbachia, Dresden, **17**: 66; fig. 15 (male head and pronotum), 16 (female forceps). – Terra typica: Western Australia (Type female: Természettudományi Múz., Budapest).

Description: Female general colour reddish-brown; antennae and legs brown or light brown; head yellow with brown spot at vertex. Head longer than wide, tumid, smooth, and rounded posteriorly. Postfrontal and coronal sutures distinct. Eyes relatively large, about as long as the scape. Antennae 16-jointed, first joint small, slightly shorter than the distance between antennal bases, second one very short. Pronotum quadrate or a little transverse; all margins straight, and all angles rounded; lateral margins convex. Median longitudinal furrow visible. Tegmina and wings entirely absent. Mesonotum broad, posterior margin straight. Metanotum transverse. Abdomen expanded to middle, depressed; tergites without carinae laterally. Ultimate tergite broad, typical. Penultimate sternite rounded posteriorly. Forceps (Fig. 189) characteristic, slender, relatively long, straight, somewhat curved apically. Inner margins with a small but prominent tooth at base. – **Male** unknown. – **Length** of body with forceps: 13,5–14 mm.

Distribution: Western Australia (Perth).

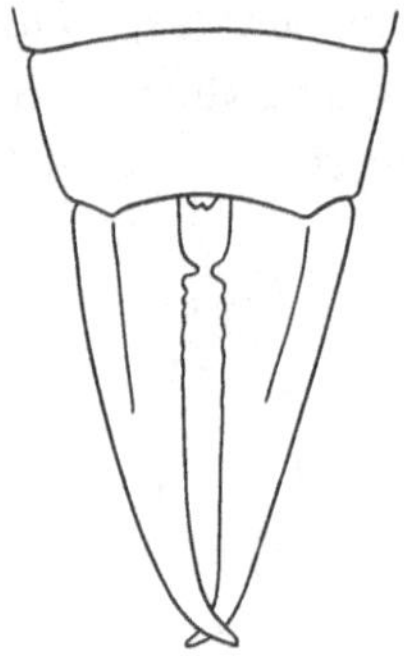

Fig. 189 Holotype ultimate tergite with forceps of *Anisolabis flavocapitata* STEINMANN. After STEINMANN.

48. Anisolabis littorea (WHITE)

1846 *Forficula littorea* WHITE, Zool. Voy. Ins.: 24, pl. VI, fig. 4 (male). – Terra typica: New Zealand (Type male: unknown locality).
1864 *Forcinella littorea* – DOHRN, Stett. ent. Ztg., **25**: 287.
1895 *Anisolabis littorea* – BRUNNER, Proc. zool. Soc., London, **1895**: 892.
1915 *Anisolabis littorea* – BURR, J. r. microsc. Soc., London, **1915**: 525; pl. X, fig. 9 (male genitalia).
1926 *Anisolabis littorea* – TILLYARD, The insects of Australia and New Zealand, Melbourne: 111, pl. 7, fig. 16 (male).
1961 *Anisolabis littorea* – GILES, Trans. r. Soc., New Zealand, Zool., Dunedin, **1**: 206; fig. 1 (male genitalia), 2 (testes and vesicula seminalis), 3 (postgenital region of male); 1: 296; fig. 1 (female reproductive organ).
1979 *Anisolabis littorea* – STEINMANN, Reichenbachia, Dresden, **17**: 66; fig. 17 (male genitalia).
Literature: BORMANS 1900; GILES 1952, 1953, 1965; HINCKS 1949; KIRBY 1904; SAKAI 1970; STEINMANN 1977, 1978.

Description: Male very large, body robust; general colour dark to black; antennae brown, all nota reddish-brown, legs yellowish-brown. Head tumid, longer than wide; postfrontal sutures and coronal suture distinct. Eyes comparatively small, shorter than the genae. Antennae 21-jointed; first joint relatively long, longer than distance between antennal bases. Pronotum more or less as long as wide; lateral margins a little convex, visibly widened posteriorly; median longitudinal furrow distinct. Tegmina and wings absent. Meso- and metanotum typical, transverse. Abdomen expanded to 6th and 7th abdominal tergites. Sides of tergites 6–9 rugoso-striate, with lateral longitudinal ridges. Ultimate tergite broad, posterior margin straight, median longitudinal sulcus deep. Forceps (Fig. 190) asymmetrical, trigonal basally, and curved apically;

dorsal keels very strongly visible. Penultimate sternite rounded posteriorly. Genitalia (Fig. 191) characteristic; paramere broad, median longitudinal incision of anterior margin very deep, paramere V-shaped; genital lobes long, with virgae. External parameres well developed, elongated. – **Female** similar to male, but forceps symmetrical, cylindrical, tapering. – **Length** of body with forceps: male: 26–31 mm, female: 25–32 mm.

Distribution: New Zealand, Tasmania, Sandwich Islands, Great Barrier Islands.

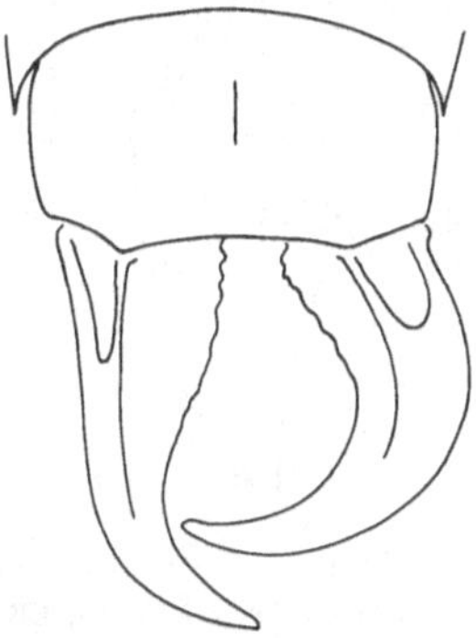

Fig. 190 Male forceps and abdominal end of *Anisolabis littorea* (WHITE). Original.

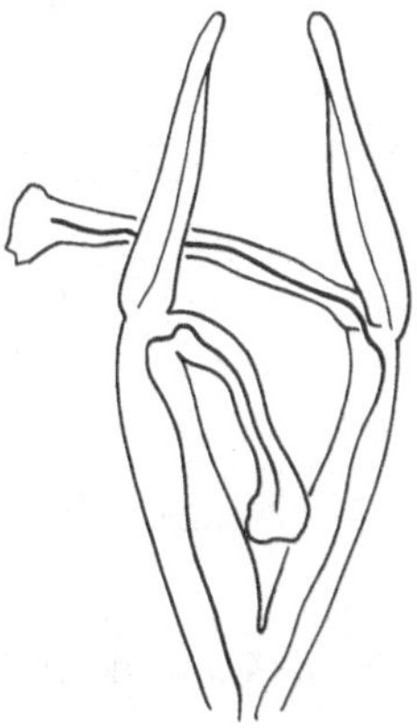

Fig. 191 Male genital armature of *Anisolabis littorea* (WHITE). Original.

49. Anisolabis westralica BURR

1908 *Anisolabis brunneri* BURR, in MICHAELSEN & HARTMEYER, Die Fauna Südwest-Australiens, Berlin, **2** (5): 71, pl. XI, fig. 6 (male). – Terra typica: West Australia (Type male: Brit. Mus. nat. Hist., London).

1911 *Anisolabis westralica* BURR, Genera Insectorum, Bruxelles, **122**: 30; pro *Anisolabis brunneri* BURR, nec DOHRN 1864 and nec BURR 1906. – Terra typica: West Australia (Type female: Zool. Inst. Mus., Hamburg).

1954 *Anisolabis westralica* – HINCKS, Lunds Univ. Årsskrift., (4) **2**: 7, fig. 1 (male genitalia).

Literature: BURR 1915; SAKAI 1970; STEINMANN 1977, 1978, 1979; WEIDNER 1964.

Description: Male dark brown or reddish-brown; legs yellowish brown. Head tumid, more or less as wide as pronotum; postfrontal sutures and coronal suture visible. Eyes small, but as long as genae, and shorter than the first antennal joint, the latter more or less as long as the distance between antennal bases. Pronotum quadrate, lateral margins a little convergent to posterior angles; median longitudinal furrow present in centre. Tegmina and wings absent. Meso- and metanotum transverse, typical. Abdomen of *Anisolabis*-type, a little expanded to abdominal tergites 6–7. Ultimate tergite normal, transverse; posterior margin straight. Forceps (Fig. 192) asymmetrical, trigonal basally, cylindrical apically;

dorsal ridge prominent. Penultimate sternite rounded posteriorly. Genitalia (Fig. 193 after HINCKS 1954) very characteristic; paramere with broad median incision medially, the incision deep and specific; genital lobes without virgae. External parameres less acuminate and more rounded distad. – **Female** similar to male, but forceps simple, symmetrical, tapering, contiguous. – **Length** of body with forceps: male: 15–16 mm, female: 16–17 mm.

Distribution: Western Australia.

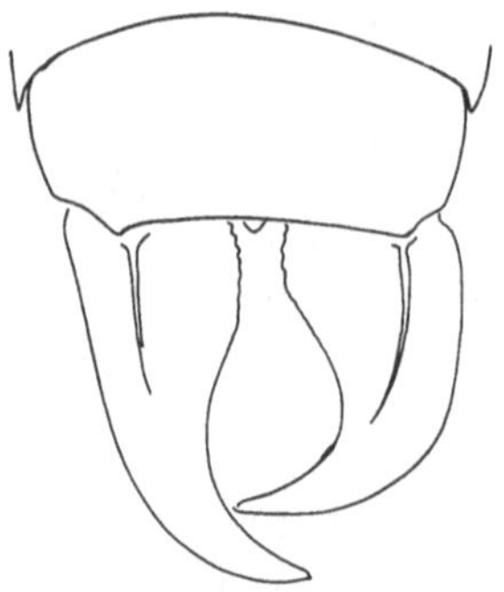

Fig. 192 Male ultimate tergite and forceps of *Anisolabis westralica* BURR. Original.

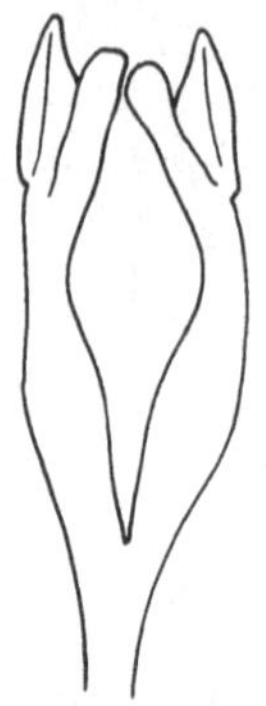

Fig. 193 Male genital armature of *Anisolabis westralica* BURR. After HINCKS.

50. Anisolabis robusta (DUBRONY)

1879 *Labidurodes robustus* DUBRONY, Ann. Mus. civ. Stor. nat., Genova, **14**: 356, fig. (male). – Terra typica: New Guinea (Type male: Mus. civ. Stor. nat., Genova).

1915 *Labidurodes robustus* – BURR, J. r. microsc. Soc., London, **1915**: 532; pl. X, fig. 3 (male genitalia).

1966 *Anisolabis robustus* – POPHAM & BRINDLE, Entomologist, London, **99**: 275. New Guinea, Burma.

1979 *Anisolabis robusta* – STEINMANN, Reichenbachia, Dresden, **17**: 59, fig. 1 (male genitalia).

Literature: BORMANS 1900; BURR 1910, 1911; GÜNTHER 1930; KIRBY 1904; SAKAI 1970; STEINMANN 1977, 1978.

Description: Male colour dark brown; antennae light brown, legs yellowish-brown; forceps reddish. Head broad, postfrontal and coronal sutures not distinct; eyes small, shorter than the scape. First antennal joint relatively long, longer than the distance between antennal bases. Posterior angles of occiput rounded. Pronotum longer than wide; all margins straight, and all angles rounded; median longitudinal furrow present. Tegmina distinct, lateral margins parallel; posterior margin truncate. Wings well developed. This species with tegmina and wings is very similar to *Carcinophora*-type, but it differs in the number of antennal segments (27-jointed), and in having well-developed glandular folds at the sides of the second and third abdominal tergites in adult specimens. Abdomen more or less widened at middle. Ultimate tergite broad; posterior margin specific, as in fig. 194; median longitudinal sulcus indistinct. Forceps (Fig. 194) characteristic; inner margins with 2–4 prominent teeth; branches more or less symmetrical, trigonal basally, cylindrical apically, not of *Anisolabis*-type. Penultimate sternite rounded posteriorly. Genitalia (Fig. 195) of *Anisolabis*-type; parameres narrow, median longitudinal incision of anterior margin deep and wide; genital lobes fully developed, but narrow, with virgae. External parameres narrow and long, obtuse apically. – **Female** similar to male, but forceps slenderer, contiguous; ultimate tergite normal, without prominent lobes posteriorly. – **Length** of body with forceps: male: 29 mm, female: 27–28 mm.

Distribution: New Guinea, Burma (after BURR 1910).

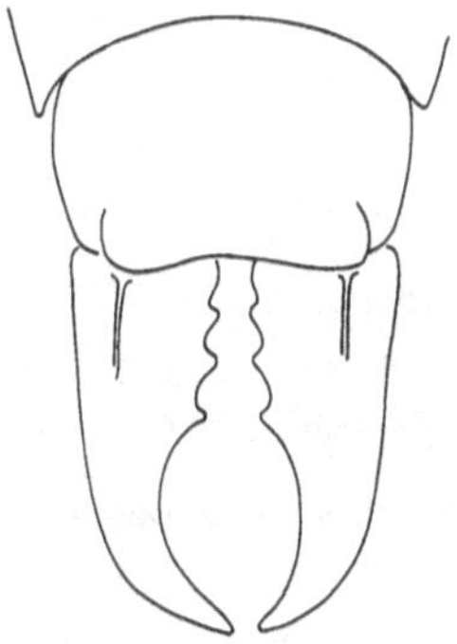

Fig. 194 Male ultimate tergite and forceps of *Anisolabis robusta* (DUBRONY). Original.

Fig. 195 Male genital armature of *Anisolabis robusta* (DUBRONY). After BURR.

51. Anisolabis egregius (GÜNTHER)

1929 *Labidurodes egregius* GÜNTHER, Mitt. Zool. Mus. Berlin, (1) **15**: 71. – Terra typica: New Guinea (Type male: Mus. Naturkunde, Berlin).
1966 *Anisolabis egregius* – POPHAM & BRINDLE, Entomologist, London, **99**: 275.
Literature: SAKAI 1970; STEINMANN 1977, 1978.

Description: Male dark reddish-brown, shiny. Head visibly longer than wide; frons tumid, occiput a little depressed; postfrontal and coronal sutures distinct. Eyes normal, shorter than the length of head behind eyes. Antennae broken, brown; first joint long, longer than eyes, but shorter than the distance between antennal bases. Pronotum quadrate; lateral margins parallel; posterior margin straight, all angles rounded; median longitudinal furrow present in centre. Tegmina and wings present, original description by GÜNTHER: "Elytren ebenfalls völlig glatt, der nach unten umgebogene Teil stark und deutlich konkav. Sehr kleines hellbraunes Scutellum. Flügelschuppen ziemlich kurz, glatt." Abdomen normal, a little expanded to middle. Ultimate tergite broad, transverse, with median longitudinal sulcus. Forceps (Fig. 196) trigonal basally, cylindrical apically, symmetrical, not of *Anisolabis*-type. Penultimate sternite rounded posteriorly. Genitalia unknown, holotype without penis. – **Female** unknown. – **Length** of body with forceps: 35,5 mm.

Distribution: New Guinea.

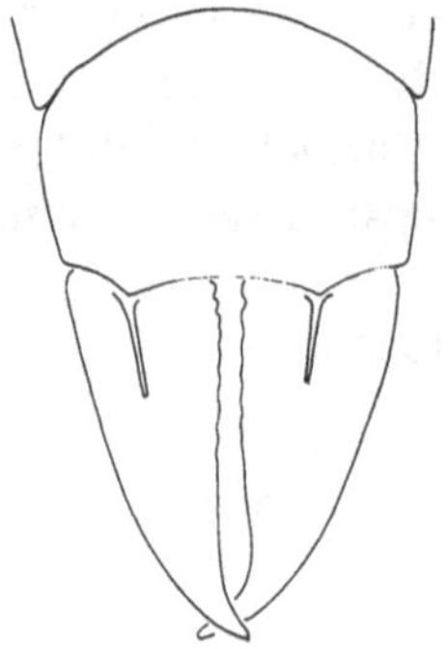

Fig. 196 Male ultimate tergite and forceps of *Anisolabis egregius* (GÜNTHER). Original.

52. Anisolabis perissa GÜNTHER

1929 *Anisolabis perissa* GÜNTHER, Mitt. Zool. Mus. Berlin, (1) **15**: 70; fig. 2 (end of abdomen with forceps of holotype), 3 (genital armature of holotype). – Terra typica: New Guinea (Type male: Mus. Naturkunde Berlin).

1979 *Anisolabis perissa* – STEINMANN, Reichenbachia, Dresden, **17**: 62, fig. 9 (male genitalia).

Literature: SAKAI 1970; STEINMANN 1977, 1978.

Description: Male colour brown, legs light brown, body shiny with fine punctations, similar to *maritima* (BONELLI), and somewhat to *horvathi* BURR, but antennae 18-jointed. Head tumid, a little longer than broad; postfrontal sutures and coronal suture distinct. Eyes small, shorter than the scape or genae. First antennal joint long, about as long as the distance between antennal bases. Pronotum wider than head, and longer than broad; lateral margins straight, all angles rounded; posterior margin truncate. Tegmina and wings absent. Meso- and metanotum typical. Abdomen slender, widened to middle; sides of abdominal tergites 6–9 rugoso-striate. Ultimate tergite broad; median longitudinal sulcus visible; posterior margin truncate, without lateral keels laterally. Forceps (Fig. 197) asymmetrical, a little elongate, trigonal basally and medially, cylindrical apically. Pygidium visible, rounded, with a little incision at posterior margin. Penultimate sternite normal, rounded posteriorly. Genitalia (Fig. 198) of *Anisolabis*-type; parameres relatively narrow, incision of anterior margin deep, narrow; genital lobes well developed, without virgae. External parameres broad basally, obtuse apically. – **Female** similar to male, but forceps simple, contiguous, tapering, symmetrical. – **Length** of body with forceps: male: 23–24 mm, female: 22–23 mm.

Distribution: New Guinea.

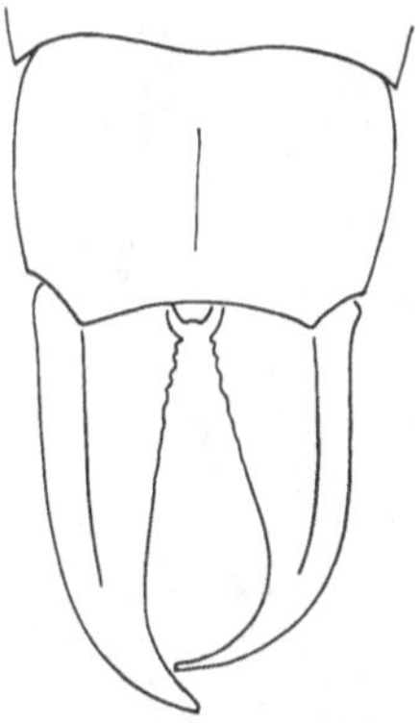

Fig. 197 Male ultimate tergite and forceps of *Anisolabis perissa* GÜNTHER. After GÜNTHER.

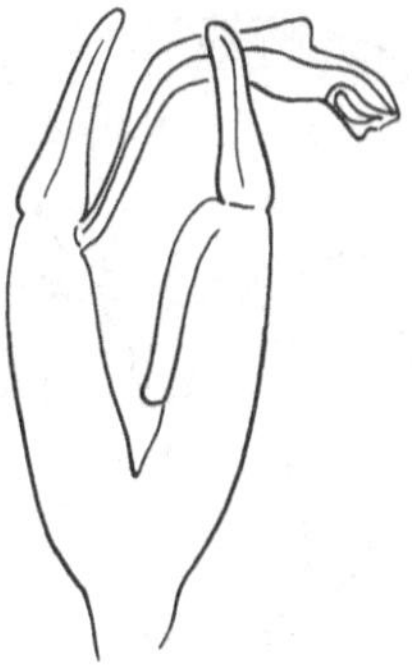

Fig. 198 Male genital armature of *Anisolabis perissa* GÜNTHER. After GÜNTHER.

53. Anisolabis horvathi BURR

1912 *Anisolabis horvathi* BURR, Ann. hist.-nat. Mus. nation. hung., Budapest, **10**: 281. – Terra typica: New Guinea (Type male: Természettudományi Múz., Budapest; but holotype destroyed in 1956).

1912 *Gonolabis sumatrana* BURR (nec BORMANS), Természetrajzi Füzetek, Budapest, **25**: 480; tab. 20, fig. 2 (male).

1979 *Anisolabis horvathi* – STEINMANN, Reichenbachia, Dresden, **17**: 60, fig. 2 (male genitalia).

Literature: SAKAI 1970; STEINMANN 1977, 1978.

Description: Male colour dark reddish-brown; antennae brown, legs yellowish-brown. Head tumid, smooth, postfrontal and coronal sutures distinct; eyes small, shorter than scape; antennae 16-jointed, first antennal joint comparatively short, shorter than distance between antennal bases. Pronotum transverse, a little wider than long; all margins straight, and all angles rounded. Median longitudinal furrow present in centre. Tegmina and wings absent. Meso- and metanotum normal, transverse, cuticle with fine punctations. Legs slender, tarsi narrow, first and second segments lond. Abdomen smooth, slightly expanded to last tergite; all abdominal tergites faintly punctate, shiny. Sides of tergites 7–8 with lateral longitudinal ridges. Ultimate tergite broad, ample, smooth; median longitudinal sulcus distinct, posterior angles with small lateral keels. Forceps (Fig. 199) asymmetrical, branches stout, trigonal, contiguous, attenuate and hooked at the apex. Penultimate sternite broad, rounded posteriorly. Genitalia (Fig. 200) specific; paramere broad, more or less parallel-sided laterally, characteristic; genital lobes long, without virgae, longer than width of median incision of anterior margin. External parameres of *Anisolabis*-type, broad basally, obtuse apically. – **Female** similar to male, but forceps slenderer. – **Length** of body with forceps: male: 9,5–10,5 mm, female: 9–10 mm.

Distribution: New Guinea.

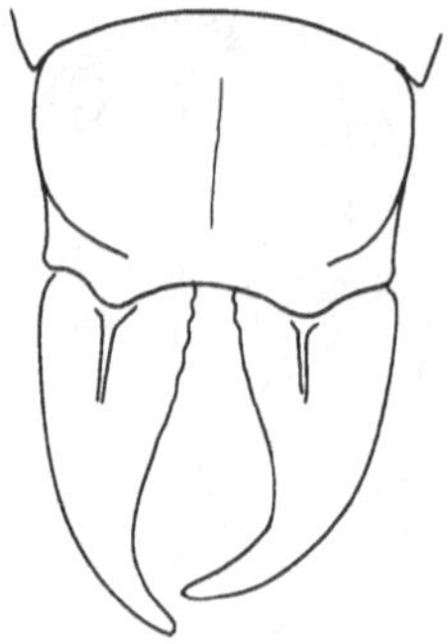

Fig. 199 Male ultimate tergite and forceps of *Anisolabis horvathi* BURR. Original.

Fig. 200 Male genitalia of *Anisolabis horvathi* BURR. After STEINMANN.

54. Anisolabis modesta STEINMANN

1979 *Anisolabis modesta* STEINMANN, Reichenbachia, Dresden, **17**: 60; fig. 3 (male forceps), 4 (genitalia of holotype; gen. prep. No. 154, det. Dr. H. STEINMANN). – Terra typica: New Guinea (Type male: Természettudományi Múz., Budapest).

Description: Male colour reddish-brown; antennae light brown, legs yellowish-brown; last tergite and forceps dark brown, body mat. Head tumid, more or less as long as wide; postfrontal sutures and coronal suture quite visible. Eyes small, shorter than first antennal joint. Antennae 13-jointed, first joint long, longer than distance between antennal bases; second joint very short. Pronotum quadrate, all margins straight, lateral margins parallel, posterior margin truncate; median

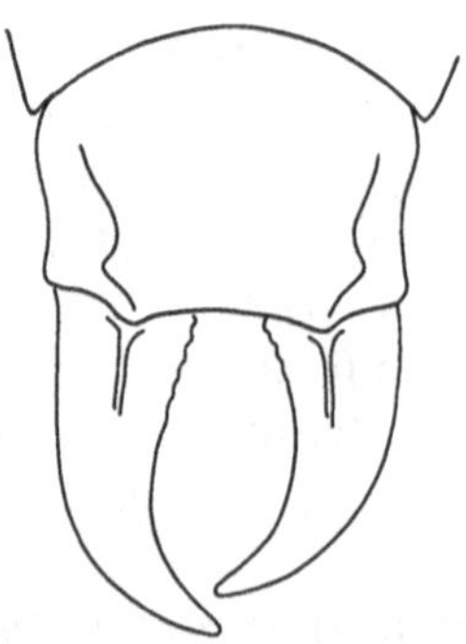

Fig. 201 Holotype ultimate tergite and forceps of *Anisolabis modesta* STEINMANN. After STEINMANN.

Fig. 202 Genitalia of holotype genital armature of *Anisolabis modesta* STEINMANN. After STEINMANN.

longitudinal furrow well marked. Tegmina and wings absent. Mesonotum transverse, posterior margin broad, metanotum typical. Abdomen widened to 5th–6th tergites. Abdominal tergites 6–9 with lateral carinae. Ultimate tergite wide, with keels laterally. Forceps (Fig. 201) asymmetrical, strongly trigonal basally, cylindrical apically. Penultimate sternite normal, rounded posteriorly. Genitalia (Fig. 202) wide; paramere oval, median incision of anterior margin deep, longer than genital lobe; the latter well developed, without virgae. External parameres of *Anisolabis*-type. – **Female** unknown. – **Length** of body with forceps: 13–14 mm.

Distribution: New Guinea.

55. Anisolabis baloghi Steinmann

1979 *Anisolabis baloghi* Steinmann, Reichenbachia, Dresden, **17**: 61; fig. 5 (male forceps), 6 (genital armature of holotype; gen. prep. No. 263, det. Dr. H. Steinmann). – Terra typica: New Guinea (Type male: Természettudományi Múz., Budapest).

Description: Male colour reddish-brown, body shiny; antennae brown, legs yellowish-brown; forceps partly light reddish. Head tumid, smooth, postfrontal sutures and coronal suture faintly distinct. Eyes relatively large, as long as first antennal joint. Antennae 17-jointed, joint 16 whitish-yellow; first antennal joint short, shorter than the distance between antennal bases. Pronotum a little wider than long, all margins straight, and all angles rounded; median longitudinal furrow visible. Femora long, posterior leg as long as the length of abdomen. Tegmina and wings entirely absent. Meso- and metanotum normal. Abdomen expanded to tergites 6–7; a little depressed; sides of tergites 6–9 not acute, but with small carinae laterally. Ultimate tergite wide; posterior margin slightly concave. Forceps (Fig. 203) more or less symmetrical, relatively long, outer and inner margins straight at middle; trigonal basally and cylindrical apically.

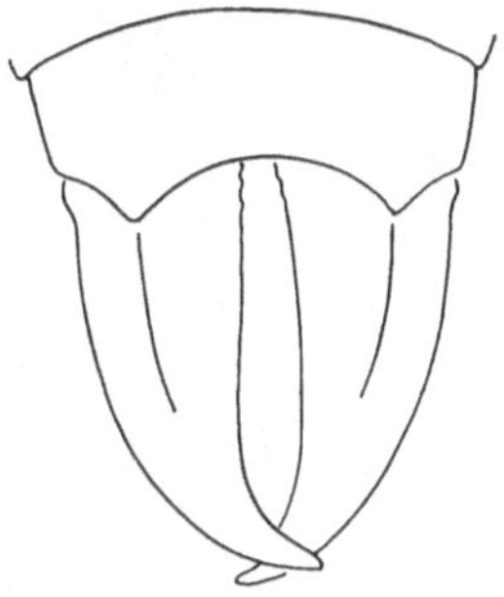

Fig. 203 Holotype ultimate tergite and forceps of *Anisolabis baloghi* Steinmann. After Steinmann.

Fig. 204 Male genital armature of *Anisolabis baloghi* Steinmann. After Steinmann.

Penultimate sternite broad, rounded at posterior margin, with yellowish hairs. Genitalia (Fig. 204) broad, specific; median incision of anterior margin of paramere deep and wide, genital lobe well developed, without virgae. External parameres characteristic, with longitudinal chitinous spine. – **Female** similar to male, but forceps slenderer, contiguous, tapering. – **Length** of body with forceps: male: 13 mm, female: 12–13 mm.

Distribution: New Guinea.

56. Anisolabis pilosa STEINMANN

1979 *Anisolabis pilosa* STEINMANN, Reichenbachia, Dresden, **17**: 61; fig. 7 (male forceps), 8 (genital armature of holotype; gen. prep. No. 268, det. Dr. H. STEINMANN). – Terra typica: New Guinea (Type male: Természettudományi Múz., Budapest).

Description: Male colour dark metallic-brown; anteclypeus yellow, antennae brown; legs light brown, last tergite and forceps reddish-brown. Head tumid, smooth, occiput a little depressed; postfrontal and coronal sutures distinct. Eyes relatively long, but shorter than the length of scape. Antennae 16-jointed, with long hairs. First antennal joint more or less as long as the distance between antennal bases. Pronotum quadrate, a little wider than long; all margins straight, and all angles rounded; median longitudinal furrow distinct. Tegmina and wings absent. Legs relatively short, posterior leg shorter than the length of abdomen without forceps. Meso- and metanotum normal. Abdomen depressed, expanded to middle; cuticle of tergites faintly punctate. Ultimate tergite very wide; posterior margin slightly concave. Abdominal tergites 6–9 with large carinae laterally. Forceps (Fig. 205) broad, curved, trigonal basally, cylindrical apically, more or less symmetrical. Penultimate sternite typical, but with yellowish hairs. Genitalia (Fig. 206) narrow; median incision of anterior margin of paramere

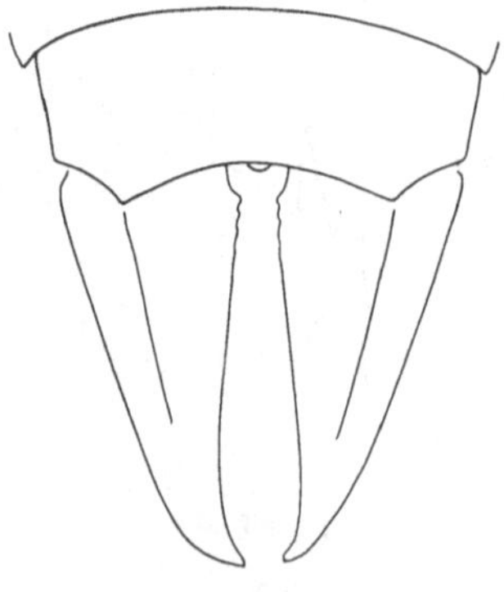

Fig. 205 Holotype ultimate tergite and forceps of *Anisolabis pilosa* STEINMANN. After STEINMANN.

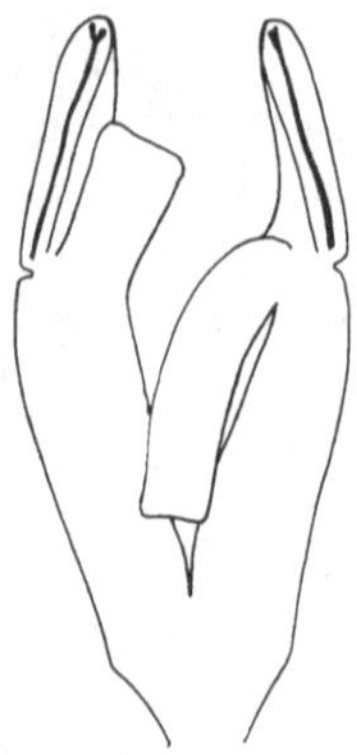

Fig. 206 Genital armature of holotype of *Anisolabis pilosa* STEINMANN. After STEINMANN.

deep, longer than genital lobe; the latter well developed, but without virgae. External parameres specific apically; apices visibly aridening. – **Female** unknown. – **Length** of body with forceps: 12 mm.

Distribution: New Guinea.

57. Anisolabis bifida BRINDLE

1970 *Anisolabis bifida* BRINDLE, Pacific Insects, Honolulu, **12**: 658; fig. 14 (male), 15 (penultimate sternite of male), 16 (genital armature of holotype). – Terra typica: Solomon Islands (Type male: Brit. Mus. nat. Hist., London).

Literature: STEINMANN 1979.

Description: Male colour blackish brown; antennae dark brown, pronotum yellowish laterally, legs yellow, forceps dark red. Head transverse, tumid, smooth, depressed medially on occiput. Postfrontal and coronal sutures indistinct. Eyes normal in size, a little shorter than the length of head behind eyes. First antennal joint short, shorter than the distance between antennal bases, second joint transverse. Pronotum transverse, all margins straight, lateral margins more or less parallel; median longitudinal furrow absent. Tegmina and wings absent. Meso- and metanotum wide, the latter with posterior margin concave. Head and all thoracal nota impunctate, smooth. Abdomen depressed, slightly widened medially, cuticle finely punctate; sides of tergites 7–9 rugoso-striate, and with lateral longitudinal ridges. Ultimate tergite transverse, posterior margin concave medially; median longitudinal sulcus indistinct. Forceps (Fig. 207) symmetrical, strongly trigonal basally, cylindrical apically, both branches narrowed distally, apices incurved, inner margins with very small denticulations. Penultimate sternite transverse, posterior margin convex, and with an excision at apex. Genitalia (Fig. 208) long and narrow; paramere with

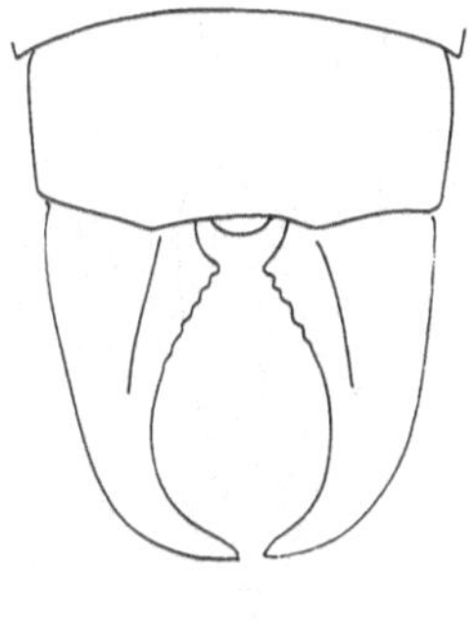

Fig. 207 Holotype ultimate tergite and forceps of *Anisolabis bifida* BRINDLE. After BRINDLE.

Fig. 208 Genital armature of holotype of *Anisolabis bifida* BRINDLE. After BRINDLE.

deep median incision between external parameres. Genital lobes well developed, with virgae. External parameres very long. – **Female** unknown. – **Length** of body with forceps: 9,5 mm.

Distribution: Solomon Islands.

58. Anisolabis verhoeffi ZACHER

1911 *Anisolabis verhoeffi* ZACHER, Zool. Jahrb., Jena, **30**: 374, fig. V (male genitalia). – Terra typica: Bismarck Archipelago (Type male: Mus. Naturkunde, Berlin).

1976 *Anisolabis verhoeffi* – BRINDLE, Rec. S. Austr. Mus., Adelaide, (13) **17**: 224; fig. 1 (male), 2 (male genitalia), 3 (external paramere of male genitalia), 4 (penultimate sternite of male). Espiritu Santo, Maewo, Aoba, Malekula, Efate, Erromanga, Aneityum.

Literature: BRINDLE 1970, 1976; BURR 1911, 1915; HINCKS 1952; POPHAM & BRINDLE 1966; SAKAI 1970, STEINMANN 1977, 1978, 1979.

Description: Male colour brown to blackish brown; antennae dark brown, sometimes with one or more distal joints pale yellow or whitish; legs yellowish. Head transverse, tumid, smooth; postfrontal and coronal sutures very distinct (or absent, after BRINDLE 1976). Eyes normal in size, a little shorter than the length of head behind eyes. First antennal joint long, but shorter than the distance between antennal bases; antennae 14-jointed, joints 11–12 white. Pronotum more or less as wide as long, all margins straight, and all angles rounded; lateral margins parallel. Tegmina and wings completely absent. Abdomen depressed, tergites 6–9 with carinae laterally; those on sixth tergite extending for distal half only. Ultimate tergite broad, with a dorso-median ridge towards each side, the ridge curved medially posteriorly, and posterior part of tergite depressed; median longitudinal sulcus present. Forceps (Fig. 209) trigonal basally, cylindrical apically. Penultimate sternite with apex concave, the plate

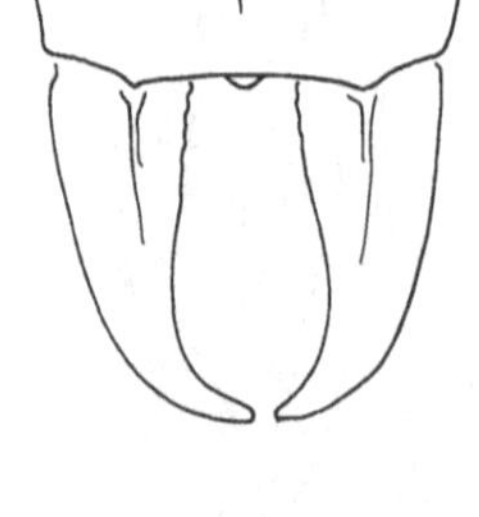

Fig. 209 Male ultimate tergite and forceps of *Anisolabis verhoeffi* ZACHER. After BRINDLE.

Fig. 210 Male genitalia of *Anisolabis verhoeffi* ZACHER. After ZACHER.

pointed. Genitalia (Fig. 210) of *Anisolabis*-type; paramere narrow, median incision of anterior margin very deep; genital lobes with long, narrow virgae. External parameres long, obtuse. – **Female** similar to male, but forceps slenderer, tapering, contiguous. – **Length** of body with forceps: male: 10,5–12 mm, female: 11–12,5 mm.

Distribution: Bismarck Archipelago, New Hebrides.

59. Anisolabis canaca BRINDLE

1976 *Anisolabis canaca* BRINDLE, Pacific Insects, Honolulu, **17** (1): 70; fig. 3 (genital armature of holotype). – Terra typica: New Caledonia (Type male: Brit. Mus. nat. Hist., London).
Literature: STEINMANN 1979.

Description: Male colour blackish, antennae dark brown or brown; legs yellow or brown, femora broadly blackish basally; body shiny, abdominal tergites smooth except for last tergite which is slightly rugoso-striate. Head transverse, tumid, but depressed medially on occiput. Postfrontal sutures and coronal suture present, but not well marked. Eyes small, shorter than genae. First antennal joint long, but a little shorter than the distance between antennal bases. Pronotum somewhat longer than wide, all angles rounded, and all margins straight; median longitudinal furrow present in centre. Tegmina and wings completely absent. Meso- and metanotum normal. Abdomen with smooth tergites. Ultimate tergite broad, posterior margin concave medially. Forceps (Fig. 211) asymmetrical, comparatively slender; trigonal basally, cylindrical apically; inner margin with pronounced teeth. Penultimate sternite typical, rounded posteriorly. Genitalia (Fig. 212) characteristic; paramere with very deep median incision at anterior margin; genital lobes well developed, virga present, specific, as in fig. 212, somewhat similar to Labidurids species. External parameres long and wide. – **Female** similar to male, but forceps straighter, the branches being more or less

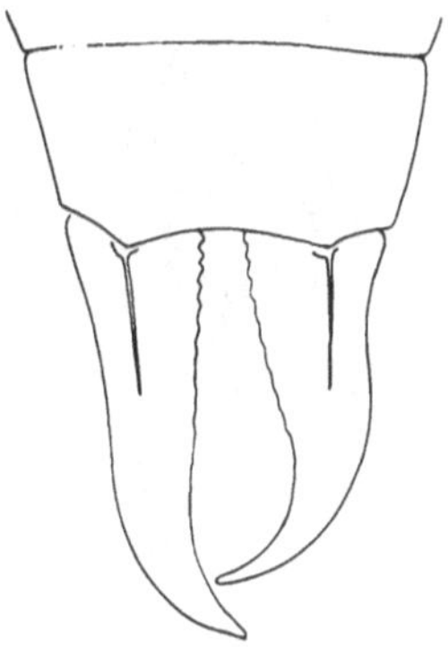

Fig. 211 Male ultimate tergite and forceps of *Anisolabis canaca* BRINDLE. Original.

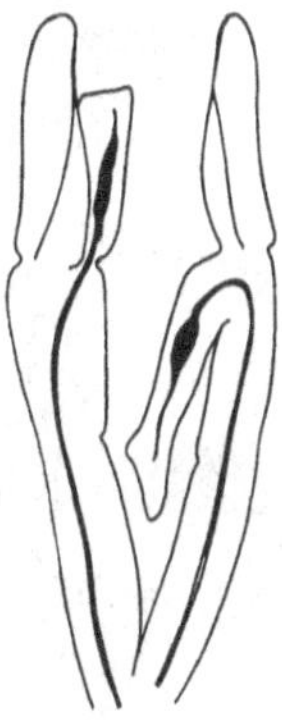

Fig. 212 Genital armature of *Anisolabis canaca*. BRINDLE. After BRINDLE.

contiguous; symmetrical. – **Length** of body with forceps: male: 17–22 mm, female: 16–21 mm.

Distribution: New Caledonia.

60. Anisolabis sarasini (BURR)

1914 *Spondax sarasini* BURR, Nova Caledonia, Wellington, **1**: 317. – Terra typica: New Caledonia (Type male: Brit. Mus. nat. Hist., London).
1966 *Anisolabis sarasini* – POPHAM & BRINDLE, Entomologist, London, **99**: 274.
1976 *Anisolabis sarasini* – BRINDLE, Pacific Insects, Honolulu, 17 (1): 68; fig. 1 (male), 2 (male genitalia).
Literature: BURR 1915; KALTENBACH 1968; SAKAI 1970; STEINMANN 1977, 1979, 1984.

Description: Male colour reddish brown to dark brown; last tergite and forceps sometimes darker; antennae, palpi and legs reddish-yellow; shiny, smooth, impunctate anteriorly. Head broad, flat dorsally, and rounded posteriorly; posterior margin concave. Postfrontal and coronal sutures distinct. Eyes moderately large, but shorter than the length of head behind eyes. First antennal joint long, about as long as the distance between antennal bases. Pronotum slightly longer than wide; lateral margins almost parallel; median longitudinal furrow absent, but plate has a transversal sulcus at posterior section. Tegmina and wings completely absent. Mesonotum large, broad; metanotum transverse. Abdomen wide, depressed, nearly parallel-sided. Ultimate tergite broad, posterior margin concave. Forceps (Fig. 213) trigonal at base, cylindrical apically; inner margin darkened but very weakly dentated, branches weakly curved and slightly asymmetrical. Penultimate sternite somewhat triangular, but with a wide blunt posterior margin. Genitalia (Fig. 214) long and narrow; paramere V-shaped, incision of anterior margin very deep; genital lobes well developed, but narrow, with specific virgae. External parameres very long, of

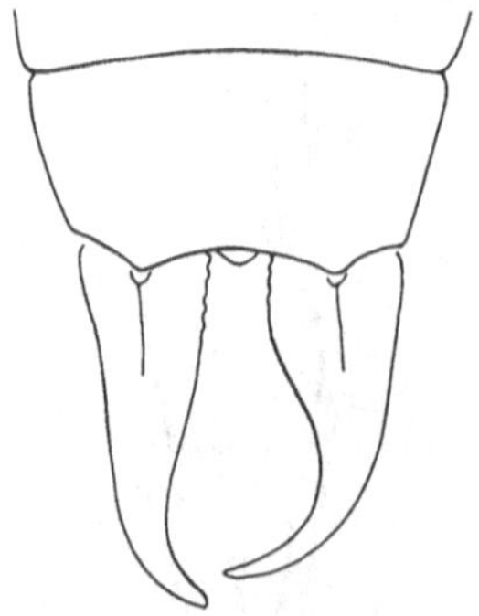

Fig. 213 Male ultimate tergite and forceps of *Anisolabis sarasini* (BURR). Original.

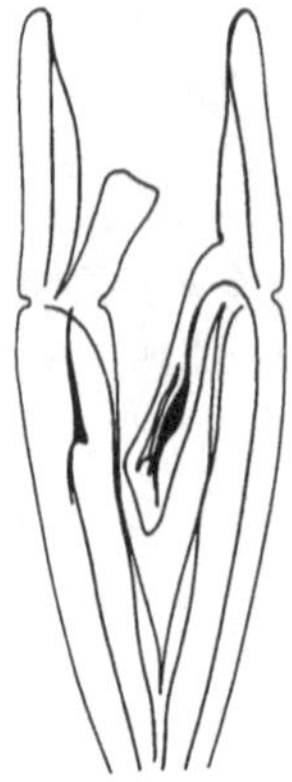

Fig. 214 Male genitalia of *Anisolabis sarasini* (BURR). After BRINDLE.

Anisolabis-type. – **Female** similar to male, but forceps more or less straight, contiguous, tapering, symmetrical. – **Length** of body with forceps: male: 25–27 mm, female: 24–28 mm.

Distribution: New Caledonia.

61. Anisolabis tegminata CAUDELL

1927 *Anisolabis tegminata* CAUDELL, Studies nat. Hist. Iowa Univ., Des Momes, **12**: 4. – Terra typica: Fiji (Type male: unknown locality).

Literature: POPHAM & BRINDLE 1966; SAKAI 1970; STEINMANN 1977, 1979, 1984.

Description: Male colour dark brown to reddish-brown; antennae and legs light brown; forceps reddish. Head more or less tumid, slightly longer than wide; postfrontal and coronal sutures distinct. Eyes relatively small, shorter than scape; the latter long, about as long as the distance between antennal bases. Pronotum quadrate, lateral margins parallel, straight; all angles rounded; posterior margin truncate; median longitudinal furrow present in anterior half. Tegmina and wings entirely absent. Meso- and metanotum typical. Abdomen slightly widened to 5th and 6th tergites; abdominal tergites punctate dorsally; sides of tergites 6–9 rugoso-striate. Ultimate tergite broad, transverse, with median longitudinal sulcus. Forceps (Fig. 215) asymmetrical, pronounced dorsal ridge in basal section, branches trigonal basally, cylindrical and curved apically. Penultimate sternite rounded posteriorly. Genitalia unknown. – **Female** similar to male, but forceps contiguous, tapering, symmetrical. – **Length** of body with forceps: male: 15–16 mm, female: 14,5–16 mm.

Distribution: Fiji.

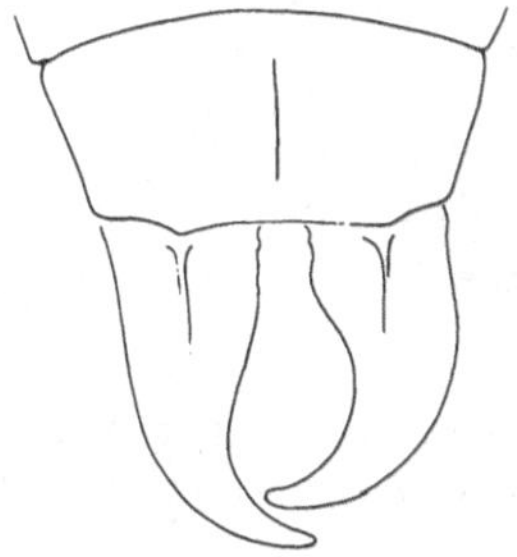

Fig. 215 Male ultimate tergite and forceps of *Anisolabis tegminata* CAUDELL. Original.

62. Anisolabis minutissima BRINDLE

1972 *Anisolabis minutissima* BRINDLE, Insects of Micronesia, Dermaptera, Honolulu, (2) **5**: 128; fig. 6a (penultimate sternite of male), 6b (male), 6c (genitalia of holotype), 7b (distal antennal segments). – Terra typica: Palau Island (Type male: U.S. nation. Mus. No. 71556, Washington).

Literature: STEINMANN 1979, 1984.

Description: Male blackish, shiny; antennae brown, legs yellow, femora darkened on basal half or more. Head transverse, tumid, postfrontal and coronal sutures absent; frons slightly depressed. Eyes small, a little shorter than first antennal joint. Scapus long, but slightly shorter than the distance between antennal bases. Pronotum slightly transverse, somewhat widened posteriorly; all margins straight, and all angles rounded. Median longitudinal furrow indistinct. Tegmina and wings absent. Meso- and metanotum transverse, the latter with posterior margin concave. Abdomen depressed, narrowed towards base, tergites finely punctate and pubescent, hairs short, sparse and yellow; tergites 7–9 rugoso-striate laterally. Ultimate tergite transverse, median longitudinal sulcus present; posterior margin concave medially. Forceps (Fig. 216) asymmetrical, strongly trigonal basally, and cylindrical apically. Penultimate sternite transverse, basically triangular, posterior margin slightly excised. Genitalia (Fig. 217) very broad, robust; paramere wide, median incision of anterior margin deep and wide; genital lobes well developed, virga present, but not visible in each genital lobe; external parameres elongated, obtuse apically. – **Female** similar to male, but abdomen more finely punctate; and forceps almost straight; inner margin smooth. – **Length** of body with forceps: male: 7 mm, female: 6,5 mm.

Distribution: Micronesia: Palau Island.

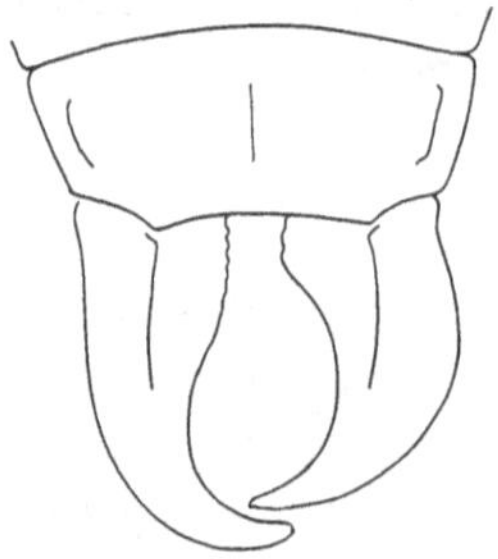

Fig. 216 Male ultimate tergite and forceps of *Anisolabis minutissima* BRINDLE. After BRINDLE.

Fig. 217 Genital armature of holotype of *Anisolabis minitissima* BRINDLE. After BRINDLE.

63. Anisolabis pacifica (ERICHSON)

1842 *Forficula pacifica* ERICHSON, Arch. Naturgesch., Berlin, **8**: 247. – Terra typica: Hawaii (Type male: unknown locality).

1891 *Anisolabis xenia* KIRBY, J. Linn. Soc. London, Zool., **23**: 517. – Terra typica: Norfolk Islands (Type male: Brit. Mus. nat. Hist., London).

1895 *Anisolabis pacifica* – BRUNNER, Proc. Zool. Soc., London, **1895**: 892.

1910 *Anisolabis perkinsi* BURR, Trans. ent. Soc. London, **1910**: 178. – Terra typica: Sandwich Islands (Type male: Brit. Mus. nat. Hist., London).

1911 *Gonolabis pacifica* – BURR, Genera Insectorum, Bruxelles, **122**: 28. Australia.

1914 *Mongolabis pacifica* – BURR, Ann. Mag. nat. Hist., London, (8) **13**: 580.
1915 *Mongolabis pacifica* – BURR, J. r. microsc. Soc., London, **1915**: 526; pl. 12, fig. 10 (male genitalia).
1933 *Gonolabis pacifica* – HEBARD, Mem. Queensland, Mus., Brisbane, **10**: 146.
1948 *Anisolabis perkinsi* – ZIMMERMANN, Insects of Hawaii, Honolulu, **2**: 201; fig. 115a (female). Hawaiian Islands.
1966 *Euborellia pacifica* – POPHAM & BRINDLE, Entomologist, London, **99**: 273. Australia.
1966 *Anisolabis perkinsi* – POPHAM & BRINDLE, Entomologist, London, **99**: 274. Polynesia.
Literature: BORMANS 1900; KIRBY 1904; SAKAI 1970; STEINMANN 1977, 1979, 1984.

Description: Male dark brown; legs yellowish brown. Head smooth and shiny. Antennae 16-jointed, first joint long and slender, apically clubbed; second transverse, third joint long and narrow, cylindrical. Postfrontal and coronal sutures not visible. Eyes comparatively small, shorter than first antennal joint. Pronotum slightly longer than wide, gently widened posteriorly, all margins straight, sides gently reflexed. Tegmina and wings absent. Meso- and metanotum transverse. Legs long and slender. Abdomen feebly dilated beyond the middle and apically attenuated. Ultimate tergite slightly narrowed apically, with median longitudinal sulcus; posterior margin subsinuate. Forceps (Fig. 218) trigonal basally, rapidly attenuate, feebly arcuate, slender, more or less symmetrical. Penultimate sternite normal, rounded posteriorly. Genitalia (Fig. 219) wide; parameral plate with median incision between external parameres. Genital lobes well developed, without virga, but with small sclerotized plates. External parameres simple. – **Female** similar to male, but forceps slenderer, tapering, contiguous; inner margin finely crenulate. – **Length** of body with forceps: male: 16–25 mm, female: 15–22 mm.

Distribution: Australia, Polynesia, Sandwich Islands, Hawaii.

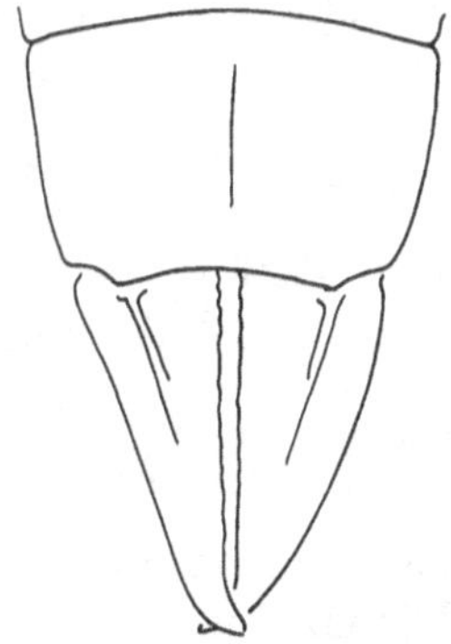

Fig. 218 Male ultimate tergite and forceps of *Anisolabis pacifica (*ERICHSON). Original.

Fig. 219 Male genital armature of *Anisolabis pacifica (*ERICHSON). After BURR.

64. Anisolabis subarmata (KIRBY)

1900 *Labia subarmata* KIRBY, Christmas Isl. Monogr., London, 134; pl. 12, fig. 7 (male). – Terra typica: Christmas Island (Type male: Brit. Mus. nat. Hist., London).
1904 *Anisolabis subarmata* – KIRBY, A Synonymic Catalogue of Orthoptera, London, **1**: 19.
Literature: BURR 1910, 1911; POPHAM & BRINDLE 1966; SAKAI 1970; STEINMANN 1978.

Description: Male dark reddish-brown; palpi, antennae light brown; legs yellowish brown. Head depressed at vertex, a little longer than wide; eyes small, shorter than first antennal base; the latter as long as distance between antennal bases. Postfrontal and coronal sutures not visible. Pronotum quadrate; all margins straight, lateral margins a little expanded posteriorly; all angles rounded. Median longitudinal furrow present in centre; posterior margin truncate. Tegmina and wings absent. Meso- and metanotum normal, broad. Abdomen more or less widened at middle; tergites finely punctate. Ultimate tergite transverse; median longitudinal sulcus indistinct; posterior margin straight, or a little concave medially. Forceps (Fig. 220) more or less symmetrical; inner margins a little lamellate, with distinct denticulations basally; branches trigonal basally, cylindrical apically. Genitalia unknown. – **Female** as male, but forceps slenderer, tapering. – **Length** of body with forceps: male: 15–17 mm, female: 15–18 mm.

Distribution: Christmas Islands.

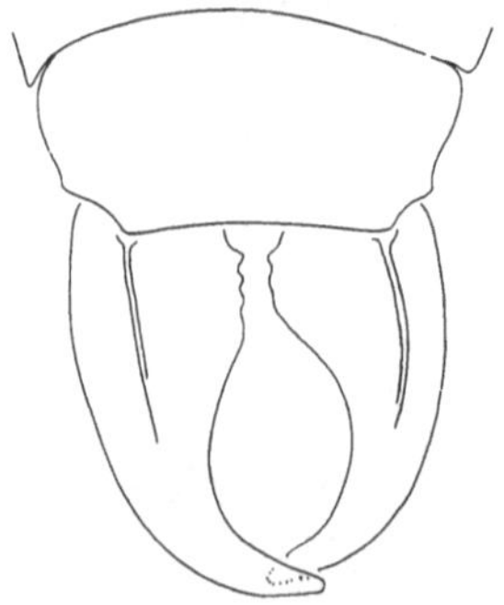

Fig. 220 Male ultimate tergite and forceps of *Anisolabis subarmata* (KIRBY). After KIRBY.

65. Anisolabis pectoralis (ESCHSCHOLTZ)

1822 *Forficula pectoralis* ESCHSCHOLTZ, Entomograph, St. Petersburg, **1**: 82. – Terra typica: Kamtschatka (Type: unknown).
1900 *Anisolabis pectoralis* – BORMANS, Das Tierreich, Berlin, **11**: 50.
Literature: BEY-BIENKO 1934; BURR 1910, 1911; KIRBY 1904; POPHAM & BRINDLE 1966; SAKAI 1970.

Original description by ESCHSCHOLTZ very short, new description by BORMANS in Das Tierreich is insufficient: "Schwarzbraun. Mundteil gelb. Antennae 15-gliedrig, braun, ausgenommen das gelbe 1. und 3. Glied, sowie das 13. und 14. ganz (oder teilweise) hellgelbe Glied. Seitenrand des Pronotum gelb, zuweilen der vordere Teil sehr breit- und der Hinterrand sehr schmal-bräunlichgelb. Prosternum blaß, die beiden folgenden Segmente braun. Beine blaßgelb. Abdominalsternite schmal braunrot gesäumt; letztes Segment sehr groß, mit runzeligem Hinterrande. Zangenarme kurz, dreikantig, an der Basis verdickt, fast gerade, nur die Spitzen nach innen und oben gebogen, Innenrand feinzähnig. Geschlecht? – L. t. 16 mm." Forceps as in fig. 221.

This species unknown to me. It is supposedly identical with *Anisolabis maritima* (BORELLI).

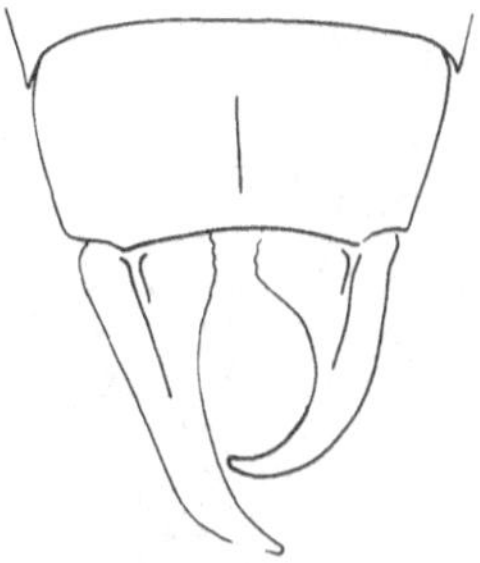

Fig. 221 Male ultimate tergite and forceps of *Anisolabis pectoralis* (ESCHSCHOLTZ). Original.

66. Anisolabis ryukyuensis (NISHIKAWA) comb. nov.

1969 *Paralabis ryukyuensis* NISHIKAWA, Kontyu, Tokyo, **37**: 44; fig. 4 (male), 5 (penultimate sternite with forceps of male), 6 (genital armature of holotype), 7 (external paramere with genital lobe). – Terra typica: Ryukyu Islands (Type male: Ent. Lab. Tokyo Univ.).

Literature: STEINMANN 1977, 1978, 1984.

Description: Male colour shiny blackish to reddish-brown; legs yellow; antennae brown, except one to four joints near the apices whitish. Head a little longer than wide, moderately tumid, slightly emarginate posteriorly, scantly punctate; postfrontal and coronal sutures distinct. Antennae 17-jointed, typically, first joint a little shorter than the distance between antennal bases. Pronotum nearly quadrate, longer than wide, widened posteriorly, as broad as head; median longitudinal furrow distinct, lateral margins parallel, anterior and posterior margins straight. Tegmina and wings entirely absent. Abdomen normal, a little expanded to middle; tergites 1–9 clearly and closely punctate, each of tergites 2–7 with a very short spine on each posterior corners, tergites 7–9 rugoso-striate laterally and ecarinate on sides. Ultimate tergite with median longitudinal sulcus. Forceps (Fig. 222) asymmetrical, trigonal basally, curved and cylindrical apically; inner margins crenulate basally. Penultimate sternite transverse, emarginate posteriorly, rugosely punctate and pubescent, with a pair of long seta near middle of lateral margin. Genitalia (Fig. 223) specific; paramere very broad, median incision of anterior margin normal; genital lobes with hyaline denticulation pads on the apices, virga clearly visible throughout. External paramere not of *Anisolabis*-type, shorter, wide medially. – **Female**

similar to male, but punctation weaker, especially on rounded sides of abdomen, ultimate tergite more narrowed posteriorly, penultimate sternite rounded posteriorly, and forceps symmetrical, almost straight. – **Length** of body with forceps: male: 17,5–22 mm; female: 17–22 mm.

Distribution: Ryukyu Islands.

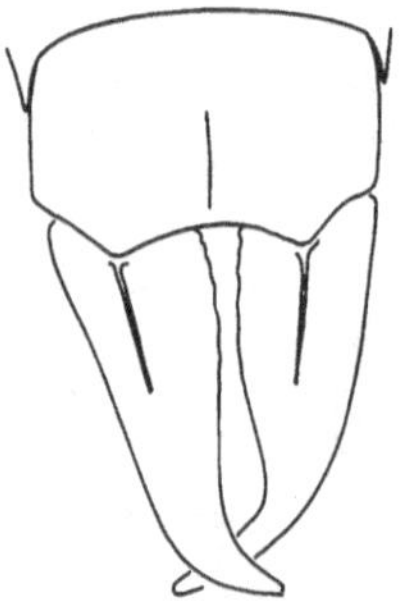

Fig. 222 Male ultimate tergite with forceps of *Anisolabis ryukyuensis* (NISHIKAWA). After NISHIKAWA.

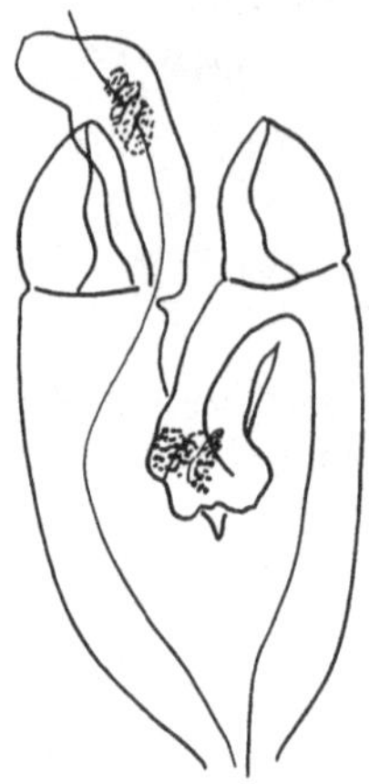

Fig. 223 Holotype genital armature of *Anisolabis ryukyuensis* (NISHIKAWA). After NISHIKAWA.

67. Anisolabis montshadskii BEY-BIENKO

1959 *Anisolabis montshadskii* BEY-BIENKO, Ent. Obozr., Moskva, **38**: 600; fig. 16 (holotype genitalia). – Terra typica: China: Yunnan (Type male: perhaps Peking Museum, 3 paratypes in Leningrad Museum).

1971 *Paralabis montshadskii* – SRIVASTAVA, Entomologist's Rec. J. Variation, London, **83**: 22 (India, China: Yunnan).

1972 *Paralabis montshadskii* – SRIVASTAVA, Rec. zool. Survey India, Calcutta, (1–4) **66**: 134; fig. 2A (distal antennal segments), 2B (penultimate sternite of male), 2C (ultimate tergite with forceps of male). 2D (male genital armature).

1975 *Paralabis montshadskii* – STEINMANN, Fol. ent. hung., Budapest, **28**: 154; fig. 12 (male genital armature of paratype).

Literature: SAKAI 1970; SRIVASTAVA 1971; STEINMANN 1977, 1978.

Description: Male stout, colour dark brown or blackish. Head normal, of *Carcinophoria*-type; eyes small, slightly shorter than scape; postfrontal and coronal sutures distinct. First antennal joint long, more or less as long as the distance between antennal bases. Pronotum with apical margin and lateral sides straight; posterior margin rounded; median longitudinal furrow present in anterior section. Tegmina and wings absent. Meso- and metanotum normal. Legs unicoloured, femora sometimes banded with black. Abdomen rugosely punctate; sides of tergites obtuse. Ultimate tergite transverse, narrowed apically; median longitudinal sulcus faint, plate with tumid elevations laterally above the bases of forceps. Forceps typical, asymmetrical, trigonal basally, cylindrical apically. Penultimate sternite with posterior margin emarginate in middle. Genitalia (Fig. 224 after SRIVASTAVA and fig. 225 after STEINMANN) characteristic; parameres with deep median incision at anterior margin; genital lobes long, visibly longer than external paramere, without virgae, with sclerotized plate apically. External parameres articulating on antero-lateral angles of paramere. – **Female** similar to male, but forceps slenderer, symmetrical, tapering. – **Length** of body with forceps: male: 12,5–17 mm, female: 12–16 mm.

Distribution: China: Yunnan, and India.

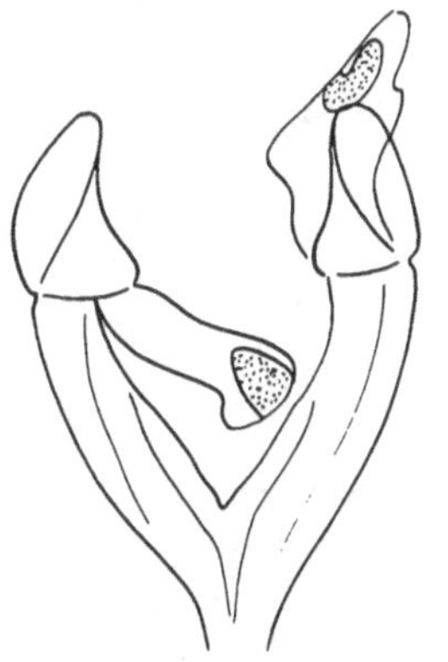

Fig. 224 Male genital armature of *Anisolabis montshadskii* BEY-BIENKO. After BEY-Bienko.

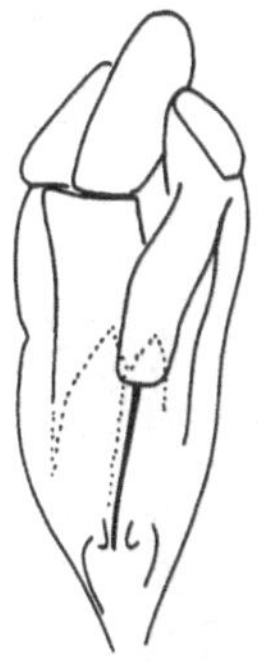

Fig. 225 Male genital armature of *Anisolabis montshadskii* BEY-BIENKO. After STEINMANN.

68. Anisolabis aborensis (BURR) comb. nov.

1913 *Euborellia aborensis* BURR, Rec. Indian Mus., Calcutta, (2) **8**: 137. – Terra typica: India: Assam (Type male: Brit. Mus. nat. Hist., London).

1968 *Paralabis aborensis* – SRIVASTAVA, Entomologist's Rec. J. Variation, London, **80**: 292.

1971 *Paralabis aborensis* – SRIVASTAVA, Entomologist's Rec. J. Variation, London, **83**: 22; fig. D (ultimate tergite with forceps of male), E (posterior margin of penultimate sternite of male), F (male genitalia).

Literature: POPHAM & BRINDLE 1966; SAKAI 1970; STEINMANN 1977, 1978, 1984.

Description: Male colour reddish chestnut with shades of yellow and black; antennae and legs yellowish-brown. Head smooth, frons tumid, occiput a little depressed; postfrontal and coronal sutures distinct. Eyes comparatively small, shorter than first antennal joint; the latter normal, about as long as the distance between antennal bases or a little shorter. Pronotum slightly longer than wide; median longitudinal furrow present but faintly marked; lateral margins straight, a little expanded posteriorly; posterior margin truncate. Tegmina present (or absent, after by SRIVASTAVA 1971, ibidem: 22). Abdomen punctate and covered with long, reddish hairs, sides of tergites 7–9 rugoso-striate, with lateral longitudinal ridges. Ultimate tergite broad, median longitudinal sulcus distinct; posterior surface with some stripes (Fig. 226); forceps asymmetrical; dorsal keels of branches typical, arms broad, strongly curved apically. Penultimate sternite rounded posteriorly. Genitalia (Fig. 227) well developed; paramere strongly V-shaped; median longitudinal incision of anterior margin very deep and wide; genital lobes with a pair chitinous denticulations apically. External parameres curved. – **Female** as male, but forceps symmetrical, slenderer, and simple; inner margins crenulate. – **Length** of body with forceps: male: 14,5–17 mm, female: 14–19 mm.

Distribution: India: Assam.

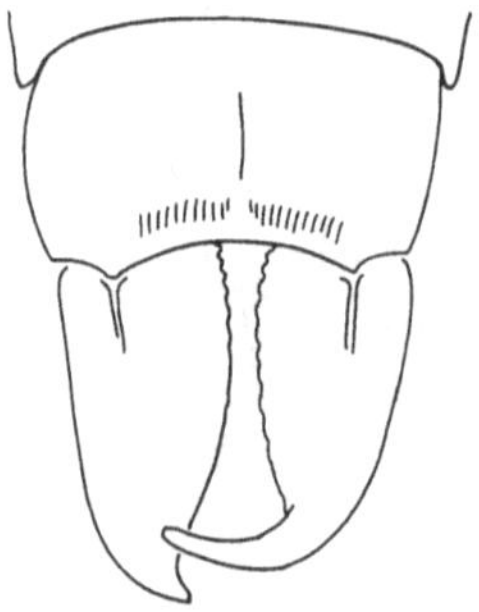

Fig. 226 Male ultimate tergite and forceps of *Anisolabis aborensis* (Burr). After Srivastava.

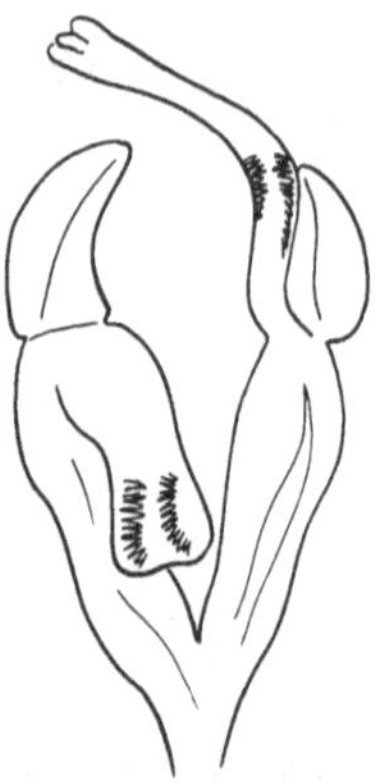

Fig. 227 Male genital armature of *Anisolabis aborensis* (Burr). After Srivastava.

69. Anisolabis antennata (Ramamurthi & David) comb. nov.

1973 *Paralabis antennatus* Ramamurthi & David, Zool. Anz., Leipzig, **190**: 441; fig. 11 (ultimate tergite and forceps of holotype), 12 (penultimate sternite), 13 (genital armature of holotype). – Terra typica: India (Type male: Ramamurthi's Collect., Univ. Mus. Madras).

Literature: Steinmann 1981.

Description: Male colour dark brown, abdomen and forceps reddish-brown, legs yellow, femora with brown bands. Head tumid, well-developed sensory bristles along entire border, caudal margin feebly emarginate mesad. Postfrontal and coronal sutures distinct, but not deep. Antennae 17-jointed; first joint a little shorter than the distance between antennal bases, second transverse, orange, rest longer than wide, dark brown, except two or three subapical joints, which are whitish. Eyes normal, shorter than the length of head behind eyes. Pronotum slightly longer than wide; lateral margins parallel, prozona separated from metazona by a transverse depression; margins truncate. Tegmina present, projecting beyond metanotum, oily and shiny, narrowed anteriorly, exposing a very small scutellum, meeting along sutural margin. Abdomen shiny, tergites finely punctate and with sparse pubescence; postero-lateral angles of tergites acute. Ultimate tergite as long as wide, median longitudinal sulcus weak; caudal margin broadly sinuate. Forceps (Fig. 228) asymmetrical, trigonal basally, sharply narrowed to apex, inner margins finely denticulate. Penultimate sternite truncate. Genitalia (Fig. 229) characteristic; central parameral plate V-shaped, median incision of anterior margin very deep; genital lobes with very specific chitinous serration, and a sclerotized plate; external parameres curved. – **Female** similar to male, but forceps straight, and subcontiguous, penultimate sternite broadly triangular. – **Length** of body with forceps: male: 14 mm, female: 15 mm.

Distribution: India.

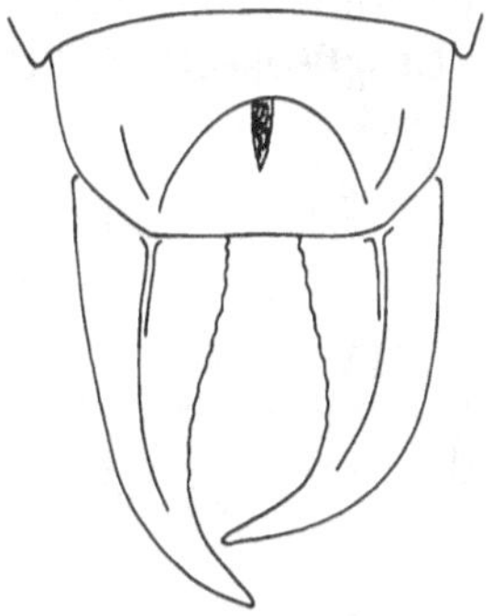

Fig. 228 Holotype ultimate tergite and forceps of *Anisolabis antennata* (RAMAMURTHI & DAVID). After RAMAMURTHI & DAVID.

Fig. 229 Holotype genital armature of *Anisolabis antennata* (RAMAMURTHI & DAVID). After RAMAMURTHI & DAVID.

70. Anisolabis lefroyi (BURR) comb. nov.

1906 *Anisolabis brunneri* BURR, (nec BURR, 1908), J. r. asiat, Soc. Bengal., Calcutta, (n. S.), **2**: 389. – Terra typica: India (Type male: Brit. Mus. nat. Hist., London).

1910 *Psalis lefroyi* BURR, Fauna Brit. India, Dermaptera, London, Bombay, 77, pl. 3, fig. 20 (male). – Terra typica: India (Type male: Brit. Mus. nat. Hist., London).

1966 *Carcinophora lefroyi* – POPHAM & BRINDLE, Entomologist, London, **99**: 275.

1967 *Euborellia lefroyi* – KAPOOR, Agra Univ. J. Res. (Sci.), **16**: 11.

1971 *Paralabis lefroyi* – SRIVASTAVA, Entomologist's Rec. J. Variation, London, **83**: 20; fig. A (ultimate tergite with forceps of male), B (posterior margin of penultimate sternite of male), C (male genitalia). India: Bihar.

Literature: BURR 1915; SAKAI 1970; STEINMANN 1977, 1978.

Description: Male blackish brown, sides of pronotum yellow, legs yellowish brown. Head smooth, about as long as wide, rounded posteriorly; eyes comparatively large, more or less as long as the length of head behind eyes.

Postfrontal and coronal sutures faintly marked; frons tumid, occiput slightly depressed. Antennae 17-jointed, blackish brown, except the joint 2nd, 13th and 14th joints, which are yellow; first joint long, but shorter than the distance between antennal bases. Pronotum slightly longer than wide, anterior margin straight, sides very faintly concave in middle; posterior margin rounded; median longitudinal furrow distinct. Tegmina present, short with posterior margin very slightly obliquely truncate. Wings absent. Abdomen gently widened medially; sides of tergites 6–9 punctate. Ultimate tergite transverse; posterior margin concave in median section; median longitudinal sulcus present. Forceps (Fig. 230) asymmetrical; branches subcontiguous at base, triangular, weakly widened apically. Penultimate sternite triangular with posterior margin briefly rounded. Genitalia (Fig. 231) typical; parameres with deep median incision between external parameres; genital lobes armed with minute chitinous teeth apically; external parameres wide at middle. – **Female** similar to male, except weakly transverse ultimate tergite narrowed posteriorly and forceps with branches straight, apices pointed and not hooked. – **Length** of body with forceps: male: 14–15,5 mm, female: 10–11,5 mm.

Distribution: India: Bihar.

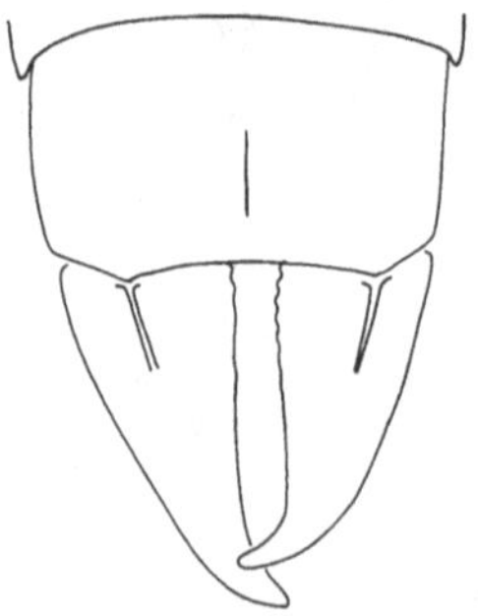

Fig. 230 Male ultimate tergite and forceps of *Anisolabis lefroyi* (BURR). After SRIVASTAVA.

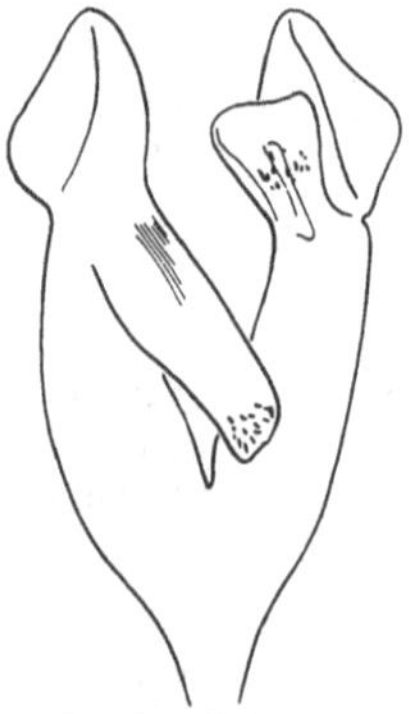

Fig. 231 Male genital armature of *Anisolabis lefroyi* (BURR). After SRIVASTAVA.

71. Anisolabis dohrni (KIRBY)

1891 *Nannopygia dohrni* KIRBY, J. Linn. Soc., London, Zool., **23**: 508. – Terra typica: Ceylon (Type female: Brit. Mus. nat. Hist., London).

1879 *Labidura femoralis* DUBRONY (nec DOHRN), Ann. Mus. civ. Stor. nat., Genova, **14**: 352. – Terra typica: Ceylon (Type male: Mus. civ. Stor. nat., Genova).

1900 *Carcinophora caeruleipennis* BORMANS, Das Tierreich, Berlin, **11**: 40. – Terra typica: Ceylon (and Northern Australia ?) (Type male: Mus. Naturkunde, Berlin). New name for *Labidura femoralis* DUBRONY.

1902 *Carcinophora dohrni* – BURR, J. Bombay nat. Hist. Soc., **14**: 328; pl. B, fig. 20 (male). New description of male.
1910 *Psalis dohrni* – BURR, Fauna Brit. India, Dermaptera, London, Bombay, 76; pl. 3, fig. 19 (male). Ceylon.
1915 *Paralabis dohrni* – BURR, J. r. microsc. Soc., London, **1915**: 540; pl. 12, fig. 1 (male genitalia).
1966 *Anisolabis dohrni* – POPHAM & BRINDLE, Entomologist., London, **99**: 274. India, Ceylon.
1971 *Carcinophora dohrni* – BRINDLE, Ent. scand., København, Suppl., **1**: 219. Ceylon.
1977 *Paralabis dohrni* – STEINMANN, Ann. hist.-nat. Mus. nation. hung., Budapest, **69**: 93. India, Ceylon, Burma, and ? North Australia.
1980 *Paralabis dohrni* – STEINMANN, Fol. ent. hung., Budapest, **41**: 337; fig. 21 (male ultimate tergite with forceps), 22 (male genitalia, gen. prep. No. 676, det. Dr. H. STEINMANN) (Ceylon).
1984 *Anisolabis dohrni* – STEINMANN, Ent. Abh. staatl. Mus. Tierk., Dresden, **48**: 74.
Literature: BURR 1901; SAKAI 1970; STEINMANN 1978.

Description: Male colour reddish to blackish brown; pronotum yellowish on sides, femora with black bands; body very shiny. Head tumid; occiput somewhat depressed; postfrontal and coronal sutures faintly distinct. Eyes normal, a little shorter than the length of head behind eyes. Antennal joints 1–3 yellowish, and 13 and extremity of 12 whitish; first antennal joint slightly shorter than the distance between antennal bases. Pronotum longer than wide, gently widening posteriorly, sides straight, posterior margin rounded; median longitudinal suture faint. Tegmina present; smooth, shiny, black or reddish with posterior margin truncate. Wings generally absent. Abdomen of *Anisolabis*-type, widened to median tergites; sides of tergites 6–9 rugoso-striate, surface punctate. Ultimate tergite transverse; posterior margin straight or somewhat concave. Forceps (Fig. 232) broad, asymmetrical, subcontiguous; strongly trigonal basally, curved and cylindrical apically; inner margin crenulate. Penultimate sternite somewhat triangular with posterior margin rounded. Genitalia (Fig. 233) oval; central parameral plate broad, median incision of anterior margin deep; genital lobes well developed, with minute chitinous teeth apically. External parameres rounded apically. – **Female** similar to male, but forceps simple, slenderer, tapering. – **Length** of body with forceps: male: 12–14,5 mm, female: 10,5–15 mm.

Distribution: Sri Lanka, India, Burma, and North Australia (after BORMANS 1900).

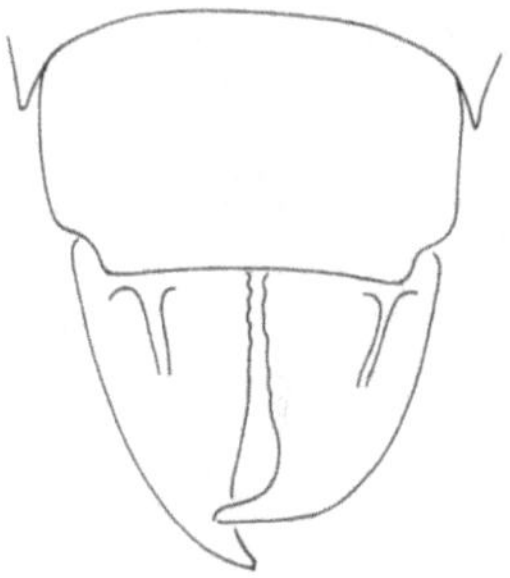

Fig. 232 Male ultimate tergite and forceps of *Anisolabis dohrni* (KIRBY). Original.

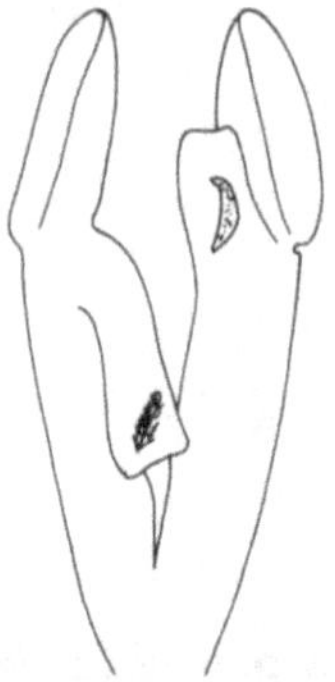

Fig. 233 Male genitalia of *Anisolabis dohrni* (KIRBY). After BURR.

72. Anisolabis orientalis (Ramamurthi)

1968 *Labidurodes orientalis* Ramamurthi, Zool. Anz., Leipzig, **181**: 130; fig. 1 (male), 2 (antennal joints), 3 (genital armature of holotype). – Terra typica: South India (Type male: Ramamurthi's Collect., Univ. Mus., Madras).

1977 *Anisolabis orientalis* – [misquote: Kapoor] – Steinmann, Ann. hist.-nat. Mus. nation. hung., Budapest, **69**: 92.

Description: Male colour dark brown, except legs, yellow. Head tumid; postfrontal and coronal sutures indistinct. Antennae 11-jointed (?); first joint thick and obliquely truncate apically, more or less as long as the distance between antennal bases; second very small, third joint a little longer than fourth. Eyes normal, shorter than the length of head behind eyes. Pronotum a little longer than broad; anterior margin not wider than head, plate slightly expanded to posterior angles. Median longitudinal furrow present. Tegmina present, but reduced to lateral flaps of mesonotum. Wings entirely absent. Metanotum broadly concave. Abdomen widening to middle, all tergites except last finely punctate. Ultimate tergite ample, transverse; median longitudinal sulcus obsolete. Forceps (Fig. 234) contiguous, short, and heavly symmetrical; trigonal basally, apex pointed and incurved. Penultimate sternite normal, broadly rounded posteriorly. Genitalia (Fig. 235) characteristic; paramere broad, incision of anterior margin not deep; genital lobes relatively short and narrow, with virgae. External parameres of *Anisolabis*-type, obtuse. – **Female** similar to male, but forceps slenderer. – **Length** of body with forceps: male: 15 mm, female: 15,5 mm.

Distribution: South India.

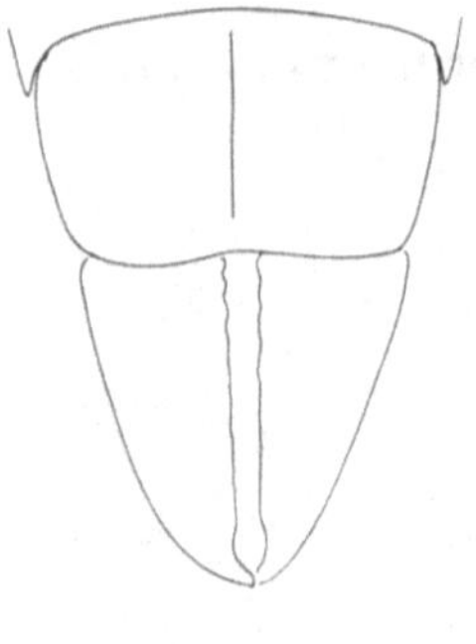

Fig. 234 Holotype ultimate tergite and forceps of *Anisolabis orientalis* (Ramamurthi). After Ramamurthi.

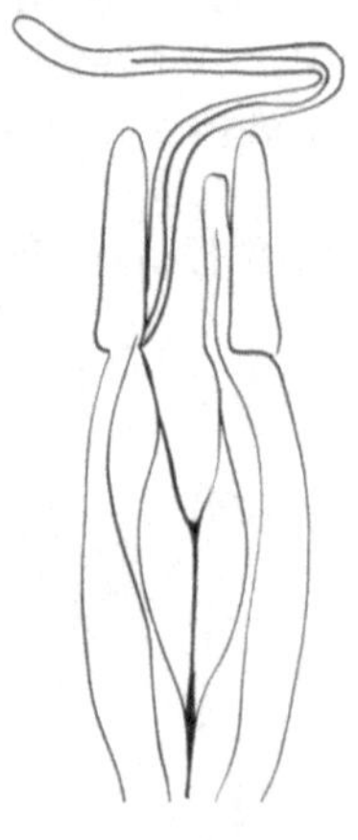

Fig. 235 Holotype genitalia of *Anisolabis orientalis* (Ramamurthi). After Ramamurthi.

73. Anisolabis mira (Bey-Bienko) comb. nov.

1959 *Plancolabis mira* Bey-Bienko, Ent. Obozr., Moskva, **38**: 608; fig. 23 (penultimate sternite of male), 29 (genital armature of holotype). – Terra typica: China: Yunnan (Type male: perhaps in Peking Museum; 3 paratypes in Leningrad Museum).
1975 *Plancolabis mira* – Steinmann, Fol. ent. hung., Budapest, **28**: 157; fig. 20 (genital armature of paratype).
1977 *Placolabis mira* – Steinmann, Ann. hist.-nat. Mus. nation. hung., Budapest, **69**: 95; fig. 24 (male genitalia).
Literature: Sakai 1970; Steinmann 1978.

Description: Male morphologically similar to *Anisolabis maritima* (Bonelli), but penultimate sternite with specific median lobe; the latter is similar to *Anisolabis cristata* Hincks, but not rounded, angular (Fig. 236). Genitalia (Fig. 237) specific; paramere broad, median longitudinal incision of anterior margin not deep; genital lobe without virga. External parameres candle-shaped, not pointed apically. – **Female** similar to *maritima* (Bonelli). – **Length** of body with forceps: male: 13–16 mm, female: 12–16 mm.
Distribution: China: Yunnan.

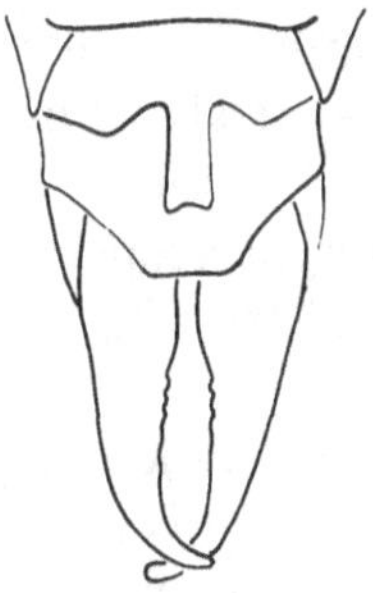

Fig. 236 Male penultimate sternite and forceps of *Anisolabis mira* (Bey-Bienko), ventrally. After Bey-Bienko.

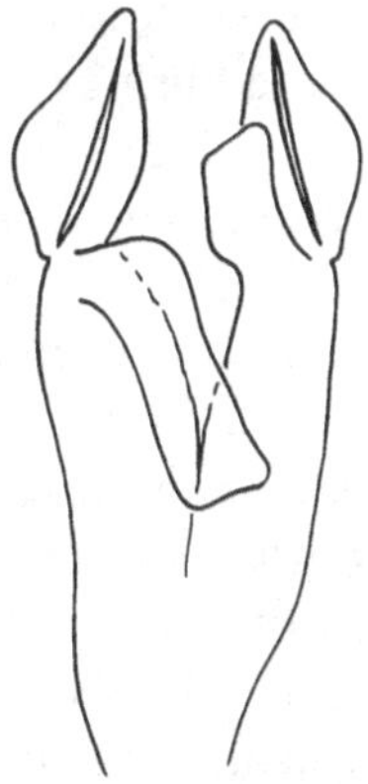

Fig. 237 Holotype genital armature of *Anisolabis mira* (Bey-Bienko). After Bey-Bienko

74. Anisolabis dubronyi Kirby

1888 *Anisolabis laeta* Bormans (nec Gerstaecker), An. Soc. esp., Hist. nat., Madrid, **17**: 435. Notes by Burr 1910: "This species is quite distinct from the E. African *Anisolabis laeta*, with which Bormans confused it."
1903 *Anisolabis dubronyi* Kirby, Ann. Mag. nat. Hist., London, (7) **11**: 68. – Terra typica: Tenasserim (Type male: Brit. Mus. nat. Hist., London).
1910 *Anisolabis dubronyi* – Burr, Fauna Brit. India, Dermaptera, London, Bombay, 85, fig. 97 (male).
1970 *Anisolabis dubronii* – Sakai, Dermapt. Cat. Prael., Tokyo, **1**: 59. Burma.
Literature: Bormans 1900; Burr 1911; Kirby 1904; Popham & Brindle 1966; Steinmann 1977.

Description: Male entirely dark reddish-black except the head with antennae, legs, yellow. Head rounded; postfrontal and coronal sutures distinct. Antennae 21-jointed, joints typical of genus; first antennal joint more or less as long as the distance between antennal bases. Eyes normal, about as long as genae. Pronotum subrectangular, longer than wide; median longitudinal furrow well marked; all margins straight, and all angles rounded. Tegmina and wings absent. Meso- and metanotum typical. Abdomen expanded medially; sides of tergites 6–9 rugoso-striate. Ultimate tergite broad, transverse; median longitudinal sulcus present. Forceps asymmetrical, strongly trigonal basally, cylindrical apically. Penultimate sternite triangular, the apex rounded. Genitalia unknown. – **Female** unknown. – **Length** of body with forceps: 24,5–25,5 mm.
Distribution: Burma.

75. Anisolabis vitalisi Burr

1917 *Anisolabis vitalisi* Burr, Ann. Soc. ent. France., Paris, **86**: 58. – Terra typica: China (Type male: Brit. Mus. nat. Hist., London).
Literature: Popham & Brindle 1966; Sakai 1970; Steinmann 1977, 1978, 1984.

Description: Male colour blackish brown; sides of pronotum, legs, yellowish brown; forceps reddish brown. Head tumid, more or less depressed at vertex; postfrontal and coronal sutures distinct, but not deep. Eyes normal, slightly shorter than the scape. Antennae 19-jointed; first joint about as long as the distance between antennal bases; second transverse. Pronotum a little longer than wide; lateral margins parallel; posterior margin truncate; all angles rounded. Median longitudinal furrow distinct. Tegmina and wings absent. Meso- and metanotum transverse, typical. Abdomen widened at middle; sides of tergites 6–9 rugoso-striate. Ultimate tergite broad, median longitudinal sulcus present; posterior margin concave at median section. Forceps asymmetrical, branches robust, slightly broad basally, and very strongly curved apically. Penultimate sternite rounded posteriorly. Genitalia unknown. – **Female** unknown. – **Length** of body with forceps: 22–24 mm.
Distribution: China, Vietnam.

76. Anisolabis penetrans Burr

1912 *Anisolabis penetrans* Burr, Ann. naturhist. Hofmus. Wien, **26**: 78. – Terra typica: Indonesia: Comoro Islands (Type male: Naturhist. Mus., Wien).
Literature: Popham & Brindle 1966; Sakai 1970; Steinmann 1977, 1978.

Description: Male colour dark reddish brown or black; shiny; antennae brown, legs yellowish-brown. Head broad, frons tumid, occiput a little depressed; postfrontal sutures finely, coronal suture normally marked; eyes comparatively small, slightly shorter than the genae or first antennal joint; the latter long, but a little shorter than the distance between antennal bases. Pronotum wider than long; lateral margins straight, concave; all angles rounded; median longitudinal furrow faintly visible. Tegmina and wings completely absent. Meso- and metanotum typical, transverse. Abdomen widened at middle, depressed, tergites faintly punctate. Ultimate tergite broad; ventro-lateral margins with keels. Pygidium not visible. Forceps trigonal basally, curved and cylindrical apically, asymmetrical, of *Anisolabis*-type. Penultimate sternite rounded posteriorly. Genitalia unknown. – **Female** unknown. – **Length** of body with forceps: 21,5 mm.

Distribution: Indonesia: Comoro Islands.

77. Anisolabis greeni Burr

1899 *Anisolabis greeni* Burr, Ann. Mag. nat. Hist., London, (7) **4**: 257. – Terra typica: Ceylon (Type male: Brit. Mus. nat. Hist., London).

1910 *Borellia greeni* – Burr, Fauna Brit. India, Dermaptera, London, Bombay, 87.

1911 *Euborellia greeni* – Burr, Genera Insectorum, Bruxelles, **122**: 30.

1915 *Paralabis greeni* – Burr, J. r. microsc. Soc., London, **1915**: 540, pl. 12, fig. 2 (male genitalia).

1966 *Anisolabis greeni* – Popham & Brindle, Entomologist, London, **99**: 274.

1971 *Paralabis greeni* – Srivastava, Entomologist's Rec. J. Variation, London, **83**: 20. India: Madras, Sri Lanka.

1971 *Anisolabis greeni* – Brindle, Ent. scand., København, Suppl., **1**: 219; fig. 16 (male genitalia), 20 (head, pro- and mesonotum). Ceylon and South India.

Literature: Hebard 1923; Sakai 1970; Steinmann 1977, 1978, 1984.

Description: Male colour varies from dark brown to blackish-brown; legs brick red; body shiny. Head tumid, postfrontal and coronal sutures distinct. Eyes relatively short, shorter than the first antennal joint; the latter normal, more or less as long as the distance between antennal bases. Pronotum rectangular, with a median impression; lateral margins straight, a little concave; posterior margin truncate. Tegmina present, narrow at base, exposing a broad, short, transverse area of the mesonotum; then suddenly and strongly dilated, so that the inner margins meet along the suture and cover the apical half of the mesonotum. Abdomen of *Anisolabis*-type, slightly expanded to median section; sides of tergites 7–9 rugoso-striate. Ultimate tergite somewhat narrowed with a faint median impression, with two faint tubercles on posterior margin. Forceps

Fig. 238 Male genital armature of *Anisolabis greeni* Burr. After Brindle.

subcontiguous, weakly asymmetrical; strongly trigonal basally, cylindrical apically, apices of branches curved; inner margins with a faint tooth at about middle or nearer to the base. Penultimate sternite triangular, its apex rounded. Genitalia (Fig. 238) simple, oval, large; parameres broad, median incision of anterior margin very deep and wide; genital lobes simple, without virgae; external parameres rounded apically. – **Female** similar to male, but forceps symmetrical, tapering, subcontiguous. – **Length** of body with forceps: male: 13,5–26,5 mm, female: 19,5–22 mm.

Distribution: Sri Lanka, India: Madras.

78. Anisolabis recurva BORELLI

1915 *Anisolabis recurva* BORELLI, Boll. Mus. Zool. Anat. comp. Univ., Torino, **30**, No. 705: 2. – Terra typica: Philippines: Luzon (Type male: Mus. Ist. Zool. sist., Torino).

Literature: BORELLI 1927; SAKAI 1970; POPHAM & BRINDLE 1966; STEINMANN 1977, 1978, 1984.

Description: Male colour dark brown; antennae light brown, legs yellowish-brown; forceps reddish-brown. Head tumid, occiput a little depressed; post-frontal and coronal sutures distinct. Eyes small, shorter than the first antennal joint. Antennae 16-jointed; first joint normal, about as long as the distance between antennal bases. Pronotum quadrate, lateral margins parallel; all margins straight, and all angles rounded; median longitudinal furrow present in centre. Tegmina and wings absent. Meso- and metanotum transverse, typical. Abdomen of *Anisolabis*-type, slightly expanded to middle; sides of tergites 6–9 rugoso-striate. Ultimate tergite broad, transverse; posterior margin concave in median section; median longitudinal sulcus finely distinct. Forceps asymmetrical, strongly trigonal basally, curved and cylindrical apically. Penultimate sternite rounded posteriorly. – Genitalia unknown (holotype without penis). – **Female** very similar to male, but forceps, as in fig. 239 straight, symmetrical; dorsal keels or ridge prominent; inner margins crenulate. – **Length** of body with forceps: male: 22–23 mm, female: 21–25,5 mm.

Distribution: Philippines.

79. Anisolabis gaudens BURR

1904 *Anisolabis gaudens* BURR, Trans. ent. Soc. London, **1904**: 291. – Terra typica: Bhutan (Type female: Mus. nation. Hist. nat., Paris).

1910 *Anisolabis gaudens* – BURR, Fauna Brit. India, Dermaptera, London, Bombay, 86. Bhutan.

Literature: BRINDLE 1975; BURR 1915; SAKAI 1970; STEINMANN 1977, 1978.

Description: Female colour black; shiny, body size medium; head red, antennae dark brownish-black, except first two joints, basally red. Head normal, a little longer than wide; postfrontal and coronal sutures distinct. Eyes relativelly short, slightly shorter than the genae or the length of head behind eyes. Antennae 15-jointed, first joint long, about as long as the distance between antennal bases. Pronotum quadrate, all margins straight, lateral margins a little expanded to posterior angles. Median longitudinal furrow present. Tegmina and wings absent. Abdomen more or less widened to middle. Ultimate tergite very broad; posterior margin concave at middle. Forceps (Fig. 239) straight, symmetrical, slightly depressed, subcontiguous; inner margins crenulate; dorsal keels prominent. – **Male** unknown. – **Length** of body with forceps: 10 mm.

Notes by BRINDLE 1975: "*Anisolabis gaudens* was described by BURR 1904 from Pedong, and the type deposited in the Paris Museum. It is a black species with a red head, although BURR 1913 provisionally assigned to the species and all female from the Abor district of Assam."

Distribution: Bhutan.

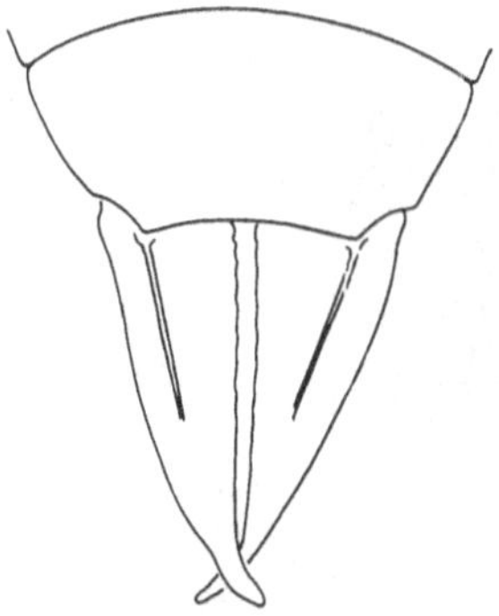

Fig. 239 Female ultimate tergite and forceps of *Anisolabis gaudens* BURR. After BURR.

80. Anisolabis maindroni (BORELLI)

1911 *Homoeolabis maindroni* BORELLI, Boll. Mus. Zool. Anat. comp. Univ., Torino, **26**, No. 640: 2. – Terra typica: India (Type male: Mus. nation. Hist. nat., Paris).

1911 *Homoeolabis maindroni* – BURR, Genera Insectorum, Bruxelles, **122**: 28, pl. 2, fig. 15 (pro-, meso- and metanotum), 15a (pro-, meso- and metasternum). Ceylon, South India.

1966 *Anisolabis maindroni* – POPHAM & BRINDLE, Entomologist, London, **99**: 274.

1971 *Homeolabis maindroni* – BRINDLE, Ent. scand., København, Suppl., **1**: 217; fig. 15 (male genitalia), 19 (head and pronotum of male).

1977 *Homoeolabis maindroni* – STEINMANN, Ann. hist.-nat. Mus. nation. hung., Budapest, **69**: 91; fig. 9 (male genitalia).

Literature: SAKAI 1970; STEINMANN 1978, 1984.

Description: Male rather large, blackish to dark reddish species, rather variable in size; legs yellow to brown, forceps of both sexes reddish. Head rounded, a little broader than pronotum; postfrontal and coronal sutures indistinct. First antennal joint normal, somewhat shorter than the distance between antennal bases. Pronotum quadrate or slightly longer than wide; median longitudinal furrow present; lateral margins parallel, posterior angles and margin rounded. Tegmina and wings absent. Mesonotum transverse, visibly wider than long. Prosternum narrow, about half as wide as metasternite; posterior margin truncate. Abdomen expanded to last tergite, a little of *Gonolabis*-type. Forceps trigonal basally, cylindrical apically. Ultimate tergite very broad; median longitudinal sulcus present; posterior margin concave in median section. Male very similar to *Euborellia stali* (DOHRN). Genitalia a little variable, characteristic; strongly sclerotized; after BURR as in fig. 240, after STEINMANN as in fig. 241, and after BRINDLE as in fig. 242. Parameres with median incision, incision deep or small; genital lobes well-developed, with or without virgae; external parameres of *Anisolabis*-type, relatively long, narrow and obtuse. – **Female** very similar to male, but forceps slenderer, narrower. – **Length** of body with forceps: male: 16–17,5 mm, female: 16–18 mm.

Distribution: South India, Sri Lanka.

Fig. 240 Male genital armature of *Anisolabis maindroni* (BORELLI). After BURR.

Fig. 241 Male genital armature of *Anisolabis maindroni* (BORELLI). After STEINMANN.

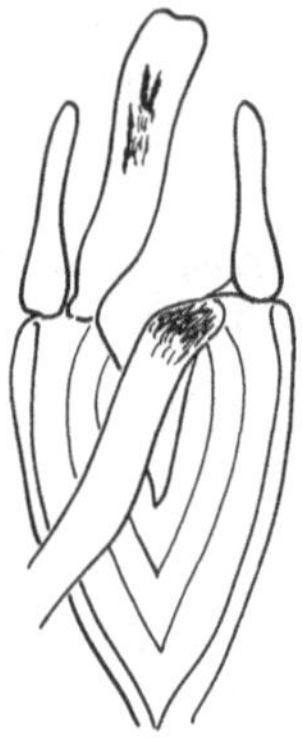

Fig. 242 Male genital armature of *Anisolabis maindroni* (BORELLI). After BRINDLE.

81. Anisolabis rubella BRINDLE

1977 *Anisolabis rubella* BRINDLE, Entomologist's Rec. J. Variation, London, **89**: 146; fig. 1 (male), 2 (genital armature of holotype). – Terra typica: Sri Lanka (Type male: Brit. Mus. nat. Hist., London).

Description: Male colour reddish-brown, legs yellow, antennae whitish basally, and brown distally. Head broad, rounded; frons tumid, occiput a little depressed. Postfrontal and coronal sutures prominent. Eyes comparatively small, slightly shorter than the length of head behind eyes. First antennal joint long, but shorter than the distance between antennal bases. Pronotum nearly as long as wide, slightly widened posteriorly; posterior margin rounded. Tegmina and wings absent. Mesonotum transverse, relatively long, metanotum short. Abdomen somewhat of *Gonolabis*-type, broadened distally; sides of tergites produced but not pointed, and those of tergites 8–9 and base of last tergite rugoso-striate but without ridges. Ultimate tergite large, median longitudinal sulcus present in centre; plate with a lateral curved ridge toward dorsum, posterior margin concave in median section. Forceps (Fig. 243) short, broad, expanded basally; asymmetrical. Penultimate sternite missing. Genitalia (Fig. 244) characteristic, not of *Anisolabis*-type; parameres short and wide; median incision of anterior margin small; genital lobes short without virgae. External parameres long, of *Anisolabis*-type. – **Female** unknown. – **Length** of body with forceps: 15,5 mm.

Distribution: Sri Lanka.

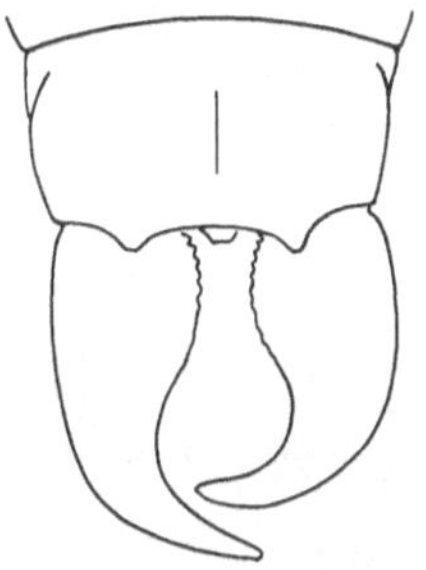

Fig. 243 Holotype ultimate tergite and forceps of *Anisolabis rubella* BRINDLE. After BRINDLE.

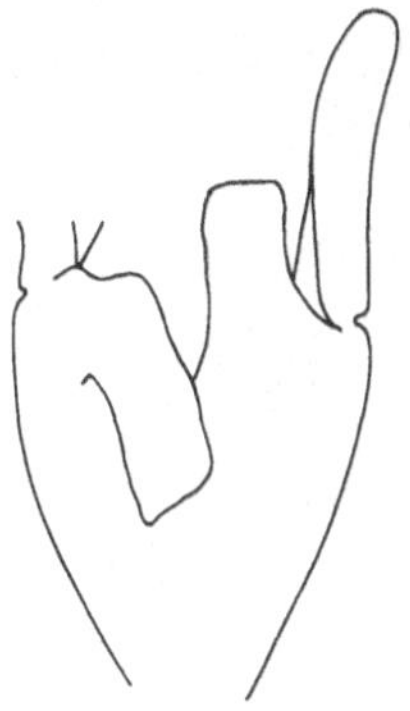

Fig. 244 Holotype genital armature of *Anisolabis rubella* BRINDLE. After BRINDLE.

82. Anisolabis kudagae BURR

1901 *Anisolabis kudagae* BURR, J. Bombay nat. Hist. Soc., **14**: 320. – Terra typica: Ceylon (Type male: Brit. Mus. nat. Hist., London).

1910 *Anisolabis kudagae* – BURR, Fauna Brit. India, Dermaptera, London, Bombay, 84; fig. 23 (male).

1971 *Euborellia kudagae* – BRINDLE, Ent. scand., København, Suppl., **1**: 220.

1977 *Anisolabis kudagae* – BRINDLE, Entomologist's Rec., J. Variation, London, **89**: 147, fig. 3 (male genitalia). Ceylon, Northen India: Kumaon.

Literature: BURR 1911, 1915; POPHAM & BRINDLE 1966; SAKAI 1970; STEINMANN 1977, 1978.

Description: Male colour blackish, legs dark brown, femora blackish at base, antennae yellowish basally, dark brown distally. Head broad, tumid; postfrontal and coronal sutures distinct. Eyes small, shorter than first antennal joint; the latter typical, a little shorter than the distance between antennal bases. Pronotum strongly transverse, lateral margins straight, and a little expanded to posterior angles; median longitudinal furrow present. Tegmina and wings totally absent. Meso- and metanotum typical. Abdomen broad, surface shiny, with exceedingly fine punctation; sides of tergites 7–9 and base of 10 rugoso-striate. Ultimate tergite with a dorso-lateral prominence; posterior margin concave. Forceps (Fig. 245) moderately short, wide; trigonal basally, curved apically; asymmetrical. Penultimate sternite with posterior margin excised medially. Genitalia (Fig. 246) elongate; paramere with very deep median incision between external parameres; genital lobes well developed, without virgae. External parameres not of *Anisolabis*-type, outer margins widened medially; apex obtuse. – **Female** similar to male, but ultimate tergite narrower, branches of forceps short, trigonal basally, almost straight and symmetrical. – **Length** of body with forceps: male: 16–17 mm, female: 15–17 mm.

Distribution: Sri Lanka, India.

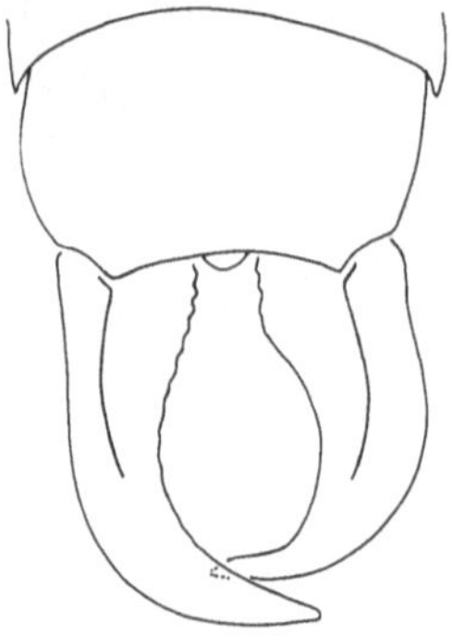

Fig. 245 Ultimate tergite with forceps of *Anisolabis kudagae* BURR. After BURR.

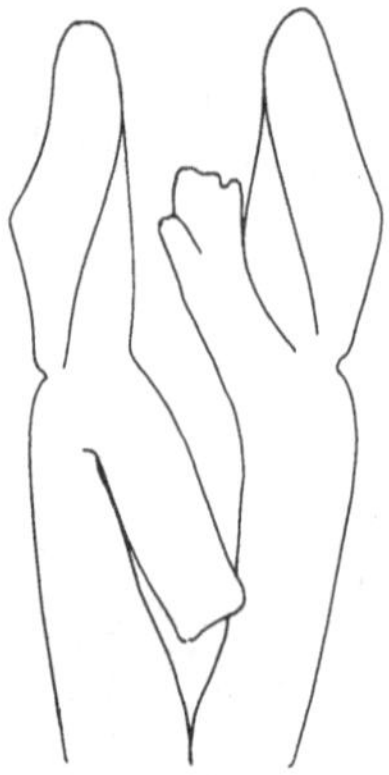

Fig. 246 Male genital armature of *Anisolabis kudagae* BURR. After BRINDLE.

Genus 14: **Foramenolabis** STEINMANN

1981 *Foramenolabis* STEINMANN, Fol. ent. hung., Budapest, **34**: 180. – Species typica: *Euborellia sisera* BURR 1914.

Description: Body similar to *Anisolabis*, but head very characteristic, trigonal, a little longer than wide, vertex specific, with a large cavity, vertical cavity about as long as first antennal joint or smaller, and a little wider than length of eye; deep. Male genitalia with very deep median incision of anterior margin of parameral plate. Genital lobes well or fully developed, elongated; apices of external parameres obtuse, rounded.

Distribution: South India, Afghanistan.

2 species.

Identification key to the species

1 Head with very large vertical cavity, length of cavity more or less as long as first antennal joint. Male genitalia resembling *Epilabis*-type, paramere wide, median incision very wide, deep (Fig. 248); external parameres very large and broad .... 1. **Foramenolabis sisera**, p. 153

1′ Head with small vertical cavity, length of cavity shorter than first antennal joint. Male genital armature with very long, but not wide external parameres (Fig. 250) .... 2. **Foramenolabis afghana**, p. 154

1. Foramenolabis sisera (BURR)

1914 *Euborellia sisera* BURR, Ann. Mus. nat. Hist., London, (8) **14**: 424. – Terra typica: South India (Type male: Brit. Mus. nat. Hist., London).
1915 *Epilabis sisera* – BURR, J. r. microsc. Soc., London, **1915**: 539; pl. XI, fig. 17 (male genitalia).
1931 *Epilabis sisera* – BORELLI, Rec. zool. Survey India, Calcutta, (1) **1**: 72; fig. 1, A (head), B (ultimate tergite of male), C (penultimate sternite of male).
1966 *Carcinophora sisera* – POPHAM & BRINDLE, Entomologist, London, **99**: 275.
1973 *Epilabis sisera* – RAMAMURTHI & DAVID, Zool. Anz., Leipzig, **190**: 435.
1981 *Foramenolabis sisera* – STEINMANN, Fol. ent. hung., Budapest, **34**, 180; fig. 13 (male genitalia, gen. prep. No. 668, det. Dr. H. STEINMANN).
Literature: SAKAI 1970; STEINMANN 1977.

Description: Male colour reddish-brown to yellowish-brown; head dark red, pronotum and thoracic nota reddish-brown, abdomen dark brown, and legs yellow. Head trigonal, a little longer than broad; vertex specific, with a large cavity, the vertical cavity about as long as the scape, and a little wider than the length of eye; deep. Postfrontal sutures distinct, but interrupted, coronal suture absent. Antennae 14-(?) jointed; first joint long, about as long as the distance between antennal bases; second transverse, third long, longer than wide, 4th joint short, but a little longer than wide. Pronotum longer than wide; lateral margins parallel, median longitudinal furrow distinct; all angles rounded. Tegmina and wings absent. Abdomen punctate dorsally, ultimate tergite broad with lateral longitudinal ridges; posterior margin straight. Forceps (Fig. 247) asymmetrical; trigonal basally, strongly curved apically, gradually tapering, left branch almost straight in basal two thirds; inner margin with a prominent tooth. Penultimate sternite triangular, slightly incrassate and obtusely rounded posteriorly and provided with a faintly raised, median stripe extending from middle to apex.

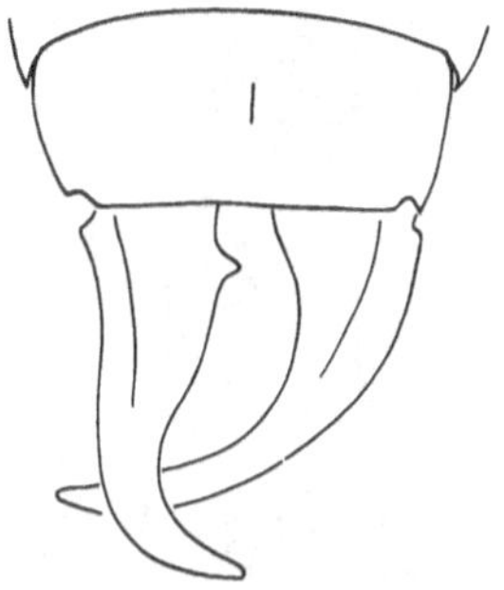

Fig. 247 Male ultimate tergite and forceps of *Foramenolabis sisera* (BURR). Original.

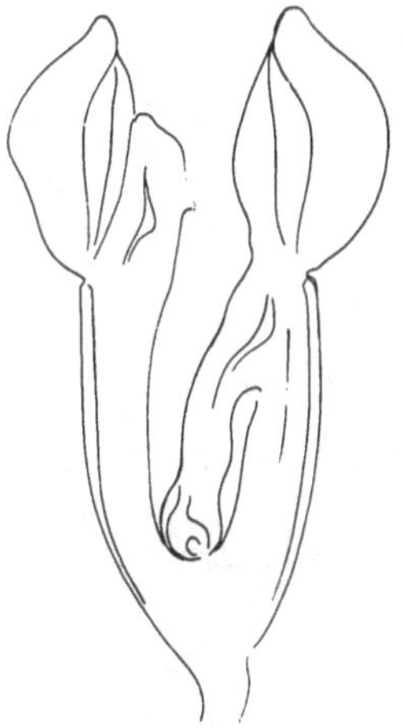

Fig. 248 Male genital armature of *Foramenolabis sisera* (BURR). Original.

Male genitalia (Fig. 248, gen. prep. No. 669, det. Dr. H. STEINMANN) characteristic; parameres relativelly short, small; median incision of anterior margin narrow but deep; genital lobes without virgae; external parameres large, strongly trigonal, expanded medially, and rounded apically. – **Female** similar to male except that the sides of abdominal tergites are convex and ecarinate, forceps symmetrical, tapering. – **Length** of body with forceps: male: 19–22 mm, female: 20–22,5 mm.

Distribution: South India.

2. Foramenolabis afghana STEINMANN

1981 *Foramenolabis afghana* STEINMANN, Fol. ent. hung., Budapest, **34**: 180; fig. 14 (head), 15 (male ultimate tergite and forceps), 16 (genital armature of holotype; gen. prep. Nr. 706, det. Dr. H. STEINMANN). – Terra typica: Afghanistan (Type male: Univ. zool. Mus., København).

Description: Male colour yellowish brown to black; head, antennae, pronotum, except posterior portion, and legs yellowish brown, meso- and metanotum, abdomen, and forceps dark brownish black. Head large, longer than wide; postfrontal sutures absent, coronal sutures small; vertical cavity moderately small, smaller than eye. Eyes small, shorter than distance between antennal bases. Antennae 13-jointed (+?), broken in holotype. First joint long, more or less as long as distance between antennal bases; second joint quadrate, third long, longer than fourth joint. Pronotum longer than wide; median longitudinal furrow distinct; lateral margins more or less parallel, all angles rounded. Tegmina and wings entirely absent. Abdomen depressed, expanded to 5th and 6th segments. Ultimate tergite broad, median longitudinal sulcus present; posterior margin concave in median section. Forceps very similar to *sisera* (BURR), but stronger (Fig. 249), slightly trigonal basally, curved and

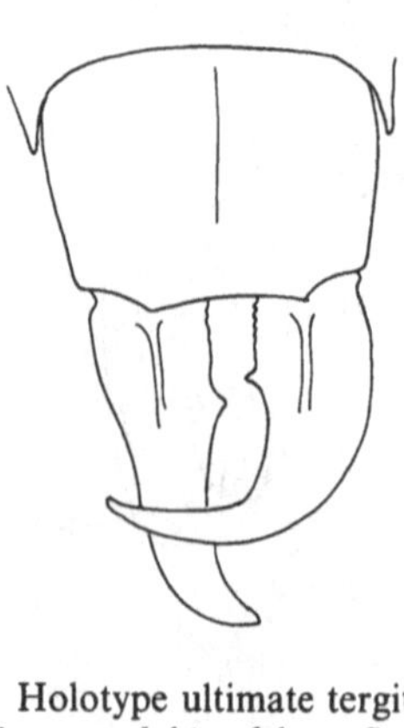

Fig. 249 Holotype ultimate tergite and forceps of *Foramenolabis afghana* STEINMANN. After STEINMANN.

Fig. 250 Holotype genital armature of *Foramenolabis afghana* STEINMANN. After STEINMANN.

cylindrical apically; asymmetrical; inner margins with one prominent tooth in median portion. Penultimate sternite simple, posterior margin rounded. Genitalia (Fig. 250) very characteristic; parameres widest basally, and narrower at anterior margin; median longitudinal incision of anterior margin of central paramere very deep; genital lobes fully developed, with sclerotized margin apically. Virga absent. External parameres elongate, apices obtuse. – **Female** unknown. – **Length** of body with forceps: 27 mm.

Distribution: Afghanistan.

Genus 15: **Aborolabis** SRIVASTAVA

1915 *Paralabis* BURR (partim), J. r. microsc. Soc., London, **1916**: 540.
1969 *Aborolabis* SRIVASTAVA, Entomologist's Rec. J. Variation, London, **81**: 246. – Species typica: *Anisolabis pervicina* BURR 1913.
1979 *Aborolabis* – STEINMANN, Eos. Rev. esp. Ent., Madrid, **53**: 213.
Literature: SRIVASTAVA 1972; STEINMANN 1977, 1978, 1984.

Description: General colour reddish-brown or brownish-black. Size medium, surface pilose. Head triangular, postfrontal and coronal sutures distinct; posterior margin of head sinuate mesad. Eyes small, shorter than genae. Pronotum quadrate or transverse; lateral margins and sides straight, gently widened posteriorly; all angles rounded. Tegmina absent or present as narrow lateral flaps on mesonotum. Wings totally absent. Abdomen gradually dilated from base to 7th abdominal tergite. Ultimate tergite broad, strongly transverse, sparsely punctate with median longitudinal sulcus. Male forceps trigonal basally, asymmetrical, female forceps slenderer, tapering. Male genitalia with external parameres three times longer than wide, tip rounded; inner margin of external paramere with characteristic triangular process; genital lobes without virgae.

Distribution: Oriental and Palae-aethiopical regions.

10 species.

Identification key to the species

1 Oriental group **2**
1′ Palaearctic or Ethiopian species **5**
2 Triangular process of the inner margin of male external parameres relatively short, about half as long as external parameres (Fig. 252); central parameral plate oval .... 1. **Aborolabis nepalensis**, p. 156
2′ Triangular process of the inner margin of external parameres long, about as long as external paramere **3**
3 Penultimate sternite emarginate posteriorly. Male genitalia (Fig. 254) strongly V-shaped, central parameral plate relatively narrow, median incision of anterior margin very deep .... 2. **Aborolabis emarginata**, p. 157
3′ Penultimate sternite entire. Male genitalia oval, median incision of anterior margin deep and wide **4**
4 Legs brownish-yellow; femora without black bands. Abdomen cylindrical, sides of the tergites 5–9 carinate. Penultimate sternite with posterior margin convex. Median incision of anterior margin of male genitalia narrow (Fig. 256) .... 3. **Aborolabis kalaktangensis**, p. 158

4′ Legs brownish, femora and tibiae slender, banded with black. Abdomen gently dilated in middle, sides of tergites 5–9 carinate. Penultimate sternite with posterior margin truncate. Median incision of anterior margin of male genitalia wide, as in fig. 258 .... 4. **Aborolabis pervicina**, p. 159
5 Tibiae and femora yellow. Triangular process of the inner margin of male genitalia wide and large basally **6**
5′ Tibiae black, femora uni- or bicolour. Inner process of external paramere of male genitalia in the middle **9**
6 Head and pronotum black. Posterior margin of penultimate sternite of male rounded. Paramere of male genitalia (Fig. 260) narrow, V-shaped, elongate; anterior margin of triangular process and inner margin of external paramere rectangular .... 5. **Aborolabis angulifera**, p. 160
6′ Head and pronotum reddish-brown, not black. Posterior margin of penultimate sternite of male excised transversely **7**
7 Posterior margin of penultimate sternite of male triangular, apex truncate. Ultimate tergite with strongly lateral keel dorsally (Fig. 261). Paramere of male genitalia broad; triangular process of the inner margin of external paramere as in fig. 262 .. 6. **Aborolabis vicina**, p. 161
7′ Posterior margin of penultimate sternite of male excised transversely **8**
8 Inner margin of male forceps crenulate basally (Fig. 263). First antennal joint dark brown; third joint one and a half times as long as fourth joint. Inner process of external paramere of male genitalia triangular (Fig. 264) .... 7. **Aborolabis cerrobarjai**, p. 162
8′ Inner margin of male forceps with prominent tooth basally (Fig. 265). First antennal joint yellowish red; third joint twice as long as fourth joint. Inner process of external paramere of male genitalia obtuse (Fig. 266) .... 8. **Aborolabis tanzanica**, p. 163
9 Femora unicolour, orange or yellowish-orange. Genital lobe relatively long and narrow; inner process of external paramere obtuse, as in fig. 268 .... 9. **Aborolabis mordax**, p. 164
9′ Femora bicolour, yellowish basally, and black apically. Genital lobe relatively short and broad; inner process of external parameres extended, as in fig. 270 .... 10. **Aborolabis mauritanica**, p. 165

1. Aborolabis nepalensis (BRINDLE)

1974 *Anisolabis nepalensis* BRINDLE, Senckenbergiana biol., Frankfurt/Main, **55**: 148; fig. 4 (male genitalia). – Terra typica: Nepal (Type male: Nat.-Mus. Senckenberg, Frankfurt am Main).
1979 *Aborolabis nepalensis* – STEINMANN, Eos. Rev. esp. Ent., Madrid, **53**: 215, fig. 1 (male genitalia). Nepal, Himalayas.

Description: Male colour generally blackish, legs yellowish brown; femora blackish except at bases and apices. Head transverse; postfrontal sutures and coronal suture fine but visible; eyes small, shorter than first antennal joint, the latter relatively short, shorter than the distance between antennal bases; second joint transverse, third joint twice as wide as long. Pronotum transverse, widened posteriorly; all margins more or less straight; median longitudinal furrow present in centre. Legs relatively short, femora glabrous but with sparse hairs. Tegmina and wings absent. Abdomen wide, visibly wider to 6th and 7th tergites. Ultimate tergite wide, posterior margin produced above the base of each branch of the forceps. Forceps (Fig. 251) strongly trigonal basally, broad, cylindrical distally; branches asymmetrical; inner margin crenulate. Penultimate sternite broadly triangular. Genitalia (Fig. 252) oval; parameres broad, median incision of anterior margin deep and wide; genital lobes well developed, without virgae.

External parameres comparatively short, triangular inner process of external parameres relatively small. – **Female** very similar to male, but branches of forceps more or less straight and almost symmetrical. – **Length** of body with forceps: male: 15–16 mm, female: 14–15,5 mm.

Distribution: Nepal, Himalayas.

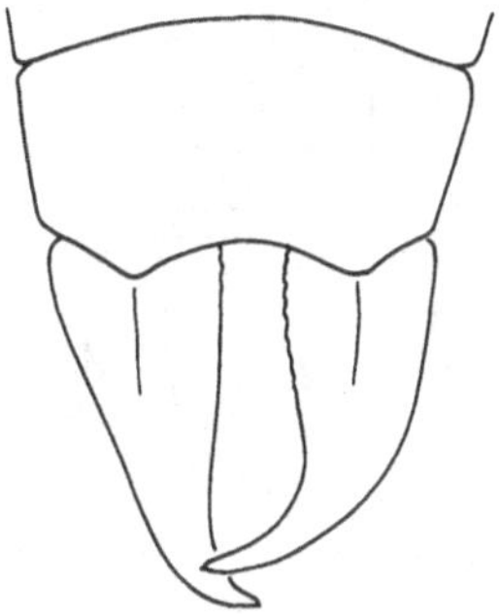

Fig. 251 Male ultimate tergite and forceps of *Aborolabis nepalensis* (BRINDLE). Original.

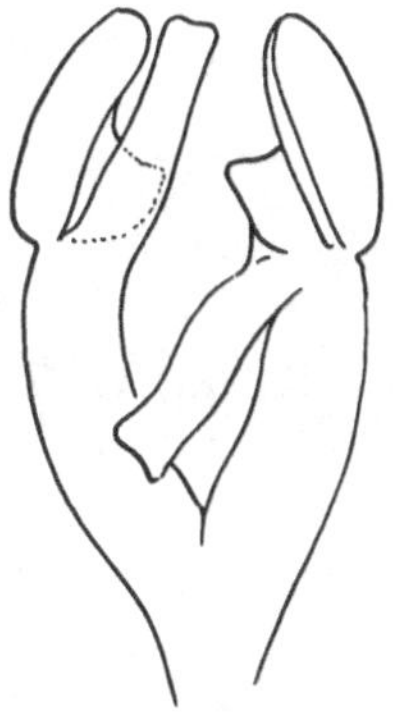

Fig. 252 Holotype genital armature of *Aborolabis nepalensis* (BRINDLE). After BRINDLE.

2. Aborolabis emarginata SRIVASTAVA

1974 *Aborolabis emarginata* SRIVASTAVA, J. zool. Soc. India, Calcutta, **26**: 59; fig. 2A (male, in text-fig.: holotype female), 2B (hind portion of penultimate sternite), 2C (male genitalia). – Terra typica: India (Type male: Nation. Collect. zool. Survey, Calcutta).

Description: Male general colour blackish brown; antennae with three basal and one or two joints in middle, yellow; sides of pronotum, and legs yellow. Head more or less as wide as long; frons tumid, occiput a little depressed. Eyes comparatively large, but shorter than the length of head behind eyes. Postfrontal and coronal sutures faintly marked. First antennal joint about as long as the distance between antennal bases. Pronotum transverse, slightly expanded posteriorly, lateral margins a little convex; median longitudinal furrow distinct in prozona. Tegmina and wings absent. Abdomen gradually depressed and expanded apically, surface finely punctate; sides of tergites 6–9 acute laterally. Ultimate tergite transverse, sparsely but finely punctate, gently narrowed posteriorly; posterior margin feebly emarginate in middle. Forceps (Fig. 253) asymmetrical, broader at base, tapering apically; trigonal basally; inner margins crenulate. Penultimate sternite concave posteriorly. Genitalia (Fig. 254) strongly V-shaped; parameral plate relatively long and narrow, median incision of anterior margin very deep; genital lobes well developed, with fine sclerotized stripes. External parameres of *Aborolabis*-type, triangular process of the inner margins relatively short, about half as long as external parameres. – **Female** similar to male, but sides of abdominal tergites rounded and carinate; penultimate sternite obtuse posteriorly and forceps simple and straight with apices gently incurved. – **Length** of body with forceps: male: 14 mm, female: 11,5 mm.

Distribution: India.

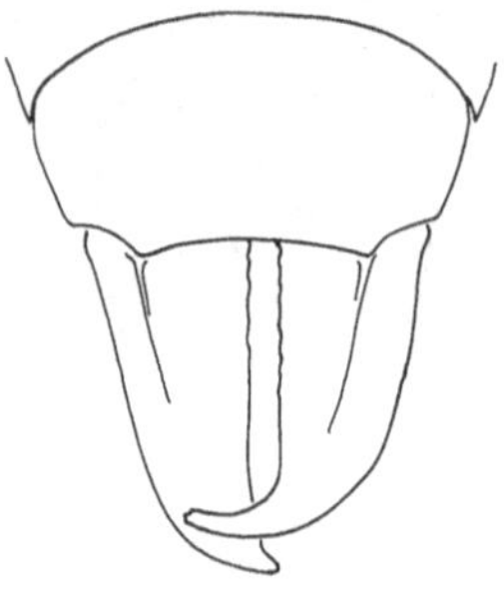

Fig. 253 Holotype ultimate tergite and forceps of *Aborolabis emarginata* SRIVASTAVA. After SRIVASTAVA.

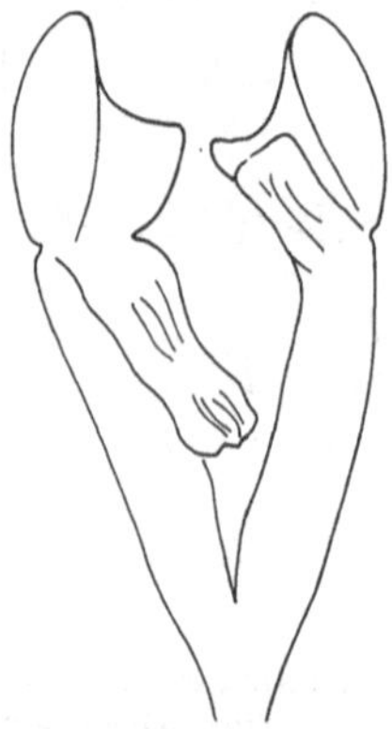

Fig. 254 Holotype genital armature of *Aborolabis emarginata* SRIVASTAVA. After SRIVASTAVA.

3. Aborolabis kalaktangensis SRIVASTAVA

1972 *Aborolabis kalaktangensis* SRIVASTAVA, Rec. zool. Survey India, Calcutta (1–4), **66**: 134; fig. C (penultimate sternite of male), D (ultimate tergite with forceps of male), E (male genitalia). – Terra typica: India (Type male: Nation. Collect. zool. Survey India, Calcutta).

Literature: SRIVASTAVA 1974; STEINMANN 1977, 1978, 1981.

Description: Male general colour brownish, with traces of black on abdomen only. Head triangular, posterior margin gently sinuate; frons tumid, occiput a little depressed. Postfrontal and coronal sutures faint; eyes small, shorter than the length of head behind eyes. First antennal joint normal, more or less as long as the distance between antennal bases; antennae 17-jointed. Pronotum almost quadrate; median longitudinal furrow well marked; anterior and posterior margins truncate, all angles rounded; lateral margins straight, but a little expanded posteriorly. Tegmina and wings absent. Legs brownish-yellow, femora proximally dark brown. Abdomen cylindrical, more or less parallel-sided; lateral sides of abdominal tergites 6–9 acute, ecarinate and without striations. Ultimate tergite transverse; median longitudinal sulcus present; posterior margin straight. Forceps (Fig. 255) trigonal at base, subcontiguous, tapering and curved, asymmetrical. Penultimate sternite rounded posteriorly. Genitalia (Fig. 256) characteristic; paramere broad, median longitudinal incision between external parameres very wide and deep; genital lobes normal, simple, without virgae. External parameres well developed, of *Aborolabis*-type, inner margin with large triangular inner process. – **Female** similar to male, but forceps slender, simple, symmetrical. – **Length** of body with forceps: male: 14,5–15 mm, female: 13,5–14,5 mm.

Distribution: India.

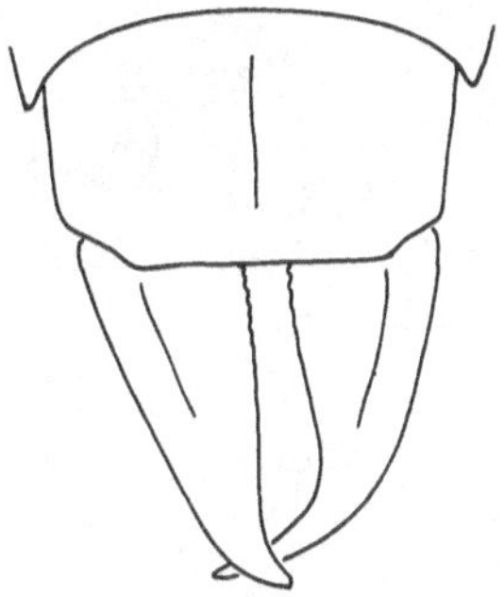

Fig. 255 Holotype ultimate tergite and forceps of *Aborolabis kalaktangensis* SRIVASTAVA. After SRIVASTAVA.

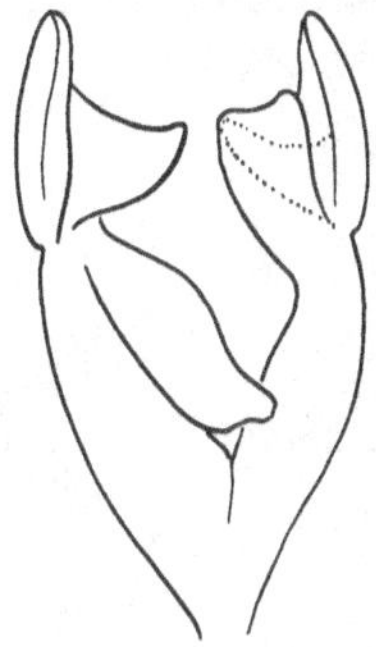

Fig. 256 Holotype genital armature of *Aborolabis kalaktangensis* SRIVASTAVA. After SRIVASTAVA.

4. Aborolabis pervicina (BURR)

1913 *Anisolabis pervicina* BURR, Rec. indian Mus., Calcutta, (2) **8**: 137. – Terra typica: Northern India (Type male: Brit. Mus. nat. Hist., London).
1915 *Paralabis pervicina* – BURR, J. r. microsc. Soc., London, **1915**: 540; pl. 12, fig. 3 (male genitalia).
1966 *Anisolabis pervicina* – POPHAM & BRINDLE, Entomologist, London, **99**: 274.
1969 *Aborolabis pervicina* – SRIVASTAVA, Entomologist's, Rec., J. Variation, London, **81**: 246; fig. 1 (male genitalia). Assam-Bhutan Frontier.
1975 *Anisolabis pervicina* – BRINDLE, Entomologia basiliensia, Basel, **1**: 18; fig. 20 (male genitalia). Bhutan.
1977 *Aborolabis pervicina* – STEINMANN, Ann. hist.-nat. Mus. nation. hung., Budapest, **69**: 94; fig. 20 (male genitalia).
Literature: SRIVASTAVA 1974; SAKAI 1970; STEINMANN 1978, 1981.

Description: Male general colour shiny black; legs brownish, femora banded with black. Head triangular, postfrontal and coronal sutures faint but marked; frons tumid, eyes normal, shorter than the length of head behind eyes. Antennae 19-jointed; first joint normal, about as long as the distance between antennal bases. Pronotum more or less quadrate; median longitudinal furrow not well marked, prozona slightly tumid and metazona flat; posterior angles rounded. Tegmina and wings absent. Legs slender. Abdomen shiny, punctate, depressed in the middle, sides of tergites 6–9 acute, rugulose and carinate. Ultimate tergite broad, transverse, less punctate with a median longitudinal sulcus; posterior margin concave medially. Forceps (Fig. 257) asymmetrical, widest and trigonal basally, strongly curved apically. Penultimate sternite normal, rounded posteriorly. Genitalia (Fig. 258) broad; paramere large, median incision of anterior margin very deep and wide; genital lobes large, without virgae. External parameres typical, triangular process very large, transverse. – **Female** similar to male, but pronotum slightly wider posteriorly, punctation of the body obscure; forceps simple, contiguous, tapering, symmetrical. – **Length** of body with forceps: male: 15–16,5 mm, female: 9–19 mm.

Distribution: India: Assam, Abor, and Bhutan.

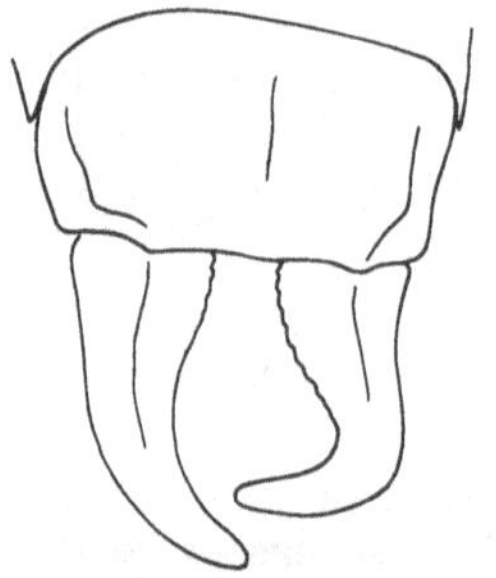

Fig. 257 Male ultimate tergite and forceps of *Aborolabis pervicina* (BURR). Original.

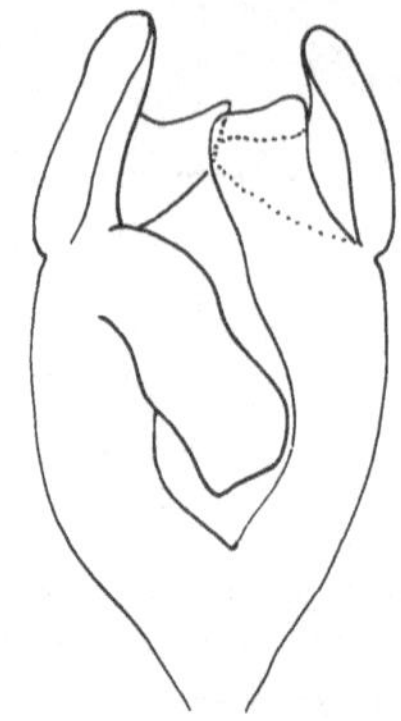

Fig. 258 Male genital armature of *Aborolabis pervicina* (BURR). After SRIVASTAVA.

5. Aborolabis angulifera (DOHRN)

1864 *Brachylabis angulifera* DOHRN, Stettiner ent. Ztg., **25**: 294. – Terra typica: S. Thomé (Type male: Inst. zool., Warszawa).
1900 *Anisolabis angulifera* – BORMANS, Das Tierreich, Berlin, **11**: 51.
1978 *Anisolabis angulifera* – BRINDLE, Dermaptera of Africa, Tervuren, **2**: 138.
1978 *Aborolabis angulifera* – STEINMANN, Fol. ent. hung., Budapest, **31**: 183; fig. 1 (male genitalia). São Tomé Island, Morocco, Algeria.
1979 *Aborolabis angulifera* – STEINMANN, Fol. ent. hung., Budapest, **32**: 154; fig. 9 (end of abdomen with forceps of holotype), 10 (genital armature of holotype, gen. prep. No. 352, det. Dr. H. STEINMANN). Sao Thomé.

Literature: BURR 1910, 1911, 1915; REHN 1905; SAKAI 1970; STEINMANN 1977, 1978, 1984.

Description: Male general colour reddish-brown; legs yellowish or light brown. Head a little longer than wide; eyes relatively small, shorter than first antennal joint; postfrontal and coronal sutures distinct. Antennae 15-jointed; first antennal joint about as long as the distance between antennal bases. Pronotum transverse, black; lateral margins expanded to posterior angles; all angles rounded, posterior margin truncate; median longitudinal furrow present in prozona. Tegmina and wings absent. Mesonotum broad, metanotum very transverse. Abdomen cylindrical, widened to 5th abdominal tergite. Ultimate tergite short and wide, at least twice as wide as long; posterior margin sharply truncate at lateral angles; median longitudinal sulcus distinct. Penultimate sternite truncate posteriorly. Forceps (Fig. 259) asymmetrical; trigonal basally, cylindrical apically; inner margin a little crenulate basally. Genitalia (Fig. 260) very characteristic, immediately recognizable; median incision of anterior margin very deep between external parameres, as a result parameres V-shaped, incision wide. External parameres medium, basally wide, then gradually and evenly attenuating, apically rounded. Genital lobes of holotype digitiform, both proclinate. – **Female** similar to male, but forceps more or less symmetrical, broad and trigonal basally, tapering, contiguous. – **Length** of body with forceps: male: 13–13,5 mm, female: 13–15 mm.

Distribution: São Tomé Island, Morocco, Algeria.

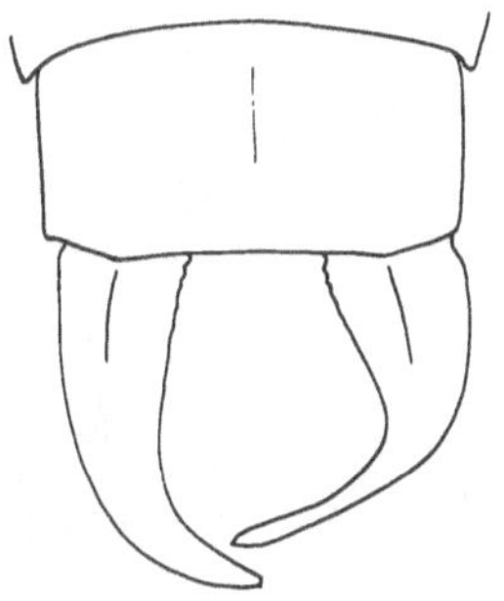

Fig. 259 Holotype ultimate tergite and forceps of *Aborolabis angulifera* (DOHRN). After STEINMANN.

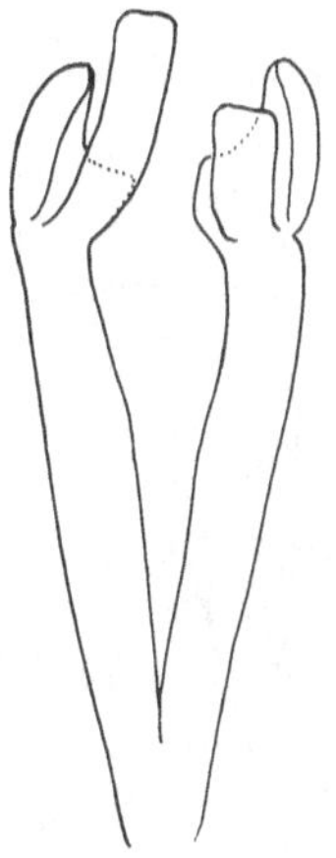

Fig. 260 Holotype genital armature of *Aborolabis angulifera* (DOHRN). After STEINMANN.

6. Aborolabis vicina (BURR)

1911 *Anisolabis vicina* BURR, Ann. Mag. nat. Hist., London, (8) **8**: 42. – Terra typica: Ethiopia (Type male: Brit. Mus. nat. Hist., London).
1966 *Carcinophora vicina* – POPHAM & BRINDLE, Entomologist, London, **99**: 275.
1978 *Anisolabis vicina* – BRINDLE, Dermaptera of Africa, Tervuren, **2**: 137; fig. 158 (male), 159 (penultimate sternite of male), 281 (external paramere of male genitalia).
1981 *Aborolabis vicina* – STEINMANN, Fol. ent. hung., Budapest, **34**: 180, fig. 12 (male genitalia).
Literature: BURR 1915; SAKAI 1970; STEINMANN 1977, 1978.

Description: Male colour various, head, meso- and metanotum, abdomen and forceps very dark reddish brown, pronotum and legs yellow. Head a little longer than wide, rounded posteriorly, postfrontal and coronal sutures distinct; eyes normal, slightly shorter than the length of head behind eyes. Antennae typical, first joint long, but a little shorter than the distance between antennal bases.

Pronotum quadrate, lateral margins a little expanded to posterior angles; median longitudinal furrow present. Tegmina and wings absent. Abdomen very strongly widened and depressed in larger males, less so in smaller males. Ultimate tergite broad; median longitudinal sulcus well marked, dorsal surface with longitudinal lateral depressions; posterior margin concave medially. Forceps (Fig. 261) asymmetrical, strongly trigonal basally, dorsal keels prominent basally, cylindrical and curved apically. Penultimate sternite simple. Genitalia (Fig. 262) elongated; paramere very long, V-shaped, median longitudinal incision of anterior margin between external parameres very deep; genital lobes well developed, without virgae. External parameres typical, triangular process on inner margin moderately small. – **Female** similar to male, but forceps symmetrical. – **Length** of body with forceps: male: 19–20 mm, female: 19–21 mm.

Distribution: Ethiopia.

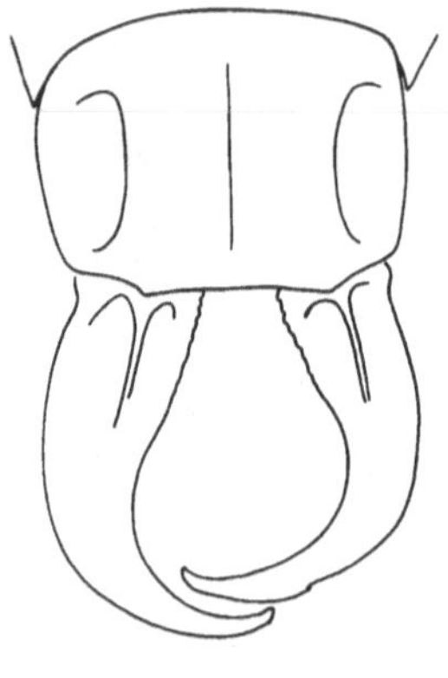

Fig. 261 Male ultimate tergite and forceps of *Aborolabis vicina* (BURR). Original.

Fig. 262 Male genital armature of *Aborolabis vicina* (BURR). After STEINMANN.

7. Aborolabis cerrobarjai STEINMANN

1979 *Aborolabis cerrobarjai* STEINMANN, EOS, Rev. esp. Ent., Madrid, **53**: 219; fig. 5 (genital armature of holotype). – Terra typica: Spain (Type male: Természettudományi Múz., Budapest).

Description: Male colour reddish-brown, antennae dark brown, legs yellow, unicolour. Head wide somewhat wider than pronotum; postfrontal sutures and coronal suture distinct; eyes small, shorter than first antennal joint. Antennae broken in holotype, first joint long but shorter than distance between antennal bases; third longer than fourth, latter quadrate. Pronotum transverse, but more or less quadrate, all margins straight, and all angles rounded; median longitudinal furrow present. Tegmina and wings entirely absent. Meso- and metanotum normal, simple. Abdomen cylindrical, widest medially; abdominal tergites 5–9 with carinae laterally. Ultimate tergite broad, simple. Forceps (Fig. 263) well developed, large, trigonal basally, cylindrical and curved apically; asymmetrical, the distance between branches of forceps large. Penultimate sternite broad, posterior margin excised transversely. Genitalia (Fig. 264; gen. prep. No. 578, det. Dr. H. STEINMANN) large, median incision of anterior margin of parameral plate deep; genital lobes fully developed, without virgae; anterior margin of triangular process and inner margin of external parameres obtuse-angled. – **Female** similar to male, but forceps more or less symmetrical, broad and trigonal basally, tapering, contiguous; ultimate tergite and forceps red, legs reddish-yellow. – **Length** of body with forceps: male: 23 mm, female: 18 mm.

Distribution: Spain: Sierra Nevada.

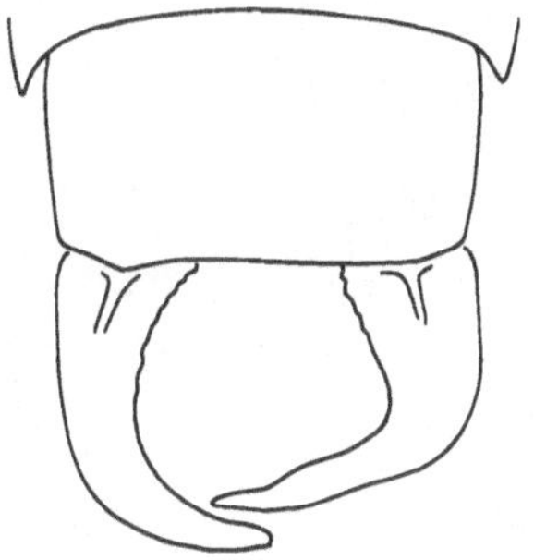

Fig. 263 Holotype ultimate tergite and forceps of *Aborolabis cerrobarjai* STEINMANN. After STEINMANN.

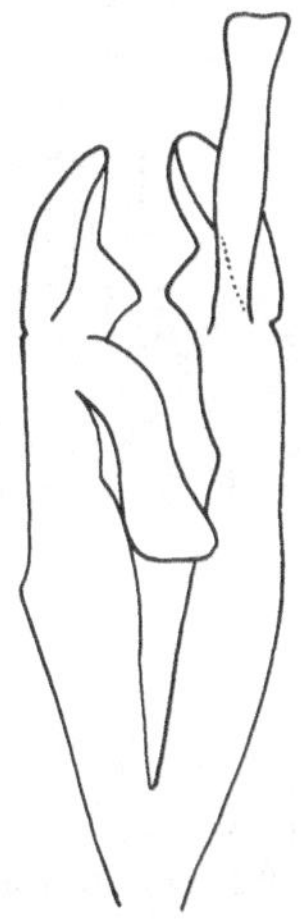

Fig. 264 Holotype genital armature of *Aborolabis cerrobarjai* STEINMANN. Original.

8. **Aborolabis tanzanica** STEINMANN

1981 *Aborolabis tanzanica* STEINMANN, Acta. zool. Acad. Sci. hung., Budapest, **27**: 192; fig. 16 (male ultimate tergite with forceps of holotype), 17 (genital armature of holotype). – Terra typica: Tanzania (Type male: Inst. taxonom. Zoöl., Amsterdam).

Description: Male general colour dark brown, brownish red to black. Head red, legs yellowish-brown. Head rounded, tumid; postfrontal and coronal sutures distinct. Eyes normal, but a little shorter than first antennal joint. Antennae 21-jointed, first joint well developed, a little longer than eye, second transverse, third long, about twice as long as fourth. Pronotum somewhat longer than wide and a little expanded posteriorly; median longitudinal furrow fine; lateral margins straight, posterior margin truncate. Tegmina and wings absent. Meso- and metanotum broad, of *Anisolabis*-type. Abdomen broad, visibly widest at 6th and 7th abdominal tergites. Ultimate tergite wide, posterior margin produced above base of branches of forceps; median longitudinal sulcus quite visible. Forceps (Fig. 265) widest and trigonal basally, cylindrical apically; branches asymmetrical; dorsal keel very prominent. Penultimate sternite normal. Genitalia (Fig. 266; gen. prep. No. 674, det. Dr. H. STEINMANN) characteristic; paramere elongated, V-shaped, median incision of anterior margin very deep and wide; genital lobes well developed, but without virgae. Triangular process of external parameres relatively small. – **Female** similar to male, but branches of forceps more or less straight and almost symmetrical. – **Length** of body with forceps: male: 18 mm, female: 21 mm.

Distribution: Tanzania: Kilimandjaro.

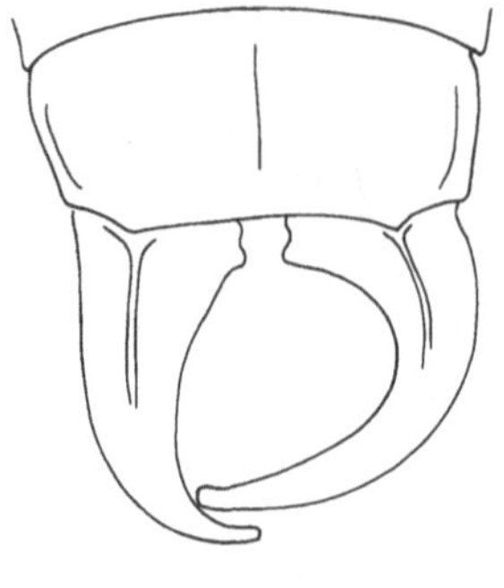

Fig. 265 Holotype ultimate tergite and forceps of *Aborolabis tanzanica* STEINMANN. After STEINMANN.

Fig. 266 Holotype genital armature of *Aborolabis tanzanica* STEINMANN. Original.

9. Aborolabis mordax STEINMANN

1978 *Aborolabis mordax* STEINMANN, Fol. ent. hung., Budapest **31**: 183; fig. 2 (genital armature of holotype). – Terra typica: Tunisia (Type male: Természettudományi Múz., Budapest).

1979 *Aborolabis mordax* – STEINMANN, EOS, Rev. esp. Ent., Madrid, **53**: 220, fig. 6 (male genitalia). Spain: Andalusia.

Description: Male very dark, dark brown to black, except dark reddish-brown antennae, femora orange or yellow, apically without dark brown or black bands, and middle abdominal tergites reddish-black. Head rounded, tumid; postfrontal and coronal sutures visible; posterior angles rounded. Eyes very small, shorter than first antennal joint. Antennae 22-jointed, first relatively short, shorter than eyes, and shorter than distance between antennal bases. Pronotum pitchy, transverse, all margins straight, and all angles rounded; median longitudinal furrow deep. Tegmina and wings totaly absent. Meso- and metanotum broad. Femora conspicuous, yellowish orange, unicolorous, tibiae black. Abdomen expanded to middle. Abdominal tergites 5–9 with lateral carinae. Ultimate tergite broad, posterior margin straight, a little concave medially; longitudinal keels present laterally; median longitudinal sulcus distinct. Forceps (Fig. 267) well developed, but short; trigonal basally, dorsal keel prominent, and cylindrical apically; asymmetrical. Penultimate sternite typical of *Aborolabis*. Genitalia (Fig. 268; gen. prep. No. 570, det. Dr. H. STEINMANN) characteristic; paramere narrow but elongate; median longitudinal incision of anterior margin between external parameres very deep and wide; genital lobes without virgae, relatively long and narrow. External parameres broad, their inner triangular process large, apices obtuse. – **Female** similar to male, but forceps symmetrical, contiguous, tapering. – **Length** of body with forceps: male: 26–28 mm, female: 25–29 mm.

Distribution: Tunisia; Spain: Andalusia.

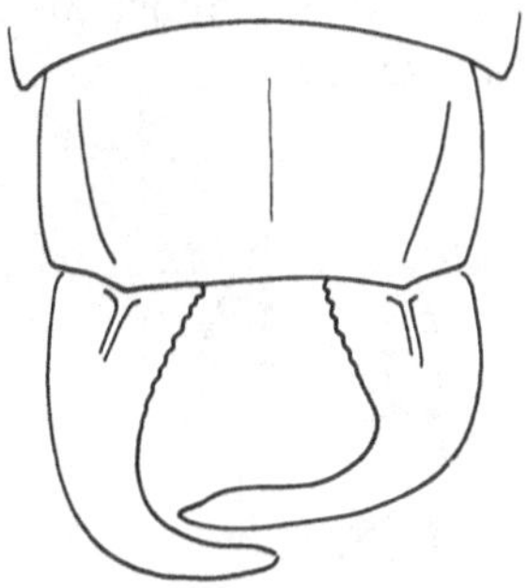

Fig. 267 Holotype ultimate tergite and forceps of *Aborolabis mordax* STEINMANN. After STEINMANN.

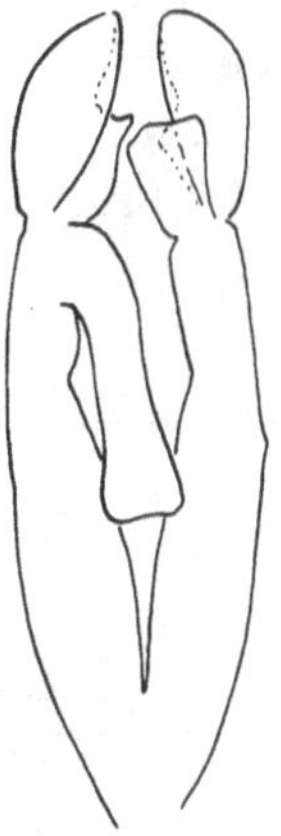

Fig. 268 Holotype genital armature of *Aborolabis mordax* STEINMANN. Original.

10. Aborolabis mauritanica (LUCAS)

1846 *Forficesila mauritanica* LUCAS, Explor. Alger., Paris, **3**: 4, pl. 1, fig. 1 (male). – Terra typica: Mauritania (Type male: unknown locality).
1864 *Brachylabis mauritanica* – DOHRN, Stettiner ent. Ztg., **25**: 292.
1900 *Anisolabis mauritanica* – BORMANS, Das Tierreich, Berlin, **11**: 45, Algeria, Tunisia.
1915 *Anisolabis mauritanica* – BURR, J. r. microsc. Soc., London, **1915**: 528; pl. 9, fig. 2 (male genitalia).
1978 *Aborolabis mauritanica* – STEINMANN, Fol. ent. hung., Budapest, **31**: 183, fig. 3 (male genitalia). Mauritania, Morocco, Tunisia, Spain.

Literature: BOESEMAN 1954; BURR 1910, 1911; KIRBY 1904; POPHAM & BRINDLE 1966; REHN 1906; SAKAI 1970; STEINMANN 1977, 1978, 1981, 1984.

Description: Male colour variable, very dark castaneous to blackish-brown, or reddish-brown. Head tumid, postfrontal and coronal sutures very deep. Eyes small, rounded, shorter than first antennal joint. Antennae 23-jointed, first joint short, shorter than distance between antennal bases. Pronotum somewhat wider than long, lateral margins parallel, convex to posterior margin; median longitudinal furrow present. Tegmina and wings entirely absent. Mesonotum transverse, posterior margin straight, metanotum cut out posteriorly. Femora bicolour, yellow basally, and black apically. Tibiae black, shiny. Abdomen widened to middle, abdominal tergites 5–9 with carinae on caudal sides. Ultimate tergite transverse, median longitudinal furrow distinct; dorsal surface of plate with two medio-lateral depressions, posterior margin straight. Forceps (Fig. 269) asymmetrical, trigonal basally, strongly curved and cylindrical apically; inner margin crenulate at base. Penultimate sternite rounded posteriorly. Genitalia (Fig. 270; gen. prep. No. 440, det. Dr. H. STEINMANN) well developed, large, paramere broad, median incision of anterior margin very deep and wide; genital lobes without virgae. External parameres of *Gonolabis*-type, a little curved apically; triangular process of inner margin extended at middle section. – **Female** similar to male, but forceps simple, symmetrical, contiguous, tapering. – **Length** of body with forceps: male: 23–26,5 mm, female: 20–29 mm.

Distribution: Mauritania, Morocco, Algeria, Tunisia, Spain.

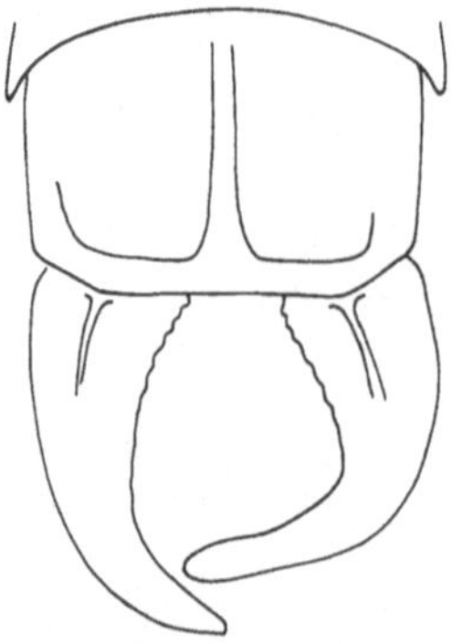

Fig. 269 Male ultimate tergite and forceps of *Aborolabis mauritanica* (LUCAS). Original.

Fig. 270 Male genital armature of *Aborolabis mauritanica* (LUCAS). Original.

Genus 16: **Ornatolabis gen. nov.**

Species typica: *Anisolabis externa* BEY-BIENKO 1959.

Description: Body of medium size, similar to *Gonolabis*. Head rounded, a little longer than wide; pronotum subtransverse, tegmina and wings totally absent.

Abdomen elongated, slightly expanded to middle; ultimate tergite broad, transverse, with asymmetrical forceps in male, e. g. strongly trigonal basally, and cylindrical apically. Male genitalia very characteristic: external parameres with ventral lobe, and apex incised.

Distribution: China.

Single species.

1. Ornatolabis externa (Bey-Bienko) comb. nov.

1959 *Anisolabis externa* Bey-Bienko, Ent. Obozr., Moskva, **38**: 604; fig. 27 (genital armature of holotype). – Terra typica: China: Yunnan (Type male: perhaps in Peking Mus.; paratype male: Zool. Inst. Ak. Nauk. Leningrad).

1975 *Euborellia externa* – Steinmann, Fol. ent. hung., Budapest, **28**: 156; fig. 18 (male genitalia).

Literature: Sakai 1970; Steinmann 1977, 1978.

Description: Male general colour blackish-brown, legs reddish-yellow. Head tumid, occiput a little depressed; postfrontal and coronal sutures distinct, well marked. Eyes relatively short, shorter than the first antennal joint or the length of head behind eyes. Antennae 16-jointed, first joint long, but shorter than the distance between antennal bases. Pronotum more or less transverse, lateral margins more or less parallel, and margins with reddish-yellow stripes; median longitudinal furrow present, but only finely visible. Tegmina and wings totally absent. Meso- and metanotum typical, normal, simple. Abdomen depressed, slightly expanded to median segments; sides of tergites 6–9 with lateral longitudinal carinae. Ultimate tergite transverse, median longitudinal sulcus present in middle section; posterior margin straight, concave in median portion. Forceps asymmetrical, of *Anisolabis*-type, strongly trigonal basally, with prominent dorsal keels; and cylindrical and curved apically. Penultimate sternite rounded posteriorly. Genitalia (Fig. 271) characteristic; paramere comparatively short, but visibly longer than wide, basally strongly attenuating. Median incision of anterior margin not reaching middle of plate. External parameres articulating antero-laterally, about as long as wide, extremely characteristic, with

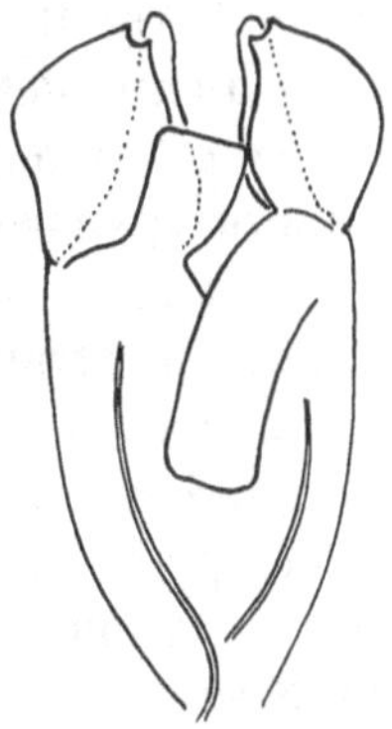

Fig. 271 Holotype genital armature of *Ornatolabis externa* (Bey-Bienko). After Bey-Bienko.

an incision on their antero-interior angles. – **Female** unknown. – **Length** of body with forceps: 14–15,5 mm.

Distribution: China: Yunnan.

Genus 17: **Anisolabella** ZACHER

1911 *Anisolabella* ZACHER, Zool. Jahrb., Jena **30**: 378. – Species typica: *Anisolabella braueri* ZACHER 1911.
1966 *Anisolabella* POPHAM & BRINDLE, Entomologist, London, **99**: 272 [= *Gonolabis* BURR 1910].
1975 *Anisolabella* – STEINMANN, Acta. zool. Acad. Sci. hung., Budapest, **21**: 205.
Literature: SAKAI 1970; STEINMANN 1975, 1977, 1978.

Description: External characteristics very similar to *Euborellia* or *Gonolabis*-types. Head rounded, but a little longer than wide; occiput depressed, eyes comparatively small. Pronotum quadrate, tegmina and wings entirely absent. Abdomen widened to middle. Male genitalia very characteristic, genital lobes with slender virgae, and external parameres more broadly rounded apically.

Distribution: North Eastern Africa.

Single species.

1. **Anisolabella braueri** ZACHER

1911 *Anisolabella braueri* ZACHER, Zool. Jahrb., Jena, **30**: 379; fig. B² (genital armature of holotype). – Terra typica: Nordost Afrika (Type male: Mus. Naturkunde, Berlin).
1966 *Gonolabis braueri* – POPHAM & BRINDLE, Entomologist, London, **99**: 273. Africa.
1978 *Anisolabis braueri* – BRINDLE, Dermaptera of Africa, Tervuren, **2**: 125, fig. 235 (external paramere of male genitalia).
Literature: BURR 1911, 1915; SAKAI 1970; STEINMANN 1975, 1977, 1978.

Description: Male colour dark brown; legs yellowish brown, clypeus yellow. Head tumid, a little depressed; postfrontal and coronal sutures distinct, well marked. Eyes comparatively small, shorter than the length of head behind eyes. Antennae 16-jointed; first joint long, more or less as long as the distance between antennal bases; second transverse, third short, scarcely twice as long as wide, fourth joint scarcely as long as wide; distal joints successively longer, spindle-shaped. Pronotum quadrate, lateral margins parallel, median longitudinal furrow distinct on anterior half. Tegmina and wings absent. Meso- and metanotum normal, broad, or transverse. Abdomen widest medially, sides of tergites 4–7 with lateral longitudinal ridges. Ultimate tergite rather small, wide, weakly punctate, median longitudinal sulcus weak; posterior margin concave at median section. Forceps asymmetrical, strongly trigonal basally, and cylindrical apically; left branch more or less straight, right branch curved sharply about midpoint. Penultimate sternite rounded apically. Genitalia (Fig. 272) characteristic; paramere very narrow, median incision of anterior margin between external parameres deep and narrow. Genital lobes with virgae. External parameres short, about as long as wide; apices rounded. – **Female** unknown. – **Length** of body with forceps: 17 mm.

Distribution: North-east Africa: Godat.

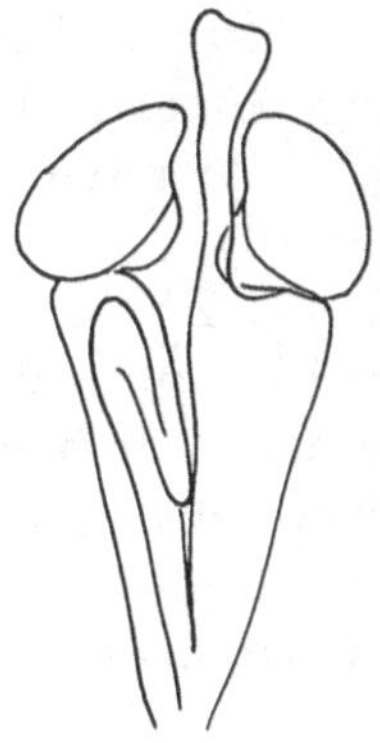

Fig. 272 Holotype genital armature of *Anisolabella braueri* ZACHER. After ZACHER.

Genus 18: **Heterolabis** BORELLI

1912 *Heterolabis* BORELLI, Bull. Mus. nation. Hist. nat., Paris, **1**: 231, **4**: 10. – Species typica: *Heterolabis brasiliensis* BORELLI 1912.
1966 *Heterolabis* – POPHAM & BRINDLE, Entomologist, London, **99**: 272 [= *Euborellia* BURR 1910].
Literature: BURR 1914, 1915; MENOZZI 1932; REICHARDT 1968; SAKAI 1970; STEINMANN 1977, 1978.

Description: Body and general characteristics very similar to *Euborellia*, but head longer than wide; postfrontal sutures finely marked, more or less straight, coronal suture distinct. Joints of antennae cylindrical. Pronotum strongly expanded posteriorly; tegmina rudimentary, represented by lateral flaps. Male forceps asymmetrical, but similar to one female-type, slender, as in fig. 273. Male genitalia specific; genital lobes with slender virgae, and virga with sclerotized vesicle basally (Fig. 274).
Distribution: Brazil.
Single species.

1. **Heterolabis brasiliensis** BORELLI

1912 *Heterolabis brasiliensis* BORELLI, Bull. Mus. nation. Hist. nat., Paris, **1**: 232, **4**: 11; fig. 1 a (genital lobe of male genitalia), 1 b (genital armature of holotype). – Terra typica: Rio de Janeiro (Type male: Természettudományi Múz., Budapest; holotype without penis).
1966 *Euborellia brasiliensis* – POPHAM & BRINDLE, Entomologist, London, **99**: 273.
Literature: BURR 1914, 1915; MENOZZI 1932; REICHARDT 1968; SAKAI 1970; STEINMANN 1975, 1977, 1978.

Description: Male colour reddish-brown; legs yellowish-brown. Head tumid, a little longer than wide; postfrontal and coronal sutures distinct. Eyes small, slightly shorter than the length of head behind eyes. Antennae broken in holotype; first joint long, but shorter than distance between antennal bases. Pronotum about as long as wide; lateral margins strongly widened posteriorly; median longitudinal furrow finely visible; posterior angles rounded. Tegmina and wings absent, tegmina represented by lateral flaps. Mesonotum broad, metanotum transverse. Abdomen of *Euborellia*-type, slender and elongate;

tergites with large, well-separated punctation. Sides of tergites without carinae or lateral longitudinal ridges. Ultimate tergite broad, posterior margin specific, as in fig. 273; plate depressed dorsally. Forceps (Fig. 273) more or less asymmetrical; trigonal basally, cylindrical apically. Penultimate sternite rounded posteriorly. Genitalia (Fig. 275) characteristic; paramere relatively narrow, median longitudinal incision of anterior margin deep; genital lobe (Fig. 274) with specific virga, the latter with basal vesicle, not of *Carcinophora*-type. – **Female** similar to male, but forceps slenderer, symmetrical, contiguous. – **Length** of body with forceps: male: 20–21 mm, female: 18–21 mm.

Distribution: Brazil.

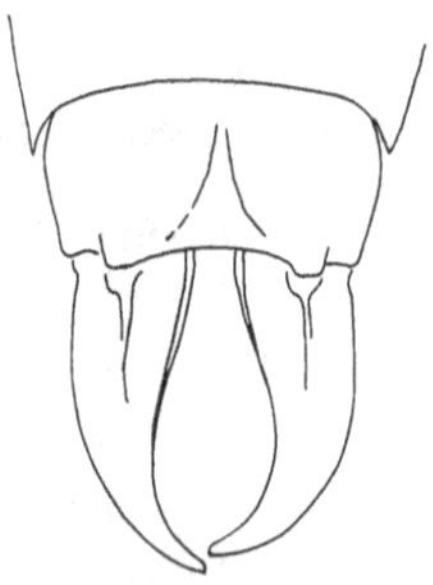

Fig. 273 Holotype ultimate tergite and forceps of *Heterolabis brasiliensis* BORELLI. Original.

Fig. 274 Genital lobe of male genitalia of *Heterolabis brasiliensis* BORELLI. After BORELLI.

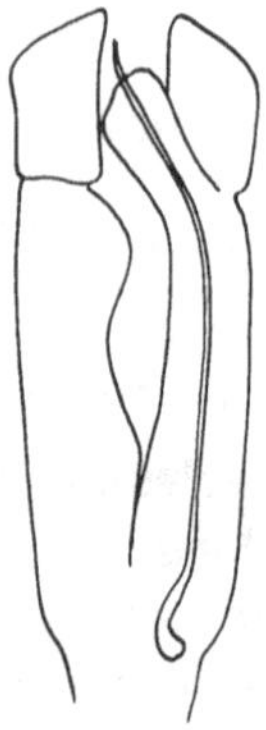

Fig. 275 Genital armature of holotype of *Heterolabis brasiliensis* BORELLI. After BORELLI.

Genus 19: **Epilabis** BURR

1915 *Epilabis* BURR, J. r. microsc. Soc., London, **1915**: 539. – Species typica: *Euborellia penicillata* BORELLI 1911.

1966 *Epilabis* – POPHAM & BRINDLE, Entomologist, London, **99**: 272 [= *Carcinophora* SCUDDER 1876].

1973 *Epilabis* s. str., RAMAMURTHI & DAVID, Zool. Anz., Leipzig, **190**: 435, and *Cryptolabis* (subgen. for *Epilabis ramachandrai* and *vallakadaiensis* spp. RAMAMURTHI & DAVID, 1973) RAMAMURTHI & DAVID.

1978 *Epilabis* – SRIVASTAVA, Bull. zool. Survey India, Calcutta, (1) **1**: 71.

Literature: SAKAI 1970; SRIVASTAVA 1978; STEINMANN 1975, 1977, 1978.

Description: Medium sized insect with tegmina covering mesonotum only and exposing a broad scutellum. Head rounded, somewhat longer than wide; postfrontal and coronal sutures well marked or finely visible; first antennal joint long, second transverse. Pronotum longer than wide, or transverse, distinctly narrowed anteriorly. Tegmina sometimes rudimentary, but meeting at the sutures. Abdomen moderately dilated just beyond the middle; abdominal sternum typical. Ultimate tergite broad, well developed; posterior margin concave medially. Male forceps more or less asymmetrical, female forceps symmetrical. Male genitalia with relatively long external parameres, which are wide and longer than half length of paramere; outer margin undulate, and inner margin straight.

Distribution: Oriental species.

9 species.

Identification key to the species

1 Genital lobes of male genitalia with virgae, which are specific, with small or larger chitinous, sclerotized vesicle basally (Figs. 277, 279, 281, 283, and 285) **2**

1′ Genital lobes of male genitalia without virgae **6**

2 Penultimate sternite without a tuft of hair in the posterior margin. Basal vesicle of the genital lobe of male genitalia small or very small. **3**

2′ Penultimate sternite with a tuft of hair in the middle of posterior margin. Basal vesicle of the genital lobe of male genitalia large and elongate **4**

3 Posterior margin of ulitmate tergite straight; forceps of male with prominent inner tooth basally (Fig. 276). Basal vesicle of male genital lobe very small, as in fig. 277. Sides of abdominal tergites acute, with a longitudinal carina on segments 5nd to 7nd.......................... 1. **Epilabis nilgiriensis**, p. 172

3′ Posterior margin of ultimate tergite concave; forceps of male without inner tooth basally (Fig. 278). Sclerotized vesicle of male genital lobe larger, as in fig. 279. Sides of tergites 7–9 obtuse, ecarinate .... 2. **Epilabis harlequin**, p. 172

4 Virga with characteristic preputial sac or vesicle smaller (Fig. 281), with a long and strongly chitinised rod measuring nearly half the length of the genital lobe. Male forceps strongly asymmetrical, as in fig. 280 .... 3. **Epilabis penicillata**, p. 173

4′ Virga with characteristic preputial sac or vesicle on basal half, as in figs. 283, and 285 ... **5**

5 Posterior margin of penultimate sternite in middle acute. Forceps of male strong and more or less symmetrical (Fig. 282). Male genitalia very similar to *analis* Ramamurthi & David, as in fig. 283 .... 4. **Epilabis emarginata**, p. 174

5′ Posterior margin of penultimate sternite rounded in middle. Forceps of male slightly asymmetrical (Fig. 284). Male genitalia very similar to *emarginata* Ramamurthi & David, as in fig. 285 .... 5. **Epilabis analis**, p. 175

6 External parameres of male genital armature obtuse apically. Male forceps asymmetrical, inner margins with or without small, but prominent, tooth basally **7**

6′ External parameres of male genital armature not rounded apically **8**

7 Male forceps with small, but prominent, inner tooth at basal section (Fig. 286). External parameres of male genitalia strongly curved medially (Fig. 287) . 6. **Epilabis punctata**, p. 177

7′ Male forceps without inner tooth basally; margins crenulate (Fig. 288). External parameres of male genitalia more or less straight medially (Fig. 289) .... 7. **Epilabis burri**, p. 178

8 External parameres of male genital armature trigonal, similar to neotropical *Metalabis*-type; genital lobes with specific chitinous prominent rod (Fig. 291) .... 8. **Epilabis ramachandrai**, p. 179

8′ External parameres of male genital armature expanded medially, not trigonal, similar to *Gonolabis*-type, as in fig. 293; genital lobes with chitinous serration.......................... 9. **Epilabis vallakadaiensis**, p. 179

1. Epilabis nilgiriensis SRIVASTAVA

1978 *Epilabis nilgiriensis* SRIVASTAVA, Bull. zool. Survey India, Calcutta, (1) **1**: 72; fig. 1D (a few basal antennal joints), 1E (thorax), 1F (posterior margin of penultimate sternite), 1G (ultimate tergite and forceps), 1H (male genitalia), 1I (ultimate tergite and forceps of female). – Terra typica: India (Type male: Mus. Zool., Ann Arbor).

Description: Male general colour blackish brown; legs testaceous. Head triangular, convex, smooth; postfrontal and coronal sutures fine but distinct. Eyes much shorter than genae. Antennae 17-jointed; first joint equal to 2nd to 4th together; third slightly longer than 4th and equal to 5th. Pronotum as long as wide, anterior and lateral margins straight, flat. Tegmina reduced, represented by lateral flaps. Wings absent. Metanota transverse, finely punctate, broadly emarginate posteriorly. Abdomen broad, finely punctate, sides of tergites acute, carinate on 5th to 7th only. Ultimate tergite sparsely punctate, transverse, gently contracted posteriorly; posterior margin straight; median longitudinal sulcus distinct in anterior half only. Forceps (Fig. 276) trigonal basally, gradually tapering, left branch almost straight in basal two thirds, afterwards curved; apices pointed; inner margin with prominent tooth. Penultimate sternite obtusely rounded posteriorly. Genitalia (Fig. 277) broad; paramere slightly V-shaped, median incision of anterior margin very deep and wide; genital lobes slender, with virgae; virga with chitinous, sclerotized vesicle basally. External parameres short, acute. – **Female** as male except that the sides of abdominal tergites are convex and ecarinate, forceps somewhat symmetrical. – **Length** of body with forceps: male: 16,5 mm, female: 18–20 mm.

Distribution: India.

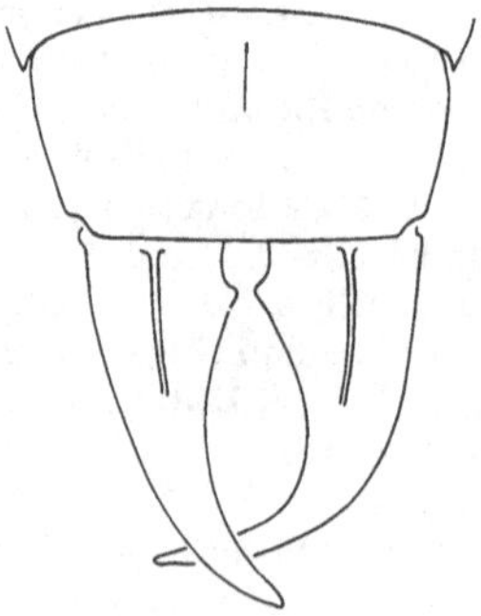

Fig. 276 Holotype ultimate tergite and forceps of *Epilabis nilgiriensis* SRIVASTAVA. After SRIVASTAVA.

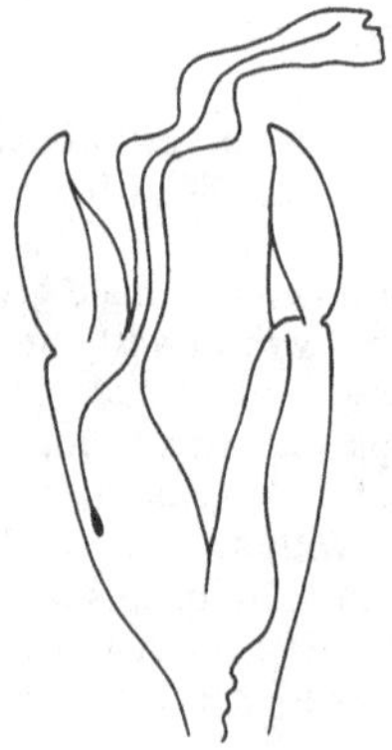

Fig. 277 Genital armature of holotype of *Epilabis nilgiriensis* SRIVASTAVA. After SRIVASTAVA.

2. Epilabis harlequin nom. nov.

Epilabis harlequin **nov. nom.** pro *Heterolabis punctata* SRIVASTAVA 1978, secondary junior homonymy of *Epilabis punctata* SRIVASTAVA 1978.

1978 *Heterolabis punctata* SRIVASTAVA, EOS, Rev. esp. Ent., Madrid, **52**: 263; fig. 11 (ultimate tergite and forceps of male), 12 (penultimate sternite and forceps of male), 13 (genital armature of holotype). – Terra typica: Philippines: Mindanao (Type male: Field Mus. nat. Hist., Chicago).

Description: Male colour dark blackish brown; frons with a dark black patch, pronotum lighter on sides, and pre-apical joints of antennae yellow, legs yellow, femora with black patch in middle. Head about as long as wide; postfrontal and coronal sutures feebly marked; eyes normal, slightly shorter than the length of head behind eyes. Antennae 16-jointed; first joint long, but a little shorter than the distance between antennal bases, second small, third long and cylindrical. Pronotum quadrate, smooth; anteriorly equal in width to head, posteriorly slightly widened; lateral margins straight, posterior margin rounded; median longitudinal furrow well marked. Tegmina and wings absent. Meso- and metanotum sparsely punctate. Abdomen very finely pubescent, a little expanded posteriorly; sides of tergites 7–9 obtuse, ecarinate, rugoso-striate. Ultimate tergite very broad, comparatively smooth, median longitudinal sulcus faint; posterior margin concave medially. Forceps (Fig. 278) weakly asymmetrical, gradually tapering, slightly widened basally; dorsal keels prominent; inner margin crenulate basally. Penultimate sternite triangular, with posterior margin rounded. Genitalia (Fig. 279) specific; paramere broad, median longitudinal incision of anterior margin wide; genital lobes well developed, virga present, with sclerotized vesicle basally. – **Female** unknown. – **Length** of body with forceps: unknown.

Distribution: Philippines: Mindanao.

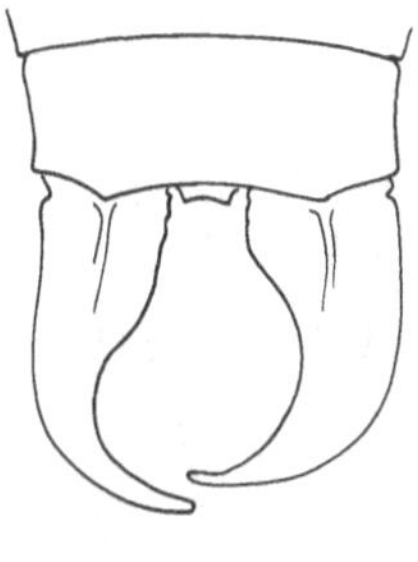

Fig. 278 Holotype ultimate tergite and forceps of *Epilabis harlequin* nom.nov. After SRIVASTAVA.

Fig. 279 Genital armature of holotype of *Epilabis harlequin* nom.nov. After SRIVASTAVA.

3. Epilabis penicillata (BORELLI)

1911 *Euborellia penicillata* BORELLI, Boll. Mus. Zool. Anat. comp. Univ. Torino, **26**, No. 640: 3. – Terra typica: India (Type male: Mus. Ist. Zool. sist. Univ. Torino; holotype without penis).

1915 *Epilabis penicillata* – BURR, J. r. microsc. Soc., London, **1915**: 539; pl. XI, fig. 15 (male genitalia).

1966 *Carcinophora penicillata* – POPHAM & BRINDLE, Entomologist, London, **99**: 275.
1973 *Epilabis penicillata* – RAMAMURTHI & DAVID, Zool. Anz., Leipzig, **190**: 435.
Literature: SAKAI 1970; SRIVASTAVA 1978; STEINMANN 1977, 1978.

Male colour reddish-black; head red or reddish-brown. Head triangular, convex, smooth; postfrontal sutures and coronal suture very distinct. Eyes much shorter than genae. Antennae 17-jointed; first joint long, more or less as long as the length of the distance between antennal bases; second transverse, third quadrate. Pronotum a little longer than wide, not broader than head anteriorly; lateral margins more or less parallel. Median longitudinal furrow present in prozona. Tegmina and wings entirely absent or mesonotum with one small lateral flap. Metanotum normal. Abdomen a little expanded to middle. Ultimate tergite broad, median longitudinal sulcus distinct; posterior margin straight; median longitudinal sulcus very finely visible. Forceps (Fig. 280) strongly asymmetrical, trigonal and broader basally, cylindrical and curved apically; dorsal keel very distinct; left branch crenulate, right branch toothed basally. Penultimate sternite with a tuft of hair in the middle of posterior margin. Genitalia (Fig. 281) broad; paramere with median incision at anterior margin; genital lobes fully developed, virgae with characteristic preputial sac or vesicle. – **Female** similar to male, but forceps symmetrical, slender, and tapering. – **Length** of body with forceps: male: 18–22 mm, female: 18,5–22 mm.

Distribution: South India, Sri Lanka.

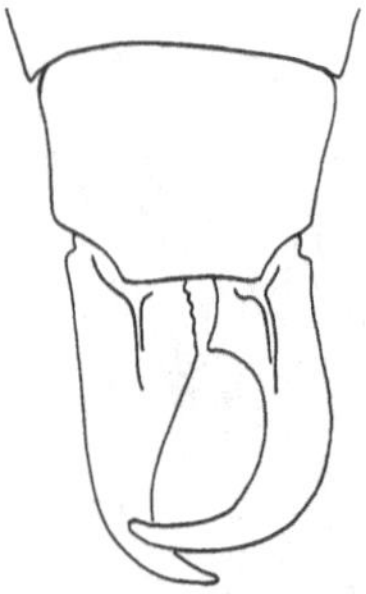

Fig. 280 Holotype ultimate tergite and forceps of *Epilabis penicillata* (BORELLI). Original.

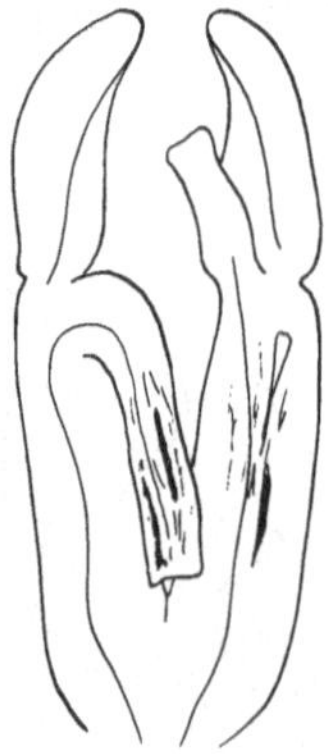

Fig. 281 Male genital armature of *Epilabis penicillata* (BORELLI). After BURR.

4. Epilabis emarginata RAMAMURTHI & DAVID

1973 *Epilabis emarginatus* RAMAMURTHI & DAVID, Zool. Anz., Leipzig, **190**: 437; fig. 4 (ultimate tergite with forceps of male), 5 (penultimate sternite of holotype), 6 (male genitalia). – Terra typica: India (Type male: RAMAMURTHI's Collect. Madras).
Literature: SRIVASTAVA 1978; STEINMANN 1978.

Description: Male general colour dark reddish-brown; legs yellow. Head tumid, posterior margin rounded; postfrontal and coronal sutures finely distinct. Eyes normal, shorter than the length of head behind eyes. Antennae broken in holotype; joints short and subconical. Pronotum longer than wide, distinctly narrowed anteriorly, all sides truncate, cephalic angles sharp, posterior angles rounded; median longitudinal furrow distinct. Tegmina and wings absent. Meso- and metanotum normal, simple. Abdomen shiny, a little expanded medially; sides of tergites 5–8 with well developed ridges above postero-lateral angle. Ultimate tergite broad, smooth; median longitudinal sulcus confined to disc; posterior margin truncate, convex medially, raised laterally; dorsal surface depressed. Forceps (Fig. 282) trigonal basally, narrowed and arched beyond 1/3 distance from base; inner margin with prominent tooth. Penultimate sternite rounded, apex acute with a tuft of well developed setae a little above apex, similar to *penicillata* (BORELLI). Genitalia (Fig. 283) characteristic; paramere very broad, V-shaped, median incision of anterior margin deep and very broad. Genital lobes fully developed, virgae with characteristic sclerotized vesicle. External parameres relatively short, broad, abtuse. – **Female** similar to male, but forceps slender, tapering, symmetrical. – **Length** of body with forceps: male: 16 mm, female: 16,5 mm.

Distribution: India.

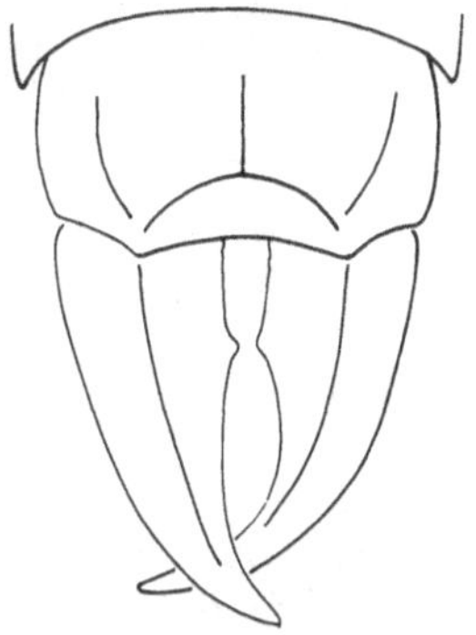

Fig. 282 Holotype ultimate tergite and forceps of *Epilabis emarginata* RAMAMURTHI & DAVID. After RAMAMURTHI & DAVID.

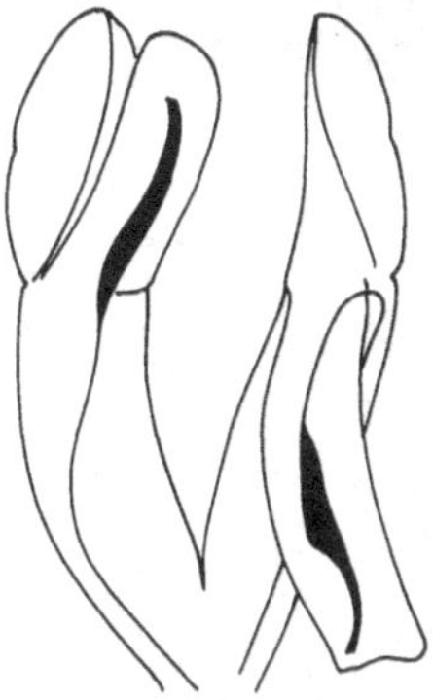

Fig. 283 Holotype genital armature of *Epilabis emarginata* RAMAMURTHI & DAVID. After RAMAMURTHI & DAVID.

5. **Epilabis analis** RAMAMURTHI & DAVID

1973 *Epilabis analis* RAMAMURTHI & DAVID, Zool. Anz., Leipzig, **190**: 436; fig. 1 (ultimate tergite and forceps), 2 (penultimate sternite of holotype), 3 (male genitalia). – Terra typica: India (Type male: RAMAMURTHI's Collect., Madras).

Literature: SRIVASTAVA 1978; STEINMANN 1977, 1978.

Description: Male colour reddish-brown; legs and antennae golden yellow. Head pear-shaped, smooth; frons tumid, and separated from depressed occiput and genae, posterior margin truncate. Eyes typical, a little shorter than the length of head behind eyes. Antennae 16– (+?) jointed, slender; first joint long, more or less as long as the distance between antennal bases; 4th and 5th equal to third, rest gradually elongated, subcylindrical. Pronotum not wider than head anteriorly, slightly widened posteriorly; prozona not distinct from metanoza and sides straight; median longitudinal furrow present. Tegmina and wings absent. Meso- and metanotum typical of the genus. Abdomen narrow anteriorly, broader across tergites 5–8, these abdominal tergites rugose laterad with postero-lateral angles acute and carinate. Ultimate tergite smooth, posterior margin convex between bases of branches of forceps. Forceps (Fig. 284) asymmetrical, subcontiguous at base and well-separated beyond due to the deeply arched right branch; inner margin with prominent tooth basally. Penultimate sternite broadly triangular due to oblique lateral sides, posterior margin rounded with a cluster of well-developed setae near apex. Genitalia (Fig. 285) specific, paramere broad, median incision between external parameres broad and deep; genital lobes with virgae, which have very long sclerotized preputial sac. External parameres short and obtuse. – **Female** similar to male, but forceps slenderer, tapering, symmetrical; abdominal tergites 5–8 blunt and smooth laterad except for a small tuberale. – **Length** of body with forceps: male: 16,5–19,5, female: 16 mm.

Distribution: India.

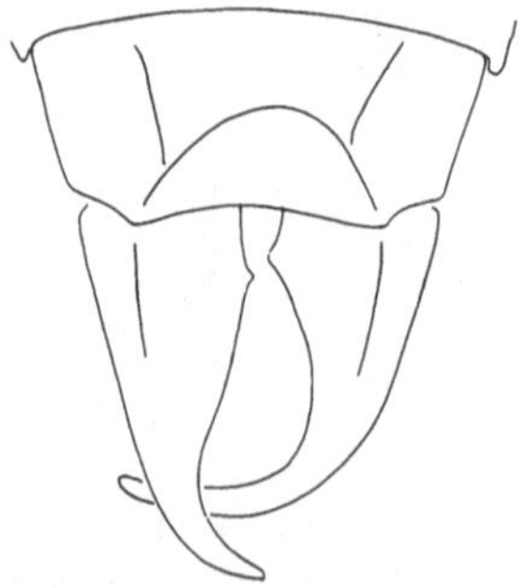

Fig. 284 Holotype ultimate tergite and forceps of *Epilabis analis* RAMAMURTHI & DAVID. After RAMAMURTHI & DAVID.

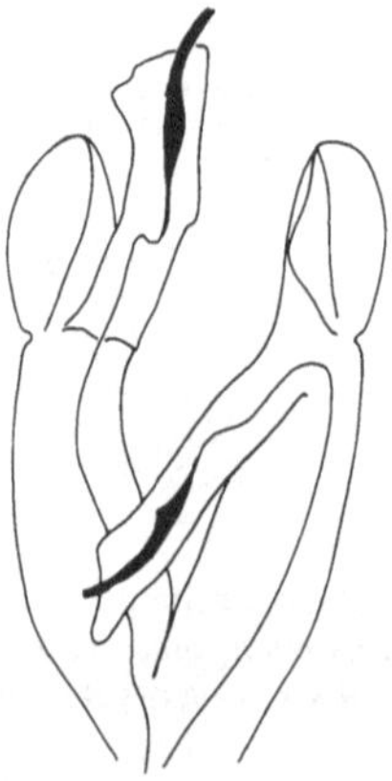

Fig. 285 Holotype genital armature of *Epilabis analis* RAMAMURTHI & DAVID. After RAMAMURTHI & DAVID.

6. Epilabis punctata SRIVASTAVA

1978 *Epilabis punctata* SRIVASTAVA, Bull. zool. Survey India, Calcutta, (1) **1**: 74; fig. 2A (basal joints of antenna), 2B (thorax), 2C (posterior margin of penultimate sternite), 2D (ultimate tergite and forceps of holotype), 2E (genital armature of holotype). – Terra typica: India (Type male: Mus. Zool., Ann Arbor).

Description: Male general colour dull blackish to reddish-brown; legs testaceous yellow. Head triangular, longer than wide; postfrontal sutures and coronal suture faint, posterior margin straight. Eyes much shorter than genae. Antennae 17-jointed, stout; first joint more or less as long as the distance between antennal bases. Pronotum finely punctate, slightly wider than long, anteriorly straight, laterally diverging posteriorly; posterior angles well-rounded and margin subtruncate; median longitudinal furrow distinct. Tegmina represented by lateral flaps. Wings absent. Metanotum very slender, strongly concave posteriorly. Abdomen convex, very slightly enlarged in middle, sides of segments obtuse, ecarinate, often rugosely punctate. Ultimate tergite transverse, sparsely punctate, median longitudinal sulcus present, deep; posterior margin concave medially. Forceps (Fig. 286) weakly asymmetrical, remote and trigonal at base with a ridge in basal one fourth only, afterwards depressed, tapering, right branch more curved than the left; inner margin small but prominent tooth present. Penultimate sternite triangular, obtusely rounded posteriorly. Genitalia (Fig. 287) of *Epilabis*-type; relatively narrow, median incision of anterior margin of central parameral plate deep and wide; genital lobes well developed, without virgae. External parameres strongly curved medially, and obtuse apically. – **Female** similar to male, except that the sides of abdominal tergites are rounded and forceps almost symmetrical. – **Length** of body with forceps: male: 18 mm, female: 17,5 mm.

Distribution: South India.

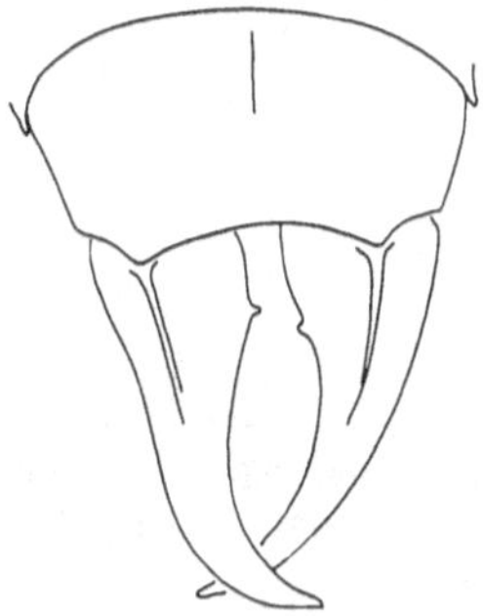

Fig. 286 Holotype ultimate tergite and forceps of *Epilabis punctata* SRIVASTAVA. After SRIVASTAVA.

Fig. 287 Holotype genital armature of *Epilabis punctata* SRIVASTAVA. After SRIVASTAVA.

7. Epilabis burri SRIVASTAVA

1970 *Epilabis burri* SRIVASTAVA, Ceylon J. Sci., biol. Sci., Colombo, (2) **8**: 46; fig. 1a (penultimate sternite of male), 1b (pro- and mesonotum), 1c (genital armature of male), 1d (ultimate tergite with forceps). – Terra typica: India (Type male:Nation. Collect. zool. Survey India, Calcutta).
Literature: SRIVASTAVA 1978; STEINMANN 1977, 1978.

Description: Male brownish black; antennae dark brown, pronotum yellowish on sides, legs light brown, femora with black bands. Head smooth, triangular, posterior margin gently emarginate in middle, frontal and coronal sutures faintly marked. Eyes relatively small, shorter than first antennal joint; the latter normal, about as long as distance between antennal bases. Pronotum slightly narrowed anteriorly, median longitudinal furrow well marked, all margins straight. Tegmina present, narrow, represented by lateral flaps on mesonotum. Wings absent. Abdomen finely punctate, narrowed anteriorly, sides of tergites 6–9 rugoso-striate. Ultimate tergite smooth, transverse; posterior margin a little specific, lamellate; median longitudinal sulcus present. Forceps (Fig. 288) strongly asymmetrical, trigonal basally, and curved and cylindrical apically; inner margin crenulate. Penultimate sternite triangular with posterior margin convex. Genitalia (Fig. 289) broad, paramere large with median incision of anterior margin deep; genital lobes with minute chitinous teeth apically. External parameres relatively short, less than half as long as paramere; obtuse apically. – **Female** similar to male, but forceps with branches contiguous, straight, and tapering apically. – **Length** of body with forceps: male: 15–15,5 mm, female: 10–12 mm.

Distribution: India.

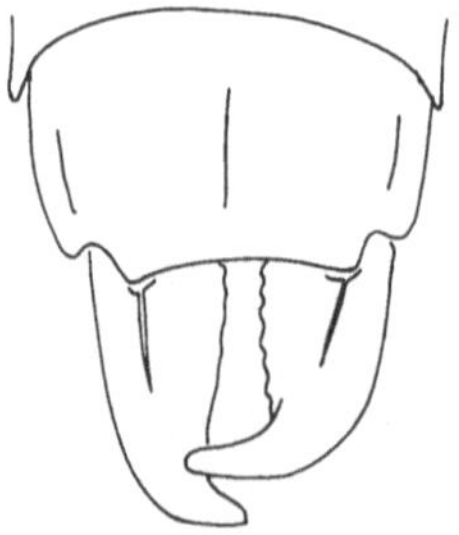

Fig. 288 Holotype ultimate tergite and forceps of *Epilabis burri* SRIVASTAVA. After SRIVASTAVA.

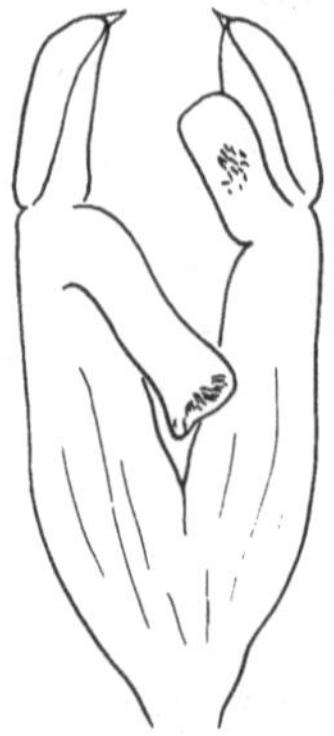

Fig. 289 Holotype genital armature of *Epilabis burri* SRIVASTAVA. After SRIVASTAVA.

8. Epilabis ramachandrai RAMAMURTHI & DAVID

1973 *Epilabis (Cryptolabis) ramachandrai* RAMAMURTHI & DAVID, Zool. Anz., Leipzig **190**: 438; fig. 7 (ultimate tergite and forceps of holotype), 8 (genital armature of holotype). – Terra typica: India (Type male: RAMAMURTHI's Collect., Madras).
1978 *Apolabis ramachandrai* – SRIVASTAVA, Bull. zool. Survey India, Calcutta, (1) **1**: 71.

Description: Male colour dark brown, antennae and legs yellow, or yellowish-brown. Head depressed, weakly rugose, only coronal suture distinct; posterior margin of head broadly truncate. Antennae broken in holotype, 4th and 5th joints short and stout, rest gradually elongated, subcylindrical. Eyes normal, shorter than the length of head behind eyes. Pronotum longer than wide, not wider than head anteriorly; lateral margins straight, parallel. Tegmina covering mesonotum only. Abdomen narrowed at base, gradually broadened to 9th tergite; surface of abdominal tergites finely punctate and setulose; lateral angles of 7, 8 and 9 with weak ridges, lateral angles blunt. Ultimate tergite wider than long, plate not separated from base and sides; median longitudinal sulcus distinct, basal region rugose; posterior margin truncate between base of forceps. Forceps (Fig. 290) more or less asymmetrical, broad at base, gradually narrowed to apex, gently arcuate, apex pointed. Penultimate sternite broadly triangular. Genitalia (Fig. 291) similar to *Metalabis*-type; trigonal, paramere V-shaped, median incision very deep. Genital lobes with distinct sclerotized plates, virga absent. – **Female** unknown. – **Length** of body with forceps: 24,5 mm.

Distribution: India.

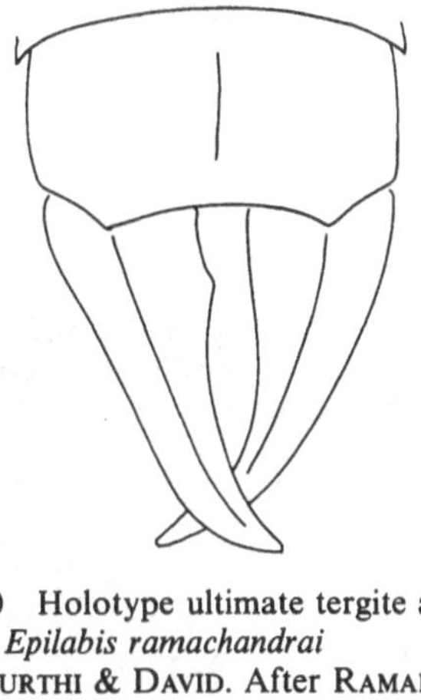

Fig. 290 Holotype ultimate tergite and forceps of *Epilabis ramachandrai* RAMAMURTHI & DAVID. After RAMAMURTHI & DAVID.

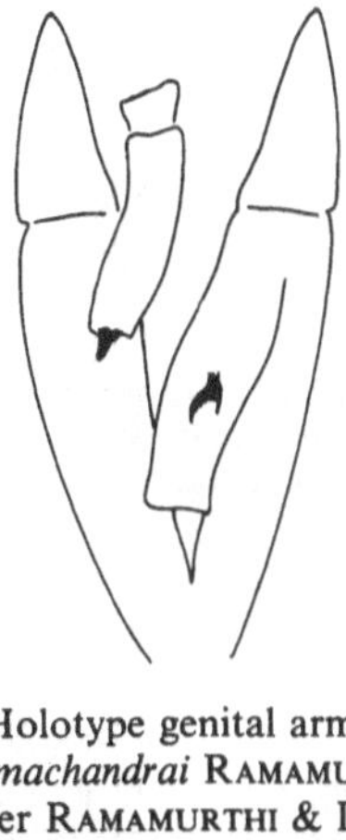

Fig. 291 Holotype genital armature of *Epilabis ramachandrai* RAMAMURTHI & DAVID. After RAMAMURTHI & DAVID.

9. Epilabis vallakadaiensis RAMAMURTHI & DAVID

1973 *Epilabis (Cryptolabis) vallakadaiensis* RAMAMURTHI & DAVID, Zool. Anz., Leipzig, **190**: 439; fig. 9 (ultimate tergite with forceps of holotype), 10 (genital armature of holotype). – Terra typica: India (Type male: RAMAMURTHI's Collect., Madras).
1978 *Gelotolabis vallakadaiensis* – SRIVASTAVA, Bull. zool. Survey India, Calcutta, (1) **1**: 71.

Description: Male colour reddish-brown; legs and antennae yellow. Head depressed, tumid, surface rugose; postfrontal and coronal sutures obsolete; occipital region with a fissure along coronal suture; posterior margin sinuate. Eyes normal, somewhat shorter than the length of head behind eyes. Antennae broken in holotype. Tegmina and wings absent. Abdomen parallel-sided, tergites punctate, finely pubescent; postero-lateral angles blunt. Ultimate tergite simple, broad, with a short median longitudinal sulcus; posterior margin concave in median section. Forceps (Fig. 292) slightly asymmetrical; trigonal basally, curved and cylindrical apically. Penultimate sternite broadly rounded posteriorly. Genitalia (Fig. 293) characteristic; paramere with median longitudinal incision between external parameres deep; genital lobes well developed, without virgae, but with minute sclerotized serration apically. External paramere of *Gonolabis*-type, not trigonal, expanded at middle portion. – **Female** unknown. – **Length** of body with forceps: unknown.

Distribution: India.

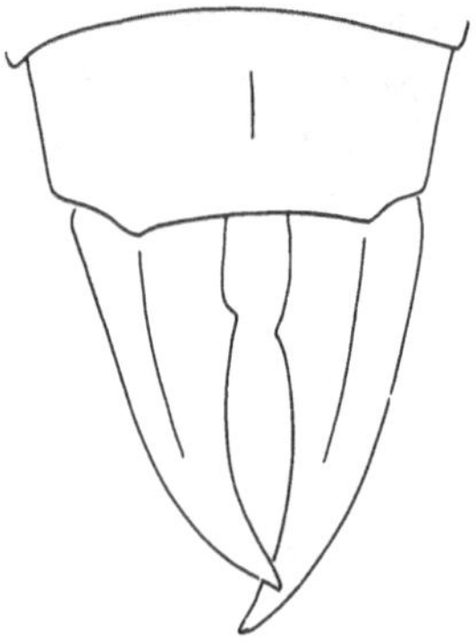

Fig. 292 Holotype ultimate tergite and forceps of *Epilabis vallakadaiensis* RAMAMURTHI & DAVID. After RAMAMURTHI & DAVID.

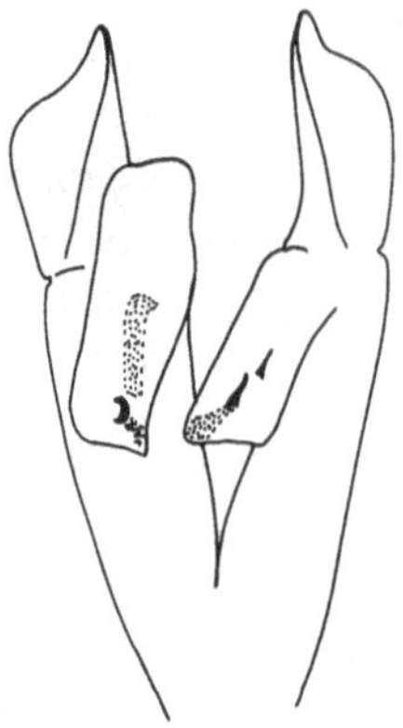

Fig. 293 Holotype genital armature of *Epilabis vallakadaiensis* RAMAMURTHI & DAVID. After RAMAMURTHI & DAVID.

Genus 20: **Metalabis** BURR

1915 *Metalabis* BURR, J. r. microsc. Soc., London, **1915**: 538. – Species typica: *Eulabis saramaccensis* ZACHER 1911.

1966 *Metalabis* – POPHAM & BRINDLE, Entomologist, London, **99**: 272 [= *Carcinophora* SCUDDER 1876].

1975 *Metalabis* – STEINMANN, Acta zool. Acad. Sci. hung., Budapest, **21**: 205.

Literature: BRINDLE 1968; HEBARD 1917; REICHARDT 1968; STEINMANN 1977, 1978.

Description: Body small or median size; apterous species; generally resembles *Anisolabis*, but forceps of male subremote and feebly asymmetrical; trigonal, very widened basally, and a little curved apically; apex of forceps obtuse. Male genitalia characteristic, virga within genital lobe not thickened.

Distribution: South America.

5 species.

Identification key to the species

1 Ultimate tergite of male with or without, short, but not keel-shaped, longitudinal ridges dorso-laterally 2

1′ Ultimate tergite of male with two long, keel-shaped longitudinal ridges dorso-laterally ... 3

2 Pronotum quadrate. Ultimate tergite with two, short longitudinal ridges, these being indicated only by a small raised portion of the cuticle (Fig. 294). Male genitalia large, broad, as in fig. 295 .... 1. **Metalabis ecarinata**, p. 181

2′ Pronotum slightly longer than wide. Ultimate tergite without longitudinal ridges laterally (Fig. 296). Male genitalia specific, both genital lobes directed forwards (Fig. 297) .... 2. **Metalabis brasiliensis**, p. 182

3 Posterior margin of ultimate tergite of male expanded apically, specific, as in fig. 298. Male genitalia very similar to *carinata* BRINDLE, as in fig. 299 ... 3. **Metalabis saramaccensis**, p. 183

3′ Posterior margin of ultimate tergite of male normal, with two dorso-lateral tubercles in the median section 4

4 Smaller species, length of body with forceps about 13 mm. Antennae 17-jointed. Male forceps strongly curved, as in fig. 300 .... 4. **Metalabis deplanata**, p. 184

4′ Larger species, length of body with forceps about 18 mm. Antennae 16-jointed. Male forceps curved, as in fig. 301 .... 5. **Metalabis carinata**, p. 185

1. Metalabis ecarinata BRINDLE

1968 *Metalabis ecarinata* BRINDLE, Stud. Fauna of Suriname and Guyanas, den Haag, **10**: 21; fig. 14 (external paramere of male genitalia), 20 (pronotum of male), 22 (ultimate tergite with forceps of male), 25 (genital armature of holotype). – Terra typica: Surinam (Type male: Rijksmus. nat. Hist., Leiden).

Literature: STEINMANN 1977, 1978.

Description: Male very similar to *carinata* BRINDLE, but with distinctive genitalia. General colour dark reddish-brown, shiny, legs reddish-yellow. Head tumid, depressed along sutures; postfrontal and coronal sutures deep. Eyes as long as genae. Antennae 13 (+?)-jointed; first joint relatively short, and broad, shorter than distance between antennal bases. Pronotum quadrate, smooth, strongly widened posteriorly; lateral margins sinuate; narrowly yellow laterally; a median longitudinal furrow present but much less strongly marked in *M. carinata* BRINDLE. Tegmina and wings absent. Mesonotum broad, transverse, with a very slight median longitudinal furrow. Metanotum short, transverse, and concave posteriorly. Abdomen slender; sides of tergites 6–9 with lateral ridges. Ultimate tergite without the dorsal longitudinal ridges, these being indicated only by a small raised portion of the cuticle. Forceps (Fig. 294) almost symmetrical; dorsal ridge present basally. Penultimate sternite rounded posteriorly. Genitalia (Fig. 295) specific; paramere broad, median longitudinal incision of anterior margin narrow; genital lobes with characteristics sclerotized serration; virga short, narrow. External parameres wide basally, narrowed

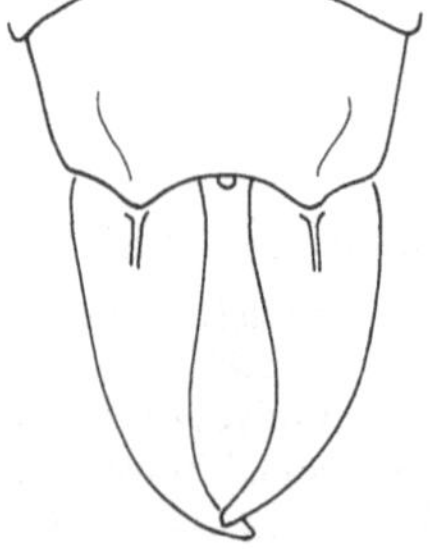

Fig. 294 Holotype ultimate tergite and forceps of *Metalabis ecarinata* BRINDLE. After BRINDLE.

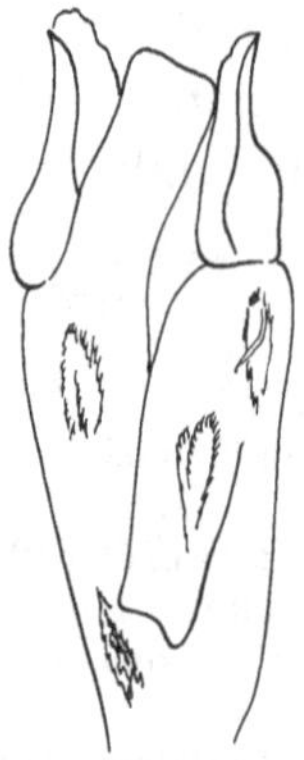

Fig. 295 Holotype genital armature of *Metalabis ecarinata* BRINDLE. After BRINDLE.

distally, inner membrane wide and extending beyond tip of paramere. – **Female** unknown. – **Length** of body with forceps: 13 mm.

Distribution: Surinam.

2. Metalabis brasiliensis BRINDLE

1979 *Metalabis brasiliensis* BRINDLE, Entomologist's month. Mag., London, **115**: 149; fig. 4 (genital armature of holotype), 5 (anterior leg), 7 (male forceps). – Terra typica: Brazil (Type male: Mus. Zool. Univ., São Paulo).

Description: Male general colour black, shiny; antennae dark brown with one (9th or 10th) distal joint pale yellow; legs yellowish-brown, femora darkened. Cuticle smooth, abdominal tergites very weakly and sparsely punctate dorsally. Head transverse, tumid; postfrontal and coronal sutures well marked; eyes large, more or less as long as the length of head behind eyes. First antennal joint long, but shorter than distance between antennal bases, second transverse, third joint twice as long as wide. Pronotum slightly longer than wide, narrowed posteriorly; lateral margins straight, and posterior margin convex; median longitudinal furrow distinct. Tegmina and wings absent. Meso- and metanotum transverse, normal. Abdomen somewhat wide, depressed and without any lateral longitudinal ridges. Ultimate tergite broad; posterior margin concave in median section. Forceps (Fig. 296) symmetrical; trigonal basally, curved and cylindrical apically. Pygidium between branches of forceps prominent, declivent, ventral part rounded. Penultimate sternite with posterior margin simply rounded. Genitalia (Fig. 297) at first sight somewhat peculiar since both genital lobes are directed the same way; paramere broad, median incision of anterior margin not deep; genital lobes with finely sclerotized serrations. External parameres comparatively small, obtuse. – **Female** unknown. – **Length** of body with forceps: 13,5 mm.

Distribution: Brazil.

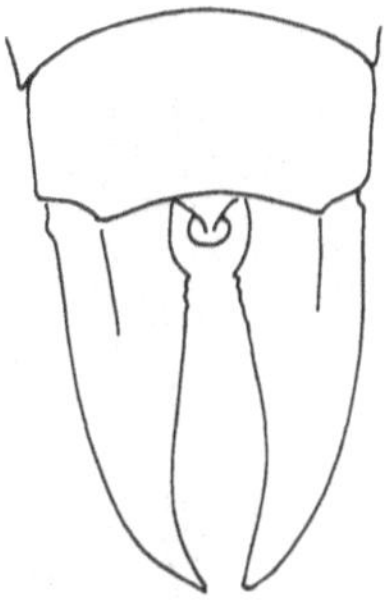

Fig. 296 Holotype ultimate tergite and forceps of *Metalabis brasiliensis* BRINDLE. After BRINDLE.

Fig. 297 Holotype genital armature of *Metalabis brasiliensis* BRINDLE. After BRINDLE.

3. Metalabis saramaccensis (ZACHER)

1911 *Eulabis saramaccensis* ZACHER, Zool. Jahrb., Jena, **30**: 378. – Terra typica: Surinam (Type male: Mus. Naturkunde, Berlin).

1917 *Metalabis saramaccensis* – HEBARD, Proc. Acad. nat. Sci. Philadelphia, **1917**: 233. Guyana, Surinam.

1966 *Carcinophora saramaccensis* – POPHAM & BRINDLE, Entomologist, London, **99**: 275. Southern America.

1968 *Metalabis saramaccensis* – BRINDLE, Stud. Fauna of Suriname and Guyanas, den Haag, **10**: 19; fig. 17 (pronotum of male), 12 (external paramere of male genitalia), 9 and 10 (abdominal end with forceps of male), 23 (male genitalia). Surinam, Guyana, Brazil.

Literature: BURR 1914, 1915; REICHARDT 1968; SAKAI 1970; STEINMANN 1977, 1978.

Description: Male colour dark brown; antennae light brown, first antennal joint yellow; legs and thoracal notum yellow. Head tumid, subpyriform; eyes moderately prominent, shorter than first antennal joint. Antennae 15-jointed; first joint comparatively short, slightly shorter than distance between antennal bases. Pronotum quadrate, as long as wide; all angles rounded; lateral margins a little convex posteriorly; median longitudinal furrow present in prozona. Tegmina and wings absent. Meso- and metanotum normal. Abdomen more or less punctate; lateral sides to abdominal tergites 4–9 with longitudinal ridges. Ultimate tergite with two longitudinal ridges dorsally, extending to posterior margin. Forceps (Fig. 298) strongly trigonal basally, curved apically; asymmetrical; inner margins crenulate at basal section. Penultimate sternite rounded posteriorly. Genitalia (Fig. 299) typical for *Metalabis*; paramere broad at anterior margin; median incision of anterior margin short and narrow; genital lobes wide with sclerotized plates, without virgae. External parameres about twice as long as wide; trigonal. – **Female** similar to male, but forceps symmetrical; contiguous, tapering. – **Length** of body with forceps: male: 9–11 mm, female: 8–11,5 mm (by ZACHER: 21–23 mm).

Distribution: Guyana, Surinam, Brazil.

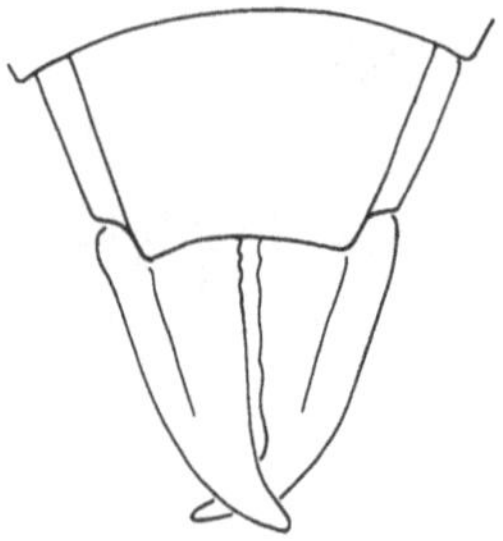

Fig. 298 Male ultimate tergite and forceps of *Metalabis saramaccensis* (ZACHER). Original.

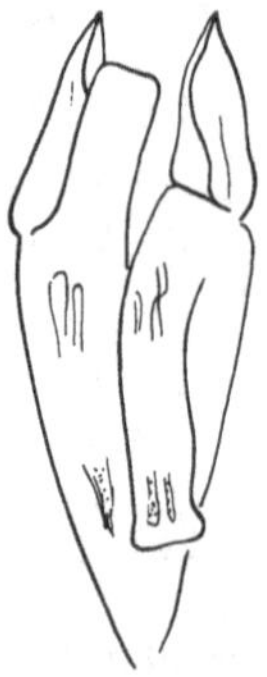

Fig. 299 Male genital armature of *Metalabis saramaccensis* (ZACHER). After BRINDLE.

4. Metalabis deplanata REHN

1932 *Metalabis deplanata* REHN, Ark. Zool., Uppsala, **24A**, 11: 3, pl. I, fig. 1 and 2 (abdomen dorsally and laterally). – Terra typica: Brazil (Type male: Naturhist. Riksmus., Stockholm).
1966 *Carcinophora deplanata* – POPHAM & BRINDLE, Entomologist, London, **99**: 275.
1978 *Metalabis deplanata* – STEINMANN, Dtsch. ent. Zeitschr., Berlin, **25**: 184.
Literature: REICHARDT 1968; SAKAI 1970.

Description: Male colour dark brown; antennae light brown, legs yellow. Head subpyriform in outline when seen from dorsum, its greatest length subequal to width across eyes; postfrontal and coronal sutures prominent, well marked. Eyes moderately large, but shorter than the length of head behind eyes. Antennae 17-jointed; first joint normal, a little shorter than distance between antennal bases; proximal joints two-thirds as long as wide. Pronotum subquadrate, greatest length very slightly more than caudal width; all angles rounded; lateral margins straight, faintly diverging posteriorly. Tegmina and wings absent. Mesonotum with posterior margin truncate. Posterior margin of metanotum broadly concave. Abdomen depressed, more or less punctate; tergites similar to *saramaccensis* (ZACHER). Ultimate tergite broad, dorsal surface with two very prominent dorso-lateral keels; posterior margin with two dorsolateral tubercles. Forceps (Fig. 300) relatively robust, short and wide, asymmetrical; strongly trigonal basally, curved and cylindrical apically; ventral surface of forceps subdeplanate. Genitalia unknown. – **Female** unknown. – **Length** of body with forceps: 13–14 mm.

Distribution: Brazil, Guyanas.

Fig. 300 Holotype ultimate tergite and forceps of *Metalabis deplanata* REHN. After REHN.

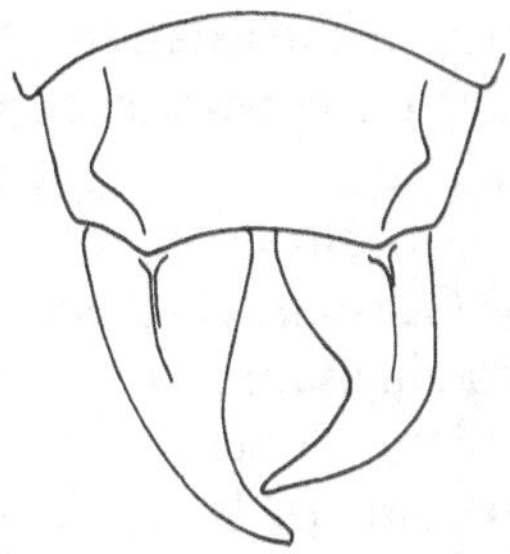

5. Metalabis carinata BRINDLE

1968 *Metalabis carinata* BRINDLE, Stud. Fauna of Suriname and Guyanas, den Haag, **10**: 20; fig. 18 (pronotum of male), 21 (ultimate tergite with forceps of male), 11 (external paramere of male genitalia), 24 (male genitalia). – Terra typica: Surinam (Type male: Rijksmus. natuurl. Hist. Leiden).
Literature: STEINMANN 1978.

Description: Male colour blackish, shiny; legs yellow. Head tumid, depressed along sutures; postfrontal and coronal sutures well marked. Eyes about as long as genae, Antennae 16-jointed; first joint short, broad, shorter than distance between antennal bases. Pronotum quadrate, smooth, widened posteriorly; lateral margins strongly sinuate; narrowly yellow laterally. Median longitudinal furrow present. Tegmina and wings absent. Meso- and metanotum typical. Femora, especially those of anterior pair, strongly dilated; anterior tibiae with numerous ventral hairs, but without spines. Abdomen reddish-black, segments 1–2 impunctate, the rest closely and strongly punctate except on actual posterior margin. Abdominal tergites 6–9 with lateral ridges, each ridge beginning about

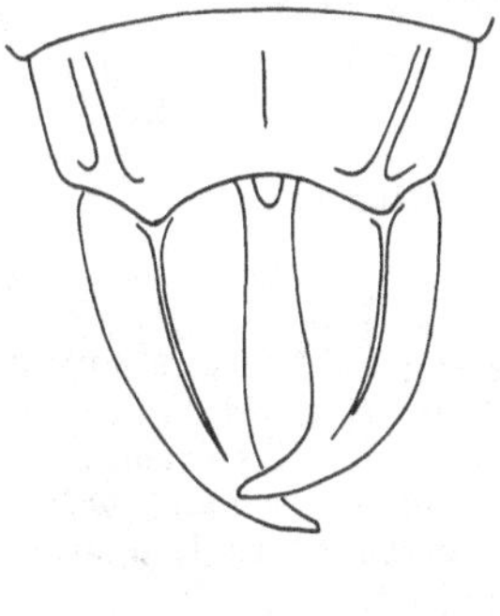

Fig. 301 Holotype ultimate tergite and forceps of *Metalabis carinata* BRINDLE. After BRINDLE.

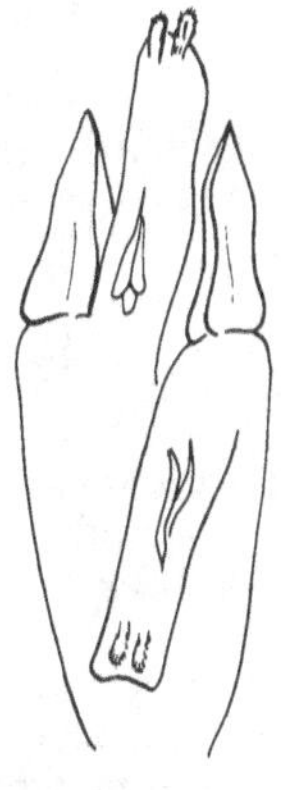

Fig. 302 Holotype genital armature of *Metalabis carinata* BRINDLE. After BRINDLE.

middle of tergite and extending to and beyond posterior margin. Ultimate tergite transverse, smooth, with median longitudinal sulcus. Forceps (Fig. 301) almost symmetrical, trigonal basally with a strong and prominent dorsal ridge extending for basal two-thirds or at apical section. Penultimate sternite rounded posteriorly. Genitalia (Fig. 302) of *Metalabis*-type, paramere broad, robust, and large; median incision comparatively small, and narrow; genital lobes wide, apex with two spinous lobes, virga specific, broadened and with accessory sclerites. External parameres long, acuminate, pointed apically, strongly trigonal. – **Female** similar to male, but no lateral ridges on the abdomen, nor ridges on ultimate tergite. – **Length** of body with forceps: male: 16–17,5 mm, female: 14,5–17 mm.

Distribution: Surinam.

Genus 21: **Gonolabis** BURR

1900 *Gonolabis* BURR, Ann. Soc. ent. belge, Bruxelles, **44**: 48. – Species typica: *Anisolabis javana* BORMANS 1883.
1902 *Gonolabidae* VERHOEFF, Zool. Anz., Leipzig, **25**: 185. – Genus typicum: *Gonolabis* BURR 1900.
1911 *Mongolabis* ZACHER, Zool. Jahrb., Jena, **30**: 385. – Species typica: *Gonolabis woodwardi* BURR 1908.
1911 *Gelotolabis* ZACHER (partim); Zool. Jahrb., Jena, **30**: 385.
1966 *Anisolabella* – POPHAM & BRINDLE, Entomologist, London, **99**: 272 [= *Gonolabis* BURR 1900].
Literature: BRINDLE 1978; BURR 1904, 1910, 1911, 1915; HANDLIRSCH 1925; REHN 1916; SAKAI 1970; STEINMANN 1975, 1977, 1978, 1980; TILLYARD 1926; TOWNES 1945.

Description: General colour dark brown to reddish-brown or yellowish-brown, rarely black. Body of *Euborellia* or *Anisolabis*-type, but abdomen strongly expanded, widening to ultimate tergite from apex to base. Head tumid. Tegmina and wings completely absent, or tegmina rudimentary. Male forceps various, male asymmetrical, and female forceps almost symmetrical; branches short, curved, widely separated at base. Male genitalia well developed, genital lobes with or without virgae; external parameres longer than wide, outer margins usually sinuate.

Distribution: African, Indo-Australian and Oriental species.

42 species.

Identification key to the species

1 Ethiopian group **2**
1′ Palearctic, Indo-Australian or Oriental species **15**
2 Ultimate tergite very wide, almost twice as wide as long (Fig. 303). Genital lobe of male genitalia with well-defined, characteristic denticulation apically (Fig. 304) .... 1. **Gonolabis electa**, p. 190
2′ Ultimate tergite broad. Genital lobe of male genitalia usually without well defined denticulated areas, although denticulations occur, and parameres relatively smaller **3**
3 Posterior margin of penultimate sternite excised at median section **4**
3′ Posterior margin of penultimate sternite not excised medially **7**
4 Pronotum quadrate **5**
4′ Pronotum transverse **6**

5 Ultimate tergite of male without dorso-lateral keel; forceps comparatively wide, inner margins more or less straight, as in fig. 305. Genital lobes of male genitalia without sclerotized plates (Fig. 306) .... 2. **Gonolabis splendida**, p. 191
5′ Ultimate tergite of male with prominent dorso-lateral keel; forceps slightly asymmetrical, inner margins undulate, as in fig. 307. Genital lobes of male genitalia with specific sclerotized plates (Fig. 308) .... 3. **Gonolabis deserta**, p. 192
6 Ultimate tergite of male very wide, transverse; male forceps moderately short, and wide, as in fig. 309. Penultimate sternite with trigonal excision at apex (Fig. 310) .... 4. **Gonolabis nitens**, p. 193
6′ Ultimate tergite of male less wide; male forceps a little longer than *nitens* (BRINDLE) as in fig. 312. Penultimate sternite with rounded excision at apex (Fig. 313) .... 5. **Gonolabis puella**, p. 194
7 Penultimate sternite with posterior margin concave **8**
7′ Penultimate sternite with posterior margin truncate or rounded **12**
8 Sides of posterior abdominal tergites with lateral longitudinal ridges. Male forceps strongly asymmetrical, apices of branches rather strongly curved (Fig. 315) .... 6. **Gonolabis silvestrii**, p. 195
8′ Sides of posterior abdominal tergites without lateral longitudinal ridges **9**
9 Sides of posterior abdominal tergites rugoso-striate. Ultimate tergite weakly punctate, punctures tending to form irregular longitudinal bands separated by smooth cuticle. Forceps as in fig. 317. Penultimate sternite with posterior part depressed and apex concave (Fig. 318) .... 7. **Gonolabis nigeriensis**, p. 196
9′ Sides of posterior abdominal tergites punctate or only slightly rugoso-striate **10**
10 Eyes fairly large. Pronotum slightly widened posteriorly; femora mainly blackish; ultimate tergite without dorso-lateral keel. Pygidium flat. External parameres of male genitalia short and strongly narrowed distally (Fig. 320) .... 8. **Gonolabis sinuata**, p. 196
10′ Eyes large. Femora yellow or only vaguely darkened medially. Pygidium protruding **11**
11 Head less transverse, eyes rather shorter than length of head behind eyes; pronotum slightly transverse. Ultimate tergite with dorso-lateral ridge, represented by a convex and narrow swelling. External parameres (Fig. 321) with broader apices, median membrane narrow... .... 9. **Gonolabis laevigata**, p. 197
11′ Head more transverse and eyes larger; pronotum strongly transverse. Ultimate tergite without dorso-lateral ridge. External parameres (Fig. 322) with narrower apices, median membrane wide .... 10. **Gonolabis transversalis**, p. 198
12 Penultimate sternite with apex truncate **13**
12′ Penultimate sternite with apex rounded **14**
13 Male forceps strongly curved and asymmetrical (Fig. 323), somewhat similar to species of *Karschiella*. External parameres of male genitalia narrowed towards apices (Fig. 324)..... .... 11. **Gonolabis hincksi**, p. 199
13′ Male forceps more or less symmetrical, inner margins with prominent tooth at base (Fig. 325). External parameres of male genitalia wide, angular (Fig. 326)......................... .... 12. **Gonolabis tanganyikae**, p. 200
14 Eyes large; pronotum strongly punctate; abdominal tergites 7–9 rugoso-striate at sides. Male forceps as in fig. 327. External parameres of male genitalia comparatively short and wide (Fig. 328) .... 13. **Gonolabis umbrosa**, p. 201
14′ Eyes fairly large; pronotum impunctate or almost so. Branches of forceps short and wide, as in fig. 329. External parameres comparatively long and narrow (Fig. 330) .... 14. **Gonolabis whellani**, p. 202
15 Palearctical species **16**
15′ Indo-Australian or Oriental species **18**
16 Distributed in Madeira and Canary Islands. Pronotum with a prominent longitudinal median impressed line extending almost over entire length. Central parameral plate of male genitalia elongate, slightly V-shaped (Fig. 332) .... 15. **Gonolabis maxima**, p. 203
16′ Asian species **17**
17 Large species, length of body with forceps 18–30 mm. Male genitalia very similar to *distincta*

(NISHIKAWA), paramere broad, oval, genital lobes with characteristic denticulation pads (Fig. 335) .... 16. **Gonolabis marginalis**, p. 204
17′ Smaller species, length of body with forceps 14–16 mm. Male genitalia very similar to *marginalis* (DOHRN), paramere broad, oval, genital lobes with specific chitinous serration (Fig. 338) .... 17. **Gonolabis distincta**, p. 205
18 Indo-Australian species **19**
18′ Oriental species **27**
19 Genital lobes of male genitalia without virgae **20**
19′ Genital lobes of male genitalia with virgae **22**
20 External parameres of male genitalia expanded at middle portion, appendicular apex bifurcate (Fig. 340); genital lobes with very distinct reticulum lobi genitalis apically. Male forceps strongly asymmetrical, trigonal basally, and curved apically (Fig. 339). .... 18. **Gonolabis bidens**, p. 206
20′ External parameres of male genitalia expanded at middle portion but apex not bifurcate . **21**
21 External paramere of male genitalia very wide at middle portion (Fig. 342), paramere with very wide median incision between external parameres .... 19. **Gonolabis gilesi**, p. 207
21′ External parameres of male genitalia normal at middle portion; paramere simple, V-shaped, genital lobes with specific; characteristic sclerotized plates, as in fig. 344 .... 20. **Gonolabis dentata**, p. 208
22 Ultimate tergite very wide, almost twice as wide as long (Fig. 303). Genital lobe of male genitalia with well-defined, characteristic denticulation apically (Fig. 304) .... 1. **Gonolabis electa**, p. 190
22′ Ultimate tergite wide. Genital lobe of male genitalia usually without well-defined denticulation pads, although denticulated areas occur, and parameres relatively small **23**
23 External parameres of male genitalia of *Carcinophora*-type, apices acute; genital lobes with specific sclerotized plate (Fig. 346). Male forceps more or less symmetrical, strongly curved, as in fig. 345 .... 21. **Gonolabis javana**, p. 209
23′ External parameres of male genitalia not of *Carcinophora*-type; genital lobes not similar to *javana* (BORMANS) **24**
24 External parameres of male genitalia expanded apically, as in figs. 348 and 350 .... **25**
24′ External parameres of male genitalia obtuse apicaly, as in figs. 352 and 353 .... **26**
25 Male forceps more or less symmetrical; ultimate tergite smooth, without dorso-lateral keels, as in fig. 347. Male genitalia with comparatively short central parameral plate (Fig. 348) .. .. 22. **Gonolabis forcipata**, p. 210
25′ Male forceps strongly asymmetrical; ultimate tergite with prominent and asymmetrical dorso-lateral kells, as in fig. 349. Male genitalia similar to *forcipata* BURR, but central parameral plate elongate (Fig. 350) .... 23. **Gonolabis tasmanica**, p. 211
26 External parameres of male genitalia comparatively small, short, and very broad medially; genital lobes very long, as in fig. 352 .... 24. **Gonolabis blanchardi**, p. 212
26′ External parameres of male genitalia comparatively large, a little expanded apically; genital lobes well developed, but slightly shorter than *blanchardi* (GUILLON) (Fig. 353) .... 25. **Gonolabis woodwardi**, p. 213
27 Genital lobes of male genitalia with virgae **28**
27′ Genital lobes of male genitalia without virgae **32**
28 External parameres of male genitalia triangular, apices not elongate **29**
28′ External parameres of male genitalia oval, apices a little elongate and obtuse **30**
29 External parameres of male genitalia strongly expanded at basal section; central parameral plate slightly oval, median incision of anterior margin very wide (Fig. 354) .... 26. **Gonolabis sechuana**, p. 214
29′ External parameres of male genitalia broad, but not very expanded at basal section; central parameral plate V-shaped, median longitudinal incision of anterior margin narrow (Fig. 355) .... 27. **Gonolabis kuekenthali**, p. 215
30 Large species, length of body with forceps 18–30 mm. Male genitalia typical, paramere broad, median incision of anterior margin deep; genital lobes with characteristical denticulation pads which are slightly sclerotized (Fig. 335) .... 16. **Gonolabis marginalis**, p. 204
30′ Smaller species, length of body with forceps 14–16 mm **31**

31 Posterior margin of penultimate sternite of male with median incision (Fig. 337). Male genitalia very similar to *marginalis* (DOHRN) and to *cavaleriei* (BORELLI) as in fig. 338 .... 17. **Gonolabis distincta**, p. 205
31' Posterior margin of penultimate sternite of male without median incision. Male genitalia very similar to *distincta* (NISHIKAWA), and to *marginalis* (DOHRN) as in fig. 357 28. **Gonolabis cavaleriei**, p. 216
32 External paramere very specific, egg-shaped, oval; genital lobes with chitinous granules (Fig. 359). Male ultimate tergite more or less quadrate, forceps small, short (Fig. 358) .... 29. **Gonolabis panayica**, p. 217
32' External parameres of male genitalia not egg-shaped, but slightly elongate. Ultimate tergite of male transverse **33**
33 Apices of external parameres excised transversally, as in fig. 360; paramere broad, median longitudinal incision of anterior margin relatively short .... 30. **Gonolabis fallax**, p. 218
33' Apices of external parameres not excised transversally **34**
34 External parameres of male genitalia oval medially, hardly widened, and apices a little expanded, obtuse, as in figs. 362 and 364 .... **35**
34' External parameres not oval, angular, but medially hardly widened **36**
35 Median incision of anterior margin of central parameral plate comparatively narrow; genital lobe well developed, with very small chitinous granulates, as in fig. 362 .... 31. **Gonolabis formosae**, p. 219
35' Median incision of anterior margin of central parameral plate comparatively wide; genital lobe fully developed, with specific chitinous and sclerotized granulates (Fig. 364) .... 32. **Gonolabis mystica**, p. 220
36 Apex of external paramere of male genitalia very elongated, characteristic, trigonal, as in fig. 365; genital lobes without virgae .... 33. **Gonolabis hwangi**, p. 221
36' Apex of external parameres of male genitalia not elongated, normal **37**
37 External paramere relatively large, long, longer than wide. Genital lobes with specific sclerotizated plates (Fig. 367) **38**
37' External parameres relatively short, visibly shorter than wide. Genital lobes without sclerotized plates **39**
38 Male forceps comparatively long, strongly asymmetrical, as in fig. 366. External parameres slightly broadened medially; genital lobe with characteristic chitinous serration, as in fig. 367 .... 34. **Gonolabis insidiata**, p. 222
38' Male forceps comparatively short, more or less symmetrical, as in fig. 368. External parameres not slightly broadened medially; genital lobe with specific sclerotized plates as in fig. 369.. .... 35. **Gonolabis insulana**, p. 223
39 Genital lobes fully developed, elongate, longer than median incision between external parameres **40**
39' Genital lobes well developed, but shorter than the length of median incision **42**
40 Apex of external parameres short; inner margin and outer margin more or less parallel basally and medially (Fig. 371); ultimate tergite of male very transverse, forceps broken medially (Fig. 370) .... 36. **Gonolabis acuta**, p. 224
40' Apex of external paramere elongated; inner and outer margins convex **41**
41 Central parameral plate of male genitalia broad, oval, U-shaped; external parameres strongly expanded medially, as in fig. 372 .... 37. **Gonolabis undata**, p. 225
41' Central parameral plate of male genitalia narrow, V-shaped; external parameres lemon-shaped, as in fig. 373 .... 38. **Gonolabis kirbyi**, p. 226
42 Central parameral plate of male genitalia with very narrow median incision between external parameres; external parameres comparatively large, as in fig. 375........ 39. **Gonolabis darevskyi**, p. 227
42' Central parameral plate of male genitalia with wide median incision between external parameres; the latter comparatively small **43**
43 External paramere of male genitalia very strongly expanded medially; median incision narrower than *panfilovi* (BEY-BIENKO), as in fig. 376 .... 40. **Gonolabis magna**, p. 228
43' External parameres of male genitalia normally expanded medially; median incision wider than *magna* (BEY-BIENKO), as in fig. 377 .... 41. **Gonolabis panfilovi**, p. 229

— Male genitalia unknown, therefore not included in the identification key. Male forceps more or less symmetrical, similar to *insulana* (BRINDLE), as in fig. 368 . 42. **Gonolabis minor**, p. 230

1. Gonolabis electa BURR

1910 *Gonolabis electa* BURR, Fauna Brit. India. Dermaptera, London, Bombay, 79, fig. 21 (male). – Terra typica: Ceylon (Type male: Brit. Mus. nat. Hist., London).

1954 *Gonolabis electa* – BOESEMAN, Zool. Verh., Leiden, **21**: 32 [= *Gonolabis minor* BORELLI 1926]. Java.

1961 *Gonolabis calas* FERNANDO, Ceylon J. Sci., biol. Sci., Colombo, **4**: 42; fig. 6 (male), 7 (penultimate sternite and forceps), 8 (antenna). – Terra typica: Ceylon (Type male: Univ. Colombo, Ceylon). – **syn. nov.**

Literature: BORELLI 1926; BRINDLE 1978; BURR 1911, 1915; HEBARD 1927; HINCKS 1954; POPHAM & BRINDLE 1966; SAKAI 1970; STEINMANN 1975, 1977, 1981.

Description: Male colour reddish-brown to blackish; antennae brown to dark brown, basal joints sometimes paler; sides of pronotum yellow; legs yellowish, and femora darkened. Head tumid, smooth, a little longer than wide, but wider than pronotum. Postfrontal and coronal sutures faintly distinct. Eyes small, shorter than first antennal joint; the latter comparatively small, slightly shorter than the distance between antennal bases. Pronotum more or less quadrate, all margins straight, and all angles rounded; median longitudinal impressed line visible. Tegmina and wings absent. Meso- and metanotum normal, typical, transverse. Abdomen of *Gonolabis*-type, expanded gradually to last tergite; tergites finely punctate, sides of tergites 7–9 more heavily punctate and slightly rugoso-striate. Ultimate tergite rectangular and about twice as wide as long, smooth and shiny; with a faint median depression; posterior margin truncate. Forceps (Fig. 303) remote at the base, very stout and trigonal basally, the upper keel corresponding to the slight tumidity of the ultimate tergite. Penultimate sternite ample, very broadly rounded. Genitalia (Fig. 304) characteristic; paramere broad, oval, relatively short; median incision of anterior margin of central parameral plate deep and wide; genital lobe with virgae, genital lobe with

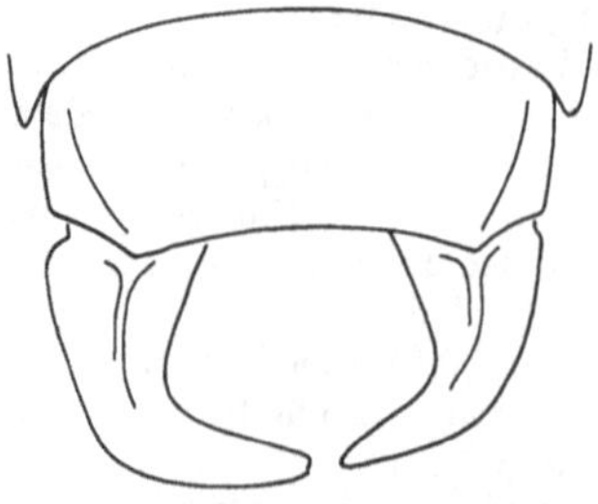

Fig. 303 Male ultimate tergite and forceps of *Gonolabis electa* BURR. Original.

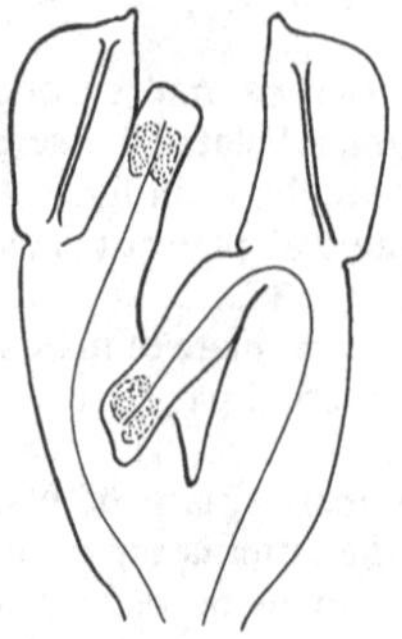

Fig. 304 Male genital armature of *Gonolabis electa* BURR. Original.

well-defined characteristic denticulation arranged apically. External parameres easily recognizable, broad, with little tip obtuse. – **Female** similar to male, but abdomen widest medially, narrowing distally; sides of tergites 7–9 less strongly punctate; both branches of forceps relatively slender, and somewhat sinuate, inner margin slightly denticulated. – **Length** of body with forceps: male: 7,5–10 mm, female: 6,5–8,5 mm.

Distribution: Oriental Region, and adventions in Africa: São Tomé and West Africa coast.

2. Gonolabis splendida STEINMANN

1978 *Gonolabis splendida* STEINMANN, Fol. ent. hung., Budapest, **31**: 194; fig. 8 (genital armature of holotype, gen. prep. No. 114, det. Dr. H. STEINMANN). – Terra typica: Congo Republic (Type male: Természettudományi Múz., Budapest).

Description: Male light brown to reddish brown; head red, antennae brown, legs yellow; femora with brown bands. Head tumid, smooth; postfrontal sutures finely, coronal suture visibly marked. Eyes comparatively large, but shorter than scape. Antennae 18-jointed; first joint long, but more or less shorter than distance between antennal bases; second joint short, transverse, third longer than 4th. Pronotum quadrate, all margins straight, and all angles rounded; lateral margins a little widened posteriorly; median longitudinal furrow present. Tegmina and wings absent. Mesonotum transverse, posterior margin straight; metanotum broad. Legs normal, femora with light brown bands. Abdomen finely punctate, somewhat widened to last tergite, of *Euborellia*-type. Ultimate tergite typical; abdominal tergites 6–8 carinate dorso-laterally. Forceps (Fig. 305) asymmetrical, broad and trigonal at basal half, cylindrical apically; inner margins crenulate. Penultimate sternite rounded posteriorly. Genitalia (Fig. 306) very

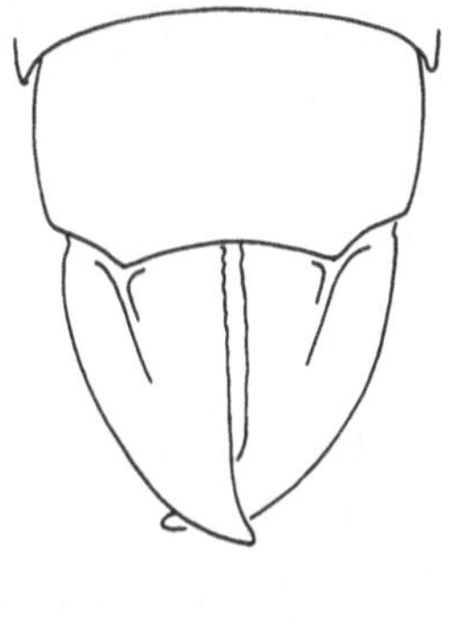

Fig. 305 Male ultimate tergite and forceps of *Gonolabis splendida* STEINMANN. Original.

Fig. 306 Holotype genital armature of *Gonolabis splendida* STEINMANN. After STEINMANN.

simple, robust, paramere broad, median incision of anterior margin not deep, but wide; genital lobes normal, without virgae; external parameres of *Anisolabis*-type. – **Female** similar to male, but abdominal tergites 8–9 without carinae dorso-laterally; forceps symmetrical, contiguous, tapering. – **Length** of body with forceps: male: 13–15,5 mm, female: 14–16 mm.

Distribution: Congo Republic.

3. Gonolabis deserta STEINMANN

1981 *Gonolabis deserta* STEINMANN, Acta zool. Acad. Sci. hung., Budapest, 27: 193; fig. 20 (ultimate tergite with forceps of male), 21 (genital armature of holotype). – Terra typica: Savanna Kamp (Tanzania ?) (Type male: Inst. taxonom. Zoöl., Amsterdam).

Description: Male colour reddish brown, shiny, legs yellowish brown. Head tumid, broad. Postfrontal sutures and coronal suture distinct. Antennae 11 (+?) jointed, broken in holotype; first antennal joint long, but scape shorter than distance between antennal bases; second joint small, transverse, third longer than 4th. Eyes comparatively large, about as long as length of head behind eyes. Pronotum quadrate, about as long as wide, but a little expanded posteriorly; lateral margins not parallel; median longitudinal furrow present in prozona. Tegmina and wings entirely absent. Meso- and metanotum normal. Abdomen punctate, shiny; sides of tergites 6–8 carinate dorso-laterally. Ultimate tergite broad, almost smooth, with very distinct keels at dorsolateral surface. Forceps (Fig. 307) asymmetrical; trigonal basally, abruptly becoming cylindrical, base broadened but without marked tooth. Penultimate sternite rounded posteriorly. Genitalia (Fig. 308; gen. prep. No. 675, det. Dr. H. STEINMANN) characteristic; paramere narrow, V-shaped, median incision of anterior margin very deep; genital lobes well developed, long, with specific paired sclerotized plates apically,

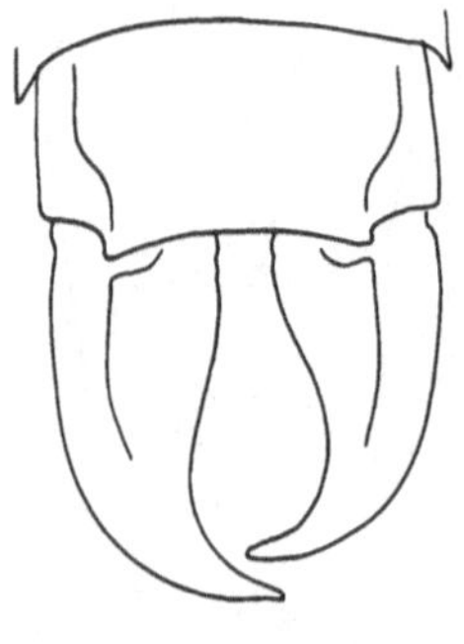

Fig. 307 Holotype ultimate tergite and forceps of *Gonolabis deserta* STEINMANN. After STEINMANN.

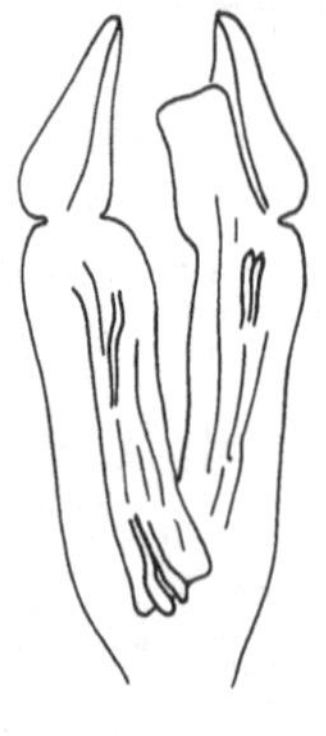

Fig. 308 Holotype genital armature of *Gonolabis deserta* STEINMANN. Original.

but without virgae. External parameres typical, abotuse. – **Female** unknown. – **Length** of body with forceps: 14 mm.

Distribution: Savanna Kamp (Tanzania ?).

4. Gonolabis nitens (BRINDLE) comb. nov.

1978 *Anisolabis nitens* BRINDLE, Dermaptera of Africa, Tervuren, **2**: 110; fig. 130 (male), 132 (penultimate sternite), 249 (external paramere of male genitalia). – Terra typica: Cameroon (Type male: Mus. nation. Hist. nat., Paris).

Description: Male general colour black, shiny; head reddish, antennae dark brown, basal joints slightly paler; legs reddish-yellow to yellowish-brown, femora vaguely darkened dorsally towards apices. Head transverse, convex; postfrontal and coronal sutures well marked; eyes small, slightly shorter than the length of head behind eyes. First antennal joint moderately small, shorter than distance between antennal bases. Pronotum transverse, widened posteriorly; lateral margins slightly sinuate; median longitudinal furrow distinct, but narrow. Tegmina and wings absent. Meso- and metanotum typical. Abdomen gradually widened distally and depressed; sides of tergites 8–9 rugoso-striate; but without lateral longitudinal ridges. Ultimate tergite smooth, very broad; median longitudinal sulcus present. Pygidium small, triangular, apex rounded and excised. Forceps (Fig. 309) short, asymmetrical, weakly curved; inner margins strongly crenulate. Penultimate sternite triangular, margins sinuate, apex excised as in fig. 310. Genitalia (Fig. 311) with very specific external parameres. Central parameral plate broad, median incision of anterior margin deep. External parameres wide with a small round apex, median membrane wide. – **Female:** similar to male; abdominal punctation almost absent, tergites 8–9 not rugoso-striate; forceps almost straight, dorsal ridge less prominent, inner margins with 4–5 relatively large crenulations on basal half. – **Length** of body with forceps: male: 15–18 mm, female: 15,5–18 mm.

Distribution: Cameroon.

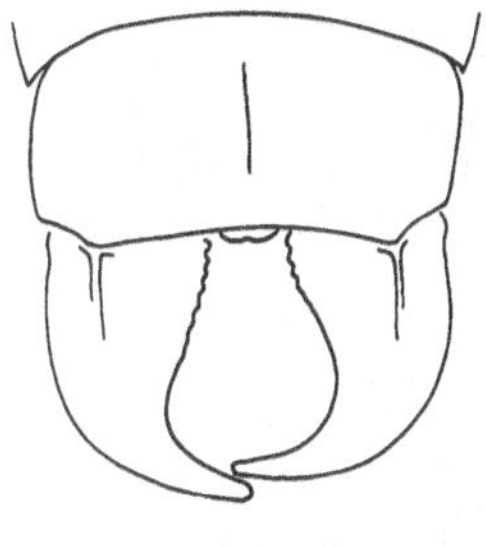

Fig. 309 Male ultimate tergite and forceps of *Gonolabis nitens* (BRINDLE). Original.

Fig. 310 Posterior margin of penultimate sternite of *Gonolabis nitens* (BRINDLE). After BRINDLE.

Fig. 311 External paramere of male genitalia of *Gonolabis nitens* (BRINDLE). After BRINDLE.

5. Gonolabis puella STEINMANN

1978 *Gonolabis puella* STEINMANN, Fol. ent. hung., Budapest, **31**: 195; fig. 10 (genital armature of holotype). – Terra typica: Ghana (Type male: Természettudományi Múz., Budapest).

Description: Male shiny blackish red or blackish brown; antennae dark brown, pronotum blackish, but with reddish posterior margin; legs light brown or yellowish brown, femora banded brown; forceps reddish brown. Head broad, broader than pronotum, postfrontal and coronal sutures distinct. Eyes relatively large, about as long as scape. Antennae broken in holotype; first joint short, shorter than distance between antennal bases, second transverse, third joint longer than 4th. Pronotum more or less transverse, a little wider than long, all margins straight, and all angles rounded; median longitudinal furrow distinct. Tegmina and wings entirely absent. Mesonotum broad, with straight posterior margin, punctate. Metanotum transverse. Abdomen visibly expanded at abdominal tergites 7–8; sides of tergites not carinate dorso-laterally. Ultimate tergite broad, posterior margin concave in median section. Forceps (Fig. 312) asymmetrical, widened and trigonal basally, tapering apically. Penultimate sternite with rounded excision at posterior margin as in fig. 313. Genitalia (Fig. 314; gen. prep. No. 241, det. Dr. H. STEINMANN) specific; central parameral plate broad, strongly V-shaped, median incision of anterior margin very deep; genital lobes well developed, without virgae. External parameres widened in basal section. – **Female** very similar to male, but general colouration lighter and forceps symmetrical. – **Length** of body with forceps: male: 9,5–10,5 mm, female: 9–10 mm.

Distribution: Ghana.

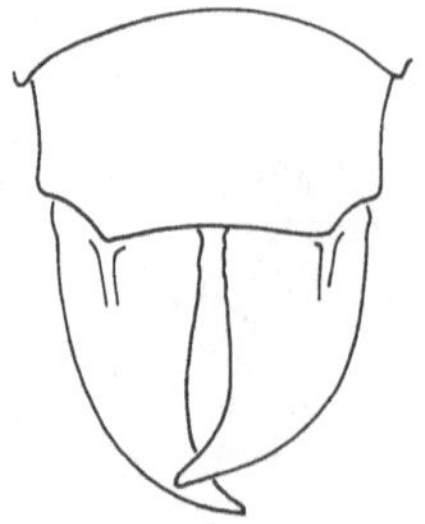

Fig. 312 Male ultimate tergite and forceps of *Gonolabis puella* STEINMANN. Original.

Fig. 313 Posterior margin of penultimate sternite of *Gonolabis puella* STEINMANN. Original.

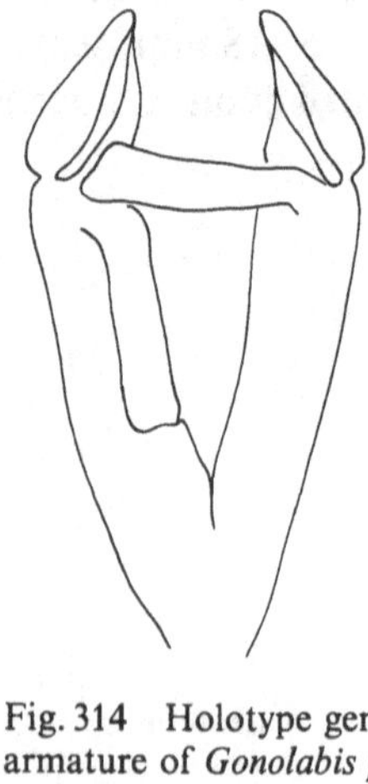

Fig. 314 Holotype genital armature of *Gonolabis puella* STEINMANN. Original.

6. Gonolabis silvestrii (Borelli) comb. nov.

1914 *Anisolabis silvestrii* Borelli, Boll. Lab. Zool. gen. agr., Protici, **8**: 261. – Terra typica: Nigeria (Type male: Mus. Ist. Zool. sist., Torino, holotype without penis).
1978 *Anisolabis silvestrii* – Brindle, Dermaptera of Africa, Tervuren, **2**: 98; fig. 148 (penultimate sternite), 151 (male forceps), 255 (external paramere of male genitalia).
Literature: Burr 1915; Popham & Brindle 1966; Sakai 1970; Steinmann 1977, 1978.

Description: Male general colour black; antennae yellowish-brown distally, segments 4–6 brown, basal two joints yellow and joint 3 yellow at base; sides of pronotum yellowish, and legs yellowish-brown, femora slightly darkened on sides. Head tumid, transverse, convex; postfrontal and coronal sutures weak; eyes comparatively large, more or less as long as the length of head behind eyes. First antennal joint long, but slightly shorter than distance between antennal bases. Pronotum transverse, widened posteriorly; median longitudinal furrow well marked. Tegmina and wings absent. Meso- and metanotum typical; thoracal nota almost smooth but with isolated punctures. Abdomen widened medially and depressed, tergites closely and rather strongly punctate; sides of tergites 6–9 rugoso-striate laterally, each with smooth well-defined lateral longitudinal ridges. Ultimate tergite broad; median longitudinal sulcus present; posterior margin straight. Forceps (Fig. 315) strongly asymmetrical, both branches of forceps short, apices rather strongly curved; inner margins crenulate. Penultimate sternite closely and finely punctate at base, distal part with small, well-separated punctures, posterior margin concave. Genitalia with paramere short, median incision of anterior margin deep; genital lobes normal. External parameres (Fig. 316) short, broad, external margin convex, inner membrane slightly sinuate, median membrane wide medially. – **Female** unknown. – **Length** of body with forceps: 16–18 mm.

Distribution: Nigeria.

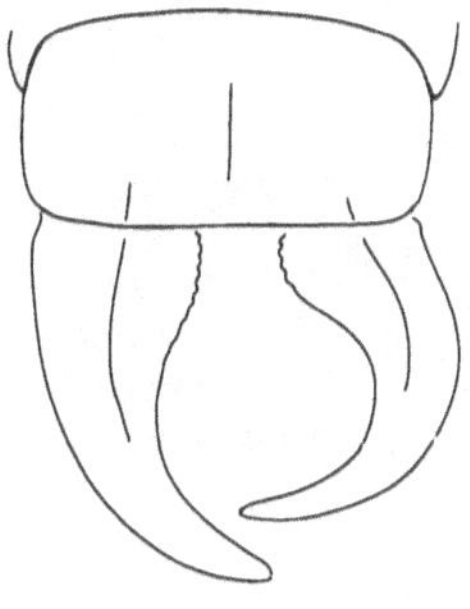

Fig. 315 Male ultimate tergite and forceps of *Gonolabis silvestrii* (Borelli). After Brindle.

Fig. 316 External paramere of male genitalia of *Gonolabis silvestrii* (Borelli). After Brindle.

7. Gonolabis nigeriensis (BRINDLE) comb. nov.

1978 *Anisolabis nigeriensis* BRINDLE, Dermaptera of Africa, Tervuren, **2**: 117; fig. 141 (male), 146 (penultimate sternite of male), 252 (external paramere of male genitalia). – Terra typica: Nigeria (Type male: Mus. r. Afr. centr., Tervuren).

Description: Male general colour black; sides of head reddish, and sides of pronotum yellow; legs yellowish, femora somewhat darkened; antennae dark brown, three basal joints paler. Head transverse, convex; postfrontal and coronal sutures faintly distinct; eyes small, slightly shorter than the length of head behind eyes; first antennal joint short, shorter than distance between antennal bases. Pronotum transverse, widened posteriorly, lateral margins slightly sinuate; median longitudinal furrow not clearly defined but depressed anteriorly. Tegmina and wings absent. Mesonotum weakly punctate, metanotum punctate. Abdomen weakly widened and depressed; sides of tergites 8–9 rugoso-striate but without lateral longitudinal ridges. Ultimate tergite broad; weakly punctate, punctures tending to irregular longitudinal bands separated by smooth cuticle. Forceps (Fig. 317) short, more or less symmetrical; branches strongly trigonal basally, and cylindrical apically. Penultimate sternite closely and finely punctate at base, posterior part depressed and apex concave (Fig. 318). Genitalia typical, paramere broad, median incision of anterior margin deep; genital lobes well developed. External parameres with outer margin strongly angular, margin of inner membrane sinuate, median membrane widened (Fig. 319). – **Female** unknown. – **Length** of body with forceps: 15 mm.

Distribution: Nigeria.

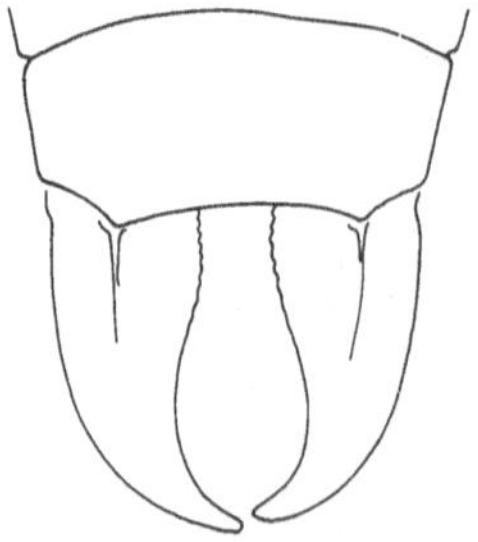

Fig. 317 Holotype ultimate tergite and forceps of *Gonolabis nigeriensis* (BRINDLE). After BRINDLE.

Fig. 318 Posterior margin of male penultimate sternite of *Gonolabis nigeriensis* (BRINDLE). After BRINDLE.

Fig. 319 External paramere of male genitalia of *Gonolabis nigeriensis* (BRINDLE). After BRINDLE.

8. Gonolabis sinuata (BRINDLE) comb. nov.

1978 *Anisolabis sinuata* BRINDLE, Dermaptera of Africa, Tervuren, **2**: 108; fig. 145 (penultimate sternite of male), 254 (external paramere of male genitalia). – Terra typica: Zaire (Type male: Mus. r. Afr. centr., Tervuren).

Description: Male colour brown to black; head and pronotum blackish, the pronotum narrowly yellow at sides; antennae dark brown, first joint black. Head transverse; postfrontal sutures indistinct, coronal suture present. Eyes fairly large, but a little shorter than the length of head behind eyes. First antennal joint long, but shorter than distance between antennal bases. Pronotum transverse, slightly widened posteriorly, cuticle coriaceous; median longitudinal furrow weak, all angles rounded. Tegmina and wings absent. Meso- and metanotum similar to pronotum. Abdomen weakly widened and depressed, sides of tergites 8–9 rugoso-striate, but without lateral longitudinal ridges. Ultimate tergite weakly punctate; median longitudinal sulcus distinct; posterior margin concave. Forceps short, inner margin with a very narrow smooth flange, apex incurved, branches weakly curved and somewhat asymmetrical. Penultimate sternite broad, rather strongly transverse, narrowed distally, apex concave and somewhat sinuate. Genitalia typical of species of genus; paramere broad, median incision of anterior margin deep; genital lobes well developed. External parameres (Fig. 320) broad basally, sharply narrowed distally, apex rounded, median membrane short and basal. – **Female** unknown. – **Length** of body with forceps: 10,5 mm.

Distribution: Zaire.

Fig. 320 External paramere of male genitalia of *Gonolabis sinuata* (BRINDLE). After BRINDLE.

9. Gonolabis laevigata (BRINDLE) comb. nov.

1978 *Anisolabis laevigata* BRINDLE, Dermaptera of Africa, Tervuren, **2**: 100; fig. 142 (head and pronotum), 145 (penultimate sternite of male), 251 (external paramere of male genitalia). – Terra typica: Guinea (Type male: Mus. nation. Hist. nat., Paris).

Description: Male colour reddish-black or dark reddish-brown, basal three joints of antennae dark yellow except apex of third blackish; legs yellow, femora vaguely darkened dorsally. Head transverse, rounded posteriorly; postfrontal and coronal sutures well marked; eyes large, but a little shorter than the length of head behind eyes. First antennal joint short, slightly shorter than distance between antennal bases. Pronotum transverse, weakly widened posteriorly;

lateral margins straight, all angles rounded; median longitudinal furrow depressed anteriorly, cuticle coriaceous, smooth, shiny. Tegmina and wings absent. Mesonotum smooth except for a triangular area anteriorly, metanotum smooth, faintly punctate basally. Abdomen nearly parallel-sided; sides of tergites not rugoso-striate, but tergites 6–8 more strongly punctate. Ultimate tergite broad, punctate, punctures variable in size and tending to form irregular longitudinal bands; median longitudinal sulcus deep in centre. Forceps short, typical, branches almost symmetrical. Penultimate sternite rounded, apex concave. Genitalia large, paramere typical, median incision of anterior margin deep; genital lobes well developed. External parameres (Fig. 321) short, external margins rounded, inner membrane with straight edge, median membrane narrow. – **Female** similar to male, but punctation finer; penultimate sternite rounded distally; branches of forceps somewhat straighter. – **Length** of body with forceps: male: 16,5–17,5 mm, female: 17–18 mm.

Distribution: Guinea.

Fig. 321 External paramere of holotype genitalia of *Gonolabis laevigata* (BRINDLE). After BRINDLE.

10. Gonolabis transversalis (BRINDLE) comb. nov.

1978 *Anisolabis transversalis* BRINDLE, Dermaptera of Africa, Tervuren, **2**: 101; fig. 143 (head and pronotum), 145 (penultimate sternite of male), 246 (external paramere of male genitalia). – Terra typica: Cameroon (Type male: Mus. nation. Hist. nat., Paris).

Description: Male colour dark reddish-brown, shiny; antennae brown to dark brown, basal joints yellow; lateral sides of pronotum yellow, legs yellow. Head very strongly transverse, convex; postfrontal and coronal sutures distinct; eyes large, slightly longer than length of head behind eyes. First antennal joint short, slightly shorter than distance between antennal bases. Pronotum strongly transverse, slightly widened posteriorly; median longitudinal furrow present in anterior half; surface coriaceous, more or less smooth, but with isolated punctures posteriorly. Tegmina and wings absent. Meso- and metanotum coriaceous. Abdomen almost parallel-sided, not of *Gonolabis*-type; tergites 1–4 closely and finely punctate towards bases, tergites 5–7 more strongly but irregularly punctate, sides not rugoso-striate laterally. Ultimate tergite smooth medially with a few small punctures, median longitudinal sulcus indistinct, lateral longitudinal ridges small, but not dorso-lateral ridge. Forceps short, broad basally, dorsal ridge weak, gradually narrowed distally, apices short and incurved, inner margins almost smooth, branches almost symmetrical. Penultimate sternite finely and closely punctate at base; posterior margin concave. Genitalia normal, central parameral plate broad, median incision of anterior margin deep; genital lobe simple, well developed. External parameres (Fig. 322) wide basally, narrowed distally, apex rounded, median membrane wide. – **Female** similar to male, but abdominal punctation rather weaker, lateral longitudinal ridge on ultimate tergite almost absent; branches of forceps straight. – **Length** of body with forceps: male: 14–15,5 mm, female: 14–16 mm.

Distribution: Cameroon.

Fig. 322 External paramere of holotype genital armature of *Gonolabis transversalis* (BRINDLE). After BRINDLE.

11. Gonolabis hincksi (BRINDLE)

1964 *Gelotolabis hincksi* BRINDLE, Ann. Mag. nat. Hist., London, (13) **7**: 753; fig. 10 (genitalia of holotype). – Terra typica: Kenya (Type male: Manchester Mus.).

1978 *Anisolabis hincksi* – BRINDLE, Dermaptera of Africa, Tervuren, **2**: 130; fig. 186 (head and pronotum of male), 189 (penultimate sternite of male), 188 (male forceps), 259 (external paramere of male genitalia).

1978 *Gonolabis hincksi* – STEINMANN, Fol. ent. hung., Budapest, **31**: 196; fig. 12 (external paramere of male genitalia).

Literature: SAKAI 1970; STEINMANN 1977, 1980.

Description: Male general colour black; antennae yellowish-brown, legs reddish-yellow, femora slightly darkened. Large species, surface impunctate on head, thorax, and basal abdominal segments. Head typical, tumid and broad; postfrontal and coronal sutures distinct. Eyes comparatively small, shorter than the length of head behind eyes. Antennae with first joint short, shorter than distance between antennal bases. Pronotum transverse, slightly widened posteriorly; lateral margins straight, all angles rounded; median longitudinal furrow present in anterior part of pronotum. Tegmina and wings absent. Meso- and metanotum typical. Abdomen dilated to last tergite; sides of tergites 7–9 rugoso-striate, but without lateral longitudinal ridges. Ultimate tergite very broad, transverse, with a very deep median longitudinal sulcus; posterior margin concave at median section. Forceps (Fig. 323) strongly asymmetrical; branches trigonal basally, cylindrical and slightly curved apically. Penultimate sternite broadly triangular, apex truncate. Genitalia of *Gonolabis*-type; paramere broad, median incision of anterior margin wide and deep; genital lobes fully developed. External parameres (Fig. 324) specific; rather narrow, strongly wider at base. – **Female** similar to male, but ultimate tergite less rugose; forceps straight, almost symmetrical, inner margins denticulate. – **Length** of body with forceps: male: 32–33 mm, female: 29–35 mm.

Distribution: Kenya.

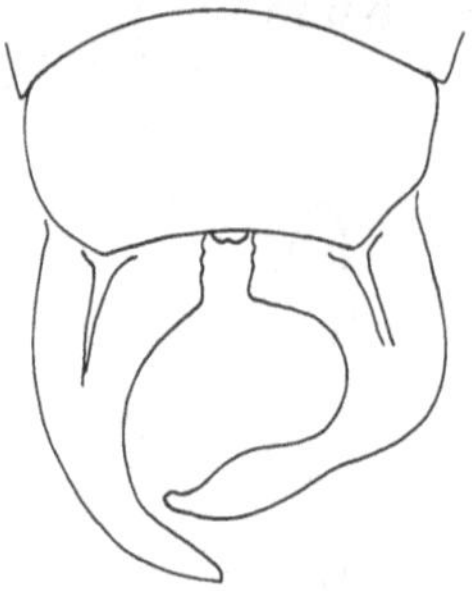

Fig. 323 Male ultimate tergite of *Gonolabis hincksi* (BRINDLE). Original.

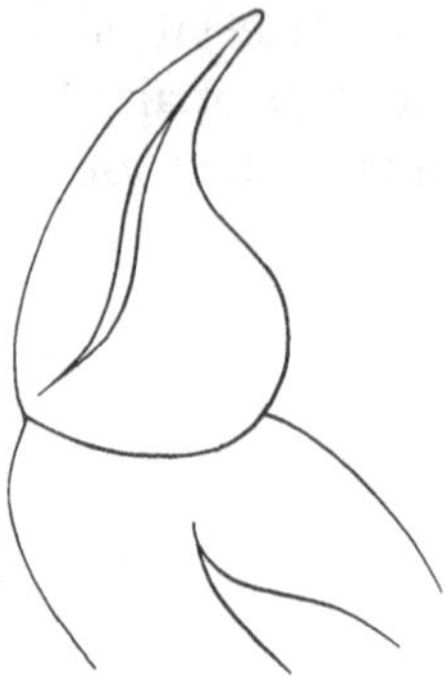

Fig. 324 External paramere of holotype genitalia of *Gonolabis hincksi* (BRINDLE). After BRINDLE.

12. Gonolabis tanganyikae (BRINDLE)

1964 *Gelotolabis tanganyikae* BRINDLE, Ann. Mag. nat. Hist., London, (13) **7**: 754; fig. 15 (male genitalia). – Terra typica: Tanganyika (Type male: Manchester Museum).
1970 *Euborellia tanganyikae* – SAKAI, Dermapt. Cat. Prael., Tokyo, **1**: 46.
1978 *Anisolabis tanganyikae* – BRINDLE, Dermaptera of Africa, Tervuren, **2**: 136.
1978 *Gonolabis tanganyikae* – STEINMANN, Fol. ent. hung., Budapest, **31**: 193; fig. 6 (male genital armature).
Literature: STEINMANN 1977, 1978.

Description: Male shiny blackish, with red head and brown antennae; femora black. Head transverse, tumid; postfrontal and coronal sutures distinct. Eyes comparatively small, shorter than scape. Antennae 20-jointed; first joint moderately short, shorter than distance between antennal bases. Pronotum more or less quadrate, impunctate, with a short median longitudinal impressed line near anterior margin. Tegmina and wings absent. Meso- and metanotum transverse, the latter with concave posterior margin. Abdomen greatly dilated to last tergite; almost impunctate except the lateral margins of tergites 5–8 which are rugose. Ultimate tergite quadrate, or transverse, almost smooth, with an impressed median longitudinal line. Forceps (Fig. 325) of *Gonolabis*-type, asymmetrical, trigonal basally, abruptly becoming cylindrical, base broadened and with well-marked tooth. Penultimate sternite rounded posteriorly. Genitalia (Fig. 326) typical, paramere slightly V-shaped, median longitudinal incision of anterior margin very wide and deep; genital lobes well developed, but without virgae. External parameres widened medially. – **Female** as male, but abdomen smoother, though rugose laterally on segments 5–8; forceps symmetrical, tapering. – **Length** of body with forceps: male: 21–21,5 mm, female: 20–24 mm.

Distribution: Tanzania.

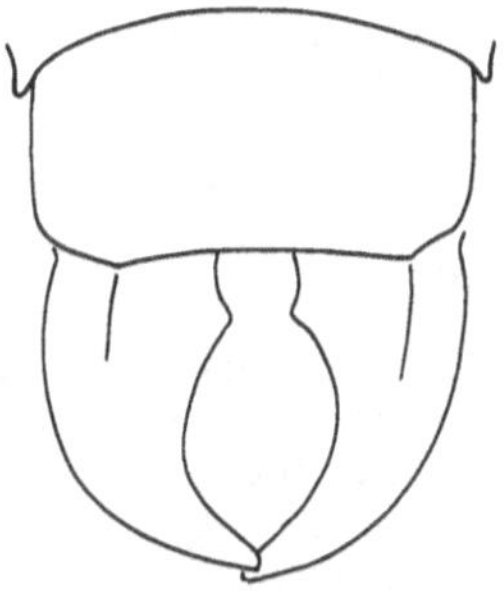

Fig. 325 Male ultimate tergite and forceps of *Gonolabis tanganyikae* (BRINDLE). Original.

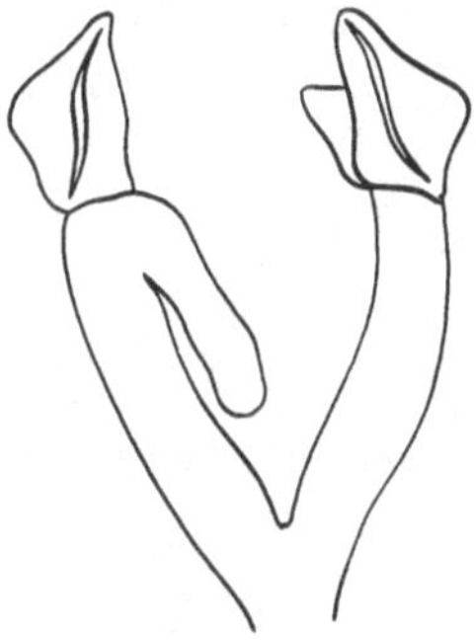

Fig. 326 Male genital armature of *Gonolabis tanganyikae* (BRINDLE). After BRINDLE.

13. Gonolabis umbrosa (BRINDLE)

1964 *Gelotolabis umbrosa* BRINDLE, Ann. Mag. nat. Hist., London, (13) **7**: 752; fig. 5 (male forceps), 13 (male genitalia). – Terra typica: Tanganyika (Type male: Manchester Mus.).

1978 *Anisolabis umbrosa* – BRINDLE, Dermaptera of Africa, Tervuren, **2**: 96; fig. 224 (penultimate sternite of male), 223 (male forceps), 271 (external paramere of male genitalia).

1978 *Gonolabis umbrosa* – STEINMANN, Fol. ent. hung., Budapest, **31**: 195; fig. 10 (male genitalia).

Literature: SAKAI 1970; STEINMANN 1977, 1978.

Description: Male colour blackish, including head; antennae and legs brown; femora with dark brown bands. Head deeply punctate; postfrontal and coronal sutures distinct; eyes about as long as genae, but shorter than first antennal joint; the latter black, the rest being brown except for distal two (joints 13–14), which are yellow; first joint long, but shorter than distance between antennal bases. Pronotum quadrate, all margins straight, and all angles rounded; median longitudinal furrow present in prozona. Tegmina and wings absent. Meso- and metanotum normal, transverse. Abdomen slightly dilated to last tergite; abdominal punctation very strong; surface shiny. Ultimate tergite rugose, transverse, with posterior margin slightly concave in median section. Forceps (Fig. 327) trigonal basally, with very strong dorsal keel, and cylindrical apically; branches more or less symmetrical; inner margins crenulate. Penultimate sternite typical. Genitalia characteristic; paramere with very deep and wide median incision between external parameres; genital lobes well developed; external parameres (Fig. 328) with apices obtuse. – **Female** unknown. – **Length** of body with forceps: 20–22 mm.

Distribution: Tanzania.

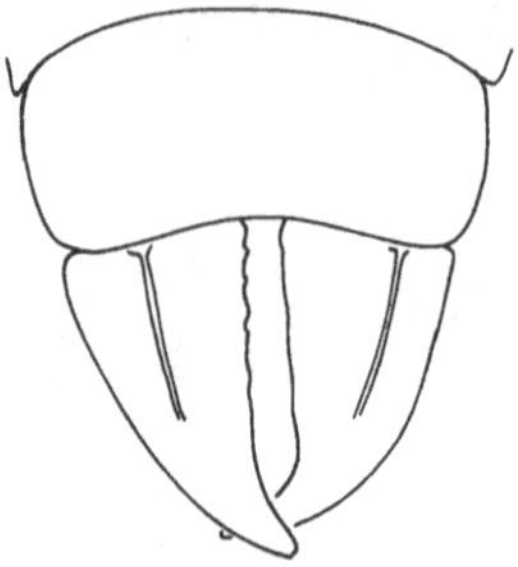

Fig. 327 Holotype ultimate tergite and forceps of *Gonolabis umbrosa* (BRINDLE). After BRINDLE.

Fig. 328 External paramere of holotype genital armature of *Gonolabis umbrosa* (BRINDLE). After BRINDLE.

14. Gonolabis whellani (BRINDLE) comb. nov.

1978 *Anisolabis whellani* BRINDLE, Dermaptera of Africa, Tervuren, **2**: 118; fig. 215 (male forceps), 216 (penultimate sternite), 279 (external paramere of male genitalia). – Terra typica: Malawi (Type male: Brit. Mus. nat. Hist., London).

Description: Male black, shiny: antennae blackish-brown, some distal joints yellowish (11 and 12 or 10 and 11, or with apices of preceding joints); legs yellow, femora vaguely darkened on anterior sides or apically. Head strongly transverse, lateral and posterior margins rounded; postfrontal and coronal sutures distinct. Eyes fairly large, but a little shorter than the length of head behind eyes. First antennal joint shorter than the distance between antennal bases. Pronotum slightly transverse, almost parallel-sided, cuticle rather strongly but not closely punctate; median longitudinal furrow distinct and depressed near anterior margin. Tegmina and wings absent. Meso- and metanotum typical. Abdomen weakly widened medially and depressed, tergites 1–2 fairly weakly punctate, rest of tergites strongly but not closely punctate basally, sides of tergites 7–9 rugoso-striate, but there are no lateral longitudinal ridges. Ultimate tergite broad, with large shallow punctures medially, with longitudinal smooth bands between, more closely punctate laterally, median sulcus deep in centre. Forceps (Fig. 329) short, slightly asymmetrical; inner margins almost simple, weakly curved; dorsal keel present in basal half. Penultimate sternite evenly rounded. Genitalia of *Gonolabis*-type; paramere with very wide, deep median incision, genital lobes well developed; external parameres (Fig. 330) short and broad, median membrane wide. – **Female** unknown. – **Length** of body with forceps: 15–16 mm.

Distribution: Malawi.

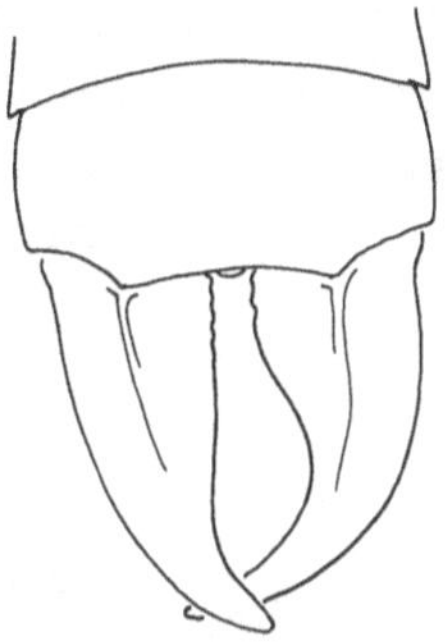

Fig. 329 Holotype ultimate tergite and forceps of *Gonolabis whellani* (BRINDLE). After BRINDLE.

Fig. 330 External paramere of holotype genital armature of *Gonolabis whellani* (BRINDLE). After BRINDLE.

15. Gonolabis maxima (BRULLÉ)

1838 *Forficula maxima* BRULLÉ, in: WEBB & BERTHELOT, Hist. nat. Canar. Ent., Madrid, **2**: 74. – Terra typica: Canary Islands (Type male: unknown locality).
1838 *Forficula major* BULLÉ, in: WEBB & BERTHELOT, Hist. nat. Canar. Ent., Madrid, **2**: 74. – Terra typica: Canary Islands (Type male: unknown locality).
1864 *Forcinella maxima* – DOHRN, Stettiner ent. Ztg., **25**: 288.
1893 *Anisolabis maxima* – BOLIVAR, Ann. Soc. esp., Hist. nat., Madrid, **22**: 46.
1912 *Euborellia maxima* – BURR, Ann. naturhist. Hofmus. Wien, **26**: 77.
1915 *Gelotolabis maxima* – BURR, J. r. microsc. Soc., London, **1915**: 541.
1964 *Gelotolabis maxima* – BRINDLE, Ann. Mag. nat. Hist., London, (13) **7**: 751; fig. 11 (male genitalia).
1966 *Gonolabis maxima* – POPHAM & BRINDLE, Entomologist, London, **99**: 273.
1985 *Canarilabis maxima* – STEINMANN, Fol. ent. Mus., Budapest, **46**: 171, fig. 4 (male genitalia, gen. prep. No. 891, det. Dr. H. STEINMANN).

Literature: BORMANS 1900; BRINDLE 1968; BURR 1910, 1911; KIRBY 1904; SAKAI 1970; STEINMANN 1977, 1978; ZACHER 1911.

Description: Male general colour dark brown. Head tumid and broad; lateral and posterior margins rounded; postfrontal and coronal sutures distinct. Antennae with first joint long but shorter than distance between antennal bases; second joint quadrate, third long, cylindrical. Eyes comparatively small, shorter than first antennal joint. Pronotum more or less quadrate, usually with a short median longitudinal impressed line; lateral margin more or less parallel, straight, all angles rounded. Tegmina and wings absent. Mesonotum transverse, metanotum normal. Abdomen of *Gonolabis*-type, slightly expanded to last tergite. Head, thoracal nota and abdominal tergites with finely, sometimes weakly marked punctures. Ultimate tergite broad, posterior margin truncate; median longitudinal sulcus present. Forceps (Fig. 331) asymmetrical, various, strongly widened basally and cylindrical apically; branches curved at apex; inner margin of right branch with or without large prominent tooth. Penultimate sternite rounded posteriorly. Genitalia (Fig. 332, gen. prep. No. 699, det. Dr. H. STEINMANN) elongate; paramere large, strongly V-shaped, median longitudinal incision of anterior margin very deep; genital lobes well developed, without virgae. External paramere (Fig. 333) strongly expanded medially, obtuse apically. – **Female** similar to male, but forceps symmetrical, more or less straight, trigonal basally, cylindrical apically, contiguous, tapering. – **Length** of body with forceps: male: 26–28 mm, female: 23–27 mm.

Distribution: Canary Islands, Madeira.

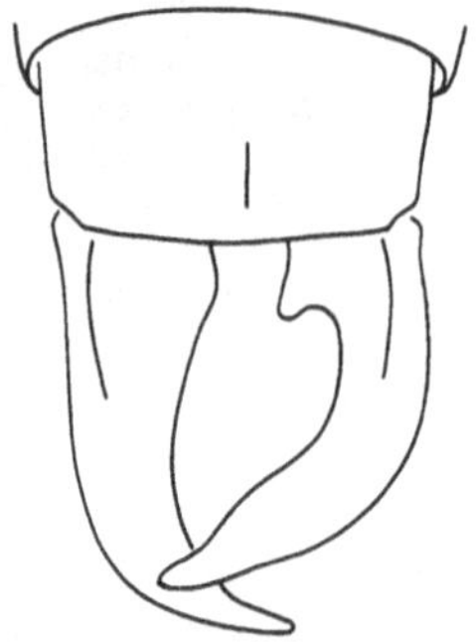

Fig. 331 Male ultimate tergite and forceps of *Gonolabis maxima* (BRULLÉ). Original.

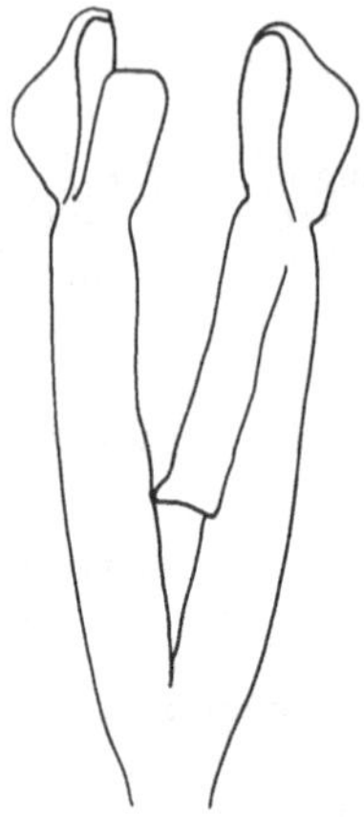

Fig. 332 Male genital armature of *Gonolabis maxima* (BRULLÉ). Original.

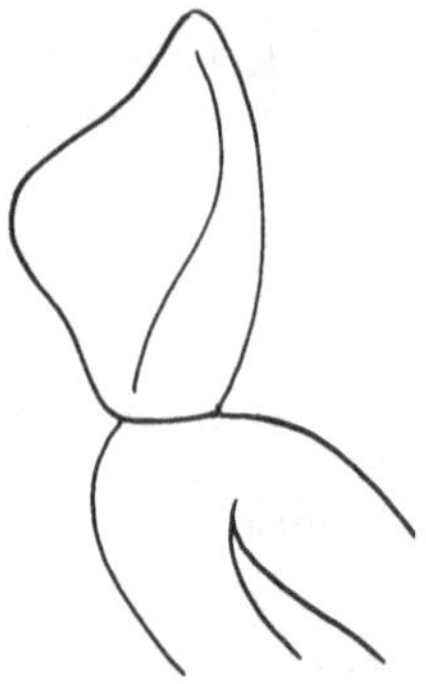

Fig. 333 External paramere of male genitalia of *Gonolabis maxima* (BRULLÉ). After BRINDLE.

16. Gonolabis marginalis (DOHRN)

1864 *Forcinella marginalis* DOHRN, Stettiner ent. Ztg., **25**: 288. – Terra typica: Japan (Type male: unknown locality).

1900 *Anisolabis marginalis* – BORMANS, Das Tierreich, Berlin, **11**: 50. Japan, Batavia.

1900 *Anisolabis marginalis* – KRAPELIN, Jahrb. hamb. wiss. Anstalten, 18: 2. Hamburg (!): "aus Japan mit Pflanzen".

1915 *Apolabis marginalis* – BURR, J. r. microsc. Soc., London, **1915**: 527; pl. XI, fig. 14 (male genitalia).

1966 *Carcinophora marginalis* – POPHAM & BRINDLE, Entomologist, London, **99**: 275.

1969 *Gelotolabis marginalis* – NISHIKAWA, Kontyû ent. Soc., Nippon, Tokyo, (1) **37**: 46; fig. 8 (penultimate sternite with forceps of male), 9 (male genitalia). Japan, Formosa, China, Java.

1970 *Carcinophora marginalis* – SAKAI, Dermapt. Cat. Prael., Tokyo, **1**: 82. Japan, Korea, Formosa, China, Deutschland.

1979 *Gonolabis marginalis* – STEINMANN, Acta zool. Acad. Sci. hung., Budapest, **25**: 167; fig. 2 (male genitalia). Japan, Formosa, Korea, China, Hong Kong to Java.

Literature: BORELLI 1927; BURR 1910, 1911; HEBARD 1923; HIGEZIRO-HASAMI 1938; KIRBY 1904; MATSUMURA 1904; SAKAI 1970; SHIRAKI 1928; STEINMANN 1977, 1978; ZACHER 1911.

Description: Male general colour dark brown; frontal area of head and legs yellowish, except dark bands near femora and tibiae; mouthparts reddish. Antennae with one to four joints near apices whitish, rest somber brown. Head tumid, occiput a little depressed; postfrontal and coronal sutures visible. Eyes relatively rather small, shorter than first antennal joint. Antennae 15-jointed; first joint long, but a little shorter than distance between antennal bases. Pronotum transverse, all margins straight, and all angles rounded; lateral margins a little concave; median longitudinal furrow present. Tegmina and wings absent. Meso- and metanotum normal. Abdomen not strongly expanded to last tergite; sides of tergites 6–9 acute but not strongly carinate; each of tergites 2–5 with a very short spine on each posterior corner. Ultimate tergite transverse, rather cristate at sides, slightly emarginate posteriorly. Forceps (Fig. 334) asymmetrical, trigonal basally, cylindrical and curved apically; dorsal keels well marked. Penultimate sternite emarginate posteriorly. Genitalia (Fig. 335) oval; paramere broad, median incision of anterior margin deep, genital lobe with virgae and with hyaline denticulation pads apically. External parameres a little longer than wide, inner margins strongly chitinized. – **Female** similar to male, but sides of abdomen rounded; ultimate tergite somewhat narrower; penultimate sternite rounded posteriorly; forceps symmetrical, almost straight. – **Length** of body with forceps: male: 18–30 mm, female: 17,5–29 mm.

Distribution: Japan, Taiwan, China, Korea, Hong Kong to Java.

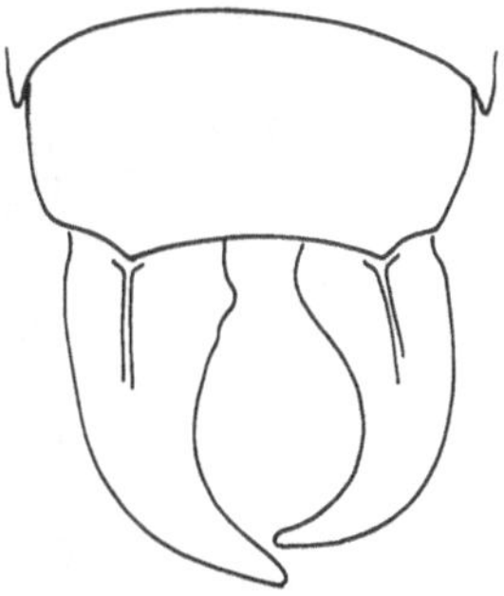

Fig. 334 Male ultimate tergite and forceps of *Gonolabis marginalis* (DOHRN). Original.

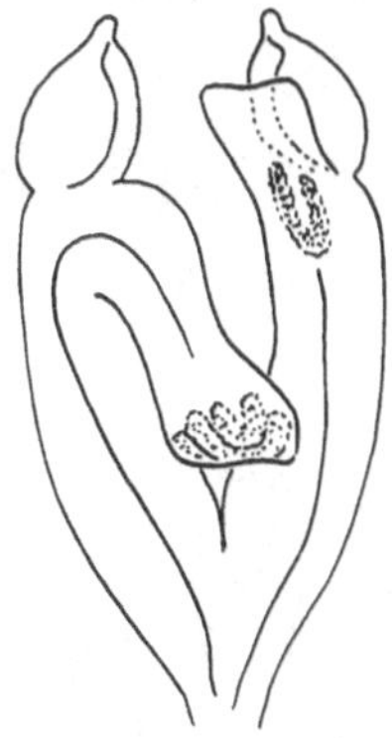

Fig. 335 Male genital armature of *Gonolabis marginalis* (DOHRN). Original.

17. Gonolabis distincta (NISHIKAWA)

1969 *Gelotolabis distincta* NISHIKAWA, Kontyû ent. Soc. Nippon, Tokyo, (1) **37**: 47; fig. 10 (male), 11 (apex of abdomen from ventral view), 12 (genital armature of holotype). – Terra typica: Chabana Island (Type male: Ent. Lab. Univ. Agriculture, Tokyo).

1979 *Gonolabis distincta* – STEINMANN, Acta zool. Acad. Sci. hung., Budapest, **25**: 168; fig. 4 (posterior margin of penultimate sternite), 5 (male genitalia). Okinawa.

Description: Male colour shiny blackish to reddish-brown; palpi, labrum, apical area of clypeus, two proximal antennal joints and legs yellow. Head longer than wide, rather flat, finely punctate, slightly emarginate posteriorly. Post-frontal and coronal sutures distinct. Eyes rather small, shorter than first antennal joint. Antennae 16-jointed; first joint shorter than distance between antennal bases. Pronotum a little wider than long at posterior margin, which is a little wider than head; all margins straight, and all angles rounded; median longitudinal furrow well marked. Tegmina and wings absent. Meso- and metanotum typical. Abdomen elongate, slightly expanded to middle; tergites punctate and pubescent; sides of tergites 6–9 rugoso-striate laterally. Ultimate tergite transverse, tumid at posterior portions, slightly cristate on sides, and emarginate posteriorly. Forceps (Fig. 336) asymmetrical, trigonal basally, cylindrical apically; dorsal keel prominent. Penultimate sternite transverse, broadly emarginate posteriorly, as in fig. 337. Genitalia (Fig. 338) characteristic; central parameral plate oval, median incision of anterior margin deep; genital lobes with slightly sclerotized denticulation pads apically, virga clearly visible throughout. – **Female** similar to male, but abdomen more slender, subparallel-sided, rugose, sides of tergites punctate, rounded; ultimate tergite not tumid, not cristate, somewhat narrower; penultimate sternite rounded posteriorly; forceps symmetrical, tapering. – **Length** of body with forceps: male: 14,5–16 mm, female: 13–16 mm.

Distribution: Okinawa.

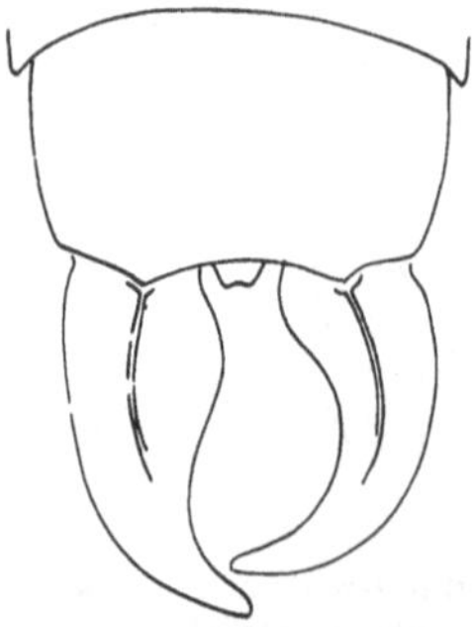

Fig. 336 Holotype ultimate tergite and forceps of *Gonolabis distincta* (NISHIKAWA). After NISHIKAWA.

Fig. 337 Posterior margin of male penultimate sternite of *Gonolabis distincta* (NISHIKAWA). After NISHIKAWA.

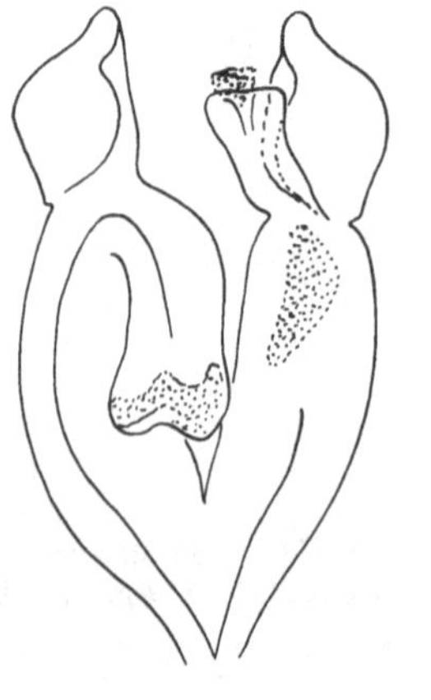

Fig. 338 Holotype genital armature of *Gonolabis distincta* (NISHIKAWA). After NISHIKAWA.

18. Gonolabis bidens STEINMANN

1981 *Gonolabis bidens* STEINMANN, Fol. ent. hung., Budapest, (34) **42**: 188; fig. 3 (genital armature of holotype). – Terra typica: New Guinea (Type male: Természettudományi Múz., Budapest).

Description: Male colour reddish-brown; antennae brown, legs yellow, ultimate tergite blackish red. Head long, longer than wide, rounded posteriorly; postfrontal sutures and coronal suture, but visible. Eyes small, shorter than first antennal joint. Antennae 15-jointed, first joint a little shorter than distance between antennal bases; second joint short, third longer than 4th. Pronotum quadrate; lateral margins a little concave, all angles rounded; median longitudinal furrow distinct. Tegmina and wings entirely absent. Meso- and metanotum normal, transverse. Abdomen not of *Gonolabis*-type, abdominal tergites punctate strongly. Ultimate tergite broad, with relatively large dorsolateral keels; posterior margin concave medially. Forceps (Fig. 339) with branches asymmetrical, very strongly trigonal basally, and broad, cylindrical apically. Penultimate sternite truncate at posterior margin. Genitalia (Fig. 340; gen. prep. No. 160, det. Dr. H. STEINMANN) very characteristic; paramere narrow, median incision of anterior margin of central parameral plate deep; the latter V-shaped; genital lobes without virgae, but armed with chitinous spirals apically. External paramere broader on basal third, and outer margin with process apically; apices of external parameres bidentate. – **Female** unknown. – **Length** of body with forceps: 12–14 mm.

Distribution: New Guinea: Papua.

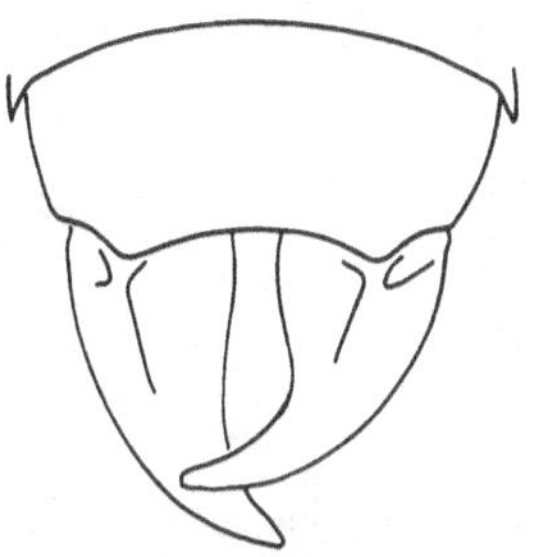

Fig. 339 Holotype ultimate tergite and forceps of *Gonolabis bidens* STEINMANN. After STEINMANN.

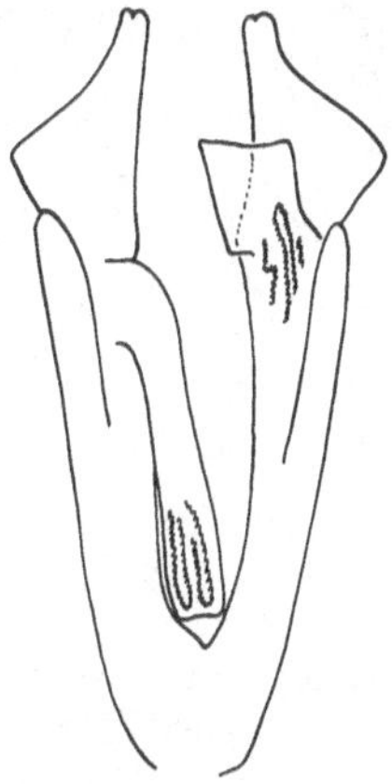

Fig. 340 Holotype genital armature of *Gonolabis bidens* STEINMANN. Original.

19. Gonolabis gilesi STEINMANN

1981 *Gonolabis gilesi* STEINMANN, Fol. ent. hung., Budapest, (34) **42**: 188; fig. 7 (genital armature of holotype). – Terra typica: Australia (Type male: Természettudományi Múz., Budapest).

Description: Male reddish-black; head black, shiny, antennae brown, legs yellow. Head longer than wide, tumid, depressed frontally, postfrontal and coronal sutures not visible. Antennae 14-jointed in holotype; first joint short, shorter than distance between antennal bases. Eyes normal, a little shorter than the length of head behind eyes. Pronotum more or less transverse; lateral margins visibly concave posteriorly; dark brownish black with light brown stripe laterally; median longitudinal furrow present. Tegmina and wings absent. Mesonotum broad, metanotum transverse, normal. Abdomen of *Gonolabis*-type, expanded to last tergite; faintly punctate dorsally, sides of tergites not acute. Ultimate tergite broad, posterior margin with two small dorso-lateral keels; posterior margin concave medially. Forceps (Fig. 341) more or less symmetrical, branches with specific blunt tooth basally; branches of forceps trigonal basally, cylindrical apically. Penultimate sternite rounded posteriorly. Genitalia (Fig. 342; gen. prep. No. 299, det. Dr. H. STEINMANN) characteristic; paramere broad, median incision of anterior margin wide and deep; plate U-shaped. Genital lobes fully developed, without virgae. External parameres broadest medially, apices a little curved. – **Female** unknown. – **Length** of body with forceps: 11,5 mm.

Distribution: Australia.

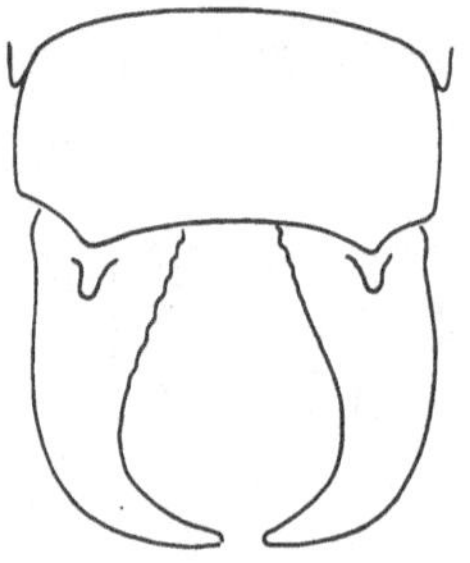

Fig. 341 Holotype ultimate tergite and forceps of *Gonolabis gilesi* STEINMANN. After STEINMANN.

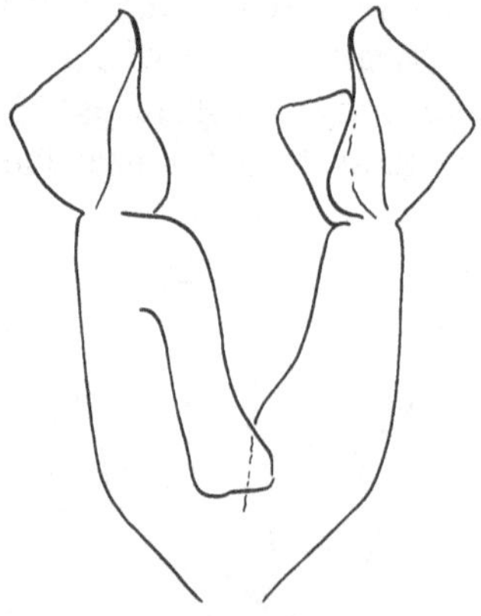

Fig. 342 Holotype genital armature of *Gonolabis gilesi* STEINMANN. Original.

20. Gonolabis dentata STEINMANN

1981 *Gonolabis dentata* STEINMANN, Fol. ent. hung., Budapest (34) **42**: 190; fig. 9 (genital armature of holotype). – Terra typica: Australia (Type male: Természettudományi Múz., Budapest).

Description: Male shiny; head black, mesonotum to forceps reddish black, legs yellowish or light brown, pronotum brown. Head rounded, depressed frontally, postfrontal sutures and coronal suture faintly distinct. Antennae 13-jointed; first joint moderately short, slightly shorter than distance between antennal bases;

second joint small, shorter than third, and third joint long, longer than fourth. Eyes relatively long, about as long as first antennal joint. Pronotum quadrate, all margins straight; lateral margins a little concave. Tegmina and wings absent. Meso- and metanotum normal, transverse. Abdomen of *Gonolabis*-type, expanded to last tergite; lateral sides of 8th–9th abdominal tergites acute. Forceps (Fig. 343) almost symmetrical; trigonal basally, and with large tooth dorso-basally; cylindrical apically. Ultimate tergite broad, posterior margin straight. Penultimate sternite broad, posterior margin truncate. Genitalia (Fig. 344; gen. prep. No. 206, det. Dr. H. STEINMANN) similar to *electa* BURR, but paramere very broad; genital lobe with characteristic sclerotized plate; median incision of anterior margin of central parameral plate very deep. External paramere broad medially; inner membrane narrow. – **Female** unknown. – **Length** of body with forceps: 12–13,5 mm.

Distribution: Australia: Darwin.

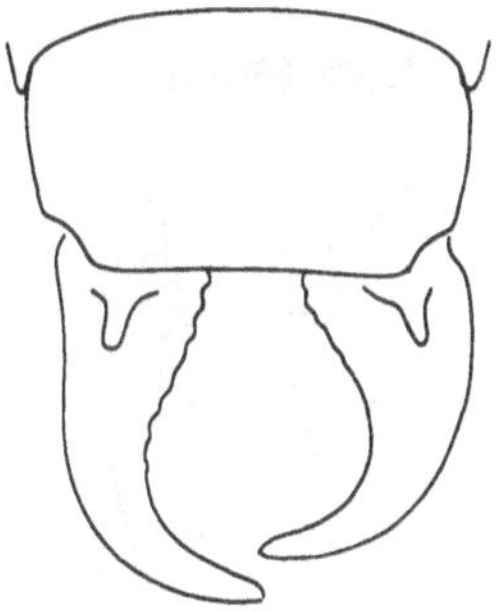

Fig. 343 Holotype ultimate tergite and forceps of *Gonolabis dentata* STEINMANN. After STEINMANN.

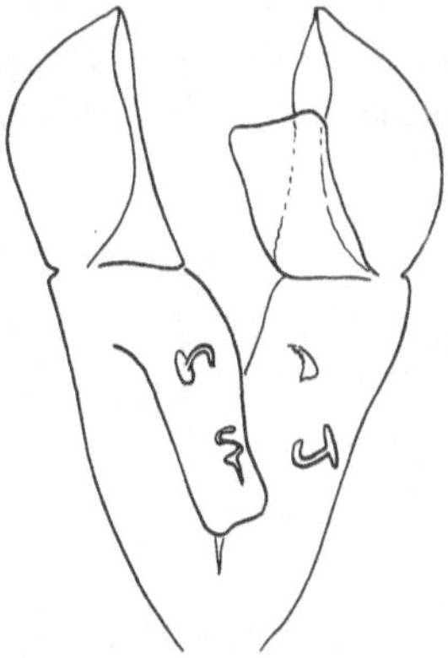

Fig. 344 Holotype genital armature of *Gonolabis dentata* STEINMANN. Original.

21. Gonolabis javana (BORMANS)

1883 *Anisolabis javana* BORMANS, Ann. Soc. ent. Belge., Bruxelles, **27**: 63; pl. 2, fig. 4 (male). – Terra typica: Java (Type male: Mus. Naturkunde, Berlin).

1900 *Gonolabis javana* – BORMANS, Ann. Mus. civ. Stor. nat., Genova, (2) **20**: 451.

1900 *Gonolabis sumatrana* BORMANS, Ann. Mus., civ. Stor. nat., Genova, **20**: 452. – Terra typica: Sumatra (Type male: Rijksmus. natuurl. Hist., Leiden, gen. prep. No. 683, det. Dr. H. STEINMANN).

1911 *Gonolabis javana* – ZACHER, Zool. Jahrb., Jena, **30**: 371; fig. H (male genitalia).

1915 *Gonolabis kükenthali* – BURR, J. r. microsc. Soc., London, **1915**: 580 [= *Gonolabis javana* BORMANS 1883)].

Literature: BOESEMAN 1954; BORELLI 1926; BURR 1911; GÜNTHER 1930; KIRBY 1904; POPHAM & BRINDLE 1966; SAKAI 1970; STEINMANN 1977, 1978, 1981.

Description: Male colour reddish brown to dark brown; antennae brown, legs yellow, femora with dark brown spot. Head tumid, broad, eyes normal; more or less as long or a little shorter than the length of head behind eyes; postfrontal and coronal sutures distinct. First antennal joint long, but a little shorter than distance between antennal bases. Pronotum more or less quadrate, lateral margins parallel; all angles rounded; median longitudinal furrow distinct in anterior half. Tegmina and wings absent. Meso- and metanotum typical of genus. Abdomen of *Gonolabis*-type; expanded to last tergite; sides of tergites 6–9 rugoso-striate, but without lateral ridges. Ultimate tergite very broad, posterior margin concave medially; median longitudinal sulcus deep and well marked. Forceps asymmetrical, sometimes symmetrical (Fig. 345); branches strongly curved; dorsal keels prominent. Penultimate sternite rounded posteriorly. Genitalia (Fig. 346) broad; central parameral plate large, median incision of anterior margin very deep; genital lobes with virgae, and with sclerotized plates medially and apically. External parameres broad at basal half; obtuse apically. – **Female** similar to male, but forceps slenderer, tapering. – **Length** of body with forceps: male: 12–14 mm, female: 12,5–15 mm.

Distribution: Java, Sumatra, Malay Archipelago, New Guinea.

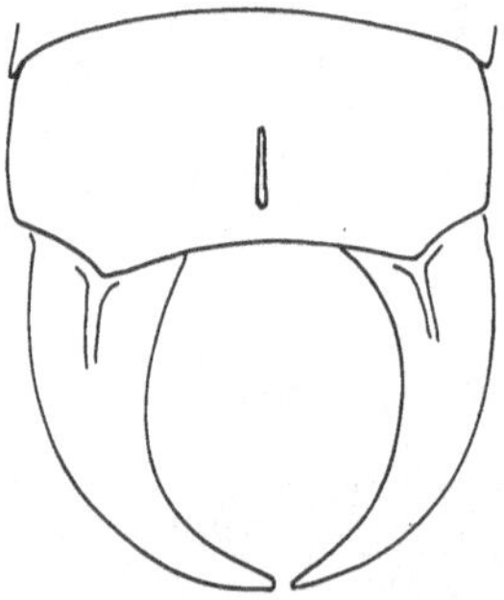

Fig. 345 Male ultimate tergite and forceps of *Gonolabis javana* (BORMANS). Original.

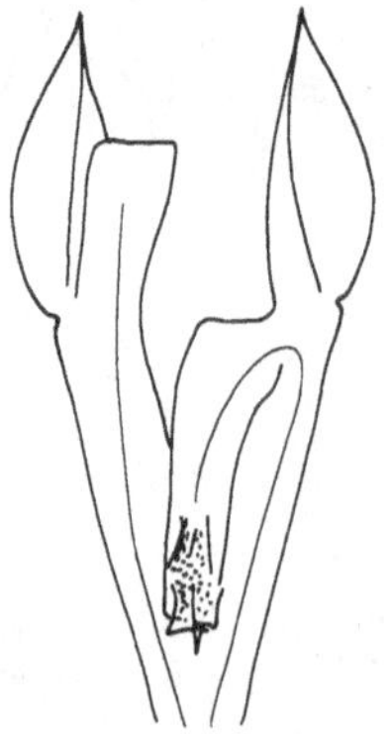

Fig. 346 Male genital armature of *Gonolabis javana* (BORMANS). After ZACHER.

22. Gonolabis forcipata BURR

1908 *Gonolabis forcipata* BURR, Fauna S.W. Australia, Dermapt.: 75. – Terra typica: Western Australia (Type male: Brit. Mus. nat. Hist., London).

1911 *Mongolabis forcipata* – ZACHER, Zool. Jahrb., Jena, **30**: 385.

1966 *Euborellia forcipata* – POPHAM & BRINDLE, Entomologist, London, **99**: 273.

1970 *Euborellia forcipata* – SAKAI, Dermapt. Cat. Prael., Tokyo, **1**: 44. West Australia; Japan: Kyoto?.

1977 *Mongolabis forcipata* – STEINMANN, Ann. hist.-nat. Mus. nation. hung., Budapest, **69**: 94.

1981 *Gonolabis forcipata* – STEINMANN, Fol. ent. hung., Budapest, (34) **42**: 192; fig. 6 (male genitalia).

Literature: BURR 1915; BORELLI 1926; SAKAI 1970; STEINMANN 1978, 1981.

Description: Male colour brownish black or reddish black; antennae brown, legs yellowish brown. Head tumid, occiput a little depressed; postfrontal and coronal sutures faintly distinct. Eyes comparatively small, shorter than the length of head behind eyes. Antennae 15-jointed; first joint long, about as long as third joint, second joint quadrate. Pronotum about as long as wide, all angles rounded; lateral margins more or less parallel; median longitudinal furrow distinct. Tegmina and wings absent. Meso- and metanotum normal, broad, transverse. Abdomen of *Gonolabis*-type, slightly expanded to last tergite; sides of tergites 6–9 rugoso-striate. Ultimate tergite very broad, a little punctate, and median longitudinal sulcus short. Forceps (Fig. 347) more or less symmetrical; branches with slightly dorsal keels basally; or a little asymmetrical. Penultimate sternite normal, simple. Genitalia (Fig. 348) narrow, paramere V-shaped, median incision of anterior margin very deep; genital lobes well developed, with virgae. External parameres broad, slightly expanded medially, apex obtuse. – **Female** similar to male, but forceps conical, contiguous, straight, and curved at apices. – **Length** of body with forceps: male: 14–16 mm, female: 14,5–17 mm.

Distribution: Western Australia, and Japan?

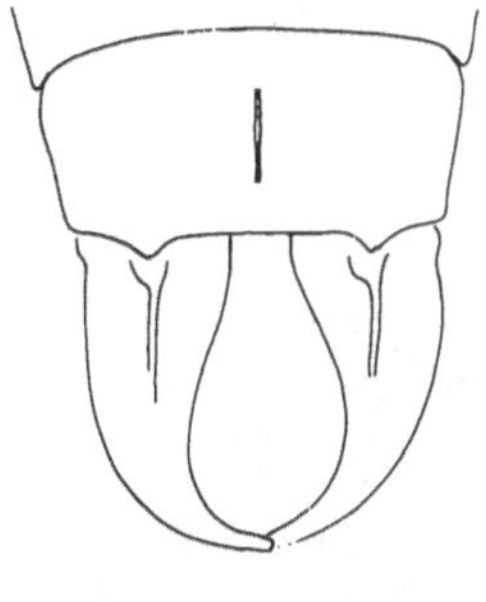

Fig. 347 Male ultimate tergite and forceps of *Gonolabis forcipata* BURR. Original.

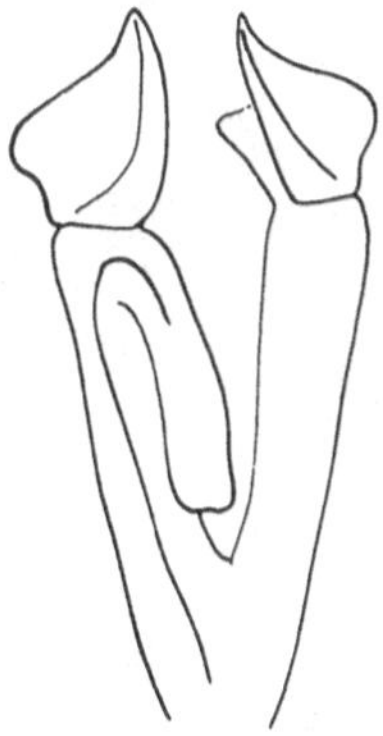

Fig. 348 Male genital armature of *Gonolabis forcipata* BURR. After STEINMANN.

23. Gonolabis tasmanica (BORMANS)

1880 *Anisolabis tasmanica* BORMANS, Bull. Soc. ent. Belge, Bruxelles, **1880**: 69; fig. 1 (male). – Terra typica: Tasmania (Type male: Mus. Naturkunde, Berlin).
1911 *Euborellia tasmanica* – BURR, Genera Insectorum, Bruxelles, **122**: 31.
1933 *Gonolabis tasmanica* – HEBARD, Mem. Queensland Mus., Brisbane, **10**: 145. Australia.
Literature: BORMANS 1900; BURR 1910, 1915; POPHAM & BRINDLE 1966; SAKAI 1970; STEINMANN 1977, 1978, 1984.

Description: Male colour dark reddish brown; antennae yellowish brown, sides of pronotum yellowish laterally. Head tumid, occiput a little depressed;

postfrontal and coronal sutures distinct. Eyes very small, slightly shorter than the length of head behind eyes. First antennal joint long, but a little shorter than distance between antennal bases; second quadrate, third longer than fourth. Pronotum transverse, lateral margins straight and slightly expanded posteriorly; median longitudinal furrow present, posterior margin truncate. Tegmina and wings absent. Mesonotum with well-marked lateral flaps; metanotum normal. Abdomen shiny, widened medially; tergites punctate dorsally; sides of tergites 6–9 without lateral ridges. Ultimate tergite transverse, slightly depressed medially, and lateral margins with asymmetrical keels. Forceps (Fig. 349) strongly asymmetrical; dorsal keel well marked, inner margins crenulate; branches strongly trigonal basally, and cylindrical apically. Penultimate sternite simple, rounded posteriorly. Genitalia (Fig. 350; gen. prep. No. 766, det. Dr. H. STEINMANN) specific; paramere elongate, large, median longitudinal incision of anterior margin very deep; genital lobes fully developed, with virgae. External parameres strongly expanded medially. – **Female** unknown to me. – **Length** of body with forceps: 17 mm.

Distribution: Tasmania, South Australia.

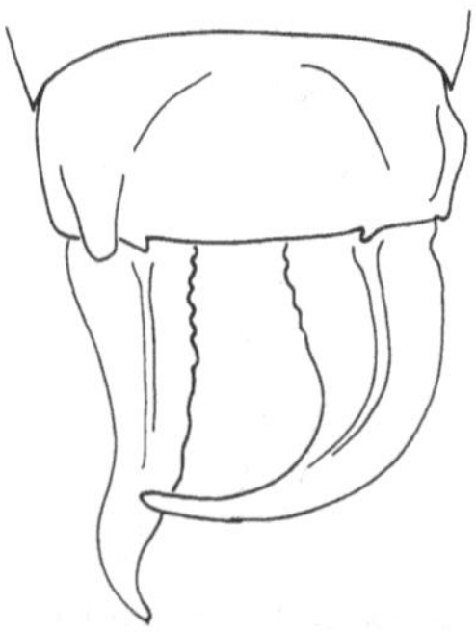

Fig. 349 Male ultimate tergite and forceps of *Gonolabis tasmanica* (BORMANS). After STEINMANN.

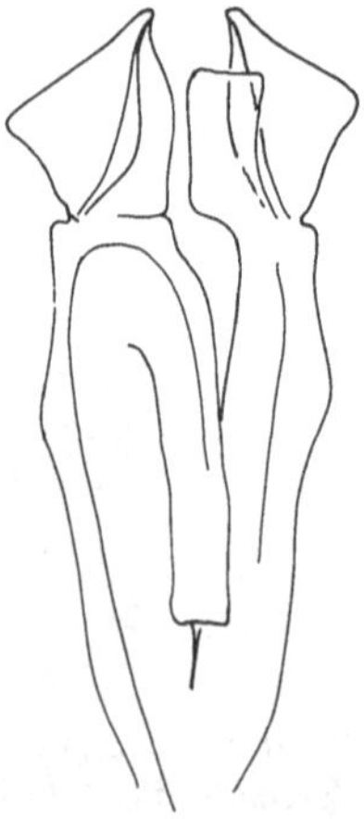

Fig. 350 Male genital armature of *Gonolabis tasmanica* (BORMANS). Original.

24. Gonolabis blanchardi (GUILLON) comb. nov.

1841 *Forficula blanchardi* GUILLON, Rev. Zool., Bot. afr., Bruxelles, **14**: 292. – Terra typica: Polynesia (Type male: unknown locality).

1876 *Anisolabis blanchardi* – SCUDDER, Proc. Boston Soc., nat. Hist., **18**: 302.

Literature: BORMANS 1900; BURR 1910, 1911, 1915; KIRBY 1904; POPHAM & BRINDLE 1966; SAKAI 1970; STEINMANN 1978, 1981.

Description: Male colour reddish yellowish-brown; abdomen and forceps dark reddish-brown. Head tumid, rounded posteriorly; postfrontal and coronal sutures not distinct. Eyes small, shorter than genae or first antennal joint. First antennal joint narrow and long, but shorter than distance between antennal bases. Pronotum visibly wider than long; all margins straight, and all angles rounded; median longitudinal furrow distinct. Tegmina and wings absent, but tegmina represented by lateral flaps. Mesonotum transverse, metanotum typical. Abdomen expanded to last tergite, tergites not acute laterally. Ultimate tergite broad, without median longitudinal sulcus, but longitudinal keels distinct laterally. Forceps (Fig. 351) trigonal basally, cylindrical apically; branches slightly asymmetrical; dorsal ridge prominent basally. Penultimate sternite broad, rounded at posterior margin. Genitalia (Fig. 352; gen. prep. No. 375, det. Dr. H. STEINMANN) specific; central parameral plate narrow, median incision of anterior margin deep; genital lobes fully developed, with virgae. External parameres broad and short, expanded at middle section. – **Female** unknown. – **Length** of body with forceps: 14 mm.

Distribution: Polynesia.

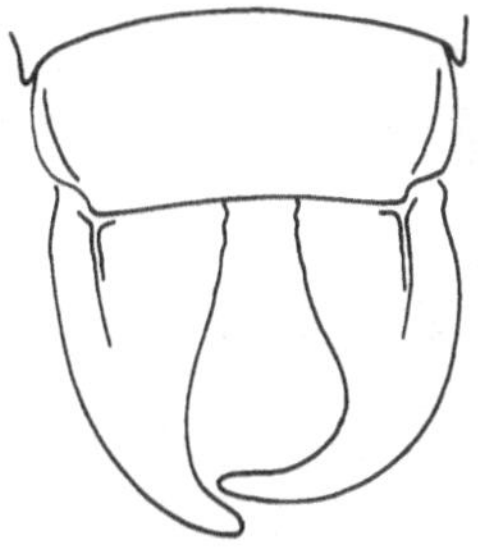

Fig. 351 Male ultimate tergite and forceps of *Gonolabis blanchardi* (GUILLON). Original.

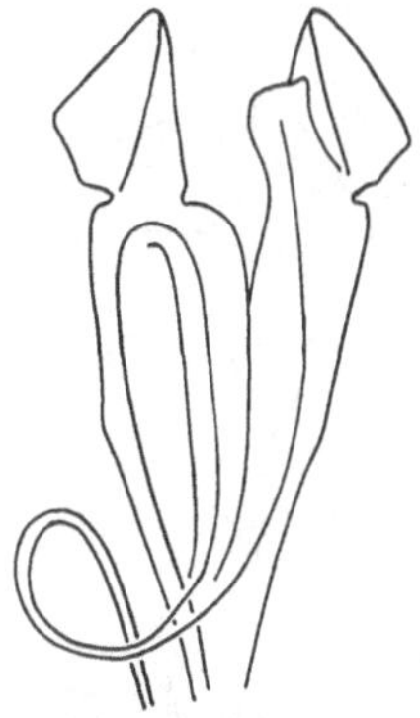

Fig. 352 Male genital armature of *Gonolabis blanchardi* (GUILLON). Original.

25. Gonolabis woodwardi BURR

1908 *Gonolabis woodwardi* BURR, Fauna S. W. Australia, Dermapt.: 75, pl. 1 (male). – Terra typica: Western Australia (Type male: Brit. Mus. nat. Hist., London).

1911 *Mongolabis woodwardi* – ZACHER, Zool. Jahrb., Jena, **30**: 370; fig. N² (male genitalia).

1913 *Anisolabis pilosiventris* MJÖBERG, Ent. Tidskr., Stockholm, **34**: 28. – Terra typica: Australia (Type male: unknown locality).

1966 *Euborellia woodwardi* – POPHAM & BRINDLE, Entomologist, London, **99**: 273.

1970 *Euborellia woodwardi* – SAKAI, Dermapt. Cat. Prael., Tokyo, **1**: 44. Australia, Japan?.

1977 *Mongolabis woodwardi* – STEINMANN, Ann. hist.-nat. Mus. nation. hung., Budapest, **69**: 94.

Literature: BURR 1910, 1911, 1915; HINCKS 1954; SAKAI 1970; SJÖSTEDT 1933; STEINMANN 1978, 1984.

Description: Male colour reddish brown; antennae brown, except antennal joints 13–15, white; pronotum brown, but lateral margins yellowish-brown; legs yellow with brown spot on femora. Head tumid, broad; postfrontal sutures and coronal suture distinct; eyes relatively small, shorter than distance between antennal bases. Antennae 15-jointed; first joint long, but shorter than frons. Pronotum a little longer than wide; lateral margins more or less parallel; posterior margin truncate; median longitudinal furrow on anterior half. Tegmina and wings absent. Mesonotum normal, metanotum typical. Abdomen of *Gonolabis*-type, slightly expanded to last tergite; tergites 6–9 rugoso-striate laterally. Ultimate tergite broad, with median depression; posterior margin concave medially. Forceps trigonal basally, curved and cylindrical apically. Penultimate sternite rounded posteriorly. Genitalia (Fig. 353) normal; paramere long, median incision of anterior margin between external parameres deep; genital lobes well developed, with virgae. External parameres relatively short, small, obtuse. – **Female** similar to male, but forceps slenderer, tapering, contiguous. – **Length** of body with forceps: male: 14–15,5 mm, female: 14,5–16 mm.

Distribution: Western Australia, and Japan?

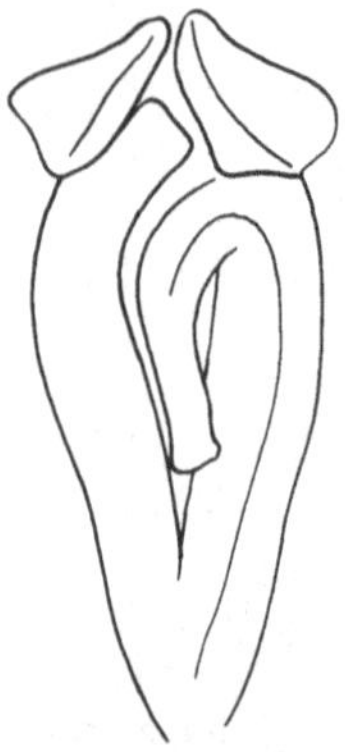

Fig. 353 Male genital armature of *Gonolabis woodwardi* BURR. After ZACHER.

26. Gonolabis sechuana (BEY-BIENKO)

1934 *Anisolabis sechuana* BEY-BIENKO, Ann. Mag. nat. Hist., London, (10) **13**: 402; fig. 1 (male genitalia), 2 (penultimate sternite of male). – Terra typica: China: Sechwan (Type male: perhaps in Peking Mus.).

1966 *Euborellia sechuana* – POPHAM & BRINDLE, Entomologist, London, **99**: 273.

1977 *Anisolabis sechuana* – STEINMANN, Ann. hist.-nat. Mus. nation. hung., Budapest, **69**: 92.

1979 *Gonolabis sechuana* – STEINMANN, Acta zool. Acad. Sci. hung., Budapest, **25**: 167, fig. 1 (male genitalia).

Literature: SAKAI 1970; STEINMANN 1978.

Description: Male general colour dark brown; shiny. Head as broad as long, a little wider than anterior margin of pronotum; rounded posteriorly. Postfrontal and coronal sutures indistinct. Eyes rather small, a little shorter than the length of head behind eyes. Antennae with short joints; first joint long, but shorter than distance between antennal bases; second joint quadrate, third distinctly shorter than eyes, subcylindrical, practically as long as fourth and fifth joints together. Pronotum a little wider than long; anterior margin truncate; median longitudinal furrow fine but distinct; lateral margins slightly expanded posteriorly. Tegmina and wings completely absent. Meso- and metanotum distinctly wider than long. Abdomen without tubercles on third and fourth segments, with very fine and dense punctation; sides of tergites 6–9 rugoso-striate, but without lateral longitudinal ridges. Ultimate tergite narrowed to apex, about twice as wide as long, with very distinct median longitudinal sulcus, posterior margin truncate. Forceps short, asymmetrical; expanded basally, curved apically; dorsal keel prominent. Penultimate sternite with a sparse and indistinct punctation. Genitalia (Fig. 354) well developed, parameres oval, median incision of anterior margin very deep and wide; genital lobe with virga; external parameres a little more than half as long as wide, acute, but not pointed on apices. – **Female** very similar to male, but forceps symmetrical, contiguous, tapering; penultimate sternite narrowed apically, broadly triangular, with rounded apex. – **Length** of body with forceps: male: 11,5–13 mm, female: 10–13,5 mm.

Distribution: China: Sechwan.

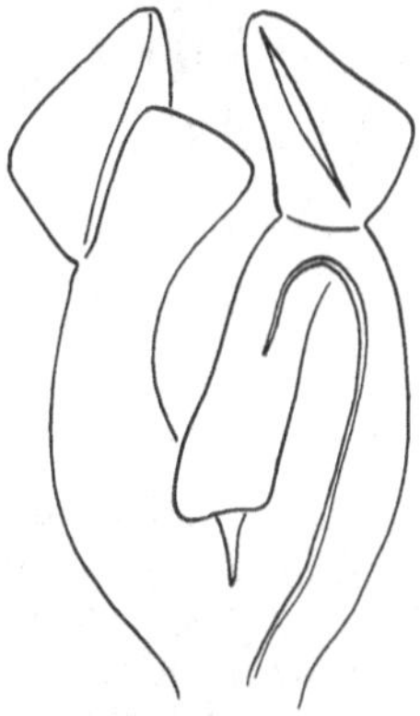

Fig. 354 Holotype genital armature of *Gonolabis sechuana* (Bey-Bienko). After Bey-Bienko.

27. Gonolabis kuekenthali Zacher

1911 *Gonolabis kuekenthali* Zacher, Zool. Jahrb., Jena, **30**: 383; fig. J² (genital armature of male). – Terra typica: Malay Archipelago (Type male: Mus. Breslau).

1915 *Gonolabis kuekenthali* – Burr, J. r. microsc. Soc., London, **1915**: 580 [= *Gonolabis javana* (Bormans 1883)].

Literature: Boeseman 1954; Burr 1914; Sakai 1970.

Description: Male colour reddish-brown to dark brownish-black; legs yellowish-brown, femora with dark brownish spot. Head tumid, broad, rounded posteriorly; postfrontal and coronal sutures distinct, well marked. Eyes comparatively small, shorter than the length of head behind eyes. Antennae with cylindrical joints; first joint long, but shorter than distance between antennal bases, second joint quadrate; joint 15 white. Pronotum more or less quadrate; all margins straight, and all angles rounded; median longitudinal furrow present in prozona. Tegmina and wings absent. Meso- and metanotum simple, typical. Abdomen of *Gonolabis*-type, slightly expanded to last tergite; sides of tergites 5–9 rugoso-striate, but without lateral ridges. Ultimate tergite broad, median longitudinal sulcus distinct; posterior margin concave in median section. Forceps asymmetrical, trigonal basally, cylindrical apically. Penultimate sternite rounded. Genitalia (Fig. 355) broad; central parameral plate large, median incision of anterior margin wide; genital lobes well developed, with virgae and sclerotized plates specific. External parameres broad medially, obtuse apically. – **Female** unknown. – **Length** of body with forceps: 20 mm.

Distribution: Malay Archipelago.

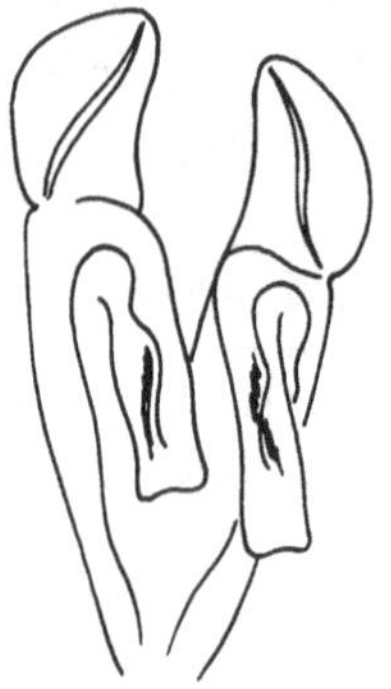

Fig. 355 Holotype genital armature of *Gonolabis kuekenthali* (ZACHER). After ZACHER.

28. Gonolabis cavaleriei (BORELLI)

1921 *Anisolabis cavaleriei* BORELLI, Bull. Mus. nation. Hist. nat., Paris, **1921**: 10. – Terra typica: China: Sechwan (Type male: Mus. nation. Hist. nat., Paris).

1927 *Anisolabis cavaleriei* – BORELLI, Bol. Lab. zool. gen. agr. Portici, **20**: 68; fig. III/1 (abdomen with forceps of male), III/2 (penultimate sternite of male), III/3 (male genitalia).

1979 *Gonolabis cavaleriei* – STEINMANN, Acta zool. Acad. sci. hung., Budapest **25**: 169; fig. 6 (posterior margin of penultimate sternite), 7 (male genitalia). China: Sechwan, Yunnan.

Literature: BORELLI 1932; SAKAI 1970; STEINMANN 1977, 1978.

Distributio: Male colour reddish brown, head blackish brown, two proximal antennal joints and legs yellow. Head a little longer than wide; postfrontal sutures and coronal suture distinct; eyes relatively small, shorter than distance between antennal bases. Antennae 16-jointed; first joint shorter than distance between antennal bases, but a little longer than eye. Pronotum quadrate, somewhat concave laterally; all angles rounded, median longitudinal furrow visible. Tegmina and wings entirely absent. Meso- and metanotum typical. Abdomen expanded to last tergite; abdominal tergites finely punctate. Ultimate tergite broad, transverse, with median longitudinal sulcus. Forceps (Fig. 356) slightly asymmetrical, trigonal basally, cylindrical and curved apically. Penultimate sternite wider than long, without emargination posteriorly. Genitalia (Fig. 357) large; paramere broad, median longitudinal incision of anterior margin wide and deep; genital lobes with virgae, similar to *distincta* (NISHIKAWA). but denticulation pads sporadic. – **Female** similar to male, but forceps with symmetrical branches, slenderer and narrower, tapering, contiguous. – **Length** of body with forceps: male: 13,5–15,5 mm, female: 13–15,5 mm.

Distribution: China: Sechwan and Yunnan.

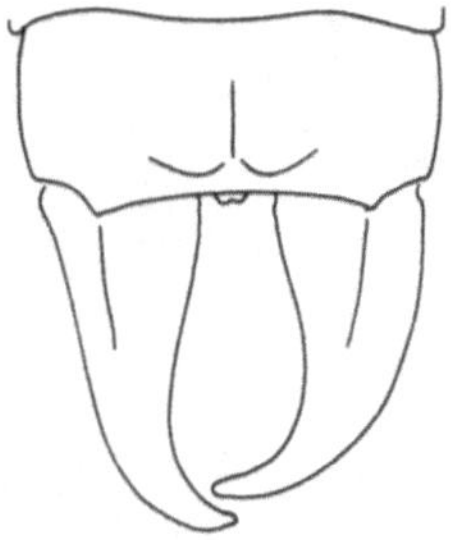

Fig. 356 Male ultimate tergite and forceps of *Gonolabis cavaleriei* (BORELI). After STEINMANN.

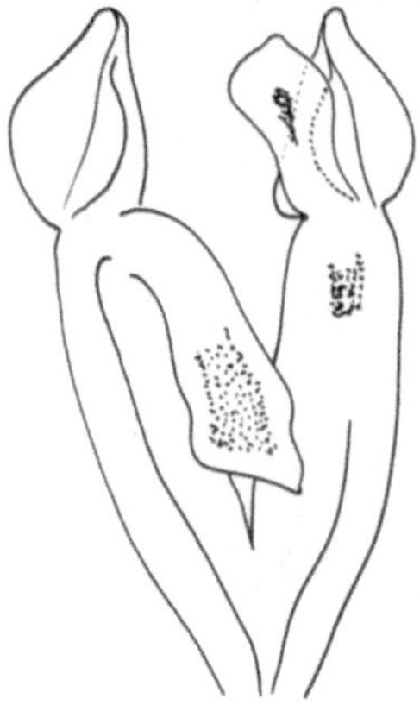

Fig. 357 Male genital armature of *Gonolabis cavaleriei* (BORELLI). After STEINMANN.

29. Gonolabis panayica STEINMANN

1981 *Gonolabis panayica* STEINMANN, Reichenbachia, Dresden, **19**: 139; fig. 5 (ultimate tergite and forceps of holotype), 6 (genital armature of holotype). – Terra typica: Panay Island (Type male: Mus. Naturkunde, Dresden).

Description: Male colour reddish dark brown; mouthparts and antennae brown, head shiny, red, pronotum and ultimate tergite shining reddish-brown, abdomen coriaceous, reddish; legs yellowish brown but femora with dark brown bands. Head tumid, postfrontal and coronal sutures indistinct. Eyes normal, shorter than the length of head behind eyes. Antennae broken in holotype; first joint relatively long, but a little shorter than distance between antennal bases; second joint short, transverse, third long, longer than fourth. Pronotum quadrate, lateral margins parallel, all angles rounded; median longitudinal furrow indistinct; posterior margin truncate. Tegmina and wings absent.

Mesonotum and metanotum transverse. Abdomen not of *Gonolabis*-type, a little expanded to median segments; sides of tergites 6–9 rugoso-striate laterally, but without lateral ridges. Ultimate tergite large, broad, typical, shiny; posterior margin slightly concave medially; median longitudinal sulcus indistinct. Forceps (Fig. 358) stout and broad, trigonal basally, and a little curved and cylindrical apically; inner margins straight, crenulate; branches asymmetrical. Penultimate sternite rounded posteriorly. Genitalia (Fig. 359; gen. prep. No. 689, det. Dr. H. STEINMANN) very characteristic; paramere well developed, V-shaped, median incision of anterior margin deep; genital lobes without virgae, somewhat similar to *formosae* (BORELLI), but narrowed. External parameres specific, oval, not expanded to middle section. – **Female** unknown. – **Length** of body with forceps: 18 mm.

Distribution: Panay Island (Arabian Sea to South India).

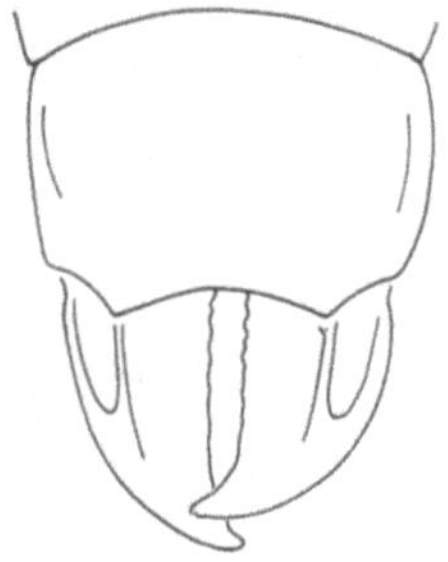

Fig. 358 Holotype ultimate tergite and forceps of *Gonolabis panayica* STEINMANN. After STEINMANN.

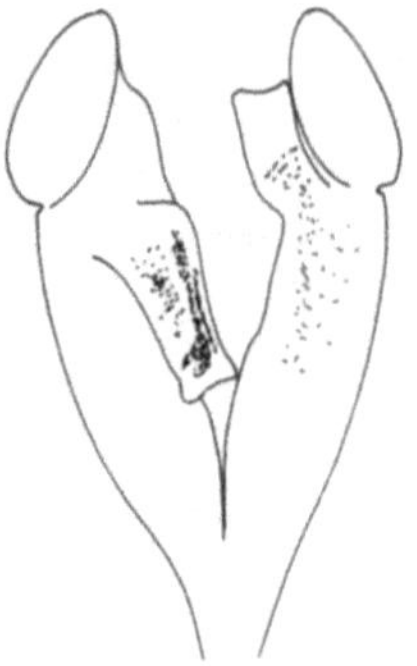

Fig. 359 Holotype genital armature of *Gonolabis panayica* STEINMANN. Original.

30. Gonolabis fallax (BEY-BIENKO)

1959 *Anisolabis fallax* BEY-BIENKO, Ent. Obozr., Moskva, **38**: 603; fig. 25 (genital armature of holotype). – Terra typica: China: Yunnan (Type male: perhaps in Peking Mus.).
1977 *Gelotolabis fallax* – STEINMANN, Ann. hist.-nat. Mus. nation. hung., Budapest, **69**: 94.
1979 *Gonolabis fallax* – STEINMANN, Acta zool. Acad. Sci. hung., Budapest, **25**: 170; fig. 8 (male genitalia).
Literature: SAKAI 1970; STEINMANN 1978.

Description: Male colour reddish blackish-brown; antennae black, except two apical joints, white or whitish. Head tumid, a little longer than wide; posterior margin rounded; postfrontal and coronal sutures distinct. Eyes comparatively small, shorter than first antennal joint. Antennae broken in holotype; first joint long, more or less as long as the distance between antennal bases. Pronotum quadrate, about as long as wide, but lateral margins convex to posterior angles; all angles rounded; median longitudinal furrow present. Tegmina and wings absent. Meso- and metanotum typical, simple. Abdomen of *Gonolabis*-type, slightly expanded to last tergite; sides of tergites 6–9 without lateral ridges, but all tergites punctate. Ultimate tergite broad with median longitudinal sulcus; posterior margin concave medially. Forceps with branches asymmetrical; trigonal basally, cylindrical and curved apically. Penultimate sternite normal, rounded posteriorly. Genitalia (Fig. 360) characteristic; paramere large, median incision of anterior margin deep; genital lobes without virgae. External parameres specific, apices excised transversally. – **Female** unknown. – **Length** of body with forceps: 22,5 mm.

Distribution: China: Yunnan.

Fig. 360 Holotype genital armature of *Gonolabis fallax* (BEY-BIENKO). After BEY-BIENKO.

31. Gonolabis formosae (BORELLI)

1927 *Anisolabis formosae* BORELLI, Boll. Lab. Zool. gen. agr. Portici, **20**: 69; fig. IV/1 (ultimate tergite with forceps of male), IV/2 (penultimate sternite), IV/3 (male genitalia), IV/4 (ultimate tergite with forceps of female). – Terra typica: Formosa (Type male: Mus. Ist. Zool. sist. Univ. Torino; holotype without penis).

1979 *Gonolabis formosae* – STEINMANN, Acta zool. Acad. Sci. hung., Budapest, **25**: 171; fig. 10 (male genitalia).

Literature: SAKAI 1970; STEINMANN 1977, 1978.

Description: Male colour brown or brownish reddish yellow, antennae, part of clypeus, legs yellowish brown or yellow. Head tumid, occiput a little depressed; vertex visibly longer than wide; postfrontal and coronal sutures distinct. Eyes relatively small, shorter than first antennal joint. Antennae 22-jointed, first joint shorter than distance between antennal bases. Pronotum rectangular, lateral margins straight, a little convex to last margin; median longitudinal furrow present in prozona. Tegmina and wings absent. Meso- and metanotum normal. Abdomen elongated, tergites punctate; ultimate tergite very broad, median longitudinal sulcus well marked, deep; posterior margin concave medially.

Forceps (Fig. 361) with asymmetrical branches, trigonal basally, curved and cylindrical apically; dorsal keels prominent on basal half. Penultimate sternite triangular, posterior margin excised posteriorly. Genitalia (Fig. 362) broad, well developed; paramere U-shaped, median longitudinal incision of anterior margin between external parameres wide and deep; genital lobes without virgae, but with sporadic denticulation pads apically. External parameres broad, outer margins expanded medially. – **Female** similar to male, but forceps symmetrical, straight, and somewhat curved at apical end. – **Length** of body with forceps: male: 26–27 mm, female: 27,5 mm.

Distribution: Taiwan.

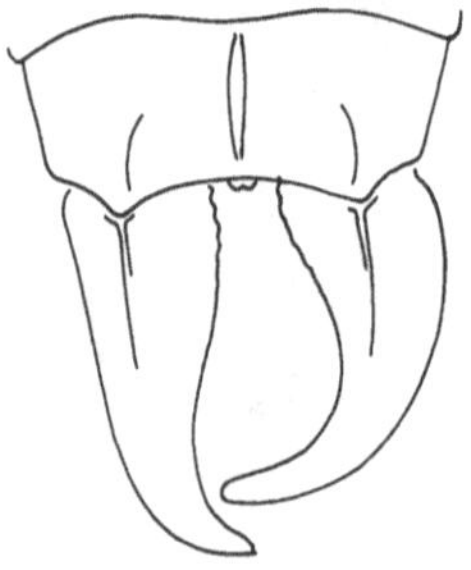

Fig. 361 Male ultimate tergite and forceps of *Gonolabis formosae* (BORELLI). Original.

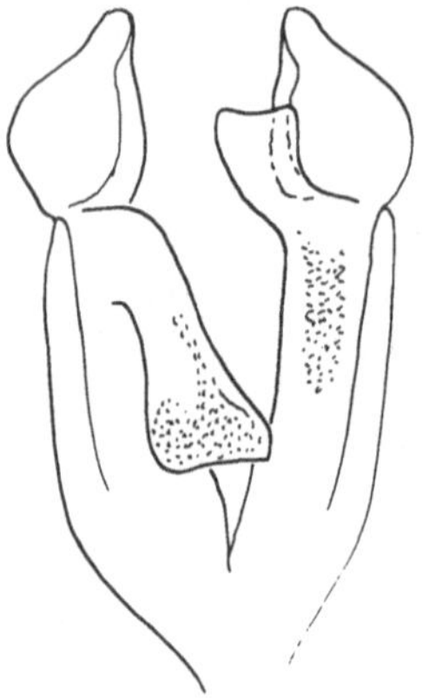

Fig. 362 Male genital armature of *Gonolabis formosae* (BORELLI). (After STEINMANN).

32. Gonolabis mystica STEINMANN

1979 *Gonolabis mystica* STEINMANN, Acta zool. Acad. Sci. hung., Budapest, **25**: 172; fig. 11 (genital armature of holotype). – Terra typica: Vietnam (Type male: Természettudományi Múz., Budapest).

Description: Male general colour dark reddish brownish black; head shiny black, antennae, thorax and abdomen brownish black, forceps reddish, legs yellowish brown. Head tumid, a little longer than wide; posterior margin rounded. Postfrontal sutures and coronal suture distinct. Eyes small, visibly shorter than distance between antennal bases. Antennae 20-jointed; first joint normal, a little shorter than distance between antennal bases; second transverse, third joint twice as long as wide; fourth shorter than third. Pronotum quadrate, as long as wide; lateral margins parallel, not convex to last margin; all angles rounded; median longitudinal furrow present. Tegmina and wings entirely absent. Meso- and metanotum normal. Abdomen elongated, a little expanded to last tergite; sides of tergites 6–9 rugoso-striate, but without lateral longitudinal ridges. Ultimate tergite broad, large; median longitudinal sulcus present, short; posterior margin concave medially. Forceps (Fig. 363) slightly asymmetrical; strongly trigonal basally, cylindrical and curved apically; dorsal keel prominent at basal half. Penultimate sternite broad, rounded posteriorly. Genitalia (Fig. 364; gen. prep. No. 580, det. Dr. H. STEINMANN) characteristic; paramere very large, V-shaped, median incision of anterior margin very deep and wide; genital lobes well developed, but without virgae; genital lobe with specific sclerotized plate and denticulated pads. External parameres oval, outer membrane somewhat chitinous. – **Female** similar to male, but colour reddish brown, forceps symmetrical, tapering, straight. – **Length** of body with forceps: male: 21,5 mm, female: 18,5–22,5 mm.

Distribution: Vietnam.

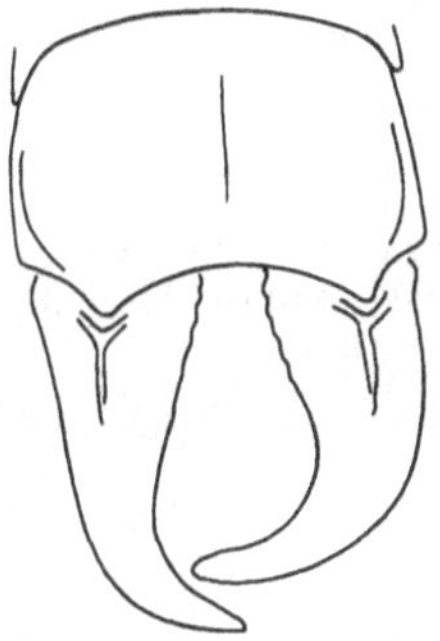

Fig. 363 Holotype ultimate tergite and forceps of *Gonolabis mystica* STEINMANN. After STEINMANN.

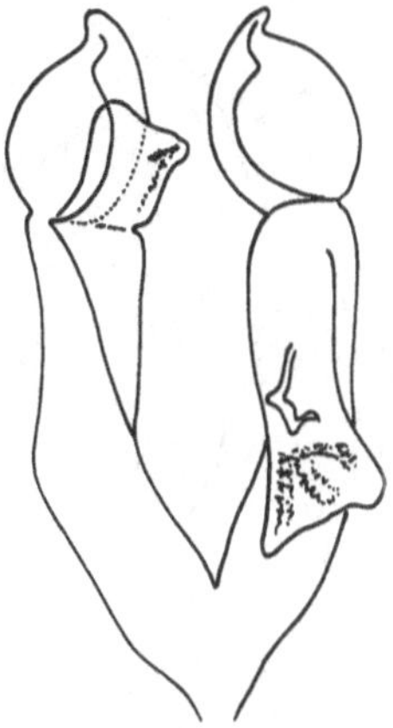

Fig. 364 Holotype genital armature of *Gonolabis mystica* STEINMANN. Original.

33. Gonolabis hwangi (BEY-BIENKO)

1959 *Anisolabis hwangi* BEY-BIENKO, Ent. Obozr., Moskva, **38**: 607; fig. 28 (genital armature of holotype). – Terra typica: China: Yunnan (Type male: perhaps in Peking Mus.).

1977 *Gelotolabis hwangi* – STEINMANN, Ann. hist.-nat. Mus. nation. hung., Budapest, **69**: 94.

1979 *Gonolabis hwangi* – STEINMANN, Acta zool. Acad. Sci. hung., Budapest, **25**: 172; fig. 12 (male genitalia).

Literature: SAKAI 1970; STEINMANN 1978.

Description: Male brownish reddish black, antennae brownish black, legs yellowish brown. Head tumid, rounded, a little longer than wide; postfrontal and coronal sutures distinct. Eyes comparatively small, shorter than scape. Antennae broken in holotype; first joint normal, but a little shorter than distance between antennal bases. Pronotum as long as wide, with distinct median longitudinal furrow; lateral margins a little convex; posterior margin rounded. Tegmina and wings absent. Mesonotum transverse, metanotum well developed, broad. Abdomen slender, expanded to last tergite; tergites punctate. Ultimate tergite broad, transverse, with median longitudinal sulcus. Forceps asymmetrical, a little expanded basally, curved and tapering apically. Penultimate sternite rounded at posterior margin. Genitalia (Fig. 365) characteristic; paramere normal, oval, median incision of anterior margin deep, but narrow; genital lobes without virgae, simple. External parameres very specific, triangular (BEY-BIENKO mentioned it as belonging to the subgenus *Cerolabis* BEY-BIENKO 1959, Ent. Obozr., Moskva, **38**: 607, of the genus *Anisolabis*). – **Female** unknown. – **Length** of body with forceps: 13 mm.

Distribution: China: Yunnan.

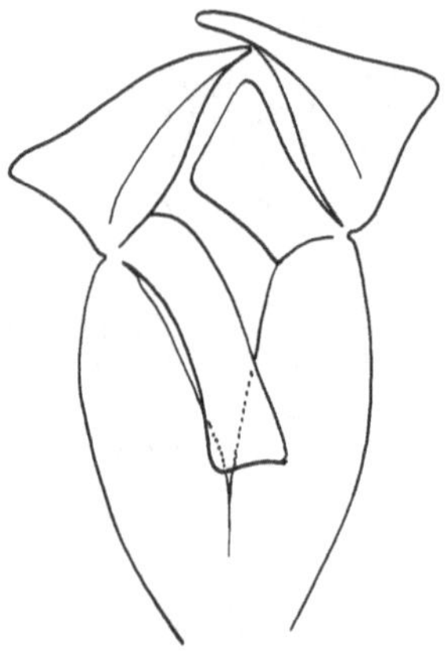

Fig. 365 Holotype genital armature of *Gonolabis hwangi* (BEY-BIENKO). After BEY-BIENKO.

34. Gonolabis insidiata STEINMANN

1979 *Gonolabis insidiata* STEINMANN, Acta zool. Acad. Sci. hung., Budapest, **25**: 173; fig. 13 (genital armature of holotype). – Terra typica: Vietnam (Type male: Természettudományi Múz., Budapest).

Description: Male large, body without forceps 24 mm. General colour reddish brown; head dark brown, scape dark brown and shiny, pedicel and flagellum yellowish brown, thorax and abdomen reddish, clypeus tarsi and tibiae yellow, femora yellow, except proximal portion, which is light brown. Head tumid, postfrontal and coronal sutures distinct. Eyes normal, a little shorter than the length of head behind eyes. Antennae 19-jointed, first joint relatively small, shorter than distance between antennal bases, but longer than eye; second joint very short, third twice as long as wide, fourth quadrate. Pronotum slightly longer than wide, lateral margins almost parallel, posterior angles rounded; median longitudinal furrow distinct. Tegmina and wings absent. Mesonotum transverse, metanotum well developed, broad. Abdomen elongate, finely punctate. Ultimate tergite broad, with very distinct median longitudinal sulcus; posterior margin concave medially. Forceps (Fig. 366) asymmetrical, of *Gonolabis*-type; trigonal basally, cylindrical and curved apically; dorsal keel very distinct at basal half. Penultimate sternite expanded, posterior margin with emargination as in *distincta* (NISHIKAWA). Genitalia (Fig. 367; gen. prep. No. 171, det. Dr. H. STEINMANN) characteristic; paramere V-shaped, with very deep median incision of anterior margin; genital lobes well developed, but shorter than length of median incision; genital lobe without virgae, but with specific sclerotized plate. External paramere expanded on basal third, the expansion quite visible on outer margins; apices obtuse. – **Female** similar to male, but general colour dark brown, head yellowish brown, vertex with a large dark brown spot, containing three yellowish speckles. Forceps symmetrical, trigonal basally. – **Length** of body with forceps: male: 28 mm, female: 19 (?)mm.

Distribution: Vietnam.

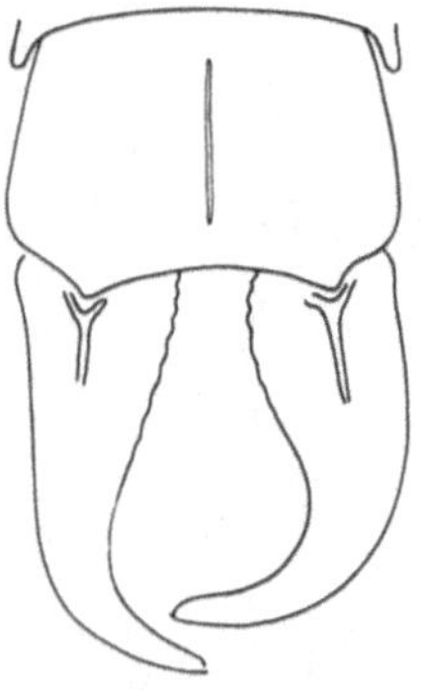

Fig. 366 Holotype ultimate tergite and forceps of *Gonolabis insidiata* STEINMANN. After STEINMANN.

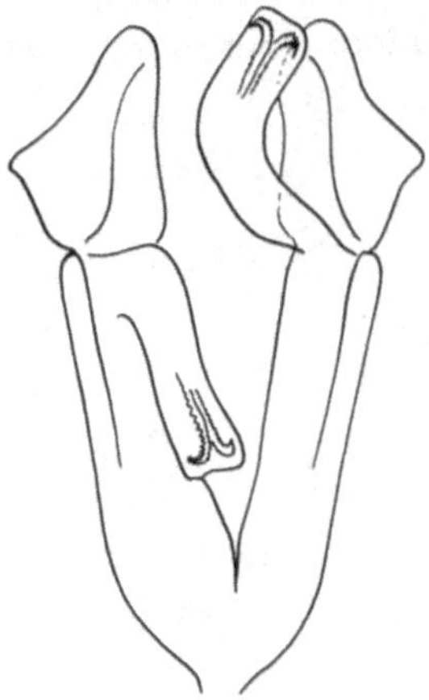

Fig. 367 Holotype genital armature of *Gonolabis insidiata* STEINMANN. Original.

35. Gonolabis insulana BRINDLE

1972 *Gonolabis insulana* BRINDLE, Insects of Micronesia, Honolulu, (2) **5**: 119; fig. 5a (male), 5b (male genitalia), 5c (penultimate sternite of male), 5d (twelfth to fourteenth antennal joints). – Terra typica: Caroline Islands (Type male: U.S. nation. Mus., Washington, No. 71555).

Literature: STEINMANN 1977, 1978.

Description: Male colour dark reddish brown; legs yellowish brown, femora darker on basal half and more; shiny. Head transverse, tumid, postfrontal sutures and coronal suture distinct; head depressed slightly on frons between the antennal bases. Eyes small, shorter than first antennal joint; latter shorter than distance between antennal bases; second joint small, quadrate, third longer than wide. Pronotum transverse, wider posteriorly, lateral margins straight, all angles rounded; median longitudinal furrow present. Tegmina and wings absent. Meso- and metanotum typical. Abdomen depressed, evenly expanded to last tergite; basal tergites impunctate, other tergites, except for last, punctate; all tergites glabrous, and lateral posterior margins of tergites evenly rounded, not acute nor rugose. Ultimate tergite transverse, depressed medially, with median longitudinal sulcus; posterior margin concave medially. Forceps (Fig. 368) more or less symmetrical, not of *Gonolabis*-type, but trigonal basally, cylindrical apically, sharply curved. Penultimate sternite transverse, without ridges, triangular, apex broad and excised medially. Genitalia (Fig. 369) similar to *electa* BURR, but external parameres shorter, and more strongly sclerotized on external margins. – **Female** similar to male, but abdomen widest medially; abdominal tergites less strongly punctate; both branches of forceps nearly straight, broader and more trigonal at base, narrower and cylindrical distally. – **Length** of body with forceps: male: 7,5–9 mm, female: 7–8,5 mm.

Distribution: Micronesia: Caroline Islands.

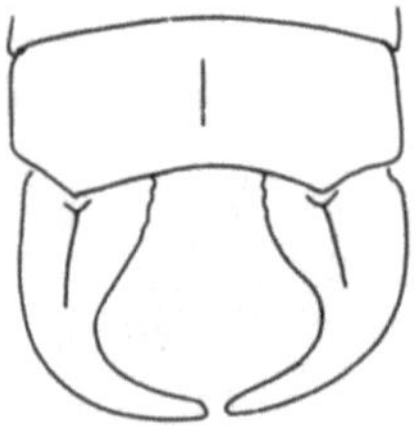

Fig. 368 Holotype ultimate tergite and forceps of *Gonolabis insulana* BRINDLE. After BRINDLE.

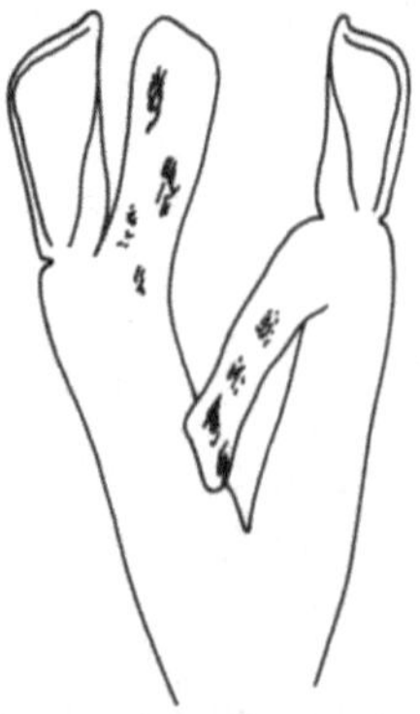

Fig. 369 Holotype genital armature of *Gonolabis insulana* BRINDLE. After BRINDLE.

36. Gonolabis acuta BOESEMAN

1954 *Gonolabis acuta* BOESEMAN, Zool. Verh., Leiden, **21**: 33; fig. 9a (male), 9b (penultimate sternite with forceps of male). – Terra typica: Java (Type male: Rijksmus. natuurl. Hist., Leiden).

1980 *Gonolabis acuta* – STEINMANN, Fol. ent. hung., Budapest, (33) **41**: 339; fig. 25 (ultimate tergite and forceps of holotype), 26 (genital armature of holotype; gen. prep. No. 682, det. Dr. H. STEINMANN).

Description: Male colour very dark, almost completely black or dark castaneous, only the distal parts of femora and tibiae, the tarsi, second, and part of fifteenth antennal joints testaceous. Head tumid, slightly longer than wide; rounded posteriorly; postfrontal sutures finely, coronal suture strongly visible. Eyes comparatively large, but a little shorter than the length of head behind eyes. Antennae broken in holotype; first joint long, more or less as long as the distance between antennal bases. Pronotum slightly longer than wide; sides slightly diverging posteriory; median longitudinal furrow distinct. Tegmina and wings absent. Mesonotum large, about half the length of pronotum, metanotum hardly wider than mesonotum. Abdomen of *Gonolabis*-type, strongly expanded to last tergite; lateral angles of abdominal tergites very acute, pointed, but with the apices slightly rounded. Ultimate tergite very broad; median longitudinal sulcus well marked; posterior margin concave medially. Forceps (Fig. 370) specific; closely resembling those in *electa* BURR, and *javana* (BORMANS); more or less symmetrical, branches characteristically medially. Penultimate sternite broad, angular, apex truncate. Genitalia (Fig. 371) of *Gonolabis*-type; paramere oval, median incision of anterior margin deep; genital lobes fully developed, and without virgae. External parameres broad medially, pointed apically. – **Female** unknown. – **Length** of body with forceps: 16 mm.

Distribution: Java.

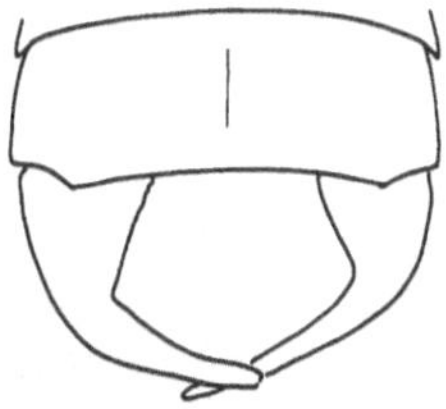

Fig. 370 Holotype ultimate tergite and forceps of *Gonolabis acuta* BOESEMAN. After BOESEMAN.

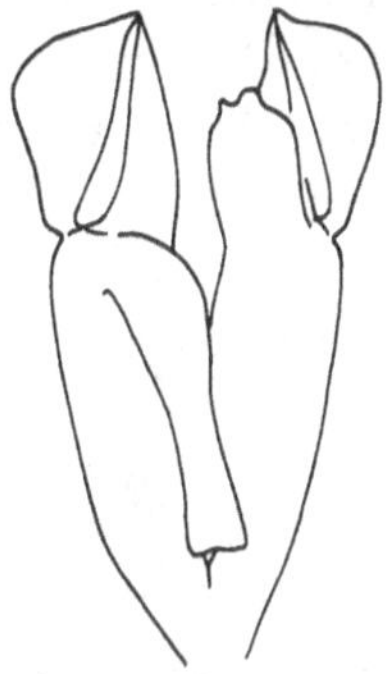

Fig. 371 Holotype genital armature of *Gonolabis acuta* BOESEMAN. After STEINMANN.

37. Gonolabis undata (BEY-BIENKO)

1959 *Anisolabis undata* BEY-BIENKO, Ent. Obozr., Moskva, **38**: 600; fig. 17 (male genitalia). – Terra typica: China: Yunnan (Type male: perhaps in Peking Mus.).

1959 *Anisolabis cheni* BEY-BIENKO, Ent. Obozr., Moskva, **38**: 602; fig. 18 (male genitalia). – Terra typica: China: Yunnan (Type male: perhaps in Peking Mus.).

1959 *Anisolabis omei* BEY-BIENKO, Ent. Obozr., Moskva, **38**: 603; fig. 24 (male genitalia). – Terra typica: China: Szechwan (Type male: perhaps in Peking Mus.).

1975 *Gelotolabis undata* – STEINMANN, Fol. ent. hung., Budapest, **28**: 154; fig. 13 (genital armature of paratype).
1975 *Gelotolabis cheni* – STEINMANN, Fol. ent. hung., Budapest, **28**: 154; fig. 14 (male genitalia).
1975 *Gelotolabis omei* – STEINMANN, Fol. ent. hung., Budapest, **28**: 156; fig. 16 (male genitalia).
1979 *Gonolabis undata* – STEINMANN, Acta zool. Acad. Sci. hung., Budapest, **25**: 170; fig. 9 (male genitalia).

Description: Male colour reddish brown, head blackish brown, antennae yellowish brown, unicolorous. Head a little longer than wide, tumid; occiput a little depressed; posterior margin rounded; postfrontal and coronal sutures fine, but distinct. Antennae 16-jointed; first joint small, slightly shorter than distance between antennal bases; about as long as length of eye; latter normal, a little shorter than the length of head behind eye. Pronotum transverse, all margins straight, and all angles rounded; lateral margins convex to posterior margin; median longitudinal furrow present. Tegmina and wings absent. Meso- and metanotum typical. Abdomen normal, abdominal tergites punctate, ultimate tergite broad, transverse, with median longitudinal sulcus; posterior margin concave medially. Forceps asymmetrical, of the *Gonolabis*-type, trigonal basally, curved and cylindrical apically. Penultimate sternite well developed, posterior margin excised transversally. Genitalia (Fig. 372) simple, oval; paramere broad, fully developed, median incision of anterior margin deep and broad; genital lobes without virgae, very long, longer than median incision; external parameres expanded medially, apices obtuse. – **Female** very similar to male, but forceps simple, contiguous, tapering, symmetrical. – **Length** of body with forceps: male: 13,5–16,5 mm, female: 13–17 mm.

Distribution: China: Yunnan and Szechwan.

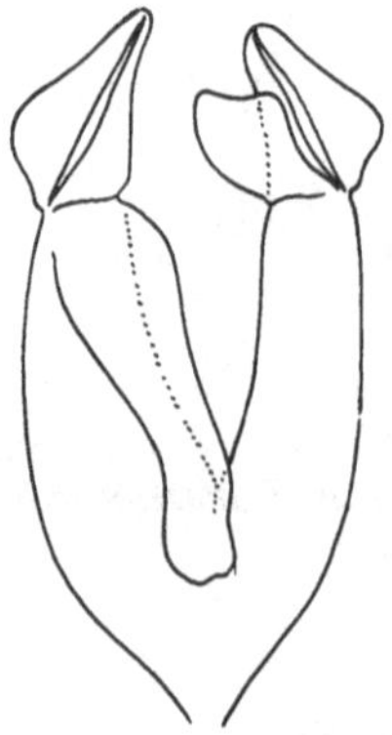

Fig. 372 Holotype genital armature of *Gonolabis undata* (BEY-BIENKO). After BEY-BIENKO.

38. Gonolabis kirbyi (Burr)

1897 *Anisolabis kirbyi* Burr, Ann. Mag. nat. Hist., London, (6) **20**: 311. – Terra typica: Java (Type male: Brit. Mus. nat. Hist., London).
1900 *Gonolabis kirbyi* – Burr, Ann. Soc. ent. belge, Bruxelles, **44**: 47.
1915 *Eulabis kirbyi* – Burr, J. r. microsc. Soc., London, **1915**: 526; pl. XI, fig. 3 (male genitalia).
1926 *Gonolabis kirbyi* – Burr, Treubia, Buitenzorg, **8**: 253.
Literature: Bormans 1900; Burr 1908, 1910, 1911; Kirby 1904; Popham & Brindle 1966; Sakai 1970; Steinmann 1977, 1978.

Description: Male colour brown, shiny; posterior part of head reddish; pronotum reddish-brown; legs yellowish brown. Head rounded; postfrontal and coronal sutures distinct; eyes normal, shorter than the length of head behind eyes. First antennal joint long, but a little shorter than distance between antennal bases. Pronotum quadrate; all margins straight, lateral margins a little concave; all angles rounded; median longitudinal furrow present in prozona. Tegmina and wings absent. Meso- and metanotum typical, broad, transverse. Abdomen elongate, somewhat expanded to last tergite; sides of tergites 6–9 rugoso-striate, but without lateral ridges. Ultimate tergite broad; median longitudinal sulcus distinct; posterior margin concave medially. Forceps asymmetrical; branches strongly trigonal basally, curved and cylindrical apically. Penultimate sternite rounded posteriorly. Genitalia (Fig. 373) specific; paramere large, median incision of anterior margin between external parameres median deep; genital lobes without virgae. External parameres lemon-shaped. – **Female** similar to male, but forceps simple, straight, contiguous. – **Length** of body with forceps: male: 15,5–16 mm, female: 15–17,5 mm.

Distribution: Java.

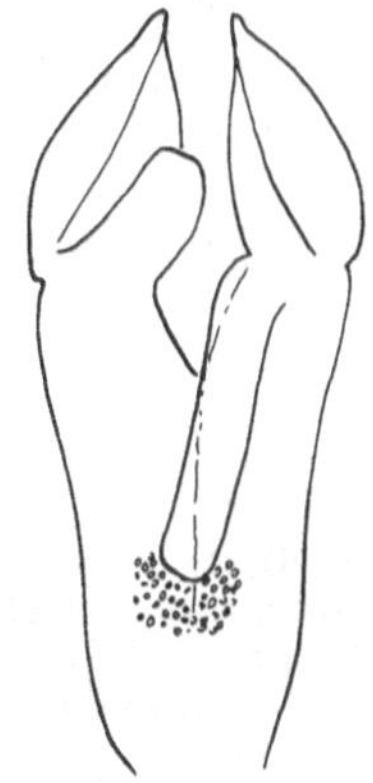

Fig. 373 Male genital armature of *Gonolabis kirbyi* (Burr). After Burr.

39. Gonolabis darevskyi Bey-Bienko

1965 *Gonolabis darevskyi* Bey-Bienko, Zool. Zshurn., Moskva, **44**: 1946; fig. 5 (genital armature of holotype), 6 (ultimate tergite with forceps of male). – Terra typica: Komodo Island (Type male: unknown locality; 4 paratypes deposited in the Leningrad Mus.).
1975 *Gonolabis darevskii* – Steinmann, Fol. ent. hung., Budapest, **28**: 154, fig. 11 (genital armature of male).
Literature: Sakai 1970; Steinmann 1977, 1978.

Description: Male dark brownish black; head brown, mouthparts and antennae lighter brown; end of abdomen and forceps reddish. Head somewhat longer than wide; posterior margin rounded; occiput a little depressed medially; postfrontal and coronal sutures present. Eyes normal, shorter than the length of head behind eyes. First antennal joint typical, shorter than distance between antennal bases. Pronotum more or less quadrate, or a little transverse; median longitudinal furrow distinct; lateral margins straight, all angles rounded. Tegmina and wings absent. Meso- and metanotum transverse. Abdomen of *Anisolabis*-type; slightly expanded to median segments; ultimate tergite broad; posterior margin concave medially. Forceps (Fig. 374) more or less symmetrical, straight, not of *Gonolabis*-type; inner margins finely crenulate; branches trigonal basally, cylindrical apically. Penultimate sternite typical. Genitalia (Fig. 375) specific; central parameral plate large, broad, robust, median incision very small, not visible; genital lobes broad, without virgae; external parameres comparatively large, broad, obtuse. – **Female** unknown. – **Length** of body with forceps: 15 mm.

Distribution: Indonesia: Komodo Island.

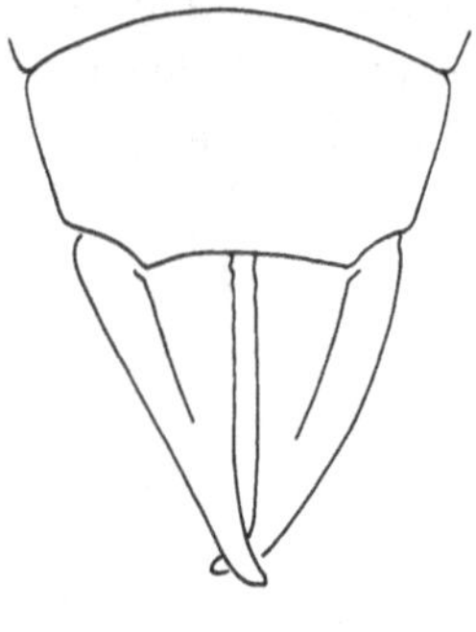

Fig. 374 Holotype ultimate tergite and forceps of *Gonolabis darevskyi* BEY-BIENKO. After BEY-BIENKO.

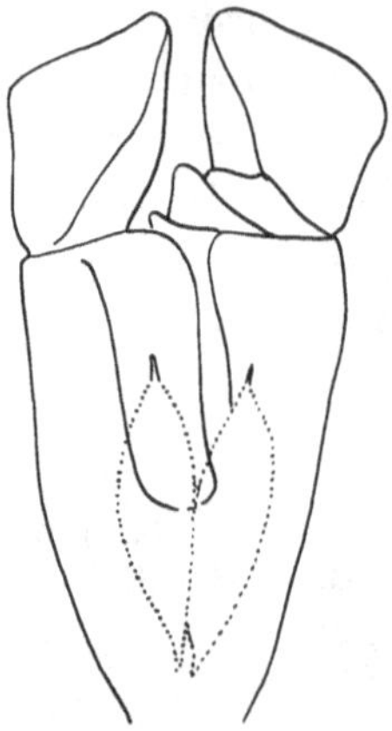

Fig. 375 Holotype genital armature of *Gonolabis darevskyi* BEY-BIENKO. After BEY-BIENKO.

40. Gonolabis magna (BEY-BIENKO)

1959 *Anisolabis (Gelotolabis) magna* BEY-BIENKO, Ent. Obozr., Moskva, **38**: 604; fig. 22 (penultimate sternite), 28 (genital armature of male). – Terra typica: China: Yunnan (Type male: perhaps in Peking Mus.).

1975 *Gelotolabis magna* – STEINMANN, Fol. ent. hung., Budapest, **28**: 156; fig. 17 (male genitalia).

1979 *Gonolabis magna* – STEINMANN, Acta zool. Acad. Sci. hung., Budapest, **25**: 174; fig. 14 (male genitalia).

Literature: SAKAI 1970; STEINMANN 1977, 1978.

Description: Male reddish brown to brownish black; clypeus yellowish brown, legs light brown. Head tumid, a little longer than wide, posterior margin rounded; postfrontal and coronal sutures distinct; eyes relatively small, shorter than the length of head behind eyes. First antennal joint long, but shorter than distance between antennal bases. Pronotum about as long as wide; relatively large, slightly widened posteriorly; median longitudinal furrow present; all angles rounded. Tegmina and wings absent. Mesonotum transverse, with well marked obtuse lateral margins. Metanotum normal. Abdomen fusiform, tergites finely punctate; sides without lateral ridges. Ultimate tergite transverse, posterior margin concave medially between bases of forceps; median longitudinal sulcus present medially. Forceps asymmetrical with branches, slightly broader at base, and narrowed distally. Penultimate sternite oval, posterior margin rounded. Genitalia (Fig. 376) strongly V-shaped; median incision between external parameres wide and deep; genital lobes without virgae; external parameres broad, obtuse; strongly widened medially. – **Female** similar to male, but forceps symmetrical, branches straight. – **Length** of body with forceps: male: 28,5 mm, female: 24 mm.

Distribution: China: Yunnan.

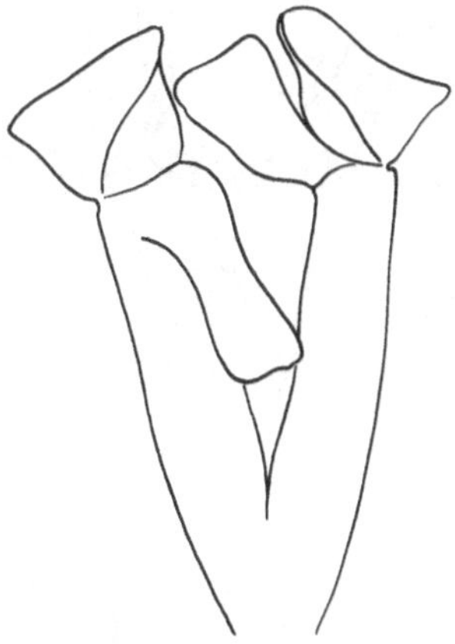

Fig. 376 Holotype genital armature of *Gonolabis magna* (BEY-BIENKO). After BEY-BIENKO.

41. Gonolabis panfilovi (BEY-BIENKO)

1959 *Anisolabis* (*Gelotolabis*) *panfilovi* BEY-BIENKO, Ent. Obozr., Moskva, **38**: 602; fig. 19 (penultimate sternite), 20 (genital armature of holotype). – Terra typica: China: Yunnan (Type male: perhaps in Peking Museum).

1975 *Gelotolabis panfilovi* – STEINMANN, Fol. ent. hung., Budapest, **28**: 156; fig. 15 (male genitalia).

1979 *Gonolabis panfilovi* – STEINMANN, Acta zool. Acad. Sci. hung., Budapest, **25**: 175; fig. 15 (male genitalia).

Literature: SAKAI 1970; STEINMANN 1977, 1978.

Description: Male colour reddish brown, much less deeply punctate and duller than rugose; shiny. Head tumid, a little transverse, finely punctate; posterior margin rounded; eyes relatively small, shorter than the length of head behind eyes. Postfrontal and coronal sutures distinct. Antennae brown, first joint long, but shorter than length of head behind eyes. Pronotum almost parallel-sided, a little longer than wide; posterior margin convex; median longitudinal furrow present. Tegmina and wings absent. Meso- and metanotum normal, typical. Abdomen elongate, tergites punctate. Ultimate tergite broad, transverse, with a visible median longitudinal sulcus. Forceps of *Gonolabis*-type, asymmetrical, widened basally, cylindrical apically. Penultimate sternite simple. Genitalia (Fig. 377) characteristic; paramere broad, median incision of anterior margin very deep and wide; genital lobes without virgae, but fully developed, as in fig. 377. External parameres broad, the outer margins specific. – **Female** similar to male, but forceps symmetrical, tapering. – **Length** of body with forceps: male: 18–21 mm, female: 19–21 mm.

Distribution: China: Yunnan.

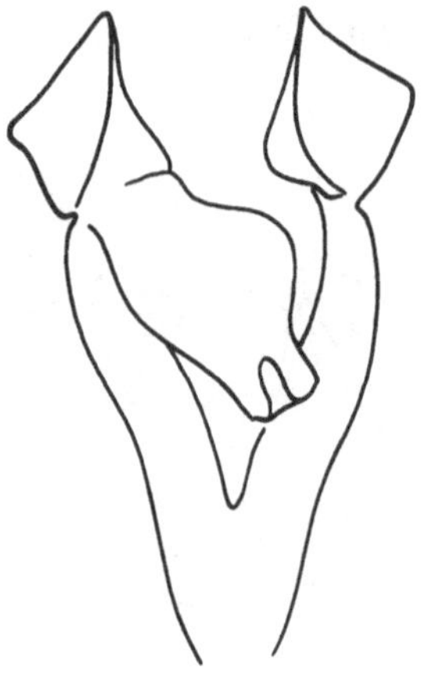

Fig. 377 Holotype genital armature of *Gonolabis panfilovi* (BEY-BIENKO). After BEY-BIENKO.

Doubtful position

Gonolabis minor BORELLI

1926 *Gonolabis minor* BORELLI, Treubia, Buitenzorg, **8**: 253. – Terra typica: Malabar (Type male: perhaps in Buitenzorg Mus.; paratype of male without penis in Mus. Ist. Zool. sist., Torino).
1954 *Gonolabis minor* – BOESEMAN, Zool. Verh., Leiden, **21**: 32 [= *Gonolabis electa* BURR 1910].
1972 *Gonolabis minor* – BRINDLE, Insects of Micronesia, Honolulu, (2) **5**: 212. Sunda Islands.
Literature: SAKAI 1970; STEINMANN 1977, 1978.

Description: Male dark brown, very small. Head a little longer than broad. Postfrontal and coronal sutures fainly visible. Eyes normal, shorter than the length of head behind eyes; first antennal joint long, but shorter than distance between antennal bases; second quadrate. Pronotum transverse.; all margins straight, and all angles rounded; median longitudinal furrow present. Tegmina and wings absent. Meso- and metanotum normal, transverse. Abdomen visibly expanded to last tergite; abdominal tergites punctate, sides of tergites without lateral ridges. Ultimate tergite very broad; dorsal surface with lateral ridges; posterior margin concave in median section. Forceps (Fig. 378) specific, very similar to *acuta* BOESEMAN, strongly curved medially; forceps more or less symmetrical; trigonal basally, cylindrical apically. Penultimate tergite normal, simple, rounded posteriorly. Genitalia unknown. – **Female** similar to male, but abdomen widest medially, and forceps slenderer, straight. – **Length** of body with forceps: male: 8,5–9,5 mm, female: 9,5–10,5 mm.

Distribution: Malabar, Java, Sunda Islands.

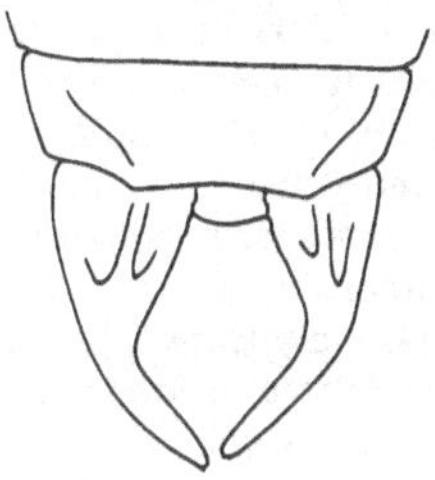

Fig. 378 Holotype ultimate tergite and forceps of *Gonolabis minor* BORELLI. Original.

Genus 22: **Euborellia** BURR

1909 *Borellia* BURR, Dtsch. ent. Zeitschr., Berlin, **1909**: 325. – Species typica: *Anisolabis moesta* GÉNÉ 1829.
1910 *Euborellia* BURR, Proc. U.S. nation. Mus., Washington, **38**: 448. Replacement name for *Borellia* BURR 1909, preoccupied by *Borellia* REHN 1906 (Orthoptera).
1966 *Mongolabis* – POPHAM & BRINDLE, Entomologist, London, **99**: 272 [= *Euborellia* BURR 1910].
1966 *Heterolabis* – POPHAM & BRINDLE, Entomologist, London, **99**: 272 [= *Euborellia* BURR 1910].
1966 *Gelotolabis* – POPHAM & BRINDLE, Entomologist, London, **99**: 272 [= *Euborellia* BURR 1910].
1966 *Horidolabis* – POPHAM & BRINDLE, Entomologist, London, **99**: 272 [= *Euborellia* BURR 1910].
Literature: BRINDLE 1970, 1978; BURR 1911, 1915; CHOPARD 1922; KAPOOR 1968; SAKAI 1970; SHIRAKI 1928; SRIVASTAVA 1976, 1977; STEINMANN 1975, 1977, 1978; TOWNES 1945; ZACHER 1911.

Description: Very similar to *Anisolabis* but smaller in size usually, and more slender. Head rounded, antennae usually more slender, moniliform and multisegmented. Eyes small to fairly large. Pronotum usually relatively narrow and often as wide as long, although sometimes longer. Tegmina well developed, long, or short; or tegmen reduced to lateral flaps on the mesonotum, sometimes absent. Wings fully developed, or shorter, concealed, sometimes absent. Abdomen elongated, a little expanded to median segments. Penultimate sternite of male rounded posteriorly, or truncate, sometimes concave. Male forceps always short, weakly curved, never strongly asymmetrical, and forceps of females more or less contiguous. Male genitalia with short external parameres, and genital lobes with or without sclerotized plates and denticulated: there are often lateral denticulated pads.

Distribution: Circumtropical throughout the World.

44 species.

Identification key to the species

1 Distributed in the Ethiopian region (after BRINDLE) **2**
1′ Neotropical, Malgassian, Indo-Australian, Palaearctic or Oriental species **19**
2 Tegmina and wings fully developed **3**
2′ Tegmina short, reduced, or absent; wings rudimentary, concealed, or entirely absent **4**
3 Colour of body blackish; both tegmina with a yellow or whitish-yellow anterior spot; wings mainly yellow or whitish-yellow, blackish apically and laterally and sometimes with a dark median spot; legs yellow, femora and tibiae strongly banded with black. Pronotum very transverse. Male forceps strongly expanded basally (Fig. 379). External parameres of male genitalia almost quadrangular; genital lobes with virgae, but without conspicuous dark denticulated areas (Fig. 380) .... 1. **Euborellia vanderbilti**, p. 237
3′ Colour of body dark brown of yellowish-brown to blackish; tegmina often paler in centre or variegated, sometimes unicolorous; wings coloured as tegmina; femora and tibiae not strongly banded with black. Pronotum longer than wide. External parameres of male genitalia as in fig. 383, rounded medially, and broad .... 2. **Euborellia cincticollis** forma *picina*, p. 238
4 Tegmina present, even though rudimentary **5**
4′ Tegmina completely absent **13**
5 Pronotum strongly transverse **6**
5′ Pronotum quadrate or longer than wide **8**
6 Pronotum strongly widened posteriorly; tegmina reduced to small lateral flaps. Abdominal tergites 7–9 of male rugoso-striate at sides, and each with a lateral longitudinal ridge. Femora vaguely darkened, not conspicuously banded with black. Male forceps strongly asymmetrical, as in fig. 384. Male genitalia (Fig. 385) V-shaped .... 3. **Euborellia moesta**, p. 239
6′ Pronotum scarcely widened posteriorly; at least femora mainly blackish or banded with black **7**
7 Tibiae black on basal half; pronotum yellow at lateral sides. Penultimate sternite of male with posterior margin weakly concave. Male forceps comparatively long (Fig. 386). External parameres of male genitalia wider, as in fig. 387 .... 4. **Euborellia flavohumeralis**, p. 240
7′ Tibiae entirely yellow; pronotum and tegmina completely black. Penultimate sternite of male with posterior margin more strongly concave. Male forceps short (Fig. 388). External parameres of male genitalia as in fig. 389 .... 5. **Euborellia ituriensis**, p. 241
8 Tegmina reduced to small lateral flaps. Pronotum long; sides of abdominal tergites 7–9 of male rugoso-striate, but without lateral longitudinal ridges. Penultimate sternite of male with posterior margin slightly truncate. Shiny black species. External parameres of male genitalia as in fig. 392. Cosmopolitan species .... 6. **Euborellia stali**, p. 242
8′ Tegmina larger **9**
9 Femora yellow, conspicuously banded with black. Tegmina black with a purplish sheen; tegmina shorter than pronotum but meeting along sutures. Male forceps as in fig. 393. External parameres of male genitalia similar to *compressa* (BORELLI), as in fig. 394........ .... 7. **Euborellia purpurea**, p. 244
9′ Femora yellowish, sometimes vaguely darkened, but not conspicuously banded with black. Tegmina without a purplish sheen **10**
10 Sides of abdominal tergites of male punctate, not rugose-striate, nor with lateral longitudinal ridges. Tegmina covering most of mesonotum. Penultimate sternite of male rounded posteriorly. Forceps of male asymmetrical and short (Fig. 395). External parameres of male genitalia as in fig. 396 .... 8. **Euborellia fulviceps**, p. 244
10′ Sides of abdominal tergites of male rugoso-striate; sometimes weakly so, and usually with lateral longitudinal ridges **11**
11 Sides of abdominal tergites 6–9 of male rugoso-striate, tergites 6–8 with blunt lateral longitudinal ridges. Penultimate sternite of male rounded apically. External parameres of male genitalia as in fig. 398 .... 9. **Euborellia feae**, p. 245
11′ Sides of abdominal tergites 7–9 rugoso-striate, tergite 7 sometimes weakly so, and tergites 7–8 usually with lateral longitudinal ridges **12**

12 Eyes smaller; tegmina smaller. Abdominal tergites 7–9 of male rugoso-striate at sides and tergites 7–8 with blunt lateral longitudinal ridges. Penultimate sternite of male rounded posteriorly. Ultimate tergite without lateral longitudinal keels, as in fig. 399............. .... 10. **Euborellia andreinii**, p. 246

12′ Eyes larger, and tegmina larger. Abdominal tergites 7–9 of male weakly or strongly rugoso-striate at sides and tergites 7–8 usually with lateral longitudinal ridges. Penultimate sternite of male with posterior margin truncate. Ultimate tergite with lateral longitudinal keels, as in fig. 381 .... 2. **Euborellia cincticollis**, p. 238

13 Posterior abdominal tergites punctate at sides, not rugoso-striate, and without lateral longitudinal ridges. Pronotum slightly transverse **14**

13′ Posterior abdominal tergites rugoso-striate at sides; larger species, body length 8 mm or more **15**

14 Postfrontal and coronal sutures at vertex indistinct. Pronotum slightly widened posteriorly. Tegmina completely absent. Male forceps elongated, as in fig. 401..................... .... 11. **Euborellia dimidiata**, p. 247

14′ Postfrontal and coronal sutures at vertex distinct. Pronotum more or less parallel-sided laterally. Tegmina represented by lateral flaps. Female forceps as in fig. 403............. .... 12. **Euborellia flava**, p. 248

15 Posterior abdominal tergites 7–9 of male rugoso-striate but without well-defined lateral longitudinal ridges, or there may be incipient ridges on tergites 7–8. Penultimate sternite of male rounded posteriorly. General colouration yellowish-brown. External paramere of male genitalia as in fig. 407 .... 13. **Euborellia compressa**, p. 249

15′ Posterior abdominal tergites of male with lateral longitudinal ridges. General colouration dark reddish-brown to black **16**

16 Legs uniformly reddish-yellow, not marked with black or darkened; lateral longitudinal ridges on sides of abdominal tergites 6–8 in male. External paramere of male genitalia as in fig. 409 .... 14. **Euborellia ugandana**, p. 250

16′ Legs yellow or yellowish, at least femora darkened medially or with a blackish ring 17

17 Eyes fairly large, ratio of length of eye to length of head behind eye about 14 : 16, legs yellow, femora usually with a conspicuous dark or blackish ring. Sides of abdominal tergites strongly rugoso-striate, and with lateral longitudinal ridges on tergites 7–9. Apex of genital lobe of male very specific, as in fig. 411. Cosmopolitan species .... 15. **Euborellia annulipes**, p. 251

17′ Eyes small, ratio of length of eye to length of head behind eye about 10 : 20 .... **18**

18 Larger species, body length 12–16 mm, and generally darker in colour. Pronotum less widened posteriorly and abdomen less irregularly punctate. External paramere of male genitalia as in fig. 413 .... 16. **Euborellia tellinii**, p. 253

18′ Smaller species, body length 8–10 mm, and generally paler in colour. Pronotum more strongly widened posteriorly and abdomen more irregularly punctate. External paramere of male genitalia as in fig. 415 .... 17. **Euborellia truncata**, p. 254

19 Neotropical species **20**

19′ Not Neotropical species **37**

20 Tegmina and wings absent **21**

20′ At least tegmina present, even through rudimentary **25**

21 Pronotum longer than wide. Abdominal tergites 7–9 rugose and with lateral longitudinal ridges. Male genitalia typical, external parameres as in fig. 417; genital lobes of *Euborellia*-type, with denticulated pads apically .... 18. **Euborellia peregrina**, p. 254

21′ Pronotum quadrate, more or less as long as wide, or transverse **22**

22 Femora usually with a conspicuous dark or blackish ring. Sides of abdominal tergites strongly rugoso-striate, and with lateral longitudinal ridges on tergites 7–9. Male genitalia (Fig. 411) strongly V-shaped, genital lobes with specific denticulated pads apically .... 15. **Euborellia annulipes**, p. 251

22′ Femora without blackish ring, yellow **23**

23 Abdominal tergites strongly punctate; basal tergites more weakly punctate. Male genitalia visibly Y-shaped (Fig. 419); median incision deep, but widened apically. External parameres short, apically obtuse .... 19. **Euborellia antoni**, p. 256

23′ Abdominal tergites mainly weakly punctate, basal tergites impunctate; male tergites with lateral longitudinal ridges on tergites 7–9 or 8–9 .... **24**

24 Pronotum as broad as long. Male with lateral longitudinal ridges on abdominal tergites 8–9. Male genitalia (Fig. 421) very similar to *annulipes* (LUCAS), genital lobes with specific denticulated pads apically. External parameres not lemon-shaped, a little longer than wide. Male forceps strongly trigonal basally, asymmetrical (Fig. 420) .... 20. **Euborellia uruguayensis**, p. 257

24′ Pronotum slightly transverse. Male with lateral longitudinal ridges on tergites 7–9. Ultimate tergite broad, median longitudinal sulcus present or absent; forceps asymmetrical, strongly curved, as in fig. 422. Male genitalia (Fig. 423) oval, median incision very broad, external parameres lemon-shaped .... 21. **Euborellia mexicana**, p. 258

25 Tegmina rudimentary, represented by lateral flaps, well separated from each other **26**

25′ Tegmina at least touching at the mid-line **31**

26 Pronotum strongly transverse; legs yellow, femora darkened; head and thoracic nota dark reddish, abdomen blackish-brown, tergites 5–9 of male with lateral longitudinal ridges. Forceps with large prominent tooth on inner margin basally (Figs. 424, and 425). Genitalia unknown .... 22. **Euborellia armata**, p. 259

26′ Pronotum quadrata or longer than wide **27**

27 Forceps with large tooth on inner margin basally, as in fig. 426. Male genitalia (Fig. 427) with very characteristic genital lobes; latter with chitinous arms apically .... 23. **Euborellia plebeja**, p. 259

27′ Forceps without large tooth on inner margin basally **28**

28 Pronotum about quadrate or almost so. Legs yellow, femora darkened. Ultimate tergite very broad, forceps (Fig. 428) strongly asymmetrical, trigonal basally, curved and cylindrical apically .... 24. **Euborellia maxima**, p. 261

28′ Pronotum longer than wide. Larger or smaller species **29**

29 Larger species, body length with forceps about 18 mm. Male forceps (Fig. 429) more or less symmetrical, straight, not of *Euborellia*-type. Male unknown .... 25. **Euborellia nitida**, p. 261

29′ Smaller species, body length with forceps 12 mm or less **30**

30 Pronotum slightly parallel-sided laterally, posterior margin of pronotum strongly convex; legs yellow, femora not ringed. Male forceps (Fig. 430) with strongly curved asymmetrical branches. Male genitalia unknown .... 26. **Euborellia flavipes**, p. 262

30′ Pronotum widened posteriorly, posterior margin of pronotum not strongly convex; legs yellow, femora with dark rings. Male forceps (Fig. 390), typical. Male genitalia (Fig. 391) very broad, genital lobes V-shaped, median incision of anterior margin very deep; genital lobes with characteristic denticulated pads, as in fig. 392. Cosmopolitan species................ .... 6. **Euborellia stali**, p. 242

31 Tegmina short, wings absent or concealed **32**

31′ Tegmina and wings normally developed **35**

32 Abdominal tergites 6–8 of male each with well-definied lateral longitudinal ridges. Forceps normal, of *Euborellia*-type, trigonal basally, curved and cylindrical apically. Male genitalia characteristic; genital lobes without prominent denticulated pads apically; external parameres (Fig. 431) with a large membranous lobe .... 27. **Euborellia boliviana**, p. 263

32′ Abdominal tergites 6–9 or 7–9 of male with lateral longitudinal ridges **33**

33 7th to 9th abdominal tergites of male with lateral longitudinal ridges. Male forceps (Fig. 432) more distinctly bowed, strongly curved. Female forceps longer and more slender......... .... 28. **Euborellia ambigua**, p. 264

33′ 6th to 9th abdominal tergites of male with lateral longitudinal ridges. Male forceps weakly bowed; female forceps shorter and heavier **34**

34 Antennae unicolorous. Femora with external faces broadly washed with brown, often weakly; internal faces similarly washed with brown, but weaker and less extensive. Male forceps relatively shorter, dorsal keel prominent but small basally (Fig. 433), dorsal surface without large blunt tooth. Male genitalia as in fig. 434 .... 29. **Euborellia caraibea**, p. 264

34′ Antennae brown or darker, with one or more distal joints whitish or yellow. Femora with median portion of external faces suffused with brown, corresponding portion of internal faces usually with a suffused patch of this color. Male forceps (Fig. 435) with a large blunt tooth on dorsal surface basally. Male genitalia of *Euborellia*-type, genital lobe with denticulated pads (Fig. 436) .... 30. **Euborellia janeirensis**, p. 265

35 Wings only with tips projecting. Female ultimate tergite with specific rounded excision at posterior margin medially, as in fig. 437. Male genitalia unknown 31. **Euborellia tatei**, p. 266
35′ Wings strongly projecting **36**
36 Outer margin of the external parameres of male genitalia sinuate, external parameres broadest medially (Fig. 382). Male ultimate tergite broad, lateral longitudinal keel strongly, developed, as in fig. 381 .... 2. **Euborellia cincticollis**, p. 238
36′ Outer margin of the external parameres of male genitalia not sinuate, but oval. Genital lobe of *Euborellia*-type, with denticulated pads apically (Fig. 439) .... 32. **Euborellia femoralis**, p. 267
37 Malgassian species **38**
37′ Not Malgassian species **43**
38 Tegmina and wings completely absent. Sides of tergites strongly rugoso-striate, and with lateral longitudinal ridges on tergites 7–9. Male genitalia (Fig. 411) V-shaped, genital lobes with specific denticulated pads apically. Cosmopolitan species .... 15. **Euborellia annulipes**, p. 251
38′ Tegmina and wings present or only tegmina rudimentary **39**
39 Tegmina rudimentary, represented by lateral flaps at mesonotum **40**
39′ Tegmina and wings present **41**
40 Forceps with large tooth on dorsal surface at inner margin basally, as in fig. 426. Male genitalia (Fig. 427) with very characteristic genital lobe; the latter with chitinous arms apically .... 23. **Euborellia plebeja**, p. 259
40′ Forceps without large tooth at dorsal surface basally; asymmetrical, but simple, of *Euborellia*-type. External parameres of male genitalia as in fig. 392. Cosmopolitan species .... 6. **Euborellia stali**, p. 242
41 Genital lobe of male genitalia without specific denticulated pads (Fig. 440), paramere broad, oval, median incision of anterior margin deep but narrow .... 33. **Euborellia insulana**, p. 269
41′ Genital lobe of male genitalia with characteristic denticulated pads apically, the latter of *Euborellia*-type **42**
42 Central parameral plate of male genitalia very broad, V-shaped; median incision of anterior margin wide; external parameres lemon-shaped, comparatively small. Genital lobe with specific denticulated pads, as in fig. 441 .... 34. **Euborellia malgacha**, p. 269
42′ Central parameral plate of male genitalia narrow, U-shaped; median incision of anterior margin broad, but narrowed as in *malgacha* Brindle. Genital lobe with specific denticulated pads, as in fig. 439 .... 32. **Euborellia femoralis**, p. 267
43 Indo-Australian species **44**
43′ Palaearctical or Oriental species **50**
44 Tegmina and wings completely absent **45**
44′ At least tegmina present, even though rudimentary **48**
45 Male forceps with very large tooth on dorsal surface basally (Fig. 442). External paramere of male genitalia characteristic, their apices with a chitinous process; external parameres very broad at median section; genital lobe without denticulated pads apically, but with sclerotized plates (Fig. 443) .... 35. **Euborellia brunneri**, p. 270
45′ Male forceps without large tooth on dorsal surface basally. External parameres of male broad, obtuse; genital lobe with denticulated pads **46**
46 Ultimate tergite of male with lateral longitudinal keels dorsally; forceps comparatively elongated, slender, as in fig. 444. External parameres of male genitalia lemon-shaped; genital lobe with denticulated pads at apical portion (Fig. 445) ..... 36. **Euborellia eteronoma**, p. 271
46′ Ultimate tergite of male without lateral longitudinal keels dorsally; forceps short. External parameres not lemon-shaped; genital lobe with denticulated pads on latero-apical portion **47**
47 Abdominal end and ultimate tergite very broad, transverse (Fig. 410). Male genitalia similar to *aporonoma* (Borelli), but external parameres comparatively shorter, and denticulated pads of genital lobe as in fig. 411 .... Cosmopolitan species .... 15. **Euborellia annulipes**, p. 251
47′ Abdominal end and ultimate tergite narrower, as in fig. 446. Male genitalia similar to *annulipes* (Lucas), but external parameres comparatively longer, and denticulated pads of genital lobe as in fig. 447 .... 37. **Euborellia aporonoma**, p. 272

48 Tegmina rudimentary, represented by lateral flaps **49**
48′ Tegmina and wings normally developed. Ultimate tergite and forceps broad, short (Fig. 438). Male genitalia (Fig. 439) large, median incision of anterior margin very deep; genital lobe with *Euborellia*-type denticulated pads. Cosmopolitan species ..... 32. **Euborellia femoralis**, p. 267
49 Male forceps without large tooth on inner margin (Fig. 390). Male genitalia very wide, genital lobe with specific denticulated pads apically, as in fig. 392 .... 6. **Euborellia stali**, p. 242
49′ Male forceps with large tooth on inner margin (Fig. 426). Male genitalia large, genital lobe with very specific chitinous serrations, as in fig. 427 .... 23. **Euborellia plebeja**, p. 259
50 Palaearctic species **51**
50′ Oriental species **52**
51 Male forceps with large tooth on dorsal surface basally (Fig. 384). Male genitalia elongated, large, median incision of anterior margin very deep and long; genital lobe with characteristic denticulated pads, as in fig. 385 .... 3. **Euborellia moesta**, p. 239
51′ Male forceps without large tooth on dorsal surface basally (Fig. 410). Male genitalia comparatively shorter, median incision of anterior margin deep, but shorter than *moesta* (GÉNÉ); genital lobe with specific denticulated pads, as in fig. 411 .... 15. **Euborellia annulipes**, p. 251
52 Tegmina and wings entirely absent **53**
52′ Tegmina and wings present, or represented by lateral flaps, or well developed **58**
53 Weak build; pronotum slightly emarginate in middle posteriorly; abdomen smooth above, almost parallel; penultimate sternite with hind margin broadly rounded. Male forceps (Fig. 448) relatively long, more or less symmetrical, not of *Euborellia*-type. Genitalia narrow, genital lobe with typical denticulated pads (Fig. 449) .... 38. **Euborellia dattai**, p. 273
53′ Stout build; pronotum emarginate posteriorly; abdomen punctate, slightly enlarged in middle; penultimate sternite somewhat triangular **54**
54 Sides of abdominal tergites carinate; external parameres of male genitalia without an inner apical concavity **55**
54′ Sides of abdominal tergites ecarinate **56**
55 Legs uniformly yellow. Posterior abdominal tergites 7–9 of male rugoso-striate, but without well-defined lateral longitudinal ridges, or there may be incipient ridges on tergites 7–8. Genital lobe with denticulated pads as in fig. 407 .... 13. **Euborellia compressa**, p. 249
55′ Legs yellow, often femora with a broad brownish or blackish ring. Sides of abdominal tergites strongly rugoso-striate, and with lateral longitudinal ridges on tergites 7–9. Genital lobe with denticulated pads, as in fig. 411 .... 15. **Euborellia annulipes**, p. 251
56 Genital lobe of male genitalia without denticulated pads apically; external parameres strongly expanded medially; paramere oval, genital lobes with specific sclerotized plates (Fig. 451). Male ultimate tergite broad, typical, forceps as in fig. 450 .... 39. **Euborellia sakaii**, p. 274
56′ Genital lobe of male genitalia with denticulated pads apically **57**
57 Ultimate tergite of male broad, robust, branches of forceps not separated basally (Fig. 452). Male genitalia narrow, central parameral plate V-shaped, denticulated pads of genital lobe as in fig. 453 .... 40. **Euborellia rajasthanensis**, p. 275
57′ Ultimate tergite of male broad, but not robust, branches of forceps well separated basally (Fig. 454). Male genitalia broad, median incision of anterior margin deep and wide; denticulated pads as in fig. 455 .... 41. **Euborellia punctata**, p. 276
58 Tegmina and wings well developed **59**
58′ Tegmina and wings reduced, tegmina represented by a lateral flaps on mesonotum; wings absent **60**
59 Forceps with large tooth on inner margin basally (Fig. 426). Genital lobe of male genitalia without denticulated pads, but with specific sclerotized serration (Fig. 427).............. .... 23. **Euborellia plebeja**, p. 259
59′ Forceps without large tooth on inner margin basally. Ultimate tergite and forceps broad, branches of forceps broadest basally, trigonal and curved apically (Fig. 438). Male genitalia (Fig. 439) large, median incision of anterior margin very deep; genital lobe of *Euborellia*-type, with characteristic denticulated pads apically .... 32. **Euborellia femoralis**, p. 267
60 Virgae within genital lobe of male genitalia present; central parameral plate with very deep and wide median incision between external parameres (Fig. 457) .... 42. **Euborellia abbreviata**, p. 277
60′ Genital lobes of male genitalia without virgae **61**

61 External parameres of male genitalia expanded medially; genital lobe with specific denticulated pads on inner margin apically (Fig. 458) ...... 43. **Euborellia annandalei**, p. 278
61′ External parameres of male genitalia normal, rounded along outer margins **62**
62 Male forceps elongated, as in fig. 459. Male genitalia comparatively narrow, median incision of anterior margin narrow, genital lobes with specific denticulated pads apically, as in fig. 460 .... 44. **Euborellia philippinensis**, p. 279
62′ Male forceps short, as in fig. 390. Male genitalia (Fig. 391) broad, median incision of anterior margin very deep and wide, genital lobe with specific denticulated pads apically, as in fig. 392 .... 6. **Euborellia stali**, p. 242

1. Euborellia vanderbilti Rehn

1936 *Euborellia vanderbilti* Rehn, Proc. Acad. nat. Sci. Philadelphia, **88**: 510; fig. 1 (male), 2 (abdomen laterally of male), 3 (penultimate sternite of male). – Terra typica: Belgian Congo (Type male: Acad. nat. Sci. Philadelphia: Type no. 5554).
1954 *Euborellia vanderbilti* – Hincks, Mém. Inst. fr. Afr. noire, Paris, Dakar, Douala, **40**: 109. Guinea.
1969 *Euborellia vanderbilti* – Brindle, Mém. Inst. fr. Afr. noire, Paris, Dakar, Douala, **31**: 64. Cameroon.
1978 *Euborellia vanderbilti* – Brindle, Dermaptera of Africa, Tervuren, **2**: 161, fig. 292 (male), 300 (male genitalia), 316 (penultimate sternite of male). West Central Africa.
Literature: Sakai 1970; Steinmann 1977, 1978.

Description: Male colour black, shiny; antennae brownish-black, but one or more distal joints white; pronotum usually whitish-yellow postero-laterally; both tegmina with a round or elliptical whitish-yellow spot; wings yellow, broadly darkened apically and sometimes with a vague dark spot medially; legs yellow, femora with a median blackish band. Head transverse, depressed transversally along postfrontal sutures; coronal suture distinct. Eyes relatively small, shorter than length of head behind eyes. Antennae 15-jointed; first joint long, but a little shorter than distance between antennal bases. Pronotum transverse, slightly widened posteriorly; lateral margins convex; median longitudinal furrow present. Tegmina and wings fully developed, extending down to third or fourth abdominal tergite. Tegmina with sharply defined oval humeral patches of mustard yellow. Abdomen of *Euborellia*-type, tergites 6–9 with a trace of a median carination. Ultimate tergite transverse. Forceps (Fig. 379) robust, slightly asymmetrical; trigonal and expanded basally, cylindrical apically. Penultimate sternite with lateral arms of distal margin straight, oblique convergent, median third of same transverse truncate. Genitalia (Fig. 380) broad; central parameral plate large; median incision of anterior margin of medium size; genital lobes well developed, virga within genital lobes long, with finely chitinous serrations apically. External parameres expanded medially, apices obtuse. – **Female** similar to male, but abdominal tergites more sparsely punctate, and abdomen narrower at apex; branches of forceps almost symmetrical, straight, and contiguous. – **Length** of body with forceps: male: 12–14,5 mm, female: 13–14 mm.

Distribution: Guinea, Cameroon, Congo Republic.

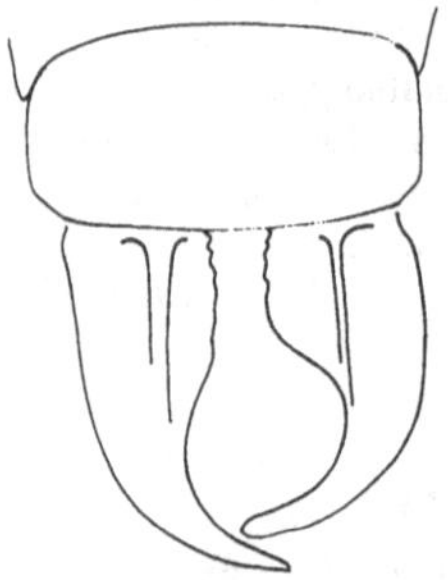

Fig. 379 Male ultimate tergite and forceps of *Euborellia vanderbilti* REHN. Original.

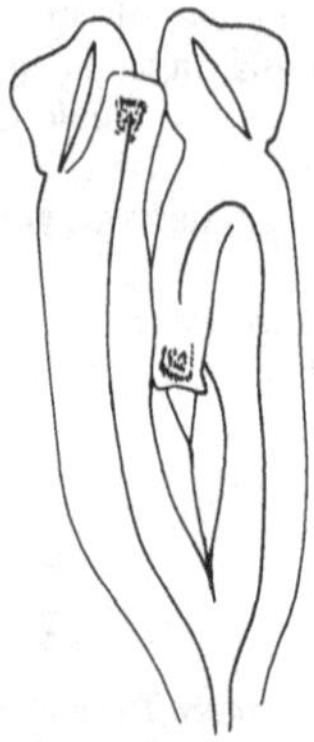

Fig. 380 Male genital armature of *Euborellia vanderbilti* REHN. After BRINDLE.

2. Euborellia cincticollis (GERSTAECKER)

1883 *Brachylabis cincticollis* GERSTAECKER, Mitt. naturwiss. Verein Neu-Vorpommern, Greifswald, **14**: 44. – Terra typica: Cameroon (Type male: Mus. Naturkunde, Berlin).
1891 *Psalis picina* KIRBY, J. Linn. Soc., London Zool., **23**: 516. – Terra typica: Gambia (Type male: Brit. Mus. nat. Hist., London.
1900 *Anisolabis cincticollis* – BORMANS, Das Tierreich, Berlin, **11**: 43.
1907 *Psalis debilis* BURR, Berliner ent. Zeitschr., **7**: 202. – Terra typica: Uganda (Type male: Brit. Mus. nat. Hist., London).
1909 *Psalis cincticollis* – BURR, Ann. Mag. nat. Hist., London, (8) **4**: 113.
1911 *Psalis cincticollis* – BURR, Genera Insectorum, Bruxelles, **122**: 32; pl. 3, fig. 7 (male). West Africa.
1915 *Euborellia debilis* – BURR, J. r. microsc. Soc., London, **1915**: 544, pl. 12, fig. 16 (male genitalia).
1924 *Euborellia cincticollis* – REHN, Bull. Amer. Mus. nat. Hist., New York **49**: 369.
Literature: BORELLI 1909, 1923; BRINDLE 1978; BURR 1909; HINCKS 1952, 1954, 1956; MENOZZI 1938; REHN 1924; SAKAI 1970; STEINMANN 1977, 1978.

Description: Male colour reddish-brown or yellowish-brown to dark brown or blackish; antennae dark brown, but basal joints yellowish; head sometimes reddish, pronotum yellow at sides, and wings often paler basally. Head transverse, convex, but sometimes slightly depressed on vertex; postfrontal and coronal sutures distinct; eyes large, but a little shorter than the length of head behind eyes. Antennae 19-jointed; first joint normal, slightly shorter than distance between antennal bases. Pronotum longer than wide, more or less parallel-sided but usually widened posteriorly; posterior margin weakly convex. Tegmina and wings fully developed or tegmina shorter and wings absent of concealed. Abdomen weakly widened medially, slightly or more strongly depressed; abdominal punctation somewhat variable, tergites 7–9 strongly punctate, and tergites 7–8 usually with lateral longitudinal ridges. Ultimate tergite very broad, with lateral longitudinal ridge usually small; there is no dorso-lateral ridge. Forceps (Fig. 381) weakly curved, asymmetrical, strongly trigonal basally, cylindrical apically. Penultimate sternite broadly triangular, apex truncate. Genitalia broad, median incision of anterior margin very deep and wide; genital lobes of *Euborellia*-type, fully developed. External parameres (Figs. 382–383) short, more or less rounded. – **Female** similar to male, but abdominal punctation, especially laterally, weaker; penultimate sternite rounded apically; forceps straight, symmetrical. – **Length** of body with forceps: male: 12–18,5 mm, female: 11,5–17 mm.

Distribution: Congo Republic, Cameroon, Gambia, Uganda, Kenya, Tanzania, Sierra Leone; (USA?, by SAKAI 1970: 32).

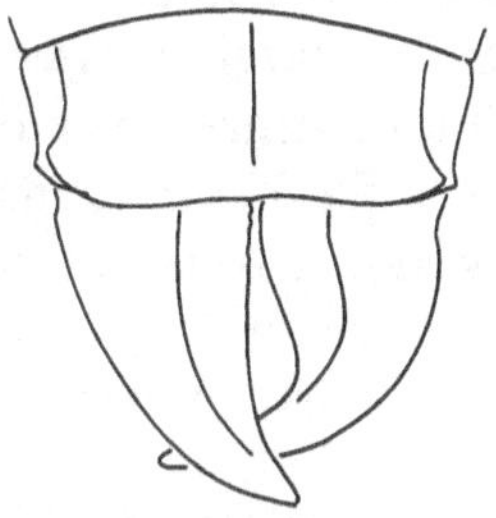

Fig. 381 Male ultimate tergite and forceps of *Euborellia cincticollis* (GERSTAECKER). Original.

Fig. 382 External paramere of male genitalia of *Euborellia cincticollis* (GERSTAECKER). After BURR.

Fig. 383 External paramere of male genitalia of *Euborellia cincticollis* (GERSTAECKER). After BRINDLE.

3. Euborellia moesta (GÉNÉ)

1839 *Forficesila moesta* GÉNÉ (apud AUDINET-SERVILLE), Hist. nat. Ins., Orthoptera, Paris, 28. – Terra typica: Sardinien (Type: sexes not cited: unknown locality).
1853 *Forficula (Labidura) moesta* – FISCHER, Orthopt. Europ., St. Petersburg, 68, pl. 6, fig. 5 (male).
1846 *Forficula taurica* MOTSCHULSKY (apud FISCHER), Ent. ross., St. Petersburg, **4**: 47. – Terra typica: South Russia (Type male: unknown locality).
1846 *Anisolabis taurica* – FISCHER, Ent. ross., St. Petersburg, **4**: 70.
1864 *Brachylabis moesta* – DOHRN, Stettiner ent. Ztg., **25**: 296.
1882 *Anisolabis moesta* – FIEBER, Lotos, Praha, **3**: 257.
1911 *Borellia moesta* – ZACHER, Zool. Jahrb., Jena, **30**: 380; fig. C^2 (male genitalia).
1911 *Euborellia moesta* – BURR, Genera Insectorum, Bruxelles, **122**: 31. South Europe.
1911 *Euborellia moesta* var. *anelytrata* VERHOEFF in M.S.: ZACHER, Zool. Jahrb., Jena, **30**: 380. Bankok, Manila, Kamerun.

Literature: BEY-BIENKO 1934; BOESEMAN 1954; BORMANS 1900; BRINDLE 1978; BURR 1910, 1915; CHOPARD 1922; HEBARD 1923; KIRBY 1904; MENOZZI 1927; POPHAM & BRINDLE 1966; SAKAI 1970; STEINMANN 1977, 1978.

Description: Male colour blackish or brownish-black; shiny, antennae dark brown, basal joints paler; pronotum yellow laterally, legs yellowish-brown, femora vaguely darkened. Head smooth, transverse; postfrontal and coronal sutures very distinct. Eyes large, more or less as long as first antennal joint; the latter comparatively small, shorter than distance between antennal bases. Pronotum a little longer than wide; lateral margins slightly expanded posteriorly;

median longitudinal furrow distinct; cuticle of pronotum coriaceous or smooth. Tegmina rudimentary, represented by lateral flaps; wings absent. Abdomen sometimes rather strongly widened and depressed; basal two segments almost smooth, the other tergites more coarsely coriaceous, and with small scattered punctures; tergites 7–9 rugoso-striate at sides. Ultimate tergite broad; lateral longitudinal ridge rugoso-striate basally, but without lateral ridges; posterior margin concave at median section. Forceps (Fig. 384) asymmetrical; strongly trigonal basally, cylindrical and curved apically; right branch curved more sharply than left. Penultimate sternite broadly rounded, apex slightly truncate. Genitalia (Fig. 385; gen. prep. No. 103, det. Dr. H. STEINMANN) characteristic; central parameral plate V-shaped, median incision of anterior margin between external parameres very deep and wide; genital lobes without virgae; apex of genital lobe of *Euborellia*-type, with denticulated pads, as in fig. 385. External parameres broad, a little extended apically. – **Female** very similar to male, but abdominal tergites almost impunctate and tergites 7–9 not rugoso-striate at sides; forceps symmetrical, with branches straight and contiguous. – **Length** of body with forceps: in both sexes: 15–22 mm.

Distribution: South Europe and East Africa.

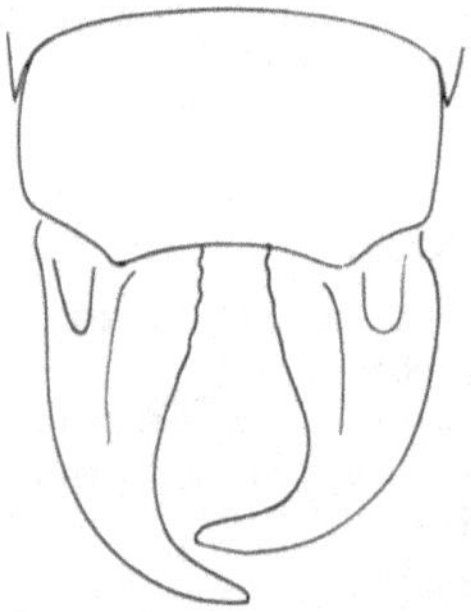

Fig. 384 Male ultimate tergite and forceps of *Euborellia moesta* (GÉNÉ). Original.

Fig. 385 Male genital armature of *Euborellia moesta* (GÉNÉ). Original.

4. Euborellia flavohumeralis BRINDLE

1978 *Euborellia flavohumeralis* BRINDLE, Dermaptera of Africa, Tervuren, **2**: 170; fig. 315 (external paramere of male genital armature). – Terra typica: Ivory Coast (Type male: Mus. r. Afr. centr., Tervuren).

Description: Male colour black, shiny; head very dark red, sides of pronotum yellow; anterior part of mesonotum yellow; and first antennal joint yellowish-brown. Head strongly transverse, and slightly depressed along epicranial sutures; the postfrontal and coronal sutures distinct. Eyes comparatively small, slightly shorter than the length of head behind eyes. First antennal joint more or less shorter than distance between antennal bases. Pronotum transverse, widened posteriorly; median longitudinal furrow present. Mesonotum with lateral rudimentary flaps; wings absent. Abdomen weakly widened medially and depressed; tergites 1–4 weakly and sparsely punctate, tergites 5–9 slightly more strongly punctate; sides of abdominal tergites 7–9 rugoso-striate laterally. Ultimate tergite broad; weakly punctate medially; plate a little depressed medially. Forceps (Fig. 386) asymmetrical; branches with dorsal keels at basal half; right branch shorter than left. Penultimate sternite with posterior margin concave. Genitalia similar to *annulipes* (LUCAS); paramere V-shaped; median incision of anterior margin very deep and wide; genital lobe typical. External parameres (Fig. 387) broad, apices well rounded. Apex of genital lobe without strongly sclerotized denticulated areas or pads. – **Female** unknown. – **Length** of body with forceps: 15,5–16 mm.

Distribution: Ivory Coast.

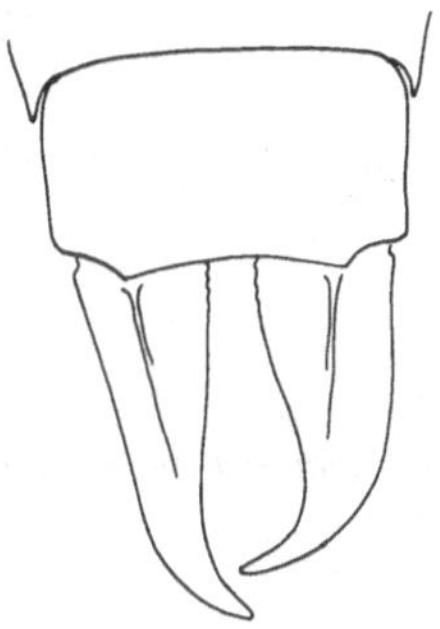

Fig. 386 Holotype ultimate tergite and forceps of *Euborellia flavohumeralis* BRINDLE. After BRINDLE.

Fig. 387 External paramere of male genitalia of *Euborellia flavohumeralis* BRINDLE. After BRINDLE.

5. Euborellia ituriensis BRINDLE

1978 *Euborellia ituriensis* BRINDLE, Dermaptera of Africa, Tervuren, **2**: 169; fig. 314 (penultimate sternite of male), 321 (external paramere of male genitalia). – Terra typica: Zaïre (Type male: Mus. r. Afr. centr., Tervuren).

Description: Male black, abdomen very dark, almost blackish-red posteriorly, forceps dark red; first antennal joint yellow; femora blackish, yellow at base and apices. Head transverse; postfrontal and coronal sutures prominent. Eyes fairly large, but a little shorter than the length of head behind eyes. First antennal joint shorter than distance between antennal bases. Pronotum strongly transverse, widened posteriorly; posterior margin truncate; all angles rounded; median longitudinal furrow present. Tegmina reduced, represented by lateral flaps; wings entirely absent. Abdomen weakly widened medially; tergites 1–2 almost impunctate, tergites 3–5 very weakly and sparsely punctate, tergites 6–8 with punctures closer and deeper but punctation irregular; sides of tergites 7–9 rugoso-striate, but sides without lateral longitudinal ridges. Ultimate tergite broad; median longitudinal sulcus present at median section; lateral longitudinal ridge present. Forceps (Fig. 388) short, asymmetrical; strongly trigonal basally, curved and cylindrical apically. Penultimate sternite narrowed distally, posterior margin concave. Genitalia of *Euborellia*-type; paramere V-shaped; median incision of anterior margin deep. Genital lobes with sclerotized denticulations at end but without lateral denticulated pads. External parameres (Fig. 389) angular externally, margin sinuate towards base, inner membrane wide. – **Female** unknown. – **Length** of body with forceps: 14,5 mm.

Distribution: Zaïre.

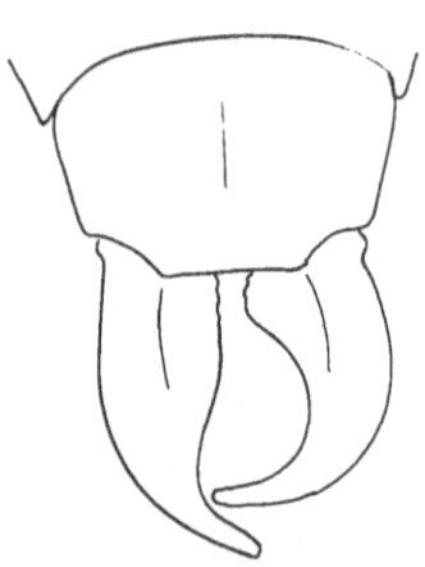

Fig. 388 Holotype ultimate tergite and forceps of *Euborellia ituriensis* BRINDLE. After BRINDLE.

Fig. 389 External paramere of holotype genitalia of *Euborellia ituriensis* BRINDLE. After BRINDLE.

6. Euborellia stali (DOHRN)

1864 *Forcinella stali* DOHRN, Stettiner ent. Ztg., **25**: 286. – Terra typica: Java (Type female: Naturhist. Riksmus., Stockholm).

1876 *Anisolabis stali* – SCUDDER, Proc. Boston Soc. nat. Hist., **18**: 308.

1900 *Anisolabis stali* – BORMANS, Das Tierreich, Berlin, **11**: 45 [= *Anisolabis femoralis* (DOHRN 1863)].

1910 *Anisolabis minuta* – BURR, Proc. U.S. nation. Mus., Washington, **38**: 447 [= *Anisolabis stali* (DOHRN 1864)].

1910 *Borellia stali* – BURR, Fauna of Brit. India, Dermaptera, London, Bombay, 88. Malay Archipelago, India: Madras, Bombay, Ceylon; Madagascar, Comoro Islands; East Africa.

1911 *Euborellia stali* – BURR, Genera Insectorum, Bruxelles, **122**: 31. Oriental Region and Madagascar.

1911 *Borellia stali* – ZACHER, Zool. Jahrb., Jena, **30**: 380; fig. D^2 (male genitalia).

1932 *Mongolabis aberrans* BORELLI, Journ. Fed. Malay, States Mus., **17**: 191. – Terra typica: Borneo (Type male: holotype unknown locality; paratype without penis: Mus. Ist. Zool. sist., Torino). – **syn. nov.**

1966 *Euborellia aberrans* – POPHAM & BRINDLE, Entomologist, London, **99**: 273.

1968 *Euborellia stali* – REICHARDT, Pap. avulsos. Zool., São Paulo, **22**: 45. Oriental Region, Madagascar, Antilles, Costa Rica, and United States: Florida.

1970 *Euborellia stali* – SAKAI, Dermapt. Cat. Prael., Tokyo, **1**: 29 [= *Euborellia plebeja* (DOHRN 1863)].

Literature: Boeseman 1954; Borelli 1927, 1932; Bormans 1900; Brindle 1978; Burr 1907, 1909, 1910, 1912, 1915; Caudell 1922; Hebard 1923; Kirby 1904; Menozzi 1931; Sakai 1970; Shiraki 1928; Srivastava 1972; Steinmann 1977, 1978.

Description: Male colour dark brown to black; shiny; antennae dark brownish, basal two joints yellow, and one or more distal joints whitish; sides of pronotum yellowish, legs yellow, femora dark brown medially or near bases. Head slightly transverse, lustrous black, sometimes brillant; posterior angles rounded; postfrontal and coronal sutures faint, but visible. Eyes moderately short, shorter than first joint. Antennae 15-jointed; first joint a little shorter than distance between antennal bases. Pronotum about as long as wide, sometimes a little longer than wide; lateral margins widened posteriorly; median longitudinal furrow distinct. Tegmina present as small ovale flaps on the side on the mesonotum. Wings absent. Abdomen faintly punctate, scarcely widened medially, somewhat depressed, basal tergites weakly and sparsely punctate, and posterior tergites more strongly punctate; sides of tergites 7–9 rugoso-striate but without lateral longitudinal ridges. Ultimate tergite transverse; posterior margin straight medially. Forceps (Fig. 390) short, trigonal in basal half, tapering and cylindrical in the apical half; sharply curved apically. Penultimate sternite rounded at posterior margin. Genitalia (Fig. 391; gen. prep. No. 812, det. Dr. H. Steinmann) similar to *plebeja* (Dohrn), but paramere narrower, median incision of anterior margin deep and very wide; Genital lobes with characteristic denticulated pads, as in fig. 392. External parameres lemon-shaped. – **Female** similar to male, but abdominal punctation weaker; forceps straighter and almost contiguous except at apices. – **Length** of body with forceps: male: 9–11 mm, female: 10–12,2 mm.

Distribution: Cosmopolitan, distributed in the Neotropical, Oriental and Malgasian faunal regions.

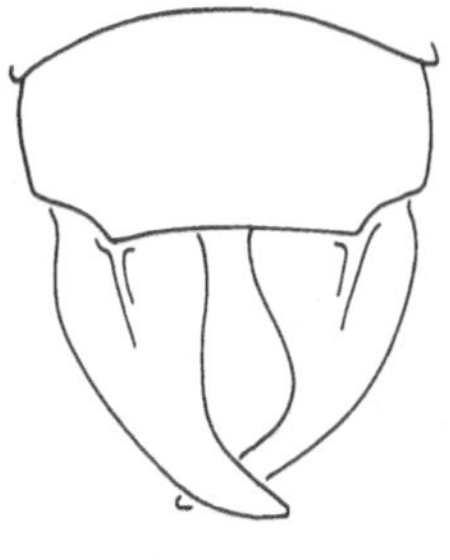

Fig. 390 Male ultimate tergite and forceps of *Euborellia stali* (Dohrn). Original.

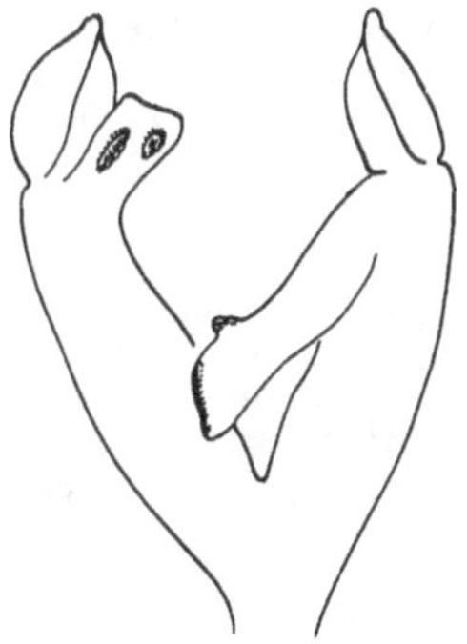

Fig. 391 Male genital armature of *Euborellia stali* (Dohrn). Original.

Fig. 392 Male genital lobe of *Euborellia stali* (Dohrn). Original.

7. Euborellia purpurea (Borelli)

1921 *Psalis purpurea* Borelli, Bull. Mus. nation. Hist. nat., Paris, **1921**: 83. – Terra typica: Belgian Congo (Type male: Mus. nation. Hist. nat., Paris).
1936 *Euborellia purpurea* – Rehn, Proc. Acad. nat. Sci. Philadelphia **88**: 512. Zaïre.
1978 *Euborellia purpurea* – Brindle, Dermaptera of Africa, Tervuren, **2**; 166.
Literature: Sakai 1970; Steinmann 1978.

Description: Male with head brillant black; antennae black, joints 1–2 and part of third yellow, sometimes eleventh joint whitish; pronotum and tegmina black, shiny, tegmina with a purplish sheen; legs yellow; abdomen weakly widened to eighth segment; ultimate tergite large, median longitudinal sulcus present on median section; posterior margin straight medially. Forceps (Fig. 393) short; asymmetrical; strongly trigonal basally, and curved apically. Genitalia similar to *compressa* (Borelli), comparatively narrow, median longitudinal incision of anterior margin deep and narrow; genital lobe (Fig. 394) specific, denticulus lobi genitalis distinct, prominent. External parameres about as long as wide, shaped like an irregular quadrilateral. – **Female** similar to male, but forceps straight, symmetrical, tapering. – **Length** of body with forceps: male: 14,5–16 mm, female: 15–17,5 mm.
Distribution: Congo Republic, Zaïre.

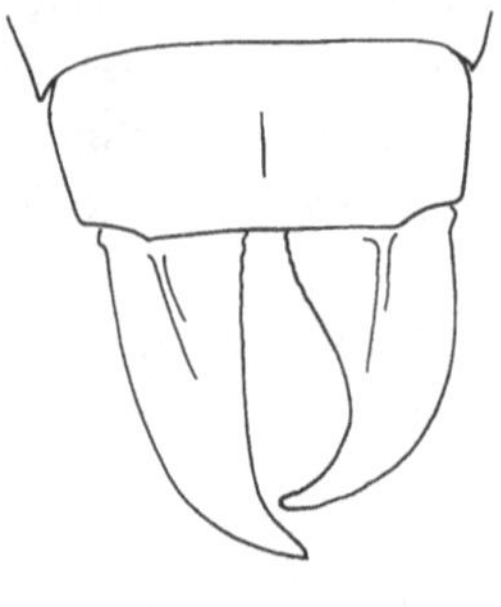

Fig. 393 Male ultimate tergite and forceps of *Euborellia purpurea* (Borelli). Original.

Fig. 394 Male genital lobe with external paramere of *Euborellia purpurea* (Borelli). Original.

8. Euborellia fulviceps Borelli

1921 *Euborellia fulviceps* Borelli, Bull. Mus. nation. Hist. nat., Paris., **1921**: 157. – Terra typica: Congo (Type male: Mus. nation. Hist. nat., Paris).
1978 *Euborellia fulviceps* – Brindle, Dermaptera of Africa, Tervuren, **2**: 167; fig. 295 (male), 307 (external paramere of male genitalia). Zaïre.
Literature: Sakai 1970; Steinmann 1978.

Description: Male general colour reddish-brown; head reddish-yellow, antennae dark brown, but basal two joints yellow, and one distal joint (14 or 15) yellowish or partly yellow; lateral sides of pronotum yellowish, legs yellow. Head rounded; postfrontal and coronal sutures deep; eyes comparatively large, but a little shorter than the length of head behind eyes. First antennal joint small, slightly shorter than distance between antennal bases. Pronotum longer than wide, widened posteriorly; median longitudinal furrow present on median section. Tegmina present but reduced, represented by lateral flaps, the latter broad, prominent. Wings absent. Abdomen weakly widened medially and depressed, tergites finely and sparsely punctate, especially on basal half of each tergite; sides of posterior tergites more strongly punctate but not rugoso-striate. Ultimate tergite more strongly punctate; transverse; median longitudinal sulcus distinct and depressed; posterior margin depressed medially and slightly rugose. Forceps (Fig. 395) strongly asymmetrical; branches trigonal basally, curved medially and apically. Penultimate sternite rounded at apex. Genitalia broad, median incision of anterior margin very deep and wide; genital lobe typical for genus, external parameres (Fig. 396) broad, rounded. – **Female** unknown. – **Length** of body with forceps: 12 mm.

Distribution: Congo Republic, and Zaïre.

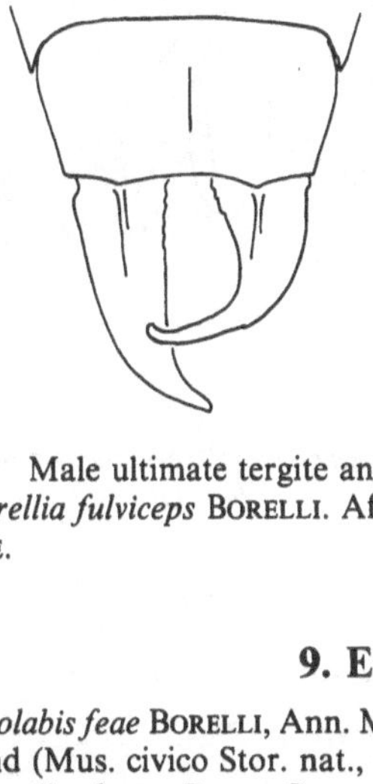

Fig. 395 Male ultimate tergite and forceps of *Euborellia fulviceps* BORELLI. After BRINDLE.

Fig. 396 External paramere of male genitalia of *Euborellia fulviceps* BORELLI. After BRINDLE.

9. Euborellia feae (BORELLI)

1907 *Anisolabis feae* BORELLI, Ann. Mus. civ. Stor. nat., Genova, **43**: 355. – Terra typica: Annobon Island (Mus. civico Stor. nat., Genova).
1911 *Euborellia feae* – BURR, Genera Insectorum, Bruxelles, **122**: 31.
1978 *Euborellia feae* – BRINDLE, Dermaptera of Africa, Tervuren, **2**: 167, fig. 304 (external paramere of male genitalia), 311 (penultimate sternite of male). Cameroon, Zaïre.
Literature: BURR 1915; POPHAM & BRINDLE 1966; SAKAI 1970; STEINMANN 1977, 1978.

Description: Male reddish-brown to blackish; head often paler in colour, antennae yellowish, legs yellow, femora vaguely darkened. Head transverse; postfrontal and coronal sutures distinct; eyes small, slightly shorter than the length of head behind eyes. First antennal joint small, shorter than distance between antennal bases. Pronotum slightly transverse, and only slightly widened posteriorly; median longitudinal furrow present. Tegmina represented by lateral flaps on mesonotum, width of each tegmina less than one third of width of mesonotum. Wings absent. Abdomen weakly widened medially and depressed; sides of tergites 6–9 rugoso-striate and 6–8 with short blunt rugoso-striate lateral longitudinal ridges. Ultimate tergite weakly punctate, posterior margin depressed between bases of forceps and somewhat; lateral longitudinal ridges well developed, but no dorso-lateral ridge. Forceps (Fig. 397) asymmetrical; right branch more sharply curved near apex. Penultimate sternite rounded apically. Genitalia of *Euborellia*-type; central parameral plate V-shaped; median longitudinal incision of anterior margin deep and wide; genital lobes typical. External parameres (Fig. 398) short, broad, outer margin angled. – **Female** similar to male, but sides of abdominal tergites not rugoso-striate; lateral longitudinal ridges on last tergite indistinct; branches of forceps straighter and more contiguous. – **Length** of body with forceps: male: 12–13,5 mm, female: 13–14 mm.

Distribution: Cameroon, Congo Republic, Zaïre.

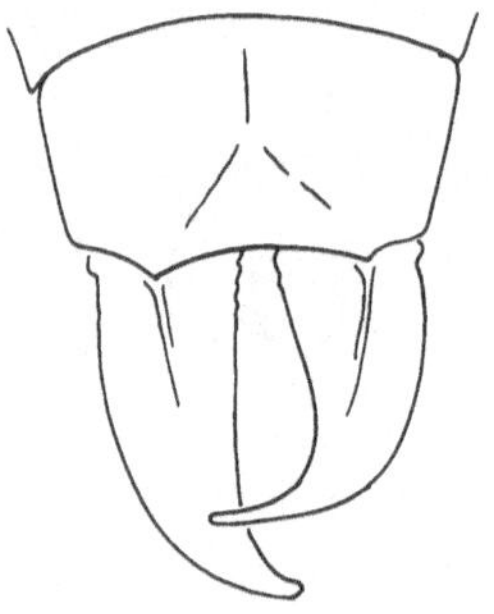

Fig. 397 Male ultimate tergite and forceps of *Euborellia feae* (BORELLI). Original.

Fig. 398 External paramere of male genitalia of *Euborellia feae* (BORELLI). After BRINDLE.

10. Euborellia andreinii (BORELLI)

1908 *Anisolabis andreinii* BORELLI, Boll. Soc. ent. ital., Firenze, Genova, **39**: 4. – Terra typica: Eritrea (Type male: unknown locality).
1908 *Borellia andreinii* – BORELLI, Boll. Mus. Zool. Anat. comp. Univ. Torino, **25**, No. 616: 1.
1911 *Euborellia andreinii* – BURR, Genera Insectorum, Bruxelles, **122**: 30.
1915 *Euborellia andreinii* – BURR, J. r. microsc. Soc., London, **1915**: 540; pl. 12, fig. 20 (male genitalia).

Literature: BURR 1912; SAKAI 1970; STEINMANN 1977, 1978.

Description: Male reddish-brown to blackish, antennae brown, basal two joints yellow, sides of pronotum and legs yellow. Head rounded and broad, postfrontal and coronal sutures weak; eyes large, but shorter than the length of head behind eyes. First antennal joint normal, a little shorter than distance between antennal bases. Pronotum about as long as wide or a little longer than wide; slightly widened posteriorly; posterior margin convex. Tegmina rudimentary, its width about one third of width of mesonotum. Wings absent. Abdomen weakly widened medially and depressed, first two tergites impunctate, other tergites more coarsely coriaceous; tergites 7–9 rugoso-striate at sides, and 7–8 with a blunt lateral longitudinal ridge. Ultimate tergite very broad, smooth or coriaceous; posterior margin rugose between the bases of forceps; median longitudinal sulcus deep. Forceps (Fig. 399) asymmetrical, trigonal basally, cylindrical apically; dorsal keel of forceps prominent. Penultimate sternite rounded apically. Genitalia (Fig. 400) very similar to *annulipes* (LUCAS), central parameral plate V-shaped, genital lobes with prominent denticulated pads, and denticulus lobi genitalis. External parameres comparatively large, oval, lemon-shaped. – **Female** similar to male, but sides of tergites 7–9 not rugoso-striate, and 7–8 without lateral longitudinal ridges; forceps less curved and branches more or less straight. – **Length** of body with forceps: male: 16–17 mm, female: 15,5–17 mm.

Distribution: Ethiopia.

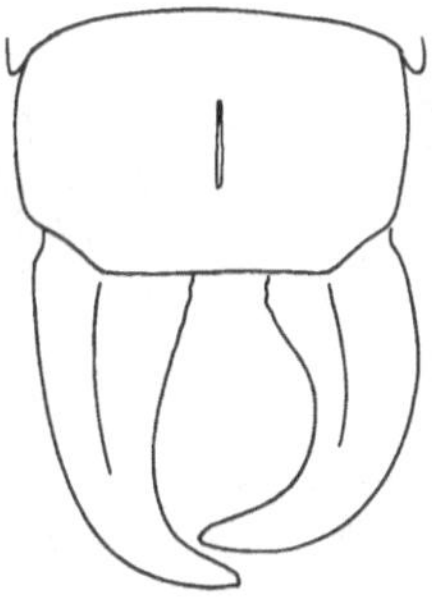

Fig. 399 Male ultimate tergite and forceps of *Euborellia andreinii* (BORELLI). Original.

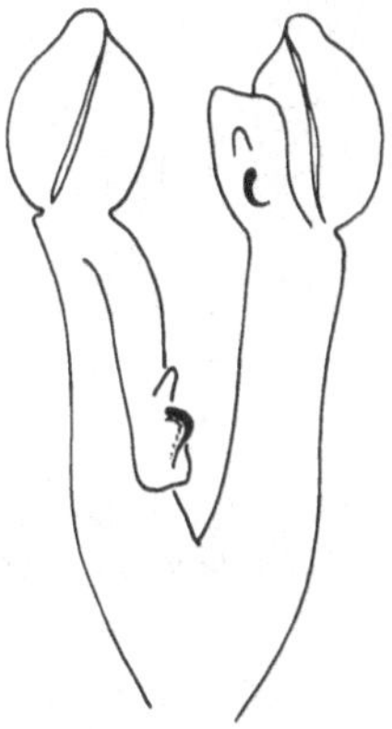

Fig. 400 Male genital armature of *Euborellia andreinii* (BORELLI). Original.

11. Euborellia dimidiata BRINDLE

1978 *Euborellia dimidiata* BRINDLE, Dermaptera of Africa, Tervuren, **2**: 181, fig. 323 (external paramere of male genitalia). – Terra typica: Zaïre (Type male: Mus. r. Afr. Centr., Tervuren).

Distribution: Male general colour black; antennae blackish, one distal joint (11 or 12) white; sides of pronotum yellow, legs yellow, femora broadly darkened, almost blackish, medially, tibiae vaguely darkened basally. Head more or less rounded, as long as wide; postfrontal and coronal sutures indistinct. Eyes comparatively small, slightly shorter than the length of head behind eyes. First antennal joint normal, a little shorter than distance between antennal bases. Pronotum transverse, widened posteriorly; median longitudinal furrow present in median section. Tegmina and wings completely absent. Abdomen scarcely widened medially, slightly depressed, tergites 1–2 almost impunctate, rest of tergites finely and sparsely punctate; sides or tergites without lateral longitudinal ridges. Ultimate tergite broad, almost smooth, punctation fine; median longitudinal sulcus indistinct. Forceps (Fig. 401) asymmetrical, branches straight medially, and curved at apical section; dorsal keel of forceps distinct. Penultimate sternite evenly rounded posteriorly. Genitalia very similar to *annulipes* (LUCAS), central parameral plate V-shaped; genital lobes without characteristic denticulus lobi genitalis and denticulated pads. External parameres (Fig. 402) rounded, obtuse. – **Female** unknown. – **Length** of body with forceps: 6,5 mm.

Distribution: Zaïre.

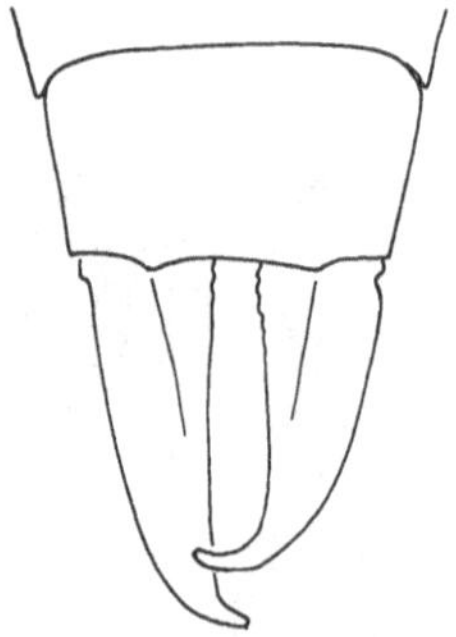

Fig. 401 Male ultimate tergite and forceps of *Euborellia dimidiata* BRINDLE. Original.

Fig. 402 External paramere of male genitalia of *Euborellia dimidiata* BRINDLE. After BRINDLE.

12. Euborellia flava STEINMANN

1981 *Euborellia flava* STEINMANN, Acta zool. Acad. Sci. hung., Budapest, **27**: 196; fig. 26 (female thorax), 27 (holotype ultimate tergite and forceps). – Terra typica: Zululand (Type female: Inst. taxonom. Zoöl., Amsterdam).

Description: Female general colour yellow and yellowish brown. Head a little reddish yellow, antennae, thoracal nota and legs yellow, abdomen and forceps yellowish brown. Head rounded, tumid, smooth. Postfrontal and coronal sutures distinct. Eyes relatively small, shorter than length of head behind eyes. Antennae 11-jointed, broken in holotype; first antennal joint of *Euborellia*-type, shorter than distance between antennal bases; second transverse, third joint longer, about as long as fourth. Pronotum transverse, a little wider than long, all angles rounded; lateral margins parallel; median longitudinal furrow distinct.

Mesonotum (Fig. 404) with lateral flaps, metanotum normal, transverse. Abdomen expanded, but not strongly widened and depressed; tergites 1–2 almost impunctate medially, slightly punctate laterally; tergite 3 more obviously punctate but punctures weak and sparse; tergites 4–9 punctate, punctation irregular, shallow and stronger laterally. Ultimate tergite broad, forceps (Fig. 403) depressed, branches short, symmetrical. – **Male** unknown. – **Length** of body with forceps: 7,5 mm.

Distribution: South Africa: Zululand.

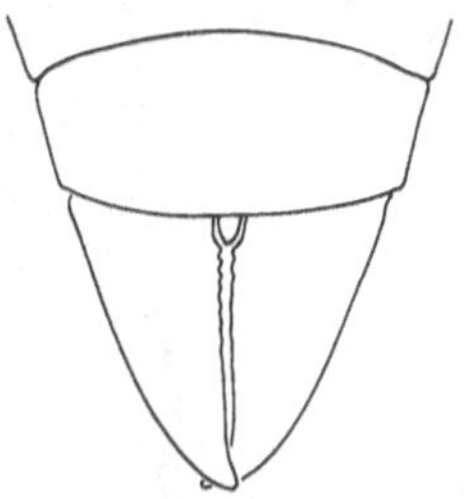

Fig. 403 Holotype ultimate tergite and forceps of *Euborellia flava* STEINMANN. After STEINMANN.

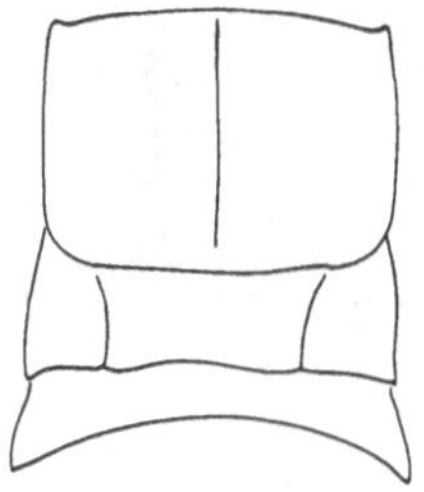

Fig. 404 Holotype thorax of *Euborellia flava* STEINMANN. After STEINMANN.

13. Euborellia compressa (BORELLI)

1907 *Anisolabis compressa* BORELLI, Boll. Mus. Zool. Anat. comp. Univ. Torino, **22**, No. 558: 3. – Terra typica: Uganda (Type male: Mus. Ist. Zool. sist., Torino).

1915 *Euborellia compressa* – BURR, J. r. microsc. Soc., London, **1915**: 542; pl. 12, fig. 18 (male genitalia).

1936 *Anisolabis compressa* – MENOZZI, Mem. Estud. Mus. zool. Univ. Coimbra, **95**: 5. Mozambique.

1978 *Euborellia compressa* – BRINDLE, Dermaptera of Africa, Tervuren, **2**: 177, fig. 309 (penultimate sternite of male), 301 (external paramere of male genitalia). Tanzania, Zaïre.

Literature: BORELLI 1909; BURR 1911; HINCKS 1938, 1954; REHN 1924; SAKAI 1970; STEINMANN 1977, 1978.

Description: Male head and thoracic nota yellowish or yellowish-brown, antennae dark brown, joints 1–2 yellow, legs yellow, abdomen and forceps reddish-brown. Head transverse, slightly depressed; postfrontal and coronal sutures indistinct. Eyes small, shorter than first antennal joint; antennae 17-jointed; first joint long, but shorter than distance between antennal bases. Pronotum a little longer than wide, convex posteriorly; all angles rounded; median longitudinal furrow distinct. Tegmina and wings absent. Abdomen weakly widened medially and depressed; sides of tergites 7–9 rugoso-striate, but without lateral longitudinal ridges. Ultimate tergite irregularly punctate with median longitudinal sulcus. Forceps (Fig. 405) somewhat asymmetrical, right branch curved more strongly than left; dorsal keel of forceps prominent. Penultimate sternite broadly rounded posteriorly. Genitalia (Fig. 406; gen. prep.

No. 562, det. Dr. H. STEINMANN) narrow; median incision of anterior margin deep but narrow; genital lobes (Fig. 407) with denticulus lobi genitalis. – **Female** similar to male, but abdominal punctation even weaker; tergites 7–9 not rugoso-striate, and forceps with straigher branches. – **Length** of body with forceps: male: 16–17 mm, female: 15,5–16,5 mm.

Distribution: Mozambique, Uganda, Tanzania, South Rhodesia, Zaïre, Congo Republic, and India (? SRIVASTAVA 1977).

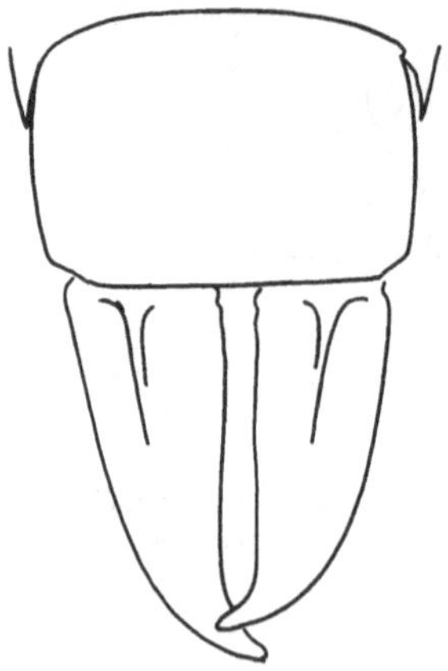

Fig. 405 Male ultimate tergite and forceps of *Euborellia compressa* (BORELLI). Original.

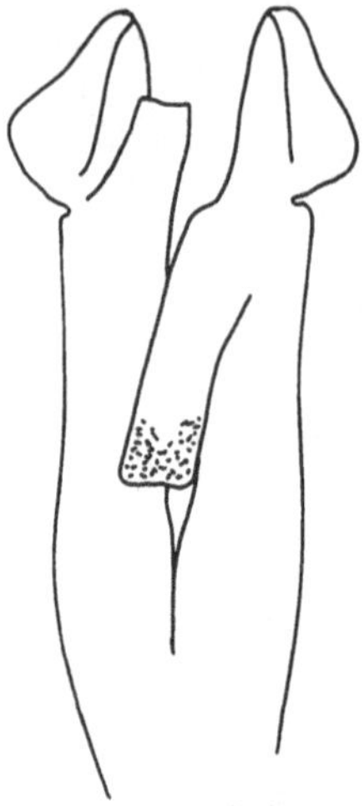

Fig. 406 Male genital armature of *Euborellia compressa* (BORELLI). Original.

Fig. 407 Male genital lobe with external paramere of *Euborellia compressa* (BORELLI). After BRINDLE.

14. Euborellia ugandana BRINDLE

1978 *Euborellia ugandana* BRINDLE, Dermaptera of Africa, Tervuren, **2**: 172, fig. 308 (external paramere of holotype genital armature). – Terra typica: Uganda (Type male: Mus. nation. Hist. nat., Paris).

Description: Male black, antennae dark brown, basal joints yellow; pronotum narrowly yellow at sides; legs reddish-yellow. Head transverse, depressed at vertex; postfrontal and coronal sutures distinct. Eyes large, but a little shorter than length of head behind eyes. First antennal joint normal, shorter than distance between antennal bases. Pronotum strongly transverse, widened posteriorly; median longitudinal furrow indistinct. Tegmina and wings entirely absent. Abdomen widened medially and depressed; tergites 1–2 almost impunctate medially, and slightly punctate laterally; sides of tergites 6–9 rugoso-striate. Ultimate tergite very broad, median longitudinal sulcus present; posterior margin concave medially. Forceps (Fig. 408) asymmetrical, almost straight except at apex, right branch slightly more curved than left. Penultimate sternite triangular, apex rounded. Genitalia of *Euborellia*-type, median incision of anterior margin of central parameral plate deep; external parameres (Fig. 409) relatively narrow, outer margins rounded. – **Female** similar to male, but abdomen with weaker punctation, and sides of tergites 6–9 not rugoso-striate; forceps straight, symmetrical. – **Length** of body with forceps: male: 15,5–17 mm, female: 16–17,5 mm.

Distribution: Uganda.

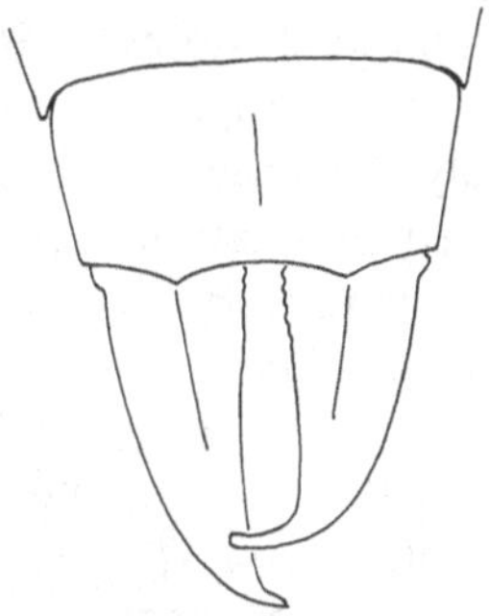

Fig. 408 Male ultimate tergite and forceps of *Euborellia ugandana* BRINDLE. Original.

Fig. 409 External paramere of male genitalia of *Euborellia ugandana* BRINDLE. After BRINDLE.

15. Euborellia annulipes (LUCAS)

1847 *Forficesila annulipes* LUCAS, Ann. Soc. ent. France, Paris, (2), **5**: 84. – Terra typica: France: Paris (Type female: unknown locality).

1851 *Anisolabis annulicornis* BLANCHARD, in GAY: Hist. nat. Chile, Zool., **6**: 10. – Terra typica: Chile, (Type female: location unknown).

1853 *Forficula* (*Labidura*) *annulipes* – FISCHER, Orth., Europaea, St. Petersburg, 69, Taf. 6, fig. 6a (male).

1863 *Anisolabis annulipes* – FUENTES, Bol. Mus. nacion. Chile, Santiago, **7**: 313.

1864 *Forcinella annulipes* – DOHRN, Stettiner ent. Ztg., **25**: 290.

1891 *Anisolabis antennata* KIRBY, J. Linn. Soc., London, Zool., **23**: 517. – Terra typica: Bermuda (Type female: Brit. Mus. nat. Hist., London).

1893 *Anisolabis bormansi* SCUDDER, Bull. Mus. comp. Zool., Harvard, **25**: 5; pl. 1, fig. 1 (female). – Terra typica: Galapagos Island (Type female: location unknown).

1906 *Anisolabis fallax* SHIRAKI, Trans. Sapporo nat. Hist., Soc., **1**: 9. – Terra typica: Japan (Type male: unknown locality).

1906 *Anisolabis piceus* SHIRAKI (partim), Trans. Sapporo nat. Hist. Soc., **1**: 9.

1908 *Anisolabis tripolitana* WERNER, Zool. Jahrb., Jena, **27**: 85, fig. 10 (male genitalia). – Terra typica: unknown.

1915 *Euborellia annulipes* – BURR, J. r. microsc. Soc., London, **1915**: 545.

1929 *Euborellia annulipes* var. *tripolitana* – CAPRA, Ann. Mus. civ. Stor. nat., Genova, **53**: 31; fig. 19b (male genitalia).

1954 *Anisolabis nana* BOESEMAN, Zool. Verh., Leiden, **21**: 39. – Terra typica: Java (Type immature female: Inst. taxonom. Zoöl., Amsterdam).

1975 *Anisolabis eteronoma* – HARZ & KALTENBACH, Orthoptera of Europe, **3**: 47 [= *Euborellia annulipes* (LUCAS 1847)].

Literature: Bey-Bienko 1934; Borelli 1909; Brindle 1978; Burr 1907, 1908, 1910, 1911, 1912; Chopard 1937; Hebard 1917, 1922; Hincks 1957; Menozzi 1939; Reichardt 1968; Sakai 1970; Scudder 1976; Steinmann 1977, 1978, 1980; Zacher 1911.

Description: Male colour reddish-brown to blackish brown; labrum, legs yellow, femora with dark brown rings; sides of pronotum yellowish. Head weakly broad; postfrontal and coronal sutures distinct. Eyes moderately short, shorter than the length of head behind eyes. First antennal joint long, but shorter than distance between antennal bases. Cuticle of head smooth. Pronotum more or less quadrate, or a little longer than wide; slightly widened posteriorly; median longitudinal furrow present. Tegmina and wings completely absent. Meso- and metanotum normal. Abdomen weakly widened posteriorly, and somewhat depressed; basal tergite weakly and rather finely punctate on basal half, impunctate posteriorly; distal tergites more completely punctate; sides of tergites 7–9 with lateral longitudinal ridges. Ultimate tergite very broad, weakly punctate, punctures rather small, well separated medially, closer laterally; median longitudinal sulcus distinct. Forceps (Fig. 410) asymmetrical, short, dorsal keels prominent; branches trigonal basally, cylindrical apically. Penultimate sternite broadly triangular, narrowed distally, sides slightly sinuate; posterior margin truncate. Genitalia (Fig. 411; gen. prep. No. 579, det. Dr. H. Steinmann) of *Euborellia*-type; central parameral plate V-shaped; median incision of anterior margin deep; genital lobe with distinct denticulated pads and denticulus lobi genitalis. External parameres short, with outer angle usually indicated but not pronounced. – **Female** similar to male, but posterior abdominal tergites more weakly punctate at sides and not rugoso-striate, nor with lateral longitudinal ridges. Forceps slender, straight, tapering, symmetrical. – **Length** of body with forceps: in both sexes: 9–13,5 mm.

Distribution: Cosmopolitan. Occurs in all faunal regions, but is restricted to south temperate, subtropical, and tropical areas.

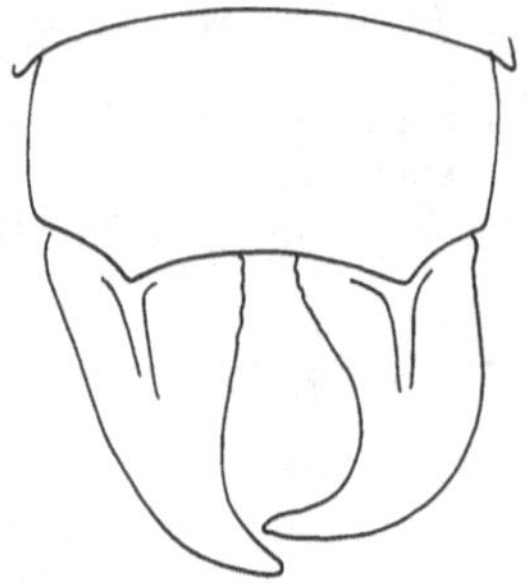

Fig. 410 Male ultimate tergite and forceps of *Euborellia annulipes* (Lucas). Original.

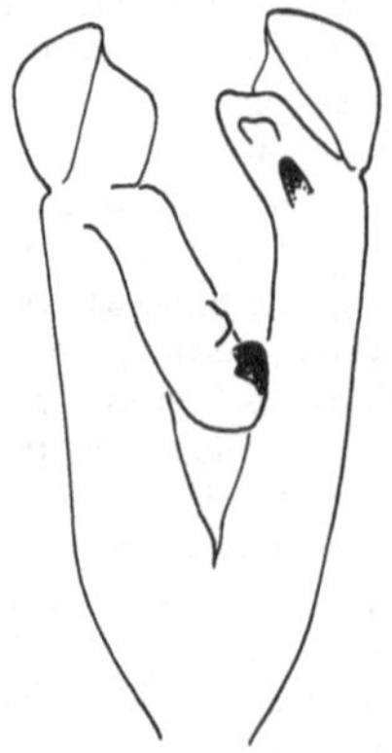

Fig. 411 Male genital armature of *Euborellia annulipes* (Lucas). Original.

16. Euborellia tellinii (BORELLI)

1908 *Anisolabis tellinii* BORELLI, Boll. Soc. ent. ital., Firenze, Genova, **39**: 6. – Terra typica: Eritrea (Type male: unknown locality).
1915 *Euborellia tellinii* – BURR, J. r. microsc. Soc., London, **1915**: 545.
Literature: BRINDLE 1978; BURR 1911, 1912; POPHAM & BRINDLE 1966; SAKAI 1970; STEINMANN 1977, 1978.

Description: Male colour dark reddish-brown to blackish; antennae dark brown, first joint rather darkened and one or more distal joints yellowish; pronotum narrowly yellow laterally; legs yellow, femora with a dark patch on both sides of femora. Head transverse, a little depressed; postfrontal and coronal sutures weel marked; eyes comparatively small, slightly shorter than the length of head behind eyes. First antennal joint typical. Pronotum transverse, widened posteriorly; lateral margins slightly sinuate; median longitudinal furrow depressed anteriorly. Tegmina and wings absent, meso- and metanotum normal. Abdomen weakly widened and depressed, tergites 2–4 weakly and irregularly punctate; sides of tergites 7–9 with lateral longitudinal ridges. Ultimate tergite broad; median longitudinal sulcus depressed postero-medially. Forceps (Fig. 412) asymmetrical; branches strongly trigonal basally, curved and cylindrical apically. Penultimate sternite rounded at sides, slightly truncate apically. Genitalia unknown to me, but external paramere (Fig. 413) short, and broad, external angles well marked, inner membrane wide. – **Female** similar to male, but ultimate tergite with lateral longitudinal ridge, branches of forceps more or less straight, evenly narrowed from base, apices curved dorsally, and almost symmetrical. – **Length** of body with forceps: male: 12–14 mm, female: 12,5–17 mm.

Distribution: Ethiopia.

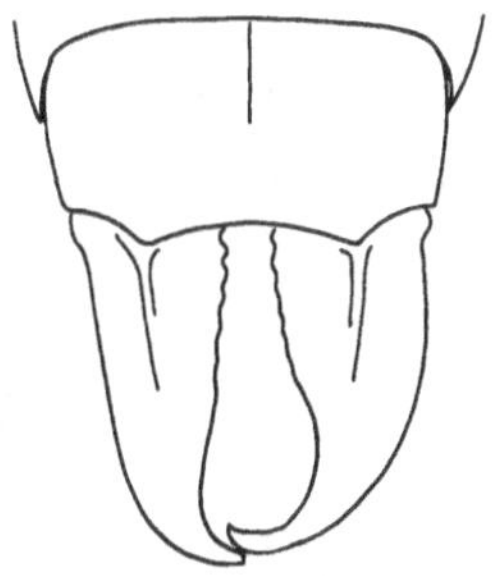

Fig. 412 Male ultimate tergite and forceps of *Euborellia tellinii* (BORELLI). Original.

Fig. 413 External paramere of male genitalia of *Euborellia tellinii* (BORELLI). After BRINDLE.

17. Euborellia truncata BRINDLE

1978 *Euborellia truncata* BRINDLE, Dermaptera of Africa, Tervuren, **2**: 175; fig. 291 (male), 322 (external paramere of male genitalia). – Terra typica: Uganda (Type male: Brit. Mus. nat. Hist., London).

Description: Male colour blackish-brown, shiny; head dark red, antennae brown, basal two joints and most of third yellow, joints 14 and 15 whitish; sides of pronotum yellow, legs yellowish-brown, femora vaguely darkened on sides. Head transverse; postfrontal and coronal sutures weak; eyes moderately small, slightly shorter than the length of head behind eyes. First antennal joint typical, shorter than distance between antennal bases. Pronotum transverse, widened posteriorly; median longitudinal furrow present. Tegmina and wings absent. Abdomen widened and depressed, tergites 1–3 very sparsely and irregularly punctate, tergites 4–5 similar but slightly stronger punctate at sides; sides of tergites 7–9 rugoso-striate, each with well-defined lateral longitudinal ridges. Ultimate tergite broad, median longitudinal sulcus present; posterior margin concave medially between the bases of forceps. Forceps (Fig. 414) asymmetrical, short; branches trigonal basally, curved and cylindrical apically. Penultimate sternite with posterior margin well rounded, but broadly truncate medially. Genitalia of *Euborellia*-type; median incision of anterior margin well marked, deep; genital lobes well developed; external parameres (Fig. 415) almost quadrate, with outer angles prominent. – **Female** similar to male, but sides of tergites 7–9 not rugoso-striate, branches of forceps straighter, and more or less contiguous. – **Length** of body with forceps: male: 10–10,5 mm, female: 10,5–11 mm.

Distribution: Uganda.

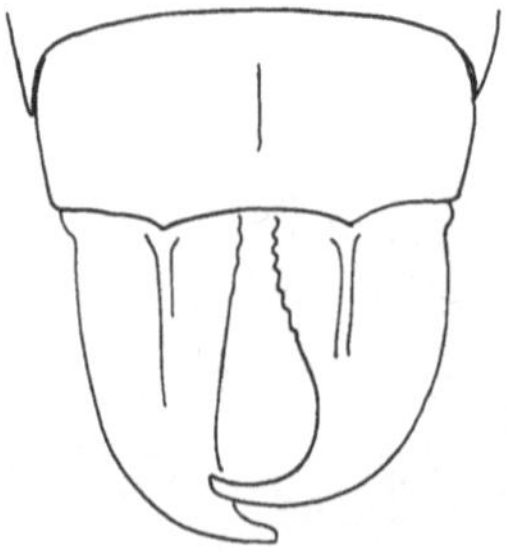

Fig. 414 Male ultimate tergite and forceps of *Euborellia truncata* BRINDLE. Original.

Fig. 415 External paramere of male genitalia of *Euborellia truncata* BRINDLE. After BRINDLE.

18. Euborellia peregrina (MJÖBERG)

1904 *Anisolabis peregrina* MJÖBERG, Ent. Tidskr., Stockholm, **25**: 131. – Terra typica: Brazil (Type female: Naturhist. Riksmus., Stockholm).

1910 *Forficula maxima* BURR, Trans. r. ent. Soc., London, **1910**: 176 [= *Anisolabis peregrina* MJÖBERG 1904].
1920 *Euborellia peregrina* – Hebard, Proc. Acad. nat. Sci. Philadelphia, **1920**: 339 (male description).
Literature: BURR 1911; POPHAM & BRINDLE 1966; REICHARDT 1968; SAKAI 1970; STEINMANN 1978.

Description: Male dark reddish-brown; shiny; pronotum yellow laterally, legs yellow. Head transverse, tumid, posterior angles rounded; postfrontal and coronal sutures distinct. Eyes small, typical; first antennal joint shorter than distance between antennal bases. Pronotum longer than wide, surface smooth, depressed just beyond midpoint transversely; median longitudinal furrow present. Tegmina and wings completely absent. Meso- and metanotum normal. Abdomen broad, depressed, slightly expanded medially. Basal tergites impunctate; distal tergites punctate; sides of tergites 7–9 rugoso-striate, but without lateral longitudinal ridges. Ultimate tergite broad, irregularly punctate, with median longitudinal sulcus; posterior margin concave medially. Forceps asymmetrical, trigonal basally, and cylindrical apically. Penultimate sternite rounded posteriorly. Genitalia typical, paramere with median incision; genital lobes of *Euborellia*-type, with denticulated pads. External parameres (Fig. 417) broad medially. – **Female** similar to male, but abdominal tergites without lateral ridges at sides; forceps (Fig. 416) well developed; branches more or less symmetrical, contiguous. – **Length** of body with forceps: male: 16–17 mm, female: 16,5–17,5 mm.

Distribution: Brazil.

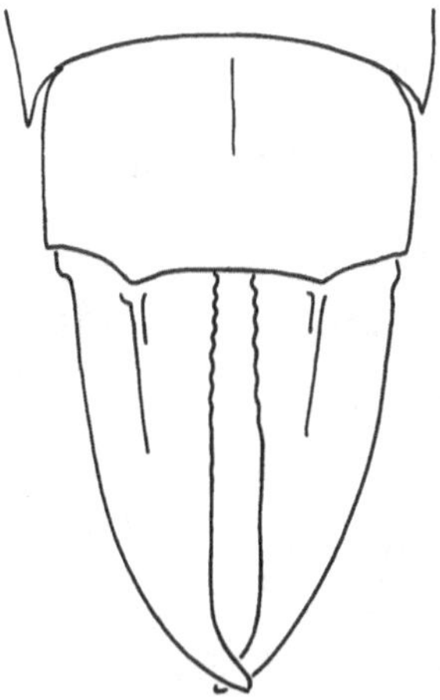

Fig. 416 Female ultimate tergite and forceps of *Euborellia peregrina* (MJÖBERG). Original.

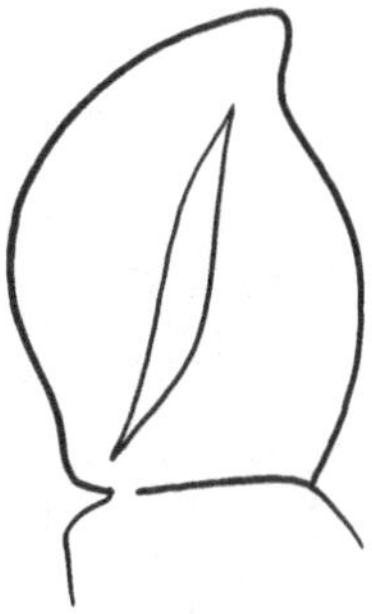

Fig. 417 External paramere of male genitalia of *Euborellia peregrina* (MJÖBERG). After BRINDLE.

19. Euborellia antoni (DOHRN)

1864 *Forcinella antoni* DOHRN, Stettiner ent. Ztg., **25**: 289. – Terra typica: Venezuela (Type male: Inst. zool., Warszawa: gen. prep. No. 364, det. Dr. H. STEINMANN).
1893 *Anisolabis antoni* – BORMANS, Biol. centr. Amer., Orth., 5, pl. 1, fig. 10 (male).
1917 *Anisolabis annulipes* – HEBARD, Ent. News, Philadelphia, **28**: 304 [= *Anisolabis antoni* (DOHRN 1864)]. Incorrect synonymization.
1924 *Euborellia antoni* – HEBARD, Proc. Acad. Sci. nat. Philadelphia, **76**: 114.
1979 *Euborellia antoni* – STEINMANN, Fol. ent. hung., Budapest, **32**: 153, fig. 7 (holotype ultimate tergite and forceps), 8 (holotype genital armature).
Literature: BURR 1911; REHN 1933; REICHARDT 1968; SAKAI 1970; STEINMANN 1977, 1978.

Description: Male colour dark brown to reddish-brown; antennae pale yellow, legs yellow. Head tumid, a little longer than wide; shining; postfrontal and coronal sutures present. Eyes small, shorter than the length of head behind eyes. First antennal joint typical, shorter than distance between antennal bases. Pronotum quadrate; lateral margins a little expanded posteriorly; median longitudinal furrow present. Tegmina and wings absent. Abdomen depressed, abdominal tergites strongly punctate, basal tergites weakly punctate. Ultimate tergite transverse; posterior margin characteristically elongated at origin of basally wide forceps, median section only slightly concave. Forceps (Fig. 418) short, broad, asymmetrical; branches trigonal except at apical end. Penultimate sternite rounded posteriorly. Genitalia (Fig. 419) specific; paramere visibly Y-shaped, median incision of anterior margin wide and deep. Genital lobes not of *Euborellia*-type, one genital lobe recurved, its end not reaching end of median incision, the other proclinate, not reaching apex of external paramere; the latter

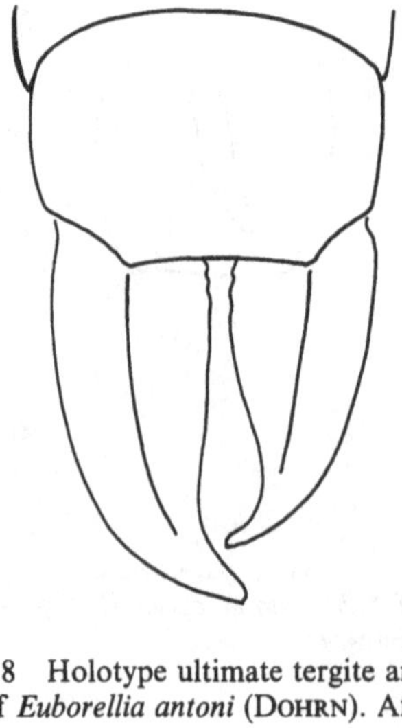

Fig. 418 Holotype ultimate tergite and forceps of *Euborellia antoni* (DOHRN). After STEINMANN.

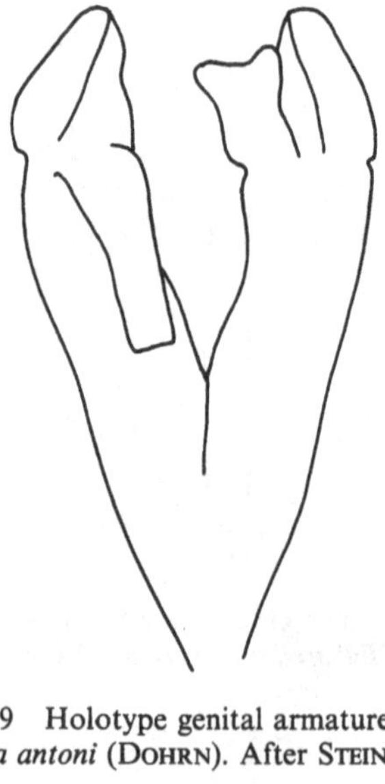

Fig. 419 Holotype genital armature of *Euborellia antoni* (DOHRN). After STEINMANN.

relatively short, apically obtuse, visibly rounded. – **Female** similar to male, but forceps symmetrical, tapering. – **Length** of body with forceps: male 15,5–17 mm, female: 16–17 mm.

Distribution: Central America: Mexico and Panama, and South America: Venezuela, French Guiana, Brazil and Chile.

20. Euborellia uruguayensis BRINDLE

1971 *Euborellia uruguayensis* BRINDLE, Ent. Tidskr., Stockholm, **92**: 10; fig. 9 (external paramere of male genitalia). – Terra typica: Uruguay (Type male: Naturhist. Riksmus., Stockholm).
Literature: STEINMANN 1977, 1978.

Description: Male colour dark reddish-brown; sides of pronotum yellowish, antennae dark brown; joints 12–13 yellow; legs yellow. Head rounded, tumid, depressed along sutures; posterior angles rounded; eyes relatively small, shorter than the length of head behind eyes. First antennal joint normal, shorter than distance between antennal bases. Pronotum quadrate, not greatly widened posteriorly; median longitudinal furrow present in median section. Tegmina and wings completely absent. Meso- and metanotum typical. Abdomen depressed, tergites punctate; sides of tergites 8–9 with lateral longitudinal ridges. Ultimate tergite large, with lateral longitudinal ridges; posterior margin concave at median section between the bases of forceps. Forceps (Fig. 420) slightly asymmetrical; strongly trigonal basally, and curved apically. Penultimate sternite evenly rounded posteriorly. Genitalia (Fig. 421; gen. prep. No. 196, det. Dr. H. STEINMANN) large; central parameral plate with very deep median incision;

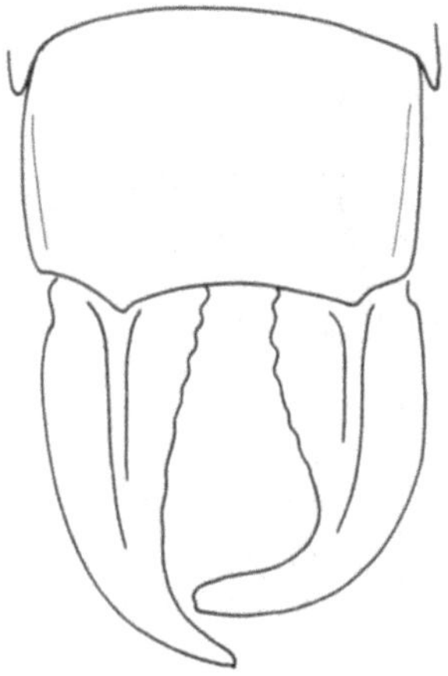

Fig. 420 Male ultimate tergite and forceps of *Euborellia uruguayensis* BRINDLE. Original.

Fig. 421 Male genital armature of *Euborellia uruguayensis* BRINDLE. Original.

genital lobes of *Euborellia*-type, with prominent denticulated pads, and denticulus lobi genitalis. External parameres broad medially. – **Female** unknown. – **Length** of body with forceps: 12–12,5 mm.

Distribution: Uruguay.

21. Euborellia mexicana Brindle

1971 *Euborellia mexicana* Brindle, Ent. Tidskr., Stockholm, **92**: 9; fig. 3 (head and pronotum of holotype), 5 (penultimate sternite of male), 4 (male forceps), 2 (male genitalia). – Terra typica: Mexico (Type male: Naturhist. Riksmus., Stockholm).

Literature: Steinmann 1977, 1978.

Description: Male colour reddish-brown; antennae dark yellowish-brown, but first joint and joints 12–13 yellow. Head rounded; tumid; postfrontal and coronal sutures distinct; eyes small, slightly shorter than the length of head behind eyes; first antennal joint small, shorter than distance between antennal bases. Pronotum more or less as long as wide, lateral margins strongly expanded posteriorly. Tegmina and wings absent. Abdomen punctate dorsally, except tergites 1–3 impunctate; tergites 7–9 rugose laterally, with a well marked lateral ridge on each side of tergites 7–8, the ridge on tergite 9 being irregular and incomplete. Ultimate tergite large, with median longitudinal sulcus. Forceps (Fig. 422) strongly trigonal basally, and curved apically; asymmetrical. Penultimate sternite truncate at apex. Genitalia (Fig. 423) oval, median incision of anterior margin wide and deep; genital lobes of *Euborellia*-type, with denticulated pads. External parameres lemon-shaped. – **Female** similar to male, but abdominal tergites less punctate, tergites not rugose laterally, nor with lateral ridges; forceps as in male, but straight, incurved only at apices. – **Length** of body with forceps: male: 12–13 mm, female: 13 mm.

Distribution: Mexico.

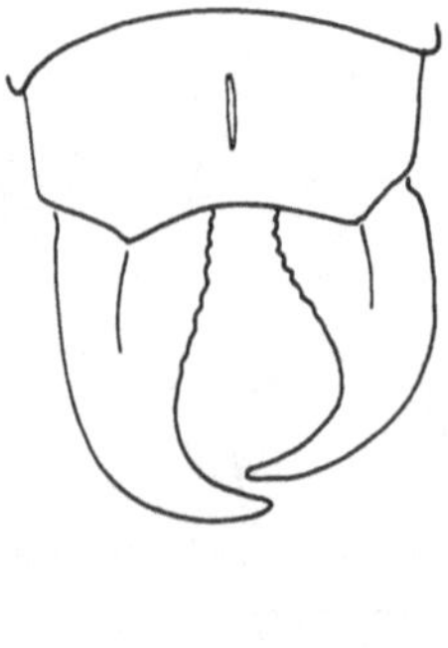

Fig. 422 Holotype ultimate tergite and forceps of *Euborellia mexicana* Brindle. After Brindle.

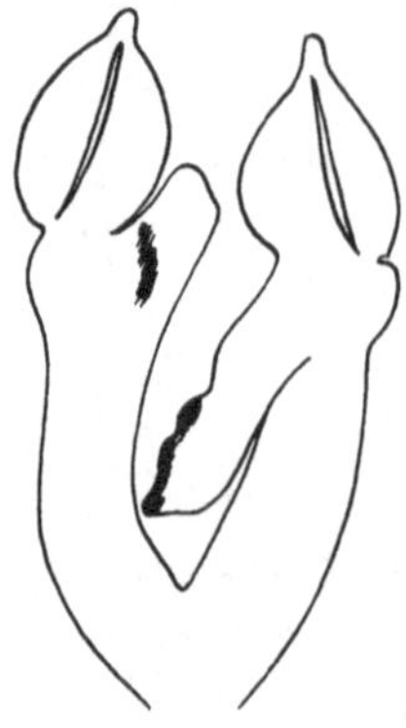

Fig. 423 Holotype genital armature of *Euborellia mexicana* Brindle. After Brindle.

22. Euborellia armata (Borelli)

1906 *Anisolabis armata* Borelli, Boll. Mus. Zool. Anat. comp. Univ. Torino, **21**, No. 531: 4, fig. a (male), b (ultimate tergite with forceps of female). – Terra typica: Costa Rica (Type male: unknown locality).
1910 *Euborellia armata* – Burr, Trans. r. ent. Soc. London, **1910**: 180, pl. 46, fig. 9 (thorax of male).
Literature: Burr 1908, 1911, 1915; Reichardt 1968; Sakai 1970; Steinmann 1977, 1978.

Description: Male colour dark brown, legs and sides of pronotum, yellow. Head tumid, vertex a little depressed; postfrontal and coronal sutures distinct. Eyes small, shorter than the length of head behind eyes. First antennal joint typical, more or less as long as distance between antennal bases. Pronotum transverse, all margins straight, and all angles rounded; median longitudinal furrow present. Tegmina rudimentary, represented by lateral flaps. Wings absent. Abdomen of *Euborellia*-type; tergites punctate, sides of tergites 6–9 normal, not acute. Ultimate tergite broad; median longitudinal sulcus present; posterior margin concave at median section. Forceps (Fig. 424) short, asymmetrical; branches with prominent inner tooth basally. Penultimate sternite rounded posteriorly. Genitalia unknown. – **Female** similar to male, but forceps (Fig. 425) straight, symmetrical, inner tooth at basal section smaller than in male. – **Length** of body with forceps: male: 23–26 mm, female: 22–25 mm.
Distribution: Costa Rica and Brazil.

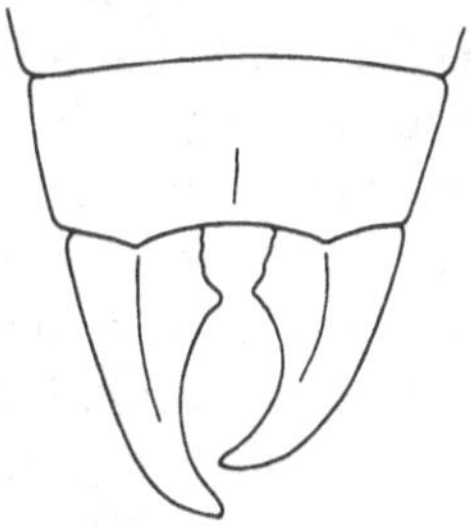

Fig. 424 Male ultimate tergite and forceps of *Euborellia armata* (Borelli). After Borelli.

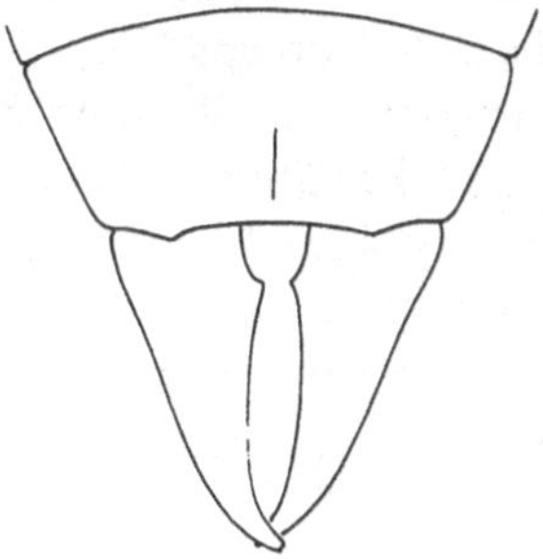

Fig. 425 Female ultimate tergite and forceps of *Euborellia armata* (Borelli). Original.

23. Euborellia plebeja (Dohrn)

1863 *Labidura plebeja* Dohrn, Stettiner ent. Ztg., **24**: 322. – Terra typica: Java (Type female: Inst. Zool., Warszawa).
1900 *Psalis plebeja* – Bormans, Das Tierreich, Berlin, **11**: 38.
1905 *Anisolabis pallipes* Shiraki, Trans. Sapporo nat. Hist. Soc., **1**: 93, fig. 3 (male). – Terra typica: Formosa (Type male: unknown locality; perhaps Formosa Mus., Taipeh).
1909 *Borellia pallipes* – Burr, Rev. russe Ent., St. Petersburg, **1909**: 340.
1911 *Euborellia pallipes* – Burr, Ann. Mag. nat. Hist., London, (8) **8**: 51.
1915 *Landex plebeja* – Burr, J. r. microsc. Soc., London, **1915**: 445.

1927 *Euborellia plebeja* – HEBARD, Proc. Acad. nat. Sci. Philadelphia, **72**: 27.
1932 *Mongolabis aberrans* BORELLI (partim), Journ. Fed. Malay States Mus., **17**: 191.
1936 *Anisolabis plebeja* – BEY-BIENKO, Fauna de l'USSR, Dermaptera, Leningrad: 97.
1937 *Euborellia kiangsuensis* HINCKS, Notes Ent. chinoise (Mus. Heude), Shanghai, **4**: 57. – Terra typica: China: Kiangsu (Type male: Brit. Mus. nat. Hist., London). – **syn. nov.**
1947 *Euborellia minuta* HINCKS, Ann. Mag. nat. Hist., London, (14) **11**: 520 [= *Euborellia plebeja* (DOHRN 1863)].
1954 *Psalis plebeja* – BOESEMAN, Zool. Verh., Leiden, **21**: 45.
1966 *Euborellia aberrans* – POPHAM & BRINDLE, Entomologist, London, **99**: 273.
1970 *Euborellia stali* – SAKAI, Dermapt. Cat. Prael., Tokyo, **1**: 29 [= *Euborellia plebeja* (DOHRN 1863)].
Literature: BORMANS 1900; BURR 1911, 1912; GÜNTHER 1930; HEBARD 1929; HINCKS 1945; NISHIKAWA 1969; STEINMANN 1977, 1978.

Description: Male colour black to blackish brown; shiny; antennae brown, except one to three joints near the apices light; frons, legs yellow; femora and tibiae sometimes darker at each base. Head tumid, rounded; postfrontal and coronal sutures faintly distinct; eyes relatively short, slightly shorter than the length of head behind eyes. First antennal joint normal, but shorter than distance between antennal bases; antennae 16-jointed. Pronotum transverse, slightly rounded posteriorly; median longitudinal furrow present medially. Tegmina rudimentary, represented by lateral flaps on mesonotum. Metasternum slightly emarginate posteriorly, sometimes almost truncate. Abdomen of *Euborellia*-type, a little expanded to median segments; abdominal tergites 7–9 acute and carinate laterally. Ultimate tergite very broad, posterior margin concave at median section. Forceps (Fig. 426) strongly asymmetrical; comparatively short, strongly trigonal basally, and curved apically; inner margin with prominent tooth basally. Penultimate sternite transverse, emarginate at posterior margin. Genitalia (Fig. 427) large, central parameral plate with deep, wide median incision between external parameres; genital lobes specific with denticulated pads, the latter similar to *femoralis* (DOHRN), but apices of genital lobes with chitinous arms. External parameres short, expanded medially, and obtuse

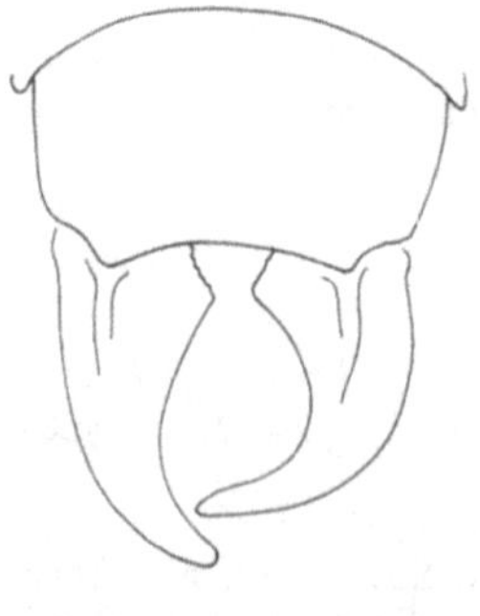

Fig. 426 Male ultimate tergite and forceps of *Euborellia plebeja* (DOHRN). After NISHIKAWA.

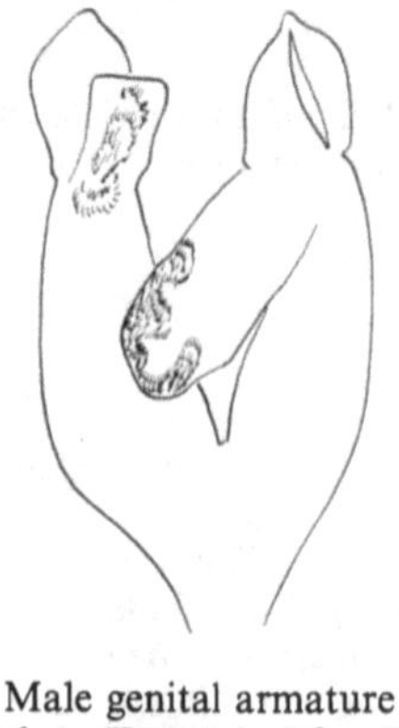

Fig. 427 Male genital armature of *Euborellia plebeja* (DOHRN). After NISHIKAWA.

apically. – **Female** similar to male, but sides of abdominal tergites not acute; ultimate tergite somewhat narrower, more narrowed posteriorly; penultimate sternite rounded posteriorly; forceps symmetrical. – **Length** of body with forceps: male: 10,5–13,5 mm, female: 12,5–14,5 mm.

Distribution: Neotropical region: Costa Rica, Jamaica, Barbados, Bahamas, Cuba, Oriental region: India to Indonesia, Japan, Formosa, Madagascar and Comoro Islands, Australia and New Guinea.

24. Euborellia maxima Moreira

1932 *Euborellia maxima* Moreira, Rev. Ent. São Paulo, **2**: 279; fig. 1 (abdomen with forceps of male), 2 (thorax of male). – Terra typica: Brazil (Type male: Departamento Zool., São Paulo.
Literature: Reichardt 1968; Sakai 1970; Steinmann 1977, 1978.

Description: Male colour dark brown; sides of pronotum and legs yellow; femora with dark brown rings. Head a little longer than wide, posterior margin rounded; postfrontal and coronal sutures distinct. Eyes relatively short, shorter than the length of head behind eyes. First antennal joint elongate, but a little shorter than distance between antennal bases. Pronotum about as long as wide; lateral margins somewhat expanded posteriorly; median longitudinal furrow present. Tegmina rudimentary, represented by lateral flaps on mesonotum. Metanotum normal. Abdomen of *Euborellia*-type, tergites faintly punctate. Ultimate tergite very broad, large; posterior margin concave in median section; median longitudinal sulcus distinct. Forceps (Fig. 428) strongly asymmetrical; branches trigonal basally, cylindrical and curved apically. Penultimate sternite rounded posteriorly. Genitalia unknown. – **Female** unknown. – **Length** of body with forceps: 19–21 mm.

Distribution: Brazil.

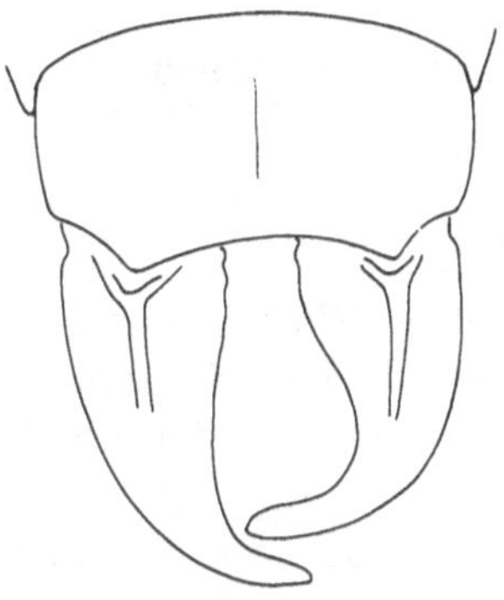

Fig. 428 Male ultimate tergite and forceps of *Euborellia maxima* Moreira. After Moreira.

25. Euborellia nitida Moreira

1932 *Euborellia nitida* Moreira, Rev. Ent. São Paulo, **2**: 279; fig. 3 (female forceps). – Terra typica: Brazil (Type female: Departamento Zool., Saõ Paulo).
Literature: Popham & Brindle 1966; Reichardt 1968; Sakai 1970; Steinmann 1977, 1978.

Description: Female general colour dark brown; sides of pronotum and legs, yellow. Head tumid, a little longer than wide; posterior margin rounded; postfrontal and coronal sutures distinct. Eyes normal, shorter than the length of head behind eyes. First antennal joint typical, shorter than distance between antennal bases. Pronotum a little longer than wide; lateral margins convex posteriorly; all margins more or less straight, and all angles rounded; median longitudinal furrow present. Tegmina rudimentary, represented by lateral flaps on mesonotum. Metanotum typical. Abdomen depressed, tergites 6–9 without carinae laterally. Ultimate tergite broad, with median longitudinal sulcus; posterior margin concave in median section. Forceps (Fig. 429) straight, symmetrical, tapering; trigonal basally and cylindrical apically. – Species very similar to *maxima* MOREIRA. – **Male** unknown. – **Length** of body with forceps: 18 mm.

Distribution: Brazil.

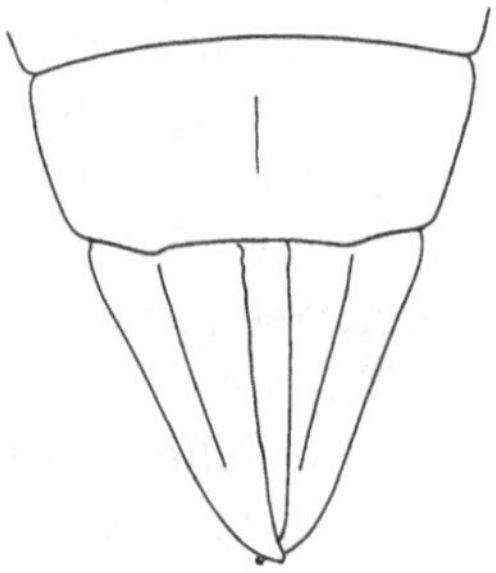

Fig. 429 Holotype ultimate tergite and forceps of *Euborellia nitida* MOREIRA. After MOREIRA.

26. Euborellia flavipes MOREIRA

1931 *Euborellia flavipes* MOREIRA, Rev. Ent. São Paulo, **1**: 258; fig. 1 (male). – Terra typica: Brazil (Type male: Inst. biol. Defesa agricola, Rio de Janeiro).

1931 *Euborellia flavipes* – RIBEIRO, Bol. Mus. nacion., Rio de Janeiro, **7**: 290 (record of female).

Literature: BRINDLE 1971; POPHAM & BRINDLE 1966; REICHARDT 1968; SAKAI 1970; STEINMANN 1977, 1978.

Description: Male colour dark brown, sides of pronotum and legs, yellow. Body very similar to *maxima* MOREIRA. Head tumid, depressed medially; postfrontal and coronal sutures faintly visible. Eyes comparatively small, slightly shorter than the length of head behind eyes. First antennal joint typical, shorter than distance between antennal bases. Tegmina rudimentary, represented by lateral flaps on mesonotum. Metanotum transverse. Femora without dark rings. Abdomen depressed, more or less parallel-sided; sides of tergites 6–9 rugoso-striate. Ultimate tergite very broad, posterior margin concave in median section. Forceps (Fig. 430) asymmetrical; trigonal basally, cylindrical and curved apically. Penultimate sternite rounded posteriorly. Genitalia unknown. – **Female**

similar to male, and similar to *nitida* MOREIRA. Forceps almost symmetrical, dorsal keels prominent; tapering, contiguous. – **Length** of body with forceps: male: 9–10 mm, female: 9–11 mm.

Distribution: Brazil.

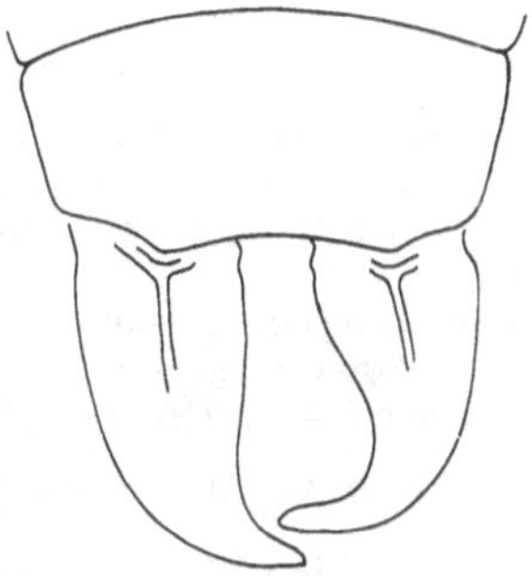

Fig. 430 Holotype ultimate tergite and forceps of *Euborellia flavipes* MOREIRA. After MOREIRA.

27. Euborellia boliviana BRINDLE

1971 *Euborellia boliviana* BRINDLE, Ent. Tidsk., Stockholm, **92**: 10; fig. 10 (external paramere of male genitalia). – Terra typica: Bolivia (Type male: Naturhist. Riksmus., Stockholm).
Literature: STEINMANN 1977, 1978.

Description: Male colour yellowish-brown anteriorly, dark brown posteriorly; shiny; legs yellow, antennae brown to dark brown, but joints 1, 12, and 13, yellow. Head tumid, depressed along postfrontal and coronal sutures; eyes small, shorter than the length of head behind eyes. First antennal joint short, much shorter than distance between antennal bases. Pronotum quadrate, slightly widened posteriorly; median longitudinal furrow almost obsolete. Tegmina very short, about half the length of pronotum, meeting in the mid-line; wings entirely absent. Abdomen of *Euborellia*-type, a little expanded medially; tergites 6–8 with a well-defined lateral longitudinal ridge on each side. Ultimate tergite with lateral ridge on each side, ridge small. Penultimate sternite broadly rounded, apex slightly truncate. Forceps trigonal basally, cylindrical apically, branches almost symmetrical and not strongly curved, nor widely separated at base; inner

Fig. 431 External paramere of holotype of *Euborellia boliviana* BRINDLE. After BRINDLE.

margin scarcely dentate. Genitalia broad, central parameral plate with median incision; genital lobe without prominent denticulated pads, not of *Euborellia*-type; external parameres (Fig. 431) with a large membranous lobe on the inner side. – **Female** unknown. – **Length** of body with forceps: 10,5 mm.

Distribution: Bolivia.

28. Euborellia ambigua (BORELLI)

1906 *Anisolabis ambigua* BORELLI, Boll. Mus. Zool. Anat. comp. Univ. Torino, **21**, No. 531: 3. – Terra typica: Costa Rica (Type female: unknown locality).

1910 *Euborellia ambigua* – BURR, Trans. r. ent. Soc. London, **1910**: 180; pl. 46, fig. 7 (thorax of female).

1921 *Euborellia ambigua* – HEBARD, Trans. amer. ent. Soc., Philadelphia, **47**: 321; pl. XIX, fig. 1 (male), 2 (abdomen laterally), 3 (ultimate tergite and forceps of female).

Literature: BOESEMAN 1954; HEBARD 1917; REICHARDT 1968; SAKAI 1970; STEINMANN 1977, 1978.

Description: Male colour dark brown; sides of pronotum, legs, yellow, its latter sometimes yellowish-brown. Head tumid, rounded, a little longer than broad; postfrontal and coronal sutures fainly visibly. Eyes comparatively small, slightly shorter than the length of head behind eyes. First antennal joint normal, a little longer than distance between antennal bases. Pronotum visibly longer than wide; lateral margins slightly expanded posteriorly; posterior margin rounded; median longitudinal furrow distinct. Tegmina present, short, wings absent. Abdomen depressed, more or less parallel-sided. Ultimate tergite broad, typical. Forceps (Fig. 432) short, strongly curved, dorsal keels prominent, branches trigonal basally, cylindrical apically. Genitalia unknown. – **Female** very similar to male, but ultimate tergite a little narrower, forceps straight, inner margins crenulate. – **Length** of body with forceps: male: 12,5–17 mm, female: 16–17,5 mm.

Distribution: United States: Florida, Bahamas; Cuba, Jamaica, Dominica, Trinidad and Costa Rica.

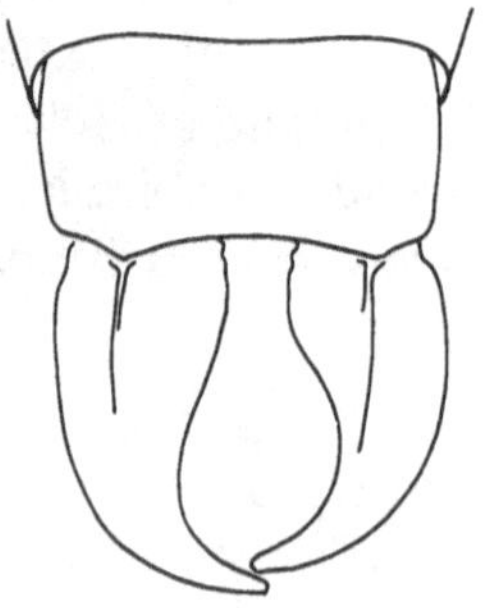

Fig. 432 Male ultimate tergite and forceps of *Euborellia ambigua* (BORELLI). After HEBARD.

29. Euborellia caraibea HEBARD

1921 *Euborellia caraibea* HEBARD, Trans. amer. ent. Soc., Philadelphia, **47**: 322; pl. XIX, fig. 4 (male), 5 (abdomen laterally), 6 (ultimate tergite and forceps of male). – Terra typica: Bahamas (Type male: HEBARD's Coll., Univ. Mus. Philadelphia).

Literature: BRINDLE 1971; KEVAN 1951; REICHARDT 1968; SAKAI 1970; STEINMANN 1977, 1978.

Description: Male very similar to *janeirensis* (DOHRN) and to *ambigua* (BORELLI); colour dark brown or shiny black; antennae brown, but basal joints paler; legs yellowish brown, femora darker on external faces. Head and pronotum similar to *stali* (DOHRN); broad. Postfrontal and coronal sutures well marked. Posterior margin of head rounded. Eyes comparatively large, more or less as long as the length of head behind eyes. Antennae 16-jointed; first joint long, but a little shorter than distance between antennal bases. Pronotum as wide as long; slightly widened posteriorly; median longitudinal furrow present in median section. Tegmina present, large but not well developed, meeting or overlapping along sutures. Wings absent or concealed; a small scutellum visible. Abdomen generally impresso-punctate, tergites 6–9 rugoso-striate laterally and each tergite with a lateral ridge on both sides. Ultimate tergite broad, with a distinct median longitudinal sulcus. Forceps (Fig. 433) asymmetrical; trigonal basally, less strongly curved and almost symmetrical. Penultimate sternite triangularly produced, with apex rather broadly truncate. Genitalia (Fig. 434) broad; median incision of anterior margin of central parameral plate very broad and deep; genital lobes not of *Euborellia*-type; simple. External parameres broad. – **Female** similar to male, but abdominal tergites 6–9 without lateral longitudinal ridges; and forceps slender, tapering, contiguous. – **Length** of body with forceps: male: 14–15 mm, female: 11–12 mm.

Distribution: Greater and Lesser Antilles.

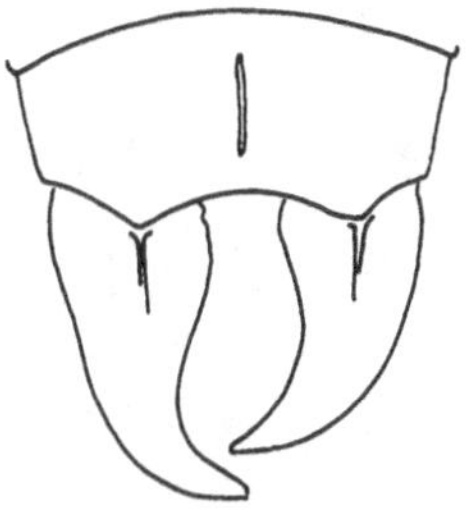

Fig. 433 Male ultimate tergite and forceps of *Euborellia caraibea* HEBARD. After HEBARD.

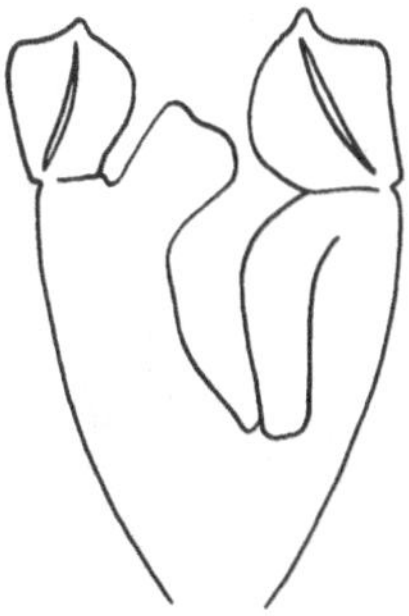

Fig. 434 Male genital armature of *Euborellia caraibea* HEBARD. After BRINDLE.

30. Euborellia janeirensis (DOHRN)

1864 *Forcinella janeirensis* DOHRN, Stettiner ent. Ztg., **25**: 285. – Terra typica: Brazil (Type: female: BRUNNER's Coll., Naturhist. Mus., Wien).

1893 *Anisolabis janeirensis* – BORMANS, Biol. centr. Amer., Orth., 5, pl. 1, fig. 8 (male).

1907 *Borellia janeirensis* – CAUDELL, Proc. U.S. nation. Mus., Washington, **33**: 168.

1910 *Euborellia janeirensis* – BURR, Proc. U.S. nation. Mus., Washington, **38**: 448.

1911 *Borellia janeireusis* – ZACHER, Zool. Jahrb., Jena, **30**: 368; fig. G² (male genitalia).
Literature: BOESEMAN 1954; BRINDLE 1971; BURR 1911, 1915; HEBARD 1921; MENOZZI 1927; REICHARDT 1968; SAKAI 1970; STEINMANN 1977, 1978.

Description: Male general colour blackish-brown; antennae with one or more distal joints whitish or yellow (according to DOHRN: „Basis des ersten und das zweite Glied rothbraun"). Head rounded, a little longer than wide; postfrontal and coronal sutures well marked; eyes normal, more or less as long as the length of head behind eyes. First antennal joint typical, shorter than distance between antennal bases. Pronotum quadrate or a little longer than wide; median longitudinal furrow present; lateral margins parallel, all angles rounded. Tegmina and wings absent. Meso- and metanotum normal. Abdomen depressed; tergites 6–9 not acute, without lateral longitudinal ridges. Ultimate tergite broad, transverse; median longitudinal sulcus present; posterior margin slightly concave at median section. Forceps (Fig. 435) strongly asymmetrical; branches very curved, trigonal basally, cylindrical apically. Penultimate sternite rounded apically. Genitalia (Fig. 436; gen. prep. No. 670, det. Dr. H. STEINMANN) broad; central parameral plate large, median incision of anterior margin deep and wide; genital lobe of *Euborellia*-type, with denticulated pads apically. External parameres broad medially. – **Female** similar to male, but forceps straight, symmetrical; tapering and contiguous. – **Length** of body with forceps: male: 12–14 mm, female: 13–14,5 mm.

Distribution: Brazil, Argentina, Paraguay, and Venezuela.

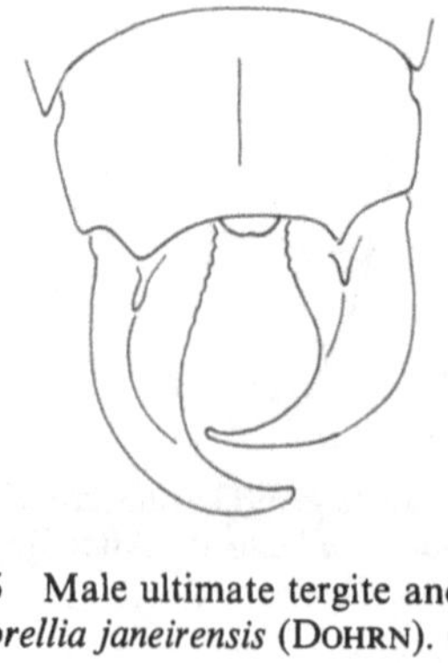

Fig. 435 Male ultimate tergite and forceps of *Euborellia janeirensis* (DOHRN). Original.

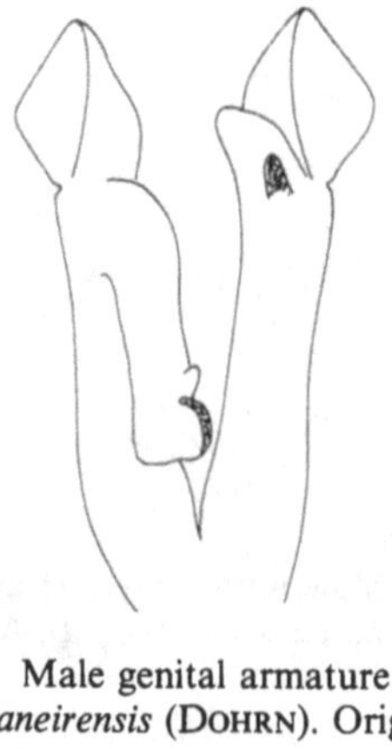

Fig. 436 Male genital armature of *Euborellia janeirensis* (DOHRN). Original.

31. Euborellia tatei HEBARD

1924 *Euborellia tatei* HEBARD, Proc. Acad. nat. Sci. Philadelphia, **76**: 114; fig. 1 (end of abdomen dorsally), 2 (abdomen laterally). – Terra typica: Ecuador (Type female: Acad. Sci., Philadelphia).
Literature: POPHAM & BRINDLE 1966; REICHARDT 1968; SAKAI 1970; STEINMANN 1977, 1978.

Description: **Female** general colour dark brown to blackish-brown; antennae light brown, legs yellowish-brown. Head smooth, rounded, and a little depressed vertically. Postfrontal and coronal sutures very faintly indicated. Eyes comparatively small, slightly shorter than the length of head behind eyes. Antennae with joints short, strongly moniliform. Pronotum weakly through distinctly transverse; lateral margins feebly diverging posteriorly; all angles rounded. Tegmina fully developed, the apices of the wings briefly projecting. Abdomen depressed, a little expanded to middle. Ultimate tergite broad, minutely acute-angulate, produced above the base of both branches of the forceps (Fig. 437). Forceps stout, straight to near apex where they are moderately incurved; inner margins with an even series of very minute, blunt teeth. Penultimate sternite with posterior margin evenly convex. – **Male** similar to female, but forceps somewhat asymmetrical; tapering, trigonal basally. Ultimate tergite with a distinct carina on each side. Genitalia unknown. – **Length** of body with forceps: female: 13–14 mm, male: 13,5–14,5 mm.

Distribution: Ecuador.

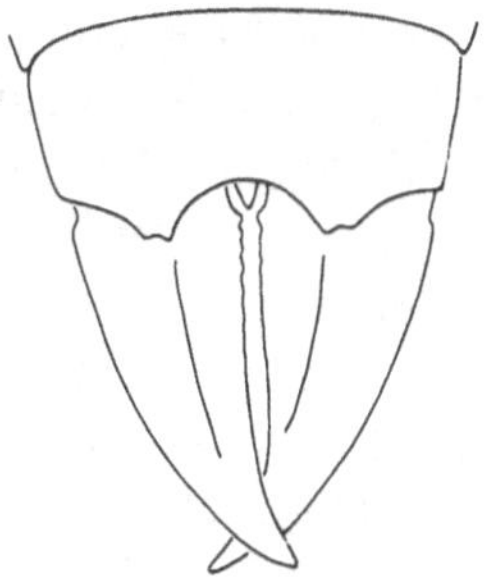

Fig. 437 Female ultimate tergite and forceps of *Euborellia tatei* HEBARD. After HEBARD.

32. Euborellia femoralis (DOHRN)

1863 *Labidura femoralis* DOHRN, Stettiner ent. Ztg., **24**: 321. – Terra typica: Ceylon (Type: female: Zool. Inst., Warszawa).

1879 *Psalis femoralis* – DOHRN, Ann. Mus. civ. Stor. nat., Genova, **14**: 353.

1900 *Anisolabis stali* – BORMANS, Das Tierreich, Berlin, **11**: 45 [= *Labidura femoralis* DOHRN 1863]. Malaysia.

1907 *Anisolabis minuta* CAUDELL, J. New York ent. Soc., **15**: 168. – Terra typica: Costa Rica (Type male: U.S. nation. Collect., Washington).

1910 *Euborellia minuta* – BURR, Proc. U.S. nation. Mus., Washington, **38**: 448.

1915 *Landex femoralis* – BURR, J. r. microsc. Soc., London, **1915**: 420; pl. 9, fig. 11 (male genitalia).

1927 *Euborellia femoralis* – HEBARD, Proc. Acad. nat. Sci. Philadelphia, **72**: 26.

1950 *Euborellia plebeja* – GURNEY, Proc. ent. Soc. Washington, **52**: (4): 202 [= *Euborellia femoralis* (DOHRN 1863)].

1959 *Anisolabis femoralis* – BEY-BIENKO, Ent. Obozr., Moskva, **38**: 606.

1968 *Euborellia kumaoensis* GANGOLA, Entomologist, London, **101**: 55; fig. 1 (head and pronotum of holotype), 2 (thorax and base of abdomen), 3 (sternal plates), 4 (forceps and last dorsal segment). – Terra typica: India (Type male: Zool. Mus. Government College, Nainital, India). – **syn. nov.**

1968 *Euborellia askotensis* GANGOLA, Entomologist, London, **101**: 56. – Terra typica: India (Type male: Zool. Mus. Th. D. S. B. Government College, Nainital, India). – **syn. nov.**

1979 *Euborellia femoralis* – STEINMANN, Fol. ent. hung., Budapest, **32**: 153 ("Holotype female, but terra typica of holotype: Java").

Literature: BOESEMAN 1954; BORELLI 1926; BRINDLE 1972; BURR 1910, 1911, 1912; CAUDELL 1922; GÜNTHER 1930; HEBARD 1929; HINCKS 1947; KIRBY 1904; MENOZZI 1927; POPHAM & BRINDLE 1966; SAKAI 1970; STEINMANN 1977, 1978; ZACHER 1911.

Description: Male colour dark brown, head black, shiny; antennae brown or a little paler; legs yellow with femora black rings; pronotum blackish-brown, but lateral margins yellowish; tegmina with reddish-brown spots, wings dark brown with inner margins yellow; forceps dark reddish-brown. Head broad, lustrous black; postfrontal sutures and coronal suture faintly distinct; eyes relatively large, as long as first antennal joint. Antennae 19-jointed; first joint about as long as distance between antennal bases. Pronotum rather longer than wide, visibly rounded broadly at posterior margin; median longitudinal furrow present. Tegmina and wings fully developed. Abdomen more or less parallel-sided, or sometimes a little expanded medially. Abdominal tergites 7–9 with lateral longitudinal ridges. Ultimate tergite broad, with distinct median longitudinal sulcus. Posterior margin of last tergite concave in median section. Forceps (Fig. 438) asymmetrical, comparatively short, strongly trigonal basally, curved and cylindrical apically. Penultimate sternite broadly rounded at posterior margin. Genitalia (Fig. 439; gen. prep. No. 297, det. Dr. H. STEINMANN) similar to *annulipes* (LUCAS), but denticulated pads within genital lobe rectangular. – **Female** similar to male, but forceps symmetrical, tapering and contiguous. – **Length** of body with forceps: male: 14,5–16,5 mm, female: 14–15,5 mm.

Distribution: Oriental region: from India, Sri Lanka eastwards through South Asia and the Sunda Islands to Borneo and the Philippines. Neotropical region: Arizona to Bahamas, and Caribbean. Madagascar, New Guinea to Micronesia.

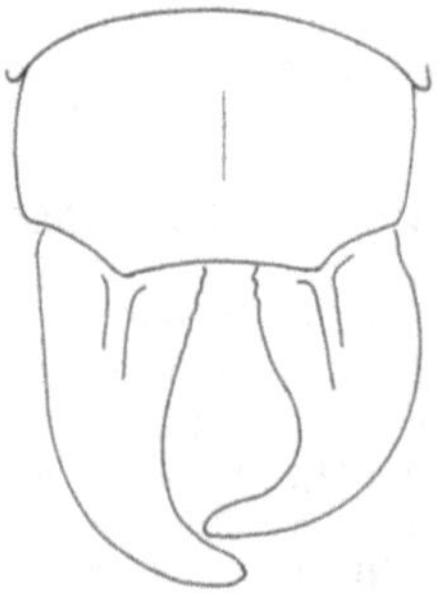

Fig. 438 Male ultimate tergite and forceps of *Euborellia femoralis* (DOHRN). Original.

Fig. 439 Male genital armature of *Euborellia femoralis* (DOHRN). Original.

33. Euborellia insulana (BORELLI)

1912 *Psalis insulana* BORELLI, Bull. Mus. nation. Hist. nat., Paris, **1912**: 225. – Terra typica: Comoro Island (Type female: unknown locality).
1915 *Euborellia insulana* – BURR, J. r. microsc. Soc., London, **1915**: 544; pl. 12, fig. 14 (male genitalia).
Literature: BURR 1914; MENOZZI 1927; POPHAM & BRINDLE 1966; SAKAI 1970; STEINMANN 1977, 1978.

Description: Male general colour brown or dark brown; pronotum brown with yellowish margins laterally; legs yellow. Head tumid, postfrontal and coronal sutures indistinct. Eyes comparatively small, slightly shorter than the length of head behind eyes. Antennae 15-jointed; first joint long, more or less as long as distance between antennal bases. Posterior angles of head rounded. Pronotum a little wider than long; lateral margins convex posteriorly; median longitudinal furrow present. Tegmina and wings well developed. Abdomen widened medially, a little depressed; tergites faintly punctate. Ultimate tergite transverse, median longitudinal sulcus present; posterior margin concave medially. Forceps strongly trigonal basally, and curved apically. Penultimate sternite typical, rounded posteriorly. Genitalia (Fig. 440) broad, central parameral plate oval, large, median incision of anterior margin deep but narrow; genital lobes well developed, without denticulated pads, but with chitinius serration apically. External parameres broad, outer margin rounded. – **Female** similar to male, but forceps straight, curved only at apical end. – **Length** of body with forceps: in both sexes: 11–12,5 mm.

Distribution: Grand Comoro Islands, and Seychelles.

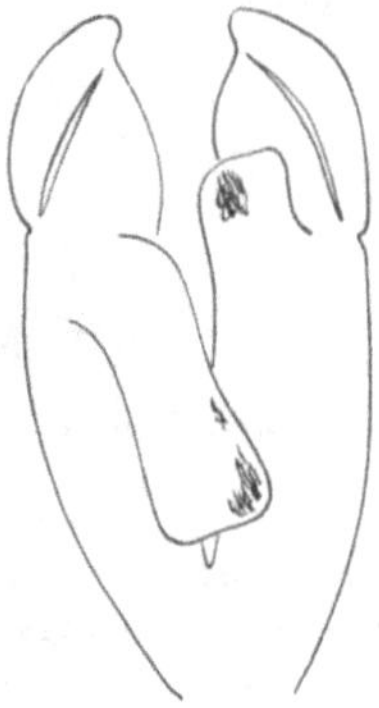

Fig. 440 Male genital armature of *Euborellia insulana* (BORELLI). After BURR.

34. Euborellia malgacha BRINDLE

1966 *Euborellia malgacha* BRINDLE, Trans. r. ent. Soc. London, **118**: 236; fig. 28 (holotype genital armature). – Terra typica: Madagascar (Type male: Inst. Recherches sci. à Madagascar).
1969 *Euborellia malgacha* – BRINDLE, Faune de Madagascar, Tananarive, **30**: 39; fig. 28 (male genitalia), 30 (male), "holotype male, deposited in the Mus. nation. Hist. nat., Paris".
Literature: SAKAI 1970; STEINMANN 1977, 1978.

Description: Male black, shiny; body and abdomen dark brown; sides of pronotum yellowish. Head tumid; postfrontal and coronal sutures faintly visible. Eyes small, shorter than the length of head behind eyes. Antennae 13-jointed; first joint normal, about as long as the distance between antennal bases. Pronotum quadrate, slightly expanded posteriorly; median longitudinal furrow present. Tegmina well developed, dark brown, finely punctate. Wings shiny. Abdomen more or less smooth, a little expanded to median segments. Ultimate tergite rugose posteriorly, transverse, with two small tubercles on posterior border. Forceps short, broad, trigonal basally, cylindrical apically, tapering, distal part narrowed and curved mesad. Penultimate sternite typical. Genitalia (Fig. 441) slightly V-shaped, median longitudinal incision between external parameres very broad and deep; genital lobes of *Euborellia*-type, with denticulated pads apically; denticulus lobi genitalis prominent. External parameres broad. – **Female** unknown. – **Length** of body with forceps: 10–11 mm.

Distribution: Madagascar.

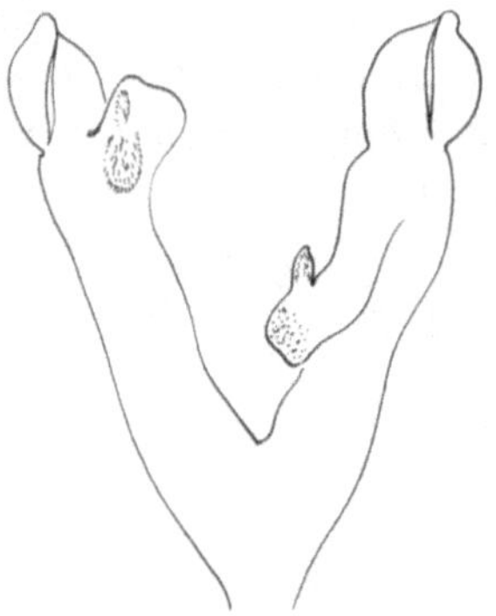

Fig. 441 Holotype genital armature of *Euborellia malgacha* BRINDLE. After BRINDLE.

35. Euborellia brunneri (DOHRN)

1864 *Forcinella brunneri* DOHRN, Stettiner ent. Ztg., **25**: 291. – Terra typica: Australia (Type female: BRUNNER's Coll., Naturhist. Mus., Wien).
1900 *Anisolabis brunneri* – BORMANS, Das Tierreich, Berlin, **11**: 48.
1911 *Gonolabis brunneri* – BURR, Genera Insectorum, Bruxelles, **122**: 28.
1914 *Mongolabis brunneri* – BURR, Ann. Mag. nat. Hist., London, (8) **13**: 581.
1915 *Gonolabis verhoeffi* – BURR, J. r. microsc. Soc., London, 1915: 528 [= *Mongolabis brunneri* (DOHRN 1864)].
1966 *Euborellia brunneri* – POPHAM & BRINDLE, Entomologist, London, **99**: 273.
Literature: BURR 1912; KIRBY 1904; SAKAI 1970; STEINMANN 1977, 1978.

Description: Male colour black or blackish-brown; clypeus and labrum brown, antennae yellowish-brown with yellow basal joints; sides of pronotum yellowish laterally. Head and body similar to *annulipes* (LUCAS). Postfrontal and coronal sutures faintly visible; eyes comparatively small, shorter than first antennal joint. Antennae 14-jointed; first joint normal, a little shorter than distance between antennal bases. Pronotum more or less quadrate; lateral margin a little expanded posteriorly; median longitudinal furrow in median section. Tegmina and wings completely absent. Meso- and metanotum typical, transverse. Abdomen slightly depressed, tergites 6–9 without lateral longitudinal ridges. Ultimate tergite very broad, posterior margin with two prominent tubercles at bases of forceps. Forceps (Fig. 442) short, strongly asymmetrical; dorsal surface with large, blunt, prominent tooth basally; branches slightly trigonal basally, curved apically. Penultimate sternite rounded posteriorly. Genitalia (Fig. 443; gen. prep. No. 700, det. Dr. H. STEINMANN) characteristic; not *Euborellia*-type; central parameral plate robust, genital lobes well developed, without denticulated pads, but with chitinous sclerotized plates. External parameres broad in median section. – **Female** very similar to male, but forceps symmetrical, tapering, contiguous. – **Length** of body with forceps: male: 15,5–16,5 mm, female: 15–17 mm.

Distribution: South Australia and Tasmania.

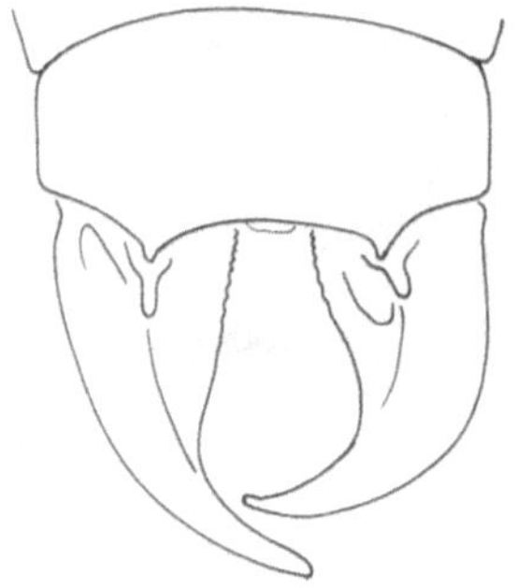

Fig. 442 Male ultimate tergite and forceps of *Euborellia brunneri* (DOHRN). Original.

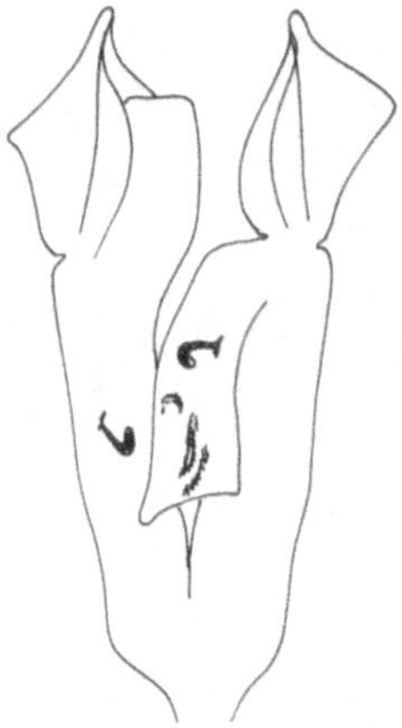

Fig. 443 Male genital armature of *Euborellia brunneri* (DOHRN). Original.

36. Euborellia eteronoma (BORELLI) comb. nov.

1909 *Anisolabis eteronoma* BORELLI, Boll. Lab. Zool. gen. agr. Portici, 3: 315. – Terra typica: Hawaii (Type male: Mus. Ist. Zool. sist., Torino; gen. prep. No. 736, det. Dr. H. STEINMANN).

Literature: BURR 1910, 1914; POPHAM & BRINDLE 1966; SAKAI 1970; STEINMANN 1977, 1978; ZIMMERMAN 1948.

Description: Male colour dark brownish red to black; antennae light brown, legs yellowish brown. Head tumid, a little depressed vertically; postfrontal and coronal sutures indistinct; head slightly rounded posteriorly. Eyes moderately small, slightly shorter than the length of head behind eyes. Antennae 18-jointed, first joint long, longer than eye, but shorter than distance between antennal bases. Pronotum quadrate, lateral margins more or less parallel, all angles rounded; median longitudinal furrow in median portion. Tegmina and wings absent. Meso- and metanotum transverse, typical. Abdomen visibly widened to abdominal tergites 5–6; sides of tergites 5–9 rugoso-striate, but without lateral longitudinal ridges. Ultimate tergite large, broad; dorso-lateral surface with faint keels laterally; posterior margin concave at median section. Forceps (Fig. 444) comparatively slender, asymmetrical, trigonal basally, cylindrical apically. Penultimate sternite normal. Genitalia (Fig. 445) large, broad, central parameral plate H-shaped, median incision between external parameres wide and deep; genital lobe not *Euborellia*-type, without denticulated pads, but with sclerotized plates apically. External paramere rounded, obtuse apically. – **Female** similar to male, but forceps slenderer, tapering. – **Length** of body with forceps: male: 19–20 mm, female: 22–24,5 mm.

Distribution: Hawaii.

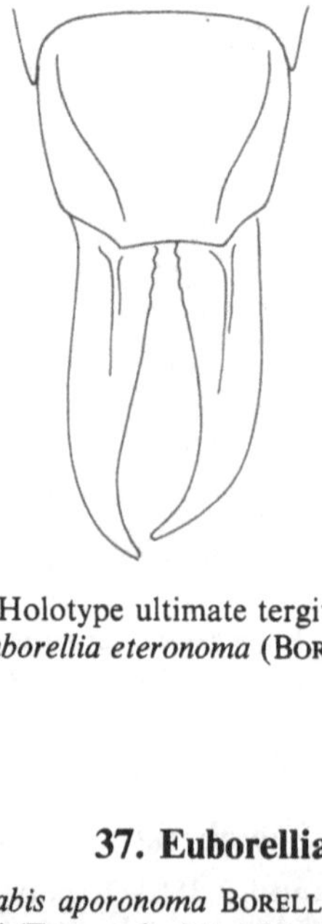

Fig. 444 Holotype ultimate tergite and forceps of *Euborellia eteronoma* (BORELLI). Original.

Fig. 445 Holotype genital armature of *Euborellia eteronoma* (BORELLI). Original.

37. Euborellia aporonoma (BORELLI) comb. nov.

1909 *Anisolabis aporonoma* BORELLI, Boll. Lab. Zool. gen. agr., Portici, **3**: 317. – Terra typica: Hawaii (Type male: Mus. Ist. Zool. sist., Torino; gen. prep. No. 737, det. Dr. H. STEINMANN).

1970 *Anisolabis aporonoma* – SAKAI, Dermapt. Cat. Prael., Tokyo, **1**: 36 [= *Euborellia annulipes* (LUCAS 1847)].

Literature: BURR 1910, 1914; POPHAM & BRINDLE 1966; SAKAI 1970; STEINMANN 1977, 1978.

Description: Male colour dark brownish red to black; antennae and abdomen, dark brown, forceps a little reddish; sides of pronotum and legs, yellowish. Head rounded, tumid, postfrontal sutures and coronal suture distinct. Eye normal, small, shorter than the length of head behind eyes. Antennae 16-jointed; first joint typical, more or less as long as distance between antennal bases. Pronotum more or less as long as wide; lateral margins a little expanded posteriorly; median longitudinal furrow present; all angles rounded. Tegmina and wings absent. Meso- and metanotum broad, normal. Abdomen similar to *annulipes* (LUCAS), somewhat expanded to median segments. Sides of tergites 6–9 rugoso-striate, but without lateral longitudinal ridges. Ultimate tergite comparatively small, elongated; posterior margin concave at median section. Forceps (Fig. 446) short, asymmetrical; branches strongly trigonal basally, and curved apically. Penultimate sternite rounded apically. Genitalia (Fig. 447) of *Euborellia*-type; central parameral plate V-shaped, large, median incision between external parameres deep and wide anteriorly; genital lobe with denticulated pads. External parameres rounded. – **Female** similar to male, but abdominal tergites 6–9 with lateral carinae; forceps straight, symmetrical. – **Length** of body with forceps: male: 14,5–18 mm, female: 15,5–23 mm.

Distribution: Hawaii.

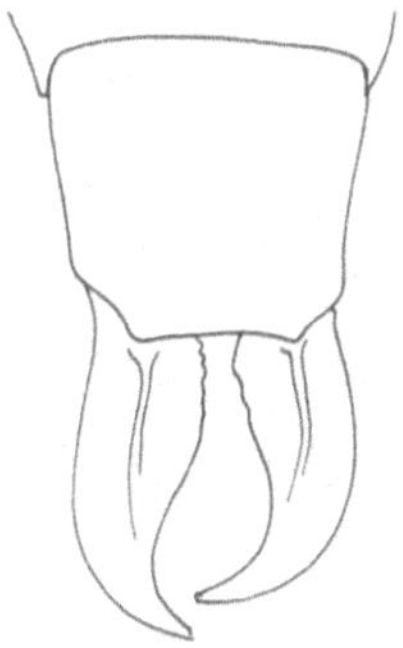

Fig. 446 Holotype ultimate tergite and forceps of *Euborellia aporonoma* (BORELLI). Original.

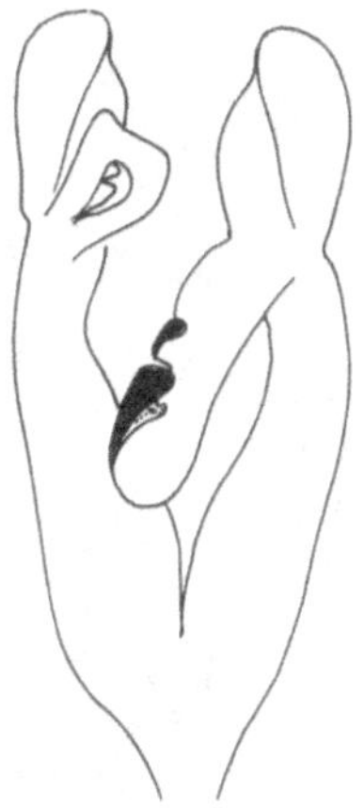

Fig. 447 Holotype genital armature of *Euborellia aporonoma* (BORELLI). Original.

38. Euborellia dattai SRIVASTAVA

1977 *Euborellia dattai* SRIVASTAVA, Oriental Insects, Delhi, **11**: 182; fig. 1e (thorax) 1f (ultimate tergite and forceps of male), 1g (posterior margin of penultimate sternite), 1h (holotype genital armature). – Terra typica: India (Type male: Zool. Survey India, Calcutta).

Description: Male general colour brownish yellow; legs yellowish with femora brown; sides of pronotum yellow; antennae with one or two joints whitish apically. Head smooth; postfrontal and coronal sutures obliterated. Eyes small, shorter than the length of head behind eyes. Antennae 17-jointed; first joint stout, gently expanded apically; a little shorter than distance between antennal bases. Pronotum a little longer than wide; lateral margins straight, but reflexed, concave in middle, slightly diverging posteriorly; median longitudinal furrow present. Tegmina and wings entirely absent. Mesonotum with straight, transverse posterior margin. Abdomen slender, scarcely enlarged in middle; sides of tergites without lateral longitudinal ridges. Ultimate tergite transverse, smooth, finely pubescent, gently sloping and narrowing backwards; median longitudinal sulcus short. Forceps (Fig. 448) more or less straight, curved only apically; trigonal basally, cylindrical apically. Penultimate sternite transverse, posterior margin broadly rounded. Genitalia (Fig. 449) narrow; median incision of anterior margin of central parameral plate deep and narrow; genital lobes well developed; lobes with denticulated pads. External parameres typical. – **Female** similar to male, but head comparatively less convex with sutures distinct and eyes less prominent. – **Length** of body with forceps: male: 10,5–11 mm, female: 9,5–10 mm.

Distribution: India.

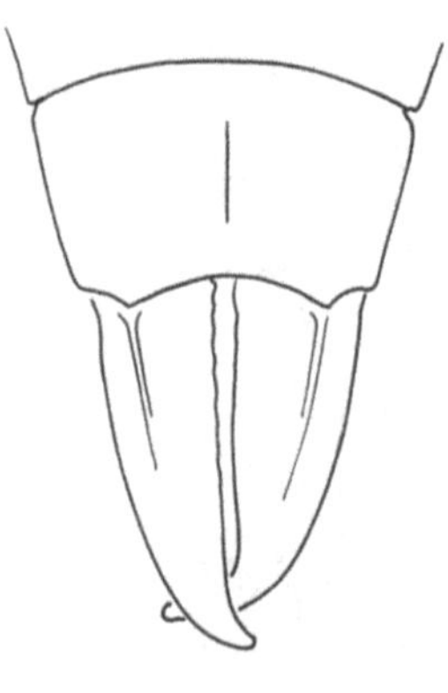

Fig. 448 Holotype ultimate tergite and forceps of *Euborellia dattai* SRIVASTAVA. After SRIVASTAVA.

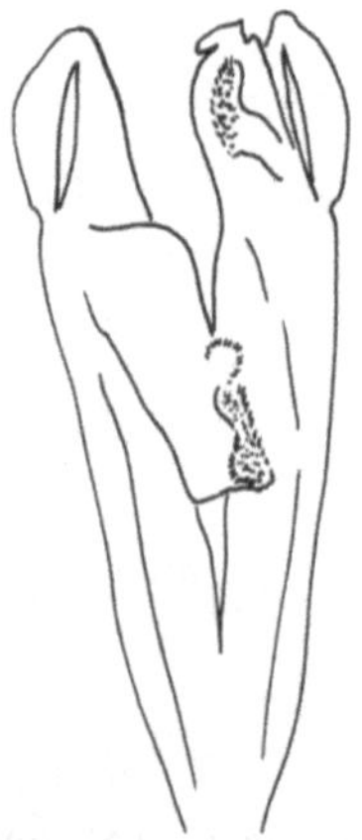

Fig. 449 Holotype genital armature of *Euborellia dattai* SRIVASTAVA. After SRIVASTAVA.

39. Euborellia sakaii STEINMANN

1978 *Euborellia sakaii* STEINMANN, Dtsch. ent. Zeitschr., Berlin, (n. F.) **25**: 186 (**nom. nov.** for *Anisolabis* (*Euborellia*) *flavipes* BEY-BIENKO, 1967, Acta ent. bohemoslov., Praha, **64**: 432; fig. 17 (male genitalia). – Terra typica: Afghanistan (Type male: Természettudományi Múz., Budapest).

Description: Male colour reddish-brown; antennae brownish, but first and second antennal joints yellowish-brown; legs yellow. Head tumid, postfrontal and coronal sutures distinct; eyes moderately small, shorter than first antennal joint, and shorter than the length of head behind eyes. First antennal joint relatively long, more or less as long as distance between antennal bases. Pronotum rather longer than wide; lateral margins widened posteriorly; median longitudinal furrow distinct; all angles rounded. Tegmina represented by lateral flaps on mesonotum. Wings absent. Metanotum typical. Abdomen of *Euborellia*-type, widened medially; all tergites punctate dorsally; sides of tergites 6–9 rugoso-striate. Ultimate tergite broad; posterior margin with two tubercles at bases of forceps. Forceps (Fig. 450) asymmetrical; branches strongly trigonal basally, curved and cylindrical apically. Penultimate sternite typical, rounded posteriorly. Genitalia (Fig. 451) characteristic; central parameral plate oval, median incision of anterior margin very deep; genital lobes specific, not *Euborellia*-type, with chitinous plates. External parameres widened medially. – **Female** similar to male, but forceps symmetrical, and body somewhat shiny. – **Length** of body with forceps: male: 9,5–11 mm, female: 9–13 mm.

Distribution: Afghanistan.

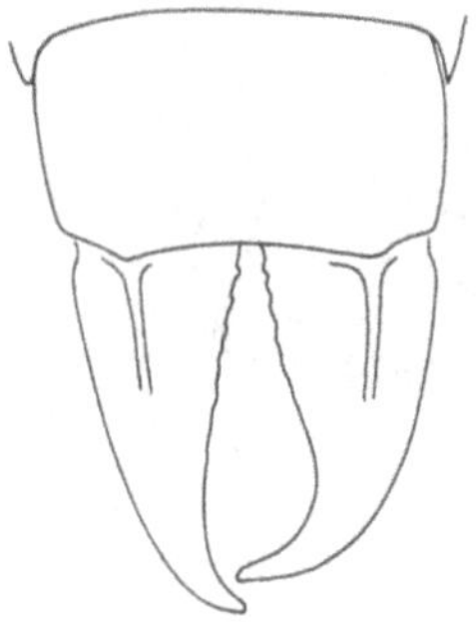

Fig. 450 Holotype ultimate tergite and forceps of *Euborellia sakaii* STEINMANN. After BEY-BIENKO.

Fig. 451 Holotype genital armature of *Euborellia sakaii* STEINMANN. After BEY-BIENKO.

40. Euborellia rajasthanensis SRIVASTAVA

1977 *Euborellia rajasthanensis* SRIVASTAVA, Oriental Insects, Delhi, **11**: 184; fig. 2a (pronotum), 2b (posterior margin of penultimate sternite), 2c (ultimate tergite and forceps of holotype), 2d (holotype genital armature), 2e (female ultimate tergite and forceps). – Terra typica: India (Type male: Zool. Survey India, Calcutta).

Description: Male general colour blackish brown, sides of pronotum and legs yellowish. Head triangular, smooth, shiny; postfrontal and coronal sutures faint; eyes comparatively small, shorter than the length of head behind eyes. Antennae (broken in holotype); first joint normal, a little shorter than distance between antennal bases. Pronotum a little longer than wide; lateral margins slightly widened posteriorly; median longitudinal furrow present at anterior half; posterior angles and margin rounded. Tegmina and wings absent. Meso- and metanotum smooth, transverse. Abdomen obscurely punctate, very slightly dilated in middle, sides of tergites 6–9 rugoso-striate, but without lateral longitudinal ridges. Ultimate tergite broad; median longitudinal sulcus present; gently contracted posteriorly. Forceps (Fig. 452) stout, trigonal in basal one third, afterwards depressed, tapering apically; branches almost straight in basal two thirds, gently curved apically. Penultimate sternite faintly punctate, narrowed apically with posterior margin scarcely emarginate in middle. Genitalia (Fig. 453) comparatively small; central parameral plate with deep, wide median incision between external parameres; genital lobes of *Euborellia*-type, with characteristic denticulated pads apically. External parameres short and broad. – **Female** similar to male, but forceps subcontiguous, apices pointed and crossing. – **Length** of body with forceps: male: 12–12,5 mm, female: 13,5–14 mm.

Distribution: India.

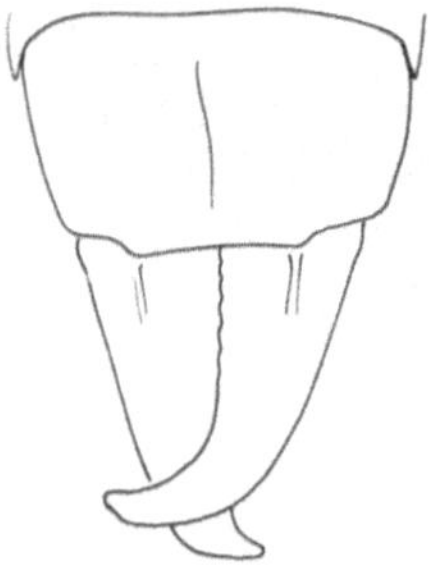

Fig. 452 Holotype ultimate tergite and forceps of *Euborellia rajasthanensis* SRIVASTAVA. After SRIVASTAVA.

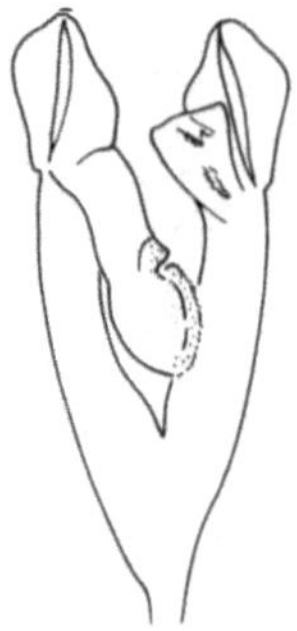

Fig. 453 Holotype genital armature of *Euborellia rajasthanensis* SRIVASTAVA. After SRIVASTAVA.

41. Euborellia punctata BORELLI

1927 *Euborellia punctata* BORELLI, Boll. Lab. Zool. gen. agr., Portici, **20**: 62; fig. 1 (abdomen with forceps of holotype), 2 (end of abdomen with forceps of female), 3 (holotype genital armature). – Terra typica: China: Nanking (Type male: Mus. Ist. Zool. sist., Torino, holotype without penis).

Literature: HEBARD 1933; SAKAI 1970; STEINMANN 1977, 1978.

Description: Male colour dark brown, with yellow legs; femora with rings, black or blackish-brown; antennae light brown; sides of pronotum yellowish-brown, laterally. Head normal, rounded posteriorly; posterior margin more or less as wide as pronotum; eyes small, a little shorter than the length of head behind eyes. Antennae 16-jointed; first joint more or less as long as distance between antennal bases. Pronotum rather longer than wide; lateral margins straight, slightly widened posteriorly; median longitudinal furrow present. Tegmina and wings completely absent. Meso- and metanotum as *annulipes* (LUCAS). Abdomen faintly punctate; tergites 7–9 with sides rugose on both sides. Ultimate tergite broad; posterior margin characteristic, strongly concave medially; posterior margin with two tubercles at bases of forceps. Forceps (Fig. 454) short, asymmetrical; branches trigonal basally, and curved apically. Penultimate sternite typical. Genitalia (Fig. 455) similar to *annulipes* (LUCAS), but genital lobes specific, as in fig. 455. – **Female** very similar to male, but forceps symmetrical. – **Length** of body with forceps: male: 12,5–14,5, female: 14–16 mm.

Distribution: China: Nanking and Hong-Kong.

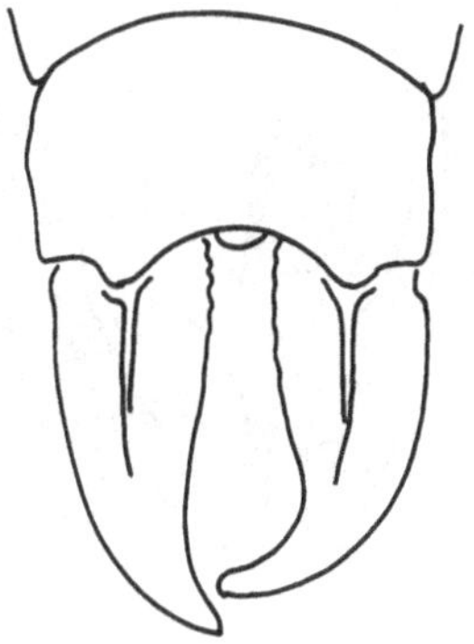

Fig. 454 Holotype ultimate tergite and forceps of *Euborellia punctata* BORELLI. After BORELLI.

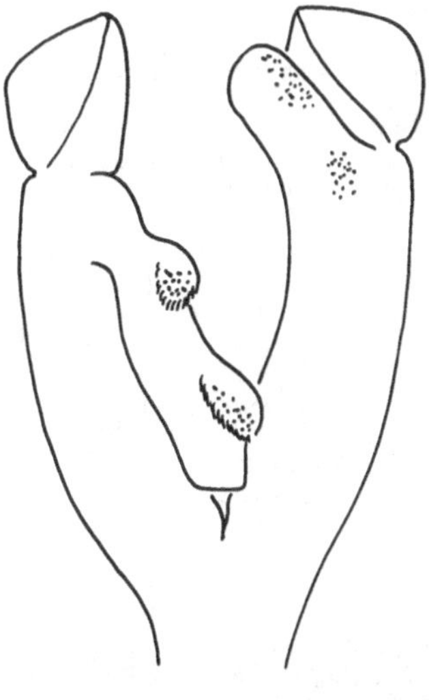

Fig. 455 Holotype genital armature of *Euborellia punctata* BORELLI. After BORELLI.

42. Euborellia abbreviata SRIVASTAVA

1977 *Euborellia abbreviata* SRIVASTAVA, Oriental Insects, Delhi, **11**: 186; fig. 2f (pronotum), 2g (posterior margin of penultimate sternite), 2h (ultimate tergite and forceps), 2i (holotype genital armature). – Terra typica: India (Type male: Zool. Survey India, Calcutta).

Description: Male colour blackish brown; sides of pronotum and legs yellowish or yellowish-brown. Head triangular, smooth; postfrontal and coronal sutures faint; posterior margin emarginate in middle. Eyes small, slightly shorter than the length of head behind eyes. First antennal joint long, more or less as long as distance between antennal bases. Pronotum a little longer than wide; lateral margins straight, but very slightly expanded posteriorly; median longitudinal furrow present. Tegmina abbreviated; narrow ovate flaps on metanotum laterally. Wings absent. Abdomen gently widened in middle, faintly punctate above, sides of tergites 6–9 rugosely punctate, acute posteriorly. Ultimate tergite broad, very slightly narrowed posteriorly, median longitudinal sulcus faint. Forceps (Fig. 456) strongly curved, asymmetrical; branches trigonal basally with strong dorsal keels, and cylindrical apically. Penultimate sternite punctate, posterior margin subtruncate in middle. Genitalia (Fig. 457) characteristic; central parameral plate V-shaped, median incision of anterior margin very deep and wide; genital lobes of *Euborellia*-type, with denticulated pads apically. External parameres short, broad, obtuse. – **Female** unknown. – **Length** of body with forceps: 12,5–13 mm.

Distribution: India

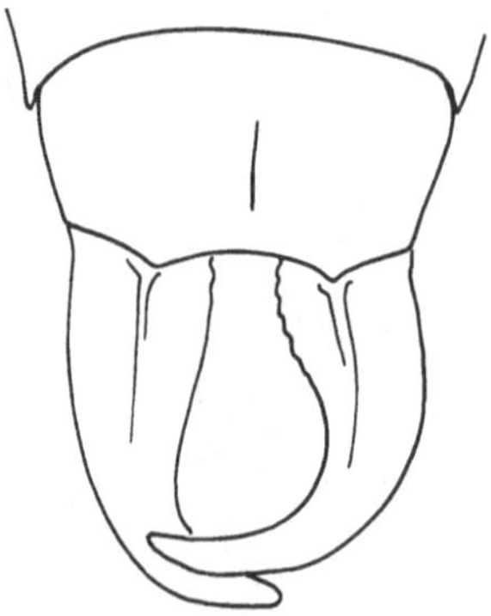

Fig. 456 Holotype ultimate tergite and forceps of *Euborellia abbreviata* SRIVASTAVA. After SRIVASTAVA.

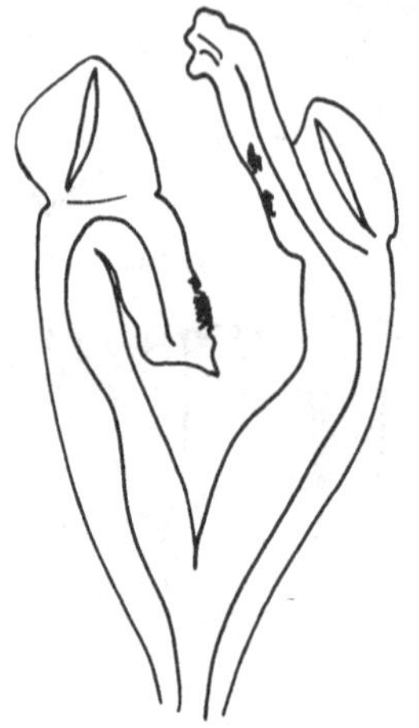

Fig. 457 Holotype genital armature of *Euborellia abbreviata* SRIVASTAVA. After SRIVASTAVA.

43. Euborellia annandalei (BURR)

1906 *Anisolabis annandalei* BURR, J. Proc. asiat. Soc. Bengal, Calcutta, (n. S.) **2**: 389. – Terra typica: Bengal (Type male: Indian Mus., Calcutta).
1910 *Borellia annandalei* – BURR, Fauna Brit. India, Dermaptera, London, Bombay, 89.
1911 *Euborellia annandalei* – BURR, Genera Insectorum, Bruxelles, **122**: 31. India.
Literature: BURR 1915; SAKAI 1970; STEINMANN 1977, 1978.

Description: Male colour reddish brown, legs yellow; antennae greyish brown; sides of pronotum yellow. Head tumid, smooth and shiny; postfrontal and coronal sutures distinct. Eyes relatively short, shorter than the length of head behind eyes. Antennae 16-jointed; first antennal joint long, slender, a little shorter than distance between antennal bases. Pronotum ample, subquadrate, somewhat wider posteriorly than anteriorly; all angles rounded; median longitudinal furrow distinct. Tegmina present, but small, elongate, oval flaps on each side of the mesonotum. Metanotum normal. Wings absent. Abdomen very finely punctate, tergites 7–9 with small longitudinal keels. Ultimate tergite ample, broad, transverse, with a deep median longitudinal sulcus and impression. Forceps asymmetrical, trigonal basally, cylindrical apically. Penultimate sternite very large, broad and ample, well rounded posteriorly. Genitalia typical, of *Euborellia*-type; central parameral plate V-shaped, median longitudinal incision between external parameres deep and wide; genital lobe (Fig. 458) with denticulated pads, and denticulus lobi genitalis. External parameres very short and broad. – **Female** similar to male, but forceps symmetrical, and sides of abdominal tergites 7–9 without longitudinal keels laterally. – **Length** of body with forceps: male: 11–13,5 mm, female: 11,5–14 mm.

Distribution: India.

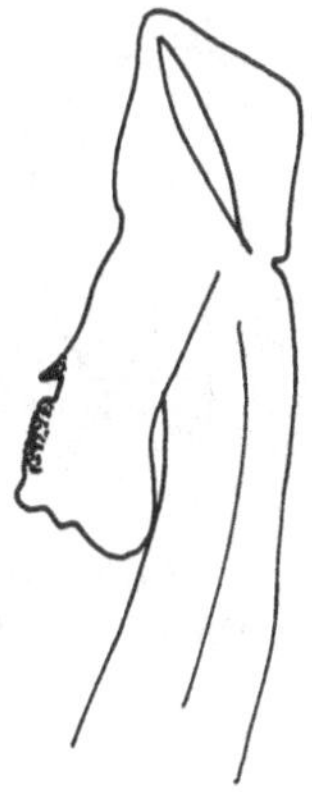

Fig. 458 External paramere with genital lobe of *Euborellia annandalei* (BURR). After BRINDLE.

44. Euborellia philippinensis SRIVASTAVA

1979 *Euborellia philippinensis* SRIVASTAVA, Bull. zool. Survey India, Calcutta (1), **2**: 49; fig. 1E (male), 1F (lateral view of abdomen), 1G (penultimate sternite) 1H (holotype genital armature). – Terra typica: Philippines: Laguna (Type male: Zool. Survey India, Calcutta).

Description: Male general colour dark brownish-black; legs clear yellow, femora with a faint brownish black spot dorsally. Head triangular, smooth, postfrontal and coronal sutures obliterate; posterior margin feebly emarginate in middle. Antennae 16-jointed; first joint slightly longer than the combined length of second and third. Pronotum longer than wide, anterior margin straight, lateral margins reflexed, slightly widened posteriorly; median longitudinal furrow faintly marked. Tegmina present as narrow ovate flaps on mesonotum with hind margin. Abdomen slender, gently enlarged in middle; sides of tergites 6–9 obtuse, smooth or sometimes with faint striations but always ecarinate. Ultimate tergite broad, median longitudinal sulcus present. Forceps (Fig. 459) asymmetrical, strongly trigonal basally, cylindrical and curved apically. Penultimate sternite with posterior margin obtusely rounded. Genitalia (Fig. 460) specific; central parameral plate with deep and wide median incision between external parameres. Genital lobes of *Euborellia*-type, with well marked denticulated pads. External parameres broad and short, obtuse apically. – **Female** similar to male, but sides of abdominal tergites 6–9 rounded, penultimate sternite comparatively narrower posteriorly and forceps simple and straight. – **Length** of body with forceps: male: 9–10,5 mm, female: 10–11,5 mm.

Distribution: Philippines.

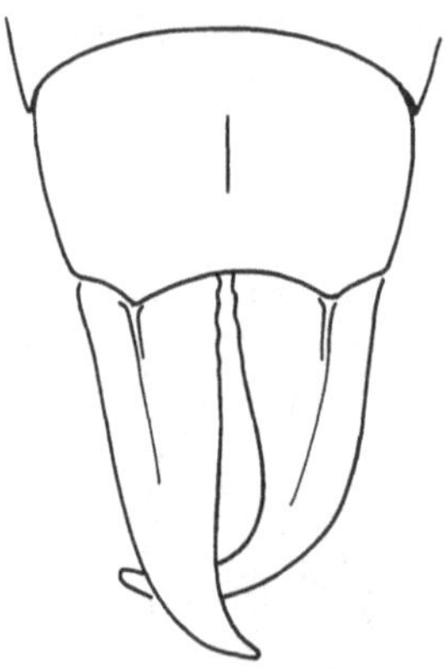

Fig. 459 Holotype ultimate tergite and forceps of *Euborellia philippinensis* SRIVASTAVA. After SRIVASTAVA.

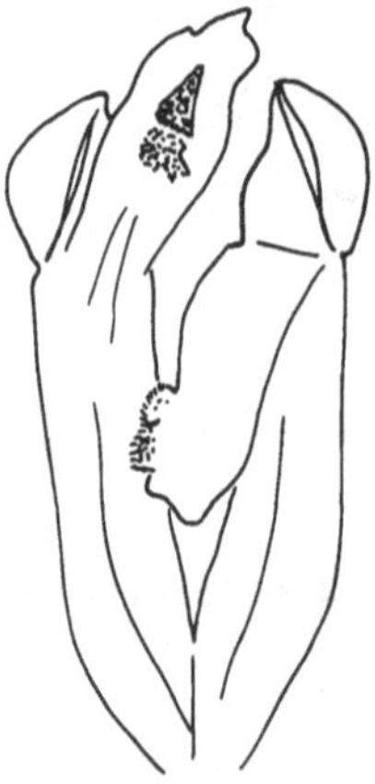

Fig. 460 Holotype genital armature of *Euborellia philippinensis* SRIVASTAVA. After SRIVASTAVA.

Subfamily 5: **Brachylabiinae** BURR

1909 *Brachylabiinae* BURR, Dtsch. ent. Zeitschr., Berlin, **1909**: 324. – Genus typicum: *Brachylabis* DOHRN 1864.
1904 *Isolabidae* (partim) VERHOEFF, Sber. Ges. naturf. Fr. Berlin,: 119.
1908 *Brachylabidae* BURR, Ann. Mag. nat. Hist., London, (8) **2**: 247.
1911 *Brachylabinae* BURR, Genera Insectorum, Bruxelles, **122**: 40.
1954 *Brachylabidae* – BRUES, MELANDER & CARPENTER, Classification of Cambridge, Insects, Mass.: 116.
1975 *Brachylabiinae* – STEINMANN, Acta zool. Acad. Sci. hung., Budapest, **21**: 206.

Literature: BOESEMAN 1954; BORELLI 1926; BORMANS 1900; BRINDLE 1978; BURR 1910, 1915; HEBARD 1923; HINCKS 1947; POPHAM & BRINDLE 1966; SAKAI 1970; STEINMANN 1977, 1978; ZACHER 1911.

Description: The species of this subfamily are characterized by the total absence of organs of flight. Body elongated, slender, legs long, tarsi, even the second segment, elongate. Head triangular; eyes very large or smaller; first antennal joint long, longer than distance between antennal bases; 4th joint long, slightly longer than wide, or quadrate. Lateral margins of mesonotum characteristic, with shorter or longer longitudinal keels. Abdomen cylindrical, tapering at the apex. Ultimate tergite discernibly excavated, excised. Forceps slender, more or less straight, or a little curved, symmetrical. Male genitalia slender, narrow, external parameres pointed, smaller or larger.

Distribution: Distributed worldwide excepting the Palaearctical region.

3 genera.

Identification key to the genera

1 Eyes in both sexes large or very large, their length equal to first antennal joint or to distance between antennal bases. Fourth antennal joint long, longer than wide, or quadrate 2

1′ Eyes in both sexes smaller, shorter than scape or shorter than distance between antennal bases .... Genus 25: **Brachylabis**, p. 299

2 Fourth antennal joint long, slightly longer than wide. Male genital armature long and narrow, genital lobes of various size .... Genus 23: **Metisolabis**, p. 281

2′ Fourth antennal joint small, short, more or less quadrate, as wide as long. Male genital armature broader and stouter .... Genus 24: **Ctenisolabis**, p. 289

Genus 23: **Metisolabis** Burr

1910 *Metisolabis* Burr, Fauna Brit., India, Dermaptera, London, Bombay, 108. – Species typica: *Brachylabis voeltzkowi* Burr 1908.

1966 *Metisolabis* – Popham & Brindle, Entomologist, London, **99**: 245 [= *Isolabis* Verhoeff 1902].

1969 *Metisolabis* – Brindle, Faune de Madagascar, Tananarive, **XXX**: 40.

Literature: Burr 1912; 1915; Hincks 1954; Steinmann 1977, 1978; Townes 1945.

Description: Agrees generally with the preceding genus, but the segments of the antennae are longer, not globular, all segments beyond the second one being at least a little longer than wide; fourth joint long, slightly longer than wide. Head rounded posteriorly; eyes very large, elliptical, and nearly the posterior angles of head. Tegmina and wings totally absent. Mesonotum with prominent lateral keels.

Distribution: Ethiopian, Malagassy and Oriental faunal regions.

7 species.

Indentification key to the species

1 Ethopian species; usually blackish in colour but dark reddish forms also occur; abdominal tergites with a distinct metallic sheen. Central parameral plate of male genitalia slightly expanded apically; virga distinct, strongly sclerotized, as in fig. 462 .... 1. **Metisolabis submetallica, p.** 282

1′ Malagasy or Oriental species 2

2 Malagasy species 3

2′ Oriental species 5

3 Tegmina present, rudimentary, represented by lateral flaps at mesonotum. Lateral keels of mesonotum acute. Virga within genital lobe of male genitalia very long (Fig. 464) .... 2. **Metisolabis milloti**, p. 283

3′ Tegmina totally absent. Lateral keels of mesonotum obtuse. Virga within genital lobe of male genitalia slightly shorter 4

4 Smaller species, body length 9–11 mm, forceps 1,5 mm. Rounded emargination of posterior margin of ultimate tergite comparatively small (Fig. 465). Male genitalia larger and broader than *malgacha* (BURR), as in fig. 466 .... 3. **Metisolabis voeltzkowi**, p. 284

4′ Larger species, body length 11–13 mm, forceps 2–2,5 mm. Rounded emargination of posterior margin of ultimate tergite comparatively large (Fig. 467). Male genitalia narrower than *voeltzkowi* (BURR), as in fig. 468 .... 4. **Metisolabis malgacha**, p. 285

5 Male forceps (Fig. 469) strongly curved, branches slightly trigonal basally, and cylindrical apically, not of *Metisolabis*-type. Sclerotized section of male virga long, as in fig. 470 .... 5. **Metisolabis bifoveolata**, p. 286

5′ Male forceps elongated, slightly straight, only curved a little at apical section; branches cylindrical, of *Metisolabis*-type **6**

6 Scape two to two and a half times longer than wide. Eyes large, as long as the length of head behind eyes. Postfrontal sutures very distinct, well marked. Posterior margin of male ultimate tergite as in fig. 471. Virga within genital lobe of male genitalia comparatively short (Fig. 472) .... 6. **Metisolabis punctata**, p. 287

6′ Scape three times longer than wide. Eyes very large, longer than the length of head behind eyes. Postfrontal sutures indistinct. Posterior margin of male ultimate tergite as in fig. 473. Virga within genital lobe of male genitalia very long (Fig. 474) .... 7. **Metisolabis fulgens**, p. 288

1. Metisolabis submetallica (BRINDLE) comb. nov.

1966 *Brachylabis submetallica* BRINDLE, Rev. Zool. Bot. afr., Bruxelles, **74**: 40; fig. 8 (head and pronotum of holotype), 14 (holotype genital armature). – Terra typica: Zaïre (Type male: Mus. r. Afr. centr., Tervuren).

1978 *Isolabis submetallica* – BRINDLE, Dermaptera of Africa, Tervuren, **2**: 50; fig. 47 (head and pronotum), 48 (male penultimate sternite), 49 (male genitalia). Burundi, Rhodesia.

Description: Male colour blackish, sides of pronotum yellow, antennae black, but second joint sometimes brown and joints 10–11 white. Cuticle rather dull on head and thorax, and on abdominal tergites 1, 2 or 3; tergites 3 or 4 to 9 shiny metallic blue, the blue sheen more conspicuous on anterior two-thirds of each tergite. Head transverse, sides rounded; posterior margin weakly convex; postfrontal and coronal sutures indistict. Eyes very large, slightly longer than the length of head behind eyes. First antennal joint long, a little longer than distance between antennal bases, and more or less equal in length to the eyes; second transverse, third one and a quarter times as long as wide; fourth joint longer than wide. Pronotum transverse, widened posteriorly; lateral margins more or less straight; median longitudinal furrow present anteriorly. Mesonotum transverse, narrowed posteriorly, each lateral margin forming a broad ridge, the keel well defined and directed dorso-laterally. Abdomen fusiform, depressed dorsally; ultimate tergite transverse, posterior margin produced above the base of each branch of forceps. Forceps (Fig. 461) cylindrical, broader towards base, narrowed distally, weakly curved. Penultimate sternite broad, posterior margin excised medially. Genitalia (Fig. 462) large; central parameral plate slightly expanded apically; genital lobes fully developed, but narrow; virga within genital lobe with strongly sclerotized section. External parameres acute. – **Female** similar to male, but branches of forceps rather less curved. – BRINDLE (1978: 51) mentioned one reddish form, named: *submetallica* "reddish form". The reddish form is structurally similar to the type form of *submetallica*, but distinguished by the colouration. Head blackish, thorax reddish, metallic sheen on abdomen less pronounced than in the type form but present and visible on most tergites. – **Length** of body with forceps: in both sexes: 6,5–9 mm.

Distribution: Zaïre, Zambia, Burundi, Rhodesia.

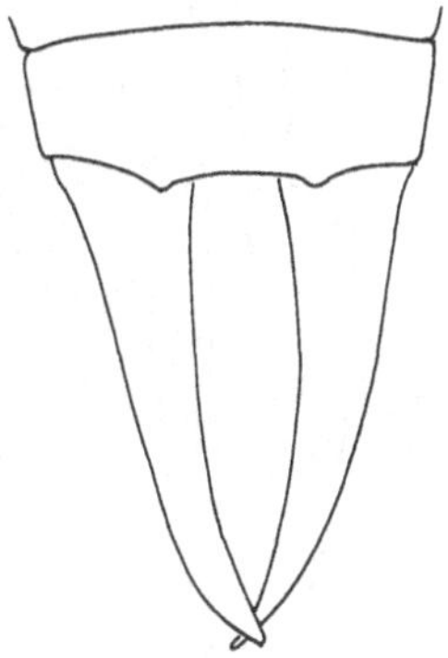

Fig. 461 Male ultimate tergite and forceps of *Metisolabis submetallica* (BRINDLE). Original.

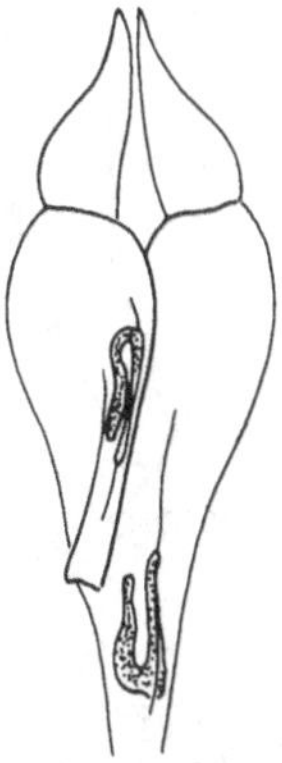

Fig. 462 Holotype genital armature of *Metisolabis submetallica* (BRINDLE). After BRINDLE.

2. Metisolabis milloti (HINCKS)

1953 *Anisolabis milloti* HINCKS, Mém. Inst. scient. Madagascar, Tananarive, (E), **4**: 365; fig. 4 (male), 5 (male genitalia). – Terra typica: Madagascar (Type male: Mus. nation. Hist. nat., Paris).
1966 *Isolabis milloti* – POPHAM & BRINDLE, Entomologist, London, **99**: 245.
1966 *Metisolabis milloti* – BRINDLE, Trans. r. ent. Soc. London, **118**: 234.
Literature: BRINDLE 1969; SAKAI 1970; STEINMANN 1977, 1978.

Description: Male colour black or blackish-brown; except antennal joints 11–12, yellow. Head tumid, about as broad as pronotum, and a little longer than wide; Eyes very large, a little longer than the length of head behind eyes. Antennae 15-jointed; first joint more or less as long as the distance between antennal bases; second transverse, third and fourth longer than wide. Pronotum well developed, longer than wide, and longer than head; convex posteriorly; median longitudinal furrow present; all angles rounded. Tegmina present, rudimentary, represented by lateral flaps. Wings entirely absent. Abdomen elongated, cylindrical; tergites punctate. Ultimate tergite broad. Forceps (Fig. 463) stout at the base, tapering; cylindrical, and curved apically. Genitalia (Fig. 464) slightly broader apically, genital lobes fully developed, virga within genital lobe very long. External parameres of *Metisolabis*-type, acute. – **Female** very similar to male, but forceps shorter, straight, tapering. – **Length** of body with forceps: in both sexes: 9 mm.

Distribution: Madagascar.

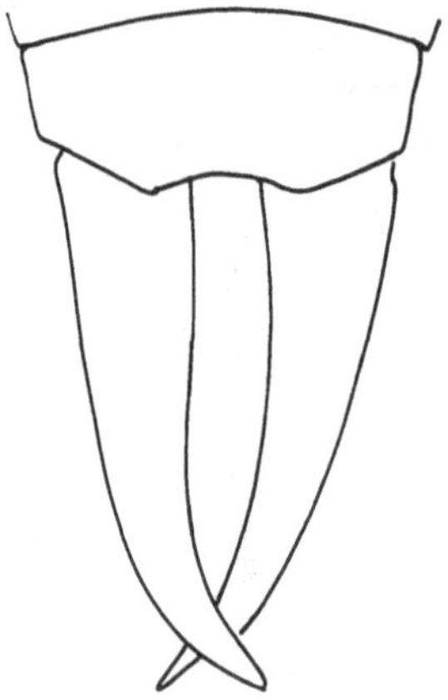

Fig. 463 Male ultimate tergite and forceps of *Metisolabis milloti* (HINCKS). After BRINDLE.

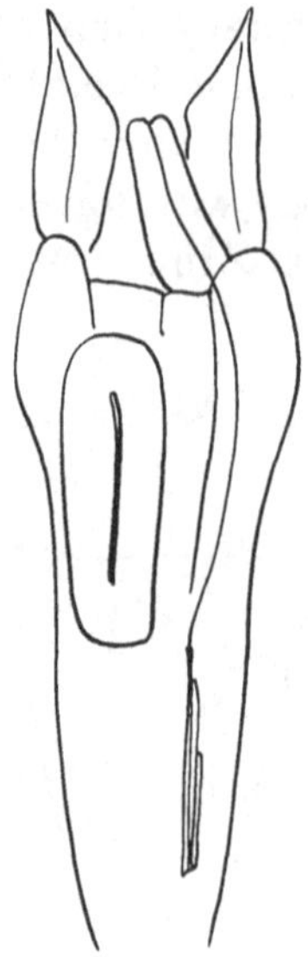

Fig. 464 Holotype genital armature of *Metisolabis milloti* (HINCKS). After HINCKS.

3. **Metisolabis voeltzkowi** (BURR)

1908 *Brachylabis voeltzkowi* BURR, Ann. Mag. nat. Hist., London, (8) **2**: 249. – Terra typica: Madagascar (Type male: Brit. Mus. nat. Hist., London).

1910 *Metisolabis voeltzkowi* – BURR, Genera Insectorum, Bruxelles, **122**: 42; pl. 4, fig. 2 (male), 2a (thorax dorsally), 2b (antenna), 2c (tarsus).

1969 *Metisolabis voeltzkowi* – BRINDLE, Faune de Madagascar, Tananarive, **30**: 44; fig. 39 (male genitalia).

Literature: BRINDLE 1966; BURR 1912, 1911; HINCKS 1953; SAKAI 1970; STEINMANN 1977, 1978.

Description: Male black, except femora at apical half, tarsi, and 10th antennal joint, yellow. Head small, about as broad as pronotum; eyes very large, longer than distance between antennal bases. Antennae 15-jointed; first joint long, but a little shorter than eye. Pronotum visibly longer than wide, about one and half times longer than wide; slightly wider posteriorly than anteriorly; median longitudinal furrow almost obsolete; all angles rounded. Mesonotum with lateral keels sharp, gently curved, obsolete just before reaching the posterior margin. Metanotum strongly sinuate posteriorly. Abdomen slender, with distinct lateral tubercle on the fourth segment only; tergites punctate dorsally. Ultimate tergite broad; posterior margin with rounded emargination medially. Forceps (Fig. 465) of *Metisolabis*-type; cylindrical; branches remote at base, gently incurved, and crossing at the apex. Genitalia (Fig. 466) specific; central parameral plate oval; virga strongly sclerotized. External parameres triangular, apices pointed. –

Female very similar to male, but forceps a little shorter, and straighter, tapering. – **Length** of body with forceps: male: 10,5–11,5 mm, female: 11–13 mm.
Distribution: Madagascar.

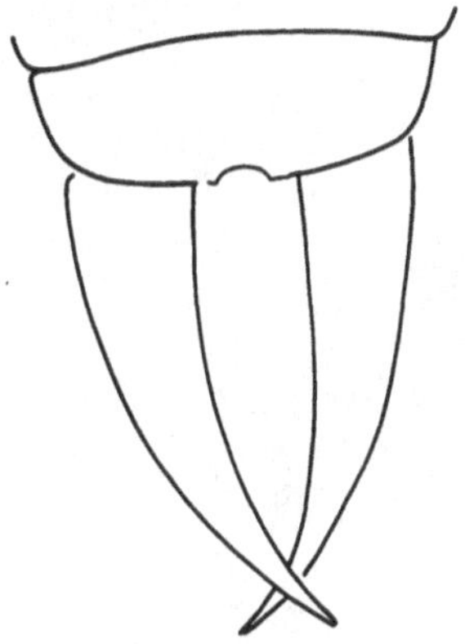

Fig. 465 Male ultimate tergite and forceps of *Metisolabis voeltzkowi* (BURR). After BURR.

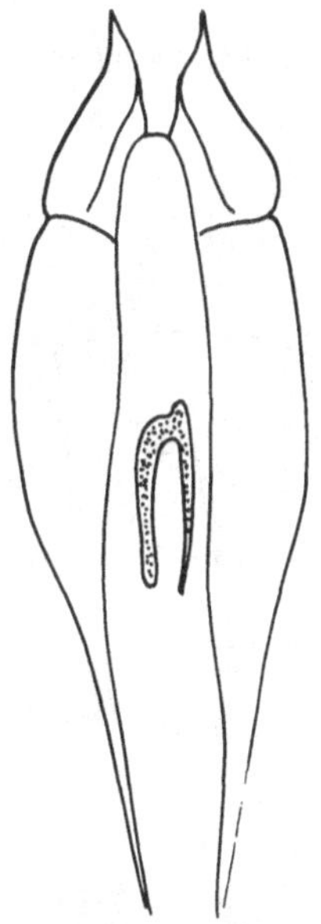

Fig. 466 Male genital armature of *Metisolabis voeltzkowi* (BURR). After BRINDLE.

4. Metisolabis malgacha (BURR)

1904 *Brachylabis malgacha* BURR, Trans. ent. Soc. London, **1904**: 292. – Terra typica: Madagascar (Type male: Brit. Mus. nat. Hist., London).
1910 *Metisolabis malgacha* – BURR, Trans. ent. Soc. London, **1910**: 196.
1966 *Isolabis malgacha* – POPHAM & BRINDLE, Entomologist, London, **99**: 245.
1969 *Metisolabis malgacha* – BRINDLE, Faune de Madagascar, Tananarive, **30**: 45; fig. 36 (male), 37 (male genitalia).
Literature: BORELLI 1906, 1932; BRINDLE 1966; BURR, 1908, 1912, 1914; HINCKS 1953; SAKAI 1970; STEINMANN 1977, 1978.

Description: Male general colour black, colour very similar to *voeltzkowi* (BURR). Head tumid, vertex a little depressed; eyes very large, visibly longer than genae, or first antennal joint. Antennae 15-jointed; first joint typical, more or less as long as distance between antennal bases. Pronotum longer than wide; all angles rounded, and all margins straight; median longitudinal furrow distinct. Lateral longitudinal keel on mesonotum sharply defined. Metanotum typical. Abdomen slender, widened to abdominal tergites 5–6. Ultimate tergite broad, posterior margin with rounded emargination medially. Forceps (Fig. 467) typical, cylindrical and curved apically. Genitalia (Fig. 468) very narrow, fully developed; virga within genital lobe long, with sclerotized section basally. External parameres acuminate at tip. – **Female** very similar to male, but forceps shorter, and straighter. – **Length** of body with forceps: male: 13–14 mm, female: 12,5–13,5 mm.

Distribution: Madagascar.

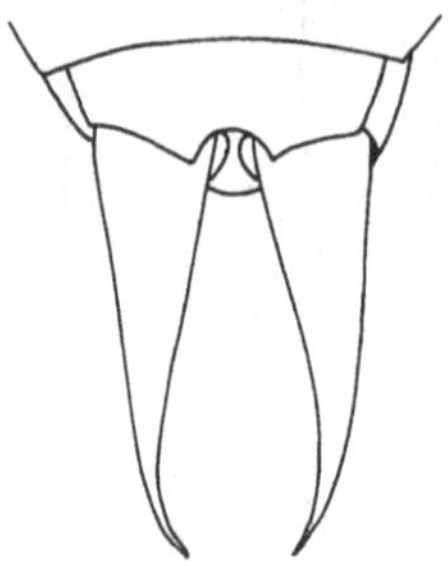

Fig. 467 Male ultimate tergite and forceps of *Metisolabis malgacha* (BURR). After BRINDLE.

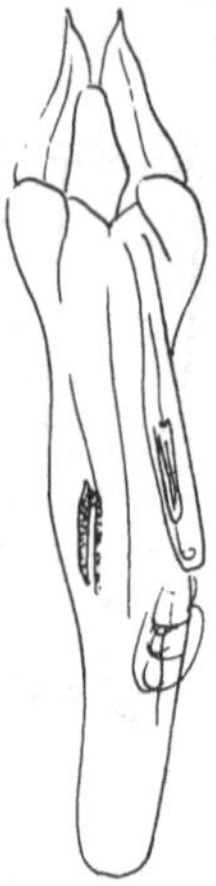

Fig. 468 Male genital armature of *Metisolabis malgacha* (BURR). After BRINDLE.

5. Metisolabis bifoveolata (BOLIVAR)

1897 *Brachylabis bifoveolata* BOLIVAR, Ann. Soc. ent. France, Paris, **66**: 285; Tab. 10, fig. 1 (male). – Terra typica: South India (Type male: Mus. nation. Hist. nat., Paris).
1911 *Ctenisolabis bifoveolata* – BURR, J. asiat. Soc. Bengal, Calcutta, **3**: 780.
1923 *Metisolabis bifoveolata* – HEBARD, Mem. Dept. Agric. India, (ent. Ser.) Pusa, **7**: 213.
1966 *Isolabis bifoveolata* – POPHAM & BRINDLE, Entomologist, London, **99**: 254.
1970 *Metisolabis bifoveolata* – SRIVASTAVA, EOS, Rev. esp. Ent., Madrid, **45**: 323, fig. 5 (male ultimate tergite and forceps), 6 (male genitalia).

Literature: BORMANS 1900; BURR 1910, 1911; KIRBY 1904; SAKAI 1970; STEINMANN 1977, 1978; VERHOEFF 1902.

Description: Male colour brownish black or blackish; antennae dark brown; punctate all over the whole body except forceps, golden yellow pubescence present. Head long, but shorter than pronotum; frons tumid, postfrontal and coronal sutures distinct. Eyes relatively large, but a little shorter than the length of head behind eyes. Antennae 14-jointed; first joint stout, narrowed at base, and expanded at apex; longer than distance between antennal bases; second transverse, third twices as long as broad, fourth longer than wide. Pronotum large, lateral margins straight, a little expanded posteriorly; median longitudinal furrow distinct. Tegmina and wings entirely absent. Mesonotum with keels laterally, strongly curving posteriorly, posterior margin subtruncate. Metanotum transverse. Abdomen strongly convex, dilated in middle; tergites finely punctate; 4th tergite with paired folds. Ultimate tergite sloping, strongly transverse; posterior margin concave with an obtuse tubercle above bases of forceps. Forceps (Fig. 469) strongly curved, not of *Metisolabis*-type; tapering, apices pointed and meeting or crossing each other. Penultimate sternite triangular, posterior margin with a triangular incision mesad. Genitalia (Fig. 470) characteristic; genital lobes fully developed; virga within genital lobe with long sclerotized section. – **Female** similar to male, but penultimate sternite with posterior margin rounded and forceps less curved. – **Length** of body with forceps: male: 15–18 mm, female: 17–18,5 mm.

Distribution: South India.

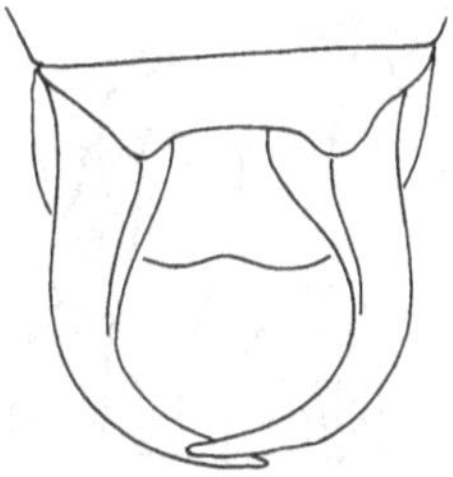

Fig. 469 Male ultimate tergite and forceps of *Metisolabis bifoveolata* (BOLIVAR). After SRIVASTAVA.

Fig. 470 Male genital armature of *Metisolabis bifoveolata* (BOLIVAR). After SRIVASTAVA.

6. Metisolabis punctata (DUBRONY)

1879 *Brachylabis punctata* DUBRONY, Ann. Mus. civ. Stor. nat., Genova, **14**: 357; fig. (female). – Terra typica: Java (Type female: Mus. civ. Stor. nat., Genova).
1908 *Brachylabis caudelli* BURR, Ann. Mag. nat. Hist., London, (8) **2**: 251. – Terra typica: Burma (Type male: Brit. Mus. nat. Hist., London). – **syn. nov.**
1910 *Metisolabis caudelli* – BURR, Fauna Brit. India, Dermaptera, London, Bombay, 109.
1911 *Leptisolabis punctata* – BURR, Genera Insectorum, Bruxelles, **122**: 43. Malay Archipelago.
1966 *Isolabis punctata* – POPHAM & BRINDLE, Entomologist, London, **99**: 245.
1978 *Isolabis caudelli* – SRIVASTAVA, Ann. Mus. civ. Stor. nat., Genova, **82**: 325; fig. 4 (anterior part of body), 5 (a portion of antenna), 6 (male ultimate tergite and forceps).

Literature: BORMANS 1900 (as *punctata* DUBRONY); BURR 1910: *Metisolabis caudelli* (BURR) proposed synonymy for *punctata* DUBRONY,: 109; HINCKS 1947; STEINMANN 1977, 1978.

Description: Male colour pitch-black; antennae with a pale ring before the apex; forceps somewhat reddish-black; legs brownish black with apex yellow. Head typical broad, tumid. Eyes very large, more or less as long as wide. Antennae 15-jointed; fourth joint more than half as long as third. Pronotum more or less quadrate; lateral margins straight, but a little expanded posteriorly; median longitudinal furrow present in anterior half. Mesonotum with the keels gently bowed at the shoulder and gently converging posteriorly. Metanotum transverse. Femora black, yellow at apex; tibiae brownish yellow, paler towards the apex; tarsi paler. Abdomen normal, expanded to abdominal tergites 5–7; ultimate tergite with two tubercles on posterior margin; posterior margin concave medially. Forceps (Fig. 471) of *Metisolabis*-type; stout at bases, rapidly tapering, gently arched, not contiguous. Genitalia (Fig. 472) very broad; central parameral plate with very short median longitudinal incision between external parameres; genital lobes comparatively small, virga within genital lobe strongly sclerotized basally. External paramere typical, pointed. – **Female** similar to male, but forceps a little shorter and contiguous. – **Length** of body with forceps: male: 12,5 mm, female: 13 mm.

Distribution: Burma, India, Java, Malayan Archipelago.

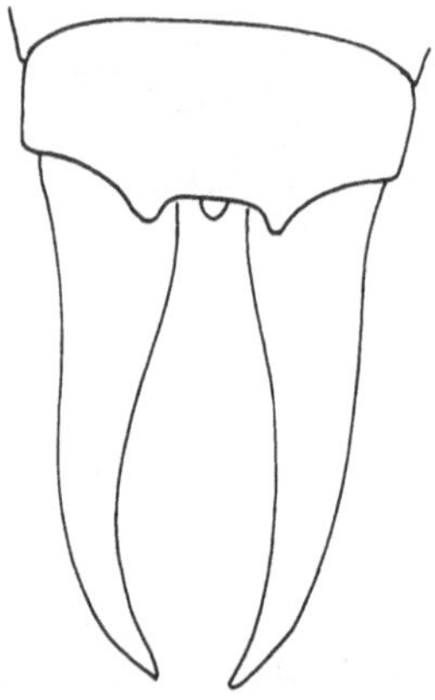

Fig. 471 Male ultimate tergite and forceps of *Metisolabis punctata* (DUBRONY). After SRIVASTAVA.

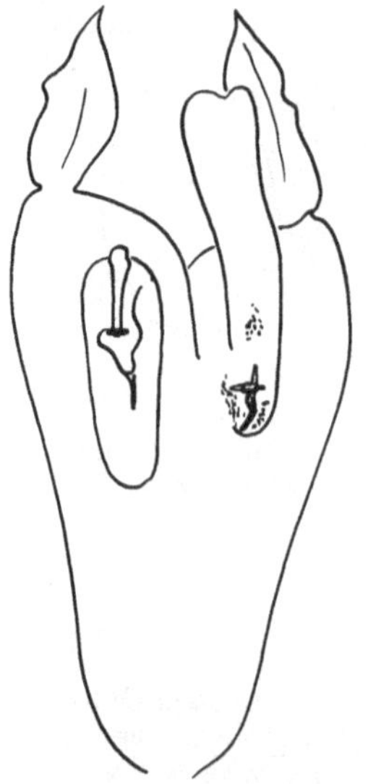

Fig. 472 Male genital armature of *Metisolabis puncatata* (DUBRONY). Original.

7. Metisolabis fulgens sp. nov.

Terra typica: Carin Cheba, 900–1100 m, V–XII. 1888, L. FEA; gen. prep. No. 588, det. Dr. H. STEINMANN (holotype), and female paratype, ditto. (Type male: Természettudományi Múz., Budapest).

Description: Male colour dark reddish-brown; legs yellow, femora dark brown at basal half; antennae brown, except joints 10–11, whitish. Head tumid;

postfrontal and coronal sutures indistinct. Antennae 15-jointed; first joint slightly longer than distance between antennal bases; second joint transverse; third and fourth joints more or less equal, longer than wide. Eyes large, longer than the length of head behind eyes. Pronotum slightly longer than wide; lateral margins straight, but expanded posteriorly; median longitudinal furrow present in anterior half. Mesonotum with prominent lateral keels; posterior margin concave. Metanotum normal. Abdomen narrowed apically as well as basally, weakly convex; lateral folds on the sides of 3rd and 4th tergites poorly developed, sides of tergites rounded. Cuticle of head, pronotum, and meso- and metanotum coriaceous, abdominal tergites punctate. Ultimate tergite narrow; posterior margin with rounded emargination. Forceps (Fig. 473) typical; basally a little trigonal, and cylindrical apically. Penultimate sternite trigonal. Genitalia (Fig. 474) characteristic; central parameral plate with deep median incision between external parameres; virga within genital lobe very long, with straight sclerotized section; external parameres comparatively small, pointed apically. – **Female** very similar to male, but forceps contiguous, straight, cylindrical. – **Length** of body with forceps: male: 12 mm, female: 18 mm.

Distribution: Thailand.

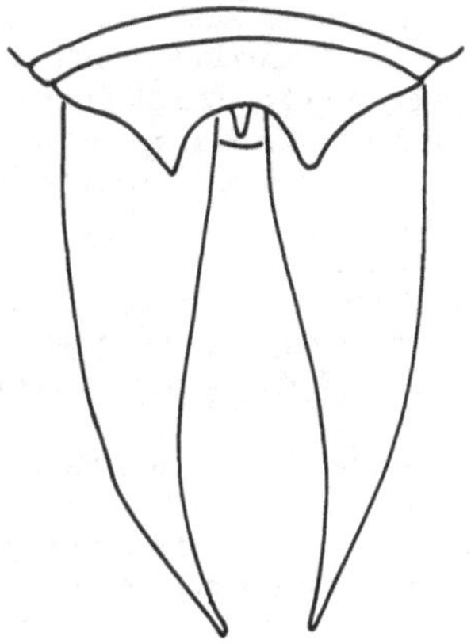

Fig. 473 Holotype ultimate tergite and forceps of *Metisolabis fulgens* sp.nov. Original.

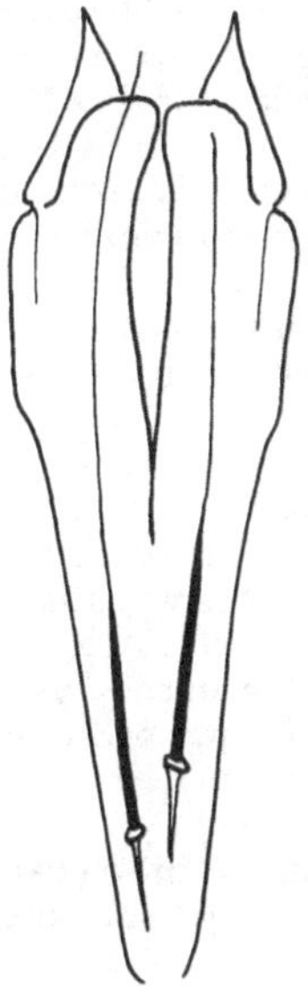

Fig. 474 Holotype genital armature of *Metisolabis fulgens* sp.nov. Original.

Genus 24: **Ctenisolabis** VERHOEFF

1902 *Ctenisolabis* VERHOEFF, Sber. Ges. naturf. Fr. Berlin, **1902**: 11. – Species typica: *Ctenisolabis togoensis* VERHOEFF 1902.

1966 *Ctenisolabis* – POPHAM & BRINDLE, Entomologist, London, **99**: 245 [= *Isolabis* VERHOEFF 1902].
1978 *Ctenisolabis* – STEINMANN, Fol. ent. hung., Budapest, **31**: 213.
Literature: BRINDLE 1978; BURR 1910, 1911, 1915; HINCKS 1953; REHN 1924; SAKAI 1970; STEINMANN 1977, 1978; TOWNES 1945; ZACHER 1911.

Description: Body similar to *Metisolabis* BURR species, but differing in the globular, antennal joints. Third joint twice longer as wide, fourth joint about as long as wide. Eyes large, elliptical situated near posterior angles of head. Tegmina absent, but mesonotum with lateral longitudinal carinae or keels; metanotum transverse. Abdomen oval, narrowed anteriorly and posteriorly, wide medially; cylindrical. Generally punctate dorsally. Male forceps simple, tapering, female forceps contiguous or subcontiguous. Male genitalia broad, and external parameres shorter than that of *Metisolabis*.

Distribution: Africa, Central America and Oriental region.

10 species.

Identification key to the species

1 Ethiopian species **2**
1′ Neotropical or Oriental species **4**
2 Eyes large, more or less as long as the length of head behind eyes **3**
2′ Eyes very large, slightly longer than the length of head behind eyes; pronotum less transverse, dark reddish-brown in colour, abdominal tergites strongly punctate. Male genitalia characteristic, as in fig. 476 .... 1. **Ctenisolabis togoensis**, p. 291
3 Head and pronotum less transverse; head black, pronotum and part of mesonotum reddish, rest of body blackish. Male genitalia with sclerotized part of virga long (Fig. 478), slender, and without a lateral appendage .... 2. **Ctenisolabis ruficollis**, p. 292
3′ Head and pronotum more transverse; head blackish, or paler, rest of body dark brown. Male genitalia with sclerotized part of virga short (Fig. 479), and with a lateral appendage 3. **Ctenisolabis traegaordhi**, p. 293
4 Neotropical species **5**
4′ Oriental species **7**
5 Antennae 12-jointed. Relatively smaller species, length of body with forceps: 9–10 mm. Abdomen strongly punctate, oval; posterior margin of ultimate tergite as in fig. 480 .... 4. **Ctenisolabis montana**, p. 294
5′ Antennae 13-jointed **6**
6 Femora black, except end near tibiae, yellow. Virga within genital lobe of male genitalia very long, but without sclerotized plate basally (Fig. 482). Posterior margin of ultimate tergite as in fig. 481 .... 5. **Ctenisolabis nigra**, p. 294
6′ Femora blackish brown basally, yellow apically. Virga within genital lobe of male genitalia very long, and with sclerotized plate basally (Fig. 484). Posterior margin of ultimate tergite as in fig. 483 .... 6. **Ctenisolabis mahunkai**, p. 295
7 Antennae 12-jointed **8**
7′ Antennae 13-jointed. Relatively large species, length of body with forceps 10–14 mm. Mesonotum ample, keels distinct but blunt **9**
8 Very small species, length of body with forceps: 6,5 mm. Cuticle of body punctate. Posterior margin of ultimate tergite with large, rounded emargination medially (Fig. 485) .... 7. **Ctenisolabis pusilla**, p. 296
8′ Relatively large species, length of body with forceps 11–14 mm. Cuticle of body aciculate. Posterior margin of ultimate tergite with two tubercles medially (Fig. 486) .... 8. **Ctenisolabis aciculata**, p. 297

9 Cuticle of body strongly punctate. Femora bicoloured, dark brown basally, and yellowish apically. Posterior margin of ultimate tergite as in fig. 487 .... 9. **Ctenisolabis fletcheri**, p. 298
9′ Cuticle of body finely punctate. Femora unicoloured, dark brown. Posterior margin of ultimate tergite as in fig. 488 .... 10. **Ctenisolabis loebli**, p. 299

1. Ctenisolabis togoensis VERHOEFF

1902 *Ctenisolabis togoensis* VERHOEFF, Sber. Ges. naturf. Fr. Berlin, **1902**: 14. – Terra typica: Togo (Type male: Mus. Naturkunde, Berlin).
1908 *Brachylabis togoensis* – BURR, Ann. Mag. nat. Hist., London, (8) **2**: 249.
1966 *Isolabis togoensis* – POPHAM & BRINDLE, Entomologist, London, **99**: 245.
1977 *Ctenisolabis togoensis* – STEINMANN, Ann. hist.-nat. Mus. nation. hung., Budapest, **69**: 96.
1978 *Isolabis togoensis* – BRINDLE, Dermaptera of Africa, Tervuren, **2**: 48; fig. 44 (male), 45 (male genitalia), 46 (posterior margin of penultimate sternite).
1978 *Ctenisolabis togoensis* – STEINMANN, Fol. ent. hung., Budapest, **31**: 214. Togo, Transvaal.
Literature: BURR 1911; KIRBY 1904; HINCKS 1953; SAKAI 1970; STEINMANN 1978.

Description: Male mat, reddish brown, metallic; legs yellowish brown. Head tumid, about as broad as pronotum; postfrontal and coronal sutures not visible. Eyes very large, slightly longer than the length of head behind eyes, and more or less as long as scape. Antennae 13-jointed; joints short, first joint more or less as long as distance between antennal bases; fourth joint about as long as wide. Pronotum quadrate or a little longer than wide; lateral margins straight, but convex; posterior margin truncate; median longitudinal furrow present. Tegmina and wings totally absent. Mesonotum with prominent keels laterally. Metanotum typical, transverse. Abdomen oval, expanded medially; tergites 3–4 with paired tubercles. Ultimate tergite transverse, with two obtuse processes posteriorly. Forceps (Fig. 475) broad basally, tapering apically; curved and pointed at apex. Penultimate sternite transverse, posterior margin rounded and with an excision medially. Genitalia (Fig. 476) characteristic; possibly somewhat

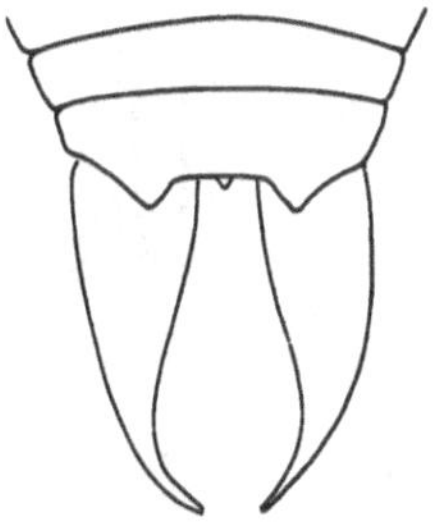

Fig. 475 Male ultimate tergite and forceps of *Ctenisolabis togoensis* VERHOEFF. After BRINDLE.

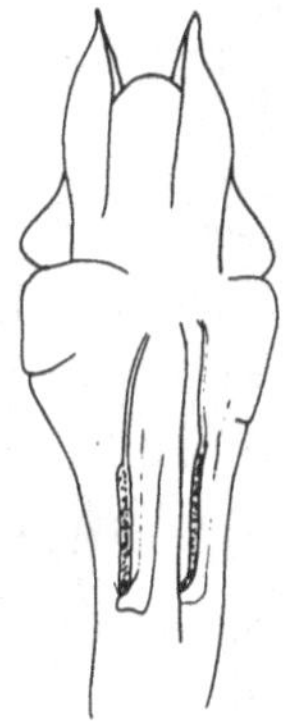

Fig. 476 Male genital armature of *Ctenisolabis togoensis*. VERHOEFF. After BRINDLE.

distorted, with membranous external parameres which are widened at base, virgal sclerites long, narrow, genital lobe with prominent denticulations. – **Female** similar to male, but colour lighter brown, and forceps slenderer. – **Length** of body with forceps: male: 8,5–9 mm, female: 9 mm.

Distribution: Togo, Transvaal.

2. Ctenisolabis ruficollis (HINCKS) comb. nov.

1957 *Brachylabis ruficollis* HINCKS, South African Animal Life, Pretoria, **4**: 69; fig. 29 (holotype genital armature), 30 (male). – Terra typica: Cape Province (Type male: Brit. Mus. nat. Hist., London).

1966 *Isolabis ruficollis* – POPHAM & BRINDLE, Entomologist, London, **99**: 245.

1978 *Brachylabis ruficollis* – BRINDLE, Dermaptera of Africa, Tervuren, **2**: 31; fig. 19 (male), 18 (male genitalia).

Literature: SAKAI 1970; STEINMANN 1977, 1978.

Description: Male colour various, head black, pronotum and anterior part of mesonotum reddish; posterior part of mesonotum and rest of body blackish; legs yellowish. Cuticle punctate and pubescent, punctation relatively fine but close; thoracic tergites with a slender longitudinal smooth line. Head transverse, rounded posteriorly; eyes large, more or less as long as the length of head behind eyes. Antennae 14-jointed; first joint a little longer than distance between antennal bases. Pronotum nearly quadrate, with lateral margins expanded to posterior angles; median longitudinal furrow present. Mesonotum broad, with prominent lateral keels. Metanotum transverse. Abdomen fusiform, contracted proximad and more strongly contracted distad. Lateral tubercles on third tergite absent, those of fourth rudimentary. Ultimate tergite transverse, posterior margin slightly produced above the base of each branch of the forceps. Forceps

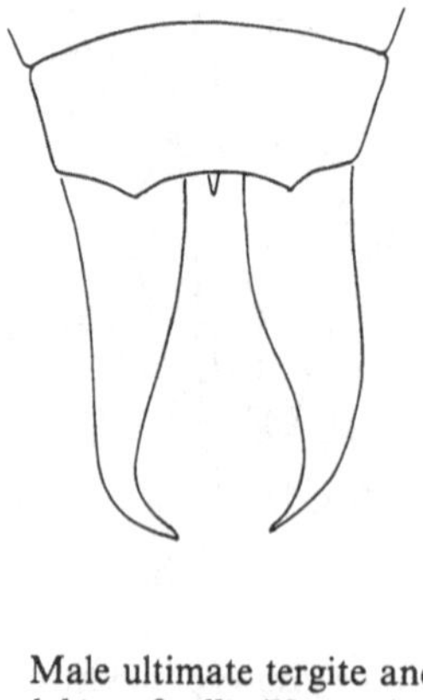

Fig. 477 Male ultimate tergite and forceps of *Ctenisolabis ruficollis* (HINCKS). After BRINDLE.

Fig. 478 Male genital armature of *Ctenisolabis ruficollis* (HINCKS). After BRINDLE.

(Fig. 477) tapering, curved distad; cylindrical. Penultimate sternite transverse, posterior margin mainly convex but slightly trisinuate. Genitalia (Fig. 478) very narrow, and long; genital lobes fully developed, virga within genital lobe very long, middle section sclerotized. External parameres pointed. – **Female** similar to male, but forceps slenderer. – **Length** of body with forceps: male: 10–11,5 mm, female: 10–12 mm.

Distribution: South Africa and Rhodesia.

3. Ctenisolabis traegaordhi (Burr) comb. nov.

1913 *Brachylabis traegaordhi* Burr, Medd. Gotesborg Mus. Zool. Afd., **2**: 3. – Terra typica: Natal (Type male: Gotesborg Museum).
1915 *Brachylabis trägardhi* – Burr, J. r. microsc. Soc., London, **1915**: 447; pl. IX, fig. 19 (male genitalia).
1966 *Isolabis traegaordhi* – Popham & Brindle, Entomologist, London, **99**: 245.
1978 *Isolabis traegaordhi* – Brindle, Dermaptera of Africa, Tervuren, **2**: 38; fig. 28 (male), 29 (male genitalia), 30 (antennal segments).
Literature: Hincks 1957; Sakai 1970.

Description: Male colour reddish brown or dark brown; antennae dark brown, joints 5 or 6 to 9 nearly blackish, and joint 10 paler in colour; legs yellow. Cuticle punctate and pubescent, rather dull; punctation relatively fine but close and strong on median abdominal tergites. Head transverse, rounded; eyes large, but not longer than the length of head behind eyes. Antennae 13-jointed; first joint broad, a little shorter than distance between antennal bases. Pronotum quadrate; lateral margins straight, but a little expanded posteriorly; all angles rounded; median longitudinal furrow present. Mesonotum transverse, with prominent lateral longitudinal keels. Metanotum narrow, transverse. Abdomen cylindrical, fusiform, depressed dorsally; lateral tubercles on third and fourth tergites rudimentary. Ultimate tergite transverse, posterior margin with small emargination medially. Forceps cylindrical, rather broader towards base, apex curved medially. Penultimate sternite with posterior margin rounded and with a very small median excision. Genitalia long and narrow; genital lobe (Fig. 479) elongated, virgal sclerite rather large, and virga visible at both ends of the sclerite.

Fig. 479 Genital lobe of male genitalia of *Ctenisolabis traegaordhi* (Burr). After Brindle.

– **Female** similar to male, but forceps less curved apically. – **Length** of body with forceps: in both sexes: 10–12 mm.

Distribution: Natal.

4. Ctenisolabis montana (BORELLI)

1909 *Brachylabis montana* BORELLI, Boll. Mus. Zool. Anat. comp. Univ. Torino, **24**: No. 611: 5. – Terra typica: Costa Rica (Type male: Mus. Ist. Zool. sist., Torino; holotype without penis).
1911 *Ctenisolabis montana* – BURR, Genera Insectorum, Bruxelles, **122**: 42; pl. 4, fig. 3 (male).
1966 *Isolabis montana* – POPHAM & BRINDLE, Entomologist, London, **99**: 245.
1977 *Ctenisolabis montana* – STEINMANN, Ann. hist.-nat. Mus. nation. hung., Budapest, **69**: 96.
Literature: STEINMANN 1978.

Description: Male colour dark brown, brownish red, except the legs and antennae, yellow; lateral margins of pronotum reddish yellow. Head broad, and short, but not broader than pronotum. Eyes large, about as long as length of head behind eyes, and as long as scape. Antennae 12-jointed; first joint comparatively short, shorter than distance between antennal bases; postfrontal and coronal sutures absent. Pronotum a little longer than wide; lateral margins almost parallel, posterior margin rounded; median longitudinal furrow present on anterior half. Tegmina and wings absent. Lateral keel of mesonotum well visible. Legs unicolourous, yellow. Abdomen oval, strongly punctate. Ultimate tergite broad; posterior margin with two tubercles medially. Forceps (Fig. 480) of *Ctenisolabis*-type, cylindrical, broad basally, curved and pointed apically. Penultimate sternite rounded posteriorly. Genitalia unknown. – **Female** very similar to male, but forceps tapering, contiguous, straight. – **Length** of body with forceps: male: 9–9,5 mm, female: 9,5–10 mm.

Distribution: Costa Rica.

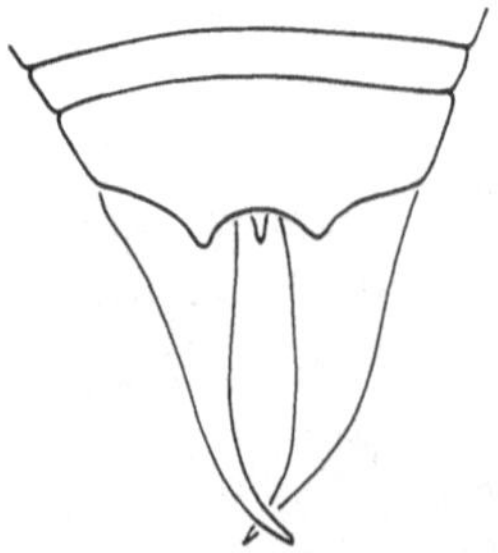

Fig. 480 Male ultimate tergite and forceps of *Ctenisolabis montana* (BORELLI). After BURR.

5. Ctenisolabis nigra (SCUDDER)

1876 *Cylindrogaster nigra* SCUDDER, Proc. Boston Soc. nat. Hist., **18**: 251. – Terra typica: Brazil (Type female: location unknown).
1900 *Brachylabis nigra* – BORMANS, Das Tierreich, Berlin, **11**: 54; fig. 22 (male).
1909 *Brachylabis fernandezi* BORELLI, Boll. Mus. Zool. Anat. comp. Univ. Torino, **24**, No. 611: 3; fig. 2 (male). – Terra typica: Costa Rica (Type male: Mus. Ist. sist., Torino; holotype without penis). – **syn. nov.**

1911 *Ctenisolabis nigra* – BURR, Genera Insectorum, Bruxelles, **122**: 42.
1911 *Ctenisolabis fernandezi* – BURR, Genera Insectorum, Bruxelles, **122**: 42.
1966 *Isolabis nigra* – POPHAM & BRINDLE, Entomologist, London, **99**: 245.
1966 *Isolabis fernandezi* – POPHAM & BRINDLE, Entomologist, London, **99**: 245.
1977 *Ctenisolabis nigra* – STEINMANN, Ann. hist.-nat. Mus. nation. hung., Budapest, **69**: 96 [= *Ctenisolabis fernandezi* (BORELLI 1909)]; fig. 27 (male genitalia).
Literature: BURR 1915 (indication of a possibly misidentification of *nigra* for *fernandezi*: 447); STEINMANN 1978.

Description: Male colour brown, light brown to reddish-brown, except femora, totally black except the end of femora near tibiae. Head a little longer than wide. Eyes large, longer than the length of head behind eyes, and more or less as long as scape. Antennae 13-jointed; first joint about as long as distance between antennal bases. Postfrontal and coronal sutures absent. Pronotum expanded posteriorly; median longitudinal furrow perfect. Mesonotum with lateral longitudinal keels, the latter curved. Abdomen a little expanded medially; tergites 3 and 4 with tubercles. Ultimate tergite transverse, posterior margin with two tubercles medially. Forceps (Fig. 481) comparatively broad basally, branches a little depressed, fang-shaped; pointed apically. Genitalia (Fig. 482) broad, median incision of anterior margin deep; genital lobes well developed, virga very long. – **Female** unknown to me. – **Length** of body with forceps: 13,5 mm.
Distribution: Costa Rica, Brazil.

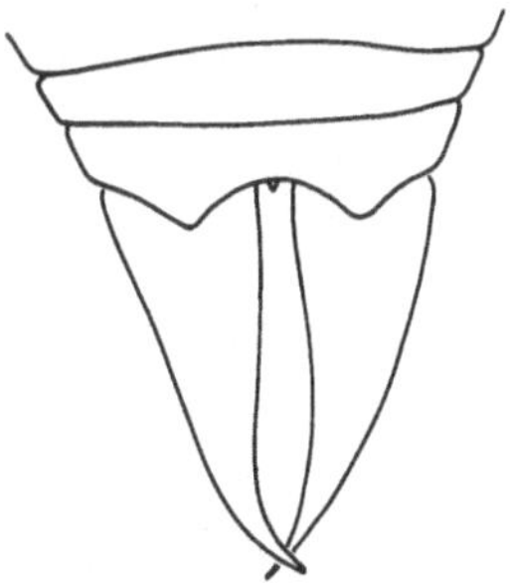

Fig. 481 Male ultimate tergite and forceps of *Ctenisolabis nigra* (SCUDDER). Original.

Fig. 482 Male genital armature of *Ctenisolabis nigra* (SCUDDER). After STEINMANN.

6. Ctenisolabis mahunkai STEINMANN

1978 *Ctenisolabis mahunkai* STEINMANN, Fol. ent. hung., Budapest, **31**: 216; fig. 2 (holotype genital armature). – Terra typica: Paraguay (Type male: Természettudományi Múz., Budapest).

Description: Male colour dark metallic brown; antennae light brown, legs brown, femora brown at the basal section and yellow apically. Body strongly punctate. Head tumid; postfrontal and coronal sutures absent. Eyes very large, slightly longer than the length of head behind eyes. Antennae 13-jointed; first joint long, but shorter than eyes, and longer than distance between antennal bases. Pronotum a little longer than wide; lateral margins straight, but convex posteriorly; all angles rounded; posterior margin straight, truncate; median longitudinal furrow distinct. Tegmina and wings entirely absent. Mesonotum with prominent lateral longitudinal keels. Metanotum typical, transverse. Abdomen oval, fourth segment with rudimentary tubercles posteriorly. Ultimate tergite of *Ctenisolabis*-type, posterior margin with two prominent tubercles. Forceps (Fig. 483) fang-shaped, broad basally, curved and pointed apically. Penultimate sternite trigonal, posterior margin rounded. Genitalia (Fig. 484; gen. prep. No. 157, det. Dr. H. STEINMANN) characteristic; central parameral plate narrow, but widened at anterior margin; median incision of anterior margin visible; genital lobes well developed, with very long virgae; virga with sclerotized section basally; external parameres relatively short, pointed. – **Female** very similar to male, but lighter brown in colour. – **Length** of body with forceps: male: 12,5 mm, female: 12 mm.

Distribution: Paraguay.

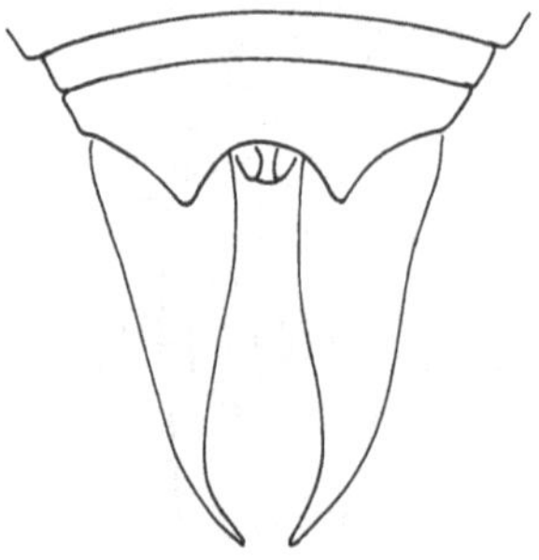

Fig. 483 Holotype ultimate tergite and forceps of *Ctenisolabis mahunkai* STEINMANN. Original.

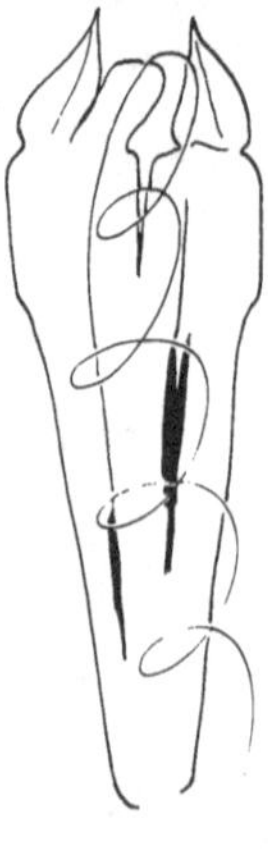

Fig. 484 Holotype genital armature of *Ctenisolabis mahunkai* STEINMANN. Original.

7. Ctenisolabis pusilla STEINMANN

1978 *Ctenisolabis pusilla* STEINMANN, Fol. ent. hung., Budapest, **31**: 216. – Terra typica: Sri Lanka (Type female: Természettudományi Múz., Budapest).

Description: Female colour dark brown, antennae and forceps light brown; femora brown basally and yellowish white apically; tibiae yellowish brown. Cuticle punctate. Head tumid, broad; postfrontal and coronal sutures quite visible; eyes very large, slightly longer than the length of head behind eyes.

Antennae 12-jointed; first very long, longer than the distance between antennal bases. Pronotum well developed, longer than wide; lateral margins straight, and expanded posteriorly; all angles rounded; posterior margin a little convex; median longitudinal furrow distinct. Tegmina and wings absent. Mesonotum with prominent lateral longitudinal keels. Metanotum typical. Abdomen broad, abdominal tergites strongly punctate; tergites three and fourth with paired tubercles. Ultimate tergite typical, posterior margin with two tubercles. Forceps (Fig. 485) cylindrical, but a little depressed, pointed apically. Penultimate sternite rounded posteriorly. – **Male** unknown. – **Length** of body with forceps: 6,5 mm.

Distribution: Sri Lanka.

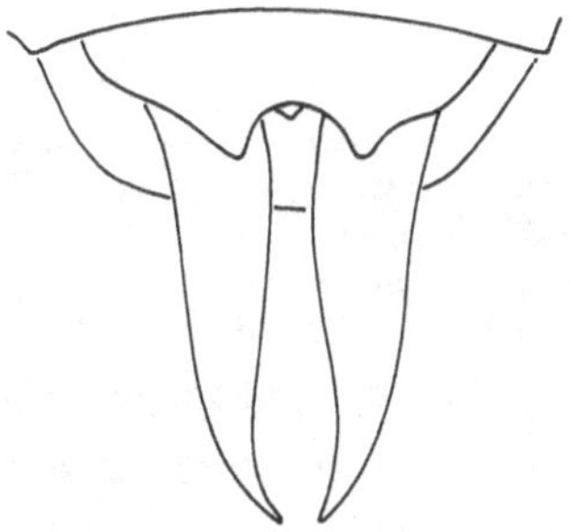

Fig. 485 Holotype ultimate tergite and forceps of *Ctenisolabis pusilla* STEINMANN. Original.

8. Ctenisolabis aciculata STEINMANN

1983 *Ctenisolabis aciculata* STEINMANN, Rev. suisse Zool., Genève, **90** (3): 545; fig. 3 (holotype ultimate tergite and forceps). – Terra typica: India (Type female: Mus. Hist. nat., Genève).

Description: Female with thoracal nota and abdominal tergites aciculate; shiny, black; head coriaceous, antennae dark brown; legs light brown, femora bicolourous, light brown apically, and brownish black basally. Head rounded, broad; postfrontal and coronal sutures finely visible; eyes large, more or less as long as the length of head behind eyes. Antennae 12-jointed; first joint long, a little longer than distance between antennal bases; second joint quadrate, third long, slightly longer than wide; fourth joint more or less as long as wide. Pronotum longer than wide; lateral margins straight, slightly expanded posteriorly; all angles rounded; posterior margin convex; median longitudinal furrow present. Tegmina and wings totally absent. Mesonotum with well-marked lateral longitudinal keels, the latter straight. Metanotum typical. Abdomen fusiform, oval; ultimate tergite broad, well developed; posterior margin with two tubercles medially. Forceps (Fig. 486) specific, broad basally, slender and curved apically; apex pointed. Penultimate sternite rounded posteriorly. – **Male** unknown. – **Length** of body with forceps: 11 mm.

Distribution: India: Meghalaya.

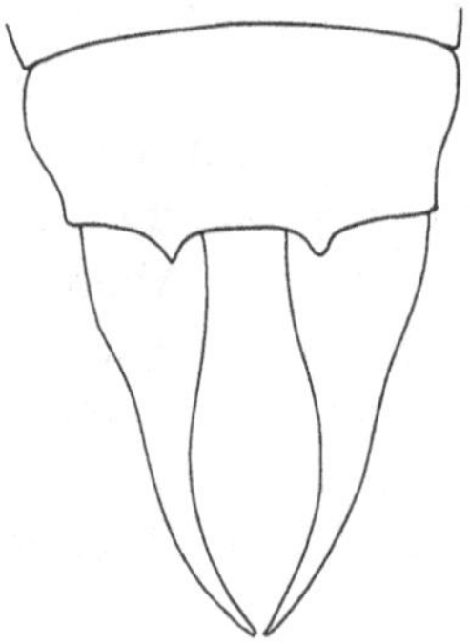

Fig. 486 Holotype ultimate tergite and forceps of *Ctenisolabis aciculata* STEINMANN. Original.

9. Ctenisolabis fletcheri BURR

1910 *Ctenisolabis fletcheri* BURR, Fauna Brit. India, Dermaptera, London, Bombay, 110, fig. 79 (female). – Terra typica: Ceylon (Type female: Brit. Mus. nat. Hist., London).
1966 *Isolabis fletcheri* – POPHAM & BRINDLE, Entomologist, London, **99**: 245.
1977 *Ctenisolabis fletcheri* – STEINMANN, Ann. hist.-nat. Mus. nation. hung., Budapest, **69**: 96.
Literature: BRINDLE 1971; BURR 1911; SAKAI 1970; SRIVASTAVA 1978; STEINMANN 1978, 1983.

Description: Female black or somewhat dark reddish black; head, thoracal nota and abdominal tergites strongly punctate; legs yellowish brown, femora bicolourous, yellowish apically, and dark brown basally. Head rounded; postfrontal and coronal sutures distinct. Eyes large, a little longer than the length of head behind eyes. Antennae 13-jointed; first joint long and wide, longer than distance between antennal bases; all joints short, third about as long as wide, fourth and fifth globular, others slightly lengthening, but apical joint is only a little longer than broad. Pronotum longer than wide; wider than head and even wider posteriorly; median longitudinal furrow prominent in anterior half. Tegmina and wings absent. Mesonotum ample, lateral longitudinal keels very distinct, a little lamellate. Metanotum normal. Abdomen fusiform, oval; tergites 3–4 with tubercles posteriorly. Ultimate tergite broad, posterior margin concave medially. Forceps (Fig. 487) typical, cylindrical, and pointed apically. – **Male** unknown. – **Length** of body with forceps: 10–11 mm.

Distribution: Sri Lanka, India: West Bengal.

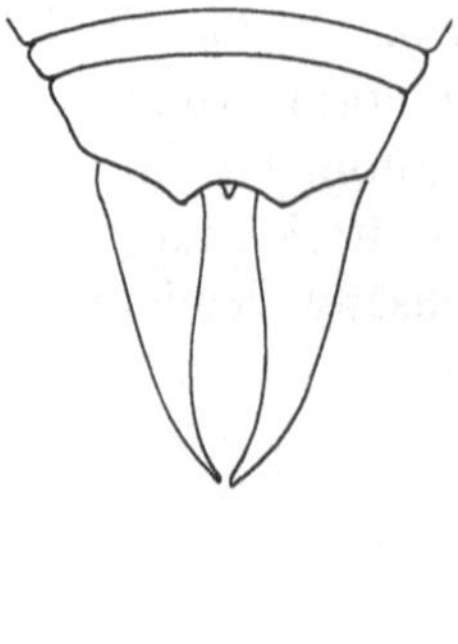

Fig. 487 Female ultimate tergite and forceps of *Ctenisolabis fletcheri* BURR. After BURR.

10. Ctenisolabis loebli STEINMANN

1983 *Ctenisolabis loebli* STEINMANN, Revue suisse Zool., Genève, **90** (3): 546; fig. 4 (holotype ultimate tergite and forceps). – Terra typica: India (Type female: Mus. Hist. nat., Genève).

Description: Female colour dark brownish black; legs unicolour, dark brown. Cuticle of head, thoracic nota and abdominal tergites strongly punctate. Head broad, tumid, rounded; postfrontal and coronal sutures indistinct; eyes large, more or less as long as the length of head behind eyes. Antennae 13-jointed; first joint long, slightly longer than distance between antennal bases; narrow basally, broader apically; third joint comparatively long, longer than wide. Pronotum longer than broad, lateral margins straight, but expanded posteriorly; posterior margin truncate; median longitudinal furrow present. Tegmina and wings absent. Mesonotum with prominent lateral longitudinal keels. Metanotum typical, transverse. Abdomen slender, fusiform, a little expanded medially. Ultimate tergite broad, posterior margin with two prominent tubercles medially. Forceps (Fig. 488) slightly broad basally, and narrower apically; branches cylindrical, and pointed. – **Male** unknown. – **Length** of body with forceps: 10,5 mm.

Distribution: India.

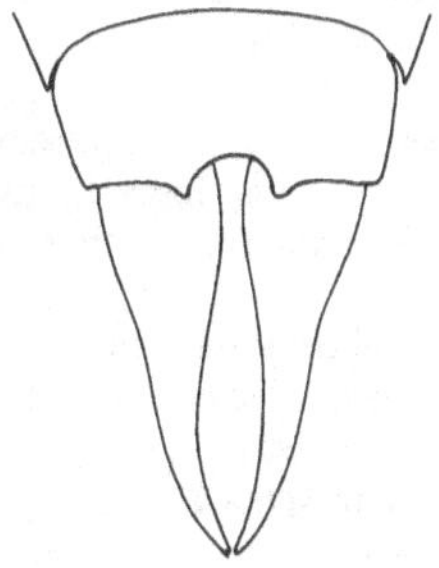

Fig. 488 Holotype ultimate tergite and forceps of *Ctenisolabis loebli* STEINMANN. Original.

Genus 25: **Brachylabis** DOHRN

1864 *Brachylabis* DOHRN, Stettiner ent. Ztg., **25**: 292. – Species typica: *Forficula chilensis* BLANCHARD 1851.

1910 *Nannisolabis* BURR, Fauna Brit. India, Dermaptera, London, Bombay, 106. – Species typica: *Nannisolabis willeyi* BURR 1910. – **syn. nov.**

Literature: BORG 1904; BORMANS 1900; BRINDLE 1970, 1971, 1976, 1978; BURR 1904, 1908, 1910, 1911, 1915; GILES 1958; HANDLIRSCH 1925; KIRBY 1891, 1904; REICHARDT 1968; SAKAI 1970; SCUDDER 1976; STEINMANN 1975, 1977, 1978; THOMAS 1945; VERHOEFF 1904; ZACHER 1911.

Description: This genus is difficult to define. Body very similar to *Metisolabis* BURR, and *Ctenisolabis* VERHOEFF, but eyes comparatively smaller, slightly shorter than the length of head behind eyes, and shorter than scape or shorter than distance between antennal bases. Cuticle of head and body punctate. Head rounded posteriorly; antennae with shorter or longer joints. Tegmina absent or rudimentary, wings totally absent. Abdomen fusiform, cylindrical. Legs moderately long, slender. Forceps simple, symmetrical. Male genitalia with narrow central parameral plate, median incision between external parameres absent or very small; genital lobes well or fully developed, virga present, sometimes very long, with or without sclerotized section. External parameres acute, pointed apically.

Distribution: Distributed in Neotropical, Ethiopian, Indo-Australian and Oriental faunal regions.

11 species.

Identification key to the species

1 Neotropical species **2**

1′ Not American species **4**

2 Very small species, body length without forceps 4 mm. Eyes very small. Found in nests of ants. First antennal joint very long .... 1. **Brachylabis lenkoi**, p. 301

2′ Larger species, body length without forceps at least 9 mm. Not associated with ants **3**

3 Pronotum as wide as long. Ultimate tergite concave posteriorly, without two prominent tubercles, as in fig. 489. Forceps strongly curved. Male genitalia similar to *coriacea* BURR, but external parameres longer, and sclerotized section of virga short (Fig. 490) .... 2. **Brachylabis chilensis**, p. 301

3′ Pronotum one and half times as long as wide; widened posteriorly. Ultimate tergite with two prominent tubercles, as in fig. 491. Forceps with straight section medially. Male genitalia similar to *chilensis* (BLANCHARD), but external parameres shorter, and sclerotized section of virga long (Fig. 492) .... 3. **Brachylabis coriacea**, p. 302

4 Ethiopian species **5**

4′ Indo-Australian or Oriental species **6**

5 Pronotum as wide as long. Smaller species, body length without forceps 4–5 mm. Virgal sclerite of male genitalia rather small and associated with denticulations (Fig. 494) .... 4. **Brachylabis carinata**, p. 303

5′ Pronotum slightly longer than wide. Larger species, body length without forceps 8–9 mm. Virga within genital lobe sclerotized for a relatively long distance and with a prominent lateral appendage (Fig. 495) .... 5. **Brachylabis fuscula**, p. 304

6 Indo-Australian species **7**

6′ Oriental species **9**

7 Genital lobe and virga of male genitalia very long (Fig. 497). Male forceps comparatively broad and short, inner margins with specific serration, as in fig. 496 .... 6. **Brachylabis yaloma**, p. 305

7′ Genital lobe and virga normal. External parameres pointed apically. Male forceps without characteristic serration at inner margins **8**

8 First antennal joint very long, slightly longer than distance between antennal bases. Posterior margin of male ultimate tergite with smaller tubercles medially (Fig. 498). Virga within genital lobe strongly sclerotized, as in fig. 499 .... 7. **Brachylabis canaca**, p. 306

8′ First antennal joint long, more or less as long as distance between antennal bases. Posterior margin of male ultimate tergite with larger tubercles medially (Fig. 500). Virga within genital lobe as in fig. 501 .... 8. **Brachylabis manawatawhi**, p. 307

9 Pronotum parallel-sided laterally **10**

9′ Pronotum slightly widened posteriorly. First antennal joint very long, rest of joints small, globular. Body reddish, pronotum red, smooth, legs dirty yellow. Male forceps strongly curved (Fig. 502) .... 9. **Brachylabis willeyi**, p. 308

10 Pronotum quadrate, more or less as long as wide. Tegmina present, but short, rudimentary, with lateral longitudinal keels. Abdominal tergites strongly punctate, shiny. Antennal joints very long and slender, fourth joint three times longer than wide .... 10. **Brachylabis tegminata**, p. 309

10′ Pronotum slightly longer than wide. Tegmina totally absent, mesonotum with lateral longitudinal keels. Abdominal tergites finely punctate, mat. Antennal joints shorter, fourth joint twice as long as wide .... 11. **Brachylabis philetas**, p. 309

1. Brachylabis lenkoi Brindle

1968 *Brachylabis lenkoi* Brindle, Pap. avulsos Zool., São Paulo, (4) **22**: 32; fig. 1 (male), 2 (male genitalia). – Terra typica: Brazil (Type male: Departamento de Zoologia, São Paulo).
Literature: Brindle 1971; Reichardt 1968.

Description: Male colour reddish brown or dark brown; antennae and legs yellowish brown. Very small species, perhaps immature. Head tumid, a little longer than wide; postfrontal and coronal sutures distinct. Eyes small, visibly shorter than genae. First antennal joint very long, longer than distance between antennal bases; fourth joint quadrate, the rest globular. Pronotum comparatively large, all angles rounded, lateral margins straight; median longitudinal furrow present. Tegmina and wings absolutely absent. Mesonotum with well-marked longitudinal lateral ridges. Metanotum transverse. Abdomen cylindrical, expanded medially, and narrower apically. Ultimate tergite simple, forceps tapering, curved apically. Found in nests of ants. – **Female** unknown. – **Length** of body with forceps: 4 mm.

Distribution: Brazil.

2. Brachylabis chilensis (Blanchard)

1851 *Forficula chilensis* Blanchard in Gay: Hist. Zool. Chile, **6**: 10; pl. 1, fig. 1 (male). – Terra typica: Chile (Type male: unknown locality; paratype male: Természettudományi Múz., Budapest: "Chile", gen. prep. No. 371, det. Dr. H. Steinmann, one male, and one female).
1851 *Forficula testaceicornis* Blanchard in Gay: Hist. Zool. Chile, **6**: 10. – Terra typica: Chile (Type female: unknown locality).
1864 *Brachylabis chilensis* – Dohrn, Stettiner ent. Ztg., **25**: 295.
1884 *Anisolabis chilensis* – Bolivar, Viajeal Pacific, Insects: 10.
1900 *Brachylabis chilensis* – Bormans, Das Tierreich, Berlin, **11**: 53. Argentina, Ecuador, Brazil.
Literature: Bormans 1883; Burr 1908, 1910, 1911, 1915; Kirby 1904; Popham & Brindle 1966; Rehn 1933; Sakai 1970; Steinmann 1977, 1978; Verhoeff 1904; Zacher 1911.

Description: Male colour reddish brown, except legs and palpi, yellow; head, pronotum, meso- and metanotum, and abdomen finely punctate. Head tumid, broad, about as broad as prozona of pronotum. Antennae 15-jointed; first joint long, longer than distance between antennal bases; fourth joint quadrate. Postfrontal and coronal sutures distinct. Eyes relatively large, about as long as genae, but somewhat shorter than the length of head behind eyes. Pronotum quadrate, as long as wide; lateral margins convex; strongly widened at posterior margin; all angles rounded; median longitudinal furrow present. Tegmina and wings entirely absent. Mesonotum normal, with prominent lateral longitudinal keels. Metanotum broad. Abdomen expanded to middle, mat. Ultimate tergite very broad; simple. Forceps (Fig. 489) a little asymmetrical, strongly curved, tapering, branches cylindrical. Genitalia (Fig. 490) narrow; central parameral plate with very narrow median incision; genital lobes well developed, virga slender, with quite visible sclerotized plate. External parameres long, pointed. – **Female** similar to male, but forceps symmetrical, contiguous, tapering. – **Length** of body with forceps: male: 13,5–15 mm, female: 13–15,5 mm.

Distribution: Chile, Argentina, Ecuador, Brazil.

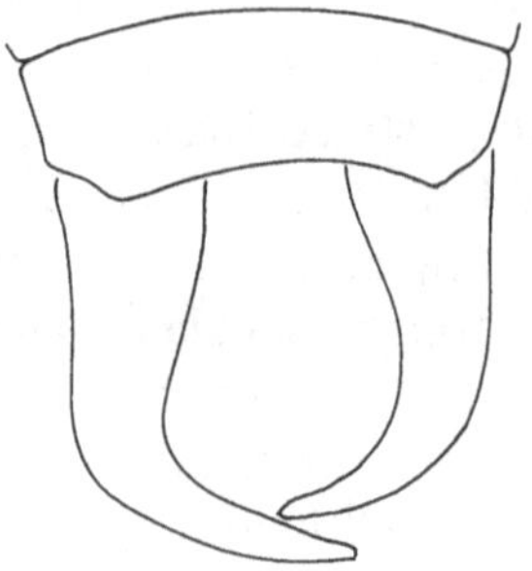

Fig. 489 Male ultimate tergite and forceps of *Brachylabis chilensis* (BLANCHARD). Original.

Fig. 490 Paratype genital armature of *Brachylabis chilensis* (BLANCHARD). Original.

3. Brachylabis coriacea BURR

1910 *Brachylabis coriacea* BURR, Trans. ent. Soc. London, **1910**: 192; pl. 47, fig. 31 (head of male). – Terra typica: Brazil (Type male: Brit. Mus. nat. Hist., London).

Literature: BURR 1911, 1915; POPHAM & BRINDLE 1966; REICHARDT 1968; SAKAI 1970; STEINMANN 1977, 1978.

Description: Male colour dark brown or reddish brown; cuticle of head fine and body strongly punctate. Head broad; tumid; postfrontal and coronal sutures finely distinct. Eyes relatively large, but slightly shorter than the length of head behind eyes. First antennal joint long, a little longer than distance between antennal bases. Pronotum long, one and half times longer than wide; widened posteriorly; median longitudinal furrow distinct. Tegmina and wings absent. Mesonotum with lateral longitudinal keels. Metanotum transverse. Abdomen cylindrical, more or less parallel-sided or a little expanded medially. Ultimate tergite broad; posterior margin with two tubercles. Forceps (Fig. 491) strong, straight basally and medially, curved apically. Penultimate sternite rounded posteriorly. Genitalia (Fig. 492) characteristic; paramere broad in anterior portion; median incision very short; genital lobes well developed; virga long, specific, with long sclerotized section basally. External parameres trigonal, very wide basally, pointed apically. – **Female** very similar to male, but forceps slenderer, tapering. – **Length** of body with forceps: male: 11–17 mm, female: 12–16 mm.

Distribution: Brazil.

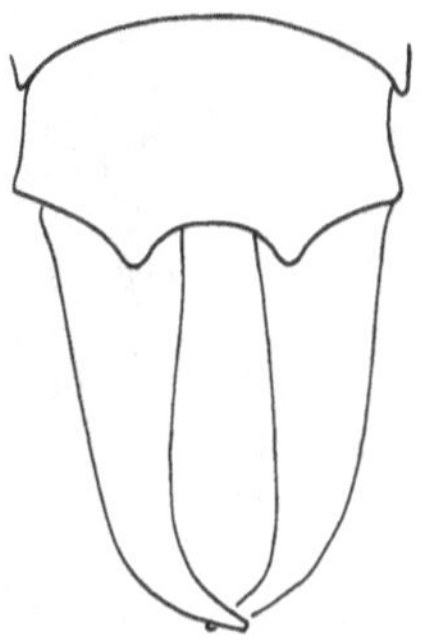

Fig. 491 Male ultimate tergite and forceps of *Brachylabis coriacea* BURR. Original.

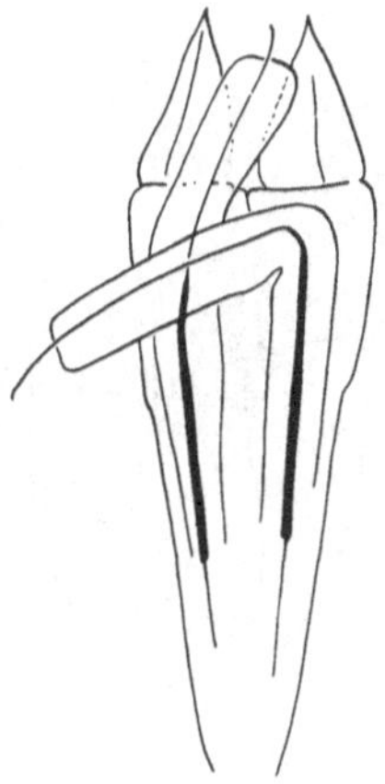

Fig. 492 Male genital armature of *Brachylabis coriacea* BURR. Original.

4. Brachylabis carinata (BRINDLE) comb. nov.

1978 *Isolabis carinata* BRINDLE, Dermaptera of Africa, Tervuren, **2**: 40; fig. 33 (holotype), 34 (holotype genitalia). – Terra typica: Rhodesia (Type male: Mus. r. Afr. centr., Tervuren).

Description: Male colour blackish-brown, antennae dark brown, legs yellowish brown, femora with basal two-thirds or more darker. Cuticle punctate and shiny. Head transverse, rounded posteriorly; postfrontal and coronal sutures indistinct. Eyes comparatively large, but slightly shorter than the length of head behind eyes. First antennal joint long, more or less as long as distance between antennal bases; second transverse, third as wide as long, fourth transverse. Pronotum strongly transverse, slightly widened posteriorly; margins almost straight; median longitudinal furrow short. Tegmina and wings absent; mesonotum transverse, with lateral longitudinal keels, the keel narrow but well defined and with an impressed line. Abdomen fusiform, somewhat depressed dorsally; ultimate tergite transverse, produced above the base of both branches of forceps. Forceps (Fig. 493) broader towards base, narrowed distally, weakly curved. Penultimate sternite transverse, posterior margin convex. Genitalia (Fig. 494) specific; central parameral plate wide apically and narrower basally; genital lobe small, virgal sclerite small and associated with denticulations. External parameres long, pointed. – **Female** very similar to male, but forceps slenderer, tapering. – **Length** of body with forceps: male: 4,5–5,5 mm, female: 4,5–6 mm.

Distribution: Rhodesia.

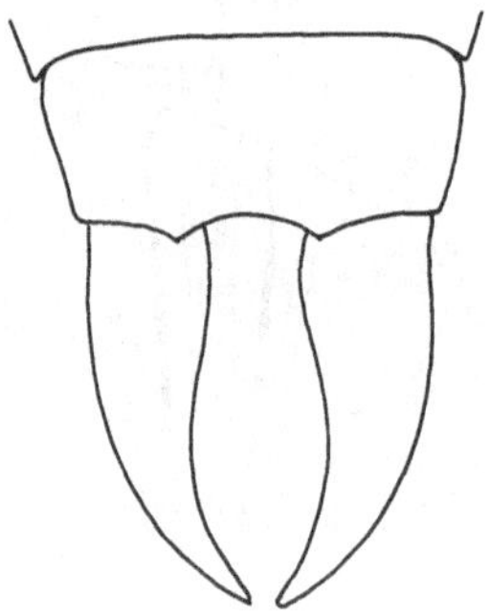

Fig. 493 Male ultimate tergite and forceps of *Brachylabis carinata* (BRINDLE). After BRINDLE.

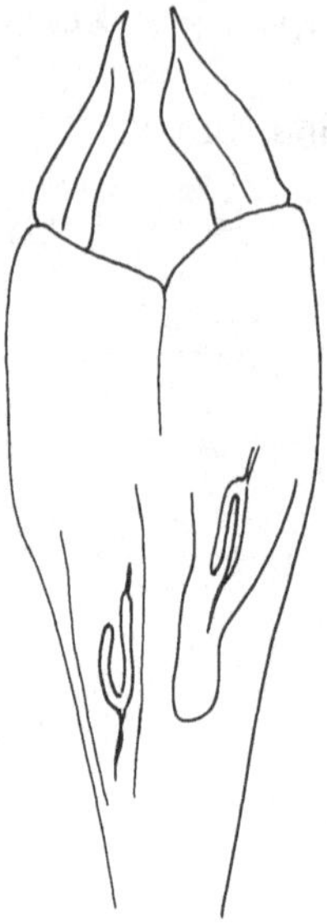

Fig. 494 Holotype genital armature of *Brachylabis carinata* (BRINDLE). After BRINDLE.

5. Brachylabis fuscula BRINDLE

1970 *Brachylabis fuscula* BRINDLE, Biologia gabonica, Paris, **6**: 15; fig. 4 (head and pronotum), 3 (penultimate sternite), 6 (holotype genital armature). – Terra typica: Gabon (Type male: Mus. nation. Hist. nat., Paris).

1978 *Isolabis fuscula* – BRINDLE, Dermaptera of Africa, Tervuren, **2**: 40; fig. 31 (male), 32 (male genitalia).

Description: Male general colour blackish; antennae dark brown, but first joint black, joint 10 light brown; sides of pronotum yellowish; legs yellowish brown, femora dark brown on basal two-thirds. Head transverse; eyes smaller than length of head behind eyes. First antennal joint more or less as long as distance between antennal bases. Postfrontal and coronal sutures indistinct. Pronotum transverse, widened posteriorly; lateral margins slightly sinuate and with a broad explanate border each side; posterior margin weakly convex. Tegmina and wings absent. Mesonotum transverse, narrowed posteriorly, lateral margins with a well-defined longitudinal ridge, the ridge oblique and passing towards the dorsum posteriorly, incomplete, present on anterior three-quarters of tergite only. Abdomen fusiform. Cuticle of head and thoracic nota finely punctate, abdominal tergites strongly punctate. Lateral tubercles on third and fourth tergites small but distinct. Ultimate tergite transverse, produced above the base of both branches of the forceps. Forceps with both branches cylindrical, slightly broader at base, narrowed distally, slightly sinuate, and with long yellow hairs. Penultimate sternite transverse, posterior margin rounded and with a small median excision. Genitalia (Fig. 495) specific; paramere broad apically, narrower basally; genital lobe well developed, virga sclerotized for a relatively long distance and with a prominent lateral appendage. External paramere relatively long, pointed. – **Female** similar to male, but forceps slender, tapering. – **Length** of body with forceps: male: 8,5–10 mm, female: 9–10,5 mm.

Distribution: Gabon.

Fig. 495 Holotype genital armature of *Brachylabis fuscula* (BRINDLE). After BRINDLE.

6. Brachylabis yaloma RAMAMURTHI

1967 *Brachylabis yaloma* RAMAMURTHI, Ent. Medd., København, **35**: 232; fig. 1 (male), 2 (antenna), 3 (abdominal end with forceps of female), 4 (abdominal end with forceps of male), 5 (holotype genitalia). – Terra typica: Bismarck Islands (Type male: Univ. zool. Mus., København).

Literature: STEINMANN 1977, 1978.

Description: Male general colour dark brown except terminal antennal joints, pale; whole body including head punctate and microsetulose. Head tumid, broad, comparatively short; posterior margin truncate. Eyes typical, more or less as long as length of head behind eyes. Antennae 11-jointed; first joint long, conical, slightly longer than distance between antennal bases; second very small, third long and slender, conical, fourth joint a little longer than wide. Pronotum longer than broad; parallel-sided laterally, median longitudinal furrow distinct. Tegmina and wings totally absent. Mesonotum with distinct lateral longitudinal keels. Metanotum transverse. Abdomen elongate, cylindrical, a little broader across tergites 5 and 6, contracted beyond. Ultimate tergite broad, simple; with smaller depression on dorsal surface. Forceps (Fig. 496) not *Brachylabis*-type, very broad basally and curved, pointed apically; very short, beak-like, inner margin crenulate. Penultimate sternite rectangular, posterior margin rounded, emarginate. Genitalia (Fig. 497) narrow, paramere long, genital lobes fully developed, virga within genital lobe very long, slender. External parameres long,

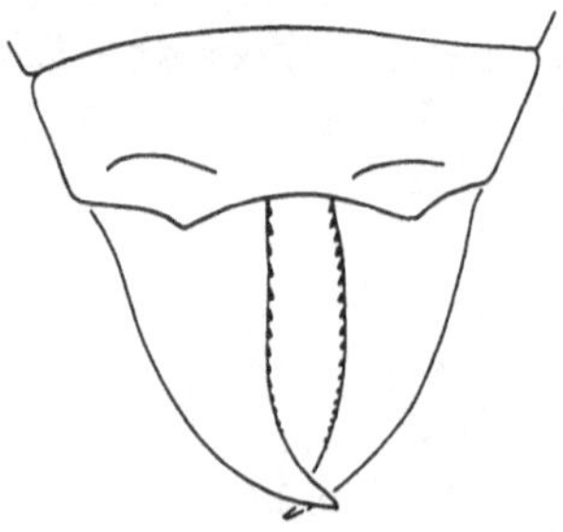

Fig. 496 Male ultimate tergite and forceps of *Brachylabis yaloma* Ramamurthi. After Ramamurthi.

Fig. 497 Holotype genital armature of *Brachylabis yaloma* Ramamurthi. After Ramamurthi.

basal and medial parts with inner and outer margins parallel; pointed apically. – **Female** as male. – **Length** of body with forceps: male: 9,5 mm, female: 8 mm.

Distribution: Bismarck Islands: New Britain.

7. Brachylabis canaca Burr

1914 *Brachylabis canaca* Burr, Nova Caledonia, Wellington, **1**: 318. – Terra typica: New Caledonia (Type male: Brit. Mus. nat. Hist., London).

1976 *Brachylabis canaca* – Brindle, Pacific Insects, Honolulu, **17**: 72; fig. 10 (head and pronotum), 11 (abdominal end with forceps), 12 (male genital armature).

Literature: Kaltenbach 1968; Sakai 1970; Steinmann 1977, 1978.

Description: Male colour dark reddish brown, with yellow longitudinal, irregular stripes of variable prominence on thoracic and abdominal tergites; antennae dark brown, legs yellowish brown to brown. Cuticle finely granulose and pubescent, dull. Head elongated, eyes comparatively small, slightly shorter than length of head behind eyes. First antennal joint very long, longer than distance between antennal bases. Pronotum long, slightly longer than wide; a little expanded posteriorly; median longitudinal furrow present in anterior half. Lateral margins straight, all angles rounded. Tegmina and wings absent. Mesonotum with well-marked lateral longitudinal keel, the ridge a little rounded. Abdomen slenderer, ultimate tergite very broad; simple. Forceps (Fig. 498) symmetrical, strongly curved, cylindrical. Penultimate sternite rounded posteriorly. Male genitalia (Fig. 499) characteristic; paramere oval, genital lobes well developed, virga very strongly sclerotized, virga within genital lobe with lateral appendage basally. – **Female** similar to male, but branches of forceps only very weakly curved. – **Length** of body with forceps: male: 11–12 mm, female: 12–13 mm.

Distribution: New Caledonia.

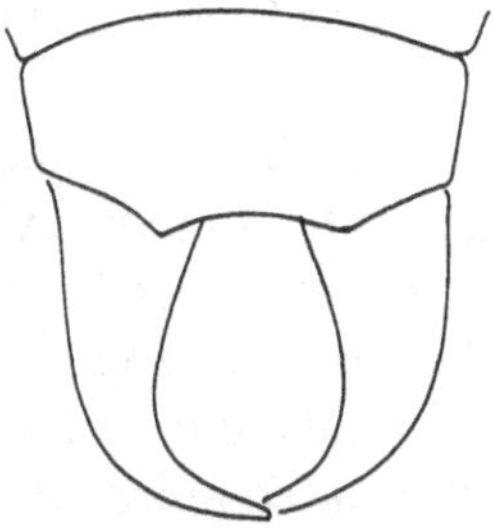

Fig. 498 Male ultimate tergite and forceps of *Brachylabis canaca* BURR. After BRINDLE.

Fig. 499 Male genitalia of *Brachylabis canaca* BURR. After BRINDLE.

8. Brachylabis manawatawhi GILES

1958 *Brachylabis manawatawhi* GILES, Rec. Auckland Inst. Mus., (1–2) **5**: 44; fig. 1 (male), 2 (penultimate sternite), (genitalia of holotype), 4 (antenna). – Terra typica: New Zealand (Type male: Auckland Inst. Mus., New Zealand).
Literature: HUDSON 1973.

Description: Male colour uniformly dark brown; head darker, legs and antennae lighter, palpi yellowish brown; nota and abdomen finely pubescent dorsally. Head small, elongate; postfrontal and coronal sutures distinct; eyes small, oval, shorter than first antennal joint. Antennae 16-jointed; first joint long, longer than distance between antennal bases; fourth joint quadrate. Pronotum as long as wide; narrower cephalad than caudad; anterior and posterior margins

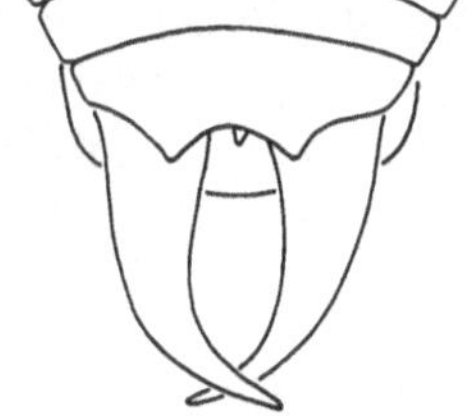

Fig. 500 Male ultimate tergite and forceps of *Brachylabis manawatawhi* GILES. After GILES.

Fig. 501 Holotype genital armature of *Brachylabis manawatawhi* GILES. After GILES.

truncate; laterally very slightly sinuate. Tegmina and wings absent. Mesonotum broad, deeply concave posteriorly; lateral longitudinal ridges very distinct. Metanotum transverse. Abdomen widest at segments 4 and 5; abdominal tergites shiny; tergite 5 with finely punctate oval areas cephalad corresponding with notches of preceding segment. Ultimate tergite broad, with blunt lobe above both forceps bases. Forceps (Fig. 500) symmetrical, curved, arcuate, cylindrical; thickened basally, tapering gradually apically, apices crossed. Genitalia (Fig. 501) characteristic; paramere flattened; genital lobes well developed, paired virgae relatively short, heavily sclerotized, expanded basally, tapering apically to fine point. External parameres relatively broad, but pointed apically. – **Female** similar to male, but forceps simple, symmetrical, nearly contiguous, tapering. – **Length** of body with forceps: male: 15–16,5 mm, female: 18,5 mm.

Distribution: New Zealand: Three Kings Islands.

9. Brachylabis willeyi (BURR) comb. nov.

1910 *Nannisolabis willeyi* BURR, Fauna of Brit. India, Dermaptera, London, Bombay, 107. – Terra typica: Ceylon (Type male: Brit. Mus. nat. Hist., London).
Literature: BRINDLE 1971, 1977; BURR 1915; SAKAI 1970; STEINMANN 1977, 1978.

Description: Male small, general colour dark reddish brown, similar to *philetas* BURR, but a little larger, and the cuticle is at most only finely punctate; head and pronotum impunctate, smooth. Head normal, rounded posteriorly; postfrontal and coronal sutures distinct. Eyes small, slightly shorter than length of head behind eyes. Antennae 13-jointed; first joint very short, fourth joint quadrate. Pronotum smooth, a little longer than wide and widened posteriorly; median longitudinal furrow present. Tegmina and wings absent. Mesonotum with a transverse depression in anterior portion, surrounded by a short raised ridge which is somewhat compressed at the anterior angles, but is not sharp. Metanotum broader than mesonotum, posterior margin concave. Legs long and slender, dirty yellow; the femora with distinct or indistinct darker shading. Abdomen with a fine golden pubescence; lateral tubercles of third and fourth segments very distinct. Ultimate tergite transverse, smooth, posterior margin nearly straight, as in fig. 502. Forceps remote at base, cylindrical, somewhat lamellate at basal section, but cylindrical and strongly curved apically. Genitalia

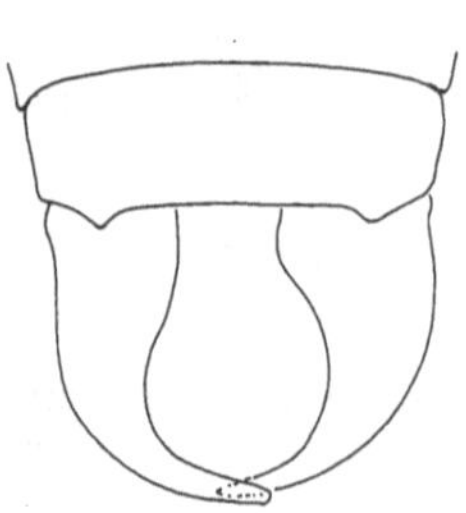

Fig. 502 Male ultimate tergite and forceps of *Brachylabis willeyi* (BURR). Original.

unknown. – **Female** similar to male, but forceps contiguous, gently arcuate from the base and overlapping. – **Length** of body with forceps: male: 9–9,5 mm, female: 8–9 mm.

Distribution: Sri Lanka.

10. Brachylabis tegminata STEINMANN

1983 *Brachylabis tegminata* STEINMANN, Rev. suisse Zool., Genève, **90** (3): 547; fig. 5 (holotype ultimate tergite and forceps). – Terra typica: India (Type female: Mus. Hist. nat., Genève).

Description: Female colour very dark brownish, somewhat reddish black; cuticle of head, pronotum and tegmina mat, metanotum and abdominal tergites strongly punctate, shiny. Antennae dark brown, sides of pronotum light brown. Head large, a little broader than pronotum; rounded posteriorly. Postfrontal and coronal sutures distinct. Eyes small, shorter than length of head behind eyes. Antennae with very long joints; first joint slightly longer than distance between antennal bases; fourth joint more or less three times as long as wide. Pronotum quadrate, lateral margins parallel; surface with transversal depression on posterior half. Tegmina present, but short, rudimentary, with well-defined lateral longitudinal keels. Wings absent. Abdomen oval, expanded medially; tergites strongly punctate, shiny black. Ultimate tergite broad, posterior margin as in fig. 503. Pygidium present. Forceps slender, a little trigonal basally, cylindrical apically; inner margins crenulate basally. – **Male** unknown. – **Length** of body with forceps: 16,5 mm.

Distribution: India: West Bengal.

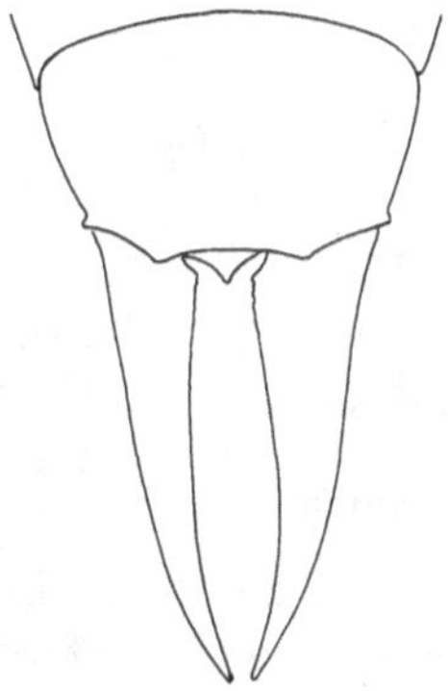

Fig. 503 Holotype ultimate tergite and forceps of *Brachylabis tegminata* STEINMANN. Original.

11. Brachylabis philetas BURR

1901 *Brachylabis philetas* BURR, J. Bombay nat. Hist. Soc., **1901**: 322; fig. 7 (male). – Terra typica: Ceylon (Type male: Brit. Mus. nat. Hist., London).

1908 *Leptisolabis philetas* – BURR, Ann. Mag. nat. Hist., London, (8) **2**: 252.

1910 *Nannisolabis philetas* – BURR, Fauna of Brit. India, Dermaptera, London, Bombay, 106.

Literature: BRINDLE 1971, 1977; BURR 1911, 1915; SAKAI 1970; STEINMANN 1977, 1978, 1983.

Description: Male colour blackish, antennae and legs dark brown. Cuticle strongly punctate. Head normal, more or less as broad as pronotum anteriorly; vertex tumid; postfrontal and coronal sutures finely distinct. Eyes small, slightly shorter than length of head behind eyes. First antennal joint long, longer than distance between antennal bases. Pronotum longer than wide; lateral margins straight; plate with transversal depression on anterior half. Tegmina and wings entirely absent. Mesonotum narrow, strongly impressed anteriorly, the posterior part being therefore slightly elevated, the lateral sides with prominent ridge longitudinally. Metanotum broader than mesonotum, posterior margin rounded, excised. Legs long, tibiae and tarsi testaceous, femora with a broad blackish band. Abdomen fusiform, shiny; ultimate tergite broad. The lateral tubercles very distinct on the second and third dorsal segments. Forceps (Fig. 504) comparatively short, simple, cylindrical. Genitalia unknown. – **Female** similar to male, but forceps straight. – **Length** of body with forceps: male: 7–7,5 mm, female: 8,5 mm.

Distribution: Sri Lanka, and West Bengal.

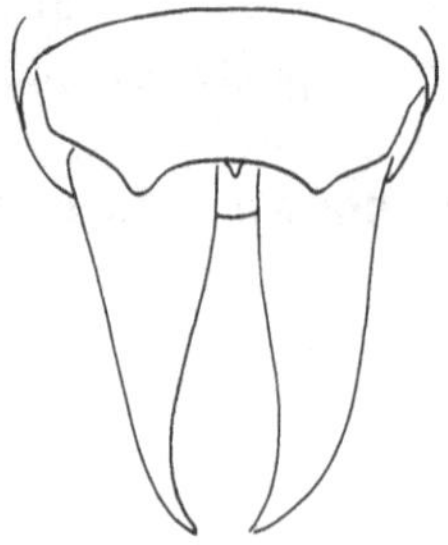

Fig. 504 Male ultimate tergite and forceps of *Brachylabis philetas* Burr. Original.

Doubtful position

? Brachylabis collinsi Brindle

1980 *Brachylabis collinsi* Brindle, Entomologist's Rec. J. Variation, London, **1980**: 172; fig. 7 (genital armature of holotype), 9 (male). – Terra typica: Borneo (Type male: Brit. Mus. nat. Hist., London).

Description: Male colour blackish-brown; shiny; antennae with some distal joints yellowish or whitish; femora pale at apices, tibiae and tarsi yellow. Cuticle very strongly, closely, and deeply punctate. Head transverse, eyes comparatively large, broad. First antennal joint normal, a little longer than distance between antennal bases. Pronotum quadrate; lateral margins parallel. Tegmina and wings absent. Mesonotum with sides parallel, punctures of mesonotum forming irregular ridges. Metanotum with sides curved. Abdomen elongate, a little expanded medially; ultimate tergite not of *Brachylabis*-type, Forceps (Fig. 505) almost cylindrical in cross section, slightly wider at base, simple. Penultimate sternite rounded posteriorly. Genitalia (Fig. 506) very specific; central parameral plate with median incision, genital lobes not visible; virga present, with specific sclerotized section medially. External parameres with inner and outer margins undulate, acute apically. – **Female** very similar to male, but punctation and rugosities less prominent. – **Length** of body with forceps: male: 8,5 mm, female: 8 mm.

Distribution: Borneo: Sarawak.

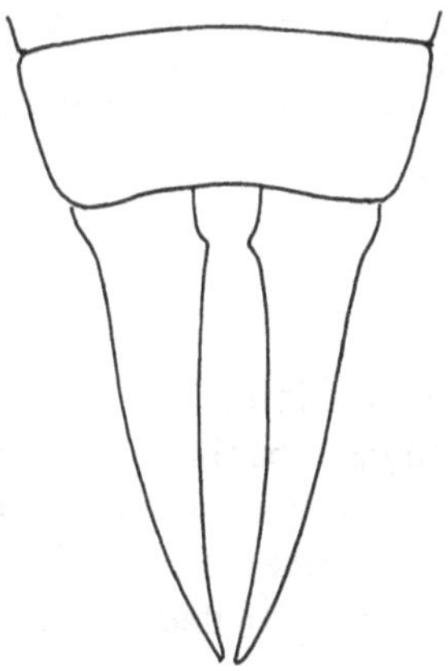

Fig. 505 Male ultimate tergite and forceps of *Brachylabis collinsi* BRINDLE. After BRINDLE.

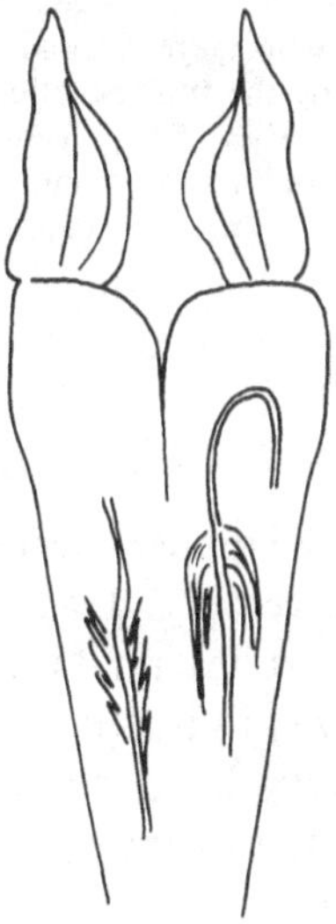

Fig. 506 Holotype genital armature of *Brachylabis collinsi* BRINDLE. After BRINDLE.

Subfamily 6: **Isolabiinae** STEINMANN

1975 *Isolabiinae* STEINMANN, Acta zool. Acad. Sci. hung., Budapest, **21**: 206. – Genus typicum: *Isolabis* VERHOEFF 1902.

1904 *Isolabiidae* (partim) VERHOEFF, Sber. Ges. naturf. Fr., Berlin, **1904**: 10.

Description: The species of this subfamily are characterized by the total absence of organs of flight, except the species of *Pterolabis villiersi* (BRINDLE), *Geracodes litus* HEBARD, and *G. labioides* HINCKS. The lateral margins of mesonotum without a ridge-like edge or keels. Head and all nota and abdomen generally brown to blackish-brown, or yellowish-brown. Eyes large, longer than length of head behind eyes, or smaller. Fourth antennal joint quadrate, as long as wide, or longer than wide. Forceps slender, more or less symmetrical, apices crossed. Male genital lobes well developed, virga present, with or without sclerotized plates; external parameres acute or obtuse.

Distribution: Distributed worldwide except the Palaearctical region.

4 genera.

Identification key to the genera

1 Eyes comparatively large, as long as length of head behind eyes **2**
1′ Eyes small, shorter than length of head behind eyes **3**
2 Winged species, tegmina well, wings fully developed. Male genitalia characteristic, similar to species of *Epilandex* HEBARD: external parameres very long (Fig. 508), central parameral plate V-shaped, median incision of anterior margin very deep and wide.. Genus 26: **Pterolabis**, p. 312
2′ Apterous species, tegmina absent or rudimentary, wings entirely absent. Male genitalia with normal external parameres .... Genus 27: **Isolabis**, p. 313
3 Both genital lobes of male genitalia with a complex arrangement of sclerites, the genital lobes not clearly distinct; eyes very small and tegmina rudimentary or eyes small and tegmina larger, meeting along the sutures, often with large visible wings .... Genus 28: **Geracodes**, p. 332
3′ Both genital lobes of male genitalia without such a complex arrangement of sclerites; eyes always very small, and tegmina represented by rudimentary flaps or completely absent ... . Genus 29: **Africolabis**, p. 336

Genus 26: **Pterolabis gen. nov.**

Species typica: *Pseudovostox villiersi* BRINDLE 1968.

Description: Smaller species; tegmina well developed, long, mainly coriaceous, wings fully developed. Eyes very large, longer than head behind eyes, and a little longer than first antennal joint. Scape normal, about as long as distance between antennal bases; fourth joint quadrate. Male genital armature specific; external parameres very long; central parameral plate with very deep, wide median incision at anterior margin.

Distribution: Ethiopian species.

Single species.

1. **Pterolabis villiersi** (BRINDLE) **comb. nov.**

1968 *Pseudovostox villiersi* BRINDLE, Bull. de l'I.F.A.N., Amsterdam, **30**: 775; fig. 1 (male), 2 (genital armature of holotype), 3 (penultimate sternite of holotype). – Terra typica: Republic Congo (Type male: Mus. nation. Hist. nat., Paris).
1970 *Geracodes villiersi* – BRINDLE, Rev. Zool. Bot. afr., Bruxelles **81**: 261; fig. 8 (male genitalia).
1978 *Pseudovostox villiersi* – BRINDLE, Dermaptera of Africa, Tervuren, **2**: 184; fig. 326 (male genitalia), 327 (penultimate sternite of male), 327 (male).
Literature: STEINMANN 1977, 1978.

Description: Male blackish or dark brown; tegmina dull yellow, blackish-brown anteriorly and laterally; cuticle of head and pronotum finely punctate and pubescent, rather shiny. Head transverse, convex, a triangular frontal area smoother and shinier; eyes very large, more or less as long as scape. First antennal joint about as long as distance between antennal bases. Pronotum relatively small, with a wide explanate lateral border; lateral margins expanded posteriorly. Tegmina well developed, long, median anterior area with minute tubercles, each of which may have originally been the base of a seta. Wings prominent. Abdomen more or less cylindrical, a little expanded medially. Ultimate tergite broad, posterior margin as in fig. 507. Forceps normal, symmetrical, cylindrical. Penultimate sternite long, punctate and pubescent, posterior margin with a median incision. Genitalia (Fig. 508) characteristic; central parameral plate with very deep, wide median incision at anterior margin, plate V-shaped; genital lobes well developed, with very long virgae. External parameres very long, similar to *Epilandex* HEBARD.

Notes by BRINDLE (1978): "This species was provisionally transferred to the genus *Geracodes* in BRINDLE (1970) pending a re-examination of the type, since the genitalia appeared to have paired penes, and the species was not included in the first part of the Dermaptera of Africa. A re-examination of the type, however, makes it clear that the species was correctly placed in *Pseudovostox* (Labiidae). The tarsal arolia are prominent, and the genitalia appear to have only a single virga, although this is broken, and a single set of virgal sclerites. On this view, the two arms of the genitalia below the parameres do not represent paired penes but a broken single penis. The present figure of the genitalia is re-drawn from the genital mount, which has cleared somewhat since the original drawing was published (BRINDLE 1968)."

A comparative anatomical examination of the genital armature of the species seems to support BRINDLE's statement, but the precence of the median incision of the central parameral plate indicates the origin of two genital lobes in spite of the fact that the preparation shows only one set of virga. Consequently, the species is relegable to Isolabiinae on the basis of its other morphological features. – **Female** unknown. – **Length** of body with forceps: 5–5,5 mm.

Distribution: Congo Republic.

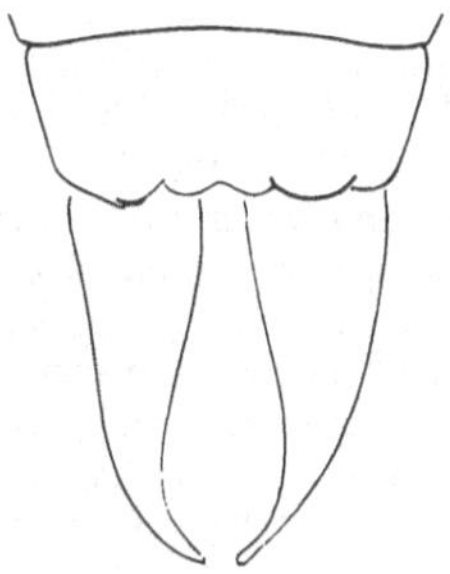

Fig. 507 Male ultimate tergite and forceps of *Pterolabis villiersi* (BRINDLE). After BRINDLE.

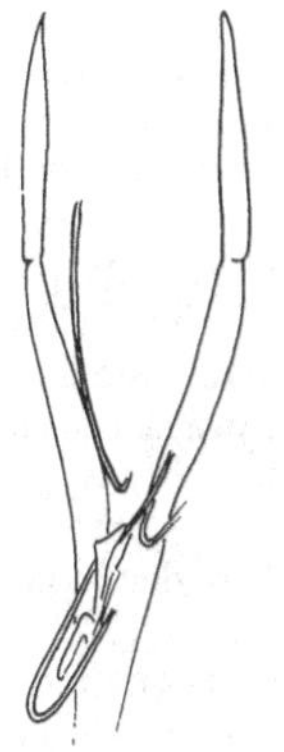

Fig. 508 Holotype genital armature of *Pterolabis villiersi* (BRINDLE). After BRINDLE.

Genus 27: **Isolabis** VERHOEFF

1902 *Isolabis* VERHOEFF, Sber. Ges. naturf. Fr. Berlin, **1902**: 11. – Species typica *Isolabis braueri* VERHOEFF, 1902.

1902 *Leptisolabis* VERHOEFF, Sber. Ges. naturf. Fr. Berlin, 1902: 12. – Species typica: *Leptisolabis usumbarana* VERHOEFF 1902.
1908 *Verhoeffia* BURR, Ann. Mag. nat. Hist., London, (8) **2**: 243. – Species typica: *Brachylabis joestedti* BORG 1904.
1910 *Arlex* BURR, Trans. ent. Soc. London, **1910**: 191 (replacement name for *Verhoeffia* BURR 1908, preoccupied by *Verhoeffia* BROLEMANN 1895 in Myriopoda).
1966 *Metisolabis* – POPHAM & BRINDLE (partim), Entomologist, London, **99**: 245 [= *Isolabis* VERHOEFF 1902].
1966 *Ctenisolabis* – POPHAM & BRINDLE, Entomologist, London, **99**: 245 [= *Isolabis* VERHOEFF 1902].
1966 *Antisolabis* – POPHAM & BRINDLE (partim), Entomologist, London, **99**: 245 [= *Isolabis* VERHOEFF 1902].
Literature: BRINDLE 1978; BURR 1910, 1911, 1915; HINCKS 1953; KIRBY 1904; REHN 1924; SAKAI, 1970; STEINMANN 1977, 1978; TOWNES 1945; ZACHER 1911.

Description: Apterous species; tegmina and wings totally absent. Mesonotum normal, without lateral longitudinal ridges or keels. First antennal joint long, more or less longer than distance between antennal bases; second joint quadrate, third twice as long as wide, and fourth joint long, slightly longer than wide. Abdomen slender, but a little expanded medially. Forceps simple, forceps of male sometimes asymmetrical, femal forceps tapering. Male genitalia narrow, central parameral plate with median incision on anterior margin; external parameres acute; genital lobes well or fully developed, with virgae.

Distribution: Ethiopian, Malgassy, Neotropic and Oriental species.

16 species.

Identification key to the species

1 Ethiopian species (after BRINDLE) **2**
1′ Malgassy, Neotropic or Oriental species **14**
2 Eyes large, more or less as long as length of head behind eyes **3**
2′ Eyes very large, longer than length of head behind eyes **6**
3 Distal antennal joints relatively slender and almost cylindrical, not wider than basal joints; pronotum relatively longer, lateral margins more or less sinuate; general colouration dark reddish-brown, legs unicolorours yellowish. Male genitalia (Fig. 509) with short broad external parameres, both plates with a long narrow tip .... 1. **Isolabis proxima**, p. 315
3′ Distal antennal joints relatively broad, somewhat conical, and wider than basal joints; pronotum relatively smaller, lateral margins straight; general colouration blackish, legs yellow with femora darkened or legs mainly dark brown **4**
4 Pronotum with relatively long conspicuous pale pubescence; posterior margin of male penultimate sternite somewhat triangular, without a median excision. External parameres of male genitalia broader basally (Fig. 511); virga stirrup-shaped basally .... 2. **Isolabis transversa**, p. 316
4′ Pronotum glabrous or almost so; posterior margin of male penultimate sternite rounded and with a small median excision **5**
5 Antennae unicoloured, blackish-brown; distal joints narrower in relation to basal joints; pronotum impunctate laterally and posteriorly. External parameres of male genitalia shorter and apices short (Fig. 513) .... 3. **Isolabis unicolor**, p. 317
5′ Antennae bicoloured, blackish-brown with joints 10–11 white or partly white; pronotum almost completely punctate. External parameres of male genitalia longer and with longer apices (Fig. 515) .... 4. **Isolabis bicolor**, p. 319
6 Rudimentary tegmina present **7**
6′ Tegmina completely absent **9**
7 Head more strongly transverse and eyes larger; pronotum parallel-sided; antennae dark

brown, joints 8–9 or 8–10 whitish, joints 10–11 or 10–13 blackish. Male genitalia (Fig. 517) very long and slender, external parameres specific, expanded medially on outer margins... .... 5. **Isolabis grandicollis**, p. 320

7′ Head less transverse and eyes smaller; pronotum widened posteriorly; antennae brown or dark brown, some distal joints paler **8**

8 Head less transverse, pronotum as wide as long, male forceps (Fig. 519) weakly curved and asymmetrical, apices not attenuated; penultimate sternite of male with posterior margin rounded .... 6. **Isolabis sjoestedti**, p. 321

8′ Head more transverse, pronotum slightly transverse, male forceps (Fig. 520) arcuate, symmetrical, apices attenuate; penultimate sternite of male with a small excision in posterior margin. Male genitalia as in fig. 521 .... 7. **Isolabis attenuata**, p. 322

9 Distal antennal joints perceptibly wider than the basal joints; antennae blackish or dark brown, some distal joints white **10**

9′ Distal antennal joints only as wide as, or less wide than the basal joints; antennae dark brown or paler, unicoloured, or with some distal joints yellowish **12**

10 Pronotum transverse, head more transverse and with posterior angles marked, posterior margin straight; third antennal joint much longer than wide. Male genitalia (Fig. 523) with undulate external parameres on outer margins .... 8. **Isolabis coiffaiti**, p. 323

10′ Pronotum almost quadrate, head less transverse and with posterior angles rounded, posterior margin slightly convex; third antennal joint slightly longer than wide **11**

11 Fourteenth antennal joint whitish, much paler than preceding segments; joints 7–8 white; general colouration dark reddish-brown. External parameres of male genitalia broader, median incision of anterior margin wide, as in fig. 525. .... 9. **Isolabis schoutedeni**, p. 324

11′ Fourteenth antennal joint blackish, concolorours with preceding joints; segments 7–8 white; general colouration blackish. External parameres of male genitalia narrower, median incision on anterior margin normal, as in fig. 527 .... 10. **Isolabis usumbarana**, p. 325

12 Distal antennal joints shorter; dark reddish-brown species with pronotum reddish; genital lobes of male genitalia well developed, virga within genital lobe with large sclerotized sac (Fig. 528) .... 11. **Isolabis rufa**, p. 326

12′ Distal antennal joints longer; blackish species; body length 10 mm or more **13**

13 Third antennal joint as long as fourth, the latter longer than wide; distal joints more slender; pronotum relatively larger; male forceps (Fig. 529) less strongly curved. Male genitalia (Fig. 530) specific, virga with sclerotized section medially straight .... 12. **Isolabis quadrata**, p. 327

13′ Third antennal joint longer than fourth, the latter only a little longer than wide; distal joints less slender; pronotum relatively smaller; male forceps (Fig. 531) more strongly curved. Male genitalia (Fig. 532) of *Isolabis*-type, virga within genital lobe with sclerotized sac .... 13. **Isolabis braueri**, p. 328

14 Malgassian species. Antennae with robust joints; pronotum longer than wide, lateral margins straight. Male genitalia narrower medially, external parameres with elongated apices (Fig. 533) .... 14. **Isolabis pauliani**, p. 329

14′ Neotropic or Oriental species **15**

15 Neotropic species. General colour jet black, dull; golden pubescence, legs black, but yellow at the base and apex; pronotum a little longer than wide, sides gently diverging posteriorly; posterior margin of ultimate tergite with short triangular lobe on each side over the forceps (Fig. 534) .... 15. **Isolabis howardi**, p. 330

15′ Oriental species. General colour dark brownish black, legs yellow with femora dark brownish black in basal two thirds; pronotum rectangular, feebly diverging posteriorly; posterior margin of ultimate tergite as in fig. 535 .... 16. **Isolabis ocellata**, p. 331

1. Isolabis proxima (Brindle)

1966 *Brachylabis proxima* Brindle, Rev. Zool. Bot. afr., Bruxelles, **74**: 42; fig. 17 (male genitalia). – Terra typica: Kivu (Type male: Mus. r. Afr. centr., Tervuren).

1978 *Isolabis proxima* – BRINDLE, Dermaptera of Africa, Tervuren, **2**: 43; fig. 37 (head and pronotum of male), 38 (male genitalia).
Literature: STEINMANN 1977, 1978.

Description: Male colour reddish-brown, rather dull; antennae dark brown, basal two joints yellowish; legs yellowish-brown. Head transverse, sides rounded; tumid, slightly rugose; eyes large, more or less as long as length of head behind eyes, but shorter than first antennal joint. Antennae 14-jointed; first long, a little longer than distance between antennal bases. Pronotum relatively large, almost quadrate, almost smooth, with sparse short yellowish hairs; posterior angles rounded but anterior angles sharply pointed; median longitudinal furrow present. Mesonotum transverse, duller, with obtuse and inconspicous lateral margins; depressed medially along a transverse line. Metanotum more transverse, with concave posterior margin. Abdomen fusiform, rather depressed, lateral tubercles on third tergite rather small, those on fourth larger; ultimate tergite transverse, posterior margin concave. Both branches of forceps cylindrical, broader on basal half, distal half narrowed and weakly curved. Penultimate sternite with posterior margin rounded and with a very small median excision. Genitalia (Fig. 509) broad apically, very narrow basally; genital lobes fully developed, with long rod-like double sclerite; external parameres long, elongated and pointed. – **Female** similar to male, but rather darker and smaller, with straight forceps. – **Length** of body with forceps: male: 11–11,5 mm, female: 10–10,5 mm.

Distribution: Ruanda-Burundi: Kivu, Tanzania and Zaïre.

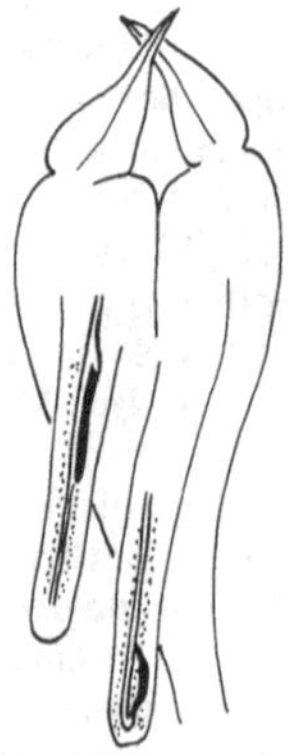

Fig. 509 Holotype genital armature of *Isolabis proxima* (BRINDLE). After BRINDLE.

2. Isolabis transversa BRINDLE

1978 *Isolabis transversa* BRINDLE, Dermaptera of Africa, Tervuren, **2**: 41; fig. 35 (head and pronotum), 36 (genital armature of holotype). – Terra typica: Zaïre (Type male: Mus. r. Afr. centr., Tervuren).

Description: Male colour black, rather dull; antennal joints 2–3 brown, and joint 10 white; femora blackish, broadly yellow at apices. Cuticle strongly punctate and pubescent. Head rounded, postero-lateral angles present, posterior margin weakly concave; eyes larger, but slightly shorter than length of head behind eyes. First antennal joint long, more or less as long as distance between antennal bases; second transverse, third slightly longer than wide. Pronotum quadrate, lateral margins slightly expanded posteriorly; posterior margin weakly convex; median longitudinal furrow present. Mesonotum transverse, weakly narrowed posteriorly, with a transverse narrow depression. Tegmina and wings absent. Abdomen fusiform, lateral tubercles small or almost absent. Ultimate tergite simple. Forceps (Fig. 510) cylindrical, broader towards base, narrowed distally, almost straight, with apices curved. Penultimate sternite with posterior margin weakly triangular, apex rounded. Genitalia (Fig. 511) narrow, median incision of anterior margin of central parameral plate distinct; genital lobes well developed, virgal sclerites rather large and associated with denticulations. External parameres broad basally, apices long. – **Female** very similar to male, but branches of forceps more slender. – **Length** of body with forceps: male: 7–7,5, female: 8 mm.

Distribution: Zaïre.

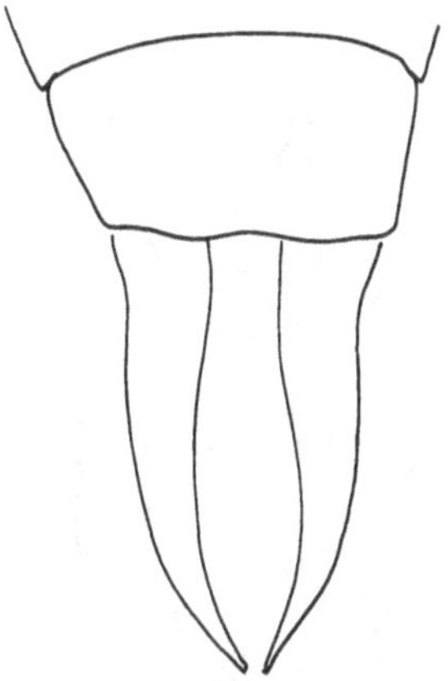

Fig. 510 Male ultimate tergite and forceps of *Isolabis transversa* BRINDLE. After BRINDLE.

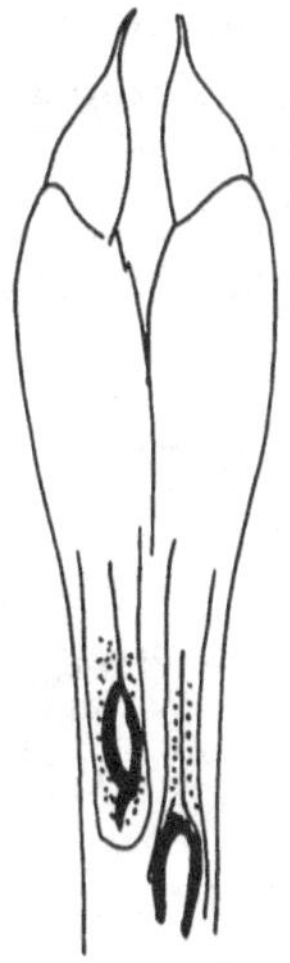

Fig. 511 Holotype genital armature of *Isolabis transversa* BRINDLE. After BRINDLE.

3. Isolabis unicolor (BRINDLE)

1975 *Brachylabis unicolor* BRINDLE, Rev. Zool. Bot. afr., Bruxelles, **89**: 684; fig. 1 (male), 2 (holotype genitalia). – Terra typica: Tanzania (Type male: Mus. r. Afr. centr., Tervuren).

1978 *Isolabis unicolor* – BRINDLE, Dermaptera of Africa, Tervuren, **2**: 44; fig. 39 (male), 40 (male genitalia). Eastern Africa.

Literature: STEINMANN 1977, 1978.

Description: Male colour blackish, antennae and legs blackish brown, tarsi paler; shiny, pronotum and head with scattered punctures, but lateral and posterior parts of the pronotum impunctate; meso- and metanotum and abdominal tergites punctate. Head short, rounded; posterior margin straight, posterior angles rounded; eyes normal, shorter than length of head behind eyes. First antennal joint comparatively long, longer than distance between antennal bases. Postfrontal and coronal sutures indistinct. Pronotum more or less quadrate, lateral margins straight, but widened posteriorly; posterior margin weakly convex. Tegmina and wings absent. Mesonotum broad with a weak lateral fold, but without lateral ridges. Abdomen fusiform and slightly depressed dorsally; lateral tubercles on third and fourth tergites weakly indicated. Ultimate tergite very broad. Forceps (Fig. 512) symmetrical, cylindrical, widened basally, acuminate apically, and inner margins crenulated basally. Penultimate sternite with a small median incision on posterior margin. Genitalia (Fig. 513) characteristic; paramere broadest at anterior margin, median incision of anterior part small; genital lobes with specific sclerotized plates, virga associated with dark sclerites and with vaguely defined areas of denticulations. – **Female** similar to male, but forceps slender, contiguous, tapering. – **Length** of body with forceps: male: 5,5–6 mm, female: 5 mm.

Distribution: Tanzania.

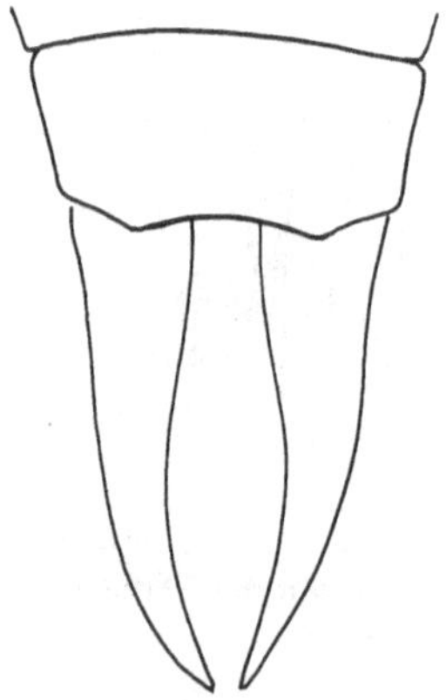

Fig. 512 Male ultimate tergite and forceps of *Isolabis unicolor* (BRINDLE). After BRINDLE.

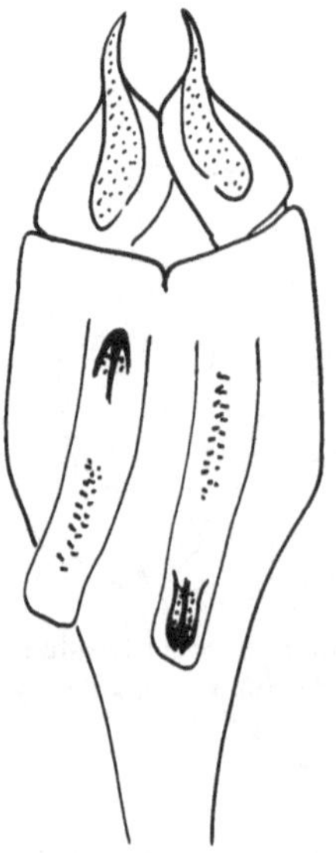

Fig. 513 Holotype genital armature of *Isolabis unicolor* (BRINDLE). After BRINDLE.

4. Isolabis bicolor (BRINDLE)

1975 *Brachylabis bicolor* BRINDLE, Rev. Zool. Bot. afr., Bruxelles, **89**: 684; fig. 3 (holotype genital armature). – Terra typica: Tanzania (Type male: Mus. r. Afr. centr., Tervuren).
1978 *Isolabis bicolor* – BRINDLE, Dermaptera of Africa, Tervuren, 2: 46; fig. 41 (male genitalia). Eastern Africa.
Literature: STEINMANN 1977,1978.

Description: Male very similar in external characters to *unicolor* (BRINDLE), but differing as follows: antennal joints 10–11 white or partly white. Head a little longer than broad, postfrontal and coronal sutures indistinct; eyes large but shorter than length of head behind eyes. First antennal joint long, slightly longer than distance between antennal bases; second transverse, fourth joint rather longer than wide. Pronotum widened posteriorly, lateral margins sinuate; posterior margin weakly convex. Pronotum almost completely punctate. Tegmina and wings absent, meso- and metanotum typical. Abdomen slightly expanded medially, ultimate tergite broad, posterior margin concave medially. Forceps (Fig. 514) slender, cylindrical. Penultimate sternite with a small median incision on posterior margin. Genitalia (Fig. 515) very similar to *unicolor* (BRINDLE), but virga within genital lobe curved, and sclerotized plate indistinct; external parameres a little longer apically. – **Female** similar to male, but forceps slenderer, and tapering. – **Length** of body with forceps: in both sexes: 5–5,5 mm.
Distribution: Tanzania.

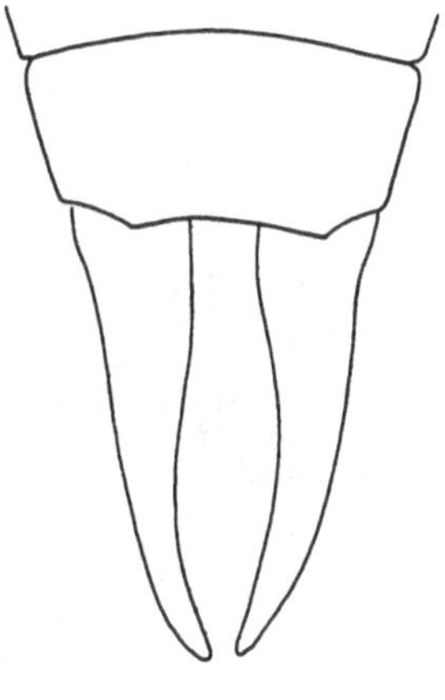

Fig. 514 Male ultimate tergite and forceps of *Isolabis bicolor* (BRINDLE). After BRINDLE.

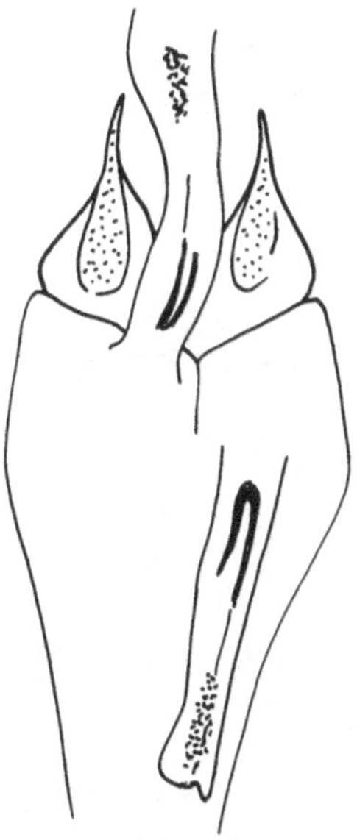

Fig. 515 Holotype genital armature of *Isolabis bicolor* (BRINDLE). After BRINDLE.

5. Isolabis grandicollis (Hincks)

1954 *Antisolabis grandicollis* Hincks, Mém. Inst. fr. Afr. noire, Paris, Dakar, **40**: 117; fig. 19 (male), 20 (holotype genital armature). – Terra typica: Guinea (Type male: Mus. nation. Hist. nat., Paris).
1966 *Isolabis grandicollis* – Popham & Brindle, Entomologist, London, **99**: 245.
1977 *Brachylabis grandicollis* – Steinmann, Ann. hist.-nat. Mus. nation. hung., Budapest, **69**: 96.
1978 *Isolabis grandicollis* – Brindle, Dermaptera of Africa, Tervuren, **2**: 46; fig. 42 (paratype female), 43 (male genitalia). Western Africa.
Literature: Sakat 1970; Steinmann 1978.

Description: Male colour blackish-brown with a reddish tint; sides of pronotum yellow, antennae dark brown at base, joints 8–9 or 8–10 whitish, joints 10–11 or 10–13 blackish; legs yellowish-brown, femora darkened on basal two-thirds or more. Cuticle punctate and pubescent, hairs rather fine and yellow. Head strongly transverse, sides rounded, posterior margin almost straight; eyes very large, slightly longer than length of head behind eyes. Antennae 13-jointed; first joint long, slightly longer than distance between antennal bases, second transverse, third one and half times as long as wide or longer. Pronotum relatively lare, rectangular, lateral margins more or less parallel, all angles rounded; median longitudinal furrow distinct, occurs on the disc with indistinct shorter anterior lines on both sides. Tegmina represented by rudimentary lateral flaps, pair of rudimentary tegmina which meet or overlap at the sutures, but which leave a wide triangular area of the mesonotum exposed in expanded specimens. Metanotum typical. Abdomen fusiform, somewhat depressed on dorsum, lateral tubercles on third and fourth tergites small. Ultimate tergite very broad, posterior margin concave. Forceps (Fig. 516) typical, cylindrical, broader towards base, and narrower distally. Penultimate sternite transverse, somewhat triangular, apex rounded. Genitalia (Fig. 517) characteristic; paramere narrow

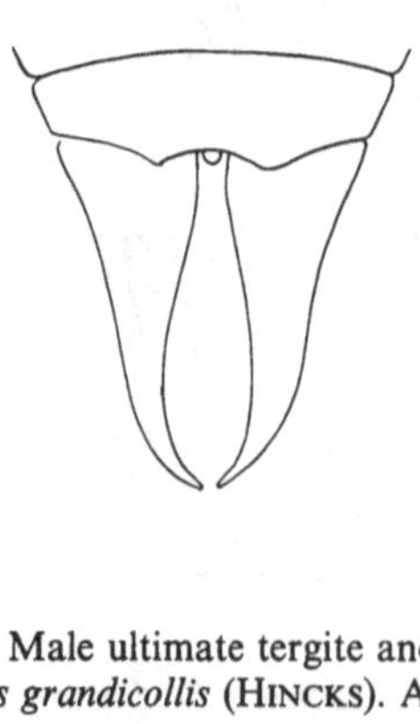

Fig. 516 Male ultimate tergite and forceps of *Isolabis grandicollis* (Hincks). After Hincks.

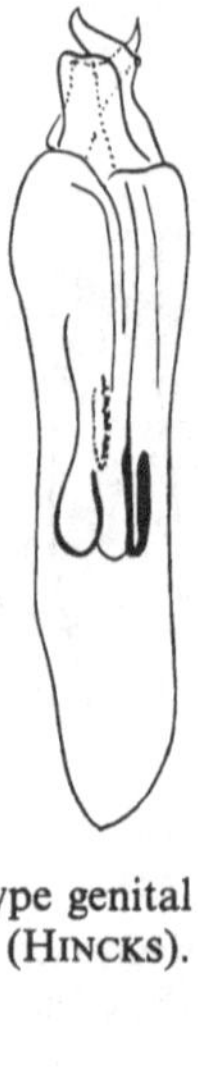

Fig. 517 Holotype genital armature of *Isolabis grandicollis* (Hincks). After Hincks.

and long, genital lobes well developed, long; virga within genital lobe elongated; the apex of each being sharply bent and with a short virgal slerite in each distal lobe. External parameres specific, expanded medially on outer margins. – **Female** similar to male, pronotum very long, longer than length of head; forceps slender, simple. – **Length** of body with forceps: male: 5,5–6,5 mm, female: 6–7 mm.

Distribution: Guinea, Bissau-Guinea, Sierra Leone.

6. Isolabis sjoestedti (BORG)

1904 *Brachylabis sjoestedti* BORG, Ark. Zool., Uppsala, 1: 568; pl. 26, fig. 2 (male). Terra typica: Cameroon (Type male: Brit. Mus. nat. Hist., London).
1908 *Verhoeffia sjoestedti* – BURR, Ann. Mag. nat. Hist., London, (8) **8**: 248.
1910 *Arlex sjoestedti* – BURR, Trans. ent. Soc. London, **1910**: 191.
1911 *Arlex sjöstedti* – BURR, Genera Insectorum, Bruxelles, **122**: 40; pl. 3, fig. 15 (head), 15a (thoracal nota). West Africa.
1978 *Isolabis sjoestedti* – BRINDLE, Dermaptera of Africa, Tervuren, **2**: 61; fig. 73 (holotype).
Literature: BURR 1912, 1915; HINCKS 1953; SAKAI 1970; STEINMANN 1977, 1978.

Description: Male colour dark reddish-brown; antennae brown, joints 12–13 yellowish; legs yellow, femora brown except at apices. Cuticle strongly punctate and pubescent, rather dull, hairs short and yellow. Head transverse, sides well rounded, posterior margin weakly convex; eyes a little longer than length of head behind eyes; antennae with normal first joint, second transverse, third joint one and half times as long as wide. Pronotum as wide as long or a little longer than wide (Fig. 518); lateral margins parallel-sided; median longitudinal furrow present on anterior section; all angles rounded. Mesonotum transverse, with rudimentary tegmina, wings absent; tegmina do not completely touch, and leave a broad expanse of the sclerite exposed; the anterior half of the mesonotum has a prominent lateral fold on both sides. Abdomen fusiform, tergites expanded to median section; lateral tubercles relatively well developed. Ultimate tergite transverse, produced above the base of both branches of forceps. Forceps (Fig. 519) not of *Isolabis*-type, asymmetrical, branches slightly broader towards base, narrower distally. Penultimate sternite with posterior margin rounded or

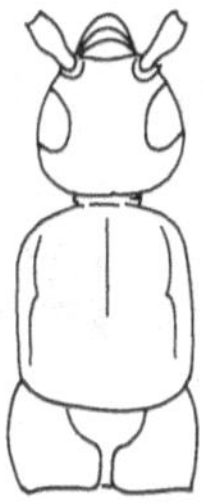

Fig. 518 Head, pro- and mesonotum of *Isolabis sjoestedti* (BORG). After BURR.

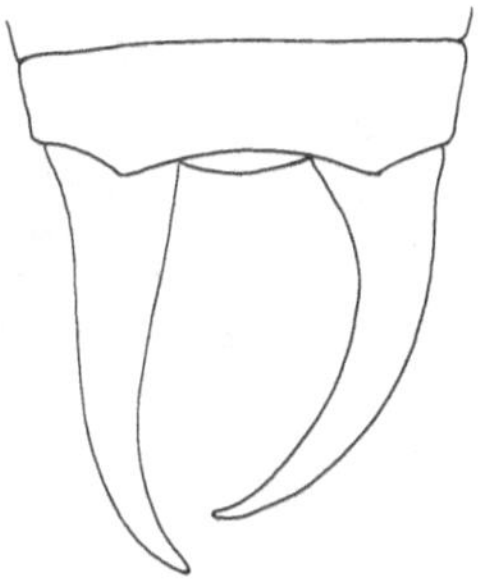

Fig. 519 Male ultimate tergite and forceps of *Isolabis sjoestedti* (BORG). After BORG.

almost so. Genitalia unknown to me. – **Female** similar to male, but branches of forceps closer together and very weakly curved, branches symmetrical and more slender, especially distally. – **Length** of body with forceps: in both sexes: 12,5–13,5 mm.

Distribution: Cameroon, and Equatorial Guinea.

7. Isolabis attenuata Brindle

1978 *Isolabis attenuata* Brindle, Dermaptera of Africa, Tervuren, **2**: 63; fig. 75 (holotype), 76 (holotype genital armature), 77 (female forceps). – Terra typica: Nigeria (Type male: Brit. Mus. nat. Hist., London).

Description: Male colour dark reddish-brown to blackish-brown; antennae brown to dark brown, distal joints paler; femora dark brown, yellow at apices or yellow for distal quarter. Cuticle of body strongly punctate and pubescent, rather dull; hairs short and yellow. Head strongly transverse, posterior angles rounded, posterior margin straight; eyes very large; first antennal joint long, more or less as long as distance between antennal bases; second quadrate, third joint between one and half and twice as long as wide. Pronotum large, slightly widened posteriorly, plate more or less quadrate; posterior margin truncate; median longitudinal furrow distinct. Mesonotum transverse, narrowed posteriorly, with rudimentary tegmina which leave a large area of the mesonotum exposed anteriorly, and which almost touch in the mid-line; laterally the mesonotum has a prominent blunt lateral fold on both sides. Abdomen fusiform, rather depressed,

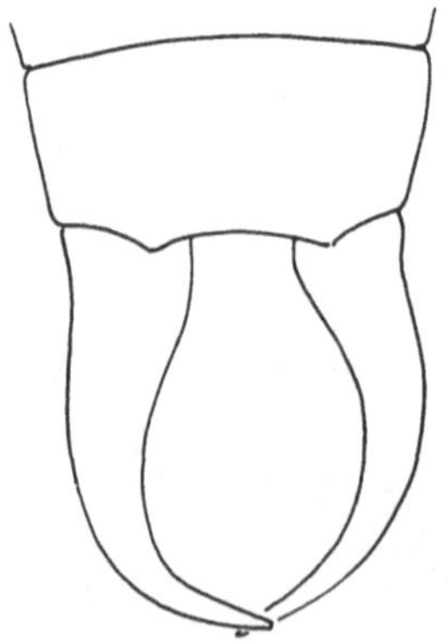

Fig. 520 Holotype ultimate tergite and forceps of *Isolabis attenuata* Brindle. After Brindle.

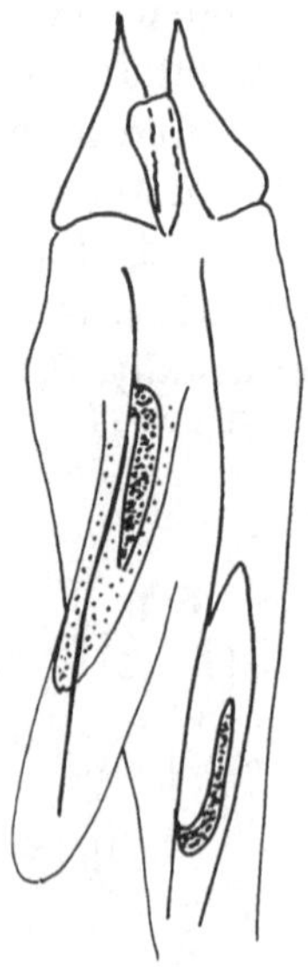

Fig. 521 Holotype genital armature of *Isolabis attenuata* Brindle. After Brindle.

lateral tubercles on third and fourth tergites small. Ultimate tergite broad, posterior margin concave medially. Forceps (Fig. 520) symmetrical, strongly curved; cylindrical, the branches conspicuously attenuated, or narrowed, immediately before apex. Penultimate sternite rounded on posterior margin, but with a small median excision. Genitalia (Fig. 521) normal, paramere with two fully developed genital lobes; virga within genital lobe long and with a lateral sclerite; denticulated areas large. External parameres trigonal, apices pointed. – **Female** similar to male, but abdomen more narrowed distally, and ultimate tergite concave between the bases of the forceps. – **Length** of body with forceps: in both sexes: 10–11,5 mm.

Distribution: Nigeria, Cameroon.

8. Isolabis coiffaiti (BRINDLE)

1970 *Brachylabis coiffaiti* BRINDLE, Biologia gabonica, Paris, (1) **6**: 14; fig. 1 (holotype), 2 (penultimate sternite of holotype), 5 (holotype genitalia). – Terra typica: Gabon (Type male: Mus. nation. Hist. nat., Paris).

1978 *Isolabis coiffaiti* – BRINDLE, Dermaptera of Africa, Tervuren, **2**: 52; fig. 50 (head and pronotum), 51 (penultimate sternite of male), 52 (male genitalia), 53 (antennal joints 1–5).

Literature: STEINMANN 1977, 1978.

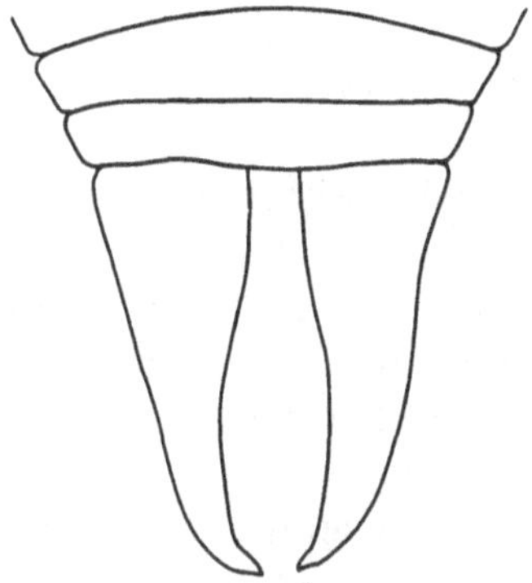

Fig. 522 Male ultimate tergite and forceps of *Isolabis coiffaiti* (BRINDLE). After BRINDLE.

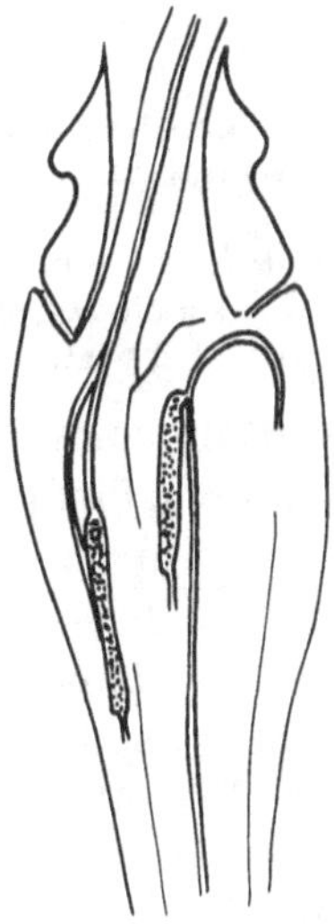

Fig. 523 Holotype genital armature of *Isolabis coiffaiti* (BRINDLE). After BRINDLE.

Description: Male colour reddish-brown, posterior tergites of abdomen blackish, antennae brown, joints 8–9 and distal half of joint 7 whitish; legs yellowish brown, and femora brown on basal two-thirds. Head transverse, short, tumid, about as wide as pronotum; eyes relatively large, but shorter than first antennal joint, which is long, longer than distance between antennal bases. Pronotum transverse, all angles rounded, lateral margins straight, but widened posteriorly; pronotum with short median longitudinal furrow and two oblique lateral lines on anterior half. Mesonotum transverse, narrowed posteriorly, with a strong lateral fold on both sides, but not actual ridges. Abdomen fusiform, lateral tubercles on fourth and third tergites only indicated by a swelling of the cuticula; abdominal tergites with very long sparse yellow hairs. Ultimate tergite mainly hidden by ninth. Forceps (Fig. 522) with each branch cylindrical, straight, broader basally, narrowed apically, sinuate. Penultimate sternite transverse, rather triangular, apex rounded. Genitalia (Fig. 523) characteristic; paramere narrow, a little wider on anterior half; genital lobes well developed, virga within genital lobe long, slender, strongly sclerotized, and with a wide lateral appendage. External parameres long, outer margins undulate. – **Female** similar to male, but branches of forceps straight, short, simple. – **Length** of body with forceps: male: 7–7,5 mm, female: 6–6,5 mm.

Distribution: Gabon.

9. Isolabis schoutedeni Hincks

1954 *Isolabis schoutedeni* Hincks, Ann. Mus. r. Congo belge, Tervuren, in-4°, Zool., **1**: 378; fig. 1 (holotype in ventral view). – Terra typica: Belgian Congo (Type male: Mus. r. Afr. centr., Tervuren).

1978 *Isolabis schoutedeni* – Brindle, Dermaptera of Africa, Tervuren, **2**: 55, fig. 61 (head and pronotum), 62 (penultimate sternite of male), 63 (male genitalia).

Literature: Sakai 1970; Steinmann 1977, 1978.

Description: Male colour dark brown, femora blackish, yellow at apices; antennae dark brown, first joint blackish, apex of joint 6, all joints 7 and 8, and base of 9 white. Cuticle strongly punctate and pubescent. Head strongly transverse, sides rounded, contracted behind eyes, as wide in widest part as pronotum at cephalic angles. Eyes very large, longer than genae. Antennae 14-jointed; first joint about as long as eye, and a little longer than distance between antennal bases. Pronotum about as long as wide; lateral margins straight, narrower cephaled than caudad; posterior margin very feebly convex; median longitudinal furrow short, on anterior half. Mesonotum about two thirds the length of pronotum; narrowed posteriorly, lateral margins forming lateral folds, but not ridges. Abdomen fusiform, lateral tubercles on third tergite not visible; those on fourth very small; abdomen contracted proximad and caudad, strongly punctate but much less so than forepart of body. Ultimate tergite transverse, pubescent, prominent on both sides above forceps. Forceps contiguous, simple, gradually tapering, slightly curved. Penultimate sternite transverse, posterior margin slightly raised near apex and with a very small excision. Genitalia very narrow (Fig. 524 after HINCKS, and fig. 525 after BRINDLE), paramere elongated, genital lobes well developed, virgal sclerites short. Genital lobes with prominent and extensive denticulations. External parameres broad basally, acute apically. – **Female** unknown. – **Length** of body with forceps: 9–10 mm.

Distribution: Zaïre, Congo Republic.

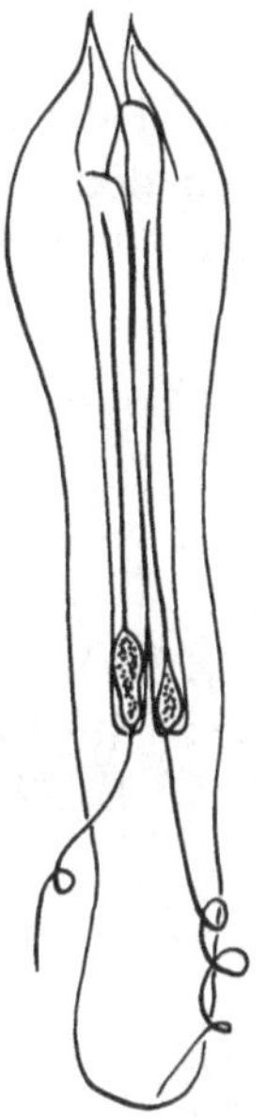

Fig. 524 Male genital armature of *Isolabis schoutedeni* HINCKS. After HINCKS.

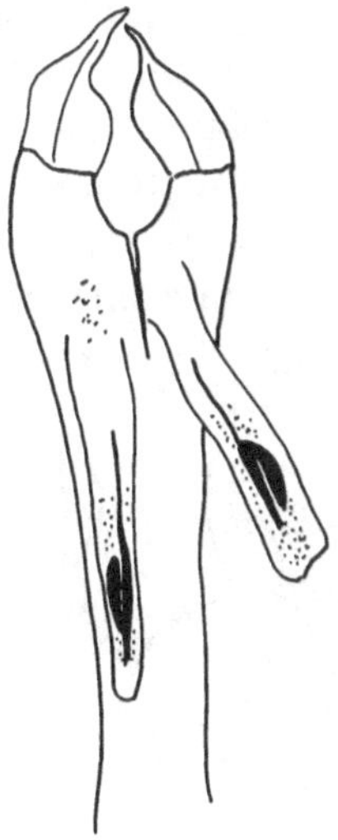

Fig. 525 Male genital armature of *Isolabis schoutedeni* HINCKS. After BRINDLE.

10. Isolabis usumbarana (VERHOEFF)

1902 *Leptisolabis usumbarana* VERHOEFF, Sber. Ges. naturf. Fr., Berlin, **1902**: 12. – Terra typica: East Africa (Type male: Mus. Naturkunde, Berlin).

1902 *Leptisolabis theoriae* VERHOEFF, Sber. Ges. naturf. Fr., Berlin, **1902**: 14. – Terra typica: East Africa (Type male: Mus. Naturkunde, Berlin).

1911 *Brachylabis usumbarana* – ZACHER, Zool. Jahrb., Jena, **30**: 388; fig. R (male genitalia).

1966 *Isolabis usumbarana* – POPHAM & BRINDLE, Entomologist, London, **99**: 245.

1966 *Isolabis theoriae* – POPHAM & BRINDLE, Entomologist, London, **99**: 245.

1978 *Isolabis usumbarana* – BRINDLE, Dermaptera of Africa, Tervuren, **2**: 57; fig. 64 (head and pronotum), 65 (posterior margin of penultimate sternite), 66 (male genitalia). Tanzania, Angola.

Literature: BRINDLE 1970; BURR 1911, 1915; HINCKS 1957; SAKAI 1970; STEINMANN 1977, 1978.

Description: Male colour blackish, rather dull; pronotum reddish-brown, legs blackish except tarsi, whitish-yellow; femora with bases and apices yellow; apex of antennal joint 6, all joint 7 and basal half of joint 8 yellowish. Head tumid, postfrontal sutures and coronal suture obsolete; eyes relatively large, larger than genae. Antennae 14-jointed; first joint long, a little longer than distance between antennal bases. Pronotum about as wide as long; lateral margins straight, widened posteriorly; median longitudinal furrow present on anterior half. Mesonotum transverse, with lateral folds well marked but rounded, and not forming lateral ridges. Metanotum normal. Abdomen depressed, fusiform, a little expanded medially; ultimate tergite transverse, simple. Forceps straight, cylindrical. Penultimate sternite with posterior margin rounded and with a median excision. Genitalia with rather broad ventral parameral plate (Fig. 526, after ZACHER, and fig. 527, after BRINDLE); genital lobes well developed, slender, virgal sclerites fairly long and associated with denticulated areas. – **Female** similar to male, but forceps simple, contiguous, tapering. – **Length** of body with forceps: in both sexes: 7,5–8 mm.

Distribution: Tanzania, Kenya, Zaïre, Angola.

Fig. 526 Male genital armature of *Isolabis usumbarana* (VERHOEFF). After ZACHER.

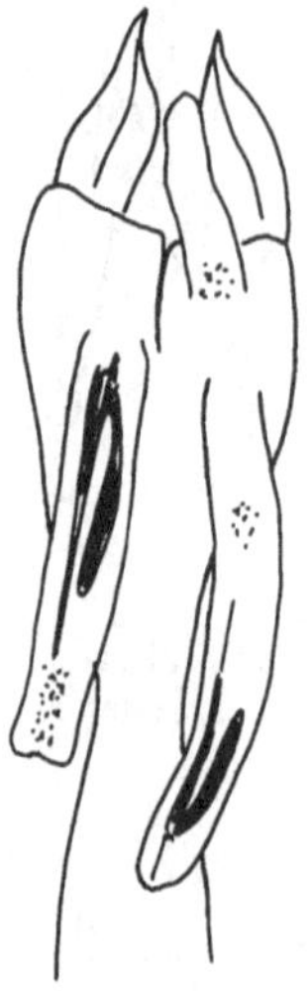

Fig. 527 Male genital armature of *Isolabis usumbarana* (VERHOEFF). After BRINDLE.

11. Isolabis rufa (BRINDLE)

1966 *Brachylabis rufa* BRINDLE, Rev. Zool. Bot. afr., Bruxelles, **74**: 41; fig. 16 (genitalia of holotype). – Terra typica: Tanzania (Type male: Mus. r. Afr. centr., Tervuren).

1978 *Isolabis rufa* – BRINDLE, Dermaptera of Africa, Tervuren, **2**: 54; fig. 54 (head and pronotum), 55 (penultimate sternite), 56 (male genitalia).
Literature: STEINMANN 1977, 1978.

Description: Male colour very dark reddish, pronotum reddish and paler, antennae dark brown, unicolorous; femora dark brown, yellow at apices. Cuticle of body strongly punctate and pubescent, hairs rather long and yellow. Head transverse, sides rounded; eyes very large, slightly longer than length of head behind eyes. Antennae 14-jointed; first joint more or less as long as distance between antennal bases. Pronotum quadrate; lateral margins straight, more or less parallel; median longitudinal furrow present on anterior half and on both sides of the line occur two shorter lines. Mesonotum transverse, depressed towards both sides, lateral folds prominent but blunt, and not forming lateral longitudinal ridges. Metanotum more transverse. Abdomen fusiform; all thoracic nota and abdominal tergites with short yellow hairs laterally; lateral tubercles on third and fourth tergites scarcely visible. Ultimate tergite transverse, posterior margin concave medially. Forceps cylindrical, broader towards base, narrowed distally, weakly curved. Penultimate sternite with posterior margin slightly sinuate and with a small median excision. Genitalia (Fig. 528) strongly narrowed basally, and broad apically; genital lobes with extensive denticulations, and virgal sclerites rather long. External parameres triangular, acuminate. – **Female** similar to male, but forceps slender and straight. – **Length** of body with forceps: in both sexes: 9–9,5 mm.

Distribution: Tanzania.

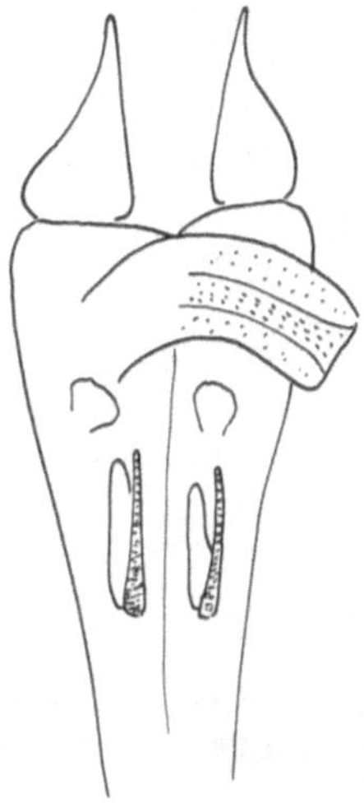

Fig. 528 Holotype genital armature of *Isolabis rufa* (BRINDLE). After BRINDLE.

12. Isolabis quadrata BRINDLE

1978 *Isolabis quadrata* BRINDLE, Dermaptera of Africa, Tervuren, **2**: 54; fig. 58 (head and pronotum), 59 (penultimate sternite of male), 59 (genital armature of holotype). – Terra typica: Zaïre (Type male: Mus. r. Afr. centr., Tervuren).

Description: Male general colour black, slightly shiny; antennae dark brown, joints 10–12 yellowish; legs yellowish-brown, femora blackish on basal two-thirds. Cuticle of body typical. Head transverse, slightly rounded posteriorly; postfrontal and coronal sutures indistinct. Eyes large, longer than length of head behind eyes. First antennal joint long, but shorter than distance between antennal bases; second transverse, third joint one and half times as long as wide; fourth as long as third but wider. Pronotum large, longer than head; lateral margins straight, but widened posteriorly; all angles rounded; median longitudinal furrow distinct and with two shorter lines on both sides, the outer one depressed posteriorly, forming a depression. Mesonotum transverse, with a lateral fold on both sides. Abdomen fusiform, rather depressed; lateral tubercles on third and fourth segments rather large and prominent. Ultimate tergite very broad, posterior margin produced above the base of both branches of the forceps. Forceps (Fig. 529) simple, tapering, contiguous, symmetrical. Penultimate sternite with posterior margin rounded and with a deep median excision. Genitalia (Fig. 530) characteristic; strongly narrowed basally, and widened apically; median incision of anterior margin small but prominent; genital lobes straight, virga within genital lobe partially sclerotized and associated with denticulations. External parameres typical. – **Female** similar to male. – **Length** of body with forceps: in both sexes: 10–12,5 mm.

Distribution: Zaïre.

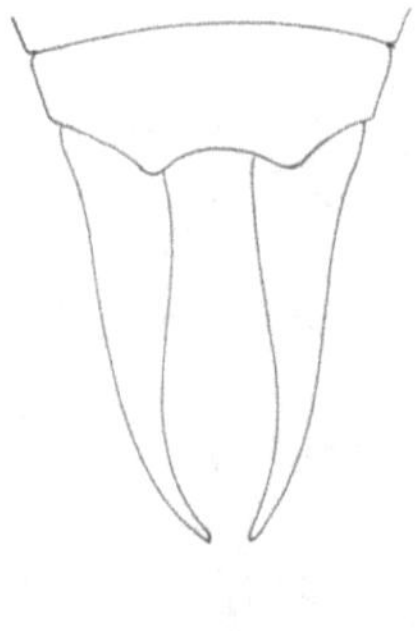

Fig. 529 Male ultimate tergite and forceps of *Isolabis quadrata* BRINDLE. Original.

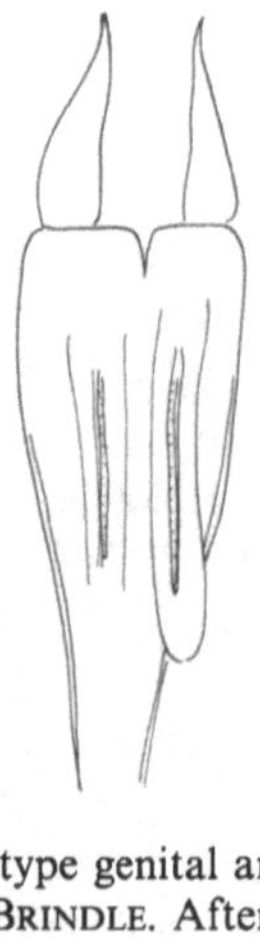

Fig. 530 Holotype genital armature of *Isolabis quadrata* BRINDLE. After BRINDLE.

13. Isolabis braueri VERHOEFF

1902 *Isolabis braueri* VERHOEFF, Sber. Ges. naturf. Fr., Berlin, **1902**: 14. – Terra typica: Guinea (Type male: Mus. Naturkunde, Berlin).

1910 *Isolabis braueri* – ZACHER, Zool. Jahrb., Jena, **30**: 388; fig. T (male genitalia).

1978 *Isolabis braueri* – BRINDLE, Dermaptera of Africa, Tervuren, **2**: 59; fig. 67 (male in ventral view), 68 (basal joints of antenna), 69 (male genitalia). Zaïre.
Literature: BRINDLE 1966; BURR 1910, 1911, 1912, 1915; HINCKS 1953; SAKAI 1970; STEINMANN 1977, 1978.

Description: Male colour blackish-brown; antennae and legs yellowish brown, femora darkened on basal half or two-thirds. Cuticle punctate and pubescent; head less strongly punctate, punctation strongest on mesonotum, metanotum and basal abdominal tergites. Head rounded; tumid, narrowed anteriorly; postfrontal and coronal sutures indistinct; eyes large, longer than length of head behind eyes. First antennal joint long, more or less as long as distance between antennal bases. Pronotum a little longer than wide; lateral margins straight, but widened posteriorly. Median longitudinal furrow distinct. Mesonotum transverse, rather large, with well marked lateral folds anteriorly, due to the plate of the sclerites being depressed, especially towards the anterior margin; metanotum typical. Abdomen broad, depressed, wider medially, lateral tubercles on third tergite rather small, those on fourth larger. Ultimate tergite transverse, posterior margin concave medially. Forceps (Fig. 531) symmetrical, cylindrical, branches relatively strongly curved. Penultimate sternite with posterior margin slightly sinuate and with a very small median excision. Genitalia (Fig. 532) broad; central parameral plate with two well-developed genital lobes, virga within genital lobe long, the latter with a lateral appendage. External parameres relatively small. – **Female** unknown. – **Length** of body with forceps: 13–13,5 mm.

Distribution: Western Africa.

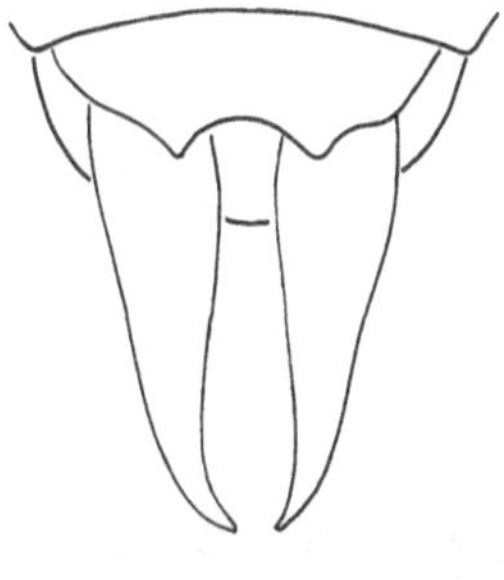

Fig. 531 Male ultimate tergite and forceps of *Isolabis braueri* VERHOEFF. After BRINDLE.

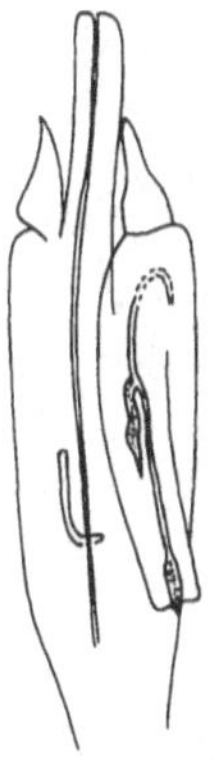

Fig. 532 Male genital armature of *Isolabis braueri* VERHOEFF. After BRINDLE.

14. Isolabis pauliani (HINCKS)

1953 *Antisolabis pauliani* HINCKS, Mém. Inst. scient. Madagascar, Tananarive, (E), **4**: 367; fig. 6 (male genitalia). – Terra typica: Madagascar (Type male: Mus. nation. Hist. nat., Paris).

1966 *Isolabis pauliani* – POPHAM & BRINDLE, Entomologist, London, **99**: 245.
1977 *Metisolabis pauliani* – BRINDLE, Trans. r. ent. Soc., London, **118**: 234.
Literature: BRINDLE 1969; SAKAI 1970; STEINMANN 1977, 1978.

Description: Male colour dark brown, shiny; palpi and proximal part of antennae yellow; sides of pronotum and legs brownish yellow. Head tumid, rounded posteriorly, a little broader than pronotum; postfrontal and coronal sutures not visible. Eyes comparatively large, longer than length of head behind eyes; first antennal joint long, normal. Pronotum slightly longer than wide: all angles rounded; lateral margins straight, but convex. Tegmina very reduced, represented by lateral flaps laterally; metanotum typical. Abdomen normal, fusiform, lateral tubercles present on third and fourth tergites. Ultimate tergite broad, posterior margin concave medially. Forceps narrow, slender, straight but a little curved apically. Penultimate sternite with posterior margin rounded. Genitalia (Fig. 533) large; central parameral plate with very small median incision on anterior margin; genital lobe with small virga, the latter with small lateral appendage. External parameres comparatively large. – **Female** similar to male, but forceps slenderer, tapering. – **Length** of body with forceps: male: 6,5 mm, female: 7,5–8,5 mm.

Distribution: Madagascar.

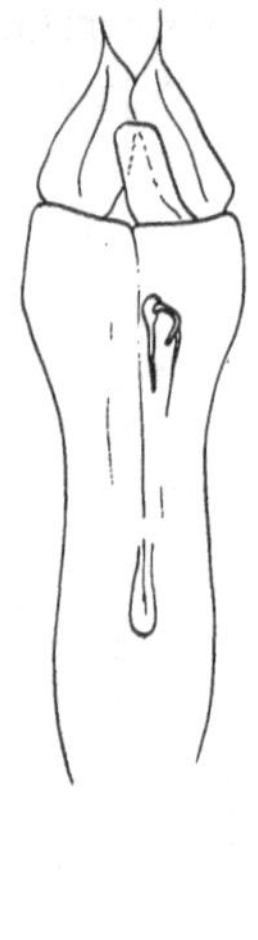

Fig. 533 Holotype genital armature of *Isolabis pauliani* (HINCKS). After HINCKS.

15. Isolabis howardi (BURR)

1907 *Brachylabis nigra* – CAUDELL, Proc. U.S. nation. Mus., Washington, **33**: 172.
1907 *Brachylabis nigra* – SAKAI, Dermapt. Cat. Prael., Tokyo, **1**: 11.
1910 *Leptisolabis howardi* BURR, Proc. U.S. nation. Mus., Washington, **38**: 449. – Terra typica: Guatemala (Type male: U.S. nation. Mus., Washington, No. 113087).
1911 *Leptisolabis aliena* BORELLI, Boll. Mus. Zool. Anat. comp. Univ. Torino, **26**, No. 644: 7. – Terra typica: Costa Rica (Type male: Mus. Ist. Zool. sist., Torino). – **syn. nov.**
1911 *Leptisolabis howardi* – BURR, Genera Insectorum, Bruxelles, **122**: pl. 4, fig. 5 (head), 5a (thoracic nota), 5b (antenna).
1966 *Isolabis howardi* – POPHAM & BRINDLE, Entomologist, London, **99**: 245.
1966 *Isolabis aliena* – POPHAM & BRINDLE, Entomologist, London, **99**: 245.
Literature: BURR 1910, 1915; REICHARDT 1968; SAKAI 1970; STEINMANN 1977, 1978.

Description: Male colour jet black, dull; body with golden pubescence; antennae brownish grey, legs black, but yellow at the base and apices. Head tumid, rounded; eyes very large, slightly longer than length of head behind eyes and first antennal joint; postfrontal and coronal sutures obsolete. Antennae 13-jointed; first joint long, more or less as long as distance between antennal bases, all antennal joint very thick, fourth joint relatively broad, more or less wider than long. Pronotum longer than wide, lateral sides gently diverging as the pronotum widens somewhat posteriorly; all angles rounded; median longitudinal furrow present on anterior half. Tegmina and wings entirely absent. Mesonotum transverse, posterior margin truncate, sides not keeled, but raised, with blunt, tumid ridges converging posteriorly. Metanotum simple, transverse. Abdomen normal, lateral tubercles on 3rd and 4th abdominal tergites faint. Ultimate tergite very short and transverse, but narrower than the abdomen; posterior margin narrower, and concave in middle, leaving a short triangular lobe on both sides projecting slightly over the forceps. Pygidium short, compressed. Forceps (Fig. 534) remote basally, cylindrical, tapering, gently arcuate towards the apex. Penultimate sternite rounded posteriorly. Genitalia unknown. – **Female** similar to male, but forceps slenderer, simple, tapering. – **Length** of body with forceps: in both sexes: 9–13,5 mm.

Distribution: Costa Rica, Guatemala.

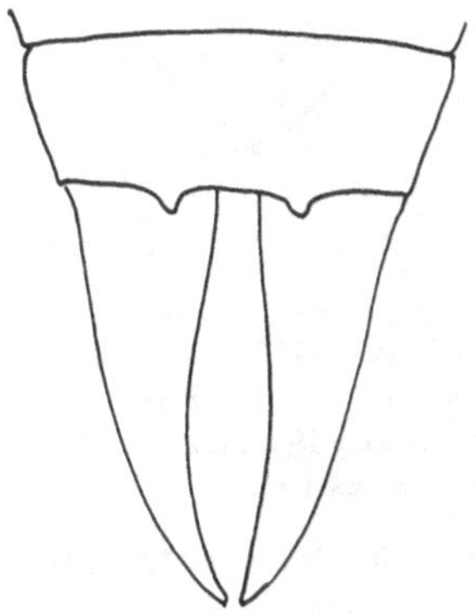

Fig. 534 Male ultimate tergite with forceps of *Isolabis howardi* (BURR). Original.

16. Isolabis ocellata SRIVASTAVA

1978 *Isolabis ocellata* SRIVASTAVA, Ann. Mus. civ. Stor. nat., Genova, **82**: 325; fig. 7 (anterior part of body), 8 (a portion of antenna), 9 (posterior margin of penultimate sternite), 10 (ultimate tergite and forceps). – Terra typica: Burma (Type male: Mus. civ. Stor. nat., Genova).

Description: Male general colour dark brownish black, lighter on the sides of pronotum and hind margin of tergites, legs yellow, femora dark brownish black in basal two thirds, antennae with one or two preapical joints yellow. Cuticle of head, thorax and abdomen punctate, forceps smooth. Head broad; postfrontal sutures distinct, coronal suture absent; eyes very large, prominent, slightly longer than length of head behind eyes. First antennal joint long, longer than distance between antennal bases; second small, third cylindrical, longer than fourth. Pronotum rectangular; lateral margins straight, gently reflexed, feebly diverging posteriorly; median longitudinal furrow short. Mesonotum transverse, sides smooth, posterior margin faintly emarginate. Metanotum normal. Abdomen narrowed apically as well as basally, weakly convex; lateral tubercles on the sides of 3rd and 4th tergites poorly developed. Ultimate tergite strongly transverse, posterior margin tri-sinuate, oblique laterally, with triangular lobe, above the bases of forceps. Forceps (Fig. 535) dilated at base, then gradually tapering, almost straight in basal two thirds, and curved apically. Penultimate sternite triangular, rounded posteriorly. Genitalia unknown. – **Female** unknown. – **Length** of body with forceps: 10,5 mm.

Distribution: Burma.

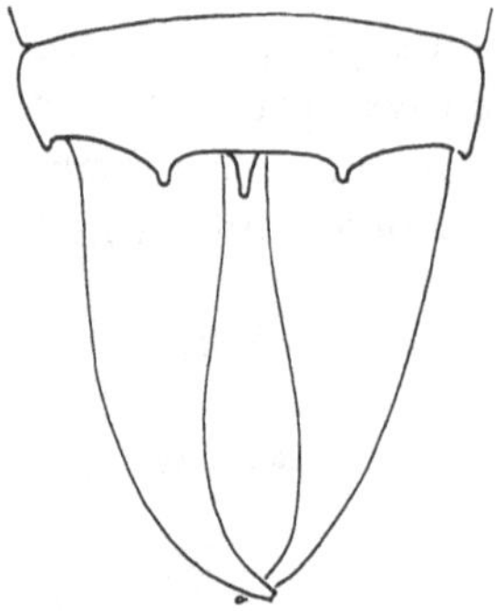

Fig. 535 Holotype ultimate tergite and forceps of *Isolabis ocellata* SRIVASTAVA. After SRIVASTAVA.

Genus 28: **Geracodes** HEBARD

1917 *Geracodes* HEBARD, Trans. amer. ent. Soc., Philadelphia, **43**: 322. – Species typica: *Geracodes litus* HEBARD 1917 (for family Labiidae).

1952 *Geracodes* – HINCKS, Proc. r. ent. Soc. London, (B) **21**: 23 (for Brachylabiinae).

Literature: BRINDLE 1978; HINCKS, 1952, 1961; REICHARDT 1968; SAKAI 1970; STEINMANN 1975, 1977, 1978; Townes 1945.

Description: Head rather long, sides rounded, eyes small, slightly shorter than length of head behind eyes, and shorter than first antennal joint. Pronotum relatively small, anterior margin narrower than hind margin; a little longer than wide, or about as long as wide. Tegmina and wings present, well developed and long, or rudimentary and wings absent. Legs long, but tarsus without arolium; the caudal tarsus is very slightly longer than the combined length of the second and third tarsal joints and has a hairy ventral surface, with two rows of widely spaced elongate, chaetiform spines. Forceps similar in both sexes. Genitalia with basal genital lobe united, distal lobes not clearly defined, virga associated with a complex arrangement of denticulated areas and sclerites.

Distribution: Neotropical and Ethiopian faunal regions.

4 species.

Identification key to the species

1 Neotropical species **2**
1′ Ethiopian species **3**
2 Tegmina and wings fully developed. Antennae 17-jointed, first antennal joint long, slightly longer than distance between antennal bases. Pygidium minute .... 1. **Geracodes litus**, p. 333
2′ Tegmina rudimentary, represented by lateral flaps; wings absent. Antennae 16-jointed, first antennal joint long, more or less as long as distance between antennal bases. Pygidium absent .... 2. **Geracodes paraguayensis**, p. 334
3 Tegmina and wings fully developed, wings often very long, sometimes shorter, and sometimes absent or concealed beneath tegmina; eyes larger than in *aptera* BRINDLE. Male genitalia as in fig. 539 .... 3. **Geracodes labioides**, p. 334
3′ Tegmina rudimentary, represented by lateral flaps; wings absent; eyes much shorter than in *labioides* HINCKS. Male genitalia as in fig. 541. .... 4. **Geracodes aptera**, p. 335

1. Geracodes litus HEBARD

1917 *Geracodes litus* HEBARD, Trans. amer. ent. Soc., Philadelphia, **43**: 323; pl. XXVI, fig. 6 (female), 7 (basal joints of antenna of female). – Terra typica: Panama (Type female: U. S. nation. Mus., Washington).

Literature: BRINDLE 1972; HINCKS 1952, HINCKS & ROY 1961; REICHARDT 1968; SAKAI 1970; STEINMANN 1977, 1978.

Description: Female colour dark brown, legs and antennae light brownish-yellow. Head with margins of cheeks almost straight, slightly converging to the sharply rounded latero-caudal angles; posterior margin straight, truncate. Eyes comparatively small, shorter than genae, or shorter than distance between antennal bases. Antennae 17-jointed; first joint very long, second quadrate, third slightly longer than wide, and fourth a little longer than wide. Pronotum about as long as broad, but very convex at lateral margins; all angles rounded; median longitudinal furrow present on anterior half. Legs relatively long, and abdomen short. Tegmina well developed, wings very long; apex of wings come close to base of forceps. Ultimate tergite very narrow, weakly produced mesad into two weakly convex portions, with posterior margin bilobate. Pygidium minute. Forceps (Fig. 536) straight, slightly depressed; inner margins straight, branches curved apically. – **Male** unknown. – **Length** of body with forceps: 5,25–5,5 mm.

Distribution: Panama.

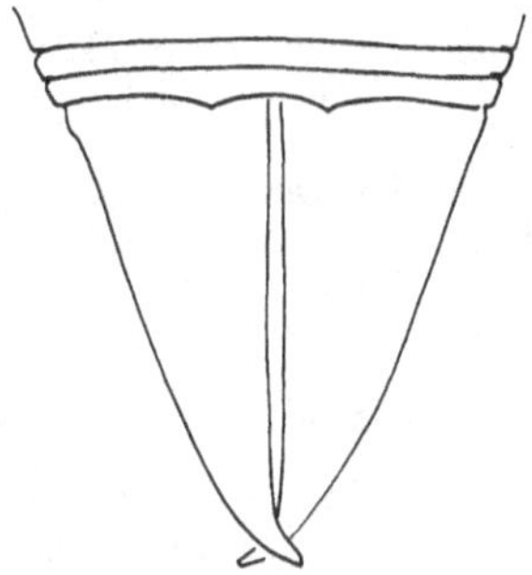

Fig. 536 Female ultimate tergite and forceps of *Geracodes litus* HEBARD. After HEBARD.

2. Geracodes paraguayensis Borelli

1932 *Geracodes paraguayensis* Borelli, Konowia, Wien, **11**: 95; fig. a (male), b (abdominal end with forceps of male), c (antenna). – Terra typica: Paraguay (Type male: Zool. Inst. Mus., Hamburg).

Literature: Reichardt 1968; Sakai 1970; Steinmann 1977, 1978.

Description: Male colour dark brown or chocolate-brown; antennae and legs yellowish brown. Head relatively long, longer than pronotum, and large, as wide as pronotum. Eyes comparatively small, shorter than first antennal joint. Antennae 16-jointed; first joint long, a little longer than distance between antennal bases; second and third small. Pronotum relatively small, all angles rounded; lateral margins convex, diverging posteriorly; posterior margin rounded; median longitudinal furrow distinct. Mesonotum with prominent but rudimentary tegmina, represented by lateral flaps. Metanotum transverse, without wings. Abdomen normal, punctate dorsally. Ultimate tergite broad, posterior margin with conceve section medially or with two tubercles. Forceps (Fig. 537) smooth, broadest basally, straight medially, and a little curved apically, tapering. Pygidium not visible. Genitalia unknown. – **Female** unknown. – **Length** of body with forceps: 5,2–5,5 mm.

Distribution: Paraguay: Cordillera.

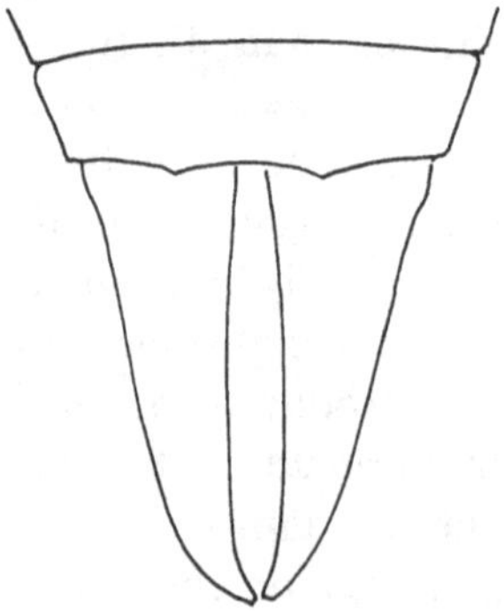

Fig. 537 Male ultimate tergite and forceps of *Geracodes paraguayensis* Borelli. After Borelli.

3. Geracodes labioides Hincks

1952 *Geracodes labioides* Hincks, Proc. r. ent. Soc. London, (B) **21**: 23; fig. 7 (male), 8 (penultimate sternite), 9 (male genitalia). – Terra typica: Sierra Leone (Type male: Brit. Mus. nat. Hist., London).

1978 *Geracodes labioides* – Brindle, Dermaptera of Africa, Tervuren, 2: 65; fig. 78 (holotype and paratype males), 79 (male genitalia), 80 (posterior margin of penultimate sternite). Sierra Leone and Ivory Coast.

Literature: Hincks & Roy 1961; Sakai 1970; Steinmann 1977, 1978.

Description: Male colour brown, but generally light, with the head and, to a lesser extent, the pronotum, sides of tegmina and abdomen, darker. Head widest across eyes, strongly contracted in front and behind; surface convex, dull, finely reticulated with decumbent pubescence. Postfrontal sutures well marked on vertex as a short depressed furrow. Eyes comparatively small, but larger than *aptera* BRINDLE. First antennal joint slightly longer than distance between antennal bases. Pronotum longer than wide. Lateral margins straight and strongly convex; all angles rounded; median longitudinal furrow present. Tegmina complete, or slightly abbreviated in flightless individuals, about one and half to two times as long as pronotum; sides parallel. Wings long, up to about one and half times as long as tegmina in fully-winged specimens. Abdomen cylindrical, moderately polished and finely punctate dorsally. Ultimate tergite very broad, simple. Forceps (Fig. 538) cylindrical, markedly pubescent, tapering to finely pointed, slightly curved distal extremity. Penultimate sternite transverse, posterior margin concave medially. Genitalia (Fig. 539) characteristic; central parameral plate oval, median incision of anterior margin small; genital lobes not clearly defined, with a complex arrangement of denticulations and sclerites. External parameres comparatively small. – **Female** similar to male; available females all with shorter tegmina, although still contiguous along sutures, and with wings absent or concealed. – **Length** of body with forceps: male: 5–6 mm, female: 6–7 mm.

Distribution: Sierra Leone, Ivory Coast.

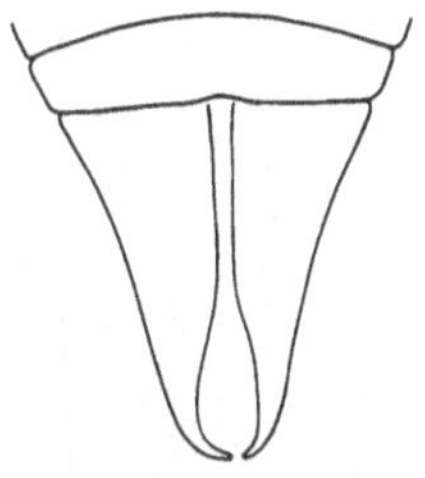

Fig. 538 Holotype ultimate tergite and forceps of *Geracodes labioides* HINCKS. After HINCKS.

Fig. 539 Holotype genital armature of *Geracodes labioides* HINCKS. After HINCKS.

4. Geracodes aptera BRINDLE

1978 *Geracodes aptera* BRINDLE, Dermaptera of Africa, Tervuren, **2**: 67; fig. 82 (male), 83 (holotype genital armature), 81 (posterior margin of penultimate sternite). – Terra typica: Guinea (Type male: Mus. nation. Hist. nat., Paris).

Description: Male colour yellowish to reddish-brown; head, mesonotum, and metanotum darker; abdomen more reddish than pronotum; antennae yellow.

Cuticle of body finely punctate and pubescent. Head slightly longer than wide, oval, sides rounded, posterior margin straight, but postero-lateral angles rounded; inter-antennal marks consisting of two short depressed lines, and vertex marked by a short furrow indicating the postfrontal sutures. Eyes comparatively small, slightly shorter than length of head behind eyes. First antennal joint very long, longer than distance between antennal bases. Pronotum slightly longer than wide; lateral margins straight, expanded posteriorly; posterior margin rounded. Tegmina present, but represented by lateral flaps; rudimentary; the flaps meet medially, but leave part of the mesonotum exposed anteriorly. Metanotum typical. Abdomen fusiform, rather broader medially; lateral tubercles present on third and fourth tergites. Ultimate tergite typical. Forceps (Fig. 540) cylindrical, broader towards base, narrowed distally, apex strongly narrowed and curved. Penultimate sternite transverse, posterior margin excised medially. Genitalia (Fig. 541) similar to *labioides* HINCKS, but central parameral plate broader and virgal sclerites differently shaped. – **Female** similar to male, but forceps slender, tapering. – **Length** of body with forceps: in both sexes: 4–4,5 mm.

Distribution: Guinea.

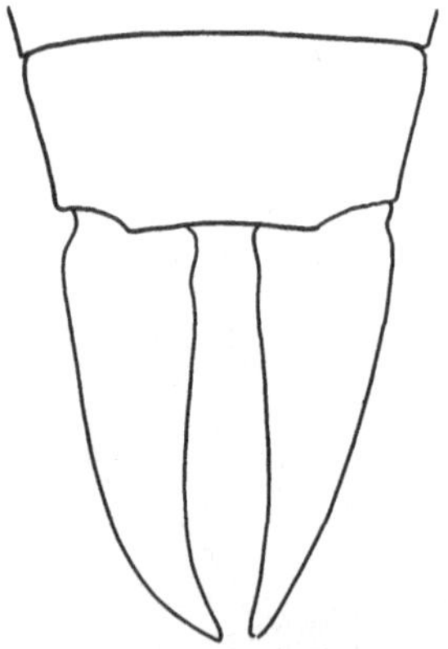

Fig. 540 Holotype ultimate and forceps of *Geracodes aptera* BRINDLE. After BRINDLE.

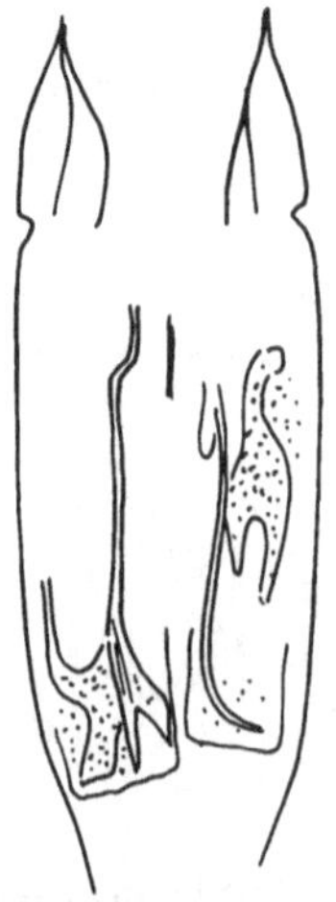

Fig. 541 Holotype genital armature of *Geracodes aptera* BRINDLE. After BRINDLE.

Genus 29: **Africolabis** BRINDLE

1978 *Africolabis* BRINDLE, Dermaptera of Africa, Tervuren, **2**: 25, – Species typica: *Africolabis pseudohumicola* BRINDLE.

Description: Small species, relatively weakly sclerotized; head slightly longer than wide; eyes very small with conspicuous facets; antennae with very short joints; first joint long, distal joints broad and short. Tegmina absent or rudimentary; wings totally absent. Abdomen somewhat depressed, widened medially, lateral tubercles on third tergite absent, those on fourth rudimentary. Forceps of both sexes simple, branches cylindrical. Male genitalia long, sclerotized part of virga short and associated with denticulated areas; genital lobes not clearly distinct.

Distribution: Central Africa.

2 species.

Identification key to the species

1 Pronotum as wide as long; tegmina present, but rudimentary, represented by lateral flaps. Distal antennal joints shorter and much paler in colour than basal joints. Virga within genital lobe of male genitalia as in fig. 543 .... 1. **Africolabis pseudohumicola**, p. 337

1′ Pronotum transverse; tegmina completely absent. Distal antennal joints longer and concolorous with the basal joints. Virga within genital lobe of male genitalia as in fig. 545 .... 2. **Africolabis humicola**, p. 338

1. Africolabis pseudohumicola Brindle

1978 *Africolabis pseudohumicola* Brindle, Dermaptera of Africa, Tervuren, **2**: 26; fig. 4 (head and thoracic tergites), 5 (holotype genital armature), 6 (distal antennal joints), 7 (male forceps). – Terra typica: Zaïre (Type male: Mus. r. Afr. centr., Tervuren).

Description: Male colour brownish, head reddish-brown or darker; legs yellowish; antennae yellowish brown basally, distal joints whitish. Cuticle of

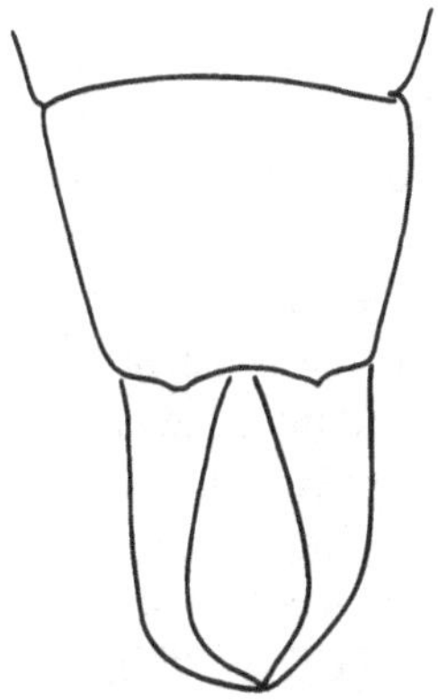

Fig. 542 Holotype ultimate tergite and forceps of *Africolabis pseudohumicola* Brindle. After Brindle.

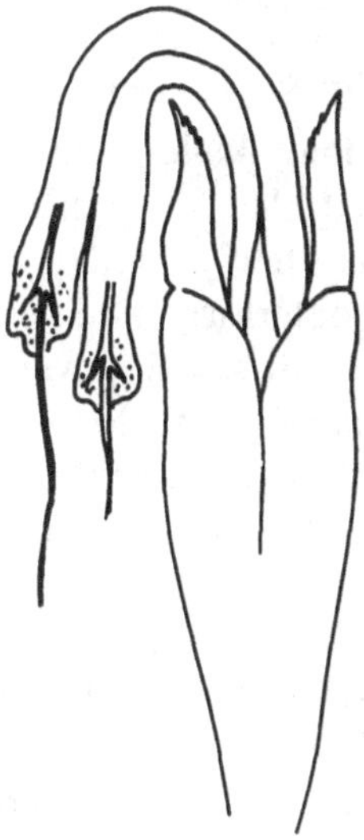

Fig. 543 Holotype genital armature of *Africolabis pseudohumicola* Brindle. After Brindle.

body finely punctate and pubescent. Head oval, longer than wide, narrowed anteriorly and less narrowed posteriorly; postfrontal and coronal sutures indistinct. Eyes very small, facets relatively large and few in number. First antennal joint very long, slightly longer than distance between antennal bases. Pronotum as long as wide; lateral margins straight, and widened posteriorly; median longitudinal furrow indistinct. Mesonotum with rudimentary tegmina, which meet or overlap at sutures but leave a wide triangular area of the tergite exposed. Metanotum transverse. Abdomen fusiform, slightly expanded medially; somewhat depressed. Ultimate tergite large, posterior margin concave medially. Forceps (Fig. 542) cylindrical, straight basally, strongly curved apically. Penultimate sternite transverse, posterior margin rounded, scarcely produced medially. Genitalia (Fig. 543) characteristic; central parameral plate narrow, median longitudinal incision of anterior margin prominent; genital lobes fully developed, virga within genital lobe rather long, partially sclerotized and associated with small sclerites and denticulated areas; external parameres comparatively large, longer than *humicola* (BRINDLE). – **Female** similar to male, but forceps with branches rather more slender and less curved. – **Length** of body with forceps in both sexes: 3,5–4,5 mm.

Distribution: Zaïre.

2. Africolabis humicola (BRINDLE)

1966 *Brachylabis humicola* BRINDLE, Rev. Zool. Bot. afr., Bruxelles, **74**: 37; fig. 19 (holotype genital armature). – Terra typica: Zaïre (Type male: Mus. r. Afr. centr., Tervuren).

1978 *Africolabis humicola* – BRINDLE, Dermaptera of Africa, Tervuren, **2**: 25; fig. 8 (holotype male), 9 (distal antennal joints), 10 (male genitalia).

Literature: HINCKS 1953; SAKAI 1970; STEINMANN 1977, 1978.

Description: Male colour dark yellowish-brown to dark brown; head sometimes darker; antennae dark brown to a little yellowish brown. Cuticle of body rather finely punctate and pubescent. Head oval, longer than wide, narrowed anteriorly and less narrowed posteriorly, convex, with indistinct postfrontal sutures. Eyes very small, facets relatively large and few in number. First antennal joint long, longer than distance between antennal bases. Pronotum transverse, widened posteriorly, lateral and posterior margins almost straight; median longitudinal furrow indistinct. Tegmina totally absent. Mesonotum and metanotum typical. Abdomen fusiform, but somewhat depressed, lateral tubercles absent or nearly so. Ultimate tergite broad, narrowed posteriorly; posterior margin concave medially. Forceps (Fig. 544) simple, cylindrical, weakly curved, branches almost contiguous at base. Penultimate sternite transverse, posterior margin produced medially to form a short rounded lobe. Genitalia (Fig. 545) narrow; central parameral plate with very small median incision at anterior margin; virga within genital lobe short and associated with sclerites and denticulated areas. – **Female** similar to male, but forceps with branches rather more slender. – **Length** of body with forceps in both sexes: 4,5–5,5 mm.

Distribution: Zaïre.

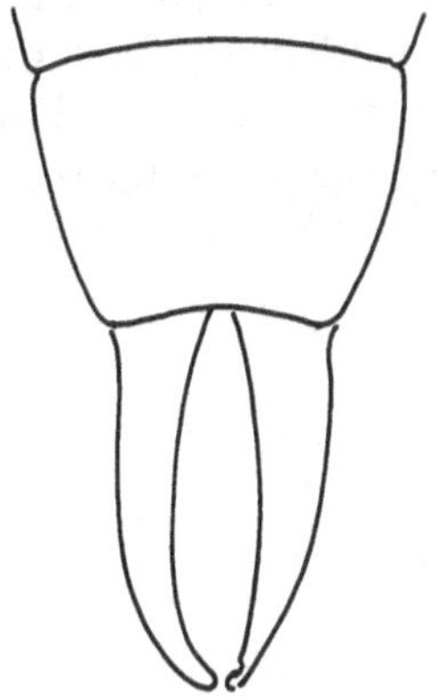

Fig. 544 Male ultimate tergite and forceps of *Africolabis humicola* (BRINDLE). After BRINDLE.

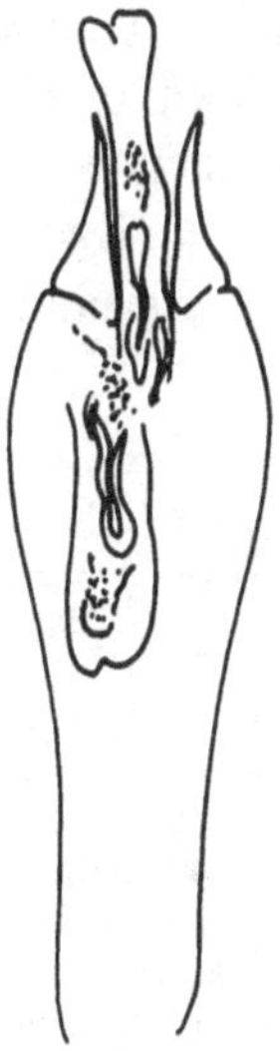

Fig. 545 Holotype genital armature of *Africolabis humicola* (BRINDLE). After BRINDLE.

Subfamily 7: **Antisolabiinae** BRINDLE

1978 Antisolabiinae BRINDLE, Dermaptera of Africa, Tervuren, **2**: 18. – Genus typicum: *Antisolabis* BURR 1911.

This subfamily has been established for the single genus *Antisolabis* BURR, which was formerly placed in the Brachylabiinae by BURR 1909 or Isolabiinae STEINMANN, 1975. The characteristic features are the long head, the short first antennal joint, and the sclerotized virga, which is unusually broad. Eyes short, slightly shorter than length of head behind eyes. Mesonotum without lateral longitudinal ridges or keel. Abdomen fusiform, a little expanded medially; ultimate tergite broad, forceps in both sexes simple.

Distribution: Ethiopian, Malgassian, Indo-Australian and Oriental faunal regions.

Single genus.

Genus 30: **Antisolabis** BURR

1911 *Antisolabis* BURR, Ann. S. Afr. Mus., Cape Town, **10**: 5. – Species typica: *Antisolabis myrmecoides* BURR 1911.

Literature: BRINDLE 1978; BURR 1912, 1915; HINCKS 1957; POPHAM & BRINDLE 1966; SAKAI 1970; STEINMANN 1975, 1977, 1978.

Description: Apterous species, tegmina and wings totally absent. Head long, oval, sides rounded; eyes very small, slightly shorter than scape or length of head behind eyes. First antennal joint very small, or longer. Mesonotum without lateral longitudinal ridges. Abdomen depressed, widened posteriorly or medially, tergites strongly pubescent. Forceps of both sexes with short branches, the branches cylindrical or almost so, symmetrical or sometimes asymmetrical. Male genitalia with narrow central parameral plate, median incision of anterior margin very small, virga within genital lobes various, gently sclerotized basally.

Distribution: Ethiopian, Malgassian, Indo-Australian, Oriental and Neotropical faunal regions.

31 species.

Identification key to the species

1 Ethiopian species **2**
1′ Not African species **8**
2 Head and pronotum coriaceous, impunctate **3**
2′ Head and pronotum punctate **4**
3 Larger species, body length 12 mm; distal antennal joints longer, three times as long as wide. Male penultimate sternite with a median longitudinal ridge near posterior margin. Virga within male genital lobe comparatively broad, as in fig. 547 .. 1. **Antisolabis sulcatipes**, p. 342
3′ Smaller species, body length 8 mm or so; distal antennal joints shorter, twice as long as wide. Male penultimate sternite simple, without a median longitudinal ridge. Virga within male genital lobe comparatively narrow, as in fig. 549 .... 2. **Antisolabis meridionalis**, p. 343
4 Fourth antennal joint longer than wide **5**
4′ Fourth antennal joint quadrate, as wide as long **7**
5 Distal antennal joints unicolorous; femora more or less unicolorous. Male genitalia with very short virgae, as in fig. 551 .... 3. **Antisolabis fusca**, p. 344
5′ Antennal joint 10 whitish, others brown or dark brown; femora dark brown, yellow on distal third **6**
6 Body colour black, with a fine pale pubescence. Pronotum transverse, gently widened posteriorly. Male genitalia (Fig. 553) characteristic, virga within genital lobe strongly sclerotized, straight .... 4. **Antisolabis myrmecoides**, p. 345
6′ Body colour dark brown, sometimes a little blackish brown. Pronotum slightly longer than wide, lateral margins almost straight. Male genitalia unknown .... 5. **Antisolabis camerunensis**, p. 346
7 Cuticle of body with very large punctures; head much more transverse, cuticle shiny. Legs dark, apices of femora and apices of tibiae yellowish. Male genitalia very similar to *minima* (Brindle), as in fig. 556 .... 6. **Antisolabis rugosa**, p. 347
7′ Cuticle of body with smaller punctures; head less transverse, cuticle only slightly shiny. Legs almost uniformly coloured. Male genitalia very similar to *rugosa* (Brindle), as in fig. 558 .... 7. **Antisolabis minima**, p. 348
8 Malagasy species **9**
8′ Indo-Australia, Oriental or Neotropical species **10**
9 Male forceps more or less symmetrical (Fig. 559). Male genitalia very similar to *seychellensis* (Brindle), but external parameres comparatively shorter. Brindle figure with three genital lobes! (Fig. 560) .... 8. **Antisolabis scotti**, p. 349
9′ Male forceps asymmetrical, not of *Antisolabis*-type. Male genitalia very similar to *scotti* (Burr), but external parameres comparatively longer ..... 9. **Antisolabis seychellensis**, p. 350
10 Indo-Australian species **11**
10′ Neotropical or Oriental species **26**

11 New Zealand species. Male forceps characteristic, not of *Antisolabis*-type, well separated basally, and with small but prominent tooth at inner margin (Fig. 563). Male genitalia similar to *myrmecoides* BURR, but virga within genital lobe with basal vesicle (Fig. 564) .... 10. **Antisolabis notabilis**, p. 351

11′ Not New Zealand species **12**

12 Australian species **13**

12′ Not Australian species **15**

13 Pronotum transverse, slightly longer than broad. Female forceps stout, strongly trigonal basally, cylindrical apically (Fig. 565), subcontiguous .... 11. **Antisolabis holdhausi**, p. 352

13′ Pronotum longer than wide **14**

14 Male genitalia large, very broad; central parameral plate with comparatively broad external parameres; virga within genital lobe with strongly sclerotized section (Fig. 567) .... 12. **Antisolabis notonoma**, p. 353

14′ Male genitalia narrow, external parameres as in fig. 569; virga within genital lobe normal. .... 13. **Antisolabis gisleni**, p. 354

15 New Guinean species **16**

15′ Polynesian, Micronesian or Melanesian group **17**

16 Pronotum quadrate. General colour dark brown. Fourth antennal joint quadrate. Male penultimate sternite with posterior margin truncate. Male genitalia elongated, genital lobes very long (Fig. 571) .... 14. **Antisolabis solitaria**, p. 355

16′ Pronotum slightly longer than wide. General colour blackish brown. Fourth antennal joint longer than wide. Female penultimate sternite with rounded posterior margin.......... .... 15. **Antisolabis venusta**, p. 356

17 Polynesian species **18**

17′ Micronesian and Melanesian species **19**

18 Pronotum quadrate. Postfrontal and coronal sutures of head indistinct. Femora unicolorous, yellow. Female forceps as in fig. 573 .... 16. **Antisolabis tongaica**, p. 357

18′ Pronotum longer than wide. Postfrontal and coronal sutures of head well marked. Femora bicolorous, brown, but apices yellow. Female forceps as in fig. 574.......... .... 17. **Antisolabis fijica**, p. 358

19 New Hebridean or Carolinean species **20**

19′ New Caledonian species **21**

20 Pronotum more strongly transverse, ratio of length to width 12,5: 9. Antennae blackish, almost unicolorous. Female forceps as in fig. 575 .... 18. **Antisolabis cordata**, p. 358

20′ Pronotum less strongly transverse, ratio of length to width 11: 9. Antennae dark brown with two or more distal joints white .... 19. **Antisolabis greensladei**, p. 359

21 Cuticle of body dull, roughened and punctate, strongly pubescent, hairs long and relatively close; larger species, body length 12 mm or more **22**

21′ Cuticle of body shiny, smooth and punctate, or granular; very weakly pubescent, hairs short and sparse; smaller species, body length less than 12 mm **23**

22 Posterior margin of head rounded; pronotum longer than wide; cuticle, especially that of abdomen, much more strongly roughened. Virga within genital lobe of male genitalia partially sclerotized, and associated with a large sclerite (Fig. 578); external parameres with a few marginal short setae .... 20. **Antisolabis rouxi**, p. 360

22′ Posterior margin of head concave medially; pronotum transverse; cuticle more uniformly and less strongly roughened. Virga within genital lobe of male genitalia simple; external parameres almost membraneous and with numerous long setae (Fig. 579).......... .... 21. **Antisolabis rouxioides**, p. 361

23 Cuticle of thoracic tergites with closely spaced raised granules; antennae brown to dark brown with distal joints white **24**

23′ Cuticle of thoracic nota smooth with large punctures, not raised granules; antennae dark brown or brown, almost unicolorous **25**

24 Pronotum quadrate, as long as wide, almost paralell-sided. Male forceps (Fig. 580) more strongly thickened basally .... 22. **Antisolabis arripiens**, p. 362

24′ Pronotum longer than wide, widened posteriorly; male forceps (Fig. 581) weakly thickened basally .... 23. **Antisolabis forficula**, p. 363

25 Pronotum strongly transverse, and more strongly widened posteriorly .... 24. **Antisolabis geniculata**, p. 364
25′ Pronotum slightly longer than wide, and weakly widened posteriorly .... 25. **Antisolabis transiens**, p. 365
26 Oriental species **27**
26′ Neotropical species. Mesonotum and metanotum more strongly punctate, the former with a lateral fold on each side for most of length, the fold curving medially. Male forceps (Fig. 585) trigonal basally, cylindrical and curved apically. Male genitalia with very sclerotized virgae, as in fig. 586 .... 26. **Antisolabis allardi**, p. 366
27 Pronotum quadrate **28**
27′ Pronotum slightly longer than wide **29**
28 Colour of body dark brown. Head smooth, shiny, thoracal and abdominal tergites with smaller punctation. Female ultimate tergite and forceps as in fig. 587 .... 27. **Antisolabis javana**, p. 367
28′ Colour of body black. Head granulose, thorax and abdomen with deep punctation. Male forceps crenulate at inner margins (Fig. 588). Male genitalia with very pointed external parameres (Fig. 589) .... 28. **Antisolabis ceylonicus**, p. 368
29 Indian species. Colour of body dark brown, cuticle of head, thoracic nota and abdominal tergites with very deep punctation. Female forceps as in fig. 590 .... 29. **Antisolabis formicoides**, p. 369
29′ Malaysian species. Colour of body blackish or yellowish-brown **30**
30 Colour of body blackish, sometimes very dark brownish-black. First antennal joint slightly longer than wide, and longer than distance between antennal bases .... 30. **Antisolabis dammermani**, p. 369
30′ Colour of body yellowish-brown. First antennal joint more or less equal to length of distance between antennal bases .... 31. **Antisolabis kelangi**, p. 370

1. Antisolabis sulcatipes Burr

1911 *Antisolabis sulcatipes* Burr, Ann. S. afr. Mus., Cape Town, **10**: 6; fig. 3a (female), 3b (antenna of female), 3c (leg), 3d (head and thorax). – Terra typica: Cape Province (Type female: S. afr. Mus., Cape Town).
1957 *Antisolabis sulcatipes* – Hincks, S. afr. Animal Life, Pretoria, **4**: 73; fig. 32 (male genitalia).
1978 *Brachylabis sulcatipes* – Brindle, Dermaptera of Africa, Tervuren, **2**: 28; fig. 14 (male genitalia), 15 (penultimate sternite in lateral view), 16 (penultimate sternite in ventral view description of male).

Literature: Brindle 1966; Burr 1911, 1915; Popham & Brindle 1966; Sakai 1970; Steinmann 1977, 1978.

Description: Male colour dark reddish-brown, almost blackish, shiny; antennae brown, the apical three or four joints whitish; legs dark brown, but tarsi yellow. Head relatively broad, narrowed anteriorly; postfrontal and coronal sutures distinct. Eyes small. Antennae 15-jointed; first joint long, more or less longer than length of eyes, and about as long as distance between antennal bases. Pronotum longer than wide; lateral margins sinuate and with a wide explanate broder; widened posteriorly. Mesonotum smooth, short, transverse, with lateral margins rounded and without lateral longitudinal ridges. Metanotum concave medially. Abdomen fusiform, depressed dorsally, lateral tubercles on third and fourth tergites represented only by swellings of the cuticle. Abdominal tergites expanded to tergites 5th and 6th. Ultimate tergite transverse, posterior margin slightly produced above the base of each branch of the forceps. Forceps (Fig. 546) symmetrical, almost cylindrical, but with inner margins flattened near base. Penultimate sternite transverse, posterior margin convex, and with a prominent raised longitudinal ridge-like process medially. Genitalia (Fig. 547) very long and narrow, genital lobes elongate, virga short, sclerotized and with associated sclerites. External parameres well developed. – **Female** similar to male, but last tergite narrowed and forceps slender, weakly curved. Penultimate sternite without ridge. – **Length** of body with forceps: in both sexes: 13–14 mm.

Distribution: South Africa.

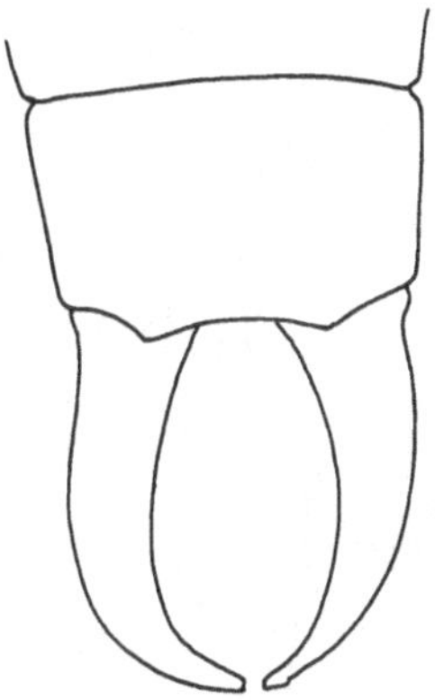

Fig. 546 Male ultimate tergite and forceps of *Antisolabis sulcatipes* BURR. Original.

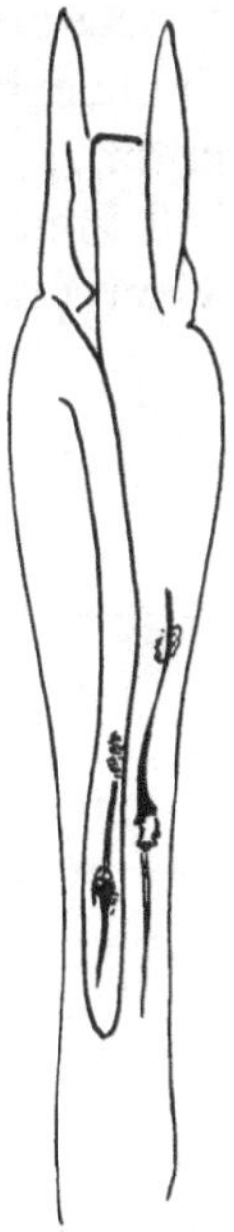

Fig. 547 Male genital armature of *Antisolabis sulcatipes* BURR. After HINCKS.

2. Antisolabis meridionalis (BRINDLE) comb. nov.

1978 *Brachylabis meridionalis* BRINDLE, Dermaptera of Africa, Tervuren, **2**: 29; fig. 11 (holotype male), 12 (posterior margin of penultimate sternite), 13 (holotype genital armature). – Terra typica: Cape Province (Type male: Brit. Mus. nat. Hist., London).

Description: Male colour dark reddish-brown, antennae brown, legs light brown. Cuticle of head and pronotum impunctate and coriaceous, meso- and metanotum and abdomen punctate and pubescent. Head slightly longer than wide, narrowed anteriorly and posteriorly, convex. Eyes very small; first antennal joint long, distal joints shorter, twice as long as wide. Pronotum longer than wide, widened posteriorly; lateral margins somewhat sinuate, posterior margin almost straight. Mesonotum transverse, without lateral longitudinal ridges; metanotum short, concave medially on posterior margin. Abdomen fusiform, depressed dorsally, lateral tubercles on third and fourth tergites small, represented by cuticular swellings. Ultimate tergite broad, posterior margin concave medially and produced above both bases of the forceps. Forceps (Fig. 548) symmetrical, strongly curved; cylindrical, inner margin flattened near base, broader towards base. Penultimate sternite transverse, posterior margin slightly concave medially. Genitalia (Fig. 549) narrow, median incision of anterior margin present; virga within genital lobe comparatively narrow, short, slender, partially sclerotized; external parameres moderately long, pointed. – **Female** similar to male, but branches of forceps slender and less curved, almost contiguous; penultimate sternite evenly rounded. – **Length** of body with forceps: in both sexes: 8–8,5 mm.

Distribution: Cape Province.

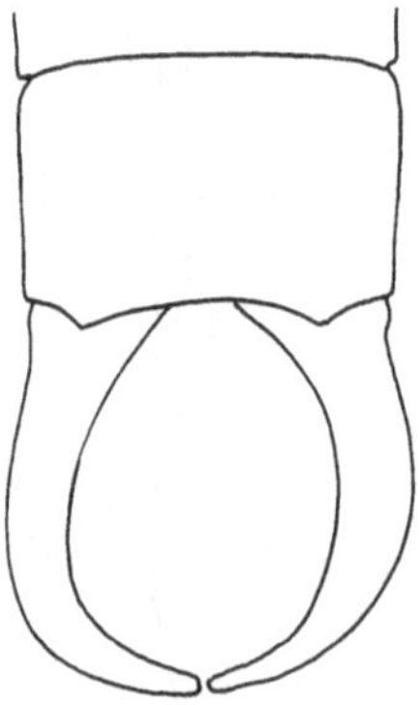

Fig. 548 Holotype ultimate tergite and forceps of *Antisolabis meridionalis* (BRINDLE). After BRINDLE.

Fig. 549 Holotype genital armature of *Antisolabis meridionalis* (BRINDLE). After BRINDLE.

3. Antisolabis fusca (BRINDLE) comb. nov.

1966 *Brachylabis fusca* BRINDLE, Rev. Zool. Bot. afr., Bruxelles, **74**: 39; fig. 13 (holotype genital armature). – Terra typica: Tanganyika (Type male: Mus. r. Afr. centr., Tervuren).

1978 *Isolabis fusca* – BRINDLE, Dermaptera of Africa, Tervuren, **2**: 36; fig. 23 (head, pronotum and mesonotum), 24 (male genitalia).

Literature: HINCKS 1953; SAKAI 1970; STEINMANN 1977, 1978.

Description: Male colour reddish-brown, much less deeply punctate and duller than *meridionalis* (BRINDLE), almost glabrous, head and pronotum coriaceous. Head transverse, sides rounded, tumid; eyes very small. Antennae 13-jointed;

first joint long, longer than distance between antennal bases, second transverse, third and fourth almost equal in length. Pronotum longer than wide, slightly widened posteriorly, margins almost straight. Mesonotum broad, depressed transversely on disc, forming a low lateral fold towards each side. Abdomen fusiform, lateral tubercles on third and fourth tergites weak; tergites punctate anteriorly, posterior margins impunctate, the margin increasing is size posteriorly. Ultimate tergite narrowed, slightly produced above the base of both branches of the forceps. Forceps (Fig. 550) cylindrical, slender, weakly curved. Penultimate sternite with posterior margin convex. Genitalia (Fig. 551) comparatively broad apically, median incision of anterior margin small, genital lobes well developed, virga within genital lobe with a short chitinous plate. External parameres very narrow, comparatively long, pointed. – **Female** similar to male, but forceps slenderer, tapering. – **Length** of body with forceps: male: 8 mm, female: 9 mm.

Distribution: Tanzania.

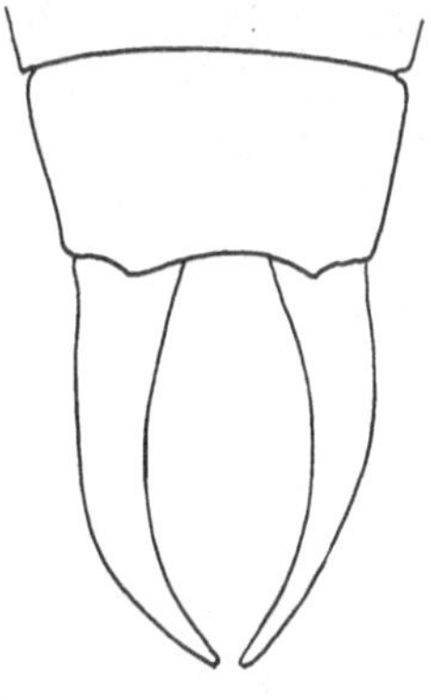

Fig. 550 Male ultimate tergite and forceps of *Antisolabis fusca* (BRINDLE). Original.

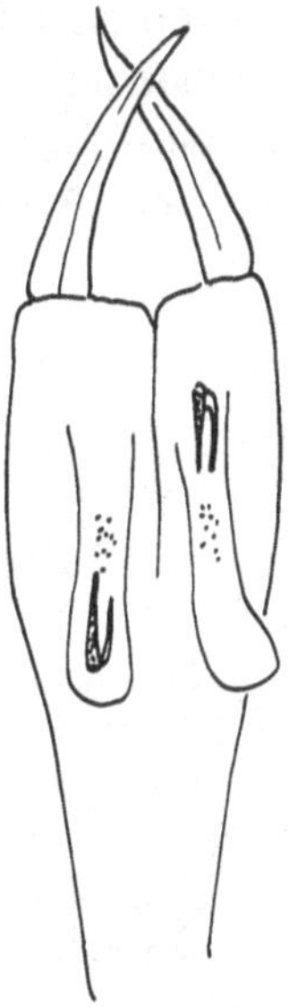

Fig. 551 Holotype genital armature of *Antisolabis fusca* (BRINDLE). After BRINDLE.

4. Antisolabis myrmecoides BURR

1911 *Antisolabis myrmecoides* BURR, Ann. S. afr. Mus., Cape Town, **10**: 5; fig. 2a (male), 2b (antenna of holotype), 2c (head and thorax of male). – Terra typica: Cape Province (Type male: South afr. Mus., Cape Town).

1957 *Antisolabis myrmecoides* – HINCKS, S. afr. Animal Life, Pretoria, **4**: 72; fig. 31 (male genitalia).

1978 *Antisolabis myrmecoides* – BRINDLE, Dermaptera of Africa, Tervuren, 2: 18; fig. 1 (paratype male), 2 (male genitalia), 3 (penultimate sternite of male).

Literature: BURR 1911, 1915; HINCKS 1953; SAKAI 1970; STEINMANN 1977, 1978.

Description: Male colour blackish, with a fine pale pubescence; sides of pronotum yellowish, legs brown but apices of femora and bases of tibiae yellowish. Cuticle mainly punctate and pubescent, head often glabrous; thoracic tergites more strongly punctate. Head rounded, tumid; eyes very small, antennae 15-jointed, first joint very small, characteristic; second joint transverse, third joint twice as long as wide, fourth one and half times as long as broad. Pronotum transverse, gently widened posteriorly; margins more or less straight; all angles rounded. Tegmina and wings absent; mesonotum transverse, narrower than pronotum, with lateral folds. Abdomen depressed, slightly widened medially. Ultimate tergite broad, posterior margin oblique laterally and concave medially. Forceps (Fig. 552) cylindrical, broader at base, branches widely separated near base, and with isolated long yellow hairs. Penultimate sternite transverse, posterior margin with a median rounded lobe. Genitalia (Fig. 553) characteristic; broad and short, median incision of anterior margin very deep; genital lobes fully developed, virga within genital lobe straight, very sclerotized and with annular thickenings, membraneous distally and basally. External parameres well developed. – **Female** similar to male, but forceps simple, contiguous, tapering. – **Length** of body with forceps: male: 5,5–6 mm, female: 6–8 mm.

Distribution: South Africa.

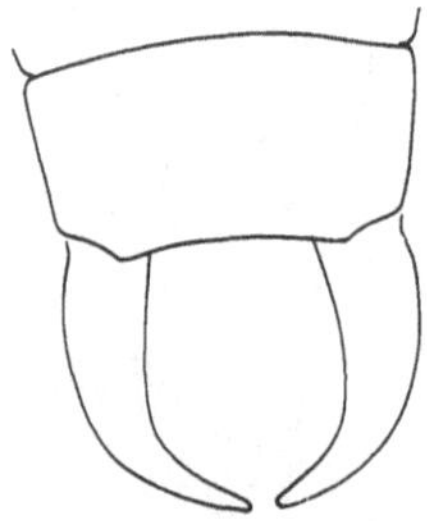

Fig. 552 Male ultimate tergite and forceps of *Antisolabis myrmecoides* Burr. Original.

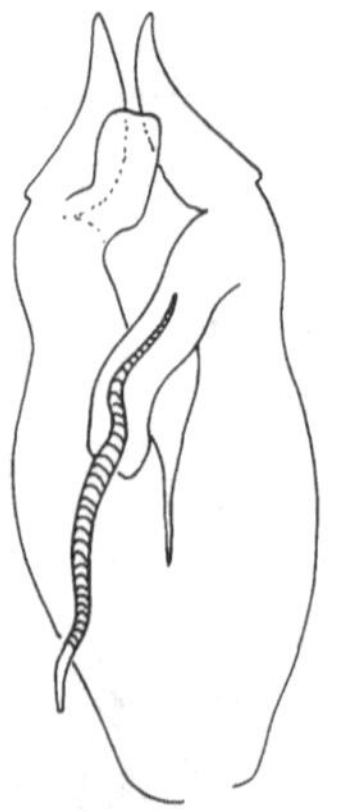

Fig. 553 Male genital armature of *Antisolabis myrmecoides* Burr. (After Brindle).

5. Antisolabis camerunensis (Borelli) comb. nov.

1914 *Nannisolabis camerunensis* Borelli, Boll. Lab. Zool. gen. agr., Portici, **8**: 270. – Terra typica: Cameroun (Type male: unknown locality).
1978 *Isolabis camerunensis* – Brindle, Dermaptera of Africa, Tervuren, **2**: 37.
Literature: Sakai 1970; Steinmann 1977, 1978.

Description: Male colour blackish, sometimes dark brown and a little blackish; antennae brown, joints 9–10 white; legs yellow, femora darkened on basal third or more; abdomen dark brown, forceps reddish-brown. Cuticle of body punctate and pubescent. Head transverse, narrowed anteriorly; eyes very small, typical. Antennae 10 (?) jointed; first joint long, a little longer than distance between antennal bases; third joint longer than broad, fourth and fifth joints longer than wide. Pronotum more or less rectangular, slightly longer than wide, margins almost straight. Mesonotum transverse, without lateral longitudinal flaps. Metanotum typical. Abdomen fusiform, lateral tubercles on third and fourth tergites present, probably small. Ultimate tergite very broad; posterior margin concave medially. Forceps (Fig. 554) simple, a little trigonal basally, curved and cylindrical apically. Penultimate sternite with posterior margin convex; pygidium small, triangular, apex truncate. Genitalia unknown. – **Female** unknown. – **Length** of body with forceps: 9,5 mm.

Distribution: Cameroon.

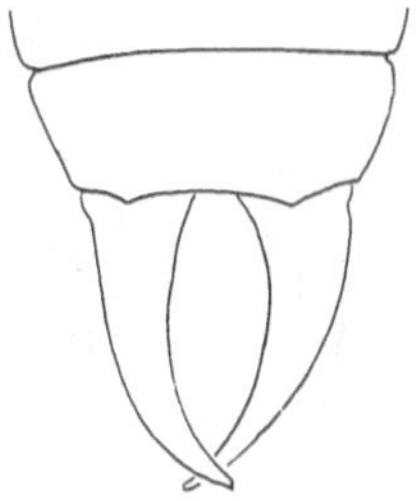

Fig. 554 Male ultimate tergite and forceps of *Antisolabis camerunensis* (BORELLI). Original.

6. Antisolabis rugosa (BRINDLE) comb. nov.

1966 *Brachylabis rugosa* BRINDLE, Rev. Zool. Bot. afr., Bruxelles, **74**: 38; fig. 12 (holotype genital armature). – Terra typica: Tanzania (Type male: Mus. r. Afr. centr., Tervuren).

1978 *Isolabis rugosa* – BRINDLE, Dermaptera of Africa, Tervuren, **2**: 34; fig. 20 (male), 21 (posterior margin of penultimate sternite of male), 22 (male genitalia).

Literature: SAKAI 1970; STEINMANN 1977, 1978.

Description: Male colour dark brown, sometimes dark reddish-brown, almost blackish; shiny; antennae blackish-brown, joint 2 paler, and joints 9–10 or 9–11 yellowish-white. Femora dark brown, yellow at apices. Cuticle of body with very large punctures, and with inconspicuous fine short yellow hairs. Head strongly transverse, margins rounded, tumid, narrower posteriorly; eyes small, slightly shorter than length of head behind eyes. Antennae 12-jointed; first joint long, longer than distance between antennal bases; second transverse, third longer than broad and widened distally; fourth joint as long as wide. Pronotum almost quadrate, slightly widened posteriorly or sometimes parallel-sided; median longitudinal furrow present on anterior half. Mesonotum transverse, narrowed posteriorly, without lateral longitudinal ridge. Metanotum typical. Abdomen with deeply punctate tergites, shiny; lateral tubercles only present on fourth tergite and very small. Ultimate tergite transverse, somewhat produced above the base of both branches of the forceps. Forceps (Fig. 555) simple, cylindrical, more or less straight, a little curved apically. Penultimate sternite with the posterior margin convex and with a median excision. Genitalia (Fig. 556) well developed; median incision of anterior margin of the central parameral plate present, but narrow; genital lobes well developed, with a short virgal sclerite and with denticulated areas. External parameres broad, acuminate, very similar to *minima* (BRINDLE). – **Female** similar to male, but body slenderer, and tergites with smaller punctation. – **Length** of body with forceps: in both sexes: 6–7 mm.

Distribution: Tanzania.

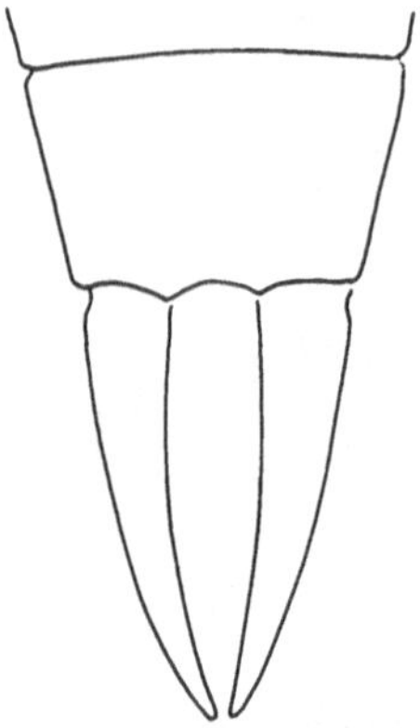

Fig. 555 Male ultimate tergite and forceps of *Antisolabis rugosa* (BRINDLE). After BRINDLE.

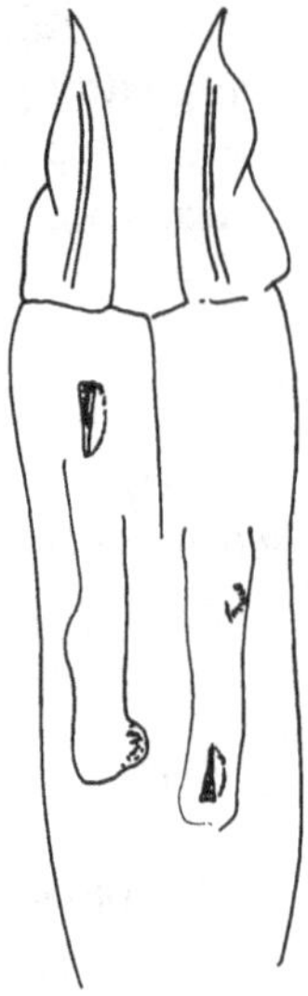

Fig. 556 Holotype genital armature of *Antisolabis rugosa* (BRINDLE). After BRINDLE.

7. Antisolabis minima (BRINDLE) comb. nov.

1966 *Brachylabis minima* BRINDLE, Rev. Zool. Bot. afr., Bruxelles, **74**: 40; fig. 15 (holotype genital armature). – Terra typica Tanganyika (Type male: Mus. r. Afr. centr., Tervuren).

1978 *Isolabis minima* – BRINDLE, Dermaptera of Africa, Tervuren, **2**: 36; fig. 26 (head, pronotum and mesonotum), 25 (male genitalia), 27 (male forceps).

Literature: HINCKS 1954; SAKAI 1970; STEINMANN 1977, 1978.

Description: Male colour dark reddish-brown to brownish-black; antennae dark brown, first joint darker and with joints 10–11 whitish; legs dark brown. Cuticle of body with smaller punctures; shiny. Head rounded; eyes small, shorter than length of head behind eyes. Antennae 13-jointed; first joint longer than distance between antennal bases; second transverse, third joint one and a quarter times as long as wide. Pronotum quadrate, very slightly widened posteriorly; lateral margins more or less straight. Mesonotum transverse, typical, plate depressed transversely; metanotum wider than mesonotum. Abdomen fusiform, slightly depressed, narrower distally, but less narrowed anteriorly; lateral tubercles rudimentary. Ultimate tergite broad, posterior margin produced above the bases of the forceps. Forceps (Fig. 557) cylindrical and slender, weakly curved apically. Penultimate sternite with posterior margin convex. Genitalia (Fig. 558) broad, large, median incision of anterior margin very small, genital lobes well developed, virga within genital lobe similar to *rugosa* (BRINDLE), with a prominent lateral sclerite and denticulated areae. – **Female** similar to male, but abdomen slender, tapering. – **Length** of body with forceps: male: 6–6,5 mm, female: 5,5–6,5 mm.

Distribution: Tanzania.

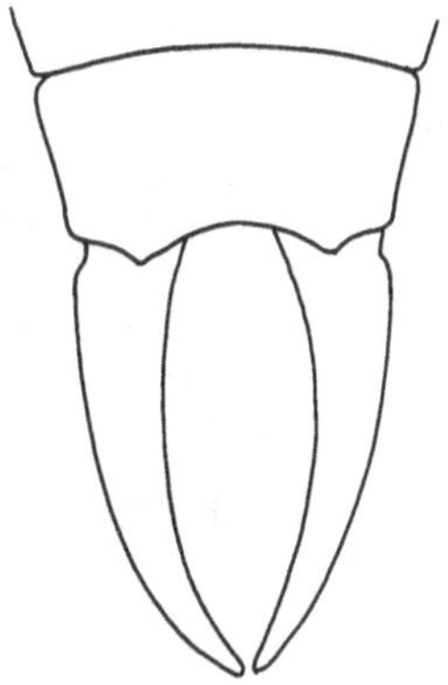

Fig. 557 Male ultimate tergite and forceps of *Antisolabis minima* (BRINDLE). After BRINDLE.

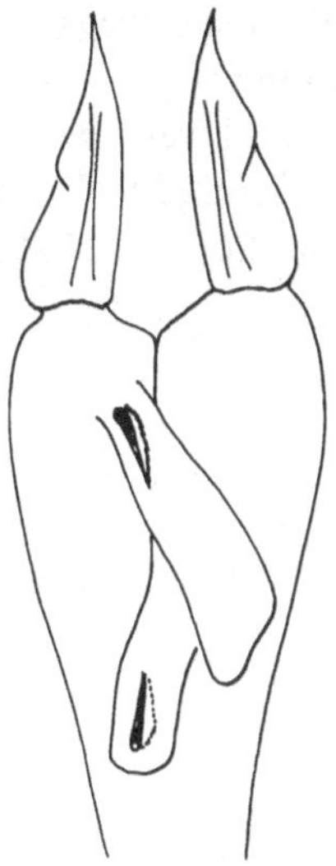

Fig. 558 Holotype genital armature of *Antisolabis minima* (BRINDLE). After BRINDLE.

8. Antisolabis scotti (BURR)

1910 *Brachylabis scotti* BURR, Trans. ent. Soc. London, **1910**: 193. – Terra typica: Seychelles (Type male: Brit. Mus. nat. Hist., London).

1966 *Antisolabis scotti* – POPHAM & BRINDLE, Entomologist, London, **99**: 244.

1970 *Chelidura geniculata* – SAKAI, Dermapt. Cat. Prael., Tokyo, **1**: 6 [= *Antisolabis scotti* (BURR 1910)]. New Caledonia and Seychelles.

1970 *Brachylabis scotti* – BRINDLE, Rev. Zool. Bot. afr., Bruxelles, **90**: 438; fig. 4 (penultimate sternite of holotype), 5 (male genitalia). Seychelles: Silhouette.

Literature: STEINMANN 1977, 1978.

Description: Male colour dark reddish-brown; antennae dark brown with one or more distal joints white; femora dark brown with yellowish apices. Cuticle of head not smooth, as described in BURR 1910, mentioned by BRINDLE 1976, but granulose, as figured in BURR; rest of body punctate and pubescent. Head about as long as wide, rounded posteriorly; postfrontal and coronal sutures obsolete. First antennal joint long, longer than distance between antennal bases, fourth joint longer than wide. Eyes small. Pronotum about as long as wide; all angles rounded, and all margins straight; median longitudinal furrow present on anterior half. Mesonotum typical, transverse and concave posteriorly. Metanotum normal. Abdomen cylindrical, a little expanded medially. Ultimate tergite broad, posterior margin concave medially. Forceps (Fig. 559) asymmetrical, a little trigonal basally, cylindrical and curved apically. Penultimate sternite rounded posteriorly. Genitalia (Fig. 560) very broad apically, median incision of anterior margin small, genital lobes well developed, BRINDLE 1976 figure with three genital lobes; virga characteristic, with sclerotized denticulated areas; external parameres pointed. – **Female** similar to male, but branches of forceps symmetrical, and very weakly curved. – **Length** of body with forceps: male: 6,5–7,5 mm, female: 6–7 mm.

Distribution: Seychelles Islands, New Caledonia.

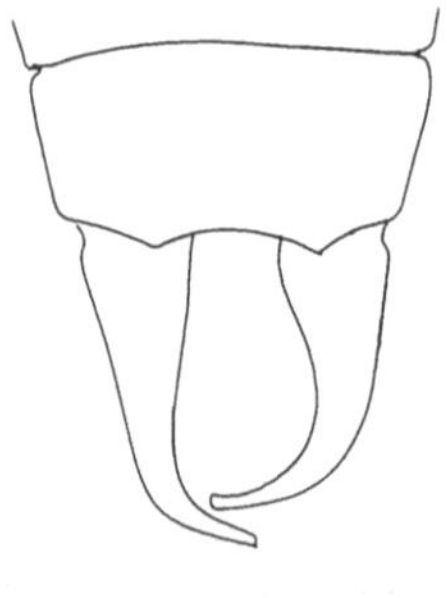

Fig. 559 Male ultimate tergite and forceps of *Antisolabis scotti* (BURR). Original.

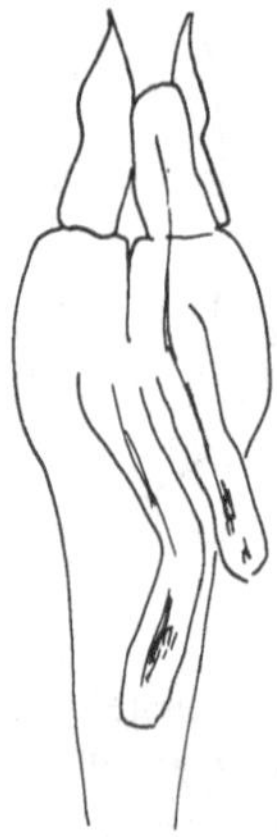

Fig. 560 Male genital armature of *Antisolabis scotti* (BURR). After BRINDLE.

9. Antisolabis seychellensis (BRINDLE) comb. nov.

1976 *Brachylabis seychellensis* BRINDLE, Rev. Zool. Bot. afr., Bruxelles, **90**: 439; fig. 1 (holotype male), 2 (penultimate sternite), 3 (holotype genital armature). – Terra typica: Seychelles (Type male: Brit. Mus. nat. Hist., London).

Description: Male colour blackish, sides of pronotum yellowish, antennae blackish-brown, one or more distal joints white; apices of femora and all tibiae yellowish. Cuticle of head granulose, pubescent and dull; pronotum partially punctate medially, rest of body closely punctate. Head transverse, tumid, postero-lateral angles with straight posterior margin; eyes fairly large, but shorter than length of head behind eyes. First antennal joint long, a little longer than distance between antennal bases; second transverse, third joint rather more than twice as long as wide. Pronotum slightly longer than wide, widened posteriorly, lateral margins sinuate; mesonotum transverse, anterior half depressed with a blunt lateral fold on both sides. Abdomen fusiform, lateral tubercles on third tergite not visible, those on fourth small. Ultimate tergite very broad, typical; posterior margin produced above the base of both branches of the forceps. Forceps (Fig. 561) strongly asymmetrical, not of *Antisolabis*-type; a little trigonal basally, cylindrical apically. Penultimate sternite transverse, posterior margin evenly convex. Genitalia (Fig. 562) (figure by BRINDLE with three genital lobes) very similar to *scotti* (BURR), but virga S-shaped. External paramere pointed. – **Female** similar to male, but forceps symmetrical. – **Length** of body with forceps: male: 5–5,5 mm, female: 5–6 mm.

Distribution: Seychelles Islands.

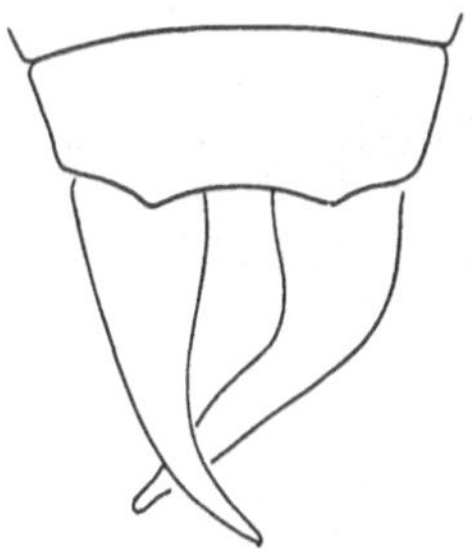

Fig. 561 Holotype ultimate tergite and forceps of *Antisolabis seychellensis* (BRINDLE). After BRINDLE.

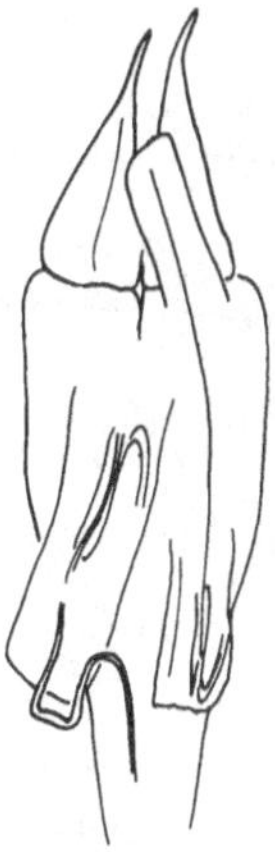

Fig. 562 Holotype genital armature of *Antisolabis seychellensis* (BRINDLE). After BRINDLE.

10. Antisolabis notabilis STEINMANN

1981 *Antisolabis notabilis* STEINMANN, Fol. ent. hung., Budapest, **34**: 184; fig. 29 (ultimate tergite and forceps of holotype), 30 (holotype genital armature). – Terra typica: New Zealand (Type male: Univ. zool. Mus., København).

Description: Male colour brownish; head reddish brown, pronotum brown but lateral margins with yellowish brown stripes, meso- and metanotum, and abdomen with forceps dark brown, legs yellowish brown. Cuticle of body punctate and pubescent. Head more or less quadrate; postfrontal sutures and coronal suture indistinct. Eyes moderately small, shorter than distance between antennal bases. Antennae 15-jointed; first joint comparatively small, shorter than distance between antennal bases; second transverse, third joint long, longer than fourth. Pronotum transverse, longer than wide; median longitudinal furrow deep; lateral margins parallel, all angles rounded. Tegmina and wings entirely absent. Mesonotum typical, transverse, posterior margin concave. Abdomen cylindrical, fusiform, punctate and shiny. Ultimate tergite very broad, posterior margin with two small tubercles above both bases of the forceps. Forceps characteristic (Fig. 563), not of *Antisolabis*-type, well-separated basally, and with small but prominent tooth at inner margin apically. Penultimate sternite simple, posterior margin rounded. Genitalia (Fig. 564; gen. prep. No. 712, det. Dr. H. STEINMANN) characteristic, paramere elongated, median longitudinal incision of anterior margin deep; genital lobes fully developed, virga very thick, basal vesicle specific, well sclerotized. External parameres trigonal, inner margin straight, apices acute. – **Female** similar to male, but forceps simple, tapering, contiguous, symmetrical. – **Length** of body with forceps: male: 12–13,5 mm, female: 12 mm.

Distribution: New Zealand.

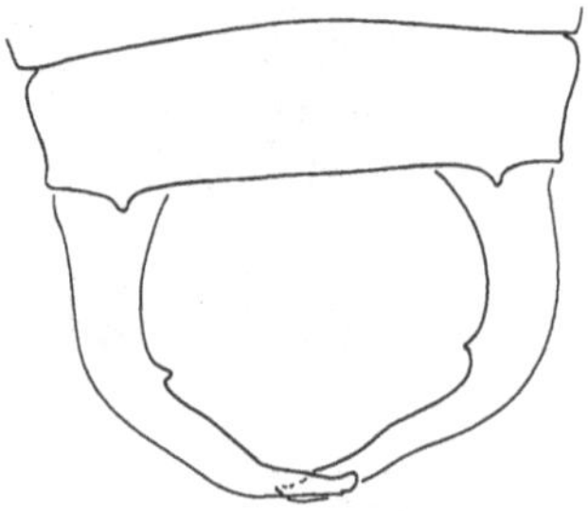

Fig. 563 Holotype ultimate tergite and forceps of *Antisolabis notabilis* STEINMANN. After STEINMANN.

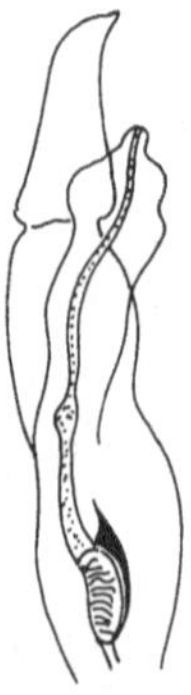

Fig. 564 Holotype genital armature of *Antisolabis notabilis* STEINMANN. Original.

11. Antisolabis holdhausi (BURR)

1910 *Nannisolabis holdhausi* BURR, Trans. ent. Soc. London, **1910**: 195; pl. 47, fig. 17 (mentioned as male, but figure female). – Terra typica: Australia (Type female: Brunner's Coll. Naturhist. Mus., Wien).

1912 *Antisolabis holdhausi* – BURR, Ann. naturhist. Hofmus., Wien, **26**: 333; fig. 4 (female).

1966 *Antisolabis holdhausi* – POPHAM & BRINDLE, Entomologist, London, **99**: 244.

Literature: SAKAI 1970; STEINMANN 1977, 1978.

Description: Female colour deep reddish brown or blackish; legs yellow, the femora and tibiae with black bands. Cuticle of body scarcely punctate. Head broad, postfrontal and coronal sutures distinct; occiput rectangular. Eyes relatively large, but shorter than length of head behind eyes. Antennae 16-

jointed; two or three prae-apical joints yellowish; first joint comparatively short, shorter than distance between antennal bases; second transverse. Pronotum broad, visibly wider than long; rectangular, all angles rounded; median longitudinal furrow short. Tegmina and wings absent. Mesonotum a little shorter and a little broader than pronotum, without lateral folds. Abdomen expanded medially, tapering apically, normal. Ultimate tergite broad, posterior margin concave medially. Forceps (Fig. 565) simple, trigonal basally, cylindrical apically; subcontiguous, curved apically. Penultimate sternite narrowly rounded. – **Male** unknown. – **Length** of body with forceps: 15 mm.

Distribution: Australia: Queensland.

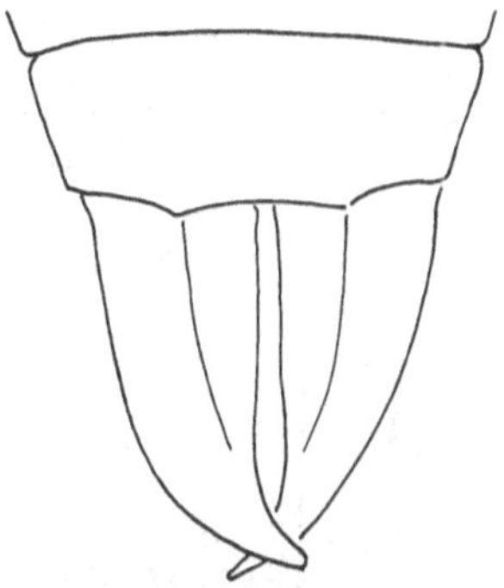

Fig. 565 Female ultimate tergite and forceps of *Antisolabis holdhausi* (BURR). After BURR.

12. Antisolabis notonoma HINCKS

1952 *Antisolabis notonoma* HINCKS, Entomologist's month. Mag., London 202, **88**: 214; fig. 1 (holotype genital armature). – Terra typica: New South Wales (Type male: Collect. Dept. Agric., Sydney).
1966 *Isolabis notonoma* – POPHAM & BRINDLE, Entomologist, London, **99**: 245.
Literature: SAKAI 1970; STEINMANN 1977, 1978.

Description: Male colour brown, legs, and antennal joints 10 to 12 dark brown. Cuticle of body closely punctate and pubescent. Head strongly contracted in front of small; eyes comparatively short; posterior margin of head truncate. First antennal joint long, as long as joints 2 to 4 together, and much stouter. Pronotum large, longer than wide; lateral margins straight; median longitudinal furrow short; plate gradually expanded posteriorly; posterior angles rounded. Tegmina and wings totaly absent; meso- and metanotum typical, normal, simple. Abdomen slightly contracted proximad and distad; lateral tubercles of tergite 3 small, of 4 distinct. Ultimate tergite transverse, sculptured as rest of abdomen, depressed mesad with short median longitudinal sulcus; posterior margin produced above both branches of the forceps. Forceps (Fig. 566) simple, branches straight medially and curved apically. Penultimate sternite transverse, posterior margin gently rounded and shallowly emarginate mesad. Genitalia

(Fig. 567) characteristic; central parameral plate broad, genital lobes well developed, virga within genital lobe very long, with smaller sclerotized plates. External parameres acuminate, hooked distad, curved outwards and closely approximated so that the space between them is completely occupied by the right one in dorsal view. – **Female** unknown. – **Length** of body with forceps: 15–18 mm.

Distribution: Australia: New South Wales.

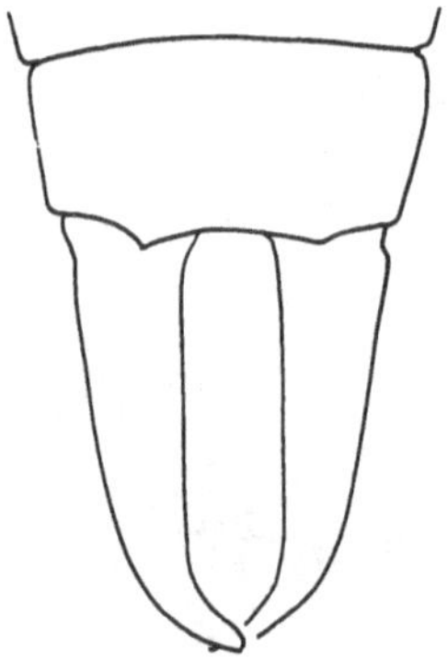

Fig. 566 Male ultimate tergite and forceps of *Antisolabis notonoma* HINCKS. Original.

Fig. 567 Holotype genital armature of *Antisolabis notonoma* HINCKS. After HINCKS.

13. Antisolabis gisleni (HINCKS) comb. nov.

1954 *Nannisolabis gisleni* HINCKS, Lunds Univ. Årsskrift, (2) **50**: 8; fig. 2 (holotype genital armature). – Terra typica: Australia (Type male: Lunds Univ. zool. Inst.).
Literature: POPHAM & BRINDLE 1966; SAKAI 1970; STEINMANN 1977, 1978.

Description: Male colour blackish, piceous; antennae and legs very dark brown, sides of pronotum light brownish-black. Cuticle of body punctate and pubescent. Head strongly convex, surface dull; postfrontal and coronal sutures feebly indicated. Eyes comparatively small, slightly shorter than length of head behind eyes. Antennae (broken in holotype) 12-jointed, first joint as long as 2nd and 5th combinated, and a little longer than distance between antennal bases; second very short, third conical, and fourth joint as long as wide. Pronotum slightly longer than wide, narrower than head at cephalic angles; median longitudinal furrow strong; lateral margins straight. Tegmina and wings absent. Mesonotum slightly more polished than pronotum, without lateral longitudinal ridge. Abdomen moderately shiny and pubescent, strongly punctate; expanded medially. Ultimate tergite transverse, posterior margin concave medially. Forceps (Fig. 568) somewhat similar to *notonoma* HINCKS, branches trigonal basally, cylindrical and curved apically; branches straight medially. Penultimate sternite rounded posteriorly. Genitalia (Fig. 569) narrow; median incision of anterior margin normal but narrow; genital lobes well developed, virga very long, with two groups of small teeth and containing a sclerotized accessory plate; external parameres acuminate at apices. – **Female** unknown. – **Length** of body with forceps: 7,5 mm.

Distribution: Western Australia.

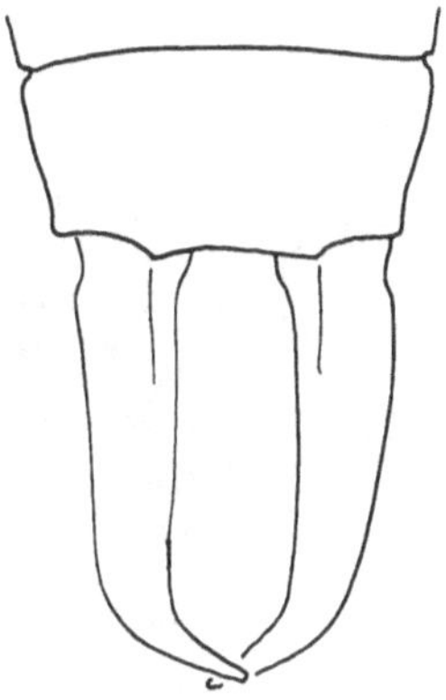

Fig. 568 Male ultimate tergite and forceps of *Antisolabis gisleni* (HINCKS). Original.

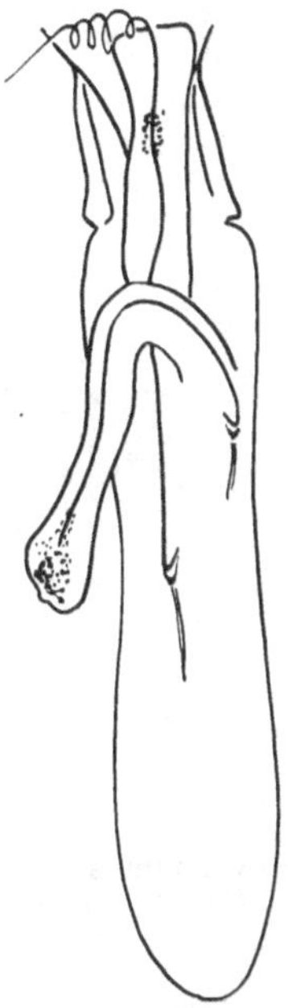

Fig. 569 Holotype genital armature of *Antisolabis gisleni* (HINCKS). After HINCKS.

14. Antisolabis solitaria sp. nov.

Terra typica: New Guinea, War Nami Creck, 21. VIII. 1968, legit: Dr. I. LOKSA, gen. prep. No. 161, det. Dr. H. STEINMANN. – (Type male: Természettudományi Múz., Budapest).

Description: Male general colour dark brown; legs yellow, but femora banded brown medially. Cuticle of head, thoracic nota and abdominal tergites strongly punctate; shiny. Head broad; posterior margin truncate; postfrontal and coronal sutures indistinct. Eyes small, slightly shorter than length of head behind eyes. First antennal joint long, longer than distance between antennal bases; second quadrate, third long, longer than fourth and fifth combined. Pronotum quadrate, lateral margins more or less parallel; median longitudinal furrow present. Mesonotum normal, broad, posterior margin straight, disc with blunt lateral ridge-like tubercle. Metanotum typical. Abdomen slender, fusiform, tubercles on fourth tergite present. Ultimate tergite broad, posterior margin concave medially. Forceps (Fig. 570) trigonal basally, curved and cylindrical apically. Penultimate sternite simple, rounded posteriorly. Genitalia (Fig. 571) very long, characteristic; central parameral plate with median incision on anterior margin; genital lobes very long, virga simple, elongate. External parameres acuminate, comparatively long. – **Female** unknown. – **Length** of body with forceps: 11 mm.

Distribution: New Guinea.

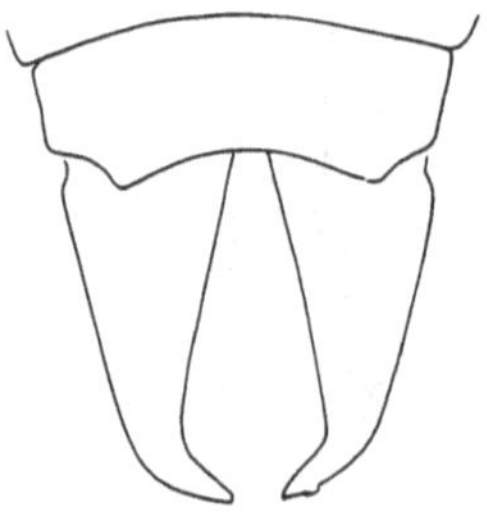

Fig. 570 Holotype ultimate tergite and forceps of *Antisolabis solitaria* sp.n. Original.

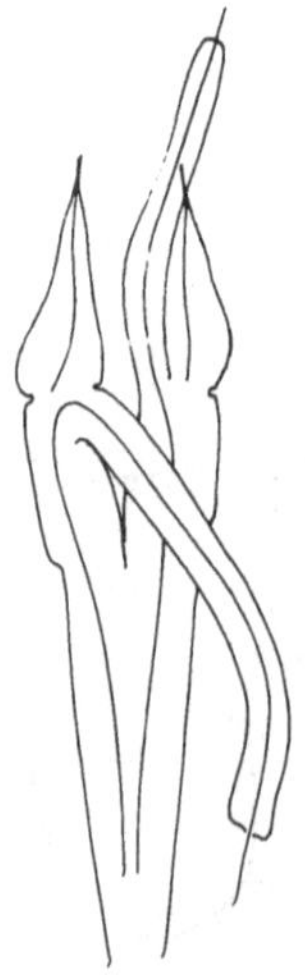

Fig. 571 Holotype genital armature of *Antisolabis solitaria* sp.n. Original.

15. Antisolabis venusta sp. nov.

Terra typica: New Guinea, Lea Bulalo, 21.VIII.1968, legit: Dr. I. LOKSA. – (Type female: Természettudományi Múz., Budapest).

Description: Female colour dark brownish black; antennae light brown, legs bicolorous, yellow except at basal half of femora. Cuticle of body punctate and pubescent. Head somewhat longer than wide; posterior margin truncate; postfrontal sutures finely, coronal suture strongly distinct. Eyes small, slightly shorter than length of head behind eyes. First antennal joint long, longer than distance between antennal bases; second quadrate, third joint longer than 4th, and fourth joint a little longer than wide. Pronotum slightly longer than wide, lateral margins more or less paralell; median longitudinal furrow present. Mesonotum typical, depressed dorsally, posterior margin straight. Metanotum normal. Abdomen fusiform, a little expanded medially. Ultimate tergite very broad, posterior margin with two tubercles above both branches of the forceps. Forceps (Fig. 572) simple, trigonal basally, cylindrical apically. Penultimate sternite broad, rounded posteriorly. – **Male** unknown. – **Length** of body with forceps: 14 mm.

Distribution: New Guinea.

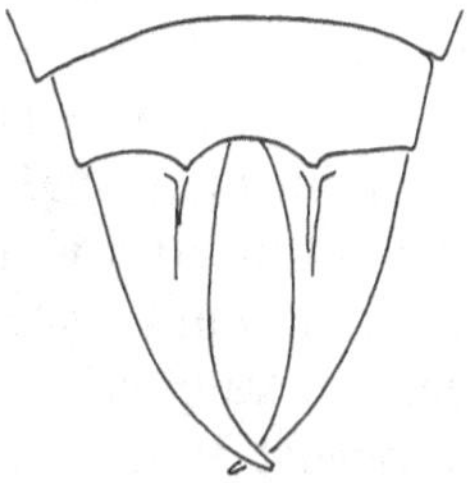

Fig. 572 Holotype ultimate tergite and forceps of *Antisolabis venusta* sp.n. Original.

16. Antisolabis tongaica sp. nov.

Terra typica: Tonga, Kiba, 8.2.40., P. 103. (Type female: Bishop Mus., Honolulu, Hawaii).

Description: Female colour dark reddish-brown; antennae light brown, legs uniformly yellow. Cuticle of body punctate with yellowish hairs. Head quadrate, more or less as long as wide; postfrontal and coronal sutures indistinct; eyes comparatively large, but shorter than the length of head behind eyes. First antennal joint long, slightly longer than distance between antennal bases; second quadrate, third joint longer than wide, fourth joint as long as wide, remaining joints conical. Pronotum quadrate, all angles rounded; lateral margins a little wided posteriorly; median longitudinal furrow present on anterior half. Tegmina and wings absent. Mesonotum transverse, with transversal depression dorsally; posterior margin truncate. Metanotum typical. Abdomen oval, expanded medially, narrowed anteriorly and apically; abdominal tergites shiny. Ultimate tergite broad, posterior margin with two blunt tubercles. Forceps (Fig. 573)

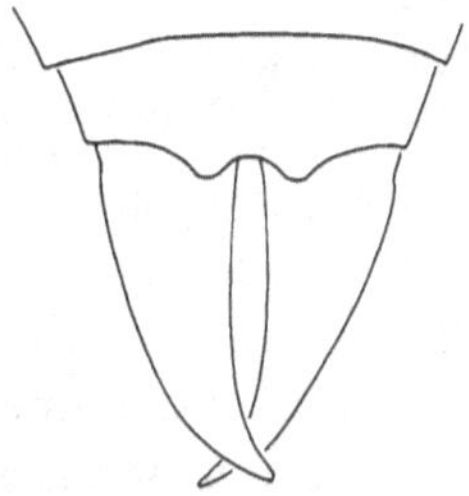

Fig. 573 Holotype ultimate tergite and forceps of *Antisolabis tongaica* sp.n. Original.

simple, cylindrical, and curved. Penultimate sternite rounded posteriorly. – **Male** unknown. – **Length** of body with forceps: 6,5 mm.

Distribution: Tonga.

17. Antisolabis fijica sp. nov.

Terra typica: Fiji Island, 4. IX. 1968, legit: Dr. J. BALOGH. (Type male: Természettudományi Múz., Budapest).

Description: Female general colour dark brown; antennae brown, but joints with narrow apical stripes, yellow; legs yellow, femora with broad brownish band medially. Cuticle of body punctate and pubescent. Head rounded, more or less as wide as long, with straight posterior margin; postfrontal and coronal sutures well marked; eyes comparatively large, but shorter than length of head behind eyes; first antennal joint long, slightly longer than distance between antennal bases; second joint quadrate, third long, longer than eyes, and fourth joint a little longer than broad. Pronotum longer than wide, lateral margins straight, more or less parallel; median longitudinal furrow distinct. Tegmina and wings absent. Mesonotum typical, transverse, posterior margin concave. Metanotum normal, simple. Abdomen fusiform, depressed; ultimate tergite very narrow, posterior margin with two tubercles. Forceps (Fig. 574) simple, somewhat trigonal basally, curved and cylindrical apically. Penultimate sternite simple, rounded posteriorly. – **Male** unknown. – **Length** of body with forceps: 10,5 mm.

Distribution: Fiji.

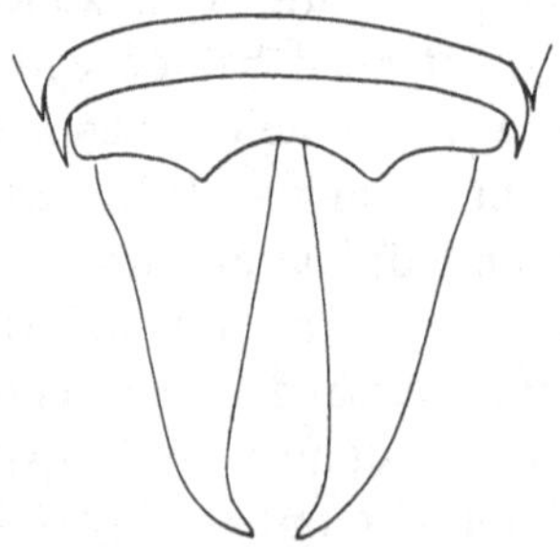

Fig. 574 Holotype ultimate tergite and forceps of *Antisolabis fijica* sp.n. Original.

18. Antisolabis cordata (BRINDLE) comb. nov.

1976 *Brachylabis cordata* BRINDLE, Rec. S.austr. Mus., Adelaide, (13) **17**: 226; fig. 5 (female). – Terra typica: New Hebrides (Type female: S. austr. Mus., Adelaide).

Description: Female general colour black, lateral margins of pronotum yellow, posterior part of abdomen with a reddish tint; antennae blackish, joint 10 paler; legs yellowish-brown, femora vaguely darkened. Cuticle of body rugose, rather shiny, with fine yellowish hairs. Head transverse, a little broader than pronotum; postfrontal and coronal sutures indistinct; eyes small, slightly shorter than length of head behind eyes. Antennae 10-jointed in type; first joint long, more or less as long as distance between antennal bases; second transverse, third joint four times as long as wide; fourth joint one and half times as long as wide. Pronotum more strongly transverse, slightly widened posteriorly. Lateral margins more or less straight, and all angles rounded; median longitudinal furrow occurs medially on anterior half. Mesonotum typical, broad, with a blunt lateral fold at base but without ridges. Metanotum normal. Abdomen fusiform, a little depressed. Ultimate tergite broad, posterior margin concave medially. Forceps (Fig. 575) simple, wider basally, cylindrical and curved apically. Penultimate sternite broad, rounded posteriorly. – **Male** unknown. – **Length** of body with forceps: 8 mm.

Distribution: New Hebrides.

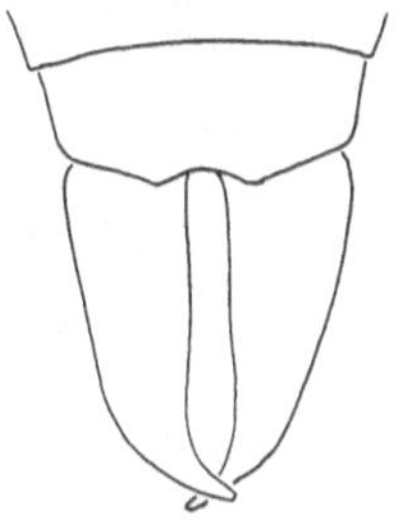

Fig. 575 Female ultimate tergite and forceps of *Antisolabis cordata* (BRINDLE). After BRINDLE.

19. Antisolabis greensladei (BRINDLE) comb. nov.

1970 *Brachylabis greensladei* BRINDLE, Pacific Insects, Honolulu, (3) **12**: 659; fig. 17 (holotype genital armature), 18 (male). – Terra typica: Solomon Islands (Type male: Brit. Mus. nat. Hist., London).

1972 *Brachylabis greensladei* – BRINDLE, Insects of Micronesa, Honolulu, (2) **5**: 132; fig. 8 (female). Palau.

Literature: BRINDLE 1976; STEINMANN 1977, 1978.

Description: Male colour blackish-brown; antennae dark brown, 2 distal joints whitish; femora dark brown basally, and brown distally. Cuticle of body strongly punctate, almost rugose and pubescent. Head transverse, tumid; posterior angles rounded; postfrontal and coronal sutures indistinct. Eyes small, slightly shorter than length of head behind eyes. Antennae broken, first joint long, longer than distance between antennal bases; second transverse, third joint three times as long as wide, fourth joint slightly longer than wide. Pronotum less strongly transverse, widened posteriorly; posterior margin weakly convex; median longitudinal furrow present. Meso- and metanotum transverse, typical. Abdomen fusiform, a little depressed. Ultimate tergite broad, simple. Forceps (Fig. 576) broadened basally, only weakly curved; cylindrical. Penultimate sternite transverse, posterior margin weakly convex. Genitalia (Fig. 577) characteristic; central parameral plate broad, median incision of anterior margin very deep, genital lobes fully developed with very long virgae. External parameres acuminate. – **Female** similar to male, but forceps contiguous, tapering. – **Length** of body with forceps: male: 5,5–7 mm, female: 6–6,5 mm.

Distribution: Solomon Islands and Palau.

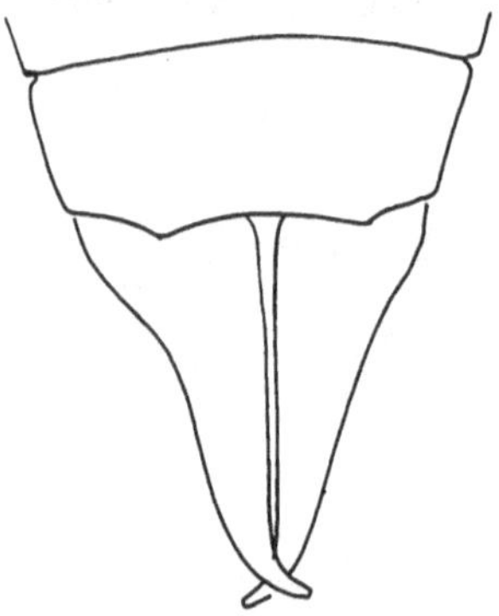

Fig. 576 Male ultimate tergite and forceps of *Antisolabis greensladei* (BRINDLE). After BRINDLE.

Fig. 577 Holotype genital armature of *Antisolabis greensladei* (BRINDLE). After BRINDLE.

20. Antisolabis rouxi BURR

1914 *Antisolabis rouxi* BURR, Nova Caledonia, Wellington, **1**: 319. – Terra typica: New Caledonia (Type male: Basel Mus.).

1915 *Antisolabis rouxi* – BURR, J. r. microsc. Soc., London, **1915**: 145; pl. IX, fig. 16 (male genitalia).

1976 *Brachylabis rouxi* – BRINDLE, Pacific Insects, Honolulu, (1) **17**: 72; fig. 8 (head and pronotum), 14 (male genitalia), 16 (penultimate sternite of male).

Literature: SAKAI 1970; STEINMANN 1977, 1978.

Description: Male colour blackish or dark reddish brown in some species; antennae dark brown, legs dark brown, with tarsi paler. Cuticle dull, roughened, that of thoracic tergites largely granular, anteriorly wider spaced; hairs arising from the granules. Abdomen strongly pubescent, the hairs long, yellowish brown. Head rounded; postfrontal sutures and coronal suture deep; eyes relatively large, length about as long as coronal suture in vertical view. First antennal joint long, more or less as long as distance between antennal bases. Pronotum longer than wide, widened posteriorly; all angles rounded, median longitudinal furrow short. Mesonotum large, wider anteriorly and swollen laterally, toward the anterior margin is a transverse depression bordered anteriorly by blunt flaps. Abdomen fusiform, a little expanded to middle. Ultimate tergite typical. Forceps cylindrical, well separated at base and broader, narrower distally; on inner margin is a small tooth of variable size. Penultimate sternite triangular posteriorly, actual apex slightly emarginate and with a circular depression on cuticle near apex. Genitalia (Fig. 578) of *Antisolabis*-type, paramere broad, genital lobes well developed; virga associated with a large sclerite; external parameres are more strongly sclerotized on outer half, with acuminate apices. – **Female** similar to male, but forceps are less curved, and relatively well-separated at base. – **Length** of body with forceps: male: 16–18,5 mm, female: 17–19 mm.

Distribution: New Caledonia.

Fig. 578 Male genital armature of *Antisolabis rouxi* BURR. After BRINDLE.

21. Antisolabis rouxiodes (BRINDLE) comb. nov.

1976 *Brachylabis rouxiodes* BRINDLE, Pacific Insects, Honolulu, (1) **17**: 74; fig. 9 (head and pronotum), 15 (holotype genital armature), 17 (penultimate sternite of male). – Terra typica: New Caledonia (Type male: Bishop Mus., Honolulu, Hawaii).

Description: Male dark reddish-brown, almost blackish brown; antennae dark brown, legs reddish yellow or yellowish brown. Cuticle dull, granular on thoracic nota, abdominal tergites less roughened than in *rouxi* BURR. All of the cuticle pubescent, hairs rather dense but generally short. Head rounded, but posterior margin concave; postfrontal and coronal sutures very distinct; eyes relatively short, shorter than length of coronal suture. First antennal joint normal, about as long as length of distance between antennal bases. Pronotum about as long as wide; lateral margins straight but widened posteriorly; median longitudinal furrow short. Mesonotum broad, wider anteriorly and swollen laterally. Abdomen normal, simple, a little expanded to middle. Ultimate tergite typical, transverse, posterior margin concave medially. Forceps cylindrical, well separated at base, tapering. Penultimate sternite without a circular depression near posterior margin medially. Genitalia (Fig. 579) characteristic; paramere more membranous, with slight median incision; genital lobe well developed, virga within genital lobe almost simple, without a large associated sclerite. External parameres specific, with numerous hairs. – **Female** similar to male, but forceps less strongly curved. – **Length** of body with forceps: male: 13,5–15 mm, female: 14–16,5 mm.

Distribution: New Caledonia.

Fig. 579 Holotype genital armature of *Antisolabis rouxiodes* (BRINDLE). After BRINDLE.

22. Antisolabis arripiens BURR

1914 *Antisolabis arripiens* BURR, Nova Caledonia, Wellington, **1**: 320. – Terra typica: New Caledonia (Type male: Basel Mus.).
1966 *Nannisolabis arripiens* – POPHAM & BRINDLE, Entomologist, London, **99**: 244.
1976 *Brachylabis arripiens* – BRINDLE, Pacific Insects, Honolulu, (1) **17**: 75; fig. 18 (male forceps).
Literature: SAKAI 1970; STEINMANN 1977, 1978.

Description: Male colour black; antennae dark brown, but joints 10–11 or 11–12 yellowish white; legs yellowish-brown. Cuticle of head and thoracic tergites closely granular, except anterior part of metanotum: of abdominal tergites sparsely punctate and pubescent; hairs and setae very sparse. Head broad, lateral margins rounded, posterior angles present. Eyes small, slightly shorter than length of head behind eyes. First antennal joint long, a little longer than distance between antennal bases. Pronotum quadrate, lateral margins more or less parallel; all angles rounded; median longitudinal furrow distinct. Mesonotum broad, swollen laterally, and without a distinct fold, but with a strong transverse median depression which has a transverse ridge bordering anterior margin of depression. Abdomen with lateral tubercles on third tergite, which is small, those on fourth larger. Ultimate tergite broad, typical. Forceps widened basally, narrowed distally, evenly and strongly curved (Fig. 580) not of

Antisolabis-type. Penultimate sternite short, transverse, posterior margin weakly rounded, and with a small median excision. Genitalia (unknown to me) of *Antisolabis*-type, paramere broad, median incision of anterior margin deep; genital lobes well developed, with virgae. Inner and outer margins of external parameres parallel. – **Female** unknown. – **Length** of body with forceps: 10–10,5 mm.

Distribution: New Caledonia.

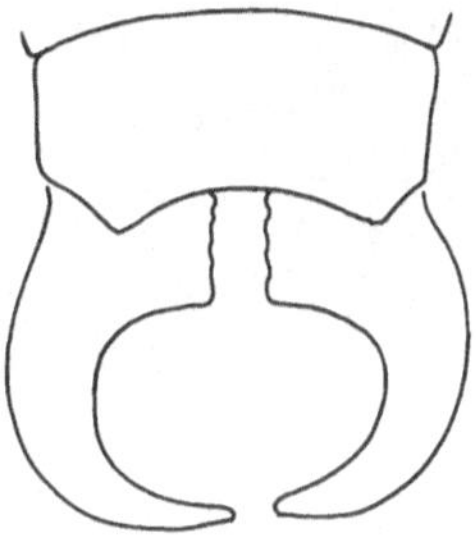

Fig. 580 Male ultimate tergite and forceps of *Antisolabis arripiens* BURR. After BRINDLE.

23. Antisolabis forficula (BURR) comb. nov.

1914 *Nannisolabis forficula* BURR, Nova Caledonia, Wellington, **1**: 322. – Terra typica: New Caledonia (Type male: Basel Mus.).

1976 *Brachylabis forficula* – BRINDLE, Pacific Insects, Honolulu, (1) **17**: 77; fig. 4 (female), 5 (male forceps), 13 (male genitalia).

Literature: SAKAI 1970; STEINMANN 1977, 1978.

Description: Male colour dark reddish-brown to blackish; antennae dark brown or blackish, joints 10–11 yellowish white; femora and tibiae largely darkened medially. Cuticle of head almost smooth anteriorly, granulose posteriorly, thoracic nota granulose, granules closest on posterior part of metanotum where they tend to form transverse irregular ridges. Abdominal tergites sparsely pubescent, hairs yellow, but forceps rather densely pubescent and hairs longer. Head tumid, postfrontal sutures and coronal suture distinct. Eyes normal, small. Antennae 12-jointed, first joint long, longer than distance between antennal bases. Pronotum longer than wide, widened posteriorly; median longitudinal furrow short. Mesonotum with a shallow transverse median depression, the depression curving anteriorly towards lateral margin and forming the inner border of an oblique lateral fold. Abdomen fusiform, with lateral tubercles on third tergite, small, those on fourth larger. Ultimate tergite broad, posterior margin straight between bases of forceps. Forceps (Fig. 581) strongly curved, microlabic form similar to *Forficula*-type, with large tooth on inner margins basally. Penultimate sternite short, transverse, posterior margin weakly convex and with a small median excision. Genitalia (Fig. 582) with large paramere, median incision of anterior margin deep; genital lobes well developed, virga present, long, more strongly sclerotized basally and associated with indistinct denticulations. External parameres with parallel margins and acuminate apices. – **Female** similar to male, but forceps straighter and with an inner dentated flange, distal part curved and directed ventro-medially as in male. – **Length** of body with forceps: male: 7,5–8,5 mm, female: 8–9,5 mm.

Distribution: New Caledonia.

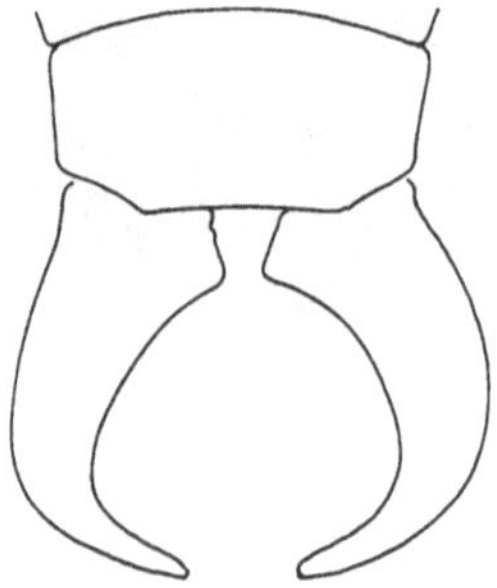

Fig. 581 Male ultimate tergite and forceps of *Antisolabis forficula* (BURR). After BRINDLE.

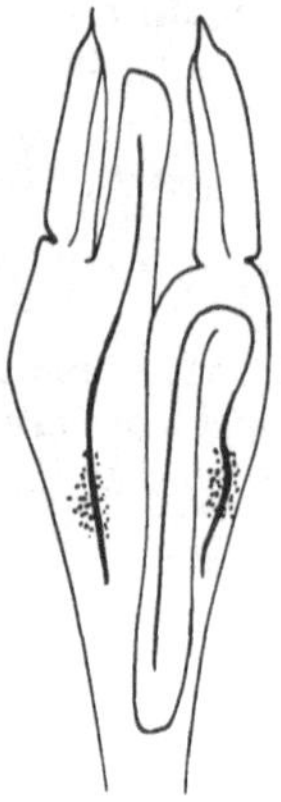

Fig. 582 Male genital armature of *Antisolabis forficula* (BURR). After BRINDLE.

24. Antisolabis geniculata (MONTROUZIER) comb. nov.

1864 *Chelidurella geniculata* MONTROUZIER, Ann. Soc. linn. Lyon, (n. S.), **11**: 222. – Terra typica: New Caledonia (Type male: Basel Mus.).
1900 *Brachylabis geniculata* – BORMANS, Das Tierreich, Berlin, **11**: 54.
1914 *Nannisolabis geniculata* – BURR, Nova Caledonia, Wellington, **1**: 321; fig. 4 (male).
1933 *Labidura geniculata* – SJÖSTEDT, Ark. Zool., Uppsala, **25 A**: 2.
1966 *Nannisolabis geniculata* – POPHAM & BRINDLE, Entomologist, London, **99**: 244.
1976 *Brachylabis geniculata* – BRINDLE, Pacific Insects, Honolulu, (1) **17**: 75; fig. 6 (head and pronotum of male).

Literature: BURR 1908, 1911; KALTENBACH 1968; KIRBY 1904; SAKAI 1970; STEINMANN 1977, 1978.

Description: Male colour reddish-brown; antennae brown to dark brown or paler basally; legs yellow, femora with dark bands of varying intensity. Cuticle shiny, very sparsely pubescent, and punctate. Head transverse, not smoothly rounded posteriorly. Eyes normal, small, shorter than length of head behind eyes. First antennal joint typical, long. Postfrontal sutures and coronal suture distinct. Pronotum strongly transverse, widened posteriorly; median longitudinal furrow short, all angles rounded. Mesonotum transverse, without lateral ridges. Metanotum broad. Abdomen fusiform, expanded to middle; ultimate tergite broad, posterior margin concave medially. Forceps (Fig. 583) cylindrical, wider at base, narrowed distally and weakly curved; branches well separated at base. Posterior margin of penultimate sternite rounded. Genitalia unknown. –

Female similar to male, but forceps closer together, scarcely curved, somewhat sinuate, rather abruptly narrowed distally. – **Length** of body with forceps: male: 8,5–10,5 mm, female: 9–11 mm.

Distribution: New Caledonia.

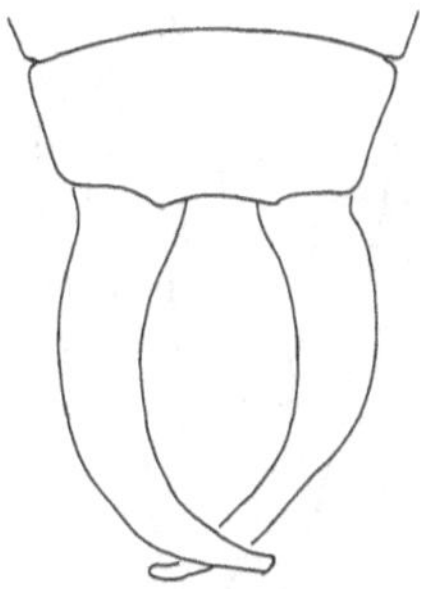

Fig. 583 Male ultimate tergite and forceps of *Antisolabis geniculata* (MONTROUZIER). After BURR.

25. Antisolabis transiens BURR

1914 *Antisolabis transiens* BURR, Nova Caledonia, Wellington, **1**: 320. – Terra typica: New Caledonia (Type male: Basel Mus.).
1915 *Antisolabis transiens* – BURR, J. r. microsc. Soc., London, **1915**: 145; pl. IX, fig. 17 (male genitalia).
1966 *Nannisolabis transiens* – POPHAM & BRINDLE, Entomologist, London, **99**: 244.
1976 *Brachylabis transiens* – BRINDLE, Pacific Insects, Honolulu, (1) **17**: 74; fig. 7 (head and pronotum), 19 (penultimate sternite of male).
Literature: SAKAI 1970; STEINMANN 1977, 1978.

Description: Male colour dark reddish-brown to almost blackish; antennae brown, legs yellow, femora largely darkened medially, and tibiae darkened basally. Cuticle shiny, very sparsely pubescent, smooth with large punctures; cuticle with very sparse but relatively short yellow hairs, more conspicuous laterally on abdomen. Head rounded; postfrontal and coronal sutures distinct. Eyes normal, moderately large, but shorter than length of head behind eyes. First antennal joint long, visibly longer than distance between antennal bases. Pronotum longer than wide; lateral margins straight and a little expanded posteriorly; median longitudinal furrow distinct in prozona; all angles rounded. Mesonotum transverse, with an oblique lateral fold on each side and a deep transverse median depression. Abdomen slightly depressed, ultimate tergite comparatively small, typical. Forceps cylindrical, widest basally, narrowed distally; tapering. Penultimate sternite somewhat triangular, with apex broad and rounded and with a small median excision. Genitalia (Fig. 584) with acuminate external parameres; median incision of anterior margin very deep; genital lobes well developed, virgae long and simple. – **Female** similar to male, but forceps much less curved, closer together. – **Length** of body with forceps: male: 10–11 mm, female: 9,5–11,5 mm.

Distribution: New Caledonia.

Fig. 584 Male genital armature of *Antisolabis transiens* BURR. After BURR.

26. Antisolabis allardi (BRINDLE) comb. nov.

1971 *Brachylabis allardi* BRINDLE, Stud. Fauna Curaçao and other Caribbean Islands, den Haag, **38**: 31; fig. 18 (male), 17 (male genitalia), 19 (penultimate sternite). – Terra typica: Hispaniola: Dominica (Type male: U.S. nation. Mus., Washington).

Description: Male colour dark reddish-brown; legs yellowish-brown, antennae and mouthparts brown. Cuticle of body more finely punctate and pubescent. Head tumid, as long as wide; lateral margins rounded, posterior margin truncate except medially, where it is slightly concave. Vertex strongly punctate. Eyes relatively small, visibly shorter than length of head behind eyes. First antennal joint long, longer than distance between antennal bases, second transverse or quadrate, third joint as long as wide, fourth joint nearly one and half times as long as wide. Pronotum longer than wide, widened posteriorly; lateral margins slightly sinuate medially; posterior margin rounded or a little straight. Mesonotum transverse, strongly and closely punctate, with a lateral fold on both sides for most of length the fold curving medially, but without lateral longitudinal ridge. Abdomen widened medially, rather depressed, strongly narrowed basally and distally; tubercles of fourth tergite small. Ultimate tergite transverse, posterior margin with concave section medially. Forceps (Fig. 585) cylindrical, well separated at base, broadened basally, strongly curved apically. Penultimate sternite transverse, posterior margin convex and with a median excision. Genitalia (Fig. 586) characteristic; somewhat of *Labidura*-type; paramere broad, median incision of anterior margin very deep and broad; genital lobes well developed, virga partially sclerotized. – **Female** unknown. – **Length** of body with forceps: 10 mm.

Distribution: Hispaniola: Dominican Republic.

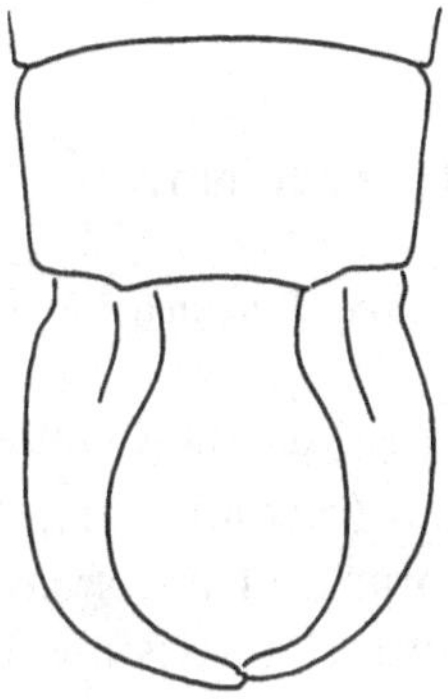

Fig. 585 Male ultimate tergite and forceps of *Antisolabis allardi* (BRINDLE). After BRINDLE.

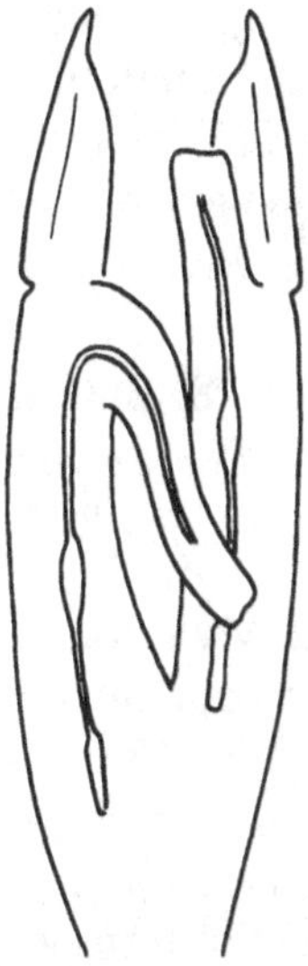

Fig. 586 Holotype genital armature of *Antisolabis allardi* (BRINDLE). After BRINDLE.

27. Antisolabis javana (BOESEMAN) comb. nov.

1954 *Nannisolabis javana* BOESEMAN, Zool. Verh., Leiden, **21**: 54; fig. 14a (female), 14b (penultimate sternite and forceps of female), 14c (antenna), 14d (thoracal sternites). – Terra typica: Java (Type female: Rijksmus. natuurl. Hist., Leiden).

Literature: SAKAI 1970; STEINMANN 1977, 1978, 1980.

Description: Female colour dark brown, almost black, covered with sparse hairs, antennae brown, joints 10–11 yellowish, femora brown, a small apical part yellowish, tibiae rather dark. Head relatively broad, tumid; postfrontal and coronal sutures distinct. Eyes small, slightly shorter than length of head behind eyes. Antennae 14-jointed; first joint more or less as long as distance between antennal bases. Pronotum quadrate, lateral margins parallel, a little concave; all angles rounded; median longitudinal furrow present. Mesonotum transverse, without lateral ridge. Abdomen slightly depressed, lateral tubercles present on

Fig. 587 Female ultimate tergite and forceps of *Antisolabis javana* (BOESEMAN). After BOESEMAN.

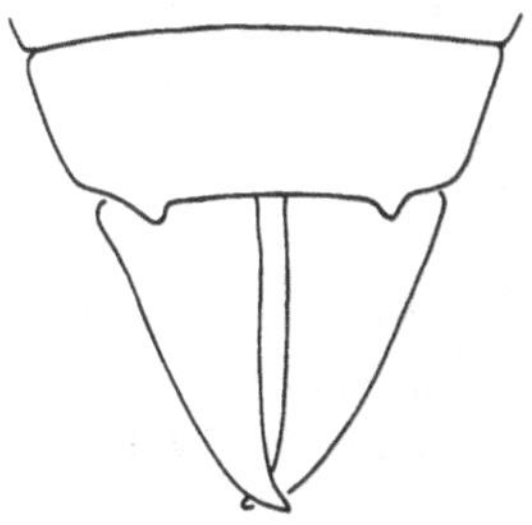

third tergite only. Ultimate tergite very broad, posterior margin with a slightly median part and small tubercles above the base of both branches of the forceps. Forceps (Fig. 587) relatively short, basally broad, with curved apices. – **Male** unknown. – **Length** of body with forceps: 10,5–12,5 mm.

Distribution: Java.

28. Antisolabis ceylonica (Brindle) comb. nov.

1977 *Nannisolabis ceylonica* Brindle, Rev. suisse Zool., Genève, (2) **84**: 456; fig. 1 (male), 2 (holotype genital armature), 3 (male antenna). – Terra typica: Ceylon (Type male: Mus. Hist. nat., Genève).

Description: Male colour blackish; antennae dark brown, legs yellow, femora dark brown except for apex (first pair), apical fifth (second pair), and apical fourth (third pair). Cuticle of head granulose, of thorax and abdomen with deep punctures. Head tumid, postfrontal and coronal sutures not visible. Eyes fairly small. Antennae 10-jointed (?); first joint as long as distance between antennal bases, second joint transverse, third about twice as long as wide, and fourth joint as long as wide. Pronotum quadrate, a little transverse, sides straight and parallel, posterior margin almost straight. Mesonotum broad, sides angled and anterior half strongly depressed. Metanotum transverse, swollen laterally. Legs relatively long. Abdomen fusiform, lateral tubercles on third and fourth tergites small. Ultimate tergite typical, posterior margin rounded, excised medially.

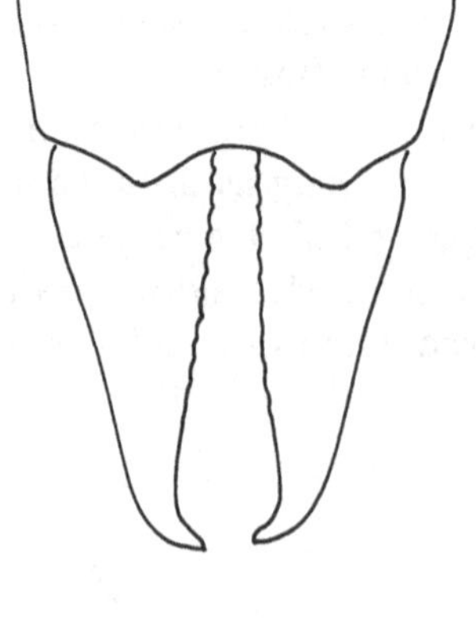

Fig. 588 Male ultimate tergite and forceps of *Antisolabis ceylonica* (Brindle). After Brindle.

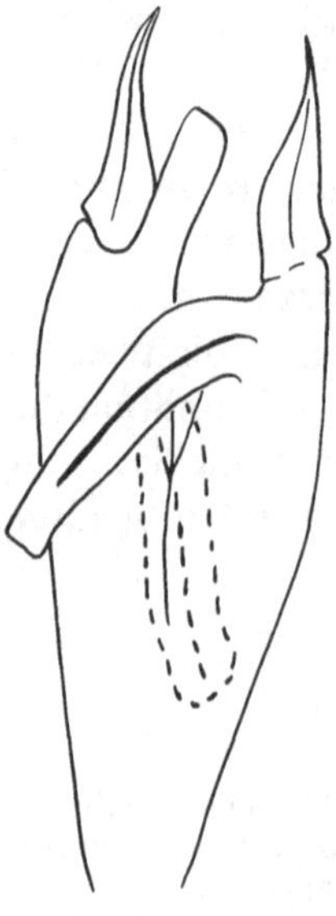

Fig. 589 Holotype genital armature of *Antisolabis ceylonica* (Brindle). After Brindle.

Forceps (Fig. 588) short, typical, evenly tapered from base, apex incurved, inner margins with small crenulations. Penultimate sternite rounded posteriorly. Genitalia (Fig. 589) broad, paramere with median incision, genital lobes well developed, with virgae slender. External parameres with pointed apices. – **Female** similar to male, but forceps slightly more slender. – **Length** of body with forceps: male: 5,5–6 mm, female: 5–6 mm.

Distribution: Sri Lanka.

29. Antisolabis formicoides (Burr)

1911 *Nannisolabis formicoides* Burr, J. asiat. Soc. Bengal, Calcutta, **7**: 781. – Terra typica: South India (Type male: Brit. Mus. nat. Hist., London).
1966 *Antisolabis formicoides* – Popham & Brindle, Entomologist, London, **99**: 244.
Literature: Burr 1914; Hebard 1923; Sakai 1970; Steinmann 1977, 1978, 1983.

Description: Male colour blackish, shiny; first antennal joint black, second yellowish, rest of joints dark brown, except joints 10–11 yellowish white; legs yellow, but femora dark brown basally. Cuticle of body strongly punctate. Head a little longer than wide; postfrontal and coronal sutures present; eyes typical, small, slightly shorter than length of head behind eyes. First antennal joint long, longer than distance between antennal bases, second and third quadrate. Pronotum longer than wide, lateral margins more or less parallel; all angles rounded; median longitudinal furrow distinct. Mesonotum typical, with transversal depression. Abdomen fusiform, abdominal tergites 2–3 with smooth transversal section anteriorly. Ultimate tergite broad. Forceps (Fig. 590) simple, cylindrical. Penultimate sternite with posterior margin rounded. Genitalia unknown. – **Female** similar to male, but forceps slenderer, tapering. – **Length** of body with forceps: in both sexes: 6,5–8,5 mm.

Distribution: South India.

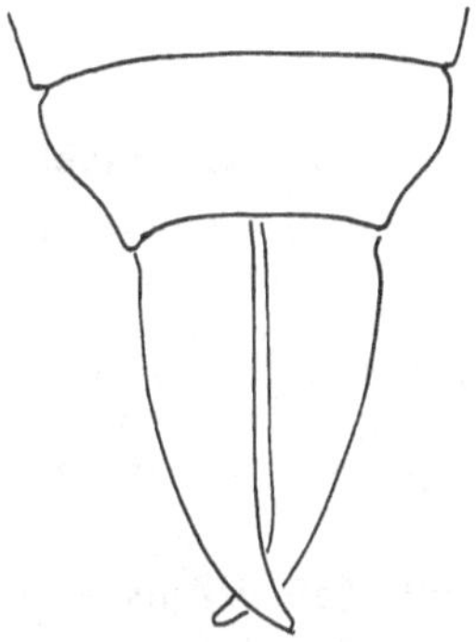

Fig. 590 Male ultimate tergite and forceps of *Antisolabis formicoides* (Burr). Original.

30. Antisolabis dammermani (Borelli) comb. nov.

1926 *Nannisolabis dammermani* Borelli, Treubia, Buitenzorg, **8**: 257. – Terra typica: Java (Type male: Mus. Ist. Zool. sist., Torino; holotype without penis).
Literature: Popham & Brindle 1966; Sakai 1970; Steinmann 1977, 1978.

Description: Male colour dark brown to blackish brown; shiny; antennae dark brown, femora yellow, banded with dark brown basally. Cuticle of body punctate and pubescent. Head broad, tumid; postfrontal and coronal sutures distinct. Eyes comparatively small, slightly shorter than length of head behind eyes. Antennae 11-jointed (?); first joint very long, longer than distance between antennal bases; second transverse, third cylindrical, a little longer than wide, and fourth oval, about as long as wide. Pronotum longer than wide, and a little longer than head, all angles rounded; lateral margins widened posteriorly; median longitudinal furrow present. Mesonotum normal, transverse. Abdomen fusiform, a little expanded medially. Ultimate tergite broad, posterior margin typical, concave medially. Forceps (Fig. 591) simple, short. Penultimate tergite rounded posteriorly. Genitalia unknown. – **Female** similar to male, but forceps slightly narrower. – **Length** of body with forceps: in both sexes: 9,5–10 mm.
Distribution: Java.

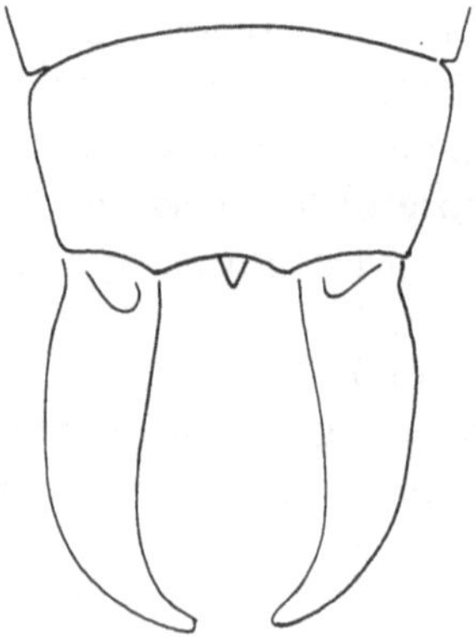

Fig. 591 Holotype ultimate tergite and forceps of *Antisolabis dammermani* (BORELLI). Original.

31. Antisolabis kelangi (BRINDLE) comb. nov.

1969 *Brachylabis kelangi* BRINDLE, Entomologist, London, **102**: 199; fig. 1 (male), 2 (holotype genital armature), 3 (external paramere of male genitalia). – Terra typica: Malaysia (Type male: Brit. Mus., nat. Hist., London).

1971 *Geracodes brincki* BRINDLE, Ent. scand., København, Suppl., **1**: 221; fig. 21 (male), 22 (holotype genital armature), 23 (male penultimate sternite). – Terra typica: Ceylon (Type male: Ent. Mus. Univ., Lund, Sweden). – **syn. nov.**

Literature: STEINMANN 1977, 1978.

Description: Male yellowish to yellowish-brown; mouthparts and legs yellow, forceps brown. Cuticle rather dull, puntured and pubescent, hairs yellow. Head oval, visibly longer than wide; lateral margins rounded and curving smoothly into posterior margin, the latter concave. Eyes very small, shorter than length of head behind eyes. Antennae relatively short; first joint long, more or less equal to the distance between antennal bases; second quadrate; all joints relatively broad, more or less cylindrical, but narrowed towards base. Pronotum longer than wide, widened posteriorly, with a median longitudinal furrow; all angles rounded; lateral margins a little sinuate. Mesonotum transverse, without lateral ridges. Legs relatively short, femora broadened; tibiae rather broad, and flattened laterally. Abdomen a little depressed, widened medially, without lateral tubercles or these are barely indicated; ultimate tergite transverse, curved apically; posterior margin produced. Forceps broad basally, cylindrical apically; inner margins irregular, but scarcely dentated. Penultimate sternite rounded posteriorly. Genitalia (Fig. 592 as *kelangi*, and fig. 583 as *brincki* by BRINDLE) very specific; paramere oval, median incision of anterior margin present; genital lobes well-developed, virga distinct, and lobes granulated blackish. – **Female** similar to male, but forceps slenderer, tapering. – **Length** of body with forceps: in both sexes: 5–5,5 mm.

Distribution: From Sri Lanka to West Malaysia.

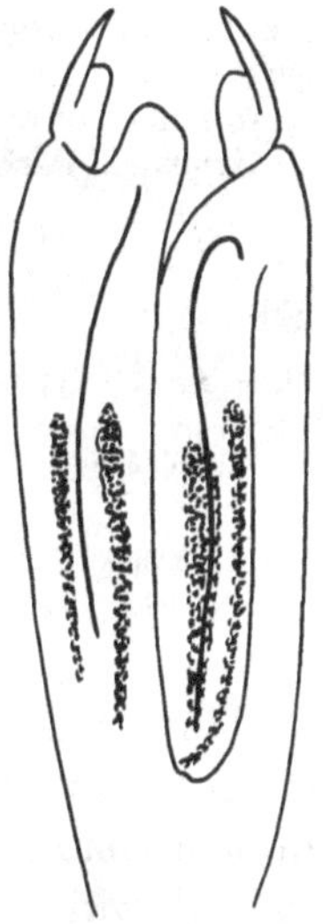

Fig. 592 Holotype genital armature of *Antisolabis kelangi* (BRINDLE). After BRINDLE.

Fig. 593 Holotype genital armature of *Geracodes brincki* BRINDLE. After BRINDLE.

Subfamily 8: **Parisolabiinae** VERHOEFF

1904 *Parisolabinae* VERHOEFF, Arch. Naturgesch., Berlin, **1904**: 120. – Genus typicum: *Parisolabis* VERHOEFF 1904.

1911 *Parisolabidae* ZACHER, Zool. Jahrb., Jena, **30**: 330.

1954 *Parisolabidae* – BRUES, MELANDER & CARPENTER, Classification of Insects, Cambridge, Mass.: 116.

1966 *Parisolabinae* – POPHAM & BRINDLE, Entomologist, London, **99**: 133.

1975 *Parisolabiinae* – STEINMANN, Acta zool. Acad. Sci. hung., Budapest, **21**: 206.

Literature: BURR 1910, 1915; HINCKS 1955, 1958; HUDSON 1973; SAKAI 1970; STEINMANN 1977, 1978.

Description: Apterous insects, tegmina and wings entirely absent. Head normal, eyes moderately large, but shorter than length of head behind eyes. Antennae multisegmented, scape short, somewhat shorter than distance between antennal bases. Pronotum and mesonotum with or without median longitudinal furrow. Legs slender, tarsi long and thin. Prosternum subparallel-sided, mesosternum and metasternum almost rectangular, truncate posteriorly. Abdomen fusiform, forceps of *Antisolabis*-type, branches well-separated at base. Male genitalia with strongly sclerotized virgae within genital lobes; median longitudinal incision of anterior margin of the central parameral plate present.

Distribution: Oriental and Indo-Australian faunal regions.

2 genera.

Identification key to the genera

1 Ultimate tergite in both sexes obtuse between posterior and lateral margins. Males with one genital lobe directed forwards. External parameres of male genitalia comparatively small, short; median incision of anterior margin hardly visible because of large lobus genitalis... .... Genus 31: **Parisolabis**, p. 372

1′ Ultimate tergite in both sexes quadrangular between posterior and lateral margins. One or both genital lobes directed backwards in male. External parameres of male genitalia larger; median incision of anterior margin clearly perceptible, reaching to middle of plate .... Genus 32: **Parisopsalis**, p. 384

Genus 31: **Parisolabis** VERHOEFF

1904 *Parisolabis* VERHOEFF, Arch. Naturgesch., Berlin, **1904**: 120. – Species typica: *Parisolabis novaezeelandiae* VERHOEFF 1904.

1908 *Pseudisolabis* BURR, Ann. Mag. nat. Hist., London (8) **2**: 254. – Species typica: *Pseudisolabis walkeri* BURR 1908.

Literature: BEY-BIENKO 1935; BORELLI 1909; BURR 1910, 1911, 1915; HEBARD 1923; HINCKS 1947; HUDSON 1973; POPHAM & BRINDLE 1966; SAKAI 1970; STEINMANN 1975, 1977, 1978; TOWNES 1945; ZACHER 1911.

Description: Colour dark brown or blackish, body shiny or dull. Apterous species. Cuticle with punctation. Head as broad as pronotum. Mesonotum without tegmina, and metanotum without wings. Abdomen dilated, fusiform, a little expanded medially. Ultimate tergite slightly narrowed, with shallow teeth above bases of forceps; posterior margin truncate. Forceps symmetrical in both sexes, moderately to strongly curved. Male genitalia with equally developed genital lobes, one directed forwards the other backwards; virga within genital lobe with strongly sclerotized basal vesicle.

Distribution: Oriental and Indo-Australian species.

12 species.

Identification key to the species

1 Oriental species **2**

1′ Indo-Australian species **5**

2 Pronotum quadrate, about as long as wide **3**

2′ Pronotum transverse, slightly wider than long **4**

3 Lateral margins of pronotum convex. Male forceps strongly curved, cylindrical, as in fig. 594. Posterior margin of male ultimate tergite straight .... 1. **Parisolabis renschi**, p. 373

3′ Lateral margins of pronotum more or less parallel. Male forceps less curved, subcylindrical, as in fig. 595. Posterior margin of ultimate sternite a little concave medially .... 2. **Parisolabis immsi**, p. 374

4 Antennae 14-jointed. Male forceps strongly curved, trigonal basally and cylindrical apically (Fig. 596) .... 3. **Parisolabis tenera**, p. 375

4′ Antennae 17-jointed. Male forceps less curved, but crossed apically; cylindrical, as in fig. 597 .... 4. **Parisolabis elegans**, p. 376

5 With semi-circular smudges of very fine setae on posterior margin of at least one abdominal segment in both sexes. Male forceps simple, subcylindrical, as in fig. 598. Male genitalia (Fig. 599) with very small external parameres .... 5. **Parisolabis setosa**, p. 376

5′ Lacking semi-circular smudges, but may have faint narrow borders of setae on posterior margins of some abdominal tergites. **6**

6 Male forceps 1,2 mm or less in length; female forceps 1,3 mm or less in length **7**

6′ Male forceps 1,3 mm or more in length; female forceps 1,4 mm or more in length **10**

7 Width of pronotum equal to or less than width of head in both sexes. Pygidium not narrowed ventrally in males. Posterior margin of ultimate tergite excised laterally (Fig. 600). Male genitalia characteristic, as in fig. 601 .... 6. **Parisolabis johnsi**, p. 377

7′ Width of pronotum greater than head at broadest point in both sexes. Pygidium narrowed ventrally in males **8**

8 Shiny species, sparsely covered with minute depressed bristles, body almost cylindrical, uniformly reddish-brown. Male forceps simple, as in fig. 602. Male genitalia as in fig. 603 . ... 7. **Parisolabis iti**, p. 378

8′ Species covered with fine hairs, body depressed, brown with muted darker pattern on abdomen and thorax in both sexes **9**

9 Male forceps strongly trigonal basally, curved and cylindrical apically (Fig. 604). Male genitalia similar to *forsteri* HUDSON, but external parameres comparatively shorter and pointed (Fig. 605) .... 8. **Parisolabis tapanuiensis**, p. 379

9′ Male forceps simple, slenderer, cylindrical or only a little trigonal basally (Fig. 606). Male genitalia with moderately longer external parameres (Fig. 607) . 9. **Parisolabis forsteri**, p. 380

10 Male forceps cylindrical, shaped as in fig. 608, usually under 2,0 mm in length; female forceps 2,1 or less in length .... 10. **Parisolabis novaezeelandiae**, p. 381

10′ Male forceps strongly trigonal basally, cylindrical or subcylindrical apically, shaped as in fig. 610 or 612, usually over 2,0 mm in length; female forceps 2,2 or more in length **11**

11 Posterior margin of male ultimate tergite straight, forceps as in fig. 610. Male genitalia very similar to *boulderensis* HUDSON, as in fig. 611 .... 11. **Parisolabis nelsonensis**, p. 382

11′ Posterior margin of male ultimate tergite concave medially, forceps as in fig. 612. Male genitalia very similar to *nelsonensis* HUDSON, as in fig. 613 .... 12. **Parisolabis boulderensis**, p. 383

1. Parisolabis renschi (GÜNTHER)

1932 *Pseudisolabis renschi* GÜNTHER, Sber. Ges. naturf. Fr., Berlin, **1932**: 472; fig. 1 (male). – Terra typica: Java (Type male: Mus. Naturkunde, Berlin).

1958 *Parisolabis renschi* – HINCKS, EOS, Rev. esp. Ent., Madrid, **34**: 132.

Literature: POPHAM & BRINDLE 1966; SAKAI 1970; STEINMANN 1977, 1978.

Description: Male colour black, shiny; labrum brown, antennae brown. Cuticle of body punctate and pubescent. Head rounded, somewhat longer than wide; postfrontal sutures and coronal suture indistinct. Eyes normal, a little shorter than length of head behind eyes. Antennae 13-jointed; first joint short, shorter than distance between antennal bases; second transverse, third joint long, about two thirds as long as scape. Pronotum quadrate, lateral margins a little convex; median longitudinal furrow absent. Mesonotum typical, transverse, wider than long, posterior margins concave. Abdomen gradually widened towards the posterior side. Ultimate tergite very broad; posterior margin straight, with a tubercle above each branch of the forceps. Forceps (Fig. 594) rounded, strongly arcuate, so as to enclose a crescent-shaped area; cylindrical, strongly separated at base. Genitalia unknown. – **Female** similar to male, but forceps less strongly curved, contiguous. – **Length** of body with forceps: male: 15–15,5 mm, female: 14,5–16 mm.

Distribution: Java.

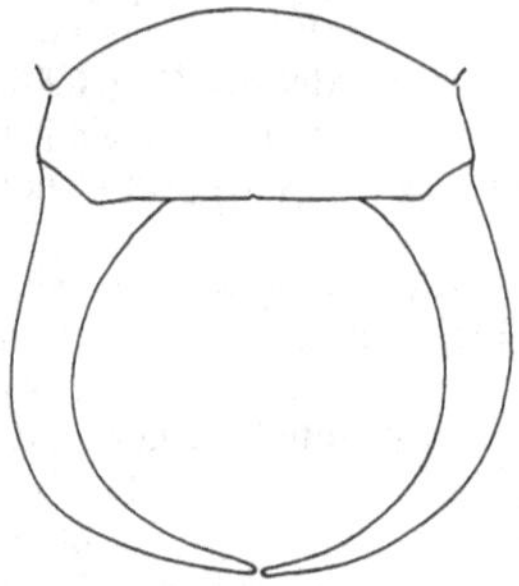

Fig. 594 Holotype ultimate tergite and forceps of *Parisolabis renschi* (GÜNTHER). After GÜNTHER.

2. Parisolabis immsi (BURR)

1913 *Pseudisolabis immsi* BURR, J. asiat. Soc. Bengal, Calcutta, **9**: 184. – Terra typica: India (Type male: Brit. Mus. nat. Hist., London).

1958 *Parisolabis immsi* – HINCKS, EOS, Rev. esp. Ent., Madrid, **34**: 132.

Literature: POPHAM & BRINDLE 1966; SAKAI 1970; STEINMANN 1977, 1978.

Description: Male colour dark brownish-red; mouthparts yellowish brown, antennae brown, legs yellow. Head tumid, rounded laterally and posteriorly; postfrontal sutures and coronal suture absent. Eyes about as long as length of head behind eyes. First antennal joint typical, more or less as long as distance between antennal bases. Pronotum quadrate, lateral margins more or less parallel; all angles rounded, posterior margin truncate; median longitudinal furrow present on anterior half. Mesonotum and metanotum normal, transverse. Abdomen fusiform, slightly widened at median segments. Ultimate tergite broad; posterior margin a little concave medially, the margin with two small tubercles above bases of the forceps. Forceps (Fig. 595) of *Parisolabis*-type, cylindrical, well separated at base, with crossed apices. Genitalia unknown. – **Female** similar to male, but forceps slenderer, and not curved strongly. – **Length** of body with forceps: male: 9,5–11 mm, female: 10–11,5 mm.

Distribution: India.

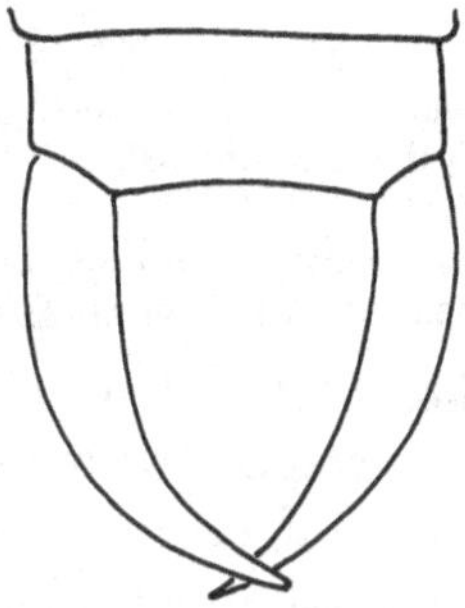

Fig. 595 Male ultimate tergite and forceps of *Parisolabis immsi* (BURR). Original.

3. Parisolabis tenera (BURR)

1910 *Pseudisolabis tenera* BURR, Fauna Brit. India, Dermaptera, London, Bombay, 104; fig. 36 (male). – Terra typica: India (Type male: Brit. Mus. nat. Hist., London).
1970 *Parisolabis tenera* – SAKAI, Dermapt. Cat. Prael., Tokyo, 1: 17.
Literature: BURR 1911; HEBARD 1917; STEINMANN 1977, 1978.

Description: Male unicolour reddish-brown; sides of pronotum and legs yellow. Head tumid, a little longer than wide; postfrontal and coronal sutures obsolete. Antennae 17-jointed; first joint clubbed, comparatively short, shorter than distance between antennal bases, second transverse, third joint cylindrical, and elongate, fourth very short, not longer than wide. Eyes normal, shorter than length of head behind eyes. Pronotum a little broader than head, and distinctly wider than long; rectangular, depressed; median longitudinal furrow distinct in fore-portion only; all angles slightly rounded; lateral margins almost parallel. Mesonotum very short and finely punctate. Sternal plates smooth, light brown, the sides parallel, transverse. Abdomen normal, subdepressed, very finely punctate, subparallel-sided, and slightly narrower apically. Ultimate tergite very broad; posterior margin truncate in middle, slightly tumid above the bases of the forceps. Forceps (Fig. 596) with the branches remote at the base, of circular cross-section; a little trigonal basally. Genitalia unknown. – **Female** unknown. – **Length** of body with forceps: 13 mm.

Distribution: India: Punjab.

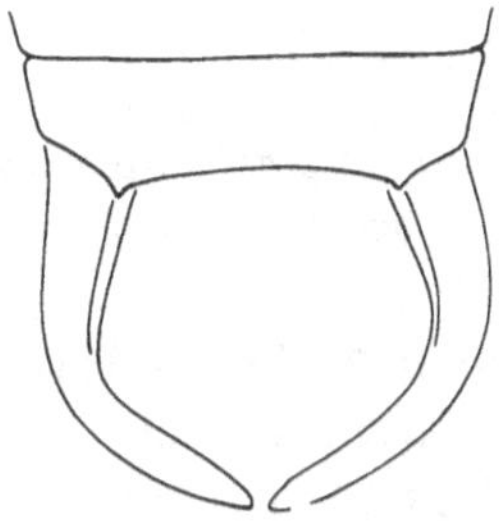

Fig. 596 Male ultimate tergite and forceps of *Parisolabis tenera* (BURR). Original.

4. **Parisolabis elegans** (HEBARD)

1917 *Pseudisolabis elegans* HEBARD, Proc. Acad. nat. Sci., Philadelphia, **69**: 236; pl. XVI, fig. 2 (male), 3 (female forceps). – Terra tyica: India (Type male: Mus. comp. Zool., Cambridge, Mass.)

1958 *Parisolabis elegans* – HINCKS, EOS, Rev. esp. Ent., Madrid, **34**: 132.

Literature: HEBARD 1923; POPHAM & BRINDLE 1966; SAKAI 1970; STEINMANN 1977, 1978.

Description: Male colour dark brown; dorsal surface of body shiny, of head and pronotum smooth with hardly and hairs, abdomen finely pubescent. Head tumid, postfrontal sutures and coronal suture absent. Eyes relatively short, shorter than length of head behind eyes. Antennae 17-jointed; first joint small, slightly shorter than distance between antennal bases; second transverse, third joint elongate, distinctly longer than sixth, the fourth joint wider than long, but varies to appreciably longer than wide. Pronotum wider than long, rectangular, with angles sharply rounded; median longitudinal furrow present on anterior half. Meso- and metanotum typical, transverse. Abdomen fusiform, a little expanded medially. Ultimate tergite very short; posterior margin concave medially. Forceps (Fig. 597) simple, cylindrical, elongate, gradually tapering and evenly curved to the aciculate apices. Genitalia unknown. – **Female** similar to male, but forceps shorter. – **Length** of body with forceps: in both sexes: 10,5–12,5 mm.

Distribution: India.

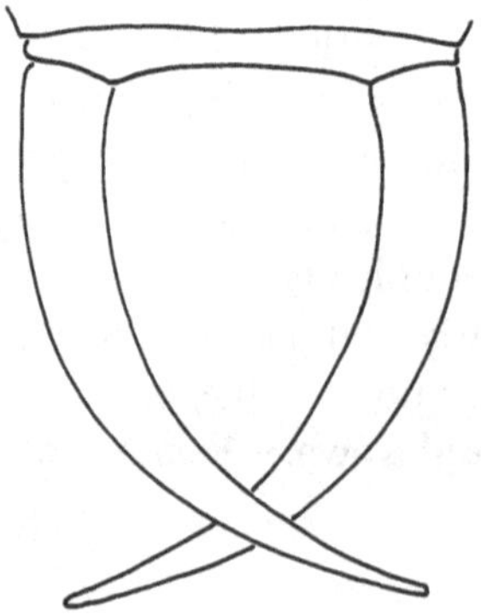

Fig. 597 Male ultimate tergite and forceps of *Parisolabis elegans* (HEBARD). After HEBARD.

5. **Parisolabis setosa** HUDSON

1973 *Parisolabis setosa* HUDSON, J. r. Soc. New Zealand, Wellington, (2) **3**: 247; fig. 13 (holotype), 40 (holotype genital armature), 62 (penultimate sternite), 80 (female forceps). – Terra typica: New Zealand (Type male: Dominion Mus., Wellington).

Description: Male colour reddish-brown; antennae paler. Head tumid, broad; posterior margin truncate; postfrontal and coronal sutures obsolete. Antennae 15-jointed; first joint short, shorter than distance between antennal bases. Eyes very small, slightly shorter than length of head behind eyes. Pronotum transverse, wider caudad than cephalad; lateral margins straight, all angles rounded; median longitudinal furrow short. Mesonotum and metanotum broad, with equal bristles. Abdomen fusiform, expanded medially, tergites punctate. Ultimate tergite very transverse, posterior margin with two tubercles above of branch of the forceps. Forceps (Fig. 598) symmetrical, cylindrical, not very remote basally, slender, unarmed, acuminate, apices crossed. Penultimate sternite rounded posteriorly. Genitalia (Fig. 599) characteristic; central parameral plate typical, genital lobes in characteristic position, both with numerous pointed tubercles. Virgae quite long, dilated basally, walls thickened in dilated region. External parameres strongly sclerotized, very short. – **Female** similar to male, but forceps closer basally than in male, cylindrical, acuminate, with minute crenulations on inner edge, apices may be crossed. – **Length** of body with forceps: in both sexes: 11–12,5 mm.

Distribution: New Zealand.

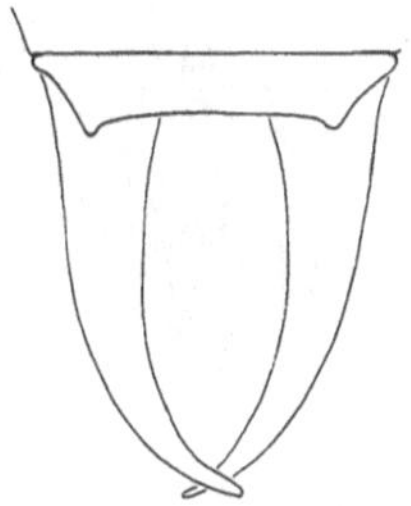

Fig. 598 Holotype ultimate tergite and forceps of *Parisolabis setosa* HUDSON. After HUDSON.

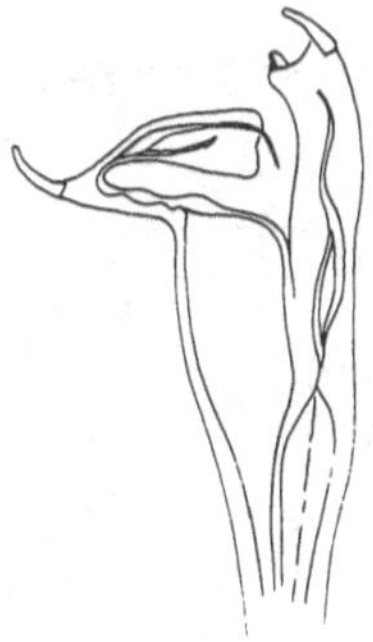

Fig. 599 Holotype genital armature of *Parisolabis setosa* HUDSON. After HUDSON.

6. Parisolabis johnsi HUDSON

1973 *Parisolabis johnsi* HUDSON, J. r. Soc. New Zealand, Wellington, (2) **3**: 244; fig. 11 (holotype), 44 (holotype genital armature), 56 (penultimate sternite), 78 (female forceps). – Terra typica: New Zealand (Type male: Dominion Mus., Wellington).

Description: Male general colour uniformly ferruginous, shining. Cuticle of body punctate and pubescent. Head a little longer than wide; posterior angles rounded, posterior margin straight; postfrontal and coronal sutures distinct. Eyes small, slightly shorter than length of head behind eyes. Antennae 15-jointed; first joint shorter than distance between antennal bases; second quadrate, third longer than fourth. Pronotum transverse, lateral margins straight, and a little expanded posteriorly; median longitudinal furrow present on anterior portion; plate with three pale areas longitudinally. Mesonotum broad, posterior margin concave, with pale markings, metanotum concave posteriorly, enclosing first abdominal tergite. Abdomen slightly dilated, depressed; posterior sternites densely haired. Ultimate tergite somewhat narrower, with shallow blunt teeth or tubercles above bases of forceps; posterior margin truncate. Forceps (Fig. 600) remote, tapering gradually, apices acuminate, subcontiguous. Penultimate sternite broad, posterior margin acute. Genitalia (Fig. 601) oval; paramere with median incision on anterior half, genital lobes well marked, typical of genus, both with numerous small, widely spaced tubercles, especially at apices. Virga within genital lobe straight, slender, slightly dilated medially, basal vesicle with 7–8 sclerotized bands and longish, bent. External parameres normal, typical of genus. – **Female** similar to male, but abdomen not depressed; tubercles on ultimate tergite closer together than in male. – **Length** of body with forceps: male: 7,5–9 mm, female: 8,5–10 mm.

Distribution: New Zealand.

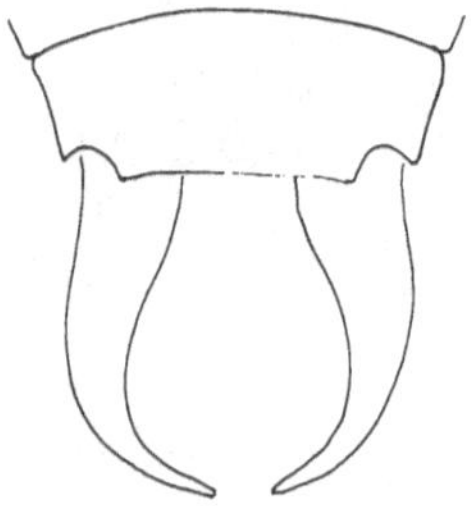

Fig. 600 Holotype ultimate tergite and forceps of *Parisolabis johnsi* HUDSON. After HUDSON.

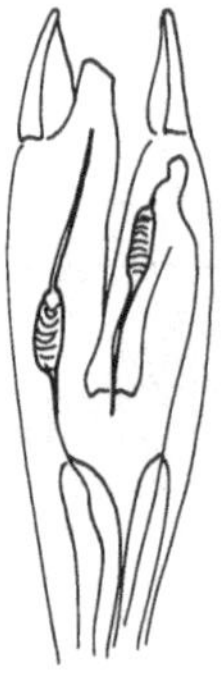

Fig. 601 Holotype genital armature of *Parisolabis johnsi* HUDSON. After HUDSON.

7. Parisolabis iti HUDSON

1973 *Parisolabis iti* HUDSON, J. r. Soc. New Zealand, Wellington, (2) **3**: 243; fig. 12 (holotype), 42 (holotype genital armature), 53 (penultimate sternite), 73 (female forceps). – Terra typica: New Zealand (Type male: Dominion Mus., Wellington).

Description: Male general colour dark brown. Head subquadrate, frons smooth, shiny, postfrontal and coronal sutures indistinct. Eyes comparatively large, but shorter than length of head behind eyes. Antennae 14-jointed; first joint long, but shorter than distance between antennal bases; all joints from fifth cylindrical, swollen. Pronotum very broad, broader than head; lateral margins straight, and strongly expanded posteriorly; median longitudinal furrow present on anterior part. Mesonotum with posterior margin rounded; caudal margin of metanotum slightly concave. Abdomen slightly dilated, not depressed, sparsely covered with minute bristles, often with faint borders of setae on some tergites. Posterior margin broad on ultimate tergite, straight. Forceps (Fig. 602) remote basally, but closer than is usual in this genus, simple, sharply tapering, extremely acuminate. Penultimate sternite subtriangular, posterior margin rounded, with distinct teeth on both sides of manubrium. Genitalia (Fig. 603) moderately broad, median incision of anterior margin small; the subequal cylindrical genital lobes pointing in opposite directions, the larger one forwards, both with numerous pointed tubercles, especially at swollen apices. Virgae short, very thick, bent, slightly dilated medially, terminating in oval basal vesicles. External parameres typical of genus. – **Female** similar to male, but forceps subcontiguous at base, simple, gradually tapering, acuminate, apices crossed. – **Length** of body with forceps: male: 7–10 mm, female: 8–12,5 mm.

Distribution: New Zealand.

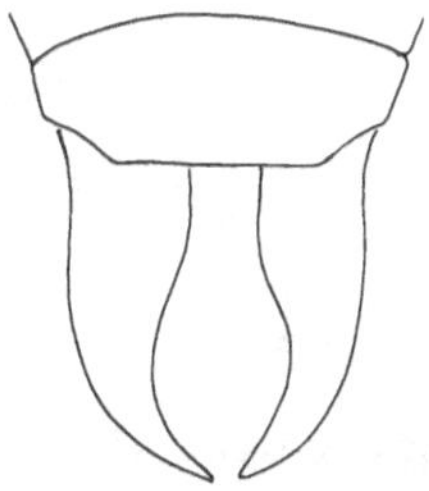

Fig. 602 Holotype ultimate tergite and forceps of *Parisolabis iti* HUDSON. After HUDSON.

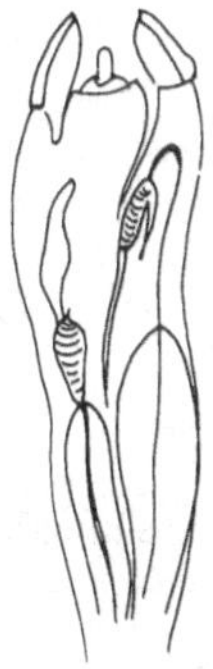

Fig. 603 Holotype genital armature of *Parisolabis iti* HUDSON. After HUDSON.

8. Parisolabis tapanuiensis HUDSON

1973 *Parisolabis tapanuiensis* HUDSON, J. r. Soc. New Zealand, Wellington, (2) **3**: 244; fig. 9 (holotype), 43 (holotype genital armature), 58 (penultimate sternite), 77 (female forceps). – Terra typica: New Zealand (Type male: Dominion Mus., Wellington).

Description: Male colour uniformly reddish-brown. Head wide, depressed; postfrontal and coronal sutures faint; eyes small, slightly shorter than length of head behind eyes. Antennae 15-jointed; first joint more or less as long as distance between antennal bases; second quadrate, third longer than wide. Pronotum subquadrate; lateral margins straight, more or less parallel; posterior margin rounded with sparse fringe of longer hairs; median longitudinal furrow present at anterior margin. Mesonotum and metanotum typical of genus, transverse. Abdomen with tergites usually vaguely patterned, depressed, dilated and pubescent. Ultimate tergite very broad; posterior margin concave medially. Forceps (Fig. 604) trigonal basally, cylindrical and curved apically. Penultimate sternite wider than long, posterior margin rounded, with small, shallow, blunt tubercles on both sides of manubrium. Genitalia (Fig. 605) elongated, oval; median incision of anterior margin present; genital lobes well developed, virgae short, thick, proximal third dilated, both capable of being everted, oval basal vesicle small, thick-walled, with 6–7 sclerotized bands. External parameres typical. – **Female** similar to male, but forceps of usual female type characteristic of genus, close together, cylindrical, acuminate. – **Length** of body with forceps: male: 7–9 mm, female: 8,5–9,5 mm.

Distribution: New Zealand.

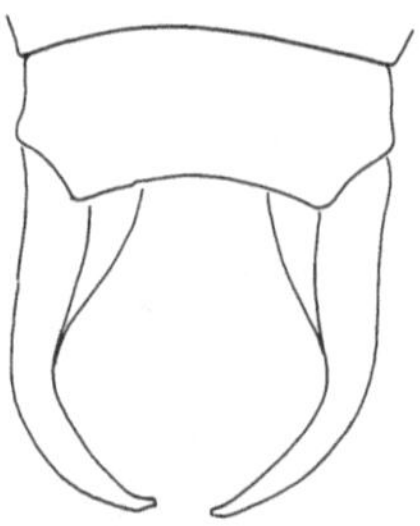

Fig. 604 Holotype ultimate tergite and forceps of *Parisolabis tapanuiensis* HUDSON. After HUDSON.

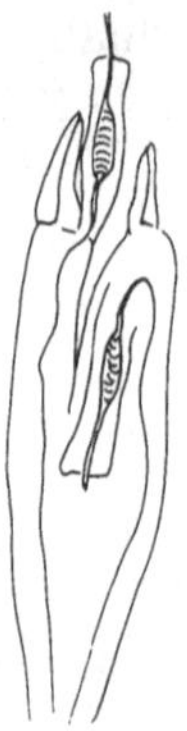

Fig. 605 Holotype genital armature of *Parisolabis tapanuiensis* HUDSON. After HUDSON.

9. Parisolabis forsteri HUDSON

1973 *Parisolabis forsteri* HUDSON, J. r. Soc. New Zealand, Wellington, (2) **3**: 246; fig. 11 (holotype), 39 (holotype genital armature), 57 (penultimate sternite), 74 (female forceps). – Terra typica: New Zealand (Type male: Dominion Mus., Wellington).

Description: Male general colour dark reddish-brown. Head broad, rectangular, posterior margin truncate; postfrontal and coronal sutures distinct. Eyes small, slightly shorter than length of head behind eyes. Antennae 14-jointed; first joint short, shorter than distance between antennal bases. Pronotum wider than long, wider posteriorly; anterior angles acute, posterior angles rounded; median longitudinal furrow present. Meso- and metanotum typical of genus; transverse. Abdomen fusiform, depressed, with long hairs on posterior margins of both tergites and sternites. Ultimate tergite very broad, with blunt tubercles over bases of forceps. Forceps (Fig. 606) very slender, cylindrical, unarmed, particularly acuminate, curvature slight, less than in *iti* HUDSON, not as remote basally as is usual in this genus and lacking characteristic arcuate shape, apices not crossed. Penultimate sternite only slightly wider than long, posterior margin convex, with distinct teeth on both sides of the manubrium. Genitalia (Fig. 607) typical of genus; paramere long, median incision of anterior margin deep; genital lobes well developed, virgae long, slender, almost straight, leading into narrow oval vesicles with numerous (11–12) narrow bands. External parameres comparatively long. – **Female** similar to male, but forceps cylindrical, armed with minute crenulations on inner margins, particularly acuminate, apices crossed. – **Length** of body with forceps: male: 9,5 mm, female: 8–10 mm.

Distribution: New Zealand.

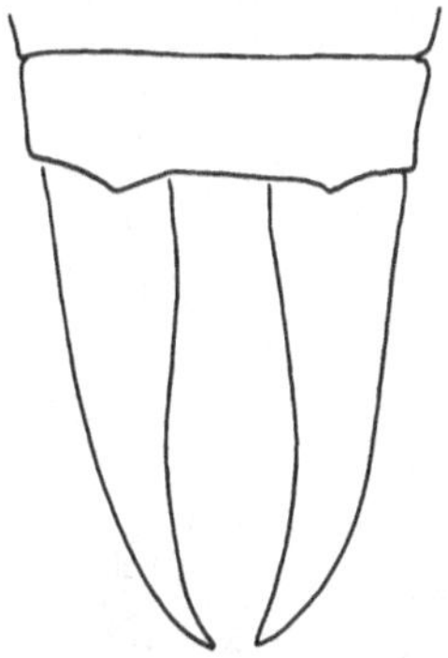

Fig. 606 Holotype ultimate tergite and forceps of *Parisolabis forsteri* HUDSON. After HUDSON.

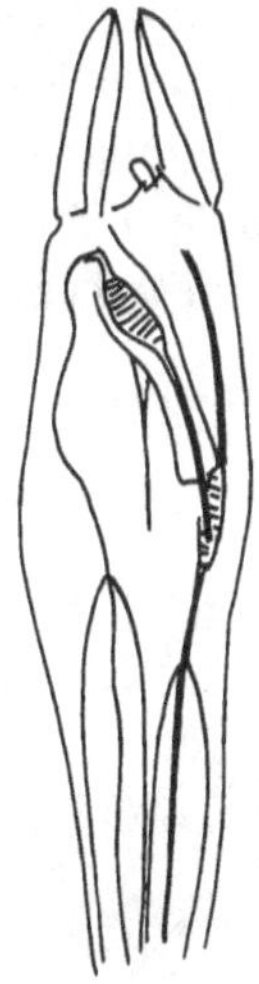

Fig. 607 Holotype genital armature of *Parisolabis forsteri* HUDSON. After HUDSON.

10. Parisolabis novaezeelandiae VERHOEFF

1904 *Parisolabis novae-zeelandiae* VERHOEFF, Arch. Naturgesch., Berlin, **1904**: 120. – Terra typica: New Zealand (Type male: Mus. Naturkunde, Berlin).
1908 *Pseudisolabis walkeri* BURR, Ann. Mag. nat. Hist., London, (8) **2**: 254. – Terra typica: New Zealand (Type male: Brit. Mus. nat. Hist., London).
1911 *Parisolabis novae-zeelandiae* – ZACHER, Zool. Jahrb., Jena, **30**: 390; fig. U² (male genitalia).
1973 *Parisolabis novaezeelandiae* – HUDSON, J. r. Soc. New Zealand, Wellington (2), **3**: 239; fig. 5 (male), 37 (male genitalia), 59 (penultimate sternite), 75 (female forceps).

Literature: BURR 1908, 1915; POPHAM & BRINDLE 1966; SAKAI 1970; STEINMANN 1977, 1978.

Description: Male colour reddish dark brown; antennae dark brown; mouthparts and legs yellowish brown. Cuticle of head, notum and abdominal tergites shiny, very strongly punctate. Head broad; postfrontal and coronal sutures not present; eyes relatively small, slightly shorter than length of head behind eyes. Antennae 17-jointed; first joint short, slightly shorter than distance between antennal bases. Pronotum a little wider than long; lateral margins straight, expanded posteriorly; median longitudinal furrow distinct. Mesonotum broad, rounded posteriorly; metanotum transverse, concave posteriorly. Abdomen comparatively broad, depressed, slightly expanded medially, dorsally covered with fine hairs, especially on posterior margin of tergites, which are themselves extended slightly caudad laterad. Ultimate tergite somewhat narrower, more densely haired, with shallow blunt tubercles above bases of forceps. Forceps (Fig. 608) symmetrical, remote, cylindrical and curved. Penultimate sternite broad, rounded posteriorly. Genitalia (Fig. 609) broad, median incision of anterior margin present; genital lobes well developed; paired virgae relatively short, thick, straight, terminating in oval vesicles with distinctive spirally thickened walls, with 7–10 sclerotized bands. External parameres typical of genus. – **Female** similar to male, but forceps a little slenderer. – **Length** of body with forceps: male: 8,5–11,5 mm, female: 9–13,5 mm.

Distribution: New Zealand.

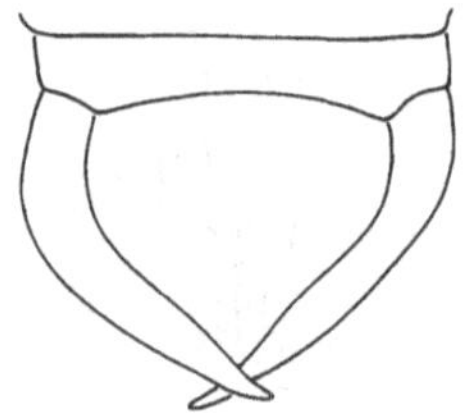

Fig. 608 Male ultimate tergite and forceps of *Parisolabis novaezeelandiae* VERHOEFF. After HUDSON.

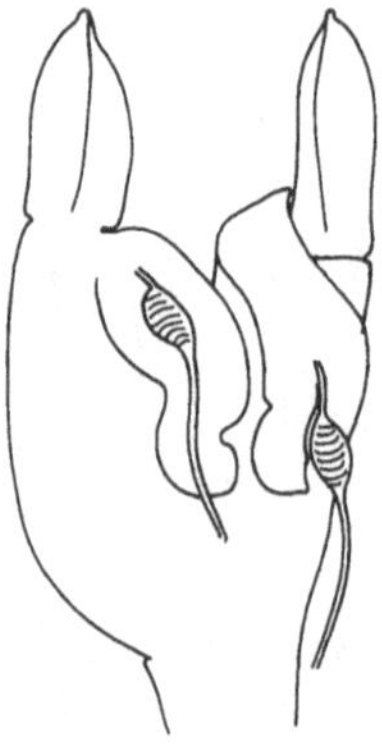

Fig. 609 Male genital armature of *Parisolabis novaezeelandiae* VERHOEFF. After ZACHER.

11. Parisolabis nelsonensis HUDSON

1973 *Parisolabis nelsonensis* HUDSON, J. r. Soc. New Zealand, Wellington, (2) **3**: 241; fig. 7 (holotype), 38 (holotype genital armature), 52 (penultimate sternite), 97 (female forceps). – Terra typica: New Zealand (Type male: Dominion Mus., Wellington).

Description: Male general colour reddish brown. Head transverse, comparatively small; eyes moderately large, but slightly shorter than length of head behind eyes. Antennae 17-jointed; first joint long, but a little shorter than distance between antennal bases. Pronotum subquadrate; lateral margins straight, wider caudad than cephalad; median longitudinal furrow present on anterior part. Mesonotum broad, rounded posteriorly; metanotum transverse, concave posteriorly. Abdomen fusiform, depressed, strongly punctate, with longer hairs on posterior margins of segments. Posterior margin of tergites produced caudad laterad, resulting in slightly sinuate margin when viewed dorsally. Ultimate tergite very broad; blunt tubercles indistinct, very wide apart, posterior margin truncate. Forceps (Fig. 610) strong, remote, trigonal basally, tapering to cylindrical shape, sharply angled, apices blunt, with very shallow tooth on inner margins apically. Penultimate sternite broad, rounded posteriorly, with distinct teeth on anterior margin. Genitalia (Fig. 611) comparatively narrow; central parameral plate with median incision of anterior margin; genital lobes well developed, paired virgae within genital lobes long, fairly straight, each basal vesicle with 9 sclerotized bands. External parameres normal, typical of genus. – **Female** similar to male, but forceps slender, cylindrical particularly acuminate, with minute crenulations on basal half at inner margins. – **Length** of body with forceps: male: 12,5–20 mm, female: 17–18,5 mm.

Distribution: New Zealand.

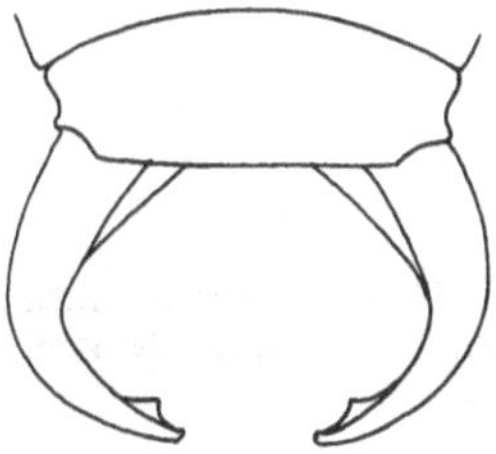

Fig. 610 Holotype ultimate tergite and forceps of *Parisolabis nelsonensis* HUDSON. After HUDSON.

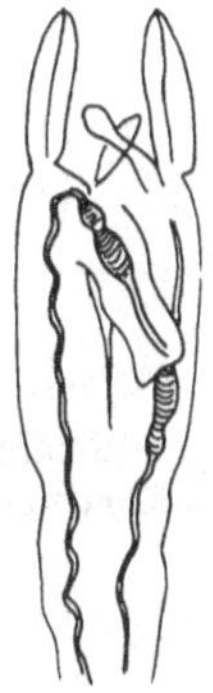

Fig. 611 Holotype genital armature of *Parisolabis nelsonensis* HUDSON. After HUDSON.

12. Parisolabis boulderensis HUDSON

1973 *Parisolabis boulderensis* HUDSON, J. r. Soc. New Zealand, Wellington, (2) 3: 242; fig. 8 (holotype), 41 (holotype genital armature), 54 (penultimate sternite). – Terra typica: New Zealand (Type male: Dominion Mus., Wellington).

Description: Male colour uniformly reddish-brown. Head comparatively small, broad, depressed. Eyes moderately large, but slightly shorter than length of head behind eyes. Antennae 16-jointed; first joint more or less as long as distance between antennal bases; postfrontal and coronal sutures obsolete. Pronotum quadrate, lateral margins more or less parallel; all angles rounded; median longitudinal furrow present on anterior half. Mesonotum broad, rounded posteriorly, metanotum transverse, concave posteriorly. Abdomen depressed, fusiform, slightly expanded medially; covered with short, fine, adpressed hairs, with longer ones on posterior margins of tergites and sternites. Ultimate tergite very broad, posterior margin concave posteriorly, with two tubercles above bases of the forceps. Forceps (Fig. 612) trigonal basally, cylindrical and curved apically. Penultimate sternite broad, gently rounded posteriorly. Genitalia (Fig. 613) typical of genus; central parameral plate with deep median longitudinal incision on anterior half; genital lobes well developed, virgae within genital lobes short, curved, oval vesicles rather narrow, with 8–9 sclerotized bands. External parameres typical. – **Female** unknown. – **Length** of body with forceps: 18–19,5 mm.

Distribution: New Zealand.

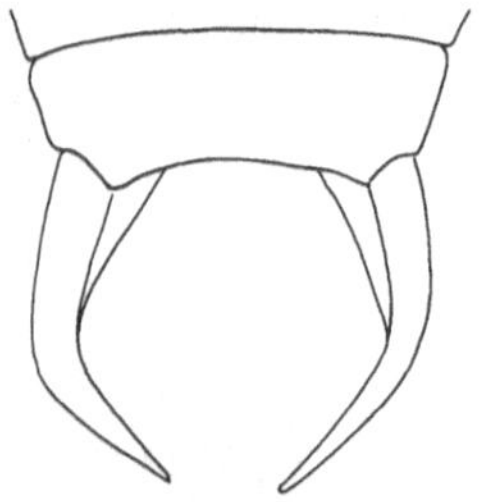

Fig. 612 Holotype ultimate tergite and forceps of *Parisolabis boulderensis* HUDSON. After HUDSON.

Fig. 613 Holotype genital armature of *Parisolabis boulderensis* HUDSON. After HUDSON.

Genus 32: **Parisopsalis** BURR

1914 *Parisopsalis* BURR, Ann. Mag. nat. Hist., London, (13) **8**: 73. – Species typica: *Parisopsalis spryi* BURR 1914.

Literature: BURR 1915; POPHAM & BRINDLE 1966; SAKAI 1970; STEINMANN 1977, 1978; TOWNES 1945.

Description: Apterous species. Tegmina and wings entirely absent. Head relatively large, more or less as long as length of pronotum. Abdomen fusiform, broad and depressed; ultimate tergite strongly depressed medially; male forceps broad and short, strongly trigonal. Male genitalia characteristic; central parameral plate comparatively small, genital lobes directed forwards; external parameres comparatively long.

Distribution: Australian species.

Single species.

1. **Parisopsalis spryi** BURR

1914 *Parisopsalis spryi* BURR, Ann. Mag. nat. Hist., London, (13) **8**: 74; pl. IV, fig. 3 (holotype), 4 (holotype genital armature). – Terra typica: Australia (Type male: BURR's Collect., Brit. Mus. nat. Hist., London).

Literature: BURR 1915; POPHAM & BRINDLE 1966; SAKAI 1970; STEINMANN 1977, 1978.

Description: Male colour reddish black, glabrous; head black, antennae blackish brown, legs yellowish brown. Head tumid, smooth, depressed, comparatively large; postfrontal and coronal sutures distinct. Antennae 15-jointed (?); first joint short, slightly shorter than distance between antennal bases. Eyes normal, shorter than length of head behind eyes. Pronotum transverse, rectangular, very gently widened posteriorly; all sides straight, and all angles rounded; median longitudinal furrow very distinct in the prozona. Mesonotum densely punctate, broad, with median longitudinal depression; posterior margin a little convex. Metanotum transverse, concave posteriorly. Abdomen depressed, strongly dilated about middle; narrowed apically. Ultimate tergite broad, unarmed, smooth, the posterior margin gently sinuate, and feebly swollen into an incipient tubercle over the bases of the forceps. The sides of each abdominal segment except the first produced into an acute strongly recurved hook, the outer edge of which is keeled, and rugulose above and below the keel. Forceps (Fig. 614) remote at base, stout, trigonal basally, tapering apically. Penultimate sternite rounded posteriorly. Genitalia (Fig. 615) characteristic; central parameral plate small, external parameres comparatively long and wide; genital lobes fully developed, directed forwards; virga within genital lobes specific, strongly sclerotized. – **Female** similar to male, but forceps slenderer, contiguous. – **Length** of body with forceps: male: 16–16,6 mm, female: 15,5–17 mm.

Distribution: Australia: Victoria.

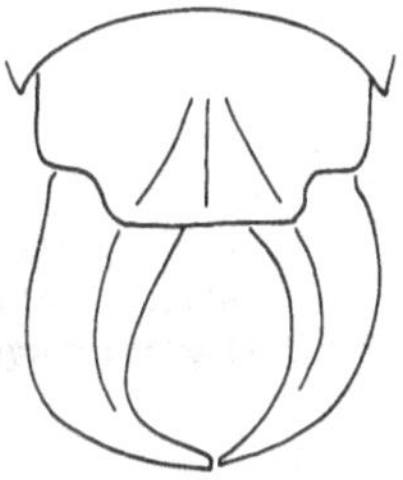

Fig. 614 Holotype ultimate tergite and forceps of *Parisopsalis spryi* BURR. After BURR.

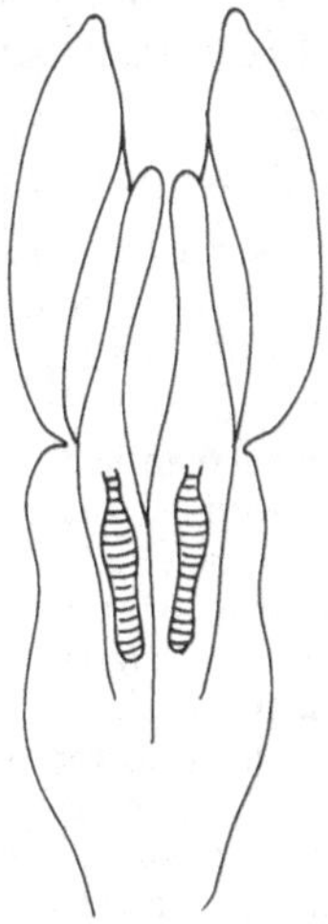

Fig. 615 Holotype genital armature of *Parisopsalis spryi* BURR. After BURR.

Subfamily 9: **Idolopsalinae** STEINMANN

1975 *Idolopsalinae* STEINMANN, Acta. zool. Acad. Sci. hung., Budapest, **21**: 206. – Genus typicum: *Idolopsalis* BORELLI 1910.

Description: Apterous insects. Tegmina and wings totaly absent. Head normal, eyes moderately large, but more or less shorter than length of head behind eyes. Antennae multisegmented, first joint relatively short, shorter than distance between antennal bases. Male genitalia characteristic, one male genital lobe more or less reduced (except for *koepckei* BRINDLE and *nigrita* BRINDLE). The group diverged phylogenetically in its evolution from the other species of the family as did the Karschiellinae in the Pygidicranidae. The reduction of the genital lobe shows a certain transition to the Labiid species with a single genital lobe. Abdomen only weakly widened, sides almost parallel.

Distribution: Neotropical faunal region.

Single genus.

Genus 33: **Idolopsalis** BORELLI

1910 *Idolopsalis* BORELLI, Bull. Mus. nation. Hist. nat., Paris, **1910**: 156. – Species typica: *Idolopsalis riveti* BORELLI 1910.

Literature: BRINDLE 1970; BURR 1911; POPHAM & BRINDLE 1966; REICHARDT 1968; SAKAI 1970; STEINMANN 1977, 1978; TOWNES 1945.

Description: Body colour dark brown to black; antennae light brown to dark, mouthparts yellowish brown. Head normal, rounded; postfrontal and coronal sutures distinct or fine; eyes relatively large, but shorter than length of head behind eyes. Pronotum quadrate, longer than wide or transverse. Abdomen fusiform, more or less depressed. Male genitalia various, virgae within genital lobe sclerotized, with or without a membranous flange on inner margin.

Distribution: Central and South America.

11 species.

Identification key to the species

1 Pronotum longer than wide. Male forceps symmetrical (Fig. 616), short, strongly curved. Male genitalia with very narrow external parameres (fig. 617). Abdominal segments 3–9 of male with lateral longitudinal ridges .... 1. **Idolopsalis borgmeieri**, p. 387

1′ Pronotum quadrate or transverse **2**

2 Pronotum about as long as wide **3**

2′ Pronotum transverse **6**

3 Abdomen in both sexes very sparsely punctate, the punctures shallow, and their diameter much less than distance between the punctures **4**

3′ Abdomen in both sexes strongly punctate, the punctures varying in size, but in general their diameter equal to or greater than the distance between the punctures **5**

4 Smaller species, length of body with forceps 9 mm. Male with lateral ridges on abdominal tergites 5–8. Apices of external parameres of male genitalia as in fig. 619 .... 2. **Idolopsalis azteca**, p. 388

4′ Larger species, length of body with forceps 19 mm. Male with lateral ridges on abdominal tergites 5–7. Apices of external parameres of male genitalia as in fig. 620 .... 3. **Idolopsalis andeana**, p. 389

5 Male forceps strongly asymmetrical (Fig. 621). External parameres of male genitalia comparatively broad, as in fig. 622 .... 4. **Idolopsalis koepckei**, p. 390

5′ Male forceps symmetrical, simple (Fig. 623). External parameres of male genitalia slightly narrow, as in fig. 624 .... 5. **Idolopsalis gracilis**, p. 391
6 External parameres of male genitalia very wide medially (Fig. 625). Male with lateral longitudinal ridges on tergites 5–9 .... 6. **Idolopsalis whymperi**, p. 391
6′ External parameres of male genitalia not very wide medially **7**
7 Male forceps (Fig. 626) of *Forficula*-type, inner margins with very large denticle. Male genitalia similar to *riveti* BORELLI, but median incision of anterior margin of the central parameral plate shorter (Fig. 627) .... 7. **Idolopsalis incerta**, p. 392
7′ Male forceps not of *Forficula*-type, without denticle on inner margins or with very small one basally. Median incision of anterior margin of the central parameral plate deep **8**
8 Male with lateral longitudinal ridges on two abdominal tergites **9**
8′ Male with lateral longitudinal ridges on more than two abdominal tergites **10**
9 Male with lateral longitudinal ridges on tergites 7–8. Body colour generally reddish-brown. Male genitalia as in fig. 629 .... 8. **Idolopsalis riveti**, p. 393
9′ Male with lateral longitudinal ridges on tergites 6–7. Body colour generally blackish. Male genitalia as in fig. 630 .... 9. **Idolopsalis nigrita**, p. 394
10 Male with lateral longitudinal ridges on tergites 7–9. Larger species, body length with forceps 14 mm. Male genitalia very similar to *parva* BRINDLE, as in fig. 631 .... 10. **Idolopsalis curta**, p. 395
10′ Male with lateral longitudinal ridges on tergites 5–8. Smaller species, body length with forceps 9 mm. Male genitalia very similar to *curta* BRINDLE, as in fig. 633 .... 11. **Idolopsalis parva**, p. 396

1. Idolopsalis borgmeieri MENOZZI

1932 *Idolopsalis borgmeieri* MENOZZI, Rev. Ent., São Paulo, **2**: 157; fig. 3 (dorsal view and male genitalia). – Terra typica: Brazil (Type male: MENOZZI's Collect., Rio de Janeiro).
Literature: REICHARDT 1968; SAKAI 1970; STEINMANN 1977, 1978.

Description: Male colour dark brown to blackish; mouthparts and palpi yellowish brown, antennae brown; legs paler yellowish. Head tumid, rounded, normal, but depressed along lines of sutures. First antennal joint normal, shorter than distance between antennal bases; second transverse, third joint slightly more than twice as long as wide; fourth joint one and half times as long as broad. Eyes typical of genus, shorter than length of head behind eyes. Pronotum longer than wide; median longitudinal furrow visible in the prozona; lateral margins straight, a little expanded posteriorly. Meso- and metanotum typical; transverse. Abdomen fusiform, a little depressed. Ultimate tergite very broad; posterior

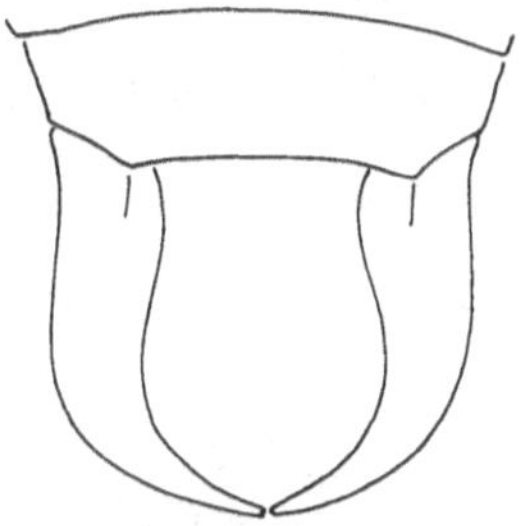

Fig. 616 Male ultimate tergite and forceps of *Idolopsalis borgmeieri* MENOZZI. Original.

Fig. 617 External paramere of male genitalia of *Idolopsalis borgmeieri* MENOZZI. After MENOZZI.

margin with two small tubercles above both branches of the forceps. Forceps (Fig. 616) symmetrical, trigonal basally, cylindrical and curved apically. Male genitalia of *Idolopsalis*-type, central parameral plate narrow, median incision of anterior margin deep; genital lobes well developed, virgae long. External parameres (Fig. 617) very narrow. – **Female** similar to male, but forceps slender, contiguous, tapering. – **Length** of body with forceps: male: 14,5 mm, female: 12 mm.

Distribution: Brazil.

2. Idolopsalis azteca (DOHRN)

1862 *Forcinella azteca* DOHRN, Stettiner ent. Ztg., **23**: 226. – Terra typica: Mexico (Type male: unknown locality).
1876 *Anisolabis azteca* – SCUDDER, Bull. U.S. geol. Survey., Jerrit, **2**: 251.
1911 *Gonolabis azteca* – BURR, Genera Insectorum, Bruxelles, **122**: 28.
1966 *Idolopsalis azteca* – BRINDLE, Proc. r. ent. Soc. London, (B) **34**: 145.
1970 *Anisolabis antennata* – SAKAI, Dermapt. Cat. Prael., Tokyo, **1**: 19 [= *Idolopsalis azteca* (DOHRN 1862)].

Literature: KIRBY 1904; STEINMANN 1977, 1978.

Description: Male colour dark brown or reddish brown; antennae brown, mouthparts and legs yellowish brown. Head and notum finely, abdomen strongly punctate, the punctures shallow, and their diameter much less than the distance between the punctures. Head tumid, slightly longer than wide, rounded posteriorly; postfrontal and coronal sutures distinct, deep; eyes relatively small, slightly shorter than length of head behind eyes. First antennal joint normal, a little shorter than distance between antennal bases. Pronotum about as long as wide; all margins straight, and all angles rounded; median longitudinal furrow absent or fine. Mesonotum normal, convex posteriorly; metanotum transverse, concave at posterior margin. Abdomen sparsely punctate, fusiform. Ultimate

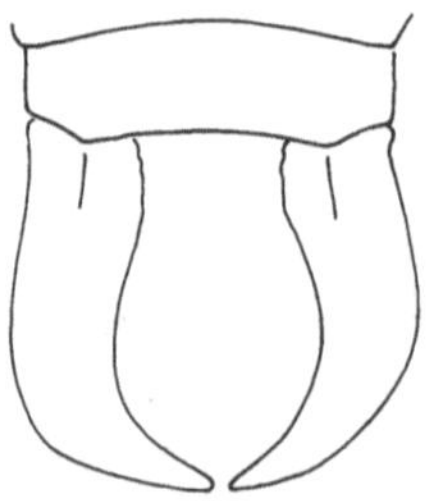

Fig. 618 Male ultimate tergite and forceps of *Idolopsalis azteca* (DOHRN). Original.

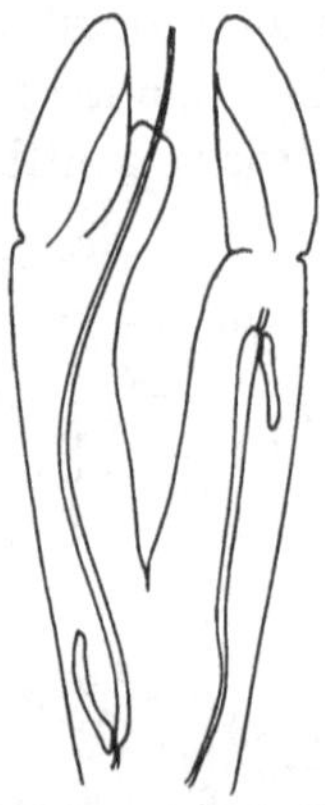

Fig. 619 Male genital armature of *Idolopsalis azteca* (DOHRN). Original.

tergite very broad, about three times as wide as long; posterior margin concave medially. Forceps (Fig. 618) symmetrical, strongly curved, cylindrical. Genitalia (Fig. 619) broad, large; central parameral plate with very deep median incision; genital lobe short, virga within genital lobes very long, with sclerotized plates basally. – **Female** similar to male, but forceps slender, and contiguous. – **Length** of body with forceps: in both sexes: 9 mm.

Distribution: Mexico, West Antilles, Peru, and Paraguay.

3. Idolopsalis andeana Burr

1914 *Idolopsalis andeana* Burr, Ann. Mag. nat. Hist., London, (14) **8**: 426. – Terra typica: Ecuador (Type male: Brit. Mus. nat. Hist., London).

1915 *Idolopsalis andeana* – Burr, J. r. microsc. Soc., London, **1915**: 446; pl. IX, fig. 14 (male genitalia).

Literature: Brindle 1966; Popham & Brindle 1966; Reichardt 1968; Sakai 1970; Steinmann 1977.

Description: Male colour brownish-black; antennae dark brown, mouthparts and tarsi yellowish brown, legs paler yellowish. Head tumid, a little depressed at vertex; postfrontal sutures and coronal suture distinct. Eyes small, slightly shorter than length of head behind eyes. First antennal joint typical, a little longer than distance between antennal bases. Pronotum as wide as long; lateral margins straight, a little expanded posteriorly; all angles rounded, and median longitudinal furrow present on anterior half. Meso- and metanotum typical, transverse. Abdomen fusiform; lateral margins of abdominal tergites 5–7 with longitudinal ridges. Ultimate tergite broad, posterior margin with two small tubercles above bases of the forceps. Forceps very similar to *azteca* (Dohrn). Genitalia (Fig. 620) large, central parameral plate comparatively short, median incision of anterior margin deep but narrow; virga with rounded sclerotized plate. External parameres specific, slightly broadened medially, outer margins rounded, apex blunt. – **Female** similar to male, but forceps slender, tapering. – **Length** of body with forceps: male: 19–19,5 mm, female: 19–20,5 mm.

Distribution: Ecuador.

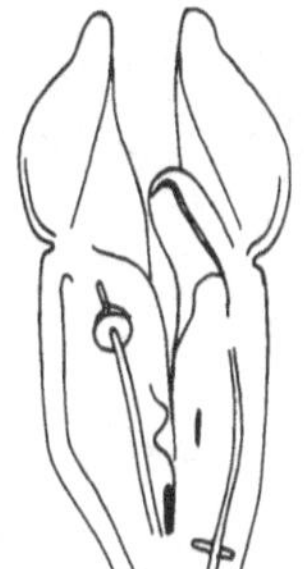

Fig. 620 Male genital armature of *Idolopsalis andeana* Burr. After Burr.

4. Idolopsalis koepckei Brindle

1966 *Idolopsalis koepckei* Brindle, Proc. r. ent. Soc. London, (B) **35**: 145; fig. 3 (male forceps), 7 (male genitalia), 14 (pronotum of male). – Terra typica: Peru (Type male: Hinck's Collect., Manchester Mus.).

Literature: Popham & Brindle 1966; Sakai 1970; Steinmann 1977, 1978.

Description: Male colour blackish, shiny; legs yellow or whitish, with black transverse bands on femora, and dark bands on tibiae. Head tumid, punctate; postfrontal sutures and coronal suture well marked; eyes comparatively small, less than length of head behind eyes. Antennae 11-jointed; first joint short, shorter than distance between antennal bases; second quadrate, third joint almost as long as first, fourth joint rather more than twice as long as wide. Pronotum quadrate, coriaceous, with median longitudinal furrow; all margins straight, and all angles rounded. Mesonotum strongly punctate, transverse. Metanotum concave posteriorly, less than half length of mesonotum medially. Abdomen strongly punctate, especially on segments 7–8; tergites 5–9 with prominent lateral longitudinal ridges. Ultimate tergite sparsely punctate, very broad, truncate, with well defined horizontal ridge between bases of forceps. Forceps (Fig. 621) characteristic, not of *Idolopsalis*-type, asymmetrical; branches cylindrical, well separated basally. Pygidium broad, slightly emarginate at tip. Genitalia (Fig. 622) with two well developed genital lobes, virgae long. External parameres elongated, broad medially. – **Female** similar to male, but without lateral longitudinal ridges on abdominal tergites; forceps short, contiguous, symmetrical. – **Length** of body with forceps: male: 13–13,5 mm, female: 13–14 mm.

Distribution: Peru.

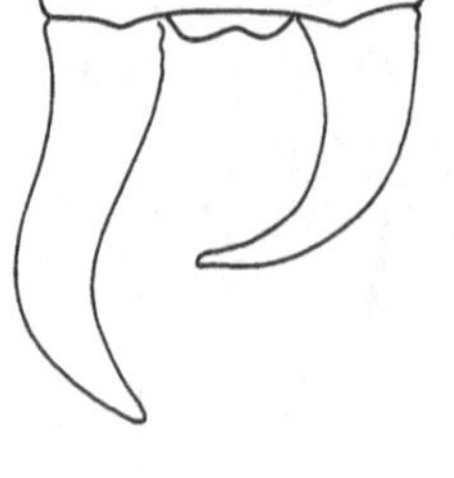

Fig. 621 Holotype ultimate tergite and forceps of *Idolopsalis koepckei* Brindle. After Brindle.

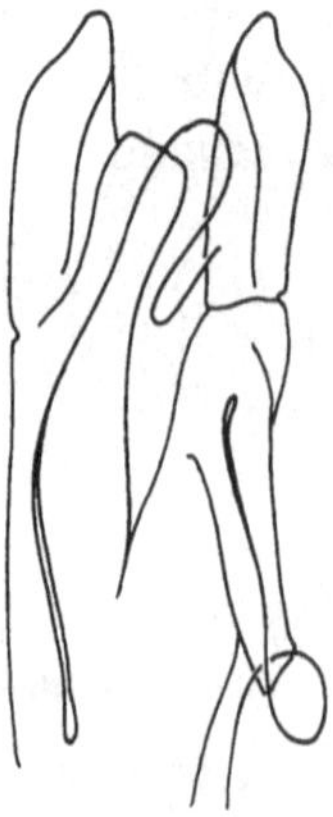

Fig. 622 Holotype genital armature of *Idolopsalis koepckei* Brindle. After Brindle.

5. Idolopsalis gracilis BRINDLE

1966 *Idolopsalis gracilis* BRINDLE, Proc. r. ent. Soc., London, (B) **35**: 146; fig. 2 (forceps of male), 12 (external paramere of holotype genitalia). – Terra typica: Ecuador (Type male: Brit. Mus. nat. Hist., London).

Literature: POPHAM & BRINDLE 1966; SAKAI 1970; STEINMANN 1977, 1978.

Description: Male colour dark reddish-brown, shiny; legs unicolour yellowish-brown. Head lightly punctate, postfrontal sutures and coronal suture distinct. Antennae 9-jointed in holotype, first joint short, slightly shorter than length of head behind eyes; second transverse, third joint shorter than first, fourth joint twice as long as wide. Pronotum quadrate, widened posteriorly, coriaceous; median longitudinal furrow well marked; lateral margins straight, posterior margin a little rounded. Mesonotum about half length of pronotum, with median depression, and with indications of lateral folds. Metanotum normal, typical of genus. Abdomen strongly punctate, widened slightly to segment 6, then narrowed; lateral ridges on tergites 6–8. Ultimate tergite broad, posterior margin truncate, closely punctate, but punctures shallow. Forceps (Fig. 623) broad, symmetrical, well separated basally. Genitalia of *Idolopsalis*-type; central parameral plate large, median incision of anterior margin deep; virga within genital lobe long. External parameres (Fig. 624) very narrow. – **Female** unknown. – **Length** of body with forceps: 11,5 mm.

Distribution: Ecuador.

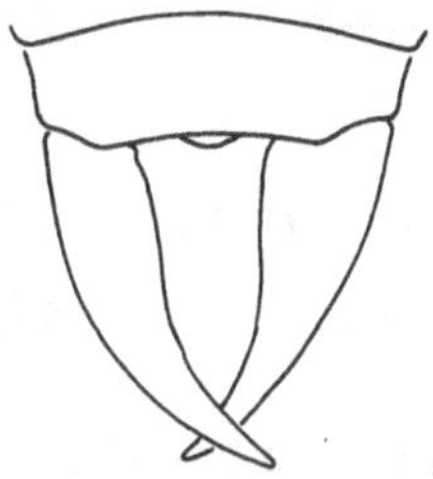

Fig. 623 Holotype ultimate tergite and forceps of *Idolopsalis gracilis* BRINDLE. After BRINDLE.

Fig. 624 External paramere of male genitalia of *Idolopsalis gracilis* BRINDLE. After BRINDLE.

6. Idolopsalis whymperi BURR

1914 *Idolopsalis whymperi* BURR, Ann. Mag. nat. Hist., London, (14) **8**: 426. – Terra typica: Ecuador (Type male: Brit. Mus. nat. Hist., London).

1915 *Idolopsalis whymperi* – BURR, J. r. microsc. Soc., London, **1915**: 446; pl. IX, fig. 2 (male genitalia).

Literature: BRINDLE 1966; POPHAM & BRINDLE 1966; SAKAI 1970; STEINMANN 1977, 1978.

Description: Male shiny blackish; antennae dark brown, mouthparts and tarsi yellowish brown, legs yellowish, femora with broad dark irregular bands. Head tumid, as broad as long, rounded posteriorly; postfrontal sutures and coronal suture well marked; broadly depressed along lines of sutures. Eyes typical of genus, shorter than length of head behind eyes. First antennal joint rather shorter than distance between antennal bases; second quadrate, third joint long, about as long as first joint; fourth joint half as long as third. Pronotum transverse, lateral margins straight, widened posteriorly; all angles rounded; median longitudinal furrow present on anterior half. Mesonotum a little shorter than pronotum, and a little longer than metanotum. Abdomen punctate dorsally, tergites 5–9 with lateral longitudinal ridges. Ultimate tergite broad, posterior margin concave medially. Forceps similar to *azteca* (DOHRN), symmetrical, widened basally, curved apically. Genitalia typical of genus, central parameral plate with median incision; virga within genital lobe long. External parameres (Fig. 625) very wide medially. – **Female** unknown. – **Length** of body with forceps: 12,5 mm.

Distribution: Ecuador.

Fig. 625 External paramere of male genitalia of *Idolopsalis whymperi* BURR. After BURR.

7. Idolopsalis incerta (BORMANS)

1903 *Anisolabis incerta* BORMANS, Ann. Mag. nat. Hist., London, (7) **11**: 234. – Terra typica: Ecuador (Type male: Brit. Mus. nat. Hist., London).

1970 *Idolopsalis incerta* – BRINDLE, Entomologist, London, **103**: 117; fig. 1 (holotype male), 2 (holotype genital armature), 3 (thoracic sterna).

Literature: BORELLI 1913; BURR 1911, 1913; HEBARD 1924; POPHAM & BRINDLE 1966; REICHARDT 1968; SAKAI 1970; STEINMANN 1977, 1978.

Description: Male colour blackish-brown, shiny, head and thoracic nota slightly reddish-tinged; legs greyish-brown. Head tumid, but depressed along lines of sutures; postfrontal and coronal sutures well marked; cuticle of head coriaceous. Eyes comparatively large, but slightly shorter than length of head behind eyes. First antennal joint short, shorter than distance between antennal bases; second joint transverse, third slightly more than twice as long as wide, and fourth joint one and half times as long as wide. Pronotum transverse, widened posteriorly, with a faint median longitudinal furrow or impressed line; lateral margins straight, posterior margin truncate. Mesonotum transverse, metanotum with concave posterior margin. Abdomen glabrous, strongly punctate, the punctures large; tubercles on fourth tergite represented only by small swellings on the cuticle. Ultimate tergite large, declivent, posterior margin thickened, and concave on posterior median section. Forceps (Fig. 626) of *Forficula*-type, somewhat asymmetrical; trigonal basally, cylindrical apically; inner margins with very large denticles. Genitalia (Fig. 627) typical of genus: virga within genital lobe very long, sclerotized, with lateral flanges. External parameres mainly membranous, but darkened along outer margins, and with a membranous flange on inner margins. – **Female** unknown. – **Length** of body with forceps: 19,5 mm.

Distribution: Ecuador.

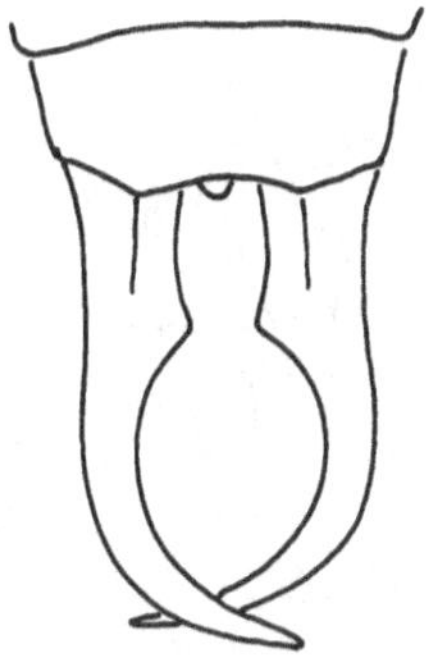

Fig. 626 Holotype ultimate tergite and forceps of *Idolopsalis incerta* (BORMANS). After BRINDLE.

Fig. 627 Holotype genital armature of *Idolopsalis incerta* (BORMANS). After BRINDLE.

8. Idolopsalis riveti BORELLI

1910 *Idolopsalis riveti* BORELLI, Bull. Mus. nation. Hist. nat., Paris, **1910**: 157. – Terra typica: Peru (Type male: Mus. nation. Hist. nat., Paris).

1915 *Idolopsalis riveti* – BURR, J. r. microsc. Soc., London, **1915**: 446; pl. IX, fig. 13 (male genitalia).

1966 *Idolopsalis riveti* – BRINDLE, Proc. r. ent. Soc. London, (B) **34**: 144; fig. 1 (male forceps), 6 (male genitalia). Ecuador.

Literature: BRINDLE 1970; BURR 1911; POPHAM & BRINDLE 1966; REICHARDT 1968; SAKAI 1970; STEINMANN 1977, 1978.

Description: Male colour dark brownish-black, shiny, head, thoracic nota and abdominal tergites punctate; pronotum reddish-brown, legs yellow, and forceps reddish-brown. Head broad, tumid; postfrontal and coronal sutures well marked, deep; eyes relatively small, slightly shorter than length of head behind eyes. Antennae 11-jointed; first joint small, shorter than distance between antennal bases; second quadrate, third joint one and half times longer than wide. Pronotum more or less transverse, lateral margins almost parallel; median longitudinal furrow long and deep. Meso- and metanotum normal, typical of genus. Abdomen with lateral longitudinal ridges on tergites 7–8. Ultimate tergite transverse, posterior margin straight, truncate. Forceps (Fig. 628) asymmetrical, widest basally, curved apically; of inner margins both branches with denticle. Genitalia (Fig. 629) of *Idolopsalis*-type; virga within genital lobe very long, sclerotized, thick; median longitudinal incision of anterior margin of central parameral plate present. External parameres approximately of equal width along most of the length. – **Female** similar to male, but forceps slender, symmetrical, tapering. – **Length** of body with forceps: male: 9,5–11,5 mm, female: 10–12,5 mm.

Distribution: Peru, Ecuador.

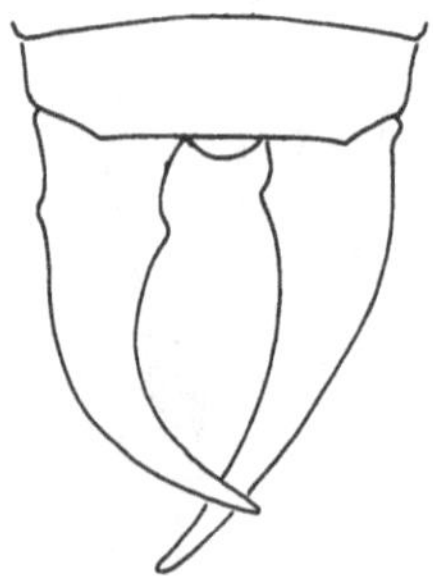

Fig. 628 Male ultimate tergite and forceps of *Idolopsalis riveti* BORELLI. After BRINDLE.

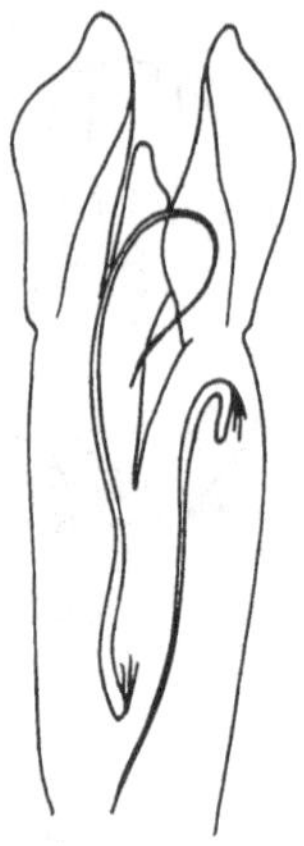

Fig. 629 Male genital armature of *Idolopsalis riveti* BORELLI. After BRINDLE.

9. Idolopsalis nigrita BRINDLE

1971 *Idolopsalis nigrita* BRINDLE, Entomologist, London, **104**: 255; fig. 9 (holotype genital armature). – Terra typica: Brazil (Type male: Mus. Zool., São Paulo).
Literature: STEINMANN 1977, 1978.

Description: Male blackish, shiny; antennae dark brown to blackish, basal three joints yellow; lateral margins of pronotum narrowly yellow; legs yellow, femora with a broad dark irregular ring medially. Head about as long as wide, rounded posteriorly; postfrontal and coronal sutures well marked; eyes small. First antennal joint small, shorter than distance between antennal bases; second quadrate, third joint two and half times as long as wide. Pronotum strongly transverse, wider posteriorly; lateral margins slightly sinuate, posterior margin straight; median longitudinal furrow distinct. Mesonotum normal, short, posterior margin concave. Metanotum typical. Abdomen fusiform, depressed, slightly widened to tergite 7, then narrower; tergite 4 sparsely punctate, tergites 5–6 more strongly punctate, tergites 7–9 strongly punctate, the punctures large. Abdominal tergites 6–9 rugose laterally, 6–7 with a lateral longitudinal ridge, 8–9 without a distinct ridge. Ultimate tergite broad, with a median longitudinal sulcus present only on middle portion. Forceps symmetrical, trigonal basally, cylindrical apically, and curved. Genitalia (Fig. 630) of *Idolopsalis*-type, median incision of anterior margin very deep, virga thick, very long, strongly sclerotized basally. External parameres broad and obtuse. – **Female** similar to male, but differing in punctation of abdominal tergites and structure of forceps. – **Length** of body with forceps: in both sexes: 12 mm.

Distribution: Brazil.

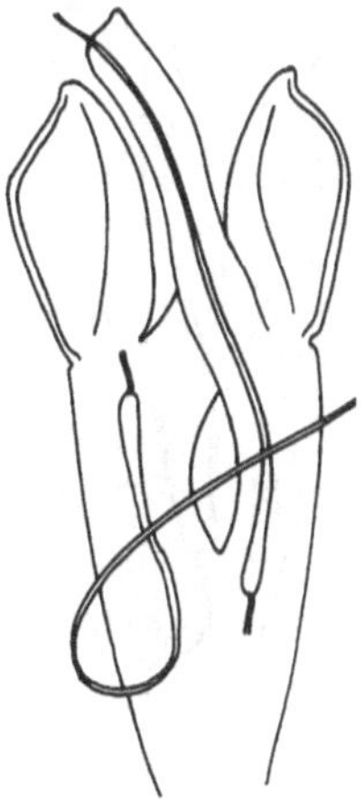

Fig. 630 Holotype genital armature of *Idolopsalis nigrita* BRINDLE. After BRINDLE.

10. Idolopsalis curta BRINDLE

1971 *Idolopsalis curta* BRINDLE, Entomologist, London, **104**: 257; fig. 10 (holotype genital armature). – Terra typica: Ecuador (Type male: California Acad. sci., San Francisco).
Literature: STEINMANN 1977, 1978.

Description: Male colour blackish-brown, thoracic nota lighter, antennae brown, legs yellow, femora with dark broad band. Cuticle mainly coriaceous and smooth, except on some abdominal tergites. Head rounded, as wide as long, broadly depressed along lines of sutures. Postfrontal and coronal sutures distinct. Eyes small, shorter than length of head behind eyes. First antennal joint rather shorter than distance between antennal bases, second transverse, third three times as long as wide, fourth joint half as long as third. Pronotum transverse, median longitudinal furrow distinct; lateral margins a little wider posteriorly. Mesonotum normal, metanotum transverse. Abdomen fusiform, depressed, weakly widened medially, tergites 2–9 punctate, the punctures large, more widely spaced on anterior margin of tergites, tergites 7–9 with punctures separated by about their own diameter and each with a lateral ridge on both sides associated with a rugose area. Ultimate tergite transverse, less strongly punctate than tergites 7–9, with a very deep median longitudinal sulcus medially, and strongly depressed near posterior margin. Forceps symmetrical, both branches with a rounded longitudinal ridge near base. Genitalia (Fig. 631) typical of genus; virga sclerotized, with lateral flanges basally; median longitudinal incision very deep. External parameres large. – **Female** unknown. – **Length** of body with forceps: 14 mm.

Distribution: Ecuador.

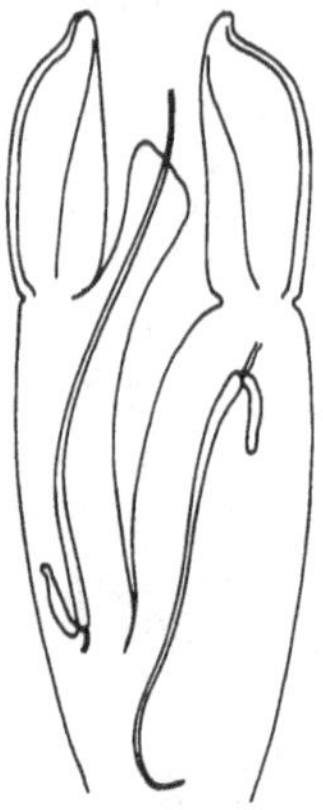

Fig. 631 Holotype genital armature of *Idolopsalis curta* BRINDLE. After BRINDLE.

11. Idolopsalis parva BRINDLE

1971 *Idolopsalis parva* BRINDLE, Entomologist, London, **104**: 256; fig. 11 (forceps of male), 12 (holotype genital armature). – Terra typica: Brazil (Type male: Mus. Zool., São Paulo).
Literature: STEINMANN 1977, 1978.

Description: Male head and thoracic nota yellowish-brown, rest of body blackish-brown; antennae yellowish-brown, but basal three joints paler; legs yellow, femora with a broad median brown band. Head rounded, about as long as wide; postfrontal sutures and coronal suture well marked; eyes small, first antennal joint short, shorter than distance between antennal bases; second quadrate, third joint two and half times as long as wide, fourth one and a quarter times as long as wide. Pronotum transverse, normal. Mesonotum and metanotum typical of genus. Abdomen fusiform, tergites strongly punctate; tergite 1 impunctate, tergite 2 slightly and sparsely punctate, tergite 3 more closely punctate, tergites 4–9 with large punctures; tergites 6–8 rugose laterally, on both sides, each tergite with a lateral longitudinal ridge. Ultimate tergite broad, with median longitudinal sulcus; posterior margin concave medially. Forceps

(Fig. 632) trigonal basally, curved and cylindrical apically. Penultimate sternite with posterior margin smoothly rounded. Genitalia (Fig. 633) with very deep median incision between external parameres; virga within genital lobe very long, with lateral flange basally. External parameres large, with sclerotized outer margins. – **Female** unknown. – **Length** of body with forceps: 8,5–9 mm.

Distribution: Brazil.

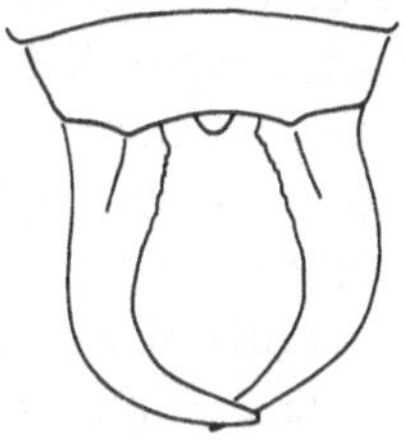

Fig. 632 Male ultimate tergite and forceps of *Idolopsalis parva* BRINDLE. After BRINDLE.

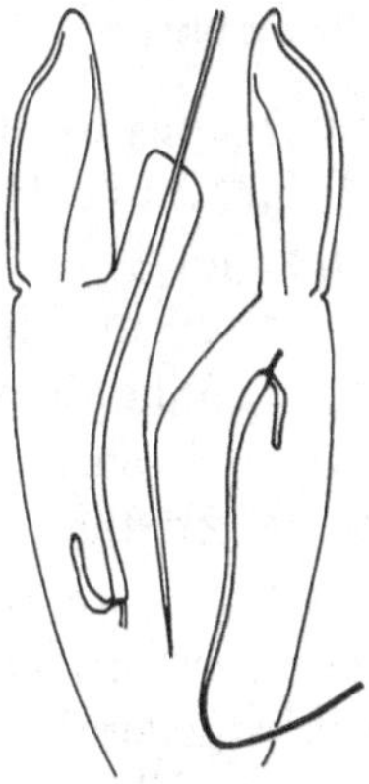

Fig. 633 Holotype genital armature of *Idolopsalis parva* BRINDLE. After BRINDLE.

Family 3: **Labiduridae** VERHOEFF

1902 *Labiduridae* VERHOEFF, Zool. Anz., Leipzig, (3) **665**: 189. – Genus typicum: *Labidura* LEACH 1815.
1911 *Labiduriales* ZACHER, Zool. Jahrb., Jena, **30**: 309.
1920 *Labidurinae* BLATCHLLEY, in HANDLIRSCH: Handbuch der Zoologie, Berlin: 475
1954 *Labiduroidea* BRUES, MELANDER & CARPENTER, Classification of Insects, Cambridge, Mass.: 115.
1975 *Labiduridae* – STEINMANN, Acta zool. Acad. Sci. hung., Budapest, **21**: 199.
Literature: BEY-BIENKO 1936; BOESEMAN 1954; BRINDLE 1973; BURR 1911; POPHAM & BRINDLE 1968; SAKAI 1970; STEINMANN 1977, 1978; TILLYARD 1926.

Description: Mainly winged species, tegmina and wings well developed or wings absent. Head and body not depressed, build generally stout; antennae multisegmented, joints 4, 5, and 6 short. Prosternum subparallel, contricted posteriorly. Pronotum long and broad, meso- and metanotum truncate posteriorly. The subfamilies Nalinae and Labidurinae are distinguished herein by the presence or absence of a longitudinal ridge along the lateral margins of the tegmina (a characteristic of subfamiliar value in every family of the order). Male genitalia of the family are distinctive, and easily separate this family from other families, genera, or species of earwigs. Abdomen parallel-sided, or a little fusiform, forceps generally remote and elongate.

Distribution: Distributed nearly all over the world.

3 subfamilies.

Identification key to the subfamilies

1 Mesosternum very narrow, its anterior portion normal, then constricting towards metasternum and narrowing to about one-half of the width of anterior end. Forceps short, about as long as width of ultimate tergite, or moderately elongated. Central parameral plate of male genitalia elongated, with more or less deep median incision; external parameres relatively wide or narrower. Basal vesicle of genital lobes appearing as a gradually widening swelling, not sharply outlined .... Subfamily 1: **Allostethinae**, p. 398

1′ Mesosternum wide, its posterior portion (towards metasternum) not constricted, normal. Forceps not short, but elongated, robust, characteristic; considerably longer than width of ultimate tergite **2**

2 Lateral margins of tegmina with a longitudinal ridge. Apices of male external parameres gradually elongated, apices evenly attenuating. Legs comparatively short, femora of hind legs not longer than pronotum .... Subfamily 2: **Nalinae**, p. 421

2′ Lateral margins of tegmina without a longitudinal ridge. Apices of male external parameres abruptly acute, thus apex sharply distinct, or apex of parameral plate widely rounded. Legs comparatively long, femora of hind leg longer than pronotum .... Subfamily 3: **Labidurinae**, p. 434

Subfamily 1: **Allostethinae** BURR

1904 *Allostethidae* VERHOEFF, Arch. Naturgesch., Berlin, **1904**: 115.
1909 *Allostethinae* BURR, Dtsch. ent. Zeitschr., Berlin, **1909**: 324. – Genus typicum: *Allostethus* VERHOEFF 1904.
1911 *Allostethidae* – ZACHER, Zool. Jahrb., Jena, **30**: 355.
1954 *Allostethidae* – BRUES, MELANDER & CARPENTER, Classification of Insects, Cambridge, Mass.: 115.

1965 *Allostethinae* – POPHAM, Entomologist, London, **98**: 134.
1965 *Allostethinae* – BRINDLE, Ann. Mag. nat. Hist., London, (13) **8**: 435.
Literature: BOESEMAN 1954; BORELLI 1932; BURR 1908, 1909, 1910, 1911, 1915; HEBARD 1927; REHN 1924; SAKAI 1970; STEINMANN 1975, 1977, 1978.

Description: The species of this subfamily are characterized by having the mesosternum strongly narrowed posteriorly (this is not narrowed in the Labidurinae, and Nalinae), and by having a more or less cylindrical body as opposed to the greatly depressed body of the Apachyidae. Body large or medium in size, between about 12 mm and 32 mm in length, and usually dark reddish brown in colour, and shiny. The tegmina may be well developed, with or without lighter markings, or may be reduced or absent. Wings fully developed, except in *Gonolabidura* ZACHER, but may vary in development in certain species. Male genitalia with two genital lobes, and with virgae; external parameres sclerotized, with acuminated apices.

Distribution: Distributed in the Oriental region.

3 genera.

Identification key to the genera

1 Tegmina and wings absent. Pronotum square, as long as wide. Male forceps strongly arcuate, gradually attenuating apicad. Third antennal joint wide, about twice as long as wide. Basal vesicle within genital lobe of male genitalia strikingly swollen Genus 1: **Gonolabidura**, p. 399

1′ Tegmina and wings developed. Pronotum not square, but slightly transversally elongated, wider than long. Third antennal joint shortened, less than twice as long as wide. Basal vesicle of male genitalia not strikingly swollen **2**

2 Pronotum only slightly wider than long. Third antennal joint very short, about as long as wide. Larger species, length of body with forceps: 25–30 mm. Male forceps moderately elongated, basally not strikingly widened. Male virga pecicular, comparatively narrow, that is, lateral margins parallel for a considerable distance .... Genus 2: **Allostethus**, p. 406

2′ Pronotum considerably wider than long. Third antennal joint moderately long, longer than wide, but not twice as long as wide. Medium-sized species, length of body with forceps about 12–21 mm. Male forceps short, basal section more or less widened. Virga of male genitalia wider .... Genus 3: **Allostethella**, p. 416

Genus 1: **Gonolabidura** ZACHER

1910 *Gonolabidura* ZACHER, Ent. Rundsch., Stuttgart, **27**: 28. – Species typica: *Gonolabidura voltzi* ZACHER 1910.
Literature: BRINDLE 1965; BURR 1911; POPHAM & BRINDLE 1968; SAKAI 1970; STEINMANN 1975, 1977; TOWNES 1945.

Description: Apterous insects, tegmina and wings totally absent. General colour in both sexes reddish or brownish to dark brownish black; mouthparts, legs moderately lighter, yellow or yellowish brown. Head broad, pronotum transverse. Abdomen fusiform, or cylindrical; forceps strongly arcuate, tapering. Male genitalia comparatively large, broad; central parameral plate with small or deep median incision; genital lobes well-developed, with virgae.

Distribution: Oriental species.

6 species.

Identification key to the species

1 Antennal joints 4–5 transverse; legs bicolorous. Female forceps (Fig. 634) almost straight, strongly trigonal, straight .... 1. **Gonolabidura javana**, p. 400
1′ Antennal joints 4–5 longer than wide; legs more or less unicolorous **2**
2 Very dark, almost black species. Male ultimate tergite simple, surface smooth though finely striate, this texture becoming more irregular near truncate caudad margin. Forceps (Fig. 635) undeveloped in holotype, male genitalia (Fig. 636) oval, undeveloped (in holotype), without genital lobes .... 2. **Gonolabidura boschmai**, p. 401
2′ Dark reddish, reddish-brown or lighter species **3**
3 Abdominal tergites sparsely covered with shallow punctures. Male ultimate tergite without lateral longitudinal ridges; both male and female forceps asymmetrical. Male genitalia of *Gonolabidura*-type; broad, virga within genital lobe sclerotized (Fig. 638) .... 3. **Gonolabidura piligera**, p. 402
3′ Abdominal tergites more deeply and closely punctate. Male ultimate tergite with well-marked lateral longitudinal ridges, both male and female forceps symmetrical **4**
4 Larger species, length of body with forceps: about 16 mm or more. Male with strongly curved forceps (Fig. 639) .... 4. **Gonolabidura astruci**, p. 403
4′ Smaller species, 15 mm in body length or less. Male with forceps less curved **5**
5 Darker and larger species; blackish-brown with abdomen lighter; median longitudinal impressed line or sulcus deep in both sexes. Posterior margin of male ultimate tergite more or less straight (Fig. 641). Male genitalia very similar to *minor* BURR, as in fig. 642 .... 5. **Gonolabidura nathani**, p. 404
5′ Lighter and smaller species; almost uniformly reddish-brown; median longitudinal sulcus of ultimate tergite almost absent in both sexes. Posterior margin of ultimate tergite in male concave (Fig. 643). Male genitalia very similar to *nathani* BRINDLE as in fig. 644 .... 6. **Gonolabidura minor**, p. 405

1. Gonolabidura javana BOESEMAN

1954 *Gonolabidura javana* BOESEMAN, Zool. Verh., Leiden, **21**: 26; fig. 6a ("habitus of male in dorsal view", but female), 6b (penultimate sternite with forceps). – Terra typica: Java (Type female, in original description as male: Rijksmus. natuurl. Hist., Leiden).

1980 *Gonolabidura javana* – STEINMANN, Fol. ent. hung., Budapest, **41**: 339; fig. 30 (holotype ultimate tergite with forceps).

Literature: POPHAM & BRINDLE 1968; SAKAI 1970; STEINMANN 1977, 1978.

Description: Female general colour dark reddish-brown, shiny; legs bicolour; antennal joints darker, one or more distal joints lighter yellow or whitish yellow. Head very broad; posterior margin straight; postfrontal and coronal sutures deep, well marked; eyes comparatively large, but shorter than length of head behind eyes. First antennal joint long, longer than distance between antennal bases; antennae generally mutilated, maximal number of joints 22. Shape of fourth and fifth joints slightly variable, transverse, distinctly shorter and thicker than the species typica. Pronotum transverse, median longitudinal furrow present; lateral margins straight, parallel; all angles rounded. Mesonotum broad with posterior margin convex; metanotum transverse, concave posteriorly. Abdomen relatively short, ultimate tergite broad with well-marked median longitudinal sulcus. Forceps (Fig. 634) broad, remote at base, straight basally but evenly curved towards apex; inner margins with one prominent tooth basally. – **Male** unknown. – **Length** of body with forceps: 17–24 mm.

Distribution: Java.

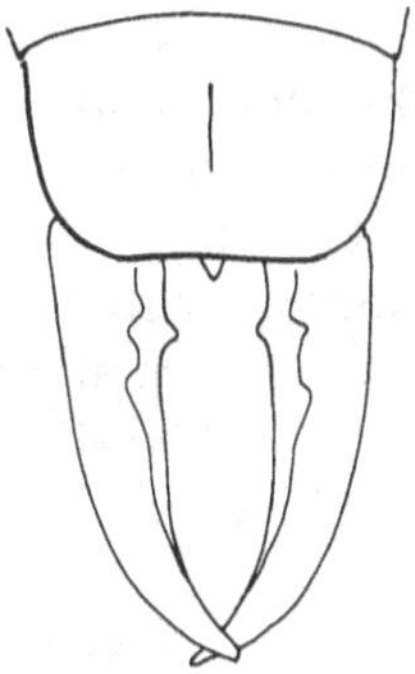

Fig. 634 Holotype ultimate tergite and forceps of *Gonolabidura javana* BOESEMAN. After STEINMANN.

2. Gonolabidura boschmai BOESEMAN

1954 *Gonolabidura boschmai* BOESEMAN, Zool. Verh., Leiden, **21**: 28; fig. 8 (male). – Terra typica: Celebes (Type male: Rijksmus. nat. Hist., Leiden. This species was presumably based on a larva, gen. prep. No. 684, det. Dr. H. STEINMANN).

1980 *Gonolabidura boschmai* – STEINMANN, Fol. ent. hung., Budapest, **41**: 341; fig. 31 (genital armature of holotype).

Literature: POPHAM & BRINDLE 1968; SAKAI 1970; STEINMANN 1977, 1978.

Description: Male body very short and broad; colour very dark, almost black; head shiny; thoracic nota and abdominal tergites punctate. Head very broad; postfrontal and coronal sutures distinct; eyes comparatively large, broad, but shorter than length of head behind eyes. Antennae broken in holotype; first antennal joint long, but shorter than distance between antennal bases. Pronotum transverse; lateral margins straight, a little expanded posteriorly: median longitudinal furrow very long and deep; posterior margin truncate. Mesonotum broad, posterior margin straight; metanotum transverse, with very concave posterior margin. Abdomen short, broad, as long the length of hind femur

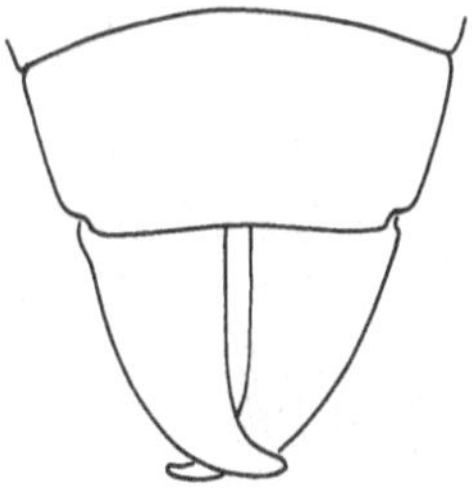

Fig. 635 Holotype ultimate tergite and forceps of *Gonolabidura boschmai* BOESEMAN. After BOESEMAN.

Fig. 636 Holotype genital armature of *Gonolabidura boschmai* BOESEMAN. After STEINMANN.

without ultimate tergite. Last tergite simple, surface smooth though finely striate, this texture becoming more irregular near the truncate posterior margin. Forceps (Fig. 635) short, stout, undeveloped in holotype, strongly tapering, round in cross-section. Penultimate sternite rounded posteriorly. Genitalia (Fig. 636) very simple in holotype; central parameral plate broad, median incision of anterior margin very small; genital lobes and virgae absent; external parameres simple, rounded apically. – **Female** unknown. – **Length** of body with forceps: 15 mm.

Distribution: Celebes.

3. Gonolabidura piligera (BORMANS)

1900 *Anisolabis piligera* BORMANS, Ann. Mus. civ. Stor. nat., Genova, **20**: 453. – Terra typica: Sumatra (Type male: Mus. civ. Stor. nat., Genova).

1910 *Gonolabidura voltzi* ZACHER, Ent. Rundsch., Stuttgart, **27**: 28. – Terra typica: Sumatra (Type male: Mus. Naturkunde, Berlin).

1911 *Gonolabidura piligera* – BURR, Genera Insectorum, Bruxelles, **122**: 25.

1965 *Gonolabidura piligera* – BRINDLE, Ann. Mag. nat. Hist., London, (13) **8**: 393; fig. 4 (basal antennal joints), 6 (pronotum), 23 (male genitalia).

Literature: BOESEMAN 1954; BORELLI 1926; KIRBY 1904; POPHAM & BRINDLE 1968; SAKAI 1970; STEINMANN 1977, 1978.

Description: Male colour dark reddish-brown, shiny; antennae brown, legs yellow or yellowish-brown. Head tumid, postfrontal and coronal sutures obsolete; eyes rather small than length of head behind eyes. Antennae with short

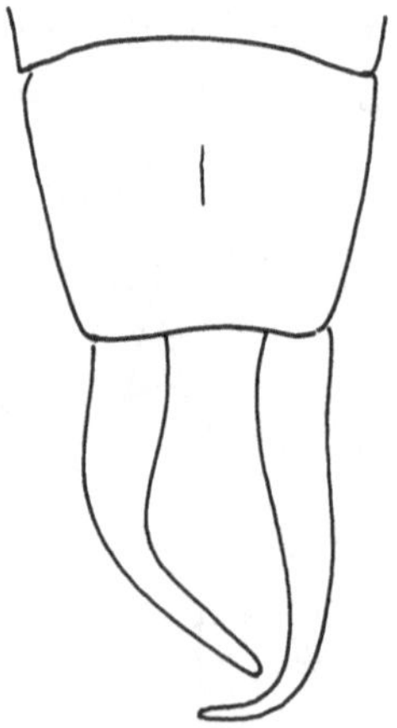

Fig. 637 Male ultimate tergite and forceps of *Gonolabidura piligera* (BORMANS). Original.

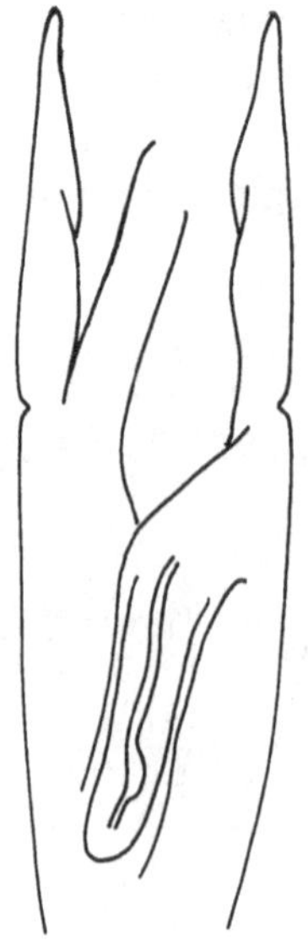

Fig. 638 Male genital armature of *Gonolabidura piligera* (BORMANS). After BRINDLE.

first joint, second joint transverse, third joint long, about twice as long as wide; joints 4–5 longer than wide. Pronotum transverse, lateral margins straight, and parallel; all angles rounded, posterior margin truncate. Mesonotum transverse, with a lateral longitudinal rounded ridge ending before posterior margin. Metanotum transverse, posterior margin concave. Abdomen short, abdominal tergites sparsely covered with shallow punctures. All thoracic nota and head impunctate. Abdomen rather dilated posteriorly. Ultimate tergite with median longitudinal sulcus only present medially, not deep. Forceps (Fig. 637) asymmetrical, widest basally, elongated apically. Penultimate sternite rounded posteriorly. Genitalia (Fig. 638) of *Gonolabidura*-type; paramere broad, median incision of anterior margin very deep, genital lobes well developed, virga strongly sclerotized. External parameres very long, narrow. – **Female** similar to male, but without lateral longitudinal ridge on ultimate tergite, and with almost straight contiguous forceps, which are broader at base. – **Length** of body with forceps: male: 19–20,5 mm, female: 19–21,4 mm.

Distribution: Sumatra.

4. Gonolabidura astruci Burr

1911 *Gonolabidura astruci* Burr, J. asiat. Soc. Bengal, Calcutta, **7**: 776. – Terra typica: South India (Type male: Brit. Mus. nat. Hist., London).

1965 *Gonolabidura astruci* – Brindle, Ann. Mag. nat. Hist., London, (13) **8**: 593; fig. 15 (male forceps), 24 (male genitalia).

Literature: Borelli 1926; Popham & Brindle 1968; Sakai 1970; Steinmann 1977, 1978.

Description: Male colour dark reddish-brown, antennae brown, legs yellowish brown, yellow basally, and darkened distally. Head tumid, broad; postfrontal and coronal sutures fairly prominent; eyes rather smaller than length of head behind eyes. First antennal joint short, shorter than distance between antennal bases. Pronotum transverse, lateral margins more or less parallel; median longitudinal furrow present at median section. Mesonotum normal, broad; metanotum transverse, concave posteriorly. Abdomen more strongly punctate; tergites expanded to median portion. Ultimate tergite broad, with lateral longitudinal ridges; posterior margin straight. Forceps (Fig. 639) symmetrical, branches strongly curved; trigonal basally, cylindrical apically; inner margin crenulate basally. Penultimate sternite typical of genus, rounded posteriorly. Genitalia (Fig. 640) typical of genus; central parameral plate broad, median incision of anterior margin very deep; plate V-shaped; genital lobes well developed, virga within genital lobes strongly sclerotized. External parameres elongated, narrow. – **Female** similar to male, but without lateral longitudinal ridge on ultimate tergite, and with almost straight contiguous forceps, which are broader at base. – **Length** of body with forceps: male: 19–20,5 mm, female: 19–21,5 mm.

Distribution: South India.

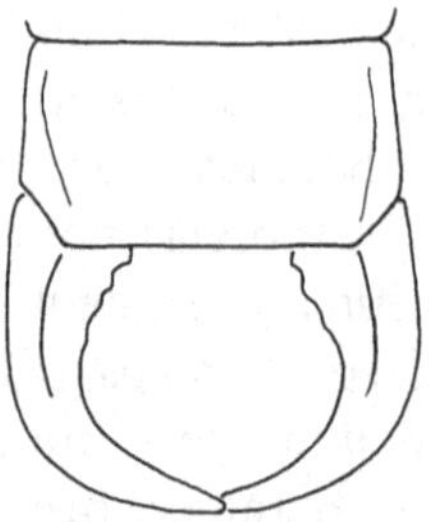

Fig. 639 Male ultimate tergite and forceps of *Gonolabidura astruci* BURR. Original.

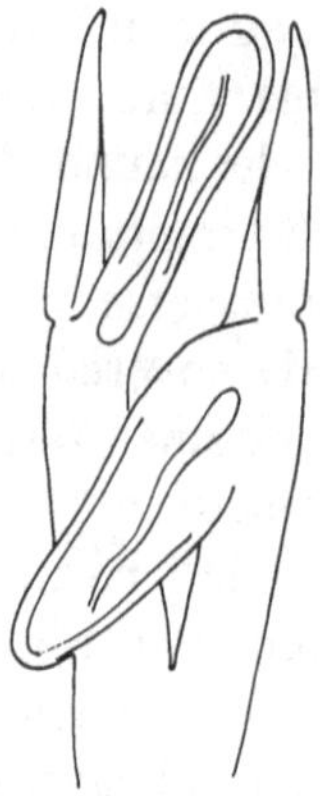

Fig. 640 Male genital armature of *Gonolabidura astruci* BURR. After BRINDLE.

5. Gonolabidura nathani BRINDLE

1965 *Gonolabidura nathani* BRINDLE, Ann. Mag. nat. Hist., London, (13) **8**: 594; fig. 21 (holotype genital armature). – Terra typica: South India (Type male: Brit. Mus. nat. Hist., London).
Literature: POPHAM & BRINDLE 1968; SAKAI 1970; STEINMANN 1977, 1978.

Description: Male head black or reddish black, pronotum shiny, blackish, abdomen reddish-brown; antennae yellow basally, darkened distally; legs yellow. Head tumid, postfrontal and coronal sutures fairly distinct; antennae 11-jointed in type; first joint shorter than distance between antennal bases, second transverse, third joint about twice as long as wide, fourth and fifth joints longer than wide. Pronotum more or less transverse, all margins straight, and all angles

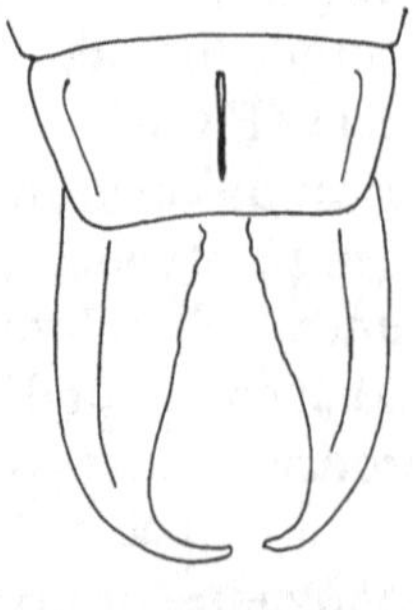

Fig. 641 Male ultimate tergite and forceps of *Gonolabidura nathani* BRINDLE. Original.

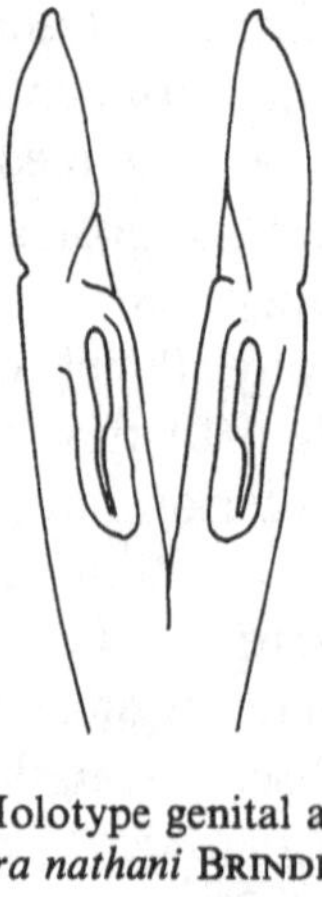

Fig. 642 Holotype genital armature of *Gonolabidura nathani* BRINDLE. After BRINDLE.

rounded; median longitudinal furrow present. Mesonotum broad, metanotum transverse. Abdomen parallel, strongly punctate. Ultimate tergite large, median longitudinal sulcus deep in middle; posterior margin more or less straight. Forceps (Fig. 641) symmetrical, trigonal basally, curved and cylindrical apically. Penultimate sternite rounded posteriorly. Genitalia (Fig. 642) of *Gonolabidura*-type, very similar to *minor* BURR; central parameral plate V-shaped, median incision of anterior margin very deep; genital lobes directed backwards, with strongly sclerotized virgae. External paramere comparatively long and broad. – **Female** similar to male, but forceps contiguous, broad at base, narrowed distally, with apices curved mesad. – **Length** of body with forceps: male: 12,5–13,5 mm, female: 13–14,5 mm.

Distribution: South India.

6. Gonolabidura minor BURR

1914 *Gonolabidura minor* BURR, Rec. Indian Mus., Calcutta, **10**: 284. – Terra typica: South India (Type male: Brit. Mus. nat. Hist., London).

1965 *Gonolabidura minor* – BRINDLE, Ann. Mag. nat. Hist., London, (13) **8**: 594; fig. 16 (male forceps), 22 (male genitalia).

Literature: POPHAM & BRINDLE 1968; SAKAI 1970; STEINMANN 1977, 1978.

Description: Male colour almost uniformly reddish-brown. Head tumid, postfrontal and coronal sutures obsolete. Antennae 15-jointed; first shorter than length of head behind eyes; joints yellowish basally, and darkened distally; second joint transverse, third about one and a half times as long as wide; fourth and fifth longer than wide. Eyes typical of genus. Pronotum more or less as long as wide; lateral margins parallel, and all angles rounded; median longitudinal furrow present. Meso- and metanotum normal. Abdomen rather dilated posteriorly, strongly punctate. Ultimate tergite with lateral longitudinal ridge, but median longitudinal sulcus absent; posterior margin concave medially.

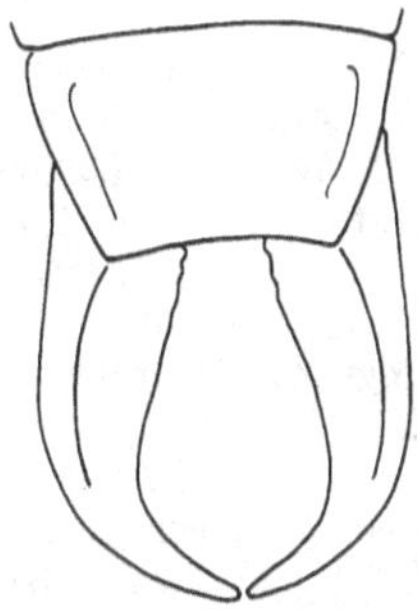

Fig. 643 Male ultimate tergite and forceps of *Gonolabidura minor* BURR. After BRINDLE.

Fig. 644 Male genital armature of *Gonolabidura minor* BURR. After BRINDLE.

Forceps (Fig. 643) symmetrical, strongly trigonal basally, curved and cylindrical apically. Penultimate sternite rounded posteriorly. Genitalia (Fig. 644) very similar to *nathani* BRINDLE; paramere V-shaped, median incision of anterior margin very deep; genital lobes directed backwards, with strongly sclerotized virgae. External parameres long and narrow. – **Female** similar to male, except for forceps which are contiguous, and that the ultimate tergite is without lateral ridge. – **Length** of body with forceps: in both sexes: 11–13 mm.

Distribution: South India.

Genus 2: **Allostethus** VERHOEFF

1904 *Allostethus* VERHOEFF, Arch. Naturgesch., Berlin, **1904**: 116. – Species typica: *Forficula indica* BURMEISTER 1838.

Literature: BOESEMAN 1954; BRINDLE 1965; BURR 1910, 1911; POPHAM & BRINDLE 1968; SAKAI 1970; STEINMANN 1977, 1978; TOWNES 1945; ZACHER 1911.

Description: Tegmina and often the wings perfectly developed. Larger species, length of body with forceps: 12–32 mm. Build powerful. Antennae multisegmented; third joint very short, about as long as wide. Prosternum and mesosternum strongly narrowed posteriorly. Pronotum and mesonotum typical, pronotum only slightly wider than long. Abdomen more or less parallel-sided or expanded medially. Forceps moderately elongated, basally not strikingly widened. Male genitalia large, median incision of anterior margin deep; virga within genital lobe feebly bowed, pecicular, comparativelly narrow, sclerotized section long.

Distribution: Oriental species.

9 species.

Identification key to the species

1 Length of body with forceps 15 mm, or less **2**

1′ Length of body with forceps at least 18 mm, usually well above **4**

2 Pronotum almost quadrate; tegmina long, about twice as long as pronotum; wings present. Female forceps (Fig. 645) with large and prominent tooth on inner margins medially .... 1. **Allostethus minus**, p. 407

2′ Pronotum transverse **3**

3 Tegmina both with yellow spot anteriorly; male forceps simple, very similar to *burri* BRINDLE, as in fig. 646. External parameres of male genitalia broader, and rather acuminate towards apices (Fig. 647) .... 2. **Allostethus philippinense**, p. 408

3′ Tegmina unicolorous, dark brown; male forceps typical, similar to *philippinense* BORELLI, as in fig. 648. External parameres of male genitalia more slender, tapered gradually towards apices, as in fig. 649 .... 3. **Allostethus burri**, p. 409

4 Tegmina unicolorous, dark reddish-brown, or almost so; wings usually present **5**

4′ Tegmina mainly yellowish, or dark brown with a distinct and well defined yellow blotch on each tegmen; wings often absent **7**

5 Wings brown; femora and tibiae dark brown, except for extreme apices; wings almost entirely dark brown; penultimate sternite of male with raised broad longitudinal ridge which is bordered by short yellow hairs. Male forceps asymmetrical, inner margins with strong tooth or teeth (Fig. 650) .... 4. **Allostethus indicum**, p. 409

5′ Wings yellow **6**

6 Male forceps asymmetrical (Fig. 652), a little lamellated and triangular in cross-section basally; inner margins with very large tooth. External parameres of male genitalia broad medially; virga within genital lobe with sclerotized preputial sac (Fig. 653) .... 5. **Allostethus setiger**, p. 411

6′ Male forceps more or less symmetrical, a little oval in cross-section basally; inner margin without or with very small tooth. External paramere of male genitalia narrow, virga within genital lobe without preputial sac (Fig. 654) .... 6. **Allostethus gracile**, p. 412

7 Forceps in both sexes short, 3 mm or so. Tegmina rather longer than pronotum; yellow with anterior and sutural margins black; wings absent. Male genitalia as in fig. 656 .... 7. **Allostethus celebense**, p. 413

7′ Forceps in both sexes longer, about 5 mm, male forceps strongly asymmetrical, as in figs. 657 or 659 .... **8**

8 Tegmina yellow with anterior and sutural margins blackish or brown, but without a metallic bluish sheen. Male forceps as in fig. 657. Male genitalia with comparatively broad external parameres, as in fig. 658 .... 8. **Allostethus lombokianum**, p. 414

8′ Tegmen brown or reddish brown, with a metallic bluish sheen. Male forceps as in fig. 659. Male genitalia with comparatively narrow external parameres, as in fig. 660 .... 9. **Allostethus anamalayanum**, p. 415

1. Allostethus minus (BORMANS)

1900 *Psalis indica* var. *minor* BORMANS, Ann. Mus. civ. Stor. nat., Genova, **20**: 447. – Terra typica: Sumatra (Type female: unknown locality).

1911 *Allostethus indicum* var. *minor* – BURR, Genera Insectorum, Bruxelles, **122**: 25. Java, Borneo.

1965 *Allostethus minor* – BRINDLE, Ann. Mag. nat. Hist., London, (13) **8**: 588; fig. 8 (forceps of female). Sumatra.

Literature: BOESEMAN 1954; BRINDLE 1968; KIRBY 1904; SAKAI 1970; STEINMANN 1977, 1978.

Description: Female reddish-brown; abdomen with reddish posterior segments. Head tumid, postfrontal sutures and coronal suture not depressed; posterior margin emarginate. Eyes relatively large, but shorter than length of head behind eyes. First antennal joint short, slightly shorter than distance between antennal bases; second transverse, third short, more or less as long as broad. Pronotum almost quadrate; shiny; lateral margins moderately parallel; median longitudinal furrow present, well marked. Tegmina long, about twice as long as pronotum. Wings almost equal in length to pronotum, yellow, darkened apically. Abdomen more or less parallel-sided, or a little expanded medially; ultimate tergite broad, posterior margin with two small tubercles; median longitudinal sulcus present. Forceps (Fig. 645) symmetrical, broad and a little

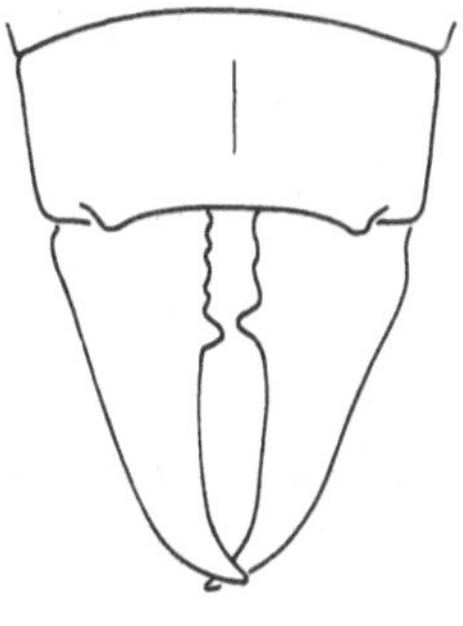

Fig. 645 Female ultimate tergite and forceps of *Allostethus minus* (BORMANS). After BRINDLE.

depressed basally, cylindrical apically; inner margin with large, prominent tooth at median section. Penultimate sternite rounded posteriorly. – **Male** unknown. – **Length** of body with forceps: 13,5–14,5 mm.

Distribution: Sumatra, Java, Borneo.

2. Allostethus philippinense Borelli

1926 *Allostethus philippinense* Borelli, Boll. Mus. Zool. Anat. comp. Univ., Torino, **1**: 62; fig. 1 (holotype genital armature). – Terra typica: Mindanao (Type male: unknown locality).
Literature: Brindle 1965; Popham & Brindle 1968; Sakai 1970; Steinmann 1977, 1978.

Description: Male colour generally reddish-brown; antennae brown, legs and mouthparts yellowish brown. Head tumid, postfrontal and coronal sutures distinct. Eyes relatively large, about as long as length of head behind eyes. Antennae 14-jointed; first joint moderately small, shorter than distance between antennal bases; second joint quadrate, third very short, about as long as wide. Pronotum transverse, lateral margins more or less parallel; all angles rounded; median longitudinal furrow distinct. Tegmina short, rather longer than pronotum, dark brown with an indistinct yellowish-brown spot towards anterior margin of each tegmen. Wings absent. Legs pale yellow, femora darkened proximally. Abdomen dark reddish, almost parallel, or a little expanded to middle. Ultimate tergite broad; posterior margin concave medially. Forceps (Fig. 646) symmetrical, simple, a little trigonal basally, cylindrical apically. Penultimate sternite rounded posteriorly. Genitalia (Fig. 647) of *Allostethus*-type, central parameral plate narrow, V-shaped, median longitudinal incision of anterior margin very deep; genital lobes well developed, virga long, broad, with small sclerotized plate at base. External parameres very long, narrow. – **Female** similar to male, but forceps approximately short, 2 mm. – **Length** of body with forceps: male: 12,5 mm, female: 15 mm.

Distribution: Philippine Islands: Mindanao.

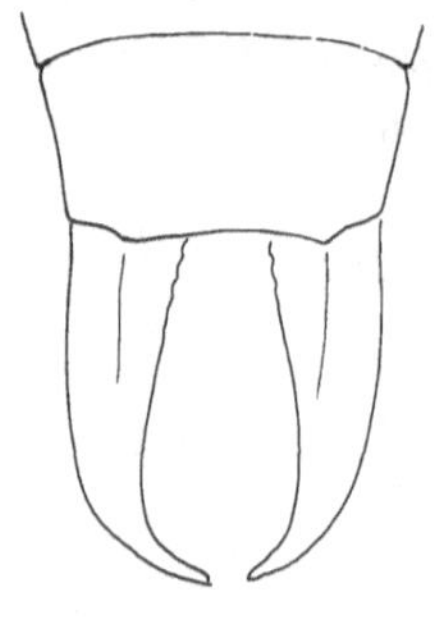

Fig. 646 Male ultimate tergite and forceps of *Allostethus philippinense* Borelli. Original.

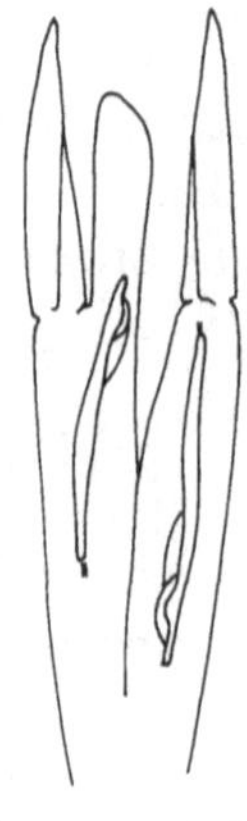

Fig. 647 Holotype genital armature of *Allostethus philippinense* Borelli. After Borelli.

3. Allostethus burri BRINDLE

1965 *Allostethus burri* BRINDLE, Ann. Mag. nat. Hist., London, (13) **8**: 589; fig. 25 (holotype genital armature). – Terra typica: Perak (Type male: BURR's Collect., Brit. Mus. nat. Hist., London).
Literature: POPHAM & BRINDLE 1968; SAKAI 1970; STEINMANN 1977, 1978.

Description: Male colour dark reddish-brown; antennae brown, last joint white, body shiny, legs yellowish brown, femora brown on basal two-thirds. Head tumid, postfrontal and coronal sutures distinct; antennae 16-jointed in type; first joint shorter than distance between antennal bases, second joint transverse, third small, only slightly longer than wide, fourth joint quadrate, fifth rather longer, sixth and succeeding joints becoming longer, tapered basally, rather truncate distally, hairy. Pronotum a little wider than long, lateral margins more or less parallel. Tegmina short, equal in length to pronotum. Scutellum visible, dark brown, triangular. Wings absent. Abdomen punctate, simple, ultimate tergite broad, typical of genus, posterior margin straight; median longitudinal sulcus short, about one-third length of tergite; lateral longitudinal ridge present on both sides of tergite, becoming well marked posteriorly, where it curves mesad to end above the base of the branches of the forceps. Forceps (Fig. 648) symmetrical, trigonal basally, cylindrical apically; broaded basally, narrowed gradually for basal two thirds, distal third curved mesad. Penultimate sternite rounded posteriorly. Genitalia (Fig. 649) large, median incision of anterior margin deep and wide; genital lobes well developed, virga thick, sclerotized. External parameres long and narrow. – **Female** unknown. – **Length** of body with forceps: 15 mm.

Distribution: Malaysia: Perak.

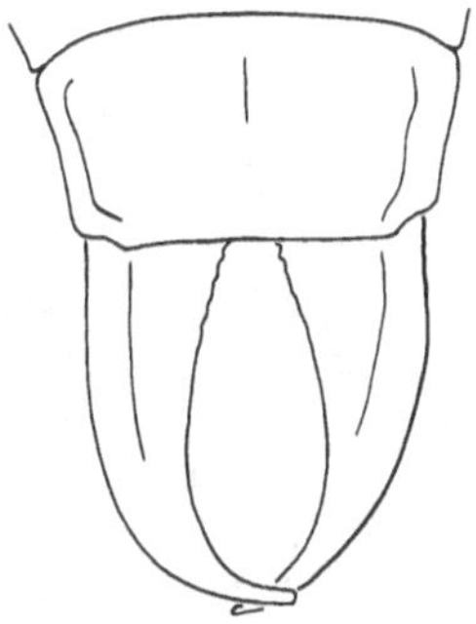

Fig. 648 Male ultimate tergite and forceps of *Allostethus burri* BRINDLE. Original.

Fig. 649 Holotype genital armature of *Allostethus burri* BRINDLE. After BRINDLE.

4. Allostethus indicum (BURMEISTER)

1838 *Forficula indica* BURMEISTER, Handb. Ent., Jena, **2**: 751. – Terra typica: Java (Type male: unknown locality).

1855 *Forficula geniculata* STÅL, Ark. Forhandl., Stockholm, **12**: 349. – Terra typica: Borneo (Type female: unknown locality).
1863 *Labidura indica* – DOHRN, Stettiner ent. Ztg., **24**: 24.
1896 *Psalis borneensis* KIRBY, J. Linn. Soc. London, Zool., **25**: 525. – Terra typica: Borneo (Type female: Brit. Mus. nat. Hist., London).
1900 *Psalis indica* – BORMANS, Das Tierreich, Berlin, **11**: 38; fig. 17 (male).
1905 *Labidurodes magnificus* REHN, Proc. U. S. nation. Mus., Washington, **29**: 504; fig. 3 (male). – Terra typica: Java (Type male: Nation. collect. U. S., Washington).
1911 *Allostethus indicum* – BURR, Genera Insectorum, Bruxelles **122**: 25; pl. 8, fig. 21 (opisthomeres of female).
1911 *Allostethus indicum* – ZACHER, Zool. Jahrb., Jena, **30**: 365; fig. R[1] (male genitalia).
1954 *Allostethus indicum* var. *maculatum* BOESEMAN, Zool. Verh., Leiden, **21**: 25. – Terra typica: Celebes (Type female: Rijksmus. natuurl. Hist., Leiden).

Literature: BORELLI 1926; BRINDLE 1965; BURR 1915; HEBARD 1927; KIRBY 1904; POPHAM & BRINDLE 1968, SAKAI 1970; STEINMANN 1977, 1978.

Description: Male almost entirely dark reddish-brown; antennae brown except for joints 15–16 which are whitish; abdomen dark blackish-brown dorsally, lighter ventrally; wings with a small yellow spot at base of sutural margins; femora and tibiae reddish-brown except at apices which are yellow. Head tumid, postfrontal and coronal sutures well marked. Antennae 18-jointed; first joint relatively small, shorter than distance between antennal bases; second quadrate, and third joint small, only slightly longer than wide. Eyes large, broad, but shorter than length of head behind eyes. Pronotum transverse, lateral margins parallel, all angles rounded. Tegmina long, with a well marked lateral longitudinal ridge, which begins ventrad to shoulders and extends almost to apex. Wings present. Abdomen normal. Ultimate tergite with median longitudinal sulcus; posterior margin with two small tubercles above bases of the forceps. Forceps (Fig. 650) more or less asymmetrical, strongly trigonal basally,

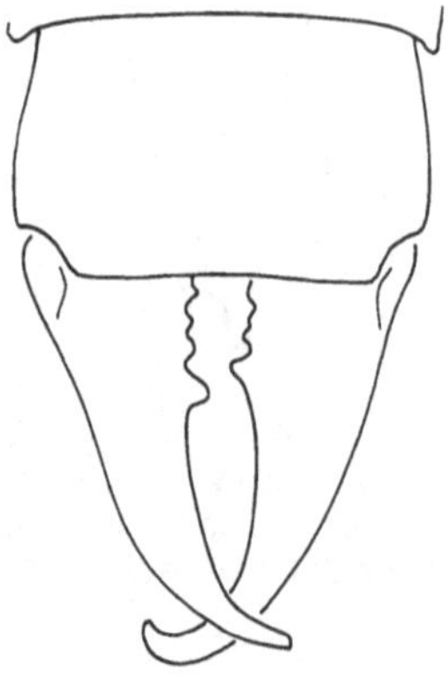

Fig. 650 Male ultimate tergite and forceps of *Allostethus indicum* (BURMEISTER). Original.

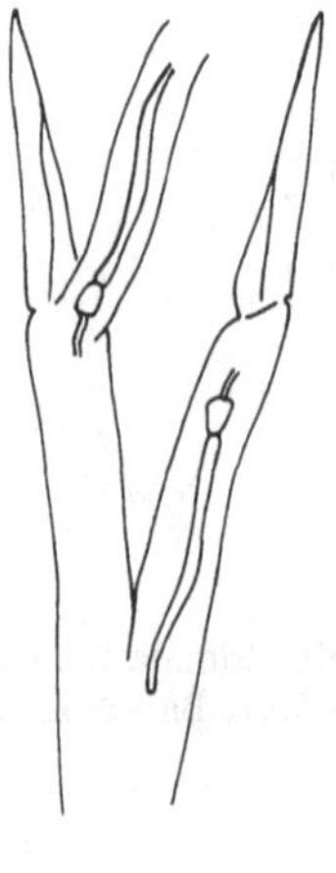

Fig. 651 Male genital armature of *Allostethus indicum* (BURMEISTER). Original.

cylindrical apically. Penultimate sternite strongly punctate, with a broad longitudinal median ridge, which extends for basal two-thirds of plate. Genitalia (Fig. 651; gen. prep. No. 734, det. Dr. H. STEINMANN) very narrow, central parameral plate with very deep and wide median incision between external parameres; genital lobes well developed, virga sclerotized. External parameres very long, acuminate. – **Female** similar to male, but penultimate sternite without ridge, and forceps slenderer, tapering. – **Length** of body with forceps: male: 24–26 mm, female: 26–30,5 mm.

Distribution: Java, Borneo, Celebes, Sumatra, Malaysia, and possibly Philippines.

5. Allostethus setiger VERHOEFF

1904 *Allostethus setiger* VERHOEFF, Arch. Naturgesch., Berlin **1904**: 117. – Terra typica: Java (Type male: unknown locality).
1904 *Allostethus martensi* VERHOEFF, Arch. Naturgesch., Berlin, **1904**: 119. – Terra typica: Sumatra (Type male: unknown locality).
1954 *Allostethus indicum* var. *setiger* – BOESEMAN, Zool. Verh., Leiden., **21**: 24. Java, Sumatra.
1965 *Allostethus setiger* – BRINDLE, Ann. Mag. nat. Hist., London, (13) **8**: 587; fig. 12 (male genitalia). Java, Thailand, Sumatra.
Literature: BURR 1911, 1915; HEBARD 1927; POPHAM & BRINDLE 1968; SAKAI 1970; STEINMANN 1977, 1978.

Description: Male colour dark, darker than *indicum* (BURMEISTER), and distinct for its yellow wings, which are only slightly darkened at the extreme costal margins, and for the femora and tibiae having the apical third yellow. Head tumid, postfrontal and coronal sutures distinct, antennal joints broader; first joint small, shorter than distance between antennal bases, second quadrate, third and fourth joints transverse. Eyes comparatively large, but a little shorter than

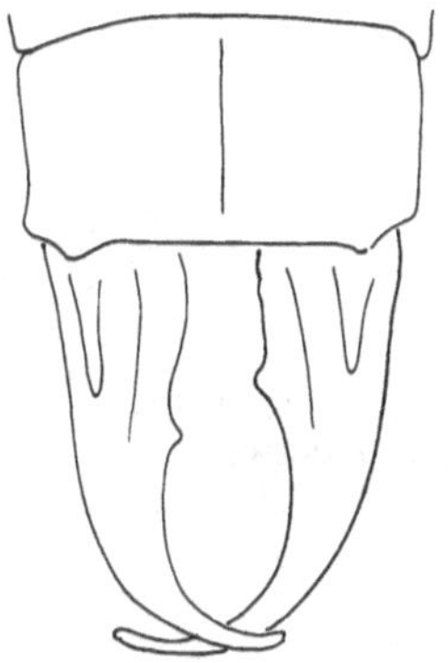

Fig. 652 Male ultimate tergite and forceps of *Allostethus setiger* VERHOEFF. Original.

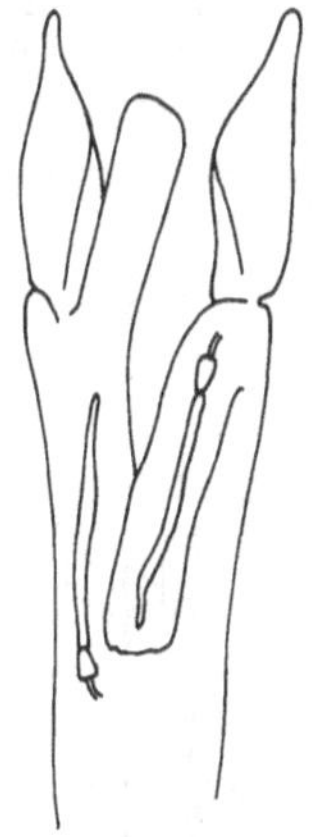

Fig. 653 Male genital armature of *Allostethus setiger* VERHOEFF. Original.

length of head behind eyes. Pronotum transverse, smooth, broader, almost as broad as tegmina; median longitudinal furrow distinct; lateral margins more or less parallel. Tegmina smooth, varying in length, of the wings varies between slightly shorter and hardly longer than the tegmina, the variation within these extremes sometimes being shown in a single species. Abdomen normal, ultimate tergite very large, broad; median longitudinal sulcus distinct; posterior margin straight, with two small tubercles. Forceps (Fig. 652) asymmetrical, strongly trigonal basally, cylindrical apically; inner margin with one very large tooth. Penultimate sternite typical of genus. Genitalia (Fig. 653; gen. prep. No. 357, det. Dr. H. STEINMANN) of *Allostethus*-type; central parameral plate with deep median incision; genital lobes well developed, relatively large, virgae sclerotized. External parameres moderately broad medially. – **Female** similar to male, but forceps slenderer, tapering. – **Length** of body with forceps: male: 23–26 mm, female: 25–32 mm.

Distribution: Thailand, Java, Sumatra.

6. Allostethus gracile BRINDLE

1965 *Allostethus gracilis* BRINDLE, Ark. Zool., Uppsala, (18) **18**: 442; fig. 6 (holotype genital armature). – Terra typica: Mindanao (Type male: Naturhist. Riksmus., Stockholm).
Literature: STEINMANN 1978.

Description: Male colour dark reddish-brown; antennae brown, legs yellowish-brown. Head broad, tumid, transverse; eyes comparatively large, equal in length to length of head behind eyes; postfrontal and coronal sutures distinct. Antennae 15-jointed in type; first joint elongate, joints 2–5 transverse to quadrate, remainder gradually lengthening distally. Pronotum transverse, impunctate; lateral margins straight and parallel, anterior margin straight, posterior margin slightly convex; median longitudinal furrow present. Tegmina long, about as long as length of pronotum, narrow, and truncate at apices. Wings

Fig. 654 Holotype genital armature of *Allostethus gracile* BRINDLE. After BRINDLE.

present, yellow, short, about one-quarter the length of tegmina. Legs typical of genus, femora darkened on basal half and at apices. Scutellum visible, small, triangular, blackish. Abdomen elongate, parallel-sided; shiny, strongly but not closely punctate, and with long sparse, brown setae. Ultimate tergite large, quadrate, narrowed apically, with median longitudinal sulcus. Forceps broadened at base, narrowed distally, triangular in cross-section basally, with dorsal ridge placed towards external margin; circular in cross-section distally. Genitalia (Fig. 654) very narrow, simple; central parameral plate with deep but narrow median incision genital lobes simple, virga characteristic, straight, a little undulate, sclerotized; external parameres very narrow, acuminate. – **Female** similar to male in almost all respects, forceps tending to be straighter. – **Length** of body with forceps: male: 19 mm, female: 19–20 mm.

Distribution: Philippines: Mindanao.

7. Allostethus celebense Burr

1911 *Allostethus celebense* Burr, Stettiner ent. Ztg., **1911**: 332. – Terra typica: Celebes (Type female: Zool. Inst., Warszawa. Type male: Természettudományi Múz., Budapest; gen. prep. No. 361, det. Dr. H. Steinmann).

1965 *Allostethus celebense* – Brindle, Ann. Mag. nat. Hist., London (13) **8**: 588; fig. 14 (male genitalia).

Literature: Burr 1911, 1912, 1915; Popham & Brindle 1968; Sakai 1970; Steinmann 1977, 1978.

Description: Male colour dark reddish-brown; tegmina yellow with anterior and sutural margins blackish or brown, but darker in adult specimens. Head, pronotum and abdomen black in mature, and brown in immature specimens. Head tumid, broad, rounded posteriorly; postfrontal and coronal sutures distinct; eyes large, equal in length to length of head behind eyes. First antennal

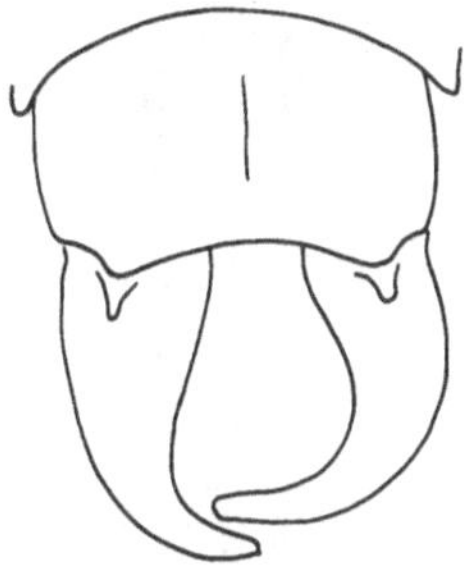

Fig. 655 Male ultimate tergite and forceps of *Allostethus celebense* Burr. After Steinmann.

Fig. 656 Male genital armature of *Allostethus celebense* Burr. Original.

joint typical of genus, more or less as long as distance between antennal bases or a little shorter. Pronotum transverse, lateral margins more or less parallel; median longitudinal furrow distinct. Tegmina rather longer than pronotum, without lateral longitudinal ridges. Wings absent, but BURR (1911) mentions specimens with yellowish-brown wings. Abdomen more or less parallel-sided, simple, elongated. Ultimate tergite very broad, large; with longitudinal impressed line deep but not reaching base of plate. Forceps (Fig. 655) short, branches broad at base, narrowed distally, almost asymmetrical. Genitalia (Fig. 656; gen. prep. No. 361, det. Dr. H. STEINMANN) very long; median incision of anterior margin of the central parameral plate very deep; genital lobes well marked, virga straight, a little sclerotized. External parameres very slender. – **Female** similar to male, but forceps tapering, somewhat trigonal basally, cylindrical apically; symmetrical. – **Length** of body with forceps: in both sexes: 18–21 mm.

Distribution: Celebes, possibly Philippine Islands.

8. Allostethus lombokianum VERHOEFF

1904 *Allostethus lombokianum* VERHOEFF, Arch. Naturgesch., Berlin, **1904**: 117. – Terra typica: Lombok (Type male: Brit. Mus. nat. Hist., London).
1911 *Allostethus lombokianum* – BURR, Genera Insectorum, Bruxelles, **122**: 25; pl. 2, fig. 9 (sternum).
1965 *Allostethus lombokianum* – BRINDLE, Ann. Mag. nat. Hist., London, (13) **8**: 588; fig. 13 (male genitalia). Lombok, Celebes, Moluccas.
Literature: BURR 1912, 1915; POPHAM & BRINDLE 1968; SAKAI 1970; STEINMANN 1977, 1978.

Description: Male colour dark, head, pronotum and abdomen dark reddish-brown, tegmina yellow with anterior and sutural margins blackish or brown; wings yellow or absent; legs dark reddish-brown with apical third or femora and tibiae yellow; all tarsi yellowish-brown. Head tumid, a little depressed;

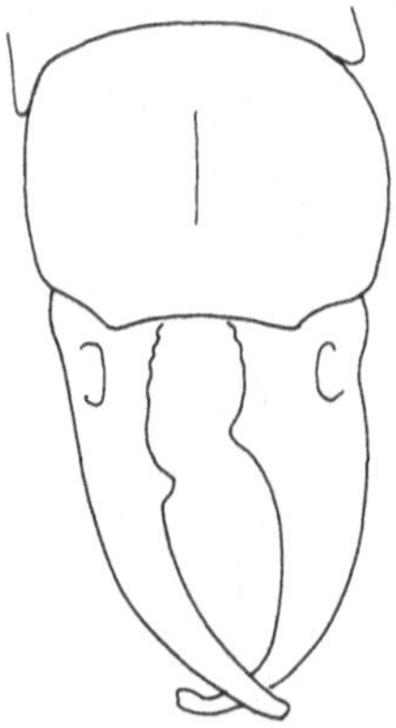

Fig. 657 Male ultimate tergite and forceps of *Allostethus lombokianum* VERHOEFF. Original.

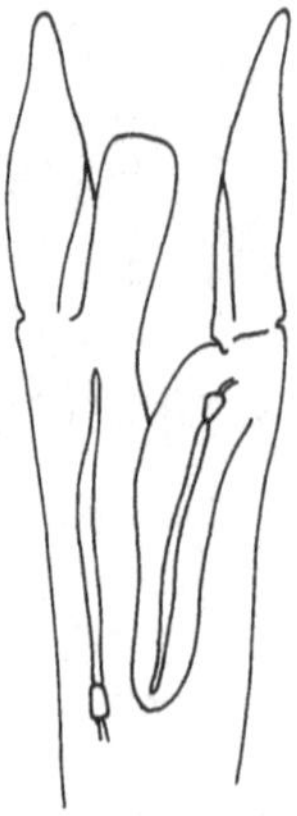

Fig. 658 Male genital armature of *Allostethus lombokianum* VERHOEFF. After BRINDLE.

postfrontal and coronal sutures distinct; eyes relatively small, slightly shorter than length of head behind eyes. First antennal joint small and broad. Pronotum transverse, lateral margins straight, more or less parallel; median longitudinal furrow present. Tegmina rather longer than pronotum, without lateral longitudinal ridge. Wings absent, but BURR (1912) mentions specimens from Moluccas with yellow wings. Abdomen fusiform, ultimate tergite very large, with median longitudinal sulcus medially. Forceps (Fig. 657) strongly asymmetrical, and curved; trigonal basally, cylindrical apically. Genitalia (Fig. 658) of *Allostethus*-type, central parameral plate with deep median incision; genital lobes well developed, virgae sclerotized, straight, thick. External parameres broad. – **Female** unknown. – **Length** of body with forceps: 21–27 mm.

Distribution: Lombok, Celebes, Molluccas.

9. Allostethus anamalayanum RAMAMURTHI

1968 *Allostethus anamalayanus* RAMAMURTHI, Zool. Anz., Leipzig, **181**: 133; fig. 5 (ultimate tergite and forceps of holotype), 6 (ultimate tergite and forceps of female), 7 (holotype genital armature). – Terra typica: India (Type male: RAMAMURTHI's Collect., Univ. Mus., Madras).
Literature: STEINMANN 1977, 1978.

Description: Male colour dark reddish-brown, head, pronotum and tegmina with a metallic bluish sheen; antenae and legs brownish, abdomen and forceps dark brown. Head tumid, distinctly longer than wide, postfrontal sutures and coronal suture moderately distinct. Antennae more than 20-jointed; first joint stouter and shorter, apical joints slender and elongate. Pronotum broadly transverse, sides depressed, median longitudinal furrow well developed, and with a shallow puncture on both sides of the prozona; all sides straight. Tegmina as

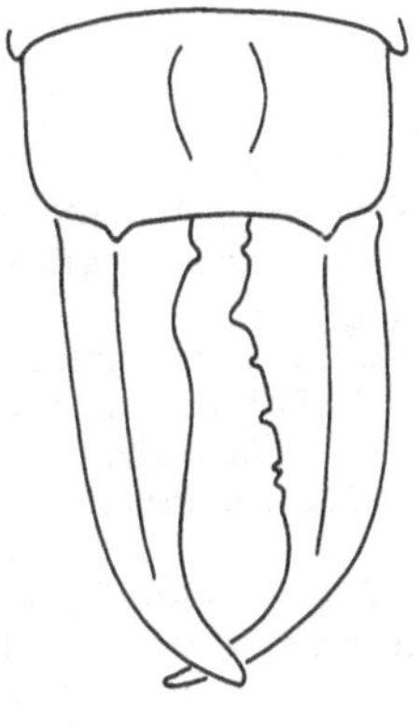

Fig. 659 Holotype ultimate tergite and forceps of *Allostethus anamalayanum* RAMAMURTHI. After RAMAMURTHI.

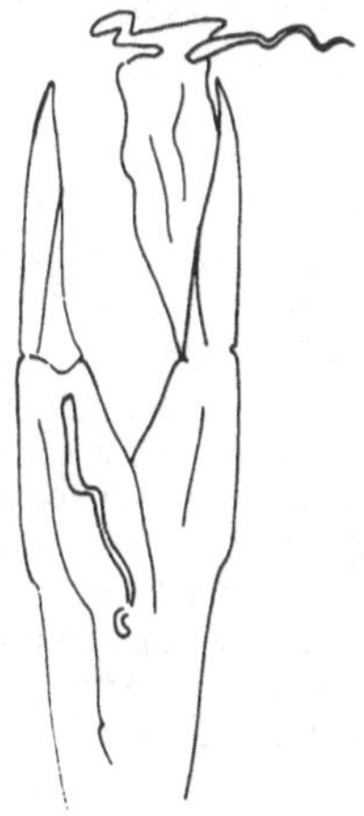

Fig. 660 Holotype genital armature of *Allostethus anamalayanum* RAMAMURTHI. After RAMAMURTHI.

long as pronotum, exposing a broadly triangular scutellum, caudal margin truncate. Abdomen widened medially, ultimate tergite transverse, anterior half sparsely, and posterior half densely, tuberculate. Forceps (Fig. 659) strongly asymmetrical, broadened at base and subcontiguous, gradually narrowed behind, arcuate; inner margins obliquely flattened, the ventral border broadly with irregularly arranged tubercles. Genitalia (Fig. 660) large; central parameral plate with two well-developed genital lobes; virgae strongly sclerotized. External parameres slender, acuminate. – **Female** similar to male, but forceps symmetrical, widened basally, curved apically, inner margins with some denticles. – **Length** of body with forceps: in both sexes: 21–24 mm.

Distribution: South India.

Genus 3: **Allostethella** ZACHER

1910 *Allostethella* ZACHER, Ent. Rundsch., Stuttgart, Species **27**: 30. – Species typica: *Allostethella malayana* ZACHER 1910.

Literature: BOESEMANN 1954; BRINDLE 1965; BURR 1911, 1915; POPHAM & BRINDLE 1968; SAKAI 1970; STEINMANN 1977, 1978; TOWNES 1945.

Description: Very similar to *Allostethus* VERHOEFF species in external characteristics, but smaller. Head broad, antennae with from eleven to eighteen joints. Lobe of metasternum pointed. Pronotum transverse, or more or less quadrate. Tegmina and wings well developed. Abdomen slender, more or less fusiform, sometimes depressed; ultimate tergite large, broad, posterior margin with two tubercles above branches of the forceps. Male forceps symmetrical or asymmetrical, branches crossed apically or gently curved. Female forceps simple, tapering. Male genitalia similar to *Allostethus*-type, but virgae within genital lobes expanded at base.

Distribution: Oriental region.

5 species.

Identification key to the species

1 Tegmina with metallic bluish-bronze sheen. Only female known; forceps (Fig. 661) stout, with inner margins crenulate .... 1. **Allostethella nitens**, p. 417

1′ Tegmina not metallic 2

2 Third antennal joint longer and more slender, as long as, but narrower than, the first; yellow spot on tegmina almost central, well defined. Male forceps (Fig. 662) asymmetrical, male genitalia (Fig. 663) with specific sclerotized plate at virgae .... 2. **Allostethella guttata**, p. 417

2′ Third antennal joint shorter and broader, obviously shorter than the first; yellow spot on tegmina anterior to centre and not well defined 3

3 Head and first ten antennal joints black; male forceps (Fig. 664) very stout, and strongly curved, symmetrical. Male genitalia without sclerotized plate in gential lobes, as in fig. 665; external parameres very long .... 3. **Allostethella doriae**, p. 418

3′ Head and first ten antennal joints brown; male forceps less strongly curved or straight medially (Fig. 666). Virga of male genital lobes with sclerotized plates (Fig. 667 and 669) 4

4 Fifth antennal joint equal in length to third. Male forceps (Fig. 666) curved, median longitudinal sulcus on ultimate tergite prominent. Male genitalia as in fig. 667 .... 4. **Allostethella sumatrana**, p. 419

4′ Fifth antennal joint shorter than third. Male forceps (Fig. 668) straight, median longitudinal sulcus on ultimate tergite very shallow and inconspicuous. Male genitalia as in fig. 669 ... . 5. **Allostethella malayana**, p. 420

1. Allostethella nitens ZACHER

1911 *Allostethella nitens* ZACHER, Zool. Jahrb., Jena, **30**: 367. – Terra typica: Indo-Malayan Archipelago (Type female: Mus. Breslau).
Literature: BRINDLE 1965; BURR 1914; POPHAM & BRINDLE 1968; SAKAI 1970; STEINMANN 1977, 1978.

Description: Female colour dark brownish, reddish-black, tegmina with metallic bluish-bronze sheen; antennae brown, legs yellowish-brown. Head tumid, broad; postfrontal and coronal sutures distinct; eyes relatively small, slightly shorter than length of head behind eyes. Antennal joints normal, brown, but joints 12–14 white. The tips of the femora and tibiae, and all the tarsi, are lighter in colour. Pronotum transverse, lateral margins straight, and more or less parallel; median longitudinal furrow present medially. Tegmina long, a little longer than pronotum. Abdomen fusiform, a little expanded medially; ultimate tergite very broad; posterior margin with two tubercles, and margin concave medially. Forceps (Fig. 661) straight, slightly trigonal basally, cylindrical apically. – **Male** unknown. – **Length** of body with forceps: 15 mm.
Distribution: Malaysia.

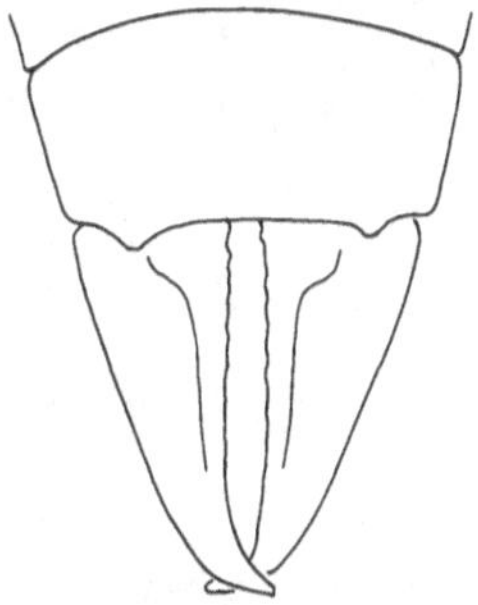

Fig. 661 Female ultimate tergite and forceps of *Allostethella nitens* ZACHER. Original.

2. Allostethella guttata (BORMANS)

1900 *Psalis guttata* BORMANS, Ann. Mus. civ. Stor. nat., Genova, **20**: 448. – Terra typica: Sumatra (Type male: unknown locality).
1911 *Psalis guttata* – BURR, Genera Insectorum, Bruxelles, **122**: 25 [= *Allostethella doridae* (DUBRONY 1869)]. BORNEO.
1965 *Allostethella guttata* – BRINDLE, Ann. Mag. nat. Hist., London, (13) **8**: 591; fig. 9 (male forceps), 18 (male genitalia). Mentawei Islands, Sumatra.
Literature: POPHAM & BRINDLE 1968; SAKAI 1970; STEINMANN 1977, 1978.

Description: Male colour dark reddish-brown; antennae joints 1–10 brown, 11 partly white, 12–13 white, 14 brown; pronotum shiny, legs yellow with femora and tibiae brown on basal two thirds. Head tumid, depressed along sutures, which are not prominent. Antennae 14-jointed, first joint rather shorter than distance between antennal bases; second transverse, third long and cylindrical, as long as but narrower than first. Pronotum transverse, lateral margins straight, or a little concave, but more or less parallel; posterior margin convex. Tegmina smooth and shiny, over twice as long as pronotum, with a well-defined large yellow spot almost centrally placed. Wings short, rather longer than half length of pronotum. Abdomen normal, strongly punctate except for ultimate tergite, which is quadrate, weakly punctate and with a median longitudinal sulcus prominent but not reaching base. Forceps (Fig. 662) straight, curved acpically, branches strongly trigonal basally. Genitalia (Fig. 663) comparatively narrow, median incision of anterior margin very deep; genital lobes well defined, virgae with broad sclerotized plate. External parameres very narrow, acuminate. – **Female** similar to male, but forceps narrower, and symmetrical. – **Length** of body with forceps: male: 12–13,5 mm, female: 12–14,5 mm.

Distribution: Sumatra, Borneo, Mentawei Islands.

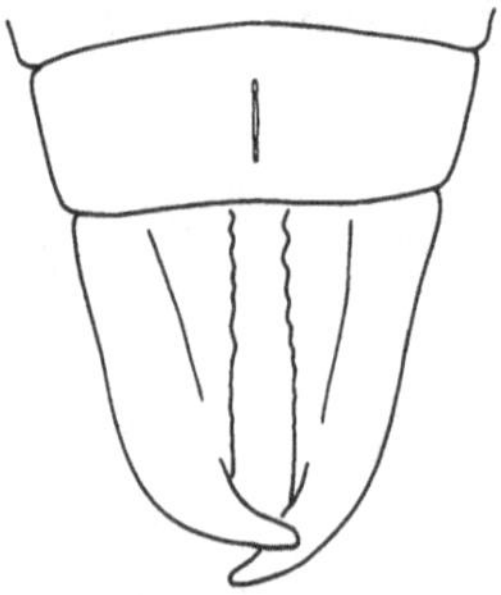

Fig. 662 Male ultimate tergite and forceps of *Allostethella guttata* (BORMANS). Original.

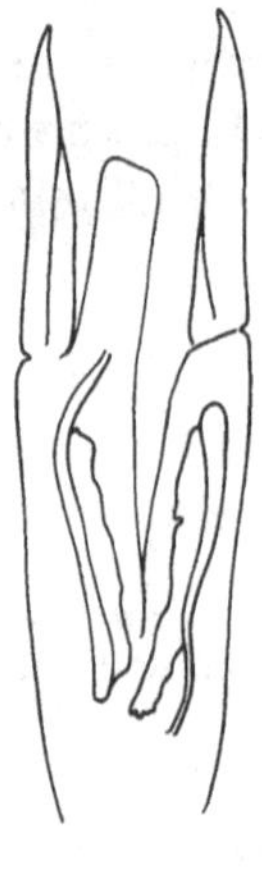

Fig. 663 Male genital armature of *Allostethella guttata* (BORMANS). After BRINDLE.

3. Allostethella doriae (DUBRONY)

1879 *Forficula doriae* DUBRONY, Ann. Mus. civ. Stor. nat., Genova, **14**: 379. – Terra typica: Borneo (Type male: unknown locality).
1900 *Apterygida doriae* – BORMANS, Das Tierreich, Berlin, **11**: 115.
1911 *Allostethella doriae* – BURR, Genera Insectorum, Bruxelles, **122**: 25; pl. 2, fig. 10 (male).
1965 *Allostethella doriae* – BRINDLE, Ann. Mag. nat. Hist., London, (13) **8**: 590; fig. 7 (male forceps).
Literature: BOESEMAN 1954; BURR 1915; KIRBY 1904; POPHAM & BRINDLE 1968; SAKAI 1970; STEINMANN 1977, 1978.

Description: Male colour reddish-brown, shiny; antennae with first ten joints black, distal three joints white; tegmina dark reddish-brown, with two orange spots near shoulders, which are sometimes indistinct or lacking; legs yellow, except the femora with dark brown bands. Head broad, tumid; postfrontal sutures and coronal suture distinct, deep; eyes moderately large, longer than length of head behind eyes. First antennal joint short, shorter than distance between antennal bases. Pronotum transverse, lateral margins straight, parallel, or more or less paralell; median longitudinal furrow distinct. Tegmina very long, longer than twice the length of pronotum. Wings present, reddish-brown. Abdomen normal, strongly punctate dorsally. Ultimate tergite broad, depressed dorsally, median longitudinal sulcus very deep. Forceps (Fig. 664) strongly curved, broad at base and rather obliquely truncate at tips. Genitalia (Fig. 665) narrow; central parameral plate with deep median incision, virga within genital lobe without sclerotized plate. External parameres narrow, acuminate. – **Female** similar to male, but forceps slightly curved, gradually tapering distally, with apices curved mesad. – **Length** of body with forceps: male: 14–15,5 mm, female: 13 mm.

Distribution: Borneo.

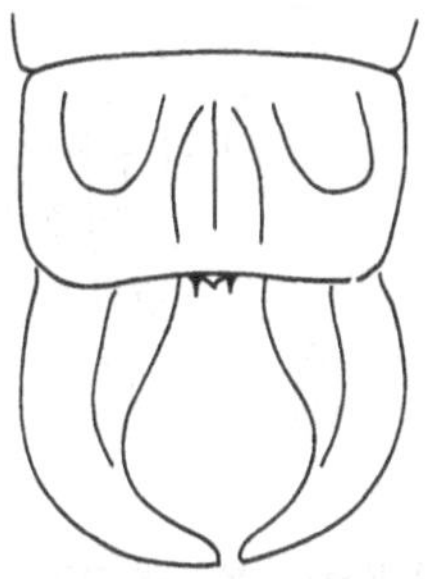

Fig. 664 Male ultimate tergite and forceps of *Allostethella doriae* (DUBRONY). After BURR.

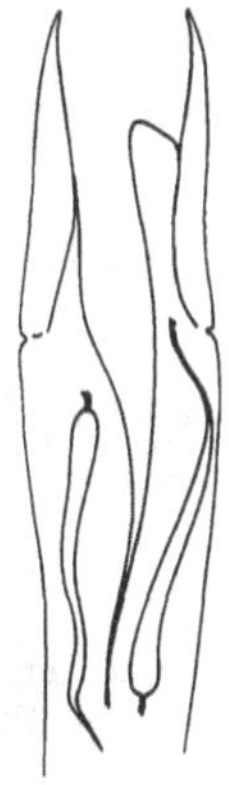

Fig. 665 Male genital armature of *Allostethella doriae* (DUBRONY). Original.

4. Allostethella sumatrana BRINDLE

1965 *Allostethella sumatrana* BRINDLE, Ann. Mag. nat. Hist., London, (13) **8**: 591; fig. 3 (antenna of male), 5 (male pronotum), 10 (holotype genital armature). – Terra typica: Sumatra (Type male: Naturhist. Riksmus., Stockholm).

Literature: SAKAI 1970; STEINMANN 1977, 1978.

Description: Male colour reddish-brown; antennae dark brown, but the last three joints whitish; body shiny, legs yellow with femora and tibiae brown on basal two thirds. Head depressed posteriorly; postfrontal sutures and coronal suture distinct. Antennae 14-jointed, first joint rather shorter than distance between antennal bases; second joint short, third small, about half length of scape; fourth joint shorter than third. Pronotum transverse, with median longitudinal furrow; lateral margins parallel. Tegmina long, longer than the length of pronotum. Wings present, brown, sutural margins lighter. Abdomen strongly punctate; ultimate tergite with lateral longitudinal ridge well marked, posterior margin flattened and concave or straight; median longitudinal sulcus strongly marked, deep, but absent from first and last quarter of plate. Forceps (Fig. 666) broad at base, narrowed distally. Genitalia (Fig. 667) broad, large; paramere with wide median incision, genital lobes well developed, virgae within genital lobes with sclerotized plates. External parameres narrow, acute at apices. – **Female** unknown. – **Length** of body with forceps: 13,5–14 mm.

Distribution: Sumatra.

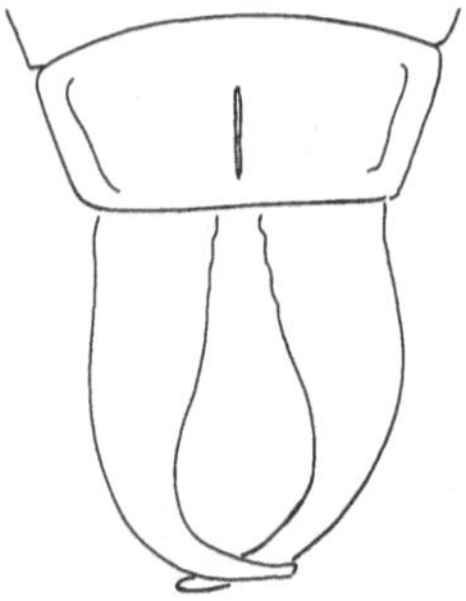

Fig. 666 Male ultimate tergite and forceps of *Allostethella sumatrana* Brindle. Original.

Fig. 667 Holotype genital armature of *Allostethella sumatrana* Brindle. After Brindle.

5. Allostethella malayana Zacher

1910 *Allostethella malayana* Zacher, Ent. Rundsch., Stuttgart, **27**: 30. – Terra typica: Sumatra (Type male: Mus. Breslau).

1911 *Allostethella malayana* – Zacher, Zool.Jahrb., Jena, **30**: 366; fig. S[1] (male genitalia).

1965 *Allostethella malayana* – Brindle, Ann. Mag. nat. Hist., London, (13) **8**: 592, fig. 9 (male ultimate tergite and forceps).

Literature: Burr 1911, 1914; Sakai 1970; Steinmann 1977, 1978.

Description: Male colour dark reddish-brown; antennae brown, but joints 11–12 white; tegmina shiny, reddish-brown, with a yellow or orange spot on anterior half of both tegmina. Wings dark brown, legs yellowish brown. Head tumid, postfrontal sutures and coronal suture distinct; eyes relatively large, a little longer than length of head behind eyes. Antennae 12-jointed (?); first joint short, much shorter than distance between antennal bases; second transverse, third short, both fourth and fifth joints shorter than third. Pronotum transverse, lateral margins straight, parallel, all angles rounded. Tegmina long, almost three times as long as pronotum. Wings short, slightly longer than pronotum; dark brown, but sometimes with a yellow spot on costa. Abdomen punctate dorsally; ultimate tergite broad, large, posterior margin straight, sometimes with lateral longitudinal ridge; median longitudinal sulcus only slightly developed. Forceps (Fig. 668) straight, specific, straight at two-third section basally, and strongly curved at apical portion. Genitalia (Fig. 669) characteristic; paramere relatively narrow; median incision of anterior margin deep, plate a little Y-shaped; genital lobes well developed, directed forwards; virga within genital lobe with sclerotized plates. External parameres narrow, and acuminate. – **Female** similar to male, but abdomen less punctate, forceps simple, tapering, contiguous. – **Length** of body with forceps: male: 15–16 mm, female: 16–17 mm.

Distribution: Sumatra.

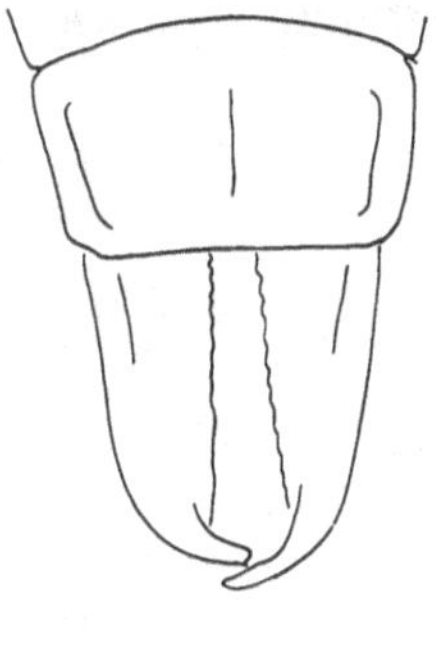

Fig. 668 Male ultimate tergite and forceps of *Allostethella malayana* ZACHER. After BRINDLE.

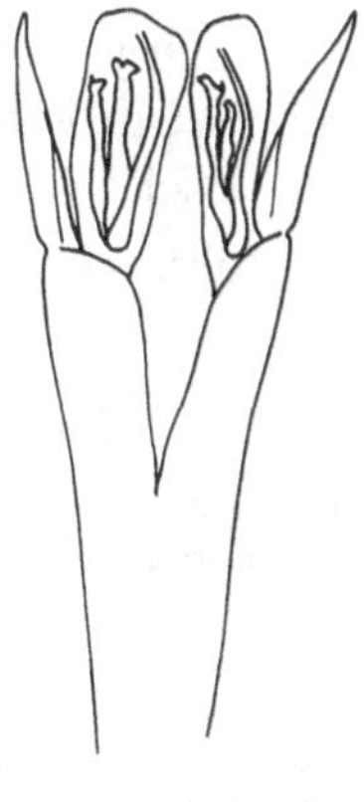

Fig. 669 Male genital armature of *Allostethella malayana* ZACHER. After ZACHER.

Subfamily 2: **Nalinae** STEINMANN

1975 *Nalinae* STEINMANN, Acta zool. Acad. Sci. hung., Budapest, **21**: 207. – Genus typicum: *Nala* ZACHER 1910.

Literature: STEINMANN 1978.

Description: Comparatively small and medium sized insects. Mesosternum wide, posterior portion (towards metasternum) not constricted, normal. Legs and tarsi short, on the hind legs neither femora nor tarsi longer than pronotum. Lateral margins of tegmina with a longitudinal ridge or keel. Forceps robust, characteristic in males, and slenderer in females; in both sexes considerably longer than width of ultimate tergite. Male genitalia of *Labidura*-type; central parameral plate with median incision between external parameres; latter gradually elongated, apices evenly attenuating.

Distribution: African, Oriental, and Southern Palaearctic species.

Single genus.

Genus 4: **Nala** ZACHER

1910 *Nala* ZACHER, Ent. Rundsch., Stuttgart, **27**: 29. – Species typica: *Forficula lividipes* DUFOUR 1829.

1910 *Paralabidura* BURR, Trans. ent. Soc. London, **1910**: 393. – Species typica: *Forficula lividipes* DUFOUR 1829.

Literature: BEY-BIENKO 1936; BRINDLE 1966; BURR 1911, 1915; CHOPARD 1922; SAKAI 1970; SHIRAKI 1928; STEINMANN 1975, 1978; TOWNES 1945.

Description: Dark species. Head transverse, tumid, sutures more or less present; eyes large or small; antennal joints elongated. Pronotum longer than wide, or transverse; posterior margin weakly or strongly convex; lateral margins more or less parallel or widest to last margin. Tegmina and wings fully developed or tegmina short and wings absent or concealed. Legs short. Abdomen more or less parallel-sided or a little expanded to middle. Both branches of male forceps arcuate, with or without a basal inner tooth. Forceps of female simple, inner margins crenulate. Virga within male genital lobes straight, well sclerotized, and with large basal vesicle.

Distribution: Ethiopian, Oriental and South Palaearctic faunal regions.

10 species.

Identification key to the species

1 Abdomen dull black or almost so, with sparse pale hairs; tegmina and wings dull brown, rarely blackish, well developed, rugose. Male forceps (Fig. 670) gently arcuate, with a small tooth basally and medially. Male genitalia as in fig. 671 .... 1. **Nala lividipes**, p. 423

1′ Abdomen coloured similarly to wings and tegmina, or abdomen not black **2**

2 African species **3**

2′ Not African species **6**

3 Body length without forceps 6 mm. Colour reddish or yellowish brown; tegmina and wings well developed. Male forceps (Fig. 672) with a large basal tooth. Male genitalia (Fig. 673) broad, paramere oval .... 2. **Nala caprea**, p. 424

3′ Body length without forceps 8 mm or longer. Colour dark reddish-brown or almost blackish; wings present or absent **4**

4 Pronotum longer than wide, slightly widened posteriorly. Both branches of male forceps (Fig. 674) with a large, lamellated inner tooth. Male genitalia as in fig. 675 .... 3. **Nala figinii**, p. 425

4′ Pronotum as wide as long or transverse. Both branches of male forceps without a large, lamellated inner tooth **5**

5 Posterior margin of pronotum strongly convex. Both branches of male forceps (Fig. 676) with isolated small teeth on inner margin. Colour reddish-brown to dark brown .... 4. **Nala intermedia**, p. 426

5′ Posterior margin of pronotum weakly convex. Both branches of male forceps (Fig. 678) with a large apical tooth towards apex. Colour blackish .... 5. **Nala saegeri**, p. 427

6 Yellowish brown; tegmina coriaceous; antennae yellowish-brown with first two joints darker; abdomen reddish-brown. Male forceps (Fig. 680) without a basal tooth, but with a small basal, and a median denticle .... 6. **Nala ornata**, p. 429

6′ Dark brown or blackish; tegmina rugose; antennae more or less unicolorous brown **7**

7 Male forceps with a large inner tooth in first third, as in fig. 681 somewhat similar to *figinii* (Burr). Male genitalia very large and broad, as in fig. 682 .... 7. **Nala nepalensis**, p. 429
7′ Male forceps with a large basal tooth or without large tooth **8**
8 Male forceps characteristic, with a moderately large tooth basally (Fig. 683). Male genitalia as in fig. 684 .... 8. **Nala timorensis**, p. 431
8′ Male forceps without large tooth on inner margins **9**
9 Male forceps with a single, comparatively small tooth medially (Fig. 685). Male genitalia as in fig. 686 .... 9. **Nala tenuicornis**, p. 432
9′ Male forceps specific, with some small teeth on inner margins (Fig. 687). Male genitalia as in fig. 688 .... 10. **Nala basalis**, p. 433

1. Nala lividipes (Dufour)

1829 *Forficula lividipes* Dufour, Ann. Sci. nat. (Zool.), Paris, **13**: 340 (replacement name for *Forficula pallipes* Dufour 1920, preoccupied by *Forficula pallipes* Fabricius 1793). – Terra typica: Spain (Type male: unknown locality).
1839 *Forficesila meridionalis* Audinet-Serville, Hist. nat. Insectes Orthopt., Suites à Buffon: 26. – Terra typica: Andalusia (Type male: unknown locality).
1839 *Forficesila castanea* Audinet-Serville, Hist. nat. Insectes Orthopt., Suites à Buffon: 26. – Terra typica: Spain (Type male: unknown locality).
1846 *Forficesila vicina* Lucas, Expl. Alger, **3**: 5. – Terra typica: Algeria (Type male: unknown locality).
1876 *Labidura dufouri* Scudder, Proc. Boston Soc. nat. Hist., **18**: 322. – Terra typica: Europe (Type male: unknown locality).
1900 *Labidura lividipes* – Bormans, Das Tierreich, Berlin, **11**: 36.
1900 *Labidura inconspicua* Kirby, in Distant, Ins. Transvaal, Pretoria, **1**: 12. – Terra typica: Pretoria (Type female: unknown locality).
1900 *Echinosoma obscurum* Kirby, in Distant, Ins. Transvaal, Pretoria, **1**: 12. – Terra typica: Pretoria (Type female: unknown locality).
1910 *Paralabidura lividipes* – Burr, Trans. ent. Soc. London, **1910**: 393.
1910 *Nala lividipes* – Zacher, Ent. Rundsch., Stuttgart, **27**: 29.
1911 *Nala lividipes* – Burr, Genera Insectorum, Bruxelles, **122**: 36; pl. 3, fig. 11 (male).
1913 *Labidura australica* Mjöberg, Ent. Tidskr., Stockholm, **1913**: 27. – Terra typica: Australia: New South Wales (Type male: unknown locality).
1915 *Nala lividipes* – Burr, J. r. microsc. Soc., London, **1915**: 444; pl. IX, fig. 10 (male genitalia).

Literature: Bey-Bienko 1936; Boeseman 1954; Borelli 1915; Brindle 1965; Burr 1911, 1915; Chopard 1922; Hebard 1923; Hincks 1955; Günther 1930; Kirby 1904; Menozzi 1927; Sakai 1970; Shiraki 1932; Steinmann 1975, 1978.

Description: Male colour blackish brown; antennae dark brown, legs yellow, femora sometimes with a median darker band; forceps dark red; tegmina and wings brown. Head transverse, cuticle strongly coriaceous, pubescent, hairs yellow; postfrontal sutures and coronal suture absent. Eyes rather large, but shorter than length of head behind eyes. Antennae 23-jointed; first joint normal, about as long as distance between antennal bases; second quadrate, third long, twice as long as fourth joint. Pronotum longer than wide, usually strongly widened posteriorly, but the pronotum is less elongated and comparatively wider in some large males. Cuticle of pronotum coriaceous or rugose posteriorly. Tegmina long, well developed, finely tuberculate, both tegmina with a lateral longitudinal keel. Wings fully developed, finely tuberculate. Legs short. Abdomen broad, depressed, sometimes widened medially or posteriorly; cuticle finely punctate and pubescent, the pubescence denser laterally. Ultimate tergite broad, posterior margin concave posteriorly. Forceps (Fig. 670) arcuate, without a large basal inner tooth, but with a very small basal and with or without a small ventro-median tooth towards apex. Penultimate sternite simple, rounded posteriorly. Genitalia (Fig. 671; gen. prep. No. 236, det. Dr. H. STEINMANN) V-shaped; median incision of anterior margin of the central parameral plate very deep and wide; genital lobes well developed, virga with large, sclerotized basal vesicle. External paramere long, broad medially, pointed apically. – **Female** similar to male, but forceps with branches slender, almost straight, and contiguous. – **Length** of body with forceps: in both sexes: 7–14 mm.

Distribution: From southern Europe to North Africa and the eastern half of Africa south of the Sahara, to Arabia, Iran and eastwards India and other parts of South-East Asia, China, Japan, Philippine Islands, and Australia.

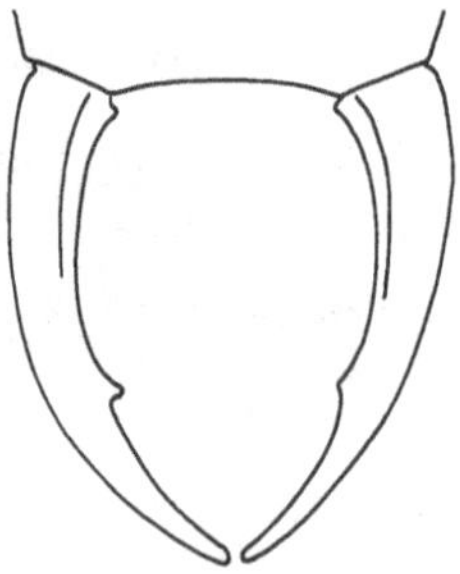

Fig. 670 Male ultimate tergite and forceps of *Nala lividipes* (DUFOUR). Original.

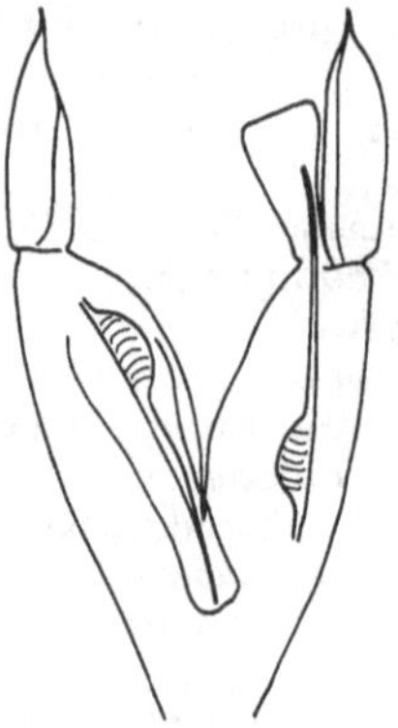

Fig. 671 Male genital armature of *Nala lividipes* (DUFOUR). Original.

2. Nala caprea MENOZZI

1929 *Nala caprea* MENOZZI, Mem. Soc. ent. Ital., Genova, **8**: 9. – Terra typica: Ivory Coast (Type male: unknown locality).

1966 *Nala caprea* – BRINDLE, Ann. Mag. nat. Hist., London, (13) **9**: 255, fig. 49 (male genitalia), 58 (male forceps). Sierra Leone, Guinea.

Literature: BRINDLE 1968, 1973; SAKAI 1970; STEINMANN 1978.

Description: Male colour reddish-brown to dark brown, rather shiny; antennae dark brown or brown, one or more distal joints pale; legs yellow, at least the posterior femora with a dark median band. Head transverse, cuticle coriaceous, pubescent; postfrontal sutures absent, but coronal suture distinct. Eyes rather large, but shorter than length of head behind eyes. Antennae similar to *lividipes* (DUFOUR). Pronotum slightly longer than wide or nearly quadrate; widened posteriorly and with posterior margin strongly convex. Tegmina and wings well developed. Tegmina long, longer than pronotum, coriaceous anteriorly, and rugose posteriorly. Wings long, rugose, finely tuberculate, shiny. Abdomen parallel-sided or almost so. Ultimate tergite transverse, posterior margin straight. Forceps (Fig. 672) trigonal basally, cylindrical apically; inner ventral margin produced into a large tooth and there is sometimes a small ventro-medial spine near apex. Genitalia (Fig. 673) large, broad; median incision of anterior margin deep, virga of *Nala*-type. External parameres broad, inner and outer margins parallel medially, apices acuminate. – **Female** similar to male, but both branches of forceps with straight margins; simple, inner margin of each branch dentated. – **Length** of body with forceps: male: 6–9 mm, female: 5–8,5 mm.

Distribution: Western Africa (from Guinea to Nigeria).

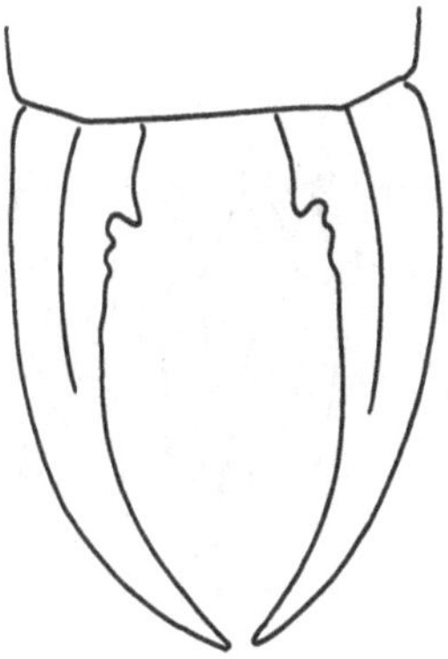

Fig. 672 Male forceps of *Nala caprea* MENOZZI. After BRINDLE.

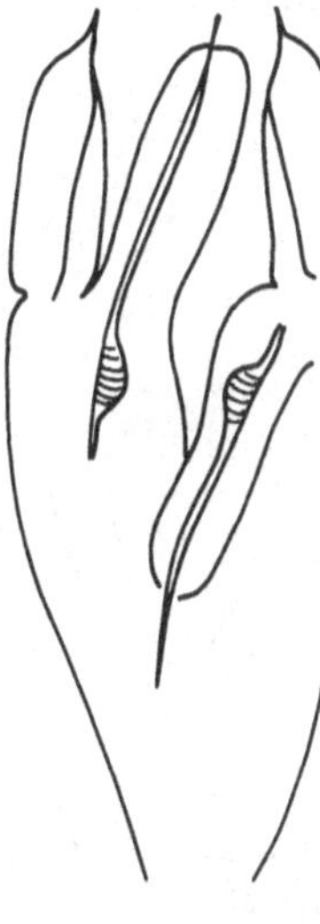

Fig. 673 Male genital armature of *Nala caprea* MENOZZI. After BRINDLE.

3. Nala figinii (BURR)

1908 *Labidura figinii* BURR, Bull. Soc. ent. Ital., Genova, **60**: 176. – Terra typica: Ethiopia (Type male: Mus. civ. Stor. nat., Genova).
1911 *Nala figinii* – BURR, Genera Insectorum, Bruxelles, **122**: 36; pl. 3, fig. 11 (male).
1966 *Nala figinii* – BRINDLE, Ann. Mag. nat. Hist., London, (13) **9**: 255; fig. 50 (male genitalia), 55 (male forceps). Africa: Guinea, Eritrea, Mozambique.

Literature: BRINDLE 1973; HINCKS 1957; MENOZZI 1936; SAKAI 1970; STEINMANN 1978.

Description: Male colour dark brown or greyish; head blackish-brown, legs yellowish-brown, femora darkened. Pronotum, tegmina and abdomen paler dorsally, and tuberculate. Head transverse, postfrontal and coronal sutures absent. Eyes relatively small, about as long as scape. Antennae 20-jointed; first joint small, shorter than distance between antennal bases; second joint quadrate, third long, longer than wide. Pronotum rounded, a little longer than wide, widened posteriorly; posterior margin convex; median longitudinal furrow present medially. Tegmina short, wings absent. Abdomen more or less parallel-sided. Ultimate tergite broad, typical of genus. Forceps (Fig. 674) characteristic; both branches of forceps long, arcuate, with a very large inner tooth, which occupies nearly half of the branch, distal half slender, but also which small teeth on inner margin. Genitalia (Fig. 675; gen. prep. No. 205, det. Dr. H. STEINMANN) typical; paramere with median incision of anterior margin; genital lobe with straight virga and with a large strongly sclerotized basal vesicle. – **Female** similar to male, but forceps with branches simple, almost straight, and contiguous. – **Length** of body with forceps: male: 9–11 mm, female: 8,5–10,5 mm.

Distribution: North-eastern and south-eastern Africa.

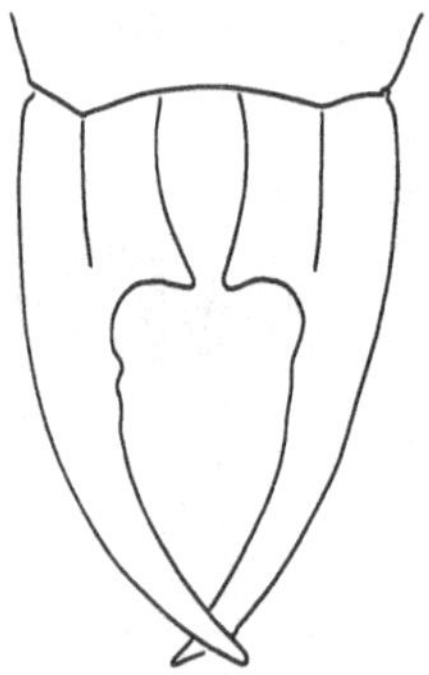

Fig. 674 Male forceps of *Nala figinii* (BURR). Original.

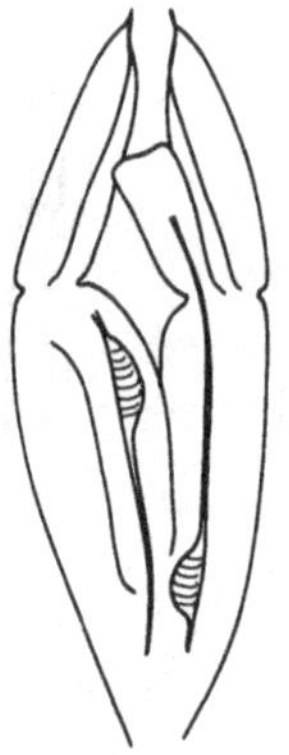

Fig. 675 Male genital armature of *Nala figinii* (BURR). Original.

4. Nala intermedia MENOZZI

1937 *Nala intermedia* MENOZZI, Rev. suisse Zool., Genève, **44**: 448. – Terra typica: Angola (Type male: unknown locality).

1966 *Nala intermedia* – BRINDLE, Ann. Mag. nat. Hist., London, (13) **9**: 255; fig. 48 (male genitalia), 54 (male forceps). Angola, South Rhodesia.

1973 *Nala intermedia* – BRINDLE, Dermaptera of Africa, Tervuren, **1**: 21; fig. 151 (male forceps), 155 (female forceps). South-west, Central and Southern Africa.

Literature: BRINDLE 1970; HINCKS 1957; MORALES-AGACINO 1951; SAKAI 1970; STEINMANN 1978.

Description: Male colour dark reddish-brown, sometimes darker; antennae brown, basal joints sometimes yellow; lateral margins of pronotum yellow, legs yellowish. Head transverse, a little tumid; postfrontal sutures faintly, coronal suture distinctly visible. Antennae 21-jointed; first joint normal, short, shorter than distance between antennal bases; second joint quadrate, third long, longer than fourth. Pronotum as wide as long or a little transverse, only slightly widened posteriorly; posterior margin convex. Tegmina short, hardly longer than pronotum. Wings absent of concealed. Abdomen blackish-brown, strongly punctate and pubescent, more strongly in female than in male; more or less parallel-sided. Ultimate tergite broad; posterior margin straight. Surface of body rugose, tegmina strongly rugose and tuberculate, both tegmina with a lateral longitudinal keel. Both branches of forceps (Fig. 676) arcuate, simple but with isolated small inner teeth, a larger one on basal third and a second larger one about distal two-thirds. Genitalia (Fig. 677) typical of genus; paramere with median incision of anterior margin; genital lobe with of *Nala*-type basal vesicle. External paramere pointed. – **Female** similar to male, but branches of forceps straight, except at apices, inner margins dentated; dorsal edge more prominent than in the previous species. – **Length** of body with forceps: male: 9–12 mm, female: 8,5–13 mm.

Distribution: Angola and South Rhodesia.

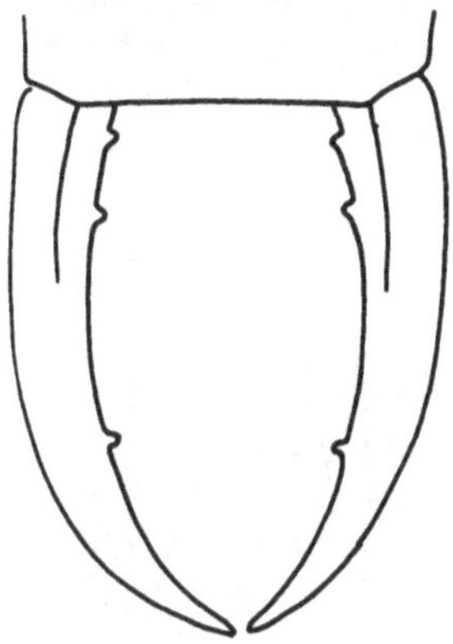

Fig. 676 Male forceps of *Nala intermedia* MENOZZI. After BRINDLE.

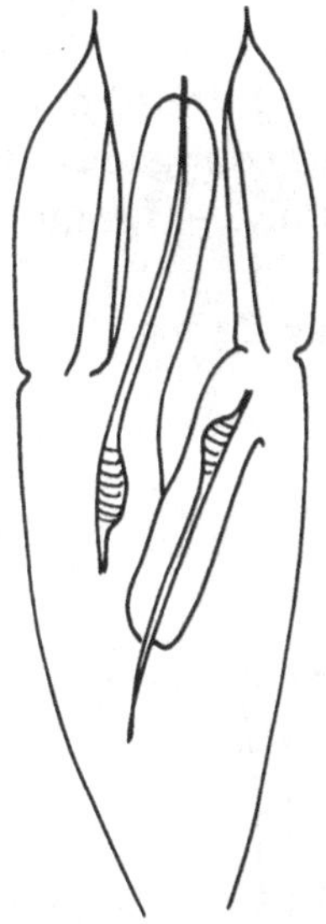

Fig. 677 Male genital armature of *Nala intermedia* MENOZZI. After BRINDLE.

5. Nala saegeri BRINDLE

1968 *Nala saegeri* BRINDLE, Explor. Parc nation. Garamba, Bruxelles, **53**: 21; fig. 11 (holotype genital armature), 12 (male forceps). – Terra typica: Zaïre (Type male: Mus. r. Afr. centr., Tervuren).

Literature: BRINDLE 1973; STEINMANN 1978.

Description: Male colour blackish, antennae dark brown, lateral margins of pronotum yellowish, legs yellowish-brown, femora darkened medially; forceps dark red to blackish. Head transverse, tumid; cuticle coriaceous, punctate and pubescent posteriorly. Eyes small, shorter than length of head behind eyes. Postfrontal and coronal sutures distinct. Antennae 17-jointed in type; first joint shorter than distance between antennal bases; second quadrate; third as long as, but narrower than, first; fourth and fifth shorter than third. Pronotum transverse, scarcely widened posteriorly, posterior margin weakly convex; coriaceous, rugose posteriorly. Tegmina short, shinier than *intermedia* MENOZZI; strongly and closely tuberculate, each tegmen with a well-marked lateral longitudinal ridge. Abdomen more or less parallel-sided, closely and finely punctate and pubescent. Ultimate tergite brightly shiny, punctate anteriorly but smooth posteriorly, with a prominent median longitudinal sulcus; posterior margin rugose between the bases of the forceps. Forceps (Fig. 678) arcuate, not of *Nala*-type, trigonal basally with a prominent dorsal ridge, cylindrical distally, inner margin with a ventro-median tooth towards apex. Genitalia (Fig. 679) typical for genus; central parameral plate comparatively large, and broad; median incision of anterior margin very deep and wide; genital lobes well developed, virgae straight with sclerotized basal vesicle. External parameres moderately long, broad medially, acuminate at apex. – **Female** similar to male, but with short contiguous forceps, which are broad at the base, both branches tapering distally. – **Length** of body with forceps: male: 7–8,5 mm, female: 7,5–9,5 mm.

Distribution: Zaïre.

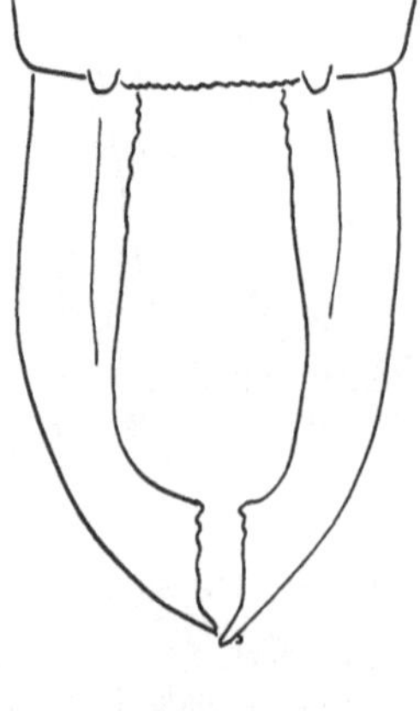

Fig. 678 Holotype forceps of *Nala saegeri* BRINDLE. After BRINDLE.

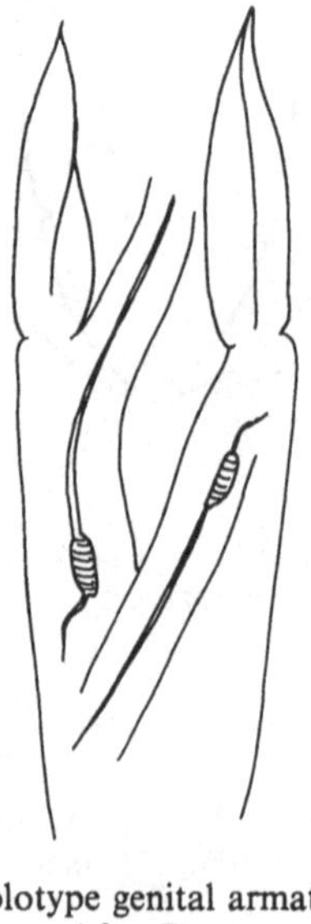

Fig. 679 Holotype genital armature of *Nala saegeri* BRINDLE. After BRINDLE.

6. Nala ornata Borelli

1932 *Nala ornata* Borelli, J. fed. Malay States Mus., Kuala Lumpur, **1932**: 182. – Terra typica: Borneo (Type male: Sarawak Mus.).
1966 *Nala ornata* – Brindle, Ann. Mag. nat. Hist., London, (13) **9**: 257; fig. 60 (male forceps).
Literature: Sakai 1970; Steinmann 1978.

Description: Male colour yellowish-brown, antennae brown, lateral margin of pronotum brownish-yellow, legs yellow, femora darkened medially, abdomen dark brown or brownish-red. Head tumid, transverse, postfrontal sutures and coronal suture distinct, but finely visible. Eyes relatively small, shorter than length of head behind eyes. First antennal joint long, but shorter than distance between antennal bases, second joint quadrate, third longer than fourth. Pronotum rounded, normal; lateral margins more or less parallel; median longitudinal furrow distinct. Tegmina long, twice as long as pronotum, tuberculate, shiny, both tegmina with a well marked lateral longitudinal ridge. Legs more or less long, longer than type of genus. Abdomen typical, almost parallel-sided. Ultimate tergite broad, with median longitudinal sulcus. Forceps (Fig. 680) specific; with two pairs of teeth on inner margins; arcuate, trigonal basally, cylindrical apically. Genitalia unknown. – **Female** similar to male, but forceps slender, contiguous, tapering. – **Length** of body with forceps: male: 11–13 mm, female: 11,5–15 mm.

Distribution: Borneo.

Fig. 680 Male forceps of *Nala ornata* Borelli. After Brindle.

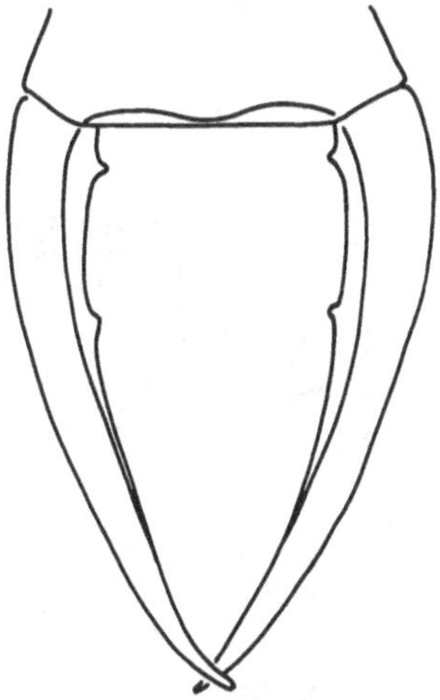

7. Nala nepalensis (Burr)

1907 *Labidura nepalensis* Burr, Rec. Indian Mus., Calcutta, **1**: 208. – Terra typica: Nepal (Type male: Brit. Mus. nat. Hist., London).
1910 *Paralabidura nepalensis* – Burr, Trans. ent. Soc. London, **1910**: 185.
1911 *Nala nepalensis* – Burr, Genera Insectorum, Bruxelles, **122**: 36.

1915 *Nala nepalensis* – Burr, J. r. microsc. Soc., London, **1915**: 444; pl. IX, fig. 9 (male genitalia).
1966 *Nala nepalensis* – Brindle, Ann. Mag. nat. Hist., London, (13) **9**: 257; fig. 52 (male genitalia), 57 (male forceps). India, Nepal, Malaysia.
Literature: Hincks 1953; Sakai 1970; Steinmann 1978.

Description: Male colour dark reddish-brown or dark brownish-black; antennae brown, legs yellowish, femora with dark brown band basally; forceps reddish-brown. Head tumid, broad; postfrontal and coronal sutures absent; cuticle of vertex coriaceous; eyes relatively small, shorter than length of head behind eyes. Antennae 21-jointed; first joint long, about as long as distance between antennal bases, second quadrate, third joint long, fourth comparatively short. Pronotum rounded; lateral margins a little convex; median longitudinal furrow present medially. Tegmina well developed, coriaceous. Wings normal, dark reddish-brown. Abdomen more or less parallel; ultimate tergite broad, simple. Forceps (Fig. 681) characteristic; branches with a basal tooth; the tooth lamelliform, broad basally, a little curved apically, and branches cylindrical apically. Genitalia (Fig. 682; gen. prep. No. 236, det. Dr. H. Steinmann) of *Nala*-type; central parameral plate broad, oval, median incision of anterior margin very deep; genital lobes broad, with virga of *Nala*-type, straight, basal vesicle strongly sclerotized. External parameres broad, obtuse. – **Female** very similar to male, but forceps slender, contiguous, tapering. – **Length** of body with forceps: male: 9–12 mm, female: 10–13 mm.

Distribution: Nepal, India and Malay Peninsula.

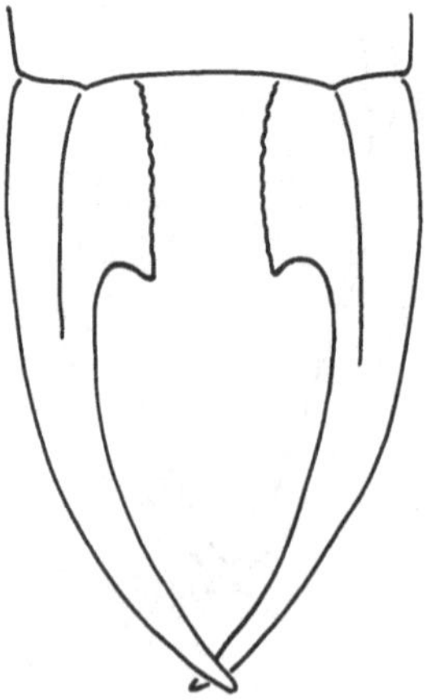

Fig. 681 Male forceps of *Nala nepalensis* (Burr). Original.

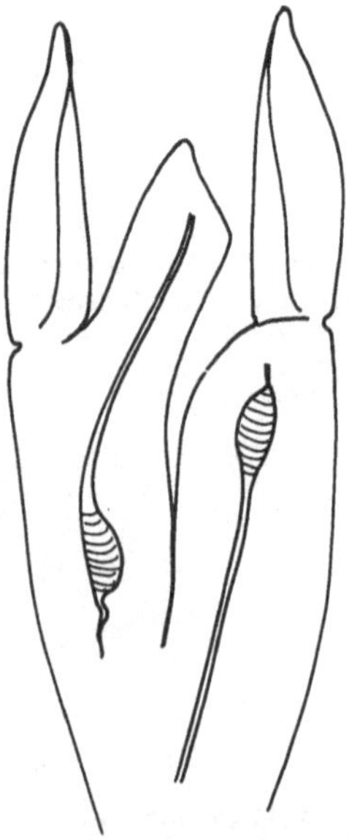

Fig. 682 Male genital armature of *Nala nepalensis* (Burr). Original.

8. Nala timorensis BRINDLE stat. nov.

1967 *Nala tenuicornis*, ssp. *timorensis* BRINDLE, Coimbra Editora, Limitada, Coimbra, **8**: 15; fig. 1 (male), 2 (female forceps), 3 (holotype genital armature). – Terra typica: Timor (Type male: Mus. Lab. zool., Univ. Coimbra).
Literature: STEINMANN 1978.

Description: Male colour reddish-brown or brownish-yellow, antennae brown, legs yellow, abdomen reddish brown, and forceps red. Head tumid, rounded posteriorly; postfrontal sutures and coronal suture absent. Eyes comparatively large, but a little shorter than length of head behind eyes. Antennae broken in type; first joint normal, a little shorter than distance between antennal bases. Pronotum longer than wide; lateral margins a little concave, posterior margin rounded. Tegmina long, but less than twice as long as pronotum; cuticle with tubercles, and with lateral longitudinal ridge. Wings absent. Abdomen normal, more or less parallel-sided anteriorly and medially, expanded posteriorly. Ultimate tergite large, posterior margin with two tubercles. Forceps (Fig. 683) arcuate, specific; trigonal basally, cylindrical apically; inner margin with a very large tooth basally. Genitalia (Fig. 684) of *Nala*-type; paramere with median incision at anterior margin; genital lobes well developed, virga within genital lobe with strongly sclerotized basal vesicle. External parameres broad basally, acuminate apically. – **Female** similar to male, but forceps simple, without inner tooth basally, but inner margin a little crenulate. – **Length** of body with forceps: male: 14 mm, female: 14,5 mm.

Distribution: Timor.

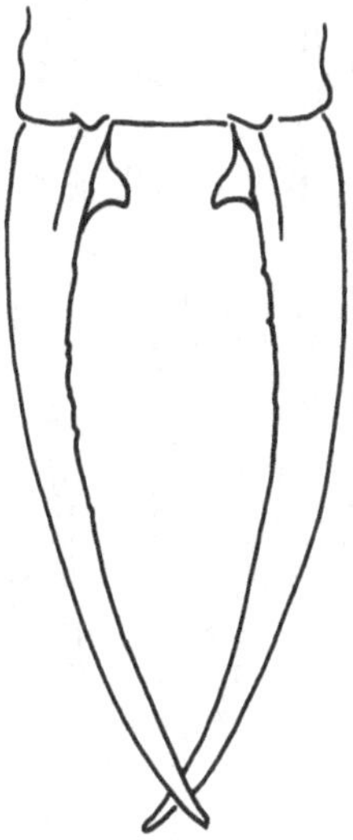

Fig. 683 Holotype forceps of *Nala timorensis* BRINDLE. After BRINDLE.

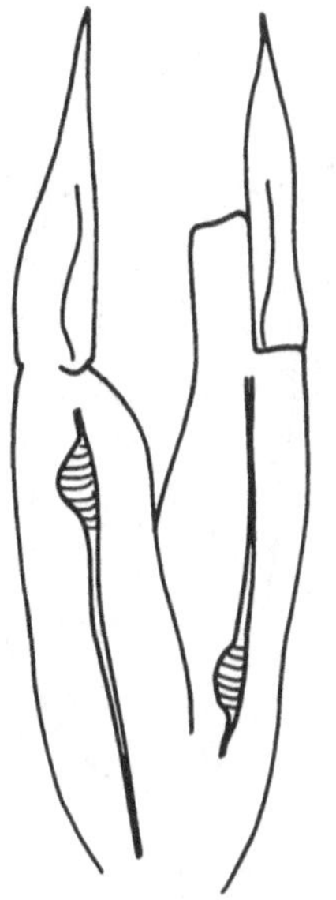

Fig. 684 Holotype genital armature of *Nala timorensis* BRINDLE. After BRINDLE.

9. Nala tenuicornis (BORMANS)

1900 *Labidura tenuicornis* BORMANS, Ann. Mus. civ. Stor. nat., Genova, (2) **20**: 446. – Terra typica: Sumatra (Type male: unknown locality).
1910 *Paralabidura tenuicornis* – BURR, Trans. ent. Soc. London, **1910**: 184.
1911 *Nala tenuicornis* – BURR, Genera Insectorum, Bruxelles, **122**: 36. Sumatra, New Guinea.
1966 *Nala tenuicornis* – BRINDLE, Ann. Mag. nat. Hist., London, (13) **9**: 257; fig. 53 (male genitalia), 59 (male forceps). Java, Sumatra, New Guinea.
Literature: BRINDLE 1967; POPHAM & BRINDLE 1968; SAKAI 1970; STEINMANN 1978.

Description: Male colour brown, antennae dark or light brown, legs yellow, femora sometimes with a median darker band; forceps reddish-brown; tegmina and wings pale yellowish-brown. Head transverse, tumid, cuticle strongly coriaceous; postfrontal sutures and coronal suture not visible. Eyes rather large, but shorter than length of head behind eyes. Antennae 21-jointed; first joint normal, a little longer than distance between antennal bases; second joint quadrate, third joint long, about twice as long as the fourth joint. Pronotum rounded posteriorly; lateral margins more or less parallel or a little concave; median longitudinal furrow present. Cuticle of pronotum coriaceous. Tegmina long, twice as long as pronotum, coriaceous or rugose posteriorly. Abdomen well developed, broad, depressed, sometimes widened medially or posteriorly. Ultimate tergite broad, simple. Forceps (Fig. 685) arcuate, without a basal tooth, but with a single tooth medially. Genitalia (Fig. 686) of *Nala*-type, paramere broad, median incision very deep; genital lobes well developed, virga typical, with a very large basal vesicle. External parameres long, acute. – **Female** similar

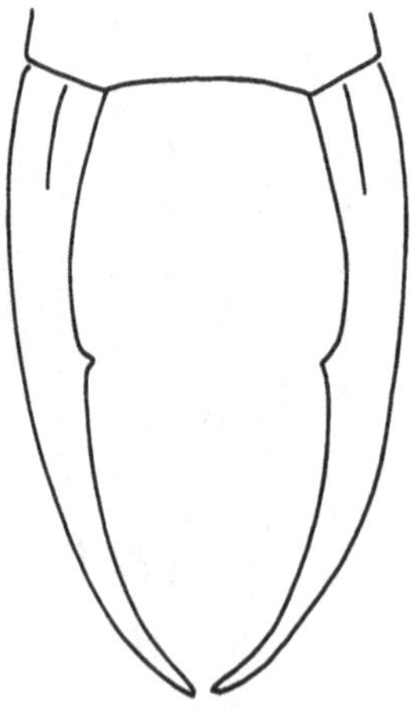

Fig. 685 Male forceps of *Nala tenuicornis* (BORMANS). Original.

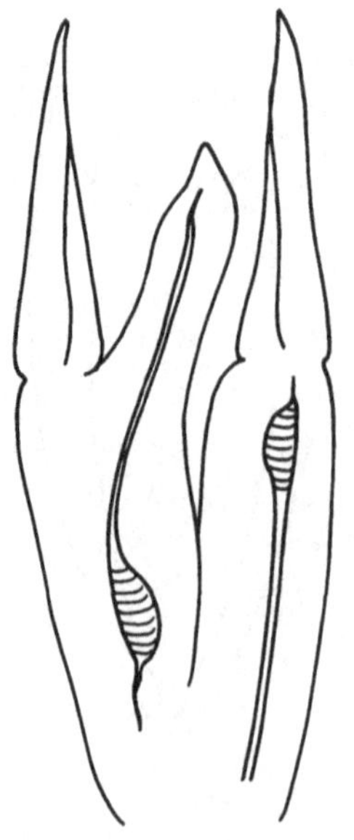

Fig. 686 Male genital armature of *Nala tenuicornis* (BORMANS). After BRINDLE.

to male, but forceps slender, tapering. – **Length** of body with forceps: male: 14–16 mm, female: 13–16,5 mm.

Distribution: Java, Borneo, Sumatra, New Guinea.

10. Nala basalis BEY-BIENKO

1970 *Nala basalis* BEY-BIENKO, Zool. Zhurnal., Moskva, **49**: 1814; fig. 9 (male), 10 (holotype genital armature), 11 (penultimate sternite of female), 12 (female forceps). – Terra typica: Western Pakistan and East Afghanistan (Type male: holotype unknown locality; paratypes: Zool. Inst. Acad. Sci., Leningrad).

1975 *Nala basalis* – STEINMANN, Fol. ent. hung., Budapest, **28**: 157; fig. 21 (male genitalia).

Literature: STEINMANN 1978.

Description: Male colour brownish-red; antennae brown, legs yellowish brown, tegmina scarcely darker. Head tumid, rounded posteriorly, about as broad as pronotum. Postfrontal sutures and coronal suture absent. Eyes relatively small, shorter than length of scape. Antennae 20-jointed; first joint long, more or less as long as distance between antennal bases; second joint quadrate, third long. Pronotum longer than the head; rounded posteriorly, lateral margins nearly convex; median longitudinal furrow present medially. Tegmina well developed, coriaceous. Wings absent. Both tegmina with a lateral longitudinal ridge. Legs long. Abdomen more or less parallel-sided. Ultimate tergite simple, broad. Forceps (Fig. 687) specific; inner margins with some small teeth, but without large tooth; arcuate, cylindrical. Genitalia (Fig. 688) oval;

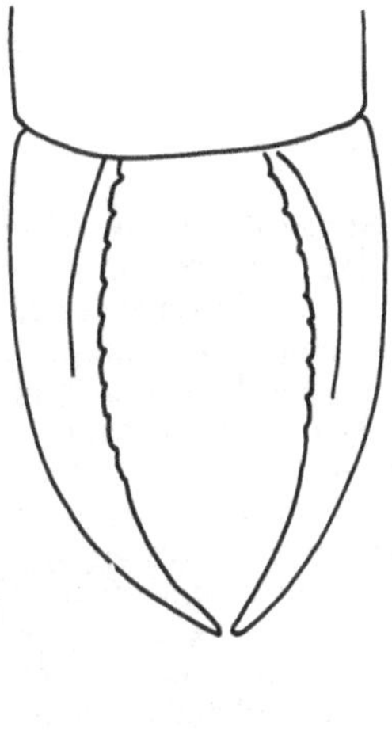

Fig. 687 Holotype forceps of *Nala basalis* BEY-BIENKO. After BEY-BIENKO.

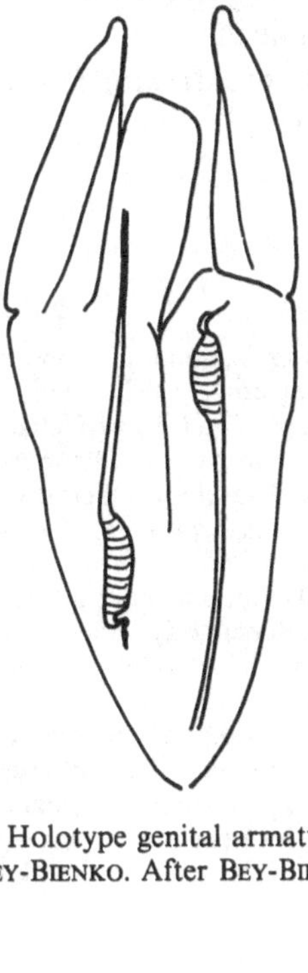

Fig. 688 Holotype genital armature of *Nala basalis* BEY-BIENKO. After BEY-BIENKO.

median incision of anterior margin narrow, genital lobes normal, virga within genital lobe with strongly sclerotized basal vesicle; virga elongated, specific. External parameres long, obtuse. – **Female** similar to male, but forceps strongly trigonal basally, cylindrical apically, simple, contiguous: inner margins with several various teeth. – **Length** of body with forceps: male: 11,5–13 mm, female: 12–14,5 mm.

Distribution: West Pakistan, and East Afghanistan.

Subfamily 3: **Labidurinae** Burr

1909 *Labidurinae* Burr, Dtsch. Ent. Zeitschr., Berlin, **1909**: 325. – Genus typicum: *Labidura* Leach 1815.

1911 *Labiduridae* Zacher, Zool. Jahrb., Jena, **30**: 306.

1954 *Labiduridae* – Brues, Melander & Carpenter, Classification of Insects, Cambridge, Mass.: 116.

Literature: Bey-Bienko 1936; Boeseman 1954; Brindle 1966; Burr 1911, 1915; Caudell 1913; Popham 1965, 1969; Sakai 1970; Steinmann 1975, 1978.

Description: Winged or wingless species. Tegmina well developed, lateral margins without lateral longitudinal ridge; posterior margin not covering dorsal shield. Pronotum quadratic, or longer than wide. Abdomen nearly parallel-sided; ultimate tergite broad; posterior margin with or without paired tubercles. Forceps fully developed for males, but simple for females. Femur of hind leg long, at least three times as long as pronotum. Male genitalia of *Labidura*-type, apices of male external parameres abruptly acute, thus apex sharply distinct or widely rounded.

Distribution: African, South American, Oriental and some nearly cosmopolitan species.

3 genus.

Identification key to the genera

1 Pronotum small, quadrate, about as long as wide. Tegmina shortened, their posterior margins not covering dorsal shield. Femur of hind leg long, at least three times as long as pronotum. Male forceps (Fig. 689) very characteristic .... Genus 5: **Tomopygia**, p. 435

1′ Pronotum longer than wide; more or less widening posteriorad, posterior margin rounded. Tegmina well developed, their posterior margin covering dorsal shield, and usually projecting beyond it. Apices of wings also projecting from beyond tegmina in most species **2**

2 Both branches of male forceps gently arcuate, relatively broad and short; sides of male abdominal tergites without tubercles; both branches of female forceps straight, without a well developed dorsal ridge and without a ventro-median tooth near apices. Apices of male external parameres widely rounded, and with a characteristic, short projecting tip (Fig. 692) .... Genus 6: **Labidura**, p. 436

2′ Both branches of male forceps usually long, and slender, sometimes short; sides of male abdominal tergites usually with tubercles, but these are absent in smaller males; both branches of female forceps almost straight, with a well-developed dorsal ridge, and with a ventro-median tooth near apex. Apices of male external parameres widely rounded, or external parameres elongated into thin, long apices .... Genus 7: **Forcipula**, p. 447

Genus 5: **Tomopygia** Burr

1904 *Tomopygia* Burr, Trans. ent. Soc. London, **1904**: 287. – Species typica: *Cylindrogaster abnormis* Bormans 1883.
Literature: Brindle 1966; Burr 1909, 1911; Popham & Brindle 1968; Sakai 1970; Steinmann 1975, 1978; Townes 1945.

Description: Wingless species. Very similar to species of *Labidura* Leach, but build more slender. Legs especially are long and thin, the posterior femora being three times as long as the pronotum, which is small and almost square. Tegmina strongly abbreviated, exposing a distinct scutellum.

Distribution: Oriental region.

Single species.

1. **Tomopygia abnormis** (Bormans)

1883 *Cylindrogaster abnormis* Bormans, Ann. Soc. ent. Belge, Bruxelles, **27**: 187; pl. 2, fig. 1 (male). – Terra typica: Java (Type male: Burr's Collect., Brit. Mus. nat. Hist., London).
1904 *Tomopygia abnormis* – Burr, Trans. ent. Soc. London, **1904**: 288.
1904 *Pygidicrana abnormis* – Kirby, A synonimic Catalogue of Orthoptera, London, **1**: 5.
1911 *Tomopygia abnormis* – Burr, Genera Insectorum, Bruxelles, **122**: 38; pl. 3, fig. 12 (male), 12a (leg), 12b (apex of abdomen), 12c (apex of female abdomen).
Literature: Boeseman 1954; Brindle 1966; Sakai 1970; Steinmann 1975, 1978.

Description: Male colour dark reddish- or blackish-brown; antennae brown, abdomen dark red, legs yellowish-brown. Head broad, rounded posteriorly. Eyes relatively large, rounded, prominent, about as long as length of head behind eyes. Postfrontal and coronal sutures absent. Antennae 12-jointed (?), broken in holotype; first joint small, slightly shorter than distance between antennal bases; second joint quadrate, third joint long, but shorter than the first. Pronotum very

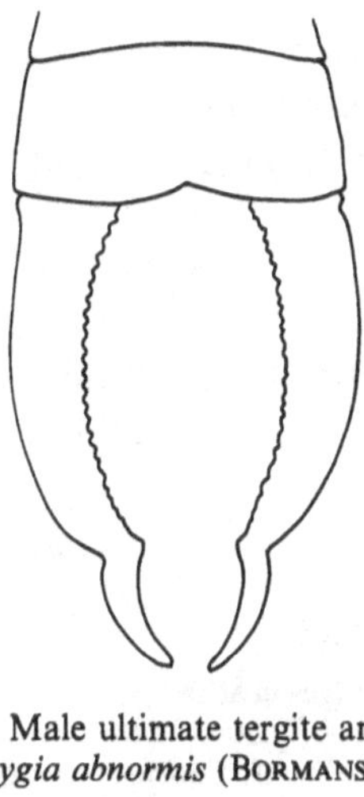

Fig. 689 Male ultimate tergite and forceps of *Tomopygia abnormis* (Bormans). After Bormans.

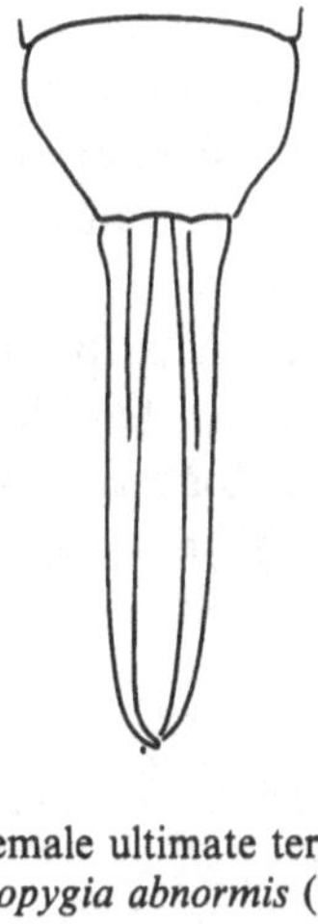

Fig. 690 Female ultimate tergite and forceps of *Tomopygia abnormis* (Bormans). After Burr.

small, more or less quadrate, shorter than the length of head in dorsal view; lateral margins parallel; median longitudinal furrow present on anterior portion; posterior margin rounded. Tegmina present, short, but a little longer than pronotum; coriaceous. Wings entirely absent. Abdomen parallel-sided at anterior and median portion, and a little expanded to last tergite. Legs with long femora and tibiae. Ultimate tergite large, broad. Forceps (Fig. 689) of *Labidura*-type, with branches remote, elongate at base, rather slender, with a single tooth at inner margin apically. Inner margins crenulate. Genitalia unknown. – **Female** similar to male, but forceps (Fig. 690) very narrow, slender, straight. – **Length** of body with forceps: in both sexes: 17–17,5 mm.

Distribution: Java.

Genus 6: **Labidura** LEACH

1815 *Labidura* LEACH, Edinbourgh Encycl., **9**: 48. – Species typica: *Forficula riparia* PALLAS 1773.

1831 *Forficesila* AUDINET-SERVILLE, Ann. Sci. nat., Paris, **12**: 32. – Species typica: *Psalis morbida* AUDINET-SERVILLE 1831.

1891 *Demogorgon* KIRBY, J. zool. Soc. London, **23**: 513. – Species typica: *Demogorgon batesi* KIRBY 1891.

Literature: BEY-BIENKO 1936; BOLIVAR 1876; BORG 1904; BORMANS 1900; BRINDLE 1966; BRUNNER 1882; BURR 1910, 1911; CAUDELL 1913; CHOPARD 1922; HANDLIRSCH 1925; KIRBY 1904; REHN, 1924; RAMME 1929; SAKAI 1970; SCUDDER 1876; STEINMANN 1975, 1978, 1979; STEPHENS 1835; TOWNES 1945; VERHOEFF 1904; ZACHER 1911.

Description: Body rather elongate and depressed. Antennae 20 to 36-jointed; third joint nearly as long as first; fourth and fifth short, as long as second. Pronotum more or less quadrate. Tegmina well developed, with a usually well developed carina, but sometimes obsolete before reaching posterior margin. Wings fully developed or rudimentary, sometimes absent. Abdomen more or less parallel-sided, a little or strongly expanded to last tergite. Abdominal tergites without lateral tubercles or spines. Ultimate tergite large, subquadratic. Penultimate sternite of male narrowed, subtruncate at apex, in female bluntly triangular. Pygidium concealed. Forceps of male with branches remote at base, elongate, rather slender, variously armed, in female subcontiguous, straight, tapering. Male genitalia specific; paramere broad, with median incision on anterior margin; genital lobes well developed, virga within genital lobe straight with large, sclerotized basal vesicle. External parameres large, lateral margins parallel for most of length, apices with a membraneous epimerite.

Distribution: Distributed in all faunal regions, but mainly tropical or subtropical, and on some oceanic islands.

7 species.

Identification key to the species

1 Abdominal tergites 6–8 with a series of short longitudinal ridges near the posterior margins, resembling the „milling" of a coin (Fig. 691). Central parameral plate of male genitalia oval, as in fig. 692 .... 1. **Labidura xanthopus**, p. 437

1′ Abdominal tergites without such ridges **2**

2 Smaller species; length of male body with forceps 16 mm or less 3
2′ Larger species; length of body with forceps of males at least 19 mm 4
3 Pronotum a little longer than wide, posterior margin rounded; wing scales about as long as pronotum, apex truncate; male forceps (Fig. 693) strongly trigonal in cross-section, with small but prominent tooth medially. Male genitalia simple, narrow, as in fig. 694 .... 2. **Labidura minor**, p. 438
3′ Pronotum quadratic; posterior margin straight; wing scales shorter than pronotum, apex obtuse. Male forceps (Fig. 695) arcuate, inner margins with two small, but prominent teeth. Male genitalia large, broad, as in fig. 696 .... 3. **Labidura dharchulensis**, p. 440
4 Wings entirely absent 5
4′ Wings present, more or less well developed, or concealed 6
5 Very large species; length of body 36–54 mm, forceps 15–24 mm. Pronotum longer than wide; lateral margins convex, posterior margin rounded. Tegmina relatively short, shorter than pronotum. Male forceps and ultimate tergite as in fig. 697 ... 4. **Labidura herculeana**, p. 441
5′ Large species; length of body 18–23 mm, forceps 6–9 mm. Pronotum quadrate; lateral margins nearly parallel, posterior margin straight. Tegmina more or less longer than pronotum. Male forceps and ultimate tergite as in fig. 699 ...... 5. **Labidura japonica**, p. 442
6 Antennal joints 8–15 long, about twice as long as wide. Ultimate tergite with two smaller or larger tubercles on posterior margin. Forceps various, but with a large paired tooth on inner margin in apical third (Fig. 701). Pronotum brown or yellowish brown .... 6. **Labidura riparia**, p. 443
6′ Antennal joints 8–15 short, only a little longer than wide. Ultimate tergite without large paired large tubercles on posterior margin. Forceps with several paired teeth on inner margins in the basal and apical thirds (Fig. 703). Pronotum black ...... 7. **Labidura orientalis**, p. 446

1. Labidura xanthopus (STÅL)

1855 *Forficesila xanthopus* STÅL, Oefvers. k. Vetensk. Akad. Förh., Stockholm, **12**: 348. – Terra typica: Brazil (Type male: Naturhist. Riksmus., Stockholm).
1860 *Forficula xanthopus* – STÅL, Eugenies Resa, Ins., Stockholm, 300.
1879 *Labidura livida* DUBRONY, Acta Soc. esp. Hist. nat., Madrid, **8**: 93. – Terra typica: Brazil (Type male: Madrid Mus.).
1891 *Demogorgon adelphus* KIRBY, J. Linn. Soc. London, Zool., **23**: 515. – Terra typica: Brazil (Type male: Brit. Mus. nat. Hist., London).
1891 *Demogorgon patagonicus* KIRBY, J. Linn. Soc., London, Zool., **23**: 515. – Terra typica: Patagonia (Type female: Brit. Mus. nat. Hist., London).
1891 *Demogorgon batesi* KIRBY, J. Linn. Soc., London, Zool., **23**: 514. – Terra typica: Santarem (Type male: Brit. Mus. nat. Hist., London).
1891 *Demogorgon bicolor* KIRBY, J. Linn., Soc., London, Zool., **23**: 514. – Terra typica: South America (Type male: Brit. Mus. nat. Hist., London).
1904 *Demogorgon longipennis* BORELLI, Boll. Mus. Zool. Anat. comp. Univ. Torino, **1904**: 4. – Terra typica: Argentina (Type male: unknown locality).
1932 *Demogorgon brasiliensis* MOREIRA, Bol. Inst. biol. Defesa agric. Rio de Janeiro, **7**: 16. – Terra typica: Brazil (Type male: Inst. biol. Defesa agric., Rio de Janeiro).
1979 *Labidura xanthopus* – STEINMANN, Acta zool. Acad. Sci. hung., Budapest, **25**: 416; fig. 1 (male forceps with ultimate tergite), 2 (male genitalia). Brazil, Argentina, Bolivia, Surinam and probably other Neotropical countries.

Literature: BRINDLE 1966; BURR 1910, 1911, 1915; HEBARD 1920; KEVAN 1951; KIRBY 1904; MACHADO, 1959; MENOZZI 1932; POPHAM & BRINDLE 1968; REHN 1933; REICHARDT 1968; SAKAI 1970; STEINMANN 1975, 1978;

Description: Male colour reddish brown or yellowish brown, usually dark brown; head reddish or darker, antennae brown or yellowish brown, tegmina often reduced and wings concealed or absent, but tegmina and wings may be fully developed. Head tumid, a little depressed; postfrontal and coronal sutures distinct. Antennae small-jointed; first joint normal, shorter than distance between antennal bases. Pronotum transverse, relatively small, lateral margins parallel; posterior margin rounded, with median longitudinal furrow. Tegmina typical, long, longer than pronotum, tuberculate. Legs long. Abdomen parallel-sided, abdominal tergites 6–8 with a series of short longitudinal ridges near posterior margins. Forceps (Fig. 691) large, both branches relatively broad, arcuate, with one inner tooth towards apex, similar to *riparia* (PALLAS). Genitalia (Fig. 692; gen. prep. No. 508, det. Dr. H. STEINMANN) of *Labidura*-type; paramere broad, median incision of anterior margin of central parameral plate very deep and wide; genital lobes well-developed, long, virga within gential lobe straight, with well-developed basal vesicle; external parameres large, inner and outer margins parallel, apex with a small epimerite. – **Female** similar to male, but forceps straight, contiguous, tapering. – **Length** of body with forceps: male: 22–26 mm, female: 20–30 mm.

Distribution: Brazil, Argentina, Bolivia, Surinam and probably other Neotropical countries.

Fig. 691 Male ultimate tergite and forceps of *Labidura xanthopus* (STÅL). Original.

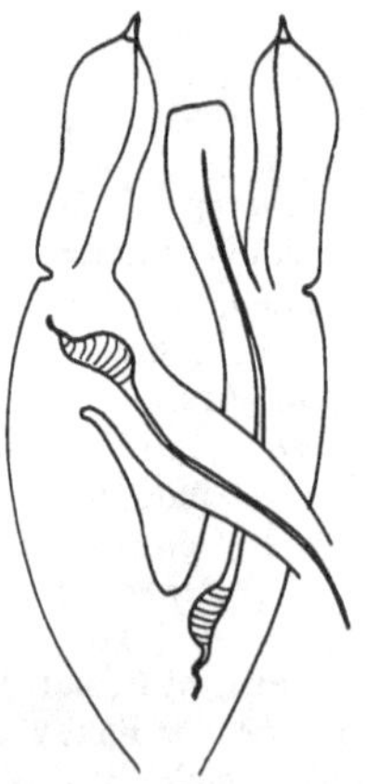

Fig. 692 Male genital armature of *Labidura xanthopus* (STÅL). Original.

2. Labidura minor BOESEMAN

1954 *Labidura minor* BOESEMAN, Zool. Verh., Leiden, **21**: 52; fig. 13a (holotype), 13b (posterior part of abdomen and proximal part of forceps of male in ventral view), 13c (antennae of male). – Terra typica: Sumatra (Type male: Rijksmus. natuurl. Hist., Leiden).

1980 *Labidura minor* – STEINMANN, Fol. ent. hung., Budapest, (33) **41**: 341; fig. 33 (ultimate tergite and forceps of holotype), 34 (holotype genital armature, gen. prep. No. 685, det. Dr. H. STEINMANN).

Literature: BRINDLE 1966; SAKAI 1970; STEINMANN 1979.

Description: Male colour yellowish brown; antennae light brown, tegmina and wings brown, texture of their surfaces slightly irregular, abdomen brown, ventral surface covered with a coat of fine yellowish hairs. Head tumid, rounded posteriorly, broad; a little broader than pronotum; postfrontal sutures and coronal suture absent. Eyes not very prominent, diameter hardly more than postorbital length of head. Antennae broken in holotype; first antennal joint conical, more or less as long as length of distance between antennal bases; second joint transverse, third long, fourth and fifth joints short, but a little longer than wide. Pronotum small, lateral margins parallel, median longitudinal furrow distinct, posterior margin broadly rounded. Both tegmina slightly narrower than pronotum and about twice its length. Wings about as long as pronotum, apex truncate. Abdomen slender, surface covered with fine impressed points. Legs long. Ultimate tergite broad, posterior margin without paired tubercles. Forceps (Fig. 693) elongate, somewhat of the *Irdex*-type, a short basal part diverging, tapering, flat inner surface forming distinct inner dorsal and ventral ridges; proximal half subcylindrical, distal half with an irregular, obtuse inner ridge beginning with a distinct blunt tooth. Genitalia (Fig. 694) very narrow, somewhat of *Nala*-type; central parameral plate with deep median incision between external parameres; genital lobes straight, virga within genital lobe of *Labidura*-type, straight, sclerotized, and with basal vesicle. External parameres long, acute. – **Female** unknown. – **Length** of body with forceps: 16 mm.

Distribution: Sumatra.

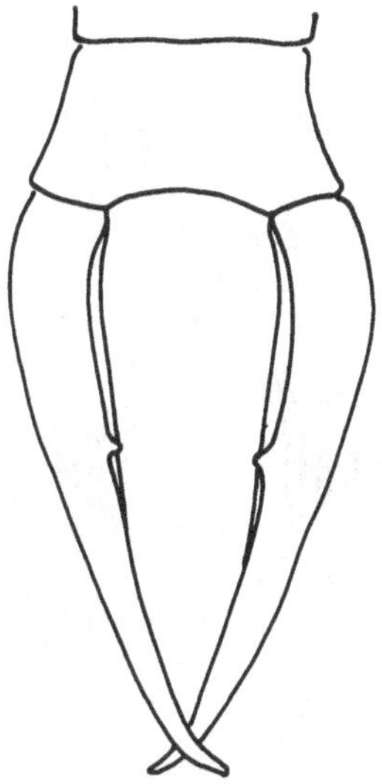

Fig. 693 Holotype ultimate tergite and forceps of *Labidura minor* BOESEMAN. Original.

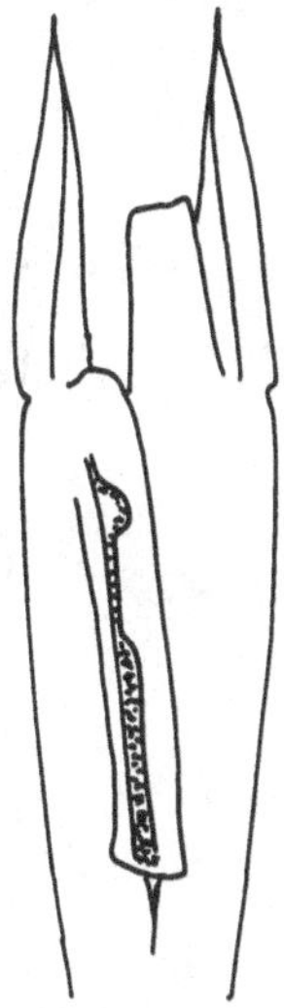

Fig. 694 Holotype genital armature of *Labidura minor* BOESEMAN. After STEINMANN.

3. Labidura dharchulensis GANGOLA

1968 *Labidura dharchulensis* GANGOLA, Entomologist, London, **101**: 59; fig. 6 (female), 7 (head and pronotum of male), 8 (pronotum and tegmen of male), 9 (male pygidium), 10 (forceps and ultimate tergite of male), 11 (male genitalia). – Terra typica: India (Type male: Government College Mus., Naini Tal, Naini, India).

1979 *Labidura dharchulensis* – STEINMANN, Acta zool. Acad. Sci. hung., Budapest, **25**: 418; fig. 4 (male ultimate tergite and forceps), 5 (genital armature).

Description: Male colour brown to dark chestnut; head light brown, pronotum yellow, but prozona with two black spots caudad and metazona lighter in shade. Tegmina blackish on sides and yellow in middle, granulated; legs yellow, abdomen brown. Head tumid, a little depressed, broad; postfrontal sutures and coronal suture not visible; eyes prominent, but small, shorter than length of head behind eyes. Antennae broken in holotype; first joint relatively small, shorter than distance between antennal bases; second transverse, third long, but shorter than length of eye. Pronotum small, quadratic, lateral margins parallel; posterior margin straight, not of the *Labidura*-type; all angles rounded. Tegmina long, longer than pronotum and longer than wide; carina sharp. Wings quite short. Abdomen expanded to 5th segment, not of the *Labidura*-type; ultimate tergite broad, rectangular, smooth, concave laterad, reddish yellow in colour, with black blunt tubercles over bases of forceps. Pygidium not visible in ventral view. Forceps (Fig. 695) cylindrical, a little curved; lower basal part flat, toothed on inner half, a dense crenulation in upper half. Genitalia (Fig. 696) characteristic;

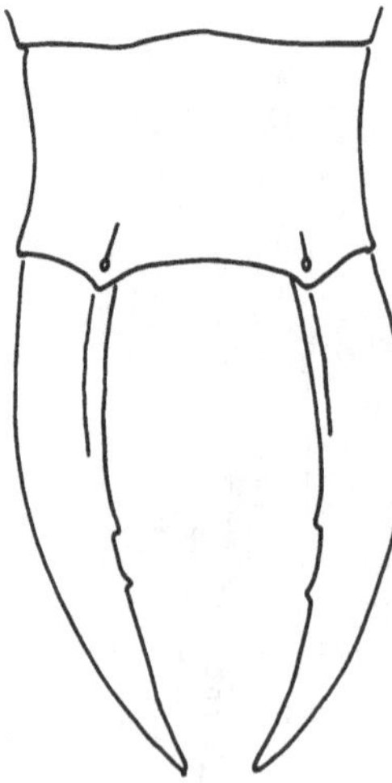

Fig. 695 Holotype ultimate tergite and forceps of *Labidura dharchulensis* GANGOLA. After GANGOLA.

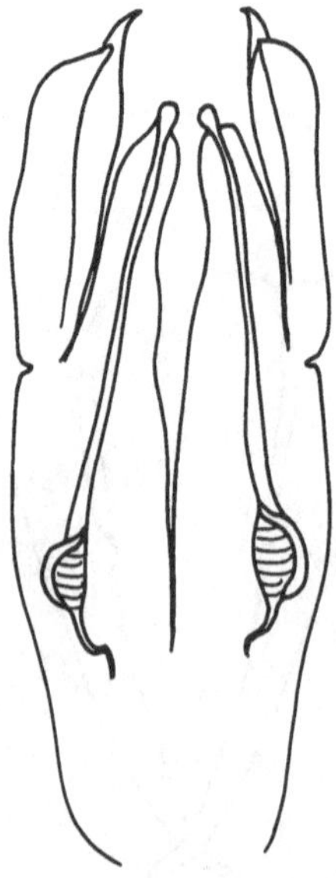

Fig. 696 Holotype genital armature of *Labidura dharchulensis* GANGOLA. After GANGOLA.

paramere broad, genital lobes well developed, with straight virgae and fully developed basal vesicle; genital lobes directed forwards. External parameres broad, epimerite distinct. – **Female** similar to male, but forceps straight. – **Length** of body with forceps: male: 16 mm, female: 15,5 mm.

Distribution: India.

4. Labidura herculeana (FABRICIUS)

1798 *Forficula herculeana* FABRICIUS, Ent. Syst. Suppl., Flensburgjiet: 185. – Terra typica: St. Helena (Type male: Univ. zool. Mus., København).

1904 *Labidura riparia* ssp. *herculeana* – KIRBY, A synonimic Catalogue, Orth., London, **1**: 11.

1911 *Labidura herculeana* – BURR, Genera Insectorum, Bruxelles, **122**: 37 [= *Labidura riparia* (PALLAS 1773)].

1970 *Labidura herculeana* – BRINDLE, Ann. Mus. r. Afr. centr., Tervuren, **181**: 221; fig. 22 (male), 23a (forceps of female), 23b (male genitalia).

1979 *Labidura herculeana* – STEINMANN, Acta zool. Acad. Sci. hung., Budapest **25**: 419; fig. 6 (male ultimate tergite and forceps), 7 (male genitalia).

Literature: REHN 1904; SAKAI 1970; STEINMANN 1978.

Description: Male black, rather shiny; antennae dark brown, first joint dark red; femora reddish, tibiae red on basal half, apical half black; cuticle rugose, except for that of slightly roughened head which displays a few scattered shallow punctures. Head rounded, longer than wide; postfrontal sutures and coronal

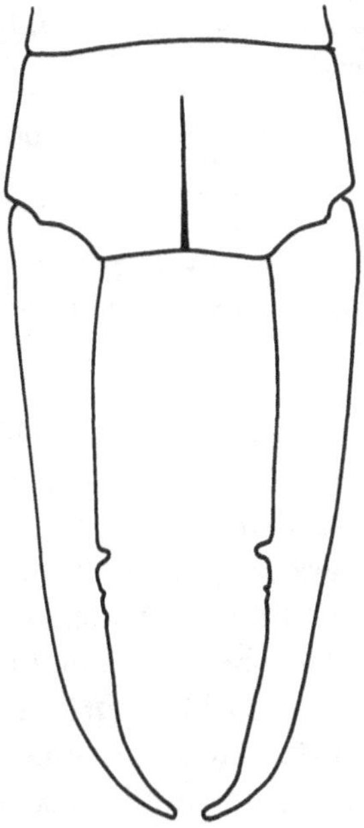

Fig. 697 Male ultimate tergite and forceps of *Labidura herculeana* (FABRICIUS). After STEINMANN.

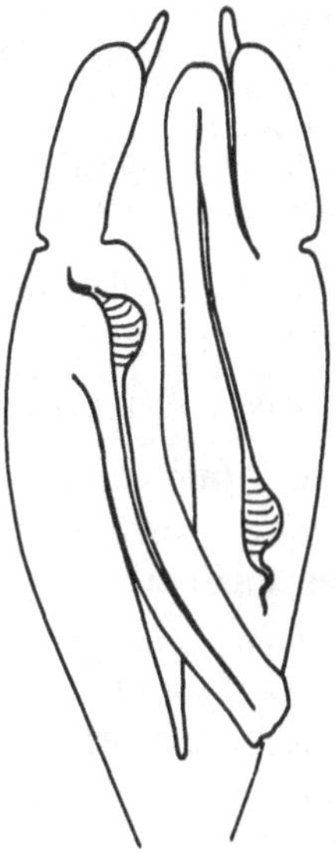

Fig. 698 Male genital armature of *Labidura herculeana* (FABRICIUS). After STEINMANN.

suture distinct, well marked; posterior margin of head transverse. Antennae 32-jointed, first joint relatively small, shorter than distance between antennal bases, second joint short, transverse, third joint about twice as long as wide; cylindrical. Eyes small, shorter than length of head behind eyes. Pronotum relatively large, longer than wide and widened posteriorly. Tegmina short, about equal in length to pronotum. Wings entirely absent. Abdomen very long, parallel-sided, or slightly widened medially; each segment with a pair of short, longitudinal, shiny lines towards each dorso-lateral border. Ultimate tergite transverse, with median longitudinal sulcus. Forceps (Fig. 697) very long, cylindrical, smooth, well separated at base, both branches arcuate narrowing distally and with one ventro-median tooth near mid-point. Genitalia (Fig. 698) of the *Labidura*-type, but paramere oval and external parameres shorter and broader than in *Labidura riparia* (PALLAS); epimerite at tip of external paramere short, broad and membranous. Virga long, dilated basally, with sinuous inner tube and a very narrow sclerite on each side. – **Female** similar to male, but forceps shorter, broad at base, narrowed distally, branches close together, both branches with a basal excavation and inner margin with isolated small tubercles. – **Length** of body with forceps: male: 45–65 mm, female: 45–70 mm.

Distribution: St. Helena.

5. Labidura japonica (de HAAN)

1842 *Forficula gigantea* var. *japonica* de HAAN, Verh. nederlandse. Overz. Bezitt., Orth., den Haag, 240. – Terra typica: Japan (Type male: Mus. Naturkunde, Berlin).

1900 *Labidura riparia* var. *japonica* – BORMANS, Das Tierreich, Berlin, **11**: 35.

1903 *Labidura icterica* var. *japonica* – Kirby, A synonymic Catalogue of Orthoptera, London, **1**: 11.

1911 *Labidura japonica* – BURR, Genera Insectorum, Bruxelles, **122**: 37.

1968 *Labidura japonica* – BRINDLE, Ann. Mag. nat. Hist., London, (13) **9**: 260 [= *Labidura riparia* (PALLAS 1773)].

1979 *Labidura japonica* – STEINMANN, Acta zool. Acad. Sci. hung., Budapest, **25**: 420; fig. 8 (male ultimate tergite and forceps), 9 (male genitalia; gen. prep. No. 639, det. Dr. H. STEINMANN). Japan, Formosa, China, India, Malaya and New Guinea).

Literature: BURR 1908, 1911; SAKAI 1970.

Description: Male colour dark reddish brown; head red or reddish brown, pronotum dark brown to blackish brown, tegmina dark brown, except at yellowish-red median part; legs yellow. Head rounded, a little broader than pronotum; postfrontal sutures and coronal suture very well marked. Eyes prominent, large, about as long as length of head behind eyes. Antennae 22 (+?) jointed; first joint relatively small, shorter than distance between antennal bases; second quadratic, third joint long, about twice as long as wide, fourth transverse or quadratic. Pronotum more or less as long as wide; lateral margins parallel, posterior margin broadly rounded; median longitudinal furrow distinct. Tegmina coriaceous, short, with a lateral longitudinal ridge. Wings absent or not discernible. Legs normal, typical of genus. Abdomen long, parallel-sided, somewhat shiny. Ultimate tergite broad; depressed at posterior margin medially.

Forceps (Fig. 699) a little trigonal basally, cylindrical apically, arcuate, inner margin with a large tooth. Genitalia (Fig. 700) specific; central parameral plate comparatively short, narrow; about as long as length of external paramere; genital lobes well developed, virga long, straight, with a large, sclerotized basal vesicle. External parameres of the *Labidura*-type, but very long, narrow, epimerite membranous. – **Female** similar to male, but colour dark brown, head, pronotum and abdomen shiny, forceps slender, inner margins crenulate. – **Length** of body with forceps: male: 24–26 mm, female: 23–25 mm.

Distribution: Japan, Taiwan, China, India, Malaysia and New Guinea.

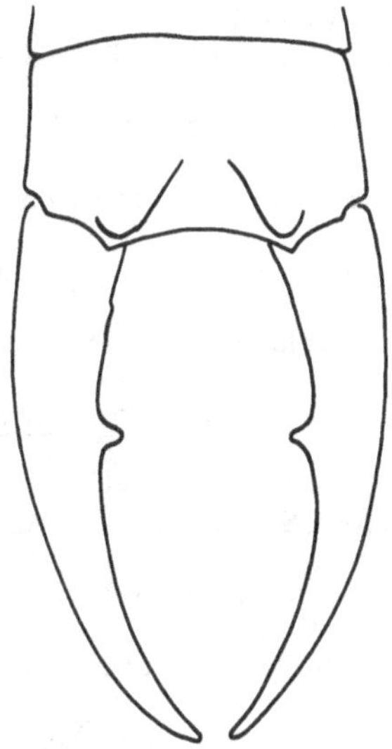

Fig. 699 Male ultimate tergite and forceps of *Labidura japonica* (DE HAAN). Original.

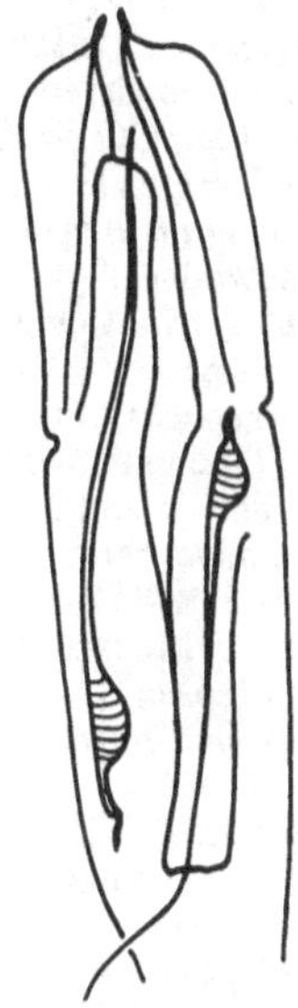

Fig. 700 Male genital armature of *Labidura japonica* (DE HAAN). Original.

6. Labidura riparia (PALLAS)

1773 *Forficula riparia* PALLAS, Reise russ. Reiche, St. Petersburg, **2**: 727. – Terra typica: Siberia: Irtysh river (Type male: Univ. zool. Mus., København).

1775 *Forficula dentata* FABRICIUS, Syst. Ent., Flensburgjiet, 270. – Terra typica: Central Africa (Type male: Univ. zool. Mus., København).

1775 *Forficula pallipes* FABRICIUS, Syst. Ent., Flensburgjiet, 270. – Terra typica: Uganda (Type male: Univ. zool. Mus., København).

1780 *Forficula maxima* VILLIERS, Linn. Ent., Paris, **1**: 427. – Terra typica: Europe (Type male: Mus. nation. Hist. nat., Paris).

1786 *Forficula bilineata* HERBST, Arch. Insektengesch., Zürich, Winterthur **7–8**: 103. – Terra typica: unknown (Type male: unknown locality).

1791 *Forficula bidens* OLIVIER, Encycl. method., Paris, **6**: 466. – Terra typica: South America (Type male: Staatl. Mus. Tierkunde, Dresden).

1791 *Forficula crenata* OLIVIER, Encycl. method., Paris, **6**: 467. – Terra typica: Central Africa (Type female: Staatl. Mus. Tierkunde, Dresden).
1793 *Forficula gigantea* FABRICIUS, Syst. Ent., Flensburgjiet, **2**: 1. – Terra typica: unknown (Type male: Univ. zool. Mus., København).
1793 *Forficula flavipes* FABRICIUS, Syst. Ent., Flensburgjiet, **2**: 1. – Terra typica: Spain (Type female: Univ. zool. Mus., København).
1793 *Forficula erythrocephala* FABRICIUS, Syst. Ent., Flensburgjiet, **2**: 4. – Terra typica: America meridionalis (Type male: unknown locality).
1805 *Forficula rufescens* PALISOT DE BEAUVOIS, Ins. rec. Afrique Amérique, Orth., Paris, 35. – Terra typica: unknown (Type male: Mus. nation. Hist. nat., Paris).
1815 *Labidura riparia* – LEACH, Edinbourgh Encycl., **9**: 48.
1831 *Psalis morbida* AUDINET-SERVILLE, Ann. Sci. nat., Paris, **22**: 35. – Terra typica: unknown (Type male: unknown locality).
1836 *Forficesila affinis* GUÉRIN-MÉNEVILLE, in RAMON DE LA SAGRA., Hist. fis. pol. nat. Isla de Cuba, Madrid, **7**: 137. – Terra typica: Cuba (Type male: unknown locality).
1838 *Forficesila gigantea* BURMEISTER nec FABRICIUS 1793, Handb. Ent., Jena, **2**: 751. – Terra typica: Antillen (Type female: unknown locality).
1838 *Forficula bivittata* KLUG (apud BURMEISTER) Handb. Ent., Jena, **2**: 751. – Terra typica: Hispaniola (Type male: unknown locality).
1838 *Forficesila suturalis* BURMEISTER, Handb. Ent., Jena, **2**: 752. – Terra typica: Colombia (Type male: unknown locality).
1839 *Forficula marginella* COSTA, Atti. Accad. Bourbon, Napoli, **4**: 50. – Terra typica: unknown (Type unknown).
1839 *Forficesila terminalis* AUDINET-SERVILLE, Hist. nat. Insectes, Orth., Paris, 25. – Terra typica: Ile de France (Type unknown).
1839 *Forficesila icterica* AUDINET-SERVILLE, Hist. nat. Insectes, Orth., Paris, 25. – Terra typica: India (Type male: Mus. nation. Hist. nat., Paris).
1846 *Forficesila riparia* – FISCHER, Waldheim Orthopt. Rossia,: 46.
1846 *Forficula bicolor* MOTSCHULSKY, in FISCHER-WALDHEIM, Horae Soc. ent. ross., St. Petersburg, **4**: 354. – Terra typica: Siberia (Type male: unknown locality).
1846 *Forficula fischeri* MOTSCHULSKY, in FISCHER von WALDHEIM, Horae Soc. ent. ross., St. Petersburg, **4**: 354. – Terra typica: Transcaucasia (Type male: unknown locality).
1858 *Forficula amurensis* MOTSCHULSKY, Bull. Soc. imp. Nat., Moscou, **32**: 499. – Terra typica: Amur (Type male: unknown locality).
1863 *Labidura bengalensis* DOHRN, Stettiner ent. Ztg., **24**: 312. – Terra typica: India (Type male: gen. prep. No. 492, det. Dr. H. STEINMANN: Zool. Inst., Warszawa).
1863 *Labidura servillei* DOHRN, Stettiner ent. Ztg., **24**: 316. – Terra typica: India: Madras (Type male: Zool. Inst., Warszawa).
1878 *Labidura auditor* SCUDDER, Proc. Boston Soc. nat. Hist., **28**: 252. – Terra typica: Brazil: Rio Grande do Norte, Natal (Type male: U.S. nation. Mus., Washington).
1882 *Labidura riparia* – var. *inermis* BRUNNER, Prodr. eur. Orth., Wien, 5. – Terra typica: Europe (Type unknown).
1891 *Labidura granulosa* KIRBY, J. Linn. Soc. London, **23**: 511. – Terra typica: Philippines (Type male: Brit. Mus. nat. Hist., London).
1891 *Labidura pluvialis* KIRBY, J. Linn. Soc. London, **23**: 512. – Terra typica: Raine Islands (Type male: Brit. Mus. nat. Hist., London).
1891 *Labidura clarki* KIRBY, J. Linn. Soc. London, **23**: 512. – Terra typica: Brazil (Type female: Brit. Mus. nat. Hist., London).
1897 *Labidura distincta* RODZIANKO, Wiener ent. Ztg., **16**: 153. – Terra typica: Europe occidentalis (Type male: Naturhist. Mus., Wien).
1897 *Labidura riparia* – var. *mixta* BOLIVAR, Ann. Sci. nat., Porto, **4**: 4. – Terra typica: unknown (Type male: unknown locality).
1900 *Labidura riparia* – ssp. *erythrocephala* – BORMANS, Das Tierreich, Berlin, **11**: 34.
1900 *Labidura riparia* – ssp. *pallipes* – BORMANS, Das Tierreich, Berlin, **11**: 34.
1900 *Labidura riparia* – ssp. *livida* – BORMANS, Das Tierreich, Berlin, **11**: 35.
1900 *Labidura riparia* – ssp. *pluvialis* – BORMANS, Das Tierreich, Berlin, **11**: 35.

1900 *Labidura riparia* – ssp. *icterica* – BORMANS, Das Tierreich, Berlin, **11**: 36.
1901 *Apterygida huseine* REHN, Proc. Acad. nat. Sci. Philadelphia, **1901**: 273. – Terra typica: Ethiopia (Type male: U.S. nation. Mus., Washington).
1903 *Labidura truncata* KIRBY, Ann. Mag. nat. Hist., London, (7) **11**: 67. – Terra typica: Australia (Type male: Brit. Mus. nat. Hist., London).
1904 *Tomopygia sinensis* BURR, Trans. r. ent. Soc. London, **1904**: 288. – Terra typica: China: Peking (Type male: BURR's Collect., Brit. Mus. nat. Hist., London).
1904 *Labidura dubronyi* BORG, Ark. Zool., Uppsala, **1**: 565. – Terra typica: unknown (Type male: unknown locality).
1904 *Labidura karschi* BORG, Ark. Zool., Uppsala, **1**: 565. – Terra typica: unknown (Type male: unknown locality).
1905 *Labidura mongolica* REHN, Proc. U.S. nation. Mus., Washington, **2**: 9: 503; fig. 2 (male). – Terra typica: Mongolia (Type male: U.S. nation. Mus., Washington).
1906 *Labidura riparia* – var. *dumonti* AZAM, Bull. Soc. ent., France, Paris,: 80. – Terra typica: France (Type male: Mus. nation. Hist. nat., Paris).
1908 *Labidura riparia* var. *herculeana* SEMENOFF, Rev. russe d'Ent., Moskva, 171. – Terra typica: South Russia (Type male: unknown locality).
1911 *Labidura riparia* – BURR, Genera Insectorum, Bruxelles, **122**: 36; pl. 8, fig. 22 (opisthomeres).
1915 *Labidura riparia* – BURR, J. r. microsc. Soc., London, **1915**: 444; pl. 9, fig. 5 (male genitalia).
1929 *Labidura confusa* CAPRA, Ann. Mus. civ. Stor. nat., Genova, **53**: 157. – Terra typica: Egypt (Type male: Mus. civ. Stor. nat., Genova).
1973 *Labidura riparia* – BRINDLE, Dermaptera of Africa, Tervuren, **1**: 108, fig. 137 (male), 138 (male genitalia), 139 (female forceps).
1979 *Labidura riparia* – STEINMANN, Acta zool. Acad. Sci. hung., Budapest, **25**: 420; fig. 10 (male ultimate tergite and forceps), 11 (male genital armature).

Literature: BEY-BIENKO 1936; BOESEMAN 1954; BORELLI 1907, 1908, 1923; BRINDLE 1966, 1968, 1970; CHOPARD 1937; DUBRONY 1879; GÜNTHER 1930; GURNEY 1950; HEBARD 1923; HINCKS 1952, 1953, 1957, 1959; HUXLEY 1968; KIRBY 1904; MATSUMURA 1935; MENOZZI 1928; RAMAMURTHI 1955; RAMME 1911; REHN 1906, 1907; SAKAI 1970; SCUDDER 1876; SHIRAKI 1905; STEINMANN 1975, 1978, 1979; ZACHER 1911.

Description: Male usually dark brown, variegated with yellow or reddish yellow; head reddish or darker, antennae brown or yellowish basally; pronotum blackish or dark brown, lateral margins yellow; tegmina dark brown, wings yellow, darkened on external margins; legs yellow. Head transverse, coriaceous; postfrontal and coronal sutures distinct, well marked. Eyes comparatively small or larger, more or less shorter than length of head behind eyes. Antennae long, multijointed; first joint relatively short, shorter than distance between antennal bases; second transverse, third joint long, about twice as long as wide. Pronotum rectangular, somewhat transverse, all angles rounded; lateral margins parallel, posterior margin weakly convex; plate with median longitudinal furrow. Tegmina long and short, rugose, rather dull, with one or more faint longitudinal or transverse irregular ridges dorsally, and both tegmina with a lateral longitudinal ridge. Wings short, usually visible, sometimes concealed or possibly absent. Legs long. Abdomen broad, elongate, depressed, often wider medially, ultimate tergite transverse, broad, with a triangular depression posteriorly and medially, posterior margin with a pair of teeth medially, teeth directed posteriorly or slightly dorso-posteriorly, or teeth occasionally absent. Forceps (Fig. 701) relatively broad, arcuate, with one inner tooth towards apex, tooth directed ventro-medially, branches widely separated at base. Genitalia (Fig. 702;

gen. prep. No. 58, det. Dr. H. STEINMANN) typical of genus; paramere broad, median incision of anterior margin of paramere broad, deep, genital lobes well developed, with broad virgae. Basal vesicle large, strongly sclerotized. External parameres parallel-sided for most of length, apices with a membranous epimerite. – **Female** very similar to male, but ultimate tergite narrower, and without paired teeth on posterior margin; both branches of forceps shorter, almost simple, straight at apex, inner margin crenulate, branches almost contiguous. Abdomen strongly punctate, especially on basal tergites, distal segments less strongly punctate. – **Length** of body with forceps: male: 17–37 mm, female: 18–38 mm.

Distribution: Cosmopolitan.

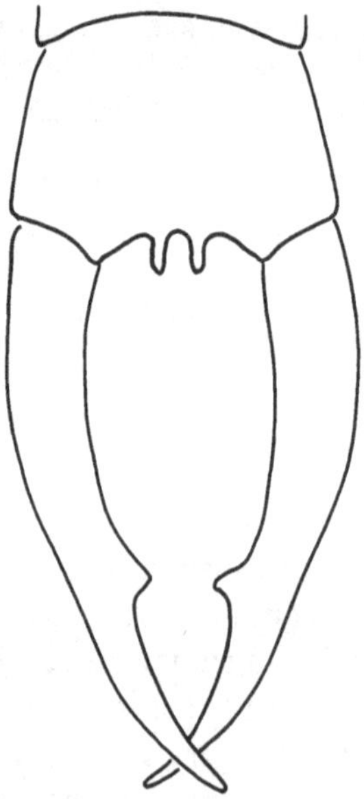

Fig. 701 Male ultimate tergite and forceps of *Labidura riparia* (PALLAS). Original.

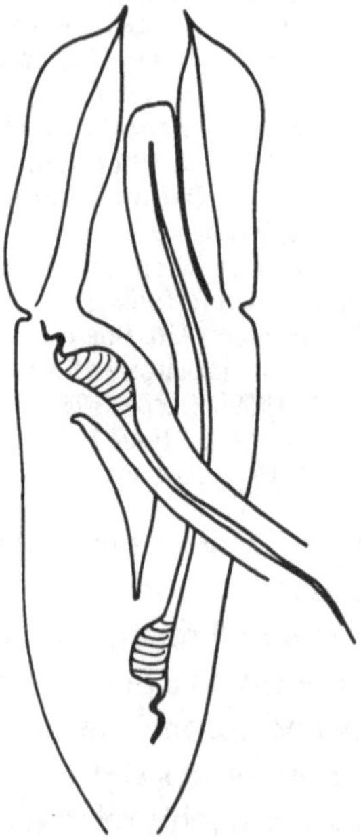

Fig. 702 Male genital armature of *Labidura riparia* (PALLAS). Original.

7. Labidura orientalis STEINMANN

1979 *Labidura orientalis* STEINMANN, Acta zool. Acad. Sci. hung., Budapest, **25**: 422; fig. 12 (ultimate tergite with forceps of holotype), 13 (holotype genital armature; gen. prep. No. 143, det. Dr. H. STEINMANN). – Terra typica: Nepal (Type male: Természettudományi Múz., Budapest).

Description: Male colour dark reddish brown; head dark red, pronotum, tegmina at lateral margins and ultimate tergite in posterior part black, legs yellow, forceps reddish brown. Head tumid, large, a little broader than pronotum; postfrontal sutures and coronal suture well marked. Eyes relatively large, but a little shorter than length of head behind eyes. Antennae broken in holotype; first joint small, shorter than distance between antennal bases; second joint quadratic, third joint twice as long as wide. Pronotum depressed, lateral margins carinate; all angles rounded, quadratic, lateral margins parallel; posterior margin rounded. Tegmina short, rugose, with a lateral longitudinal ridge. Wings yellow, apices blackish brown, coriaceous. Legs long. Abdomen a little expanded medially, strongly punctate, especially on basal tergites, distal segments less strongly punctate. Ultimate tergite transverse, with a triangular depression posteriorly; posterior margin with a pair of blunt teeth dorsally, obtuse. Forceps (Fig. 703) trigonal basally, cylindrical apically, curved, with large inner teeth on basal to apical portions. Genitalia (Fig. 704) large, paramere relatively narrow, median incision of anterior margin of paramere very deep; genital lobes well developed, virgae broad and long, basal vesicle large. External parameres of the *Labidura*-type, inner and outer margins parallel, epimerite membranous. – **Female** unknown. – **Length** of body with forceps: 37,5 mm.

Distribution: Nepal.

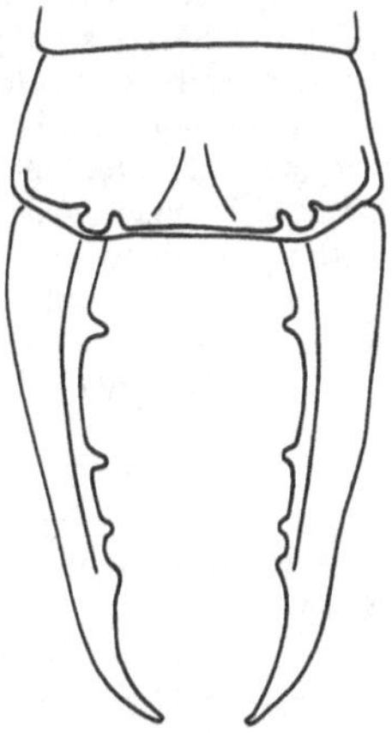

Fig. 703 Holotype ultimate tergite and forceps of *Labidura orientalis* STEINMANN. After STEINMANN.

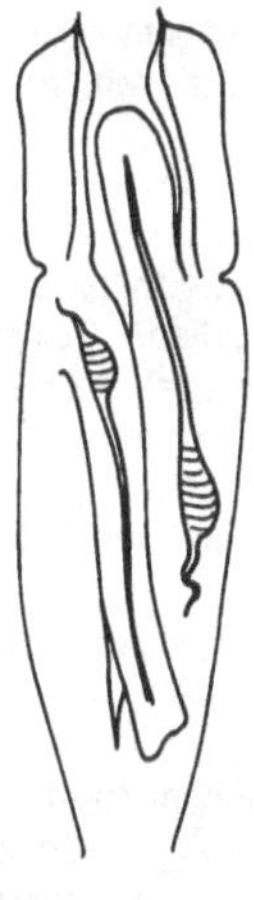

Fig. 704 Holotype genital armature of *Labidura orientalis* STEINMANN. After STEINMANN.

Genus 7: **Forcipula** BOLIVAR

1897 *Forcipula* Bolivar, Ann. Soc. ent. France, Paris, **66**: 283. – Species typica: *Labidura quadrispinosa* DOHRN 1863.

Literature: BOESEMAN 1954; BORMANS 1900; BRINDLE 1966; BURR 1904, 1910, 1911; HANDLIRSCH 1925; HEBARD 1917; SAKAI 1970; STEINMANN 1975, 1980; VERHOEFF 1902; ZACHER 1911.

Description: Medium to large species with the general characteristic of *Labidura*, but certain abdominal segments have spines, crests or ridges at the sides. The ridges, the lateral longitudinal keels of the tegmina, present at the shoulder only, not reaching the posterior margin, or almost obsolete. The forceps of male very long and slender. Tegmina and wings usually fully developed, sometimes the tegmina are short and the wings absent or concealed. Legs and abdomen relatively slender, long, depressed. Genitalia of male somewhat similar to *Labidura*, but slenderer.

Distribution: Mainly Oriental and Australian in distribution, but some species are recorded from Africa and South America.

26 species or subspecies.

Identification key to the species and subspecies

1 Neotropic species **2**

1′ Ethiopian, Palaearctic, Oriental or Indo-Australian species **4**

2 Tegmen coloured as abdomen, blackish brown, sometimes slightly lighter on shoulders; abdominal segments 4–6 of male with single lateral spine, or without spines. Male pygidium with two small teeth (Fig. 705) and forceps curved apically, with single tooth medially. Male genitalia as in fig. 706 .... 1 a. **Forcipula quelchi quelchi**, p. 450

2′ Tegmen colour mainly yellowish brown or reddish yellow, much lighter than abdomen. Male forceps without (Fig. 707) or with inner tooth or teeth (Fig. 709) **3**

3 Tegmen rugose, shiny, yellowish brown; abdominal segments 4–6 of male with single lateral spines. Male pygidium with two teeth, and forceps (Fig. 707) elongate, slender, almost straight. External parameres of male genitalia (Fig. 708) concave on inner and outer margins .... 1 b. **Forcipula quelchi boliviana**, p. 451

3′ Tegmen tuberculate dull, reddish yellow, with lateral margins and outer angles black. Abdominal segments 3–6 of male with or without crests, never with spines. Male pygidium without teeth, male forceps (Fig. 709) sinuate, typical form with two teeth on inner margin of median portion. Male genitalia (Fig. 710) narrow, median incision of anterior margin of paramere relatively short, and external parameres of male genitalia more or less parallel along inner and outer margins .... 2. **Forcipula americana**, p. 452

4 African species **5**

4′ Palaearctic, Oriental or Indo-Australian species **7**

5 Yellowish brown, eyes larger than length of head behind eyes; male forceps (Fig. 711) gently arcuate with three teeth on inner margin of each branch. Male ultimate tergite without too large vertical spines near base of forceps. Pygidium without two teeth .... 3. **Forcipula congo**, p. 453

5′ Blackish brown, eyes smaller than length of head behind eyes. Male forceps with one single tooth on inner margin. Pygidium with two teeth at posterior margin **6**

6 Male forceps strongly curved apically, as in fig. 712. Tegmina well developed, long, about twice as long as wide. Male genitalia (Fig. 713) very narrow, long; median incision of anterior margin of paramere deep .... 4. **Forcipula gariazzi**, p. 454

6′ Male forceps straight, arcuate, with crenulation on inner margin, or sometimes a little curved (Fig. 714), and with a single tooth on medial portion. Tegmina short, a little longer than wide. Wings absent or concealed. Male genitalia (Fig. 715) with short median incision; genital lobes well developed but narrow .... 5. **Forcipula tanganyikae**, p. 455

7 Legs with a broad dark transverse band on middle and posterior femora, or almost all blackish **8**

7′ Legs yellow or almost so **9**

8 Legs mainly blackish, tegmina rather longer than pronotum; wings fully developed, visible. Male forceps (Fig. 716) slender, straight, without *Forcipula*-type curvature apically .... 6. **Forcipula tarsata**, p. 456

8′ Legs yellowish, median and posterior femora with dark transverse band; tegmina short, about equal in length to pronotum; wings concealed. Male forceps (Fig. 717) with *Forcipula*-type curvature apically .... 7. **Forcipula borellii**, p. 457

9 Yellowish brown. Eyes rather large, as long as length of head behind eyes. Pronotum rather longer than wide, lateral margins curving inwards slightly; basal abdominal segments of male with lateral crests. Male forceps (Fig. 718) characteristic, both branches curved apically, with single tooth on inner margins. Male genitalia (Fig. 719) relatively short, broad, with spine-like epimerite .... 8. **Forcipula lurida**, p. 458
9′ Blackish brown, eyes smaller, lateral margins of pronotum not curving inwards **10**
10 Tegmen tuberculate or rugose **11**
10′ Tegmen punctate or with setae **15**
11 Tegmen tuberculate **12**
11′ Tegmen rugose **13**
12 Male forceps (Fig. 720) specific, of *Cranopygia*-type, crenulate on inner margin apically. External parameres of male genitalia (Fig. 721) broad, genital lobes very long .... 9. **Forcipula banksi**, p. 459
12′ Male forceps (Fig. 722) of *Forcipula*-type, with single large tooth on inner margin. External parameres of male genitalia (Fig. 723) characteristic, paramere V-shaped, median incision of anterior margin broad .... 10. **Forcipula tuberculata**, p. 460
13 Head red; postfrontal and coronal sutures faint. Pronotum about as long as wide, quadrate. Tegmina with light reddish median longitudinal stripe. Wings yellow, with a little dark brown colouring apically .... 11. **Forcipula leonardii**, p. 461
13′ Head dark brown to black; postfrontal sutures and coronal suture strongly visible, strikingly marked. Pronotum more or less longer than wide. Tegmina without median longitudinal stripe **14**
14 Male with lateral spines on abdominal tergites 3–6. Male forceps (Fig. 725) without crenulations on inner margin basally. Median incision of male genitalia (Fig. 726) short, paramere relatively broad .... 12. **Forcipula quadrispinosa**, p. 462
14′ Male with lateral spines on abdominal tergites 3–5. Forceps of male (Fig. 727) with fine crenulations on inner margin basally. Median incision of male genitalia very long, narrow, as in fig. 728 .... 13. **Forcipula simplex**, p. 463
15 Pronotum transverse, widening slightly posteriorly. Male without abdominal spines. Male forceps (Fig. 729) of *Forcipula*-type, but inner tooth on median portion characteristic. Male genitalia (Fig. 730) very narrow, long; external parameres very long .... 14. **Forcipula despinosa**, p. 464
15′ Pronotum quadrate or longer than wide **16**
16 Pronotum quadrate, about as long as wide **17**
16′ Pronotum longer than wide **21**
17 Male with single spine on each basal abdominal tergite **18**
17′ Male with two spines on each basal abdominal tergite **20**
18 Male forceps with simple but well-visible inner tooth basally **19**
18′ Male forceps without single large inner tooth basally but with some small tubercles (Fig. 731). External parameres of male genitalia (Fig. 732) relatively small, shorter than the length of incision on anterior margin of the central parameral plate ... 15. **Forcipula yunnanea**, p. 465
19 Male forceps strongly curved (Fig. 733), characteristic. Ultimate tergite with transverse, simple posterior margin. Male genitalia narrow, as in fig. 734 16. **Forcipula trispinosa**, p. 466
19′ Male forceps straight; trigonal basally, cylindrical apically (Fig. 735). Male genitalia very specific, both penes directed forwards, as in fig. 736 .... 17. **Forcipula afghana**, p. 467
20 Male forceps of *Forcipula*-type, normal, strongly curved basally but without large denticle on inner margin (Fig. 737). External parameres of male genitalia relativly short, shorter than the length of central parameral plate, as in fig. 738 ..... 18a. **Forcipula decolyi decolyi**, p. 468, 469
20′ Male forceps with very characteristic large denticle on inner margin (Fig. 739). External parameres of male genitalia relatively long, about as long as length of central parameral plate, as in fig. 740 .... 18b. **Forcipula decolyi novaguineae**, p. 470
21 Sides of head behind eyes widening posteriorly. Tegmina short, wings absent or concealed. Male with single spines on each basal abdominal tergite, those on the fifth short. Male forceps (Fig. 741) with single inner tooth basally. Male genitalia as in fig. 742 .... 19. **Forcipula aborensis**, p. 470

21′ Sides of head parallel behind eyes or narrowing. Tegmina normally developed, wings visible. Male forceps without large inner tooth basally **22**
22 Tegmina shorter, less than twice the length of pronotum **23**
22′ Tegmina longer, much longer than twice the length of pronotum **24**
23 Wings yellow or yellowish brown to brown; abdomen chestnut-brown, abdominal tergites 3–5 or 6 of male with two spines on each side. Ultimate tergite with well-marked median longitudinal sulcus (Fig. 743). Male genitalia typical of genus, as in fig. 744 .... 20. **Forcipula clavata**, p. 471
23′ Wings and tegmina blackish brown, abdomen blackish; abdominal tergites 3–5 of male with single spines. Male forceps (Fig. 745) slender, not of *Forcipula*-type, without inner tooth on basal portion. Male genitalia slender, narrow, as in fig. 746 .... 21. **Forcipula walkeri**, p. 473
24 Smaller, total length 20–21 mm; abdominal segments of male without spine. Male forceps (Fig. 747) only slightly sinuate. External parameres of male genitalia a little expanded basally, as in fig. 748 .... 22. **Forcipula vanheurni**, p. 474
24′ Larger, total length about 30–34 mm; abdominal segments 3–6 of male with one or two spines **25**
25 Abdominal tergites 3–6 of male with single spines. Male forceps (Fig. 749) sharply curved near middle. External parameres of male genitalia comparatively broad, as in fig. 750 .... 23. **Forcipula indica**, p. 475
25′ Abdominal tergites 3–6 of male with two spines. Male forceps (Fig. 751) strongly curved, with characteristic serration basally on the inner margin. External paramere comparatively narrow, as in fig. 752 .... 24. **Forcipula obscura**, p. 476

1. Forcipula quelchi BURR

1904 *Forcipula quelchi* BURR, Trans. ent. Soc. London, **1904**: 289. – Terra typica: British Guiana (Type male: BURR's Collect., Brit. Mus., nat. Hist., London).
1966 *Forcipula quelchi quelchi* – BRINDLE, Ann. Mag. nat. Hist., London, (13) **9**: 267; fig. 32 (male genitalia). Surinam.
1980 *Forcipula quelchi quelchi* – STEINMANN, Acta zool. Acad. Sci. hung., Budapest, **26**: 232; fig. 1 (male ultimate tergite and forceps), 2 (male genitalia).
Literature: BURR 1910; POPHAM & BRINDLE 1968; REICHARDT 1968; SAKAI 1970; STEINMANN 1978.

Description: Male colour blackish brown, tegmina and wings dark brown, antennae and legs yellow; forceps blackish or reddish. Head tumid, depressed, broad, a little broader than pronotum; postfrontal sutures and coronal suture distinct. Eyes large, about as long as length of head behind eyes. Antennae 21-jointed; first joint normal, short, shorter than distance between antennal bases, second transverse, third joint long, but only twice as long as wide. Pronotum more or less parallel-sided, posterior margin broadly rounded. Tegmina coriaceous, wings present. Abdomen parallel-sided, abdominal segments 4–6 with single lateral spines; ultimate tergite broad, posterior margin straight; pygidium with two small tubercles on posterior margin. Forceps (Fig. 705) curved apically, inner margin with a single very well marked tooth medially; branches trigonal basally, cylindrical apically. Genitalia (Fig. 706) characteristic; central parameral plate broad, median incision of anterior margin very deep; genital lobes well developed, virga straight, basal vesicle large, strongly sclerotized. External parameres long, inner and outer margin more or less parallel or a little convex, epimerite very long. – **Female** similar to male, but forceps straight, contiguous, tapering. – **Length** of body with forceps: male: 30–38 mm, female: 30–41 mm.

Distribution: British Guiana, Surinam.

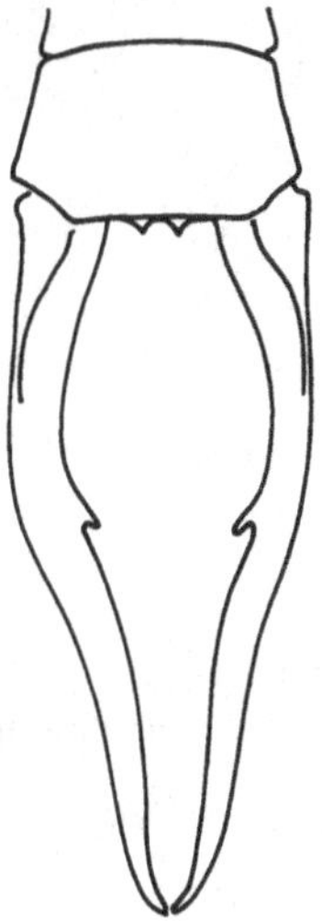

Fig. 705 Male ultimate tergite and forceps of *Forcipula quelchi quelchi* BURR. After STEINMANN.

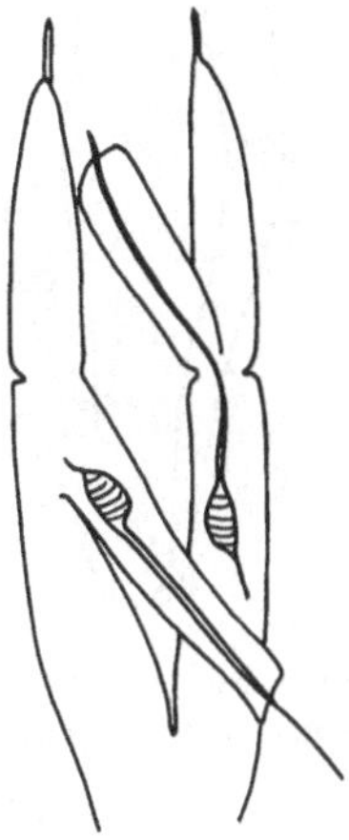

Fig. 706 Male genital armature of *Forcipula quelchi quelchi* BURR. After BRINDLE.

1a. Forcipula quelchi quelchi BURR

1904 *Forcipula quelchi* BURR, Trans. ent. Soc. London, 1904: 289, remainder see above.

1b. Forcipula quelchi boliviana BRINDLE

1966 *Forcipula quelchi* ssp. *boliviana* BRINDLE, Ann. Mag. nat. Hist., London (13) **9**: 267. – Terra typica: Bolivia (Type male: HINCK's Collect., Manchester Mus.).

1980 *Forcipula quelchi boliviana* – STEINMANN, Acta zool. Acad. Sci. hung., Budapest, **26**: 233; fig. 3 (male ultimate tergite and forceps), 4 (male genitalia).

Description: Male as typical form but tegmina reddish or yellowish brown. Lateral spines of male smaller than in typical form. Male forceps (Fig. 707) relatively short, slender, slightly sinuate; inner margin without large tooth. Male genitalia (Fig. 708) very similar to *quelchi quelchi* BURR, but genital lobes a little longer. – **Female** similar to male, but abdomen much broader than that of male, and forceps straight, tapering. – **Length** of body with forceps: male: 23,5 mm, female: 25,5 mm.

Distribution: Bolivia.

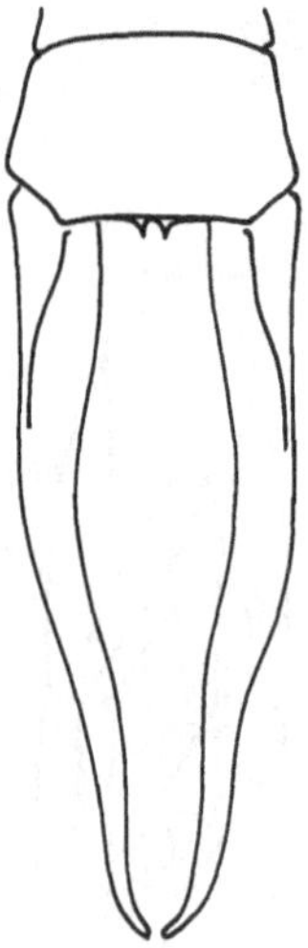

Fig. 707 Male ultimate tergite and forceps of *Forcipula quelchi boliviana* BRINDLE. After STEINMANN.

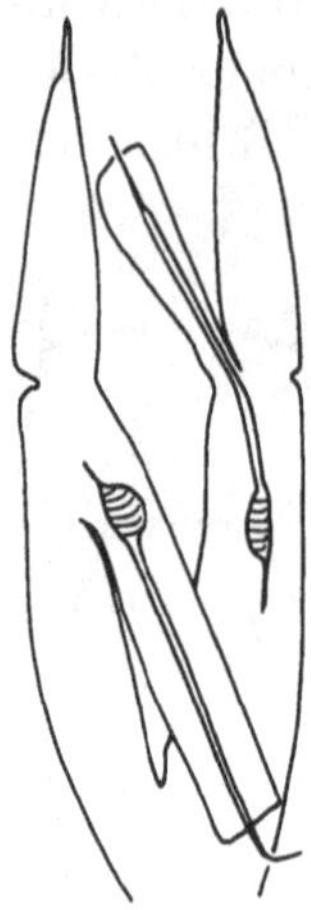

Fig. 708 Male genital armature of *Forcipula quelchi boliviana* BRINDLE. After STEINMANN.

2. Forcipula americana BORMANS

1900 *Forcipula americana* BORMANS, Ann. Mus. civ. Stor. nat., Genova, **20**: 443. – Terra typica: Bolivia (Mus. civ. Stor. nat., Genova).

1911 *Forcipula americana* – BURR, Genera Insectorum, Bruxelles, **122**: 37; pl. 8, fig. 1 (male).

1966 *Forcipula americana* – BRINDLE, Ann. Mag. nat. Hist., London, (13) **9**: 262; fig. 15 (male forceps), 44–45 (external parameres of male genitalia). Bolivia, Peru.

1980 *Forcipula americana* – STEINMANN, Acta zool. Acad. Sci. hung., Budapest, **26**: 233; fig. 5 (male ultimate tergite and forceps), 6 (male genitalia).

Literature: BURR 1910; KIRBY 1904; POPHAM & BRINDLE 1968; SAKAI 1970; STEINMANN 1975, 1978.

Description: Male colour blackish; antennae dark brown, tegmina reddish yellow; legs yellow; forceps red. Head broad, shiny; postfrontal and coronal sutures well marked. Eyes relatively large, about as long as width of head behind eyes. Antennae 21-jointed; first joint long, but shorter than distance between antennal bases; second joint transverse, third long, half as long as wide. Pronotum small, black, but lateral margins yellowish; lateral margins parallel, posterior margin rounded broadly. Tegmina long, twice as long as pronotum; black or reddish brown, coriaceous. Wings fully developed. Legs typical for genus. Abdomen parallel-sided; abdominal tergites 3–6 with an oblique serrated crest on each side. Ultimate tergite broad, with two characteristic spines on posterior margin. Forceps (Fig. 709) sinuate, of *Forcipula*-type, inner margin with one or two large teeth medially. Genitalia (Fig. 710; gen. prep. No. 248, det. Dr. H. STEINMANN) broad, median incision of anterior margin of central parameral plate normal, quite visible; genital lobes well developed, virgae typical, basal vesicle large. External parameres very slender, inner and outer margins more or less parallel; epimerite distinct. – **Female** similar to male, but forceps without ventro-medial tooth or teeth, slender, tapering, somewhat trigonal basally, slightly cylindrical apically. – **Length** of body with forceps: male: 23–31 mm, female: 21–28 mm.

Distribution: Bolivia and Peru.

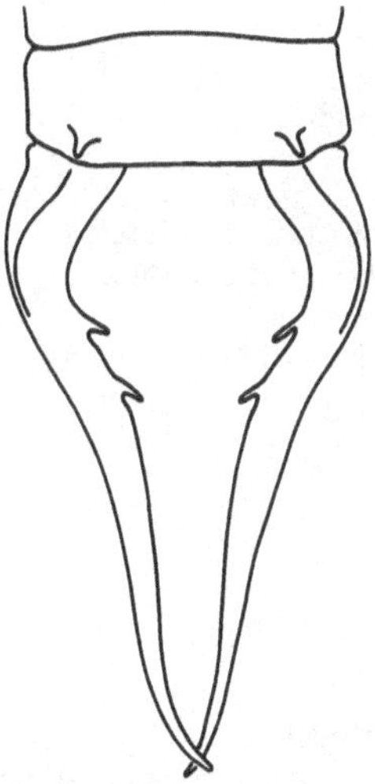

Fig. 709 Male ultimate tergite and forceps of *Forcipula americana* BORMANS. Original.

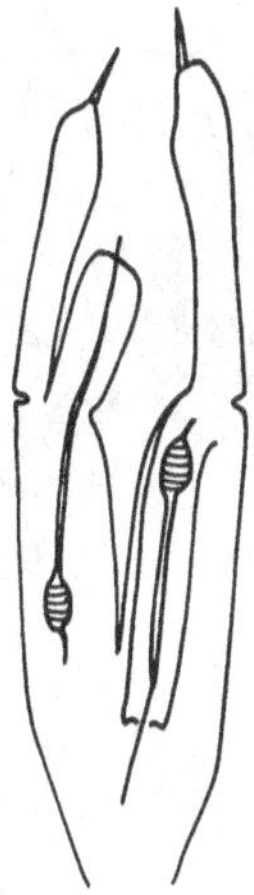

Fig. 710 Male genital armature of *Forcipula americana* BORMANS. Original.

3. Forcipula congo BURR

1900 *Forcipula congo* BURR, Ann. Soc. ent. belge, Bruxelles, **44**: 49. – Terra typica: Congo (Type male: Mus. r. Afr. centr., Tervuren).

1973 *Forcipula congo* – BRINDLE, Dermaptera of Africa, Tervuren, **1**: 116, fig. 147 (left branches of male forceps from Nigeria (left, and from Cameroon right). Nigeria, Cameroon.

1980 *Forcipula congo* – STEINMANN, Acta zool. Acad. Sci. hung., Budapest, **26**: 234; fig. 7 (male ultimate tergite and forceps). Central Africa.

Literature: BORELLI 1923; BRINDLE 1966; BURR 1911; REHN 1924; STEINMANN 1978.

Description: Male colour yellowish, tegmina rather lighter medially, wings yellow, darkened on sutural margin; legs yellow, abdomen somewhat darker, brown to dark brown. Head transverse, postfrontal sutures and coronal suture distinct. Eyes very large, about as long as first antennal joint. Scape short, shorter than distance between antennal bases. Pronotum longer than wide, posterior margin convex; lateral margins more or less parallel. Tegmina rather rugose, but without well-defined lateral longitudinal ridges; wings present, fully developed. Abdomen strongly and closely punctate; abdominal tergites 2–4 each with a single lateral spine; ultimate tergite with two very long vertical spines near both bases of forcipes. Forceps (Fig. 711) very characteristic, long, gently arcuate, with three large teeth on inner margins, the margins dentate. Genitalia unknown. – **Female** similar to male, but forceps slenderer, tapering, contiguous. – **Length** of body with forceps: male: 45–46 mm, female: 43–46 mm.

Distribution: Mainly Central Africa, extending west to Nigeria.

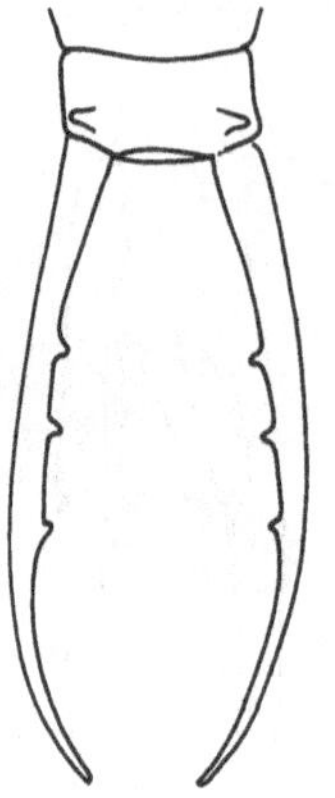

Fig. 711 Male ultimate tergite and forceps of *Forcipula congo* BURR. After STEINMANN.

4. Forcipula gariazzi BORELLI

1900 *Forcipula gariazzi* BORELLI, Boll. Mus. Zool. Anat. comp. Univ. Torino, **15**, No. 381: 1. – Terra typica: Congo (Type male: Mus. Ist. Zool. sist., Torino; without penis).
1915 *Forcipula gariazzi* – BURR, J. r. microsc. Soc., London, **1915**: 444, pl. IX, fig. 7 (male genitalia).
1966 *Forcipula gariazzi* – BRINDLE, Ann. Mag. nat. Hist., London, (13) **9**: 265, fig. 16 (male forceps), 34 (male genitalia). Tanganyika.
1973 *Forcipula gariazzi* ssp. *gariazzi* – BRINDLE, Dermaptera of Africa, Tervuren, **1**: 113; fig. 140 (male), 141 (male genitalia), 142–143 (forms of male forceps).
Literature: BORELLI 1923; BRINDLE 1968; BURR 1907, 1911; HINCKS 1957; HUXLEY 1968; REHN 1924; SAKAI 1970; STEINMANN 1980.

Description: Male colour blackish, antennae brown to dark brown, tegmina sometimes with humeral part paler in colour, or sometimes dark reddish brown; legs yellow or yellowish brown, femora often darkened medially or with dark longitudinal stripes. Head more or less transverse, postfrontal sutures and coronal suture well marked. Eyes rather large, but shorter than length of head behind eyes. First antennal joint small, shorter than the distance between antennal bases. Pronotum small, lateral margins parallel, quadrate; posterior margin convex. Cuticle of pronotum glabrous or coriaceous. Tegmina well developed, cuticle and wings tuberculate and with irregular longitudinal ridges. Abdomen long, parallel-sided or almost so; tergites finely tuberculate at each side; basal half of each tergite with a pair of longitudinal smooth stripes on each side of plate. Abdominal tergites 3–6, or occasionally 4–5, each with one large cylindrical tubercle on both sides. Pygidium with two teeth. Ultimate tergite very broad, posterior margin concave medially. Forceps (Fig. 712) well developed, long, very slender, trigonal basally, cylindrical apically, with one small inner tooth near middle. Genitalia (Fig. 713; gen. prep. No. 60, det. Dr. H. STEINMANN) narrow, a little similar to *Labidura*-type, but with the external parameres more slender and with a longer epimerite; inner tube of virga sinous only at base. – **Female** similar to male, but abdominal pubescence longer, and ultimate tergite narrower. Forceps slender, short, and inner margin crenulate. – **Length** of body with forceps: male: 19–38 mm, female: 20–35 mm.

Distribution: Congo Republic, Ghana, Chad, Sudan, Cameroon, and Tanzania.

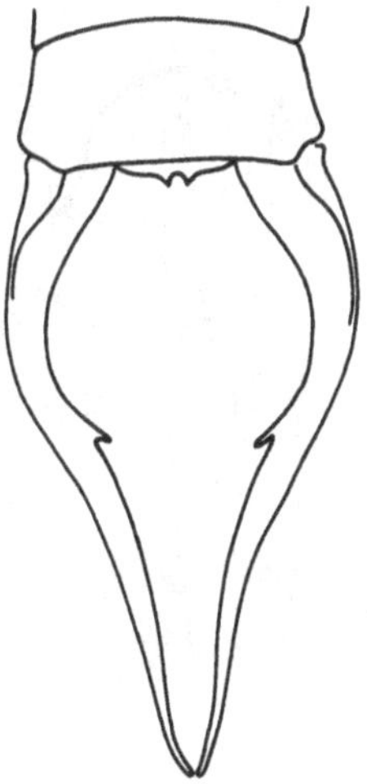

Fig. 712 Male ultimate tergite and forceps of *Forcipula gariazzi* BORELLI. Original.

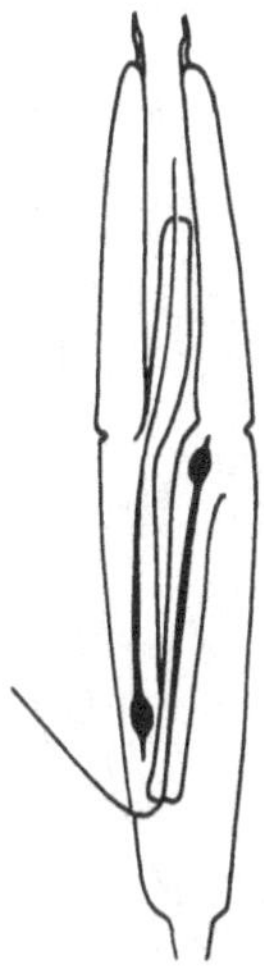

Fig. 713 Male genital armature of *Forciupula gariazzi* BORELLI. Original.

5. Forcipula tanganyikae HINCKS

1957 *Forcipula gariazzi* – var. *tanganyikae* HINCKS, Explor. Lac Tanganyika, **3**: 42. – Terra typica: Tanganyika (Type male: Brit. Mus. nat. Hist., London).
1966 *Forcipula gariazzi* – ssp. *tanganyikae* – BRINDLE, Ann. Mag. nat. Hist., London, (13) **9**: 265.
1973 *Forcipula gariazzi* – ssp. *tanganyikae* – BRINDLE, Dermaptera of Africa, Tervuren, **1**: 116; fig. 144 (male), 145 (male forceps).
1980 *Forcipula tanganyikae* – STEINMANN, Acta. zool. Acad. Sci. hung., Budapest, **26**: 235, fig. 10 (male ultimate tergite and forceps), 11 (male genitalia).

Description: Male very similar to *gariazzi* BORELLI, but tegmina short, only a little longer than wide; wings absent or concealed. Pronotum small, lateral margins more or less parallel-sided, cuticle coriaceous. Abdomen long, tergites finely tuberculate at each side; tergites 3–6, or sometimes 4–6 with one large cylindrical tubercle on each side. Ultimate tergite very broad, posterior margin concave medially. Forceps (Fig. 714) trigonal basally, and a little laminate, inner margin with several small teeth and one large tooth. Forma *macrolabia*, as in fig. 714, without small teeth, but with one single large tooth. Male genitalia (Fig. 715) narrow, median incision of anterior margin deep, genital lobes narrow, long, virgae straight, basal vesicle of *Forcipula*-type. External parameres characteristic, epimerite distinct. – **Female** similar to male, but forceps slenderer, contiguous, tapering. – **Length** of body with forceps: male: 20–35 mm, female: 22–34 mm.

Distribution: Tanzania, Tchad, Ethiopia, Cameroon, Central African Republic, Uganda, Nigeria and Liberia.

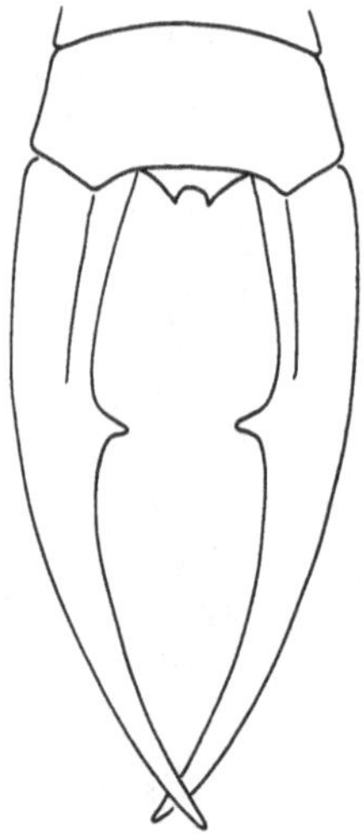

Fig. 714 Male ultimate tergite and forceps of *Forcipula tanganyikae* HINCKS. After STEINMANN.

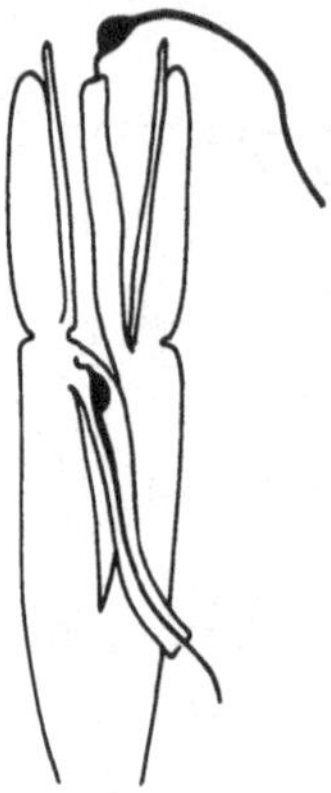

Fig. 715 Male genital armature of *Forcipula tanganyikae* HINCKS. Original.

6. Forcipula tarsata (WESTWOOD)

1857 *Forficula tarsata* WESTWOOD, Proc. zool. Soc. London, **5**: 129. – Terra typica: Philippines (Type male: unknown locality).
1904 *Forcipula tarsata* – BURR, Trans. ent. Soc. London, **1904**: 289.
1966 *Forcipula tarsata* – BRINDLE, Ann. Mag. nat. Hist., London, (13) **9**: 267.
1980 *Forcipula tarsata* – STEINMANN, Acta zool. Acad. Sci. hung., Budapest, **26**: 236; fig. 12 (male ultimate tergite and forceps).
Literature: SAKAI 1970; STEINMANN 1978.

Description: Male colour blackish brown; antennae brown to dark brown; legs dark brown, mainly blackish, tips of tibiae and tarsi yellowish; tegmina shiny, abdomen dull blackish brown. Head tumid, transverse, rounded posteriorly; postfrontal sutures and coronal suture distinct. Eyes relatively large, but shorter than length of head behind eyes. Antennae with long first joint. Pronotum small, lateral margins more or less parallel; coriaceous. Tegmina short, rather longer than pronotum; wings present but short. Abdomen long, more or less parallel-sided. Ultimate tergite broad, posterior margin concave medially. Forceps (Fig. 716) elongated, trigonal basally, cylindrical apically, without inner tooth medially. Genitalia unknown. – **Female** similar to male, but forceps tapering, slenderer. – **Length** of body with forceps: in both sexes: 25–32 mm.

Distribution: Philippines.

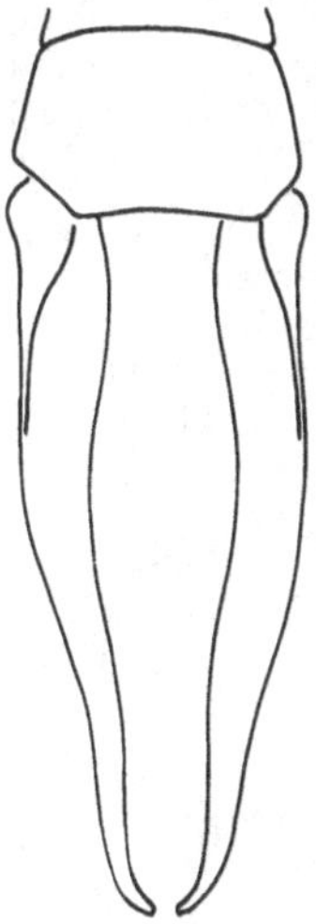

Fig. 716 Male ultimate tergite and forceps of *Forcipula tarsata* (WESTWOOD). After STEINMANN.

7. Forcipula borellii CHOPARD

1924 *Forcipula borellii* CHOPARD, Rec. Indian Mus., Calcutta, **26**: 90. – Terra typica: India (Type male: Mus. nation. Hist. nat., Paris).

1980 *Forcipula borellii* – STEINMANN, Acta zool. Acad. Sci. hung., Budapest, **26**: 236; fig. 15 (male ultimate tergite and forceps).

Literature: BRINDLE 1966; POPHAM & BRINDLE 1968; SAKAI 1970.

Description: Male colour dark brown, abdomen reddish brown; antennae brown, legs yellowish, median and posterior femora with a black or dark brown transverse band. Head tumid, depressed, postfrontal sutures and coronal suture well marked. Eyes large, about as long as the length of head behind eyes or scape. Antennae typical of genus. Pronotum small, lateral margins almost parallel, posterior margin convex; median longitudinal furrow present. Tegmina short, equal in length to pronotum, with a lateral longitudinal ridge only indicated at shoulders; wings concealed beneath tegmina. Abdominal tergites 3–5 with one spine on each side, segment 6 with a small lateral tubercle on each side. Ultimate tergite large, posterior margin concave medially. Forceps (Fig. 717) trigonal basally, cylindrical apically; with a strong narrow tooth towards base on the inner margin of both branches, the tooth directed postero-medially. Genitalia unknown. – **Female** similar to male, but forceps with a ventro-medial tooth. – **Length** of body with forceps: male: 24–25 mm, female: 22–23 mm.

Distribution: India.

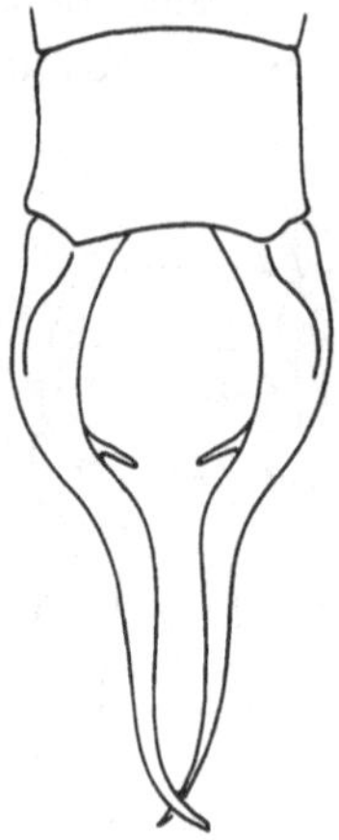

Fig. 717 Male ultimate tergite and forceps of *Forcipula borellii* CHOPARD. After STEINMANN.

8. Forcipula lurida BOLIVAR

1897 *Forcipula quadrispinosa* var. *lurida* BOLIVAR, Ann. Soc. ent. France, Paris, **66**: 283. – Terra typica: South India (Type male: Mus. nation. Hist. nat., Paris).
1900 *Forcipula quadrispinosa* ssp. *lurida* – BORMANS, Das Tierreich, Berlin, **11**: 31.
1911 *Forcipula lurida* – BURR, Genera Insectorum, Bruxelles, **122**: 38.
1966 *Forcipula lurida* – BRINDLE, Ann. Mag. nat. Hist., London, (13) **9**: 266; fig. 24 (male forceps), 40 (male genitalia).
Literature: BURR 1910; KIRBY 1904; POPHAM & BRINDLE 1968; SAKAI 1970; STEINMANN 1980.

Description: Male colour yellowish-brown, abdomen reddish, antennae and legs yellow. Head tumid, postfrontal sutures and coronal suture distinct. Eyes relatively large, but shorter than length of head behind eyes. Antennae with small joints; first joint somewhat long, but shorter than distance between antennal bases. Pronotum rather longer than wide, lateral margins curving inwards slightly. Tegmina normal. Abdomen elongated; abdominal tergites 2–5 or 3–5 with an oblique serrated crest. Ultimate tergite broad, well developed. Forceps (Fig. 718) characteristic; both branches curved apically, with single tooth on inner margin. Genitalia (Fig. 719; gen. prep. No. 687, det. Dr. H. STEINMANN) with small central parameral plate, median incision of anterior margin very deep; genital lobes fully developed, virga within genital lobes sclerotized, straight, basal vesicle normal. External parameres specific, epimerite well developed. – **Female** unknown. – **Length** of body with forceps: 45 mm.

Distribution: South India and Assam.

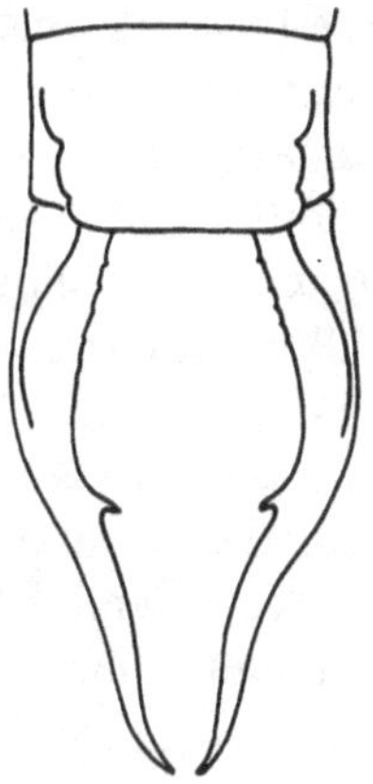

Fig. 718 Male ultimate tergite and forceps of *Forcipula lurida* BOLIVAR. Original.

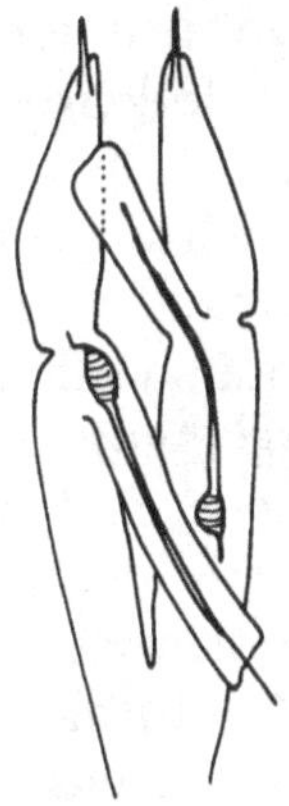

Fig. 719 Male genital armature of *Forcipula lurida* BOLIVAR. Original.

9. Forcipula banksi BORELLI

1915 *Forcipula banksi* BORELLI, Boll. Mus. Zool. Anat. comp. Univ. Torino, **30**: 3. – Terra typica: Philippines (Type male: Mus. Zool. sist., Torino; holotype without penis).
1966 *Forcipula banksi* – BRINDLE, Ann. Mag. nat. Hist., London, (13) **9**: 262; fig. 7 (pronotum of male), 22 (male genitalia).
Literature: SAKAI 1970; STEINMANN 1980.

Description: Male colour blackish or dark reddish brown; antennae dark brown, legs yellowish brown. Head tumid, depressed, postfrontal sutures and coronal suture distinct. Eyes smaller, visibly shorter than length of head behind

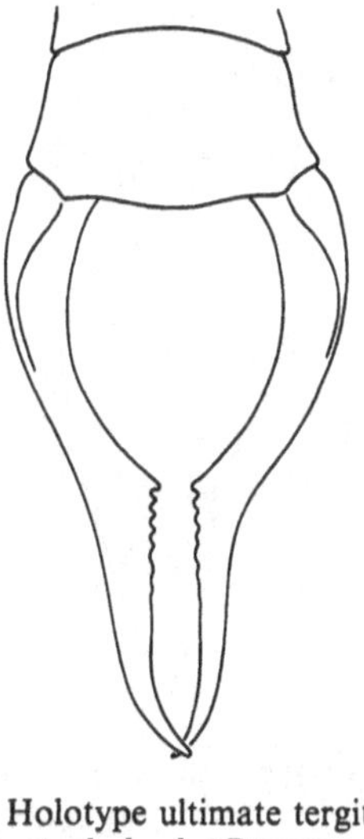

Fig. 720 Holotype ultimate tergite and forceps of *Forcipula banksi* BORELLI. Original.

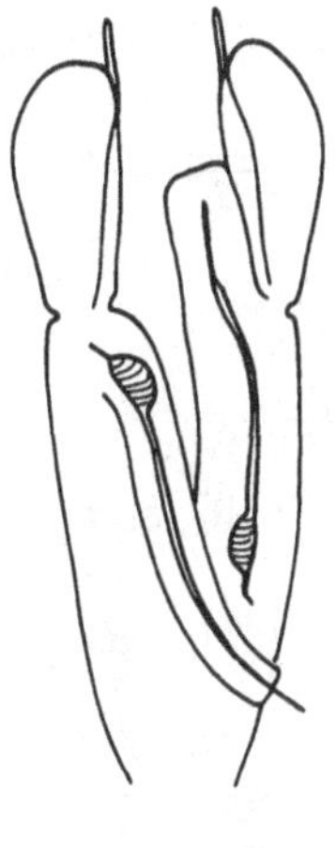

Fig. 721 Male genital armature of *Forcipula banksi* BORELLI. After BRINDLE.

eyes. Antennae small-jointed; first joint more or less twice as long as broad. Pronotum a little longer than wide, with posterior margin strongly convex; lateral margin not curving inwards. Tegmina tuberculate, well developed, but shorter, not much longer than pronotum (measured along sutures), median area reddish. Abdomen slender, abdominal tergites 3–6 with an oblique serrated crest, which ends on a short spine posterior-dorsally. Ultimate tergite very broad; posterior margin more or less straight or undulate. Forceps (Fig. 720) specific, similar to *Cranopygia*-type; trigonal basally, strongly ridged dorsally, crenulate on inner margin apically. Genitalia (Fig. 721) characteristic; paramere well developed, median incision of anterior margin very deep; genital lobes long, virgae straight, basal vesicle oval. External parameres rounded apically, epimerite long. – **Female** similar to male, but forceps short, contiguous, and without a ventro-medial tooth. – **Length** of body with forceps: male: 17–25 mm, female: 16–20 mm.

Distribution: Philippines.

10. Forcipula tuberculata SRIVASTAVA

1977 *Forcipula tuberculata* SRIVASTAVA, Dtsch. ent. Zeitschr., Berlin, **24**: 289; fig. 1 (holotype ultimate tergite and forceps), 2 (external paramere of male genitalia), 3 (holotype genital armature). – Terra typica: India (Type male: Zool. Survey India, Calcutta).
Literature: STEINMANN 1980.

Description: Male general colour shiny yellowish brown; antennae, head and pronotum darker; legs light yellow. Head tumid, slightly longer than wide, finely pubescent, gently depressed in middle at the junction of median and transverse

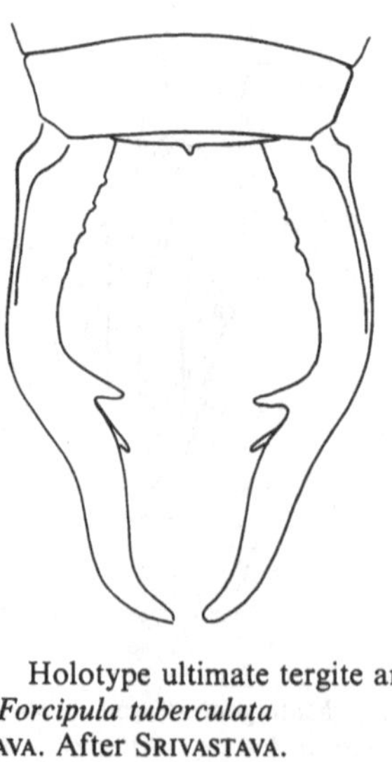

Fig. 722 Holotype ultimate tergite and forceps of *Forcipula tuberculata* SRIVASTAVA. After SRIVASTAVA.

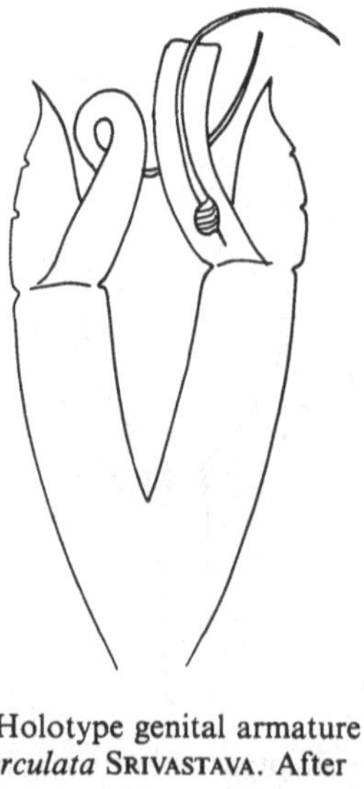

Fig. 723 Holotype genital armature of *Forcipula tuberculata* SRIVASTAVA. After SRIVASTAVA.

sutures. Eyes large, prominent. Antennae 35-jointed; first joint normal, second joint small, equal to fourth, third long, equal to fourth and fifth together, the rest long and thin. Pronotum longer than wide, lateral margins gently emarginate medially, posterior margin broadly rounded. Tegmen twice as long as pronotum, tuberculated; posterior margin truncate. Wings slightly longer than pronotum. Abdomen finely punctate, gently expanded posteriorly, sides of segments 3–5 with an oblique, triangular, serrated ridge. Ultimate tergite very broad, pygidium present, quite visible. Forceps (Fig. 722) remote, with an oblique ridge above base, curved up to a little beyond middle, then almost straight, with two pairs of teeth on inner margins. Genitalia (Fig. 723) specific, not of *Forcipula*-type; paramere V-shaped, median incision of anterior margin very deep and wide; genital lobes well developed, virgae and basal vesicle typical of genus. External parameres very characteristic, with two incisions at outer margin. – **Female** unknown. – **Length** of body with forceps: 24,5 mm.

Distribution: India.

11. Forcipula leonardii STEINMANN

1981 *Forcipula leonardii* STEINMANN, Mem. Soc. ital. Sci. nat., Mus. Milano, **22**: 160, fig. 5 (holotype ultimate tergite and forceps). – Terra typica: Isl. Andamana (Type female: Mus. civ. Stor. nat., Milano).

Description: Female general colour reddish brown, head red, shiny, antennae dark brown; pronotum black with yellowish grey lateral margins; tegmina and abdomen reddish brown, legs dark brown. Tegmina with median longitudinal light reddish stripe, wings yellow. Head relatively large, a little broader than pronotum; postfrontal sutures and coronal suture distinct but faint. Eyes large, longer than length of head behind eyes, and more or less equal in length to first antennal joint. Antennae 27-jointed; first joint long, but shorter than distance between antennal bases; second very short, third twice as long as fourth. Pronotum about as long as wide; lateral margins parallel, posterior margin

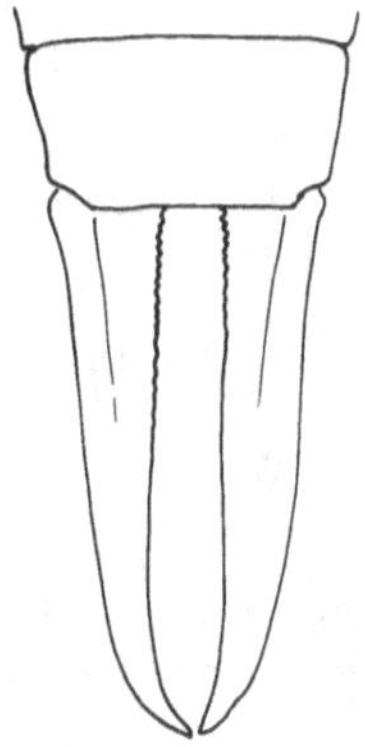

Fig. 724 Holotype ultimate tergite and forceps of *Forcipula leonardii* STEINMANN. After STEINMANN.

rounded. Tegmina rugose, well developed, wings fully developed, long, their apices dark brown coloured. Legs long, slender, normal. Abdomen punctate, ultimate tergite and forceps slender. Forceps (Fig. 724) elongated, slightly dentate on inner margins. Penultimate sternite broad, posterior margin rounded, with yellowish hairs. – **Male** unknown. – **Length** of body with forceps: 27 mm.

Distribution: Island Andamana.

12. Forcipula quadrispinosa (DOHRN)

1863 *Labidura quadrispinosa* DOHRN, Stettiner ent. Ztg., **24**: 311. – Terra typica: India (Type male: Univ. zool. Mus., København).
1891 *Labidura morosa* KIRBY, J. Linn. Soc. London, **23**: 513. – Terra typica: Hab.? (Type female: Brit. Mus. nat. Hist., London).
1897 *Forcipula quadrispinosa* – BOLIVAR, Ann. Soc. int. France, Paris, **46**: 283.
1908 *Forcipula jacobsoni* BURR, Ann. Mag. nat. Hist., London, (8) **1**: 47. – Terra typica: Java (Type male: Brit. Mus. nat. Hist., London).
1911 *Forcipula quadrispinosa* – ZACHER, Zool. Jahrb., Jena, **30**: 392; fig. Y^2 (male genitalia).
1980 *Forcipula quadrispinosa* – STEINMANN, Acta zool. Acad. Sci. hung., Budapest, **26**: 239; fig. 20 (male ultimate tergite and forceps), 21 (male genitalia). India, Bhutan, Sikkim, Ceylon, Bangladesh, Burma, Thailand, Cambodia, Laos, Vietnam, Philippines, Java, Reunion, Mauritius.

Literature: BRINDLE 1966; BORMANS 1900; BURR 1910, 1911; GANGOLA 1968; HEBARD 1917; HINCKS 1947; KAPOOE 1968; POPHAM & BRINDLE 1968; SAKAI 1970; STEINMANN 1981.

Description: Male dark brown, abdomen blackish, head dark brown to black. Head broad, depressed; postfrontal and coronal sutures distinct, fine. Antennae 32-jointed; first joint conical, long, but a little shorter than distance between antennal bases; second quadrate, third long, about as long as wide. Eyes relatively small, much shorter than length of head behind eyes. Pronotum a little

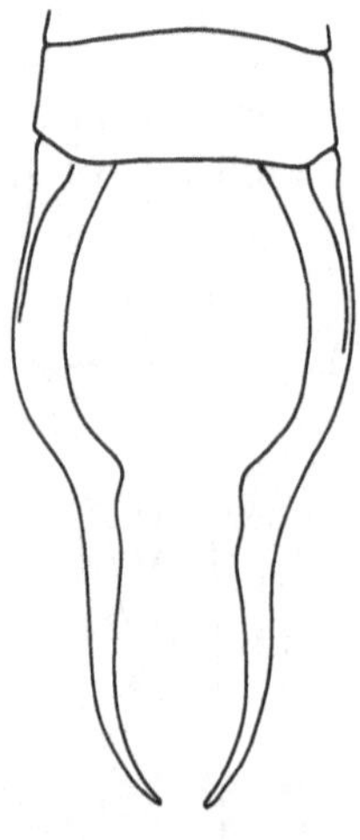

Fig. 725 Male ultimate tergite and forceps of *Forcipula quadrispinosa* (DOHRN). After STEINMANN.

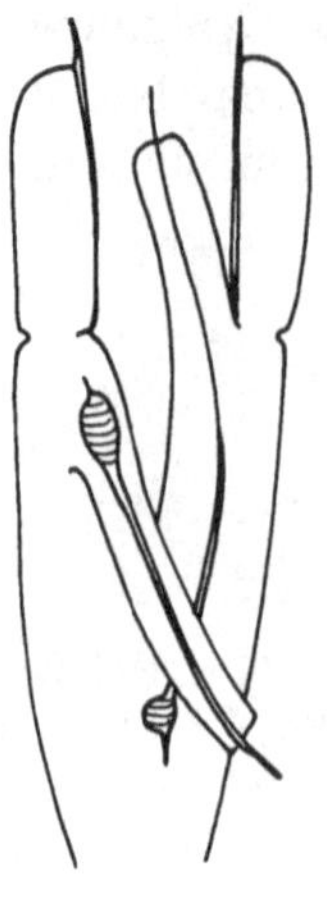

Fig. 726 Male genital armature of *Forcipula quadrispinosa* (DOHRN). Original.

longer than wide; lateral margins parallel, all angles rounded; median longitudinal furrow present. Tegmina rugose, shiny, wings well developed. Abdomen parallel-sided, abdominal tergites 3–6 with lateral spines. Forceps (Fig. 725) dark reddish, strongly arcuate basally, or in smaller males with less curvature medially; branches trigonal basally, cylindrical apically. Genitalia (Fig. 726; gen. prep. No. 368, det. Dr. H. STEINMANN) large; paramere broad, median incision of anterior margin deep; genital lobes fully developed, narrow, virgae straight, basal vesicle oval, large. External parameres long, inner and outer margins parallel, epimerite long. – **Female** very similar to male, colour dark reddish-brown, forceps straight, with ventro-medial tooth on inner margin. – **Length** of body with forceps: male: 22–34 mm, female: 24–35 mm.

Distribution: India, Bhutan, Sikkim, Sri Lanka, Bangladesh, Burma, Thailand, Vietnam, Cambodia, Laos, Philippines, Java, Reunion, and Mauritius.

13. Forcipula simplex BEY-BIENKO

1970 *Forcipula simplex* BEY-BIENKO, Zool. Zhurn., Moskva, **49**: 1815?; pl. 1 (male forceps), 2 (male genitalia). – Terra typica: North India (Type male: Zool. Staatssammlung, München).
Literature: STEINMANN 1980.

Description: Male colour brownish black, tegmina reddish dark brown, moderately rugose; antennae brown, legs yellowish. Head tumid, postfrontal sutures and coronal suture distinct. Eyes normal, shorter than length of head behind eyes. Antennae multi-jointed, first joint long, but shorter than distance between antennal bases, second quadratic, third long, about twice as long as wide. Pronotum about as long as wide; lateral margins parallel, or so. Tegmina

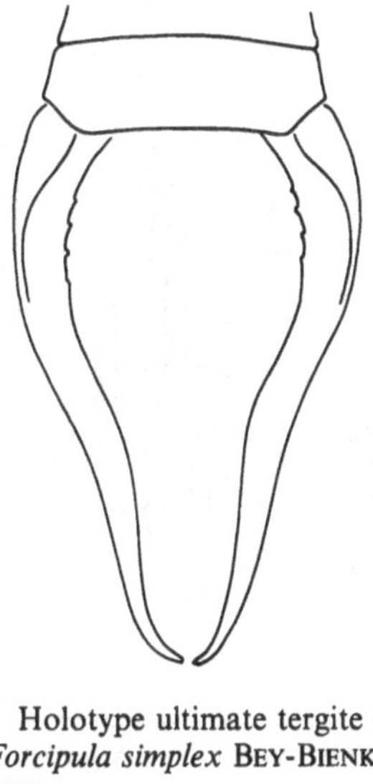

Fig. 727 Holotype ultimate tergite and forceps of *Forcipula simplex* BEY-BIENKO. After BEY-BIENKO.

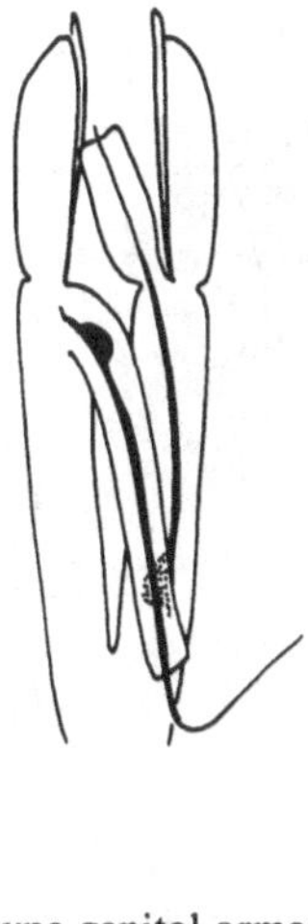

Fig. 728 Holotype genital armature of *Forcipula simplex* BEY-BIENKO. After BEY-BIENKO.

well developed, wings present. Abdomen elongated, parallel-sided, tergites 3–5 with lateral spines. Ultimate tergite very broad, depressed medially; posterior margin concave. Forceps (Fig. 727) slender, a little denteted on inner margin, but without single large tooth on median portion. Genitalia (Fig. 728) narrow, paramere with very deep median incision; genital lobes fully developed, slender, virga of *Forcipula*-type, basal vesicle large, oval. External parameres large, inner and outer margins more or less parallel, epimerite relatively small. – **Female** unknown. – **Length** of body with forceps: 28 mm.

Distribution: North India: Himalaya.

14. Forcipula despinosa Hebard

1917 *Forcipula despinosa* Hebard, Prov. Acad. nat. Sci. Philadelphia, **1917**: 234; pl. 16, fig. 1 (male). – Terra typica: North India (Type male: Hebard's Collect., No. 437, U.S. nation. Mus., Washington).

1966 *Forcipula despinosa* – Brindle, Ann. Mag. nat. Hist., London, (13) **9**: 265; fig. 19 (male forceps).

1980 *Forcipula despinosa* – Steinmann, Acta zool. Acad. Sci. hung., Budapest, **26**: 241; fig. 24 (ultimate tergite and forceps of male), 25 (male genital armature).

Literature: Popham & Brindle 1968; Sakai 1970; Steinmann 1978.

Description: Male colour dark reddish brown, legs and antennae yellow. Head tumid, broad, a little broader than pronotum. Postfrontal and coronal sutures present, but fine. Antennae with first joint normal, a little shorter than distance between antennal bases; second joint quadratic, third joint subequal in length, fourth about half again as long as wide. Pronotum small, transverse, lateral margins expanded to posterior angles; posterior margin rounded. Tegmina and

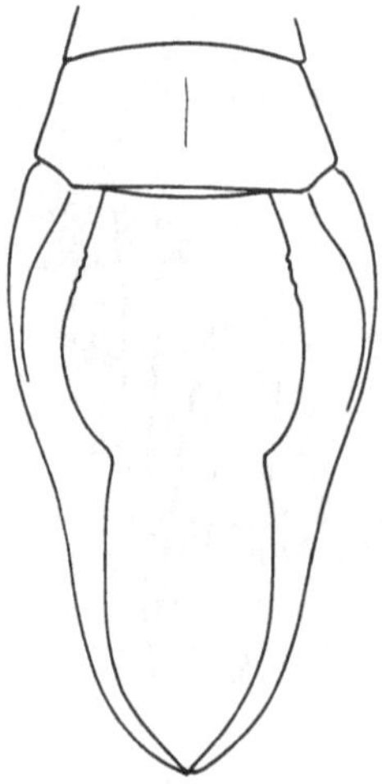

Fig. 729 Male ultimate tergite and forceps of *Forcipula despinosa* Hebard. After Steinmann.

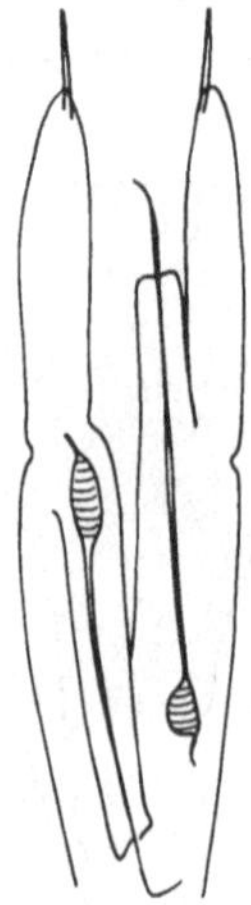

Fig. 730 Male genital armature of *Forcipula despinosa* Hebard. After Steinmann.

wings fully developed, broadly rounding lateral. Abdomen not of *Forcipula*-type, without projections. Ultimate tergite very broad; median longitudinal sulcus present. Forceps (Fig. 729) elongated, internal margins broadly concave for slightly more than proximal half, with minute, widely spaced serrations, then straight, with more numerous, slightly larger subserrations to distal third of this portion, then concave and smooth to the apex. Pygidium declivous, large. Genitalia (Fig. 730) comparatively narrow, elongate; central parameral plate with median incision on anterior margin; genital lobes well developed, virgae straight, basal vesicle oval, large. External parameres very long, inner and outer margins more or less parallel, epimerite long, thick. – **Female** unknown. – **Length** of body with forceps: 24–25 mm.

Distribution: Northern India, and Philippines.

15. Forcipula yunnanea Bey-Bienko

1970 *Forcipula yunnanea* Bey-Bienko, Zool. Zhurn., Moskva, **49**: 1816; pl. 2, fig. 3 (male forceps), 4 (holotype genital armature). – Terra typica: China: Yunnan (Type male: unknown locality; perhaps in Peking Mus.).
Literature: Steinmann 1980.

Description: Male colour brownish black, tegmina densely punctate; antennae light brown, legs yellow. Head tumid, depressed; postfrontal sutures and coronal suture well marked. Very similar to *clavata* Liu. First antennal joint conical, rather cylindrical; shorter than distance between antennal bases; second quadrate, third long, but a little shorter than first. Eyes relatively large, but

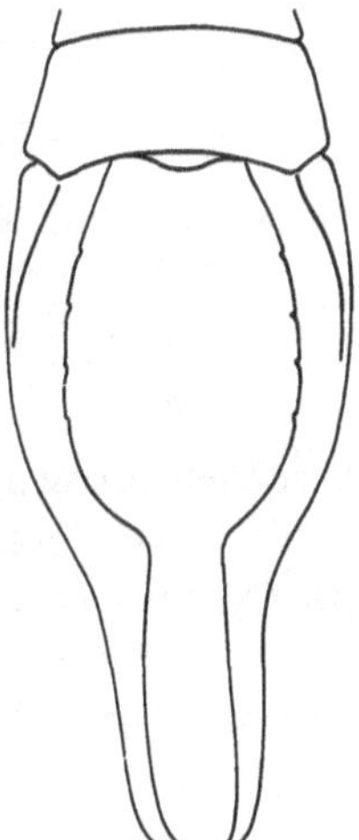

Fig. 731 Holotype ultimate tergite and forceps of *Forcipula yunnanea* Bey-Bienko. After Bey-Bienko.

Fig. 732 Holotype genital armature of *Forcipula yunnanea* Bey-Bienko. After Bey-Bienko.

shorter than length of head behind eyes. Pronotum small, quadrate; lateral margins more or less parallel, posterior margin rounded. Tegmina well developed, wings concealed. Abdomen elongate, parallel-sided, abdominal tergites 3–5 with single spines. Ultimate tergite broad; posterior margin strongly concave; pygidium simple. Forceps (Fig. 731) similar to *clavata* LIU; trigonal basally, cylindrical apically, strongly curved in basal half, and inner margin with several small teeth. Genitalia (Fig. 732) typical of genus; paramere V-shaped, median longitudinal incision of anterior margin very deep; genital lobes fully developed, elongate, with virgae straight; basal vesicle strongly sclerotized. External parameres long, obtuse at apices, epimerite long. – **Female** similar to male, but abdominal tergites without lateral spines, and forceps slenderer, contiguous, tapering. – **Length** of body with forceps: male: 25,5–31 mm, female: 26–29 mm.

Distribution: China: Yunnan.

16. Forcipula trispinosa (DOHRN)

1863 *Labidura trispinosa* DOHRN, Stettiner ent. Ztg., **24**: 310. – Terra typica: India (Type male: unknown locality; paratype male: Természettudományi Múz., Budapest).

1891 *Labidura pugnax* KIRBY, J. Linn. Soc. London, **23**: 510; pl. XII, fig. 1 (male). – Terra typica: North India (Type male: Brit. Mus. nat. Hist., London).

1891 *Labidura morosa* KIRBY (partim), J. Linn. Soc., London, **23**: 513.

1900 *Forcipula trispinosa* – BORMANS, Das Tierreich, Berlin, **11**: 30; fig. 15a (male), 15b (female forceps). Östliches Indien. Abessinien. – According to DUBRONY, this species occurs in Abyssinia, but the record refers to an African species, *F. gariazzi* BORELLI.

1910 *Forcipula trispinosa* var. *minor* BURR, Fauna of Brit. India, Dermapt., London, Bombay, 93. – Terra typica: Nepal (Type male: Naturhist. Mus., Wien).

1911 *Forcipula pugnax* – ZACHER, Zool. Jahrb., Jena, **30**: 392; fig. Z^2 (external paramere of male genitalia).

1912 *Forcipula pugnax* var. *parallela* BORELLI, Bull. Mus. nation. Hist. nat., Paris, **1912**: 14. – Terra typica: Bhutan (Type male: Mus. nation. Hist. nat., Paris).

1980 *Forcipula trispinosa* – STEINMANN, Acta zool. Acad. Sci. hung., Budapest, **26**: 242; fig. 26 (paratype male, ultimate tergite and forceps), 27 (paratype genital armature; gen.-prep. No. 377, det. Dr. H. STEINMANN).

Literature: BHARADWAJ 1961; BRINDLE 1966; BURR 1911; CHOPARD 1924; GANGOLA 1968; HEBARD 1917; KIRBY 1904; POPHAM & BRINDLE 1968; SAKAI 1970; STEINMANN 1978.

Description: Male colour blackish brown or dark reddish brown; antennae dark brown or yellowish brown, legs yellow. Head tumid, broad, shiny; postfrontal sutures and coronal suture distinct, but fine. Eyes relatively large, but shorter than length of head behind eyes. Antennae small-jointed; first joint conical, about twice as long as wide, second transverse, third joint longer than fourth. Pronotum small, longer than wide, lateral margins more or less parallel, posterior margin rounded, arch-shaped. Tegmina and wings well developed, coriaceous, shiny. Abdomen longer, tergites 3–5 with spines laterally. Abdominal tergites 3–5 with one flattened thorn-like spine on each side, which is at the end of an oblique crest. Ultimate tergite broad, median longitudinal sulcus present medially. Forceps (Fig. 733) very long, with large, acute ventro-lateral tooth at first, basal portion; branches strongly trigonal basally, and cylindrical apically. Genitalia (Fig. 734) narrow, typical of genus, paramere with deep median incision; genital lobes well developed, with straight virgae; basal vesicle large, oval. External parameres very long, with long epimerite. – **Female** similar to male, but abdominal tergites without spines, and forceps simple, straight, with ventro-medial tooth. – **Length** of body with forceps: male: 26–42 mm, female: 25–40 mm.

Distribution: India, Burma, Pakistan, Vietnam.

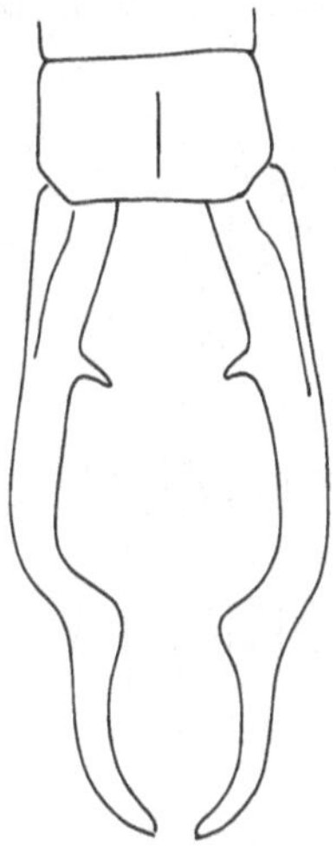

Fig. 733 Male ultimate tergite and forceps of *Forcipula trispinosa* (DOHRN). After STEINMANN.

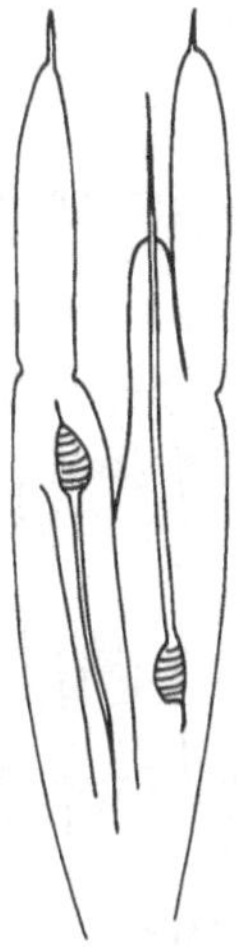

Fig. 734 Paratype genital armature of *Forcipula trispinosa* (DOHRN). Original.

17. Forcipula afghana STEINMANN

1980 *Forcipula afghana* STEINMANN, Acta zool. Acad. Sci. hung., Budapest, **26**: 243; fig. 28 (ultimate tergite and forceps of holotype), 29 (holotype genital armature). – Terra typica: Afghanistan (Type male: Természettudományi Múz., Budapest; gen. prep. No. 691, det. Dr. H. STEINMANN).

Description: Male colour reddish to dark brown; head brownish black; pronotum, tegmina and abdominal tergites dark reddish brown; forceps red, legs yellow. Head tumid, broad, more or less somewhat broader than pronotum; postfrontal sutures and coronal suture distinct, fine, but well visible. Eyes relatively small, shorter than length of head behind eyes. Antennae 28-jointed; first joint short, shorter than distance between antennal bases; second joint quadrate, third joint long, but shorter than scape. Pronotum about as long as wide; all margins and angles rounded; median longitudinal furrow quite visible.

Tegmina normal, a little longer than pronotum; wings present. Abdomen parallel-sided; abdominal tergites 2–4 with small spines on each side. Ultimate tergite broad; posterior margin with one pair of tooth; pygidium small. Forceps (Fig. 735) straight, trigonal basally, cylindrical apically; with single small, but quite visible tooth at inner margin basally. Genitalia (Fig. 736) characteristic; paramere relatively short, about as long as length of external paramere with epimerite; median incision of anterior margin narrow and not deep; genital lobes well developed with straight virgae; basal vesicle strongly sclerotized, oval. External parameres long, broadened basally, with long epimerite. – **Female** unknown. – **Length** of body with forceps: 21–22 mm.

Distribution: Afghanistan: Hindukusch and Nuristan; India, Nepal.

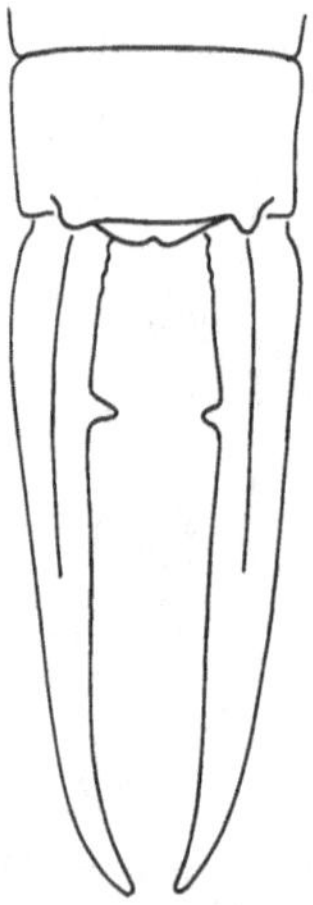

Fig. 735 Holotype ultimate tergite and forceps of *Forcipula afghana* STEINMANN. After STEINMANN.

Fig. 736 Holotype genital armature of *Forcipula afghana* STEINMANN. After STEINMANN.

18. Forcipula decolyi BORMANS

1900 *Forcipula decolyi* BORMANS, Ann. Mus. civ. Stor. nat., Genova, **20**: 444. – Terra typica: India (Type male: Mus. civ. Stor. nat., Genova).

1911 *Forcipula decolyi* – BURR, Genera Insectorum, Bruxelles, **122**: 37; pl. 7, fig. 1 (male). India, New Guinea.

1912 *Forcipula decolyi var. longipennis* BORELLI, Bull. Mus. nation. Hist. nat., Paris, **18**: 224. – Terra typica: Tonkin (Type male: Mus. nation. Hist. nat., Paris). – **syn. nov.**

1915 *Forcipula decolyi* – BURR, J. r. miscrosc. Soc., London, **1915**: 444; pl. IX, fig. 8 (male genitalia).

1966 *Forcipula decolyi decolyi* – BRINDLE, Ann. Mag. nat. Hist., London, (13) **9**: 264; fig. 28 (male forceps), 42 (external parameres of male genitalia). India, Bhutan, Sikkim, Vietnam.

Literature: BEY-BIENKO 1959; BOESEMAN 1954; BURR 1904, 1906, 1910; GÜNTHER 1930; HEBARD 1923; KIRBY 1904; POPHAM & BRINDLE 1968; SAKAI 1970; STEINMANN 1980.

Description: Male colour blackish brown, antennae and legs yellow; pronotum and tegmina shiny, coriaceous. Head tumid, postfrontal sutures and coronal suture distinct, well marked. Eyes large, but a little shorter than length of head behind eyes. Antennae small-jointed; first joint small, conical, shorter than distance between antennal bases; second quadrate, third long, about twice as long as wide. Pronotum shiny black, lateral margins yellowish, more or less quadrate, or a little longer than wide; lateral margins parallel, posterior angles rounded. Tegmina well-developed, more or less twice as long as pronotum. Wings absent, or concealed. Abdomen of *Forcipula*-type, abdominal tergites 3–6 with two spines on each side, the posterior-dorsal spine being the largest of the two; spines on segment 6 much smaller than the others. Ultimate tergite very broad; pygidium simple. Forceps (Fig. 737) slender, inner margin on basal half without tooth or denticulation; trigonal basally, cylindrical apically. Genitalia (Fig. 738) narrow; paramere relatively short, more or less as long as external paramere; median incision of anterior margin deep; genital lobes well developed; virgae straight, basal vesicle typical of genus. External parameres long, specific, as in fig. 738. – **Female** similar to male, but a little darker in colour; abdominal tergites without spines, and forceps tapering, with inner tooth on apical half. – **Length** of body with forceps: male: 32–44 mm, female: 24–31 mm.

Distribution: India, Bhutan, Sikkim, Vietnam, South-Western China.

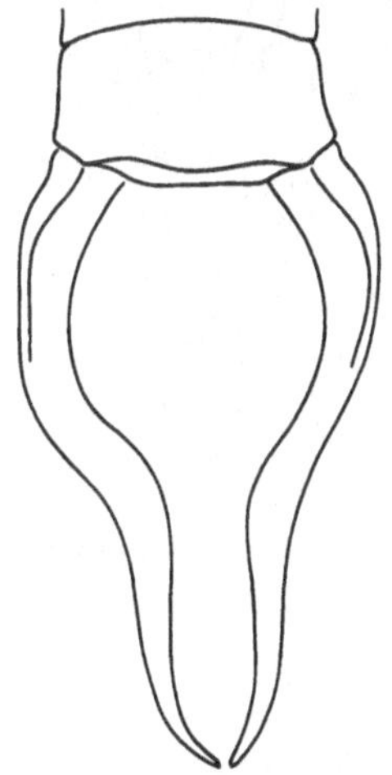

Fig. 737 Male ultimate tergite and forceps of *Forcipula decolyi decolyi* BORMANS. After STEINMANN.

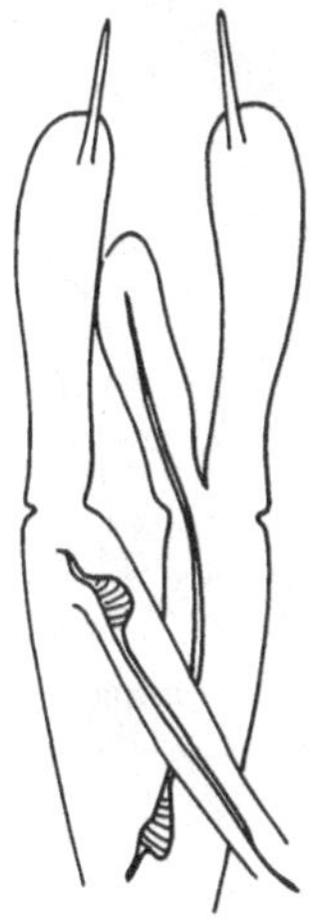

Fig. 738 Male genital armature of *Forcipula decolyi decolyi* BORMANS. After BURR.

18a. Forcipula decolyi decolyi BORMANS

1900 *Forcipula decolyi* BORMANS, Ann. Mus. civ. Stor. nat. Genova, **20**: 444. – Remainder see above.

18b. Forcipula decolyi novaguineae BRINDLE

1966 *Forcipula decolyi novaguineae* BRINDLE, Ann. Mag. nat. Hist., London, (13) **9**: 264; fig. 27 (male forceps), 43 (male genitalia). – Terra typica: New Guinea (Type male: Brit. Mus. nat. Hist., London).

1980 *Forcipula decolyi novaguineae* – STEINMANN, Acta zool. Acad. Sci. hung., Budapest, **26**: 246; fig. 35 (male ultimate tergite and forceps), 36 (male genitalia).

Description: Male characteristics very similar to *decolyi decolyi* BORMANS, but forceps (Fig. 739) with specific internal ridge extending from base end ending in a tooth; trigonal basally, cylindrical apicall. Curvature of forceps of *Forcipula*-type. Genitalia (Fig. 740) characteristic; central parameral plate slightly short, shorter than length of external paramere; median incision of anterior margin very deep; genital lobes fully developed, virgae straight, with strongly sclerotized basal vesicle. External parameres very long, inner and outer margins more or less parallel, apex obtuse, epimerite acute. – **Female** unknown. – **Length** of body with forceps: male: measurement unknown.

Distribution: New Guinea.

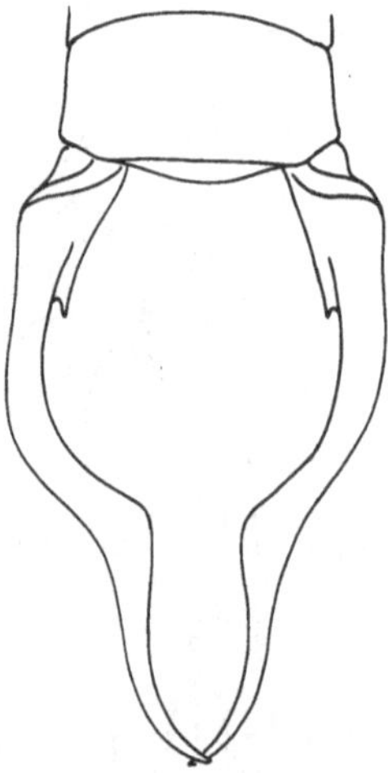

Fig. 739 Holotype ultimate tergite and forceps of *Forcipula decolyi novaeguineae* BRINDLE. After BRINDLE.

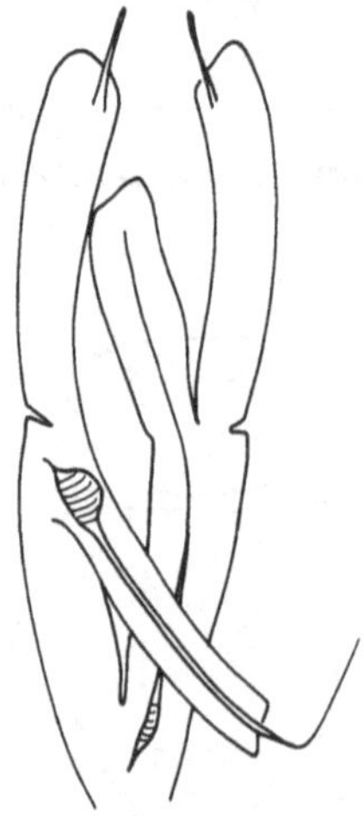

Fig. 740 Holotype genital armature of *Forcipula decolyi novaeguineae* BRINDLE. After BRINDLE.

19. Forcipula aborensis BRINDLE

1966 *Forcipula aborensis* BRINDLE (new name for *Forcipula pugnax* BURR 1910, nec KIRBY 1891), Ann. Mag. nat. Hist., London, (13) **9**: 262; fig. 25 (male forceps), 37 (male genitalia). – Terra typica: India (Type male: BURR's Collect., Brit. Mus. nat. Hist., London).

1910 *Forcipula pugnax* BURR, Rec. indian Mus., Calcutta, **8**: 139.

Literature: STEINMANN 1980.

Description: Male colour dark reddish or blackish brown, antennae brown, legs and forceps yellowish brown. Very similar to *decolyi decolyi* BORMANS, but head tumid, depressed; postfrontal sutures and coronal suture distinct; eyes large, but shorter than length of head behind eyes. Antennae small-jointed; first joint conical, rather cylindrical. Pronotum broad, rather longer than wide, all margins more or less straight, cuticle of pronotum coriaceous. Tegmina normal, wings absent or concealed. Abdomen gradually dilated posteriorly, spines thorn-like, those on segments 3–4 well-developed, those on segment 5 small. Ultimate tergite very broad; posterior margin straight. Forceps (Fig. 741) with single large inner tooth basally, a little similar to *trispinosa* (DOHRN), but basal part broadened. Genitalia (Fig. 742) narrow, specific; paramere V-shaped, median incision of anterior margin deep; genital lobes narrow and long, with virgae straight; basal vesicle large. External parameres characteristic, epimerite long. – **Female** unknown. – **Length** of body with forceps: 38,5 mm.

Distribution: India.

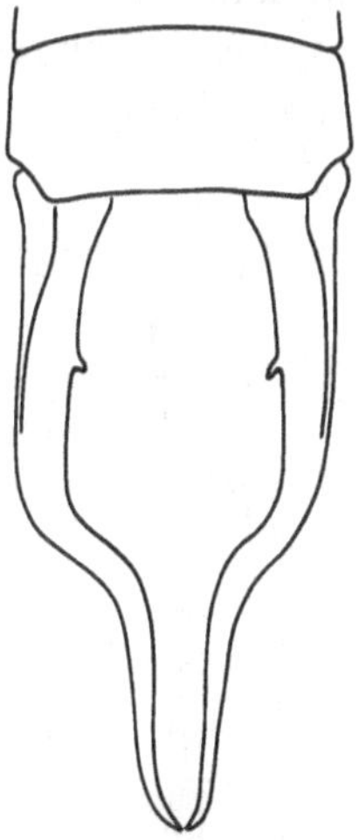

Fig. 741 Holotype ultimate tergite and forceps of *Forcipula aborensis* BRINDLE. After BRINDLE.

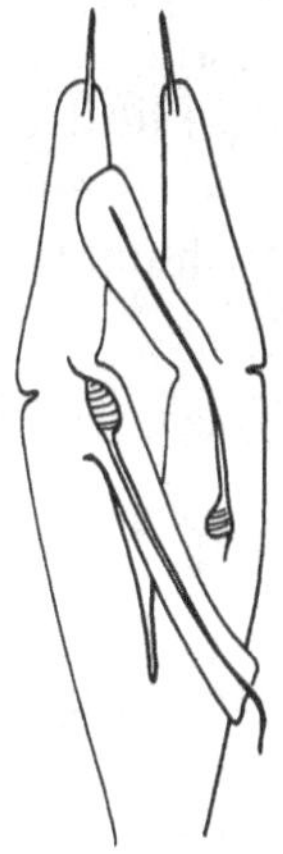

Fig. 742 Holotype genital armature of *Forcipula aborensis* BRINDLE. After BRINDLE.

20. Forcipula clavata LIU

1946 *Forcipula clavata* LIU, Jl. W. China Border Res. Soc., Shanghai, **16**: 22; pl. 2, fig. 8 (male). – Terra typica: China: Chungking (Type male: Peking Mus.).

1959 *Forcipula clavata* – BEY-BIENKO, Ent. Obozr., Moskva, **38**: 609 [= *Forcipula decolyi decolyi* BORMANS 1912].

1959 *Forcipula clavata* var. *unica* – BEY-BIENKO, Ent. Obozr., Moskva, **38**: 605.

1966 *Forcipula clavata* – BRINDLE, Ann. Mag. nat. Hist., London, (13) **9**: 263.

1980 *Forcipula clavata* – STEINMANN, Acta zool. Acad. Sci. hung., Budapest, **26**: 247; fig. 39 (male ultimate tergite and forceps), 40 (male genitalia; gen. prep. No. 285, det. Dr. H. STEINMANN).

Description: Male colour blackish or reddish brown; head black, shiny, antennae dark brown, pronotum black, tegmina and wings dark brown, abdomen reddish, legs yellow. Head tumid, wide, broader than pronotum; postfrontal sutures and coronal suture distinct; antennae 26-jointed; first joint conical, short, shorter than distance between antennal bases, second transverse, third joint about twice as long as wide; fourth joint quadrate. Pronotum small, a little longer than broad; lateral margins parallel, posterior margin rounded. Tegmina short, less than twice the length of pronotum; coriaceous, shiny. Wings present, yellowish brown to dark brown. Abdomen with strong punctation, abdominal tergites 3–5 or 6 with two spines on each side, those on segment 6 being much smaller that the others. Forceps (Fig. 743) resemble those of *decolyi decolyi* BORMANS, but there are short teeth on the inner margins towards the base and the angle at the mid-point is more rounded. Ultimate tergite broad, median longitudinal sulcus prominent. Genitalia (Fig. 744) specific; paramere narrow, V-shaped, median longitudinal incision of anterior margin deep; genital lobes well developed, with straight virgae; basal vesicle large; external parameres narrow, long, apices acute, epimerite long. – **Female** similar to male, but abdomen dark blackish brown, with hairs yellow, but without lateral spines; forceps contiguous, trigonal basally, a little curved and cylindrical apically. – **Length** of body with forceps: male: 38–42 mm, female: 36–38 mm.

Distribution: China: Chungkiang and Kiangsi.

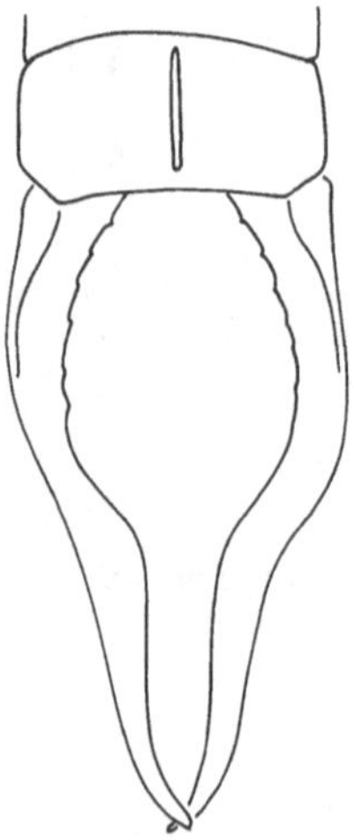

Fig. 743 Male ultimate tergite and forceps of *Forcipula clavata* LIU. After STEINMANN.

Fig. 744 Male genital armature of *Forcipula clavata* LIU. Original.

21. **Forcipula walkeri** (KIRBY)

1896 *Labidura walkeri* KIRBY, J. Linn. Soc., London, **25**: 524. – Terra typica: Hong Kong (Type male: Brit. Mus. nat. Hist., London).
1910 *Forcipula walkeri* – BURR, Trans. ent. Soc. London, **1910**: 187.
1966 *Forcipula walkeri* – BRINDLE, Ann. Mag. nat. Hist., London (13) **9**: 268; fig. 19 (male forceps), 41 (external paramere of male genitalia).
1980 *Forcipula walkeri* – STEINMANN, Acta zool. Acad. Sci. hung., Budapest, **26**: 248; fig. 41 (male ultimate tergite and forceps), 42 (male genitalia).
Literature: BORMANS 1900; BURR 1910, 1911; POPHAM & BRINDLE 1968; SAKAI 1970; STEINMANN 1978; WU 1935.

Description: Male colour blackish or dark reddish brown, head and pronotum black, shiny, palpi yellowish, antennae dark brown, legs yellow, tegmina shiny, abdomen very dull. Head rounded, postfrontal sutures and coronal suture distinct, but fine. Eyes normal, first antennal joint typical, conical, rather cylindrical. Pronotum longer than wide, with a median longitudinal furrow; lateral margins more or less parallel. Tegmina short, less than twice the length of pronotum. Wings well developed. Segments 3–5 of abdomen with a lateral spine on both sides; segment 2 with a small tubercle, segments 2–7 with a double row of short stripes in front. Ultimate tergite with a well-marked median longitudinal sulcus. Forceps (Fig. 745) almost straight, slightly curved inwards in the middle, and then again outwards, points turned rather sharply inwards at the extremities. Genitalia (Fig. 746) narrow, typical of genus; genital lobes well developed, with straight virgae; basal vesicle strongly sclerotized, oval. External parameres a little oval, but long, epimerite acute, long. – **Female** similar to male, but abdomen without lateral spines, forceps slender, tapering. – **Length** of body with forceps: male: 21–27 mm, female: 17–19 mm.

Distribution: Hong-Kong.

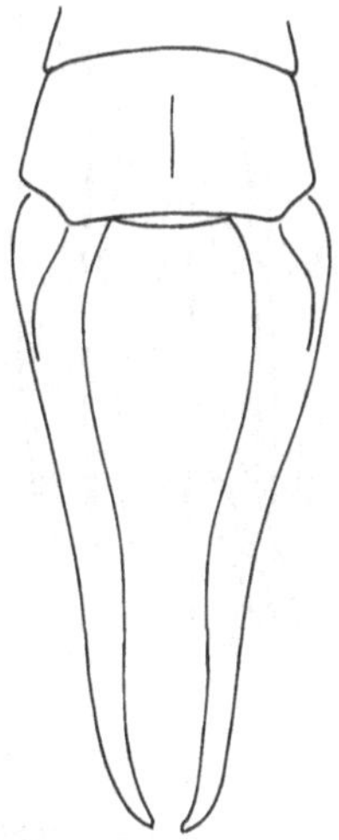

Fig. 745 Male ultimate tergite and forceps of *Forcipula walkeri* (KIRBY). Original.

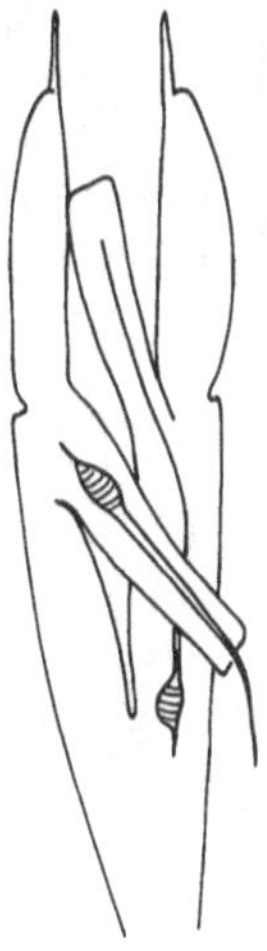

Fig. 746 Male genitalia of *Forcipula walkeri* (KIRBY). After BRINDLE.

22. Forcipula vanheurni (BOESEMAN)

1954 *Labidura vanheurni* BOESEMAN, Zool. Verh., Leiden, **21**: 49; fig. 12 (male). – Terra typica: Java (Type male: Rijksmus. natuurl. Hist., Leiden; gen. prep. No. 686, det. Dr. H. STEINMANN).
1966 *Forcipula vanheurni* – BRINDLE, Ann. Mag. nat. Hist., London, (13) **9**: 268; fig. **21** (male forceps).
1980 *Forcipula vanheurni* – STEINMANN, Acta zool. Acad. Sci. hung., Budapest, **26**: 249; fig. 43 (holotype ultimate tergite and forceps), 44 (holotype genital armature).
1980 *Forcipula vanheurni* – STEINMANN, Fol. ent. hung., Budapest, (33) **41**: 341; fig. 35 (male forceps), 36 (male genitalia).

Description: Male colour blackish, pronotum yellowish laterally, tegmina and wings dark brown; antennae and legs yellowish brown, basal three antennal joints lighter than rest. Head tumid, rounded, but longer than wide, and broader than pronotum. Postfrontal sutures and coronal suture finely visible; eyes relatively large, about equalling length of head behind eyes. Antennae 23-jointed; first joint conical, short, shorter than distance between antennal bases, second joint quadrate, third joint three times as long as wide; fourth joint small. Pronotum very small, somewhat longer than wide; lateral margins parallel, posterior margin broadly rounded. Tegmina surface with a slightly irregular texture, without hairs; coriaceous. Wings well developed, long. Abdomen subcylindrical, lateral margins parallel; an inconspicuous row of very small tubercles along the free margins of most segments; a marginal row of slightly longer whitish hairs on all but posterior dorsal segments. Ultimate tergite tumid, with median longitudinal sulcus. Forceps (Fig. 747) slender, elongate, trigonal

Fig. 747 Holotype ultimate tergite and forceps of *Forcipula vanheurni* (BOESEMAN). Original.

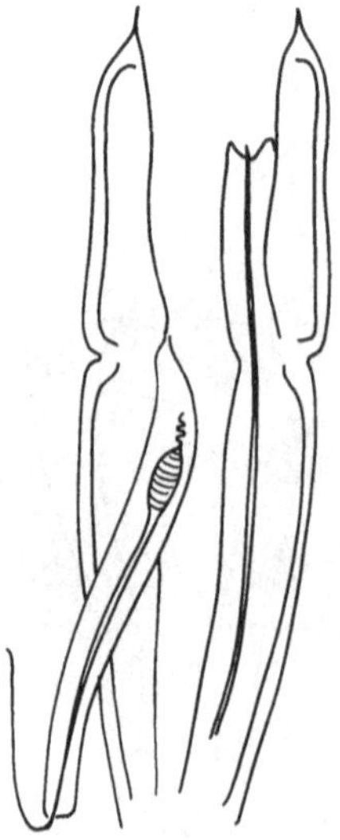

Fig. 748 Holotype genital armature of *Forcipula vanheurni* (BOESEMAN). After STEINMANN.

basally, cylindrical apically, and with a small dorsal keel on the short diverging basal part, inner margin only proximally with several distinct blunt small teeth, becoming inconspicuous more distally. Genitalia (Fig. 748) characteristic; paramere long, median incision of anterior margin very deep; genital lobes fully developed, virga elongate, with straight virgae. External parameres specific, a little expanded basally, inner and outer margins concave; epimerite normal, acute. – **Female** unknown. – **Length** of body with forceps: 27 mm.

Distribution: Java.

23. Forcipula indica Brindle

1966 *Forcipula indica* Brindle, Ann. Mag. nat. Hist., London, (13) **9**: 265; fig. 26 (left branch of male forceps). – Terra typica: North India: Punjab (Type male: Brit. Mus. nat. Hist. London).

1980 *Forcipula indica* – Steinmann, Acta zool. Acad. Sci. hung., Budapest, **26**: 250; fig. 45 (male ultimate tergite and forceps), 46 (male genitalia). Bhutan.

Description: Male colour dark reddish brown; tegmina sometimes lighter, abdomen blackish; antennae tawny, head black, pronotum black, with yellow margins. Head broad, postfrontal sutures and coronal suture distinct, well marked; eyes large, but shorter than length of head behind eyes. Antennae small-jointed; first joint conical, relatively short, significantly shorter than distance between antennal bases. Pronotum a little longer than wide, exceedingly finely punctate, pubescent. Tegmina normal, wings well developed. Abdomen long, abdominal tergites 3–6 with single lateral spines, which are thorn-like basally but tuberculate apically; largest spine on third segment, smallest on the sixth.

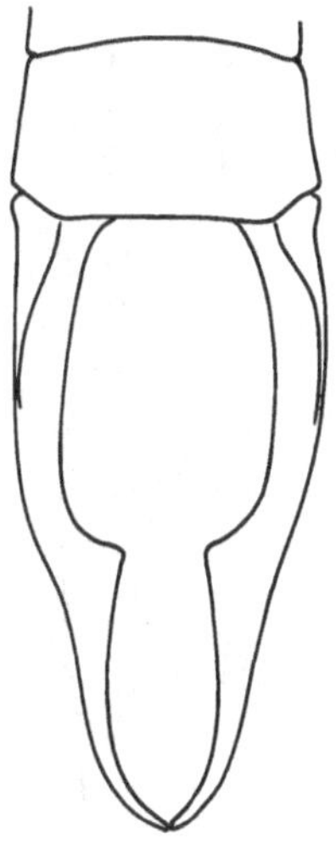

Fig. 749 Male ultimate tergite and forceps of *Forcipula indica* (Brindle). After Steinmann.

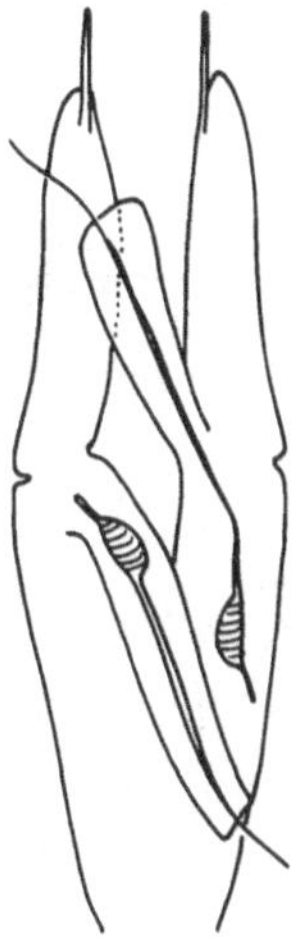

Fig. 750 Male genital armature of *Forcipula indica* Brindle. After Steinmann.

Ultimate tergite smooth, with a median impression, truncate posteriorly. Forceps (Fig. 749) with branches remote at base, trigonal, straight for about two thirds of length, then abruptly bent inwards, with a blunt tooth on inner margin at the angle of the bend, then straight, simple. Genitalia (Fig. 750) broad, typical of genus; paramere with median incision at anterior margin; genital lobes well developed, virga typical, basal vesicle large. External parameres narrow and long, epimerite long, acute. – **Female** unknown. – **Length** of body with forceps: 32–38 mm.

Distribution: North India, Bhutan.

24. Forcipula obscura Steinmann

1980 *Forcipula obscura* Steinmann, Acta zool. Acad. Sci. hung., Budapest, **26**: 250; fig. 47 (holotype ultimate tergite and forceps), 48 (holotype genital armature; gen. prep. No. 693, det. Dr. H. Steinmann). – Terra typica: Vietnam (Type male: Természettudományi Múz., Budapest).

Description: Male colour dark reddish brown to black; head and pronotum black, tegmina, wings and abdominal tergites dark reddish brown, shiny; forceps dark red; antennae dark brown, legs dark yellow. Head tumid, postfrontal sutures and coronal suture well marked. Eyes relatively large, about equal to the length of head behind eyes. Antennae broken in holotype, 16 – ?-jointed; first joint relatively large, conical, dark brown, and shorter than distance between antennal bases; second quadrate, third relatively long, but shorter than scape, the rest cylindrical. Pronotum longer than wide, lateral margins parallel, black with yellowish borders; median longitudinal furrow distinct. Tegmina long, more or

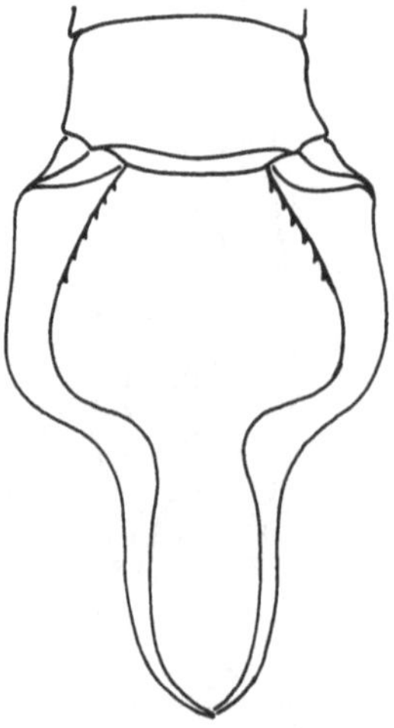

Fig. 751 Holotype ultimate tergite and forceps of *Forcipula obscura* Steinmann. After Steinmann.

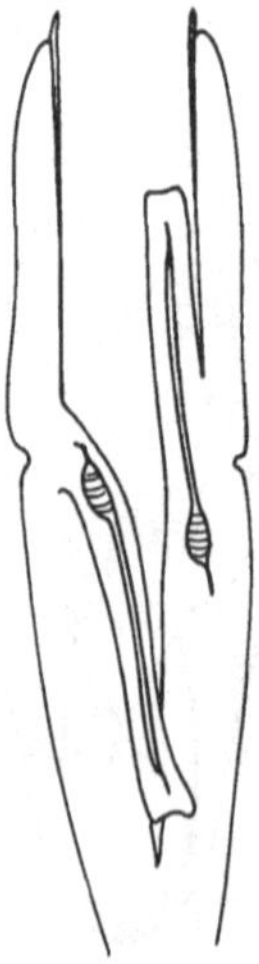

Fig. 752 Holotype genital armature of *Forcipula obscura* Steinmann. After Steinmann.

less twice as long as pronotum, wings well developed, long. Abdomen with parallel-sided tergites; abdominal tergites 3–6 with two lateral spines. The spines similar to *decolyi decolyi* BORMANS, but larger. Ultimate tergite broad, with depression at postero-dorsal portion. Forceps (Fig. 751) very specific, robust; trigonal basally and cylindrical apically, strongly curved medially; with blunt teeth at inner margin basally. Genitalia (Fig. 752) characteristic; central parameral plate comparatively short, more or less as long as the external parameres, median incision of anterior margin very deep; paramere V-shaped; genital lobes long and narrow, virga within genital lobe straight, with strongly sclerotized basal vesicle. External parameres very long, and narrow, epimerite distinct. – **Female** unknown. – **Length** of body with forceps: 34,5 mm.

Distribution: Vietnam.

Section 3: **Paradermaptera** VERHOEFF

1902 *Paradermaptera* VERHOEFF, Zool. Anz., Leipzig, **25** (665): 199. – Superfamilia typica: *Apachyoidea* STEINMANN 1975.
1911 *Labiduriales* – ZACHER (partim) Zool. Jahrb., Jena, **30**: 309.
1911 *Paradermaptera* – ZACHER, Zool. Jahrb., Jena, **30**: 333.
Literature: BOESEMAN 1954; BURR 1911, 1915; STEINMANN 1975.

Description: Body very strongly depressed. The pronotum is narrowed anteriorly and forms a distinct neck between the head and pronotum. Tegmina and wings present; well, sometimes fully, developed. Abdomen more or less parallel-sided, flattened. The last tergite is produced posteriorly to form an anal process.

Distribution: African, Oriental and Indo-Australian faunal regions.

1 superfamilia.

Superfamily 4: **Apachyoidea** STEINMANN

1975 *Apachyoidea* STEINMANN, Acta zool. Acad. Sci. hung., Budapest, **21**: 199. – Familia typica: Apachyidae VERHOEFF 1902.

Description: A very distinctive superfamily. The body is uniformly depressed. Head broad, sometimes transverse; postfrontal and coronal sutures distinct, well marked. Eyes large approximately, as long as length of head behind eyes. Antennae multi-segmented; first joint comparatively short. Tegmina weak at axillary angle, exposing ample scutellum. Abdominal segments without lateral ridges, keels or spines. Legs compressed. Forceps in both sexes symmetrical or more or less symmetrical. Anal process various.

Distribution: African, Oriental, and Indo-Australian species.

Single family.

Family 4: **Apachyidae** VERHOEFF

1902 *Apachyidae* VERHOEFF, Zool. Anz., Leipzig, **665**: 199. – Genus typicum: *Apachyus* AUDINET-SERVILLE 1831.
1915 *Apachyinae* BURR, J. r. microsc. Soc., London, **1915**: 447.
1954 *Apachyoidea* BRUES, MELANDER & CARPENTER, Classification of Insects, Cambridge (Mass.): 113.
1965 *Apachyinae* BURR, in: *Labiduridae*; BRINDLE, Ann. Mag. nat. Hist., London, (13) **8**: 435.
1975 *Apachyidae* – STEINMANN, Acta zool. Acad. Sci. hung., Budapest, **21**: 207.
Literature: BOESEMAN 1954; BURR 1911; POPHAM 1969; POPHAM & BRINDLE 1968; ZACHER 1911.

Description: Very strongly flattened species, extremely so in the dominant and typical genus *Apachyus*, less so in *Dendroiketes*. Legs with compressed femora. Head transverse, narrowed behind eyes, depressed; posterior margin concave or straight. Cuticle of head, pronotum, and tegmina gently coriaceous, wings and abdominal tergites rugosely punctate. The male genitalia have a prominent virga, which is narrower than those of the Labiduridae. The virga has no inner sinous tube, its base is slightly wider and is associated with a small sclerite.

Distribution: Ethiopian, Oriental and Indo-Australian faunal regions.

Single subfamily.

Subfamily 1: **Apachyinae** BURR

1909 *Apachyinae* BURR Dtsch. ent. Zeitschr., Berlin, **1909**: 321. – Genus typicum *Apachyus* AUDINET-SERVILLE 1831.

Literature: BOESEMAN 1954; BRINDLE 1965; BURR, 1915; REHN 1924; SAKAI 1970; STEINMANN 1975.

Description: Head broad, strongly depressed; pronotum elliptical, longer than wide, quadrate or transverse. Tegmina and wings present, well or fully developed, sometimes wings concealed. Ultimate tergite flat, ending posteriorly in a large anal process which often occupies much of the space between the branches of the forceps. Forceps similar in both sexes, with wide curved branches. Male genitalia with two genital lobes, which are with more or less sclerotized virgae.

Distribution: as in family.

2 genera.

Identification key to the genera

1 Body less strongly flattened; pronotum strongly transverse, usually smaller and less depressed species .... Genus 1: **Dendroiketes**, p. 479

1′ Body very strongly flattened; pronotum quadrate or longer than wide, larger and more depressed species .... Genus 2: **Apachyus**, p. 483

Genus 1: **Dendroiketes** BURR

1909 *Dendroiketes* BURR, Dtsch. ent. Zeitschr., Berlin, **1909**: 321. – Species typica: *Apachyus corticinus* BURR 1908.

Literature: BOESEMAN 1954; BORELLI 1923; BURR 1910, 1911, 1915; POPHAM & BRINDLE 1968; REHN 1924; SAKAI 1970; STEINMANN 1975; TOWNES 1945; ZACHER 1911.

Description: The body less depressed, as in species of *Apachyus*. Antennae with about thirty joints; first joint long and thick, second transverse, third joint long and cylindrical; fourth and fifth joints short and subconical, together not longer than third; sixth and ninth slightly longer, subconical, the rest more elongated, nearly cylindrical. Pronotum rectangular, strongly transverse; tegmina and wings perfectly developed, femora somewhat compressed. Tibiae slender, tarsi long, first and third segments about equally long. Abdomen more or less parallel-sided, depressed and smooth. Ultimate tergite of male and female ample, slightly wider than long, densely granulated; penultimate sternite of male convex, rounded and obtusely triangular, punctate; in female punctate, produced into a long, slender, sharp pointed lobe. Pygidium smoother, depressed, the margin itself finely crenulated, rounded in male, in female parallel-sided at the base. Forceps with the branches depressed, finely punctate.

Distribution: Oriental and Indo-Australian faunal regions.

3 species.

Identification key to the species

1 Tegmina brown with yellowish area on anterior half. Posterior margin of ultimate tergite with very short anal process (Fig. 753). Indo-Australian species. . . . . 1. **Dendroiketes novaeguineae**, p. 480

1′ Tegmina unicolorous or with inner margins lighter. Posterior margin of ultimate tergite with well-developed anal process (Figs. 756, and 757). Oriental species 2

2 Tegmina greyish-brown, smooth. Pronotum rectangular. Male anal process (Fig. 756) rounded, crenulate on margin, female anal process parallel-sided at base, then triangular . . . . 2. **Dendroiketes corticinus**, p. 481

2′ Tegmina dark brownish or pitchy, punctate. Pronotum transverse. Female anal process (Fig. 757) rectangular basally with parallel-sides, triangular apically . . . . 3. **Dendroiketes punctatus**, p. 482

1. Dendroiketes novaeguineae BOESEMAN

1954 *Dendroiketes novaeguineae* BOESEMAN, Zool. Verh., Leiden, **21**: 61; fig. 10b (male), 10c (penultimate sternite with forceps of male in ventral view). – Terra typica: New Guinea (Type male: Inst. taxonom. Zoöl., Amsterdam; gen. prep. No. 705, det. Dr. H. STEINMANN).

1967 *Dendroiketes similis* RAMAMURTHI, Ent. Medd., København, **35**: 247; fig. 26 (ultimate tergite with forceps of male), 27 (ultimate tergite with forceps of female), 30 (male genitalia), 28 (penultimate sternite of male), 29 (penultimate sternite of female). – Terra typica: New Britain (Type male: Univ. zool. Mus., København).

1981 *Dendroiketes novaeguineae* – STEINMANN, Acta zool. Acad. Sci. hung., Budapest, **27**: 199; fig. 36 (holotype ultimate tergite and forceps), 37 (holotype genital armature). New Guinea, Bismarck Islands.

Description: Male general colour dark brown; head yellow or brownish yellow, a dark brown band along the coronal suture, a narrow band along occipital margins and all except basal three antennal joints slightly brownish. Tegmina brown. Head broad; postfrontal and coronal sutures distinct; antennae more than 35-jointed; first antennal joint long, about as long as distance between antennal bases; second short, transverse, third joint long, longer than first; joints 4–8 short, as broad as long. Pronotum transverse, rectangular, median longitudinal furrow distinct; very dark brown with two lighter yellowish spots at about three fifths of length. Tegmina brown, dark along median margins and sides, with a lighter, almost ovate yellowish area on anterior half. Wings well developed, light brown, lateral and apical parts darker, proximal two fifths and narrow band along median margin yellow. Abdomen more or less parallel-sided; ultimate tergite fully developed, broad, with median longitudinal sulcus; posterior margin not elongate, characteristic, straight (Fig. 753), caudal half sparsely denticulated; pygidium very slightly produced, concave with a row of short yellowish hairs along ventral margin. Penultimate sternite transverse, posterior margin broadly triangular. Forceps cylindrical, very strongly curved.

Genitalia (Fig. 754) large, median incision of anterior margin very wide and deep; genital lobes comparatively small, virga specific; external parameres oval, obtuse. – **Female** similar to male, but pygidium very narrow, dorsal surface triangular and drawn into a peg; ventral surface flat, looked from above appearing like a wing on either side of the peg. – **Length** of body with forceps; male: 24,5–25 mm, female: 24–25,5 mm.

Distribution: New Guinea, Bismarck Islands.

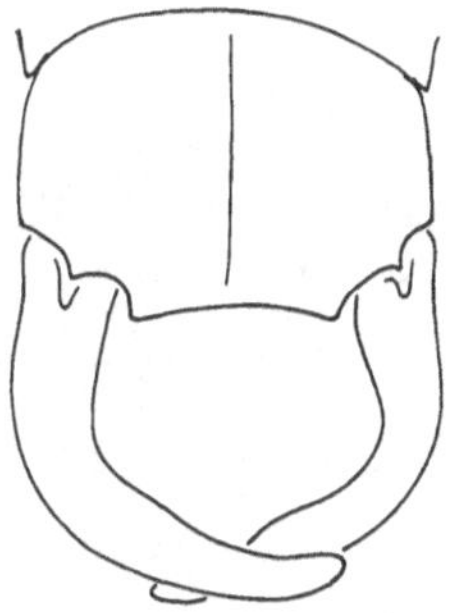

Fig. 753 Holotype ultimate tergite and forceps of *Dendroiketes novaeguineae* BOESEMAN. After STEINMANN.

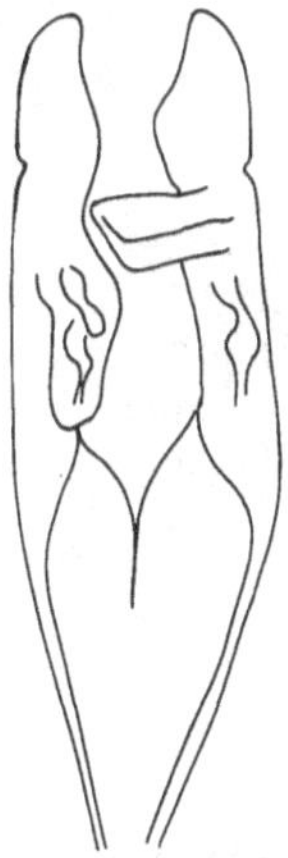

Fig. 754 Holotype genital armature of *Dendroiketes novaeguineae* BOESEMAN. After STEINMANN.

Fig. 755 Male genital armature of *Dendroiketes similis* RAMAMURTHI. After RAMAMURTHI.

2. Dendroiketes corticinus (BURR)

1908 *Apachyus corticinus* BURR, Ann. Mag. nat. Hist., London (8) **1**: 51. – Terra typica: Ceylon (Type male: Brit. Mus. nat. Hist. London).

1910 *Dendroiketes corticinus* – BURR, Fauna of Brit. India, Dermapt.: London, Bombay, 36; fig. 3 (male), 3a (penultimate sternite and forceps of male), 3b (tarsus of male).

Literature: BRINDLE 1965; BURR, 1911, 1915; POPHAM & BRINDLE 1968; SAKAI 1970; STEINMANN 1975; ZACHER 1911.

Description: Male colour greyish or dark brown; antennae with first three joints yellowish, the rest grey; head yellow, the mouthparts darker, legs yellow, femora with darker shading; ultimate tergite reddish brown, forceps red. Head smooth, tumid; postfrontal sutures and coronal suture very distinct; antennae about 30-jointed; first joint long, but shorter than distance between antennal bases; second very short, third long, about as long as scape. Pronotum rectangular, nearly square, prozona tumid, median longitudinal furrow distinct; metazona depressed; posterior margin straight. Tegmina and wings fully developed, tegmina narrow, not rounded at the shoulders, about twice as long as pronotum, smooth. Wings longer than tegmina, the scales of the same colour.

Abdomen more or less parallel-sided, depressed, smooth; ultimate tergite broad, with median longitudinal sulcus; anal process (Fig. 756) rounded, crenulate on margin. Forceps depressed, strongly curved, finely punctate. Genitalia unknown. – **Female** similar to male, but anal process parallel-sided at the base, then obtusely triangular, with rather sharp points at each side; forceps with a blunt tubercle on the underside near the base. – **Length** of body with forceps: in both sexes: 10–11,5 mm.

Distribution: Sri Lanka.

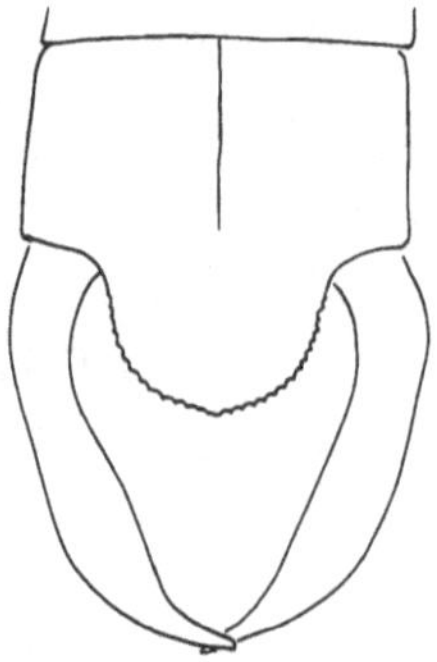

Fig. 756 Male ultimate tergite and forceps of *Dendroiketes corticinus* (BURR). After BURR.

3. Dendroiketes punctatus BORELLI

1923 *Dendroiketes punctatus* BORELLI, Boll. Mus. Zool. Anat. comp. Univ. Torino, **38**: 5. – Terra typica: Mindanao (Type female: Mus. Ist. Zool. sist., Torino).
Literature: BRINDLE 1965; POPHAM & BRINDLE 1968; SAKAI 1970.

Description: Female colour yellowish brown, head yellow, pronotum brown, tegmina blackish-brown, legs yellow; abdomen yellow, darker laterally, ultimate tergite reddish. Head tumid, depressed; postfrontal sutures and coronal suture distinct, well marked. Antennae broken in holotype, 25(?)-jointed; first joint long, about as long as distance between antennal bases; second joint transverse, third long, more or less as long as the length of scape. Eyes normal, a little shorter than length of head behind eyes. Pronotum transverse, with median longitudinal furrow; posterior margin truncate. Tegmina and wings well developed. Abdomen

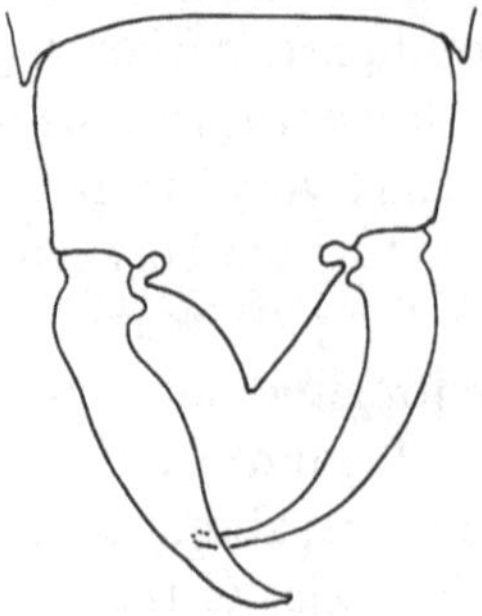

Fig. 757 Female ultimate tergite and forceps of *Dendroiketes punctatus* BORELLI. Original.

depressed, more or less parallel-sided. Ultimate tergite broad; posterior margin with triangular anal process (Fig. 757); process with parallel-sided section basally, and triangular portion apically. Forceps asymmetrical in holotype, left branch slenderer, and less curved than right one. – **Male** unknown. – **Length** of body with forceps: 12,5 mm.

Distribution: Philippines.

Genus 2: Apachyus AUDINET-SERVILLE

1831 *Apachyus* AUDINET-SERVILLE, Ann. Sci. nat., Paris, **22**: 35. – Species typica: *Forficula depressa* PALISOT de BEAUVOIS 1805.

1839 *Apachya* – AUDINET-SERVILLE, Hist. nat. Ins., Orth. Paris, 54.

1846 *Apachys* AGASSIZ, Nomencl. zool. Ind., Madrid, 27.

Literature: BOESEMAN 1954; BORG 1904; BORMANS 1900; BRINDLE 1965; BURR 1911, 1915; HANDLIRSCH 1925; KIRBY 1904; REHN 1924; SAKAI 1970; STEINMANN 1975; TOWNES 1945; VERHOEFF 1902; ZACHER 1911.

Description: Entire body remarkably flattened, and depressed. Head broad, antennae long, with 30–50 joints; first joint stout and long, second transverse of quadrate, third joint long. Tegmina ample and smooth, wings ample and long. Pronotum elliptical, quadrate, or longer than wide. Legs normal, but femora compressed, carinulate, tibiae slender, first tarsal segment short and thick, second very small, third long and slender. Abdomen depressed, lateral margins parallel; ultimate tergite well developed, anal process produced between the branches of forceps from the last dorsal tergite into a flat lobe, rounded, pentagonal or lanceolate. Forceps very similar in both sexes, flat, arcuate or incurved at an angle, without teeth. Male genitalia narrow or broader, median incision at anterior margin, genital lobes with virgae.

Distribution: Africa, New Guinea, and Tropical Asia.

11 species.

Identification key to the species

1 African species 2

1′ Oriental or Indo-Australian species 5

2 In both sexes the width of the anal process surpasses its length; species of larger size: together with the forceps body length is equal to 22 mm or more; head yellowish or darker brown; lateral margins behind eyes converging posteriorad 3

2′ Anal process of both sexes rather long, length equals or considerably surpasses width; medium sized species, of 18–20 mm in length or smaller; head yellow, lateral margins behind eyes parallel, not converging 4

3 Tegmina dark-brown; head yellowish brown or brown; the first three basal joints of antennae are yellow, the others brown; anal process of males more prominent, with well-marked serration (Fig. 758); anal process of females triangular, as in fig. 760 .... 1. **Apachyus depressus**, p. 484

3′ Tegmina black, head brown or dark brown; the first three basal joints of antennae are yellow, the others black; anal process of male shorter and wider (Fig. 761); anal process of females rather long, with slightly convex margins, as in fig. 763 .... 2. **Apachyus reichardi**, p. 485

4 Insect of medium size, length of body together with forceps: 18–20 mm; abdomen yellow, feet pale yellow. Length of the anal process approximately equals its width, that of the male (Fig. 764) is roughtly rounded off, that of the female (Fig. 766) triangular; at the point of the process the apical margins subtend a right angle .... 3. **Apachyus murrayi**, p. 486
4′ Insect of small size, together with the forceps the length of its body is 10,5 mm; abdomen for the most part brown, excepting the last tergite, which is ochre in colour; its feet are pale brown; anal process longer than wide, that of the female triangular (Fig. 767); at the apex of process margins enclose an acute angle .... 4. **Apachyus baloghi**, p. 488
5 Oriental species **6**
5′ Indo-Australian species **8**
6 Wings yellow, tegmina dark brown. Anal process of male (Fig. 768) almost pentagonal. Male genitalia with large external parameres, as in fig. 769. Anal process of female triangular (Fig. 770) .... 5. **Apachyus feae,** p. 488
6′ Wings coloured as tegmina at least darkened on lateral margins **7**
7 Wings and tegmina entirely dark; pronotum uniformly yellow; anal process of male more or less triangular apically (Fig. 771). Male genitalia very similar to *javanus* VERHOEFF, as in fig. 772. Anal process of female as in fig. 773 .... 6. **Apachyus chartaceus**, p. 490
7′ Wings and tegmina lighter medially, darkened on sutural and lateral margins; pronotum yellow, bordered with brown; anal process of male specific, excised transversally (Fig. 774). Male genitalia similar to *chartaceus* (DE HAAN). Anal process of female as in fig. 776 .... 7. **Apachyus javanus**, p. 491
8 Very large species, body length with forceps: 40–42 mm. Male forceps strongly curved, and anal process not triangular, posterior margin excised transverselly, but with a blunt tubercle medially (Fig. 777) .... 8. **Apachyus petersensi**, p. 492
8′ Smaller species; colour dark brown, uniformly coloured or variegated; anal process of male triangular **9**
9 Larger species, body length with forceps: 24 mm or more; uniformly dark reddish brown. Anal process of male (Fig. 778) triangular, wider than long. Anal process of female triangular, equilateral, as in fig. 779 .... 9. **Apachyus queenslandicus**, p. 493
9′ Smaller species, body length with forceps: 20 mm or less; tegmina with yellow discal spot, or pale yellow medially **10**
10 Male anal process elongate, slightly longer than wide (Fig. 780). Tegmina with two elongated yellow patches .... 10. **Apachyus athertonensis**, p. 493
10′ Male anal process broad, slightly wider than long (Fig. 781). Tegmina pale yellow medially, and lateral margins with broad brown stripe .... 11. **Apachyus beccari**, p. 494

1. Apachyus depressus (PALISOT DE BEAUVOIS)

1805 *Forficula depressa* PALISOT DE BEAUVOIS, Ins. rec. Afrique Amérique, Orth., Paris, 36; pl. 1, fig. 5 (male). – Terra typica: Nigeria (Type male: unknown locality).
1839 *Apachyus depressus* – AUDINET-SERVILLE, Hist. nat. Insectes, Orth., Paris: 55.
1911 *Apachyus depressus* – ZACHER, Zool. Jahrb., Jena, **30**: 395; fig. E^3 (male genitalia).
1965 *Apachys depressus* – BRINDLE, Ann. Mag. nat. Hist., London, (13) **8**: 441; fig. 1 (anal process of male), 16 (male genitalia). Central Africa.
1970 *Apachyus reichardi* [= *Apachyus depressus* PALISOT DE BEAUVOIS] – SAKAI, Dermapt. Cat. Prael., Tokyo, **1**: 27.

Literature: BOESEMAN 1954; BORG 1904; BRINDLE 1966, 1973; BURR 1907, 1910, 1911; HINCKS 1948, 1952; KIRBY 1904; MENOZZI 1928; STEINMANN 1977, 1979, 1981.

Description: Male rather variable in colour; head yellowish brown or brown; pronotum, tegmina and wings dark brown, sutural margins of wings yellow or white; legs yellow. Head transverse, narrowed behind eyes, depressed; post-frontal sutures and coronal suture well marked; posterior margin of head concave. Eyes relatively large, equal to the length of head behind eyes. First

antennal joint small, slightly shorter than distance between antennal bases. Pronotum longer than wide, strongly narrowed and rounded anteriorly, with a prominent neck separating pronotum from head; lateral margins sinuate, posterior margin straight; median longitudinal furrow distinct. Tegmina long and wide, with a large triangular scutellum exposed, wings fully developed. Pronotum coriaceous, tegmina and wings rugosely punctate except for inner pale part of wings, which is impunctate. Abdomen almost parallel-sided, rather narrowed towards base; rugosely punctate. Ultimate tergite with depressed anal process, which is rounded (Fig. 758), process short, with marginal serrations prominent. Forceps of *Apachyus*-type, strongly curved, cylindrical or subcylindrical, a little expanded apically. Genitalia (Fig. 759) broad; paramere wide, genital lobes short, virga within genital lobe long, with a basal circular vesicle associated with a small sclerite. Median incision of anterior margin of central parameral plate deep and very wide. External parameres more heavily sclerotized on lateral margins. – **Female** similar to male, but anal process (Fig. 760) short and triangular; penultimate sternite triangular but longer and much narrowed distally; apex almost pointed and extending nearly to apex of anal process ventrally. – **Length** of body with forceps: male: 20–28,5 mm, female: 24–30 mm.
Distribution: West and Central Africa.

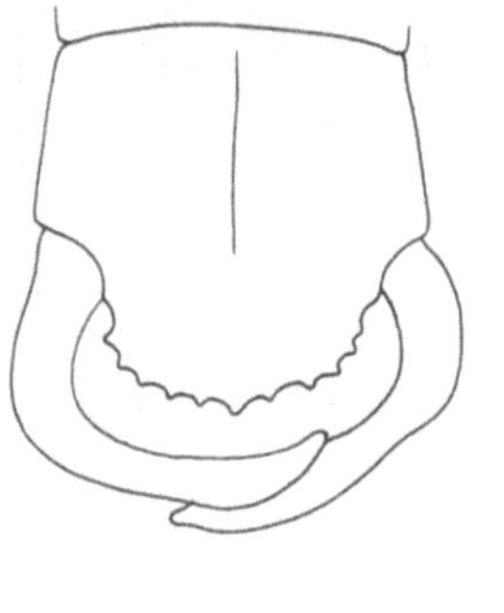

Fig. 758 Male ultimate tergite and forceps of *Apachyus depressus* (PALISOT DE BEAUVOIS). After BRINDLE.

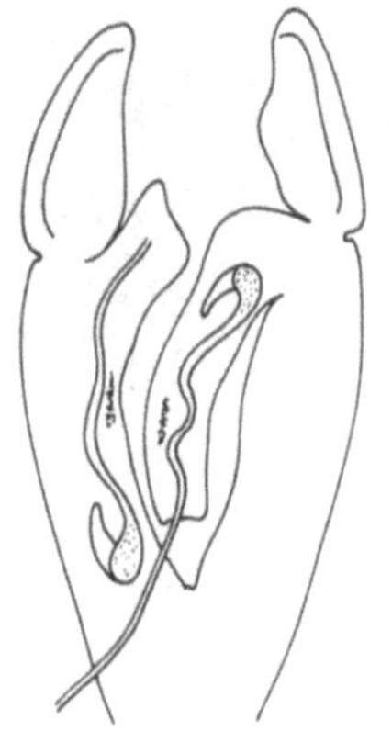

Fig. 759 Male genital armature of *Apachyus depressus* (PALISOT DE BEAUVOIS). After BRINDLE.

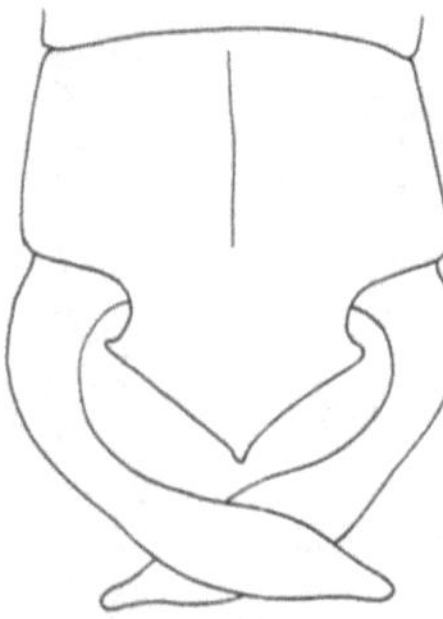

Fig. 760 Female ultimate tergite and forceps of *Apachyus depressus* (PALISOT de BEAUVOIS). After BRINDLE.

2. Apachyus reichardi KARSCH

1886 *Apachyus reichardi* KARSCH, Berliner ent. Zeitschr., **30**: 85. – Terra typica: Tanzania (Type male: Mus. Naturkunde, Berlin).
1911 *Apachyus reichardi* – BURR, Genera Insectorum, Bruxelles, **122**: 45. Africa.
1911 *Apachyus reichardi* – ZACHER, Zool. Jahrb., Jena, **30**: 396.
1911 *Apachyus murrayi* var. *reichardi* – BURR, Stettiner ent. Ztg., **1911**: 327.

1912 *Apachyus reichardi* – BURR, Ann. naturhist. Mus. Wien., **26**: 339 (Congo).
1965 *Apachys reichardi* – BRINDLE, Ann. Mag. nat. Hist., London, (13) **8**: 442; fig. 15a (external paramere of male genitalia). Central Africa, not extending to either West or East coasts.
Literature: BRINDLE 1973; BORMANS 1900; BURR 1911; KIRBY 1904; STEINMANN 1977.

Description: Male almost entirely dark reddish brown or blackish; antennae blackish, head dark brown, tegmina and wings blackish. Head broad, transverse, narrowed behind eyes; postfrontal sutures and coronal suture well marked; eyes relatively small, slightly shorter than length of head behind eyes. First antennal joint small, about as long as eye. Pronotum longer than wide; similar to *depressus* (PALISOT DE BEAUVOIS), strongly narrowed and rounded anteriorly; posterior margin straight; median longitudinal furrow present. Tegmina long, longer than pronotum; coriaceous; wings fully developed. Abdomen parallel, tergites rugosely punctate. Ultimate tergite long, with median longitudinal sulcus; anal process (Fig. 761) with marginal serration; forceps of *Apachyus*-type. Genitalia (Fig. 762, gen. prep. No. 185, det. Dr. H. STEINMANN) comparatively narrow; paramere very long, median incision of anterior margin very deep, but narrow, genital lobes relatively short, virga long, basal vesicle large. External parameres normal. – **Female** similar to male, but ultimate tergite broad; anal process (Fig. 763) triangular. – **Length** of body with forceps in both sexes: 20–30 mm.
Distribution: Centra-eastern to South-east Africa.

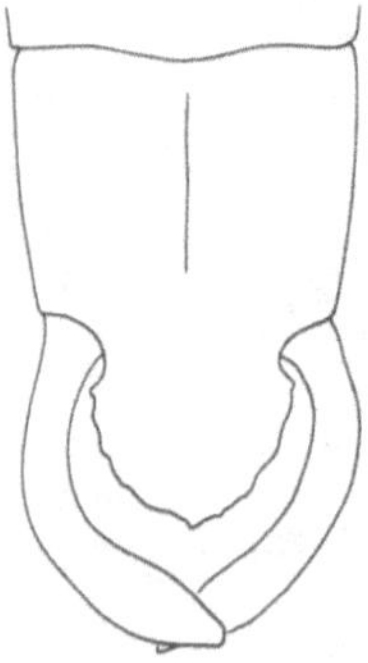

Fig. 761 Male ultimate tergite and forceps of *Apachyus reichardi* KARSCH. Original.

Fig. 762 Male genital armature of *Apachyus reichardi* KARSCH. Original.

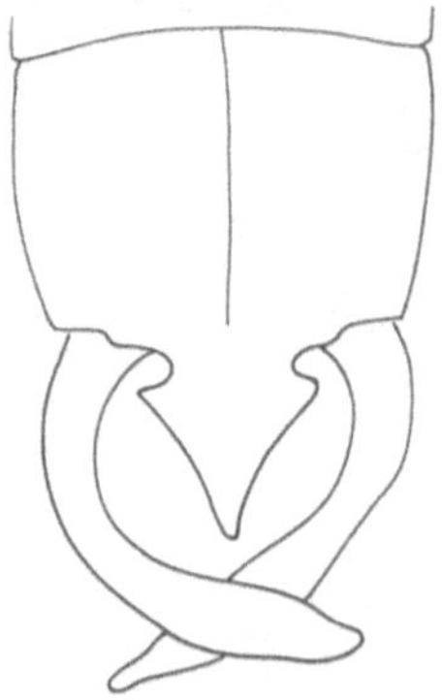

Fig. 763 Female ultimate tergite and forceps of *Apachyus reichardi* KARSCH. Original.

3. Apachyus murrayi DOHRN

1863 *Apachya murrayi* DOHRN, Stettiner ent. Ztg., **24**: 44. – Terra typica: Old Calabar (Type male: Zool. Inst., Warszawa; gen. prep. No. 385, det. Dr. H. STEINMANN).
1908 *Apachyus murrayi* – BURR, Ann. Soc. ent. belge, **52**: 34. (Congo).

1965 *Apachys murrayi* – BRINDLE, Ann. Mag. nat. Hist., London, (13) **8**: 442; fig. 13 (male anal process), 14 (female anal process), 15 (external paramere of male genitalia).
1973 *Apachyus murrayi* – BRINDLE, Dermaptera of Africa, Tervuren, **1**: 129; fig. 163 (ultimate tergite with anal process of male), 164 (ultimate tergite with anal process of female).
1979 *Apachyus murrayi* – STEINMANN, Fol. ent. hung., Budapest, **32**: 159; fig. 1 (anal process of holotype), 22 (holotype genital armature).
Literature: BOESEMAN 1954; BORELLI 1914; BRINDLE 1968, 1970; BURR 1911, 1912; KIRBY 1904; MENOZZI 1935; MORALES-AGACINO 1951; REHN 1945; SAKAI 1970; STEINMANN 1977.

Description: Male colour reddish brown to brownish yellow; head yellow or brownish, antennae dark brown; legs yellow, abdomen and forceps dark reddish brown. Cuticle coriaceous, similar to *depressus* (PALISOT DE BEAUVOIS), but punctation usually less strong. Head broad, depressed; postfrontal sutures and coronal suture distinct. Eyes normal, a little shorter than length of head behind eyes. First antennal joint relatively small, shorter than distance between antennal bases; second joint transverse, third joint long, as long as first. Pronotum longer than wide; strongly narrowed and rounded anteriorly; posterior margin straight; median longitudinal furrow well marked. Tegmina long, longer than pronotum. Wings well developed. Abdomen more or less parallel-sided, strongly depressed; ultimate tergite broad, median longitudinal sulcus very long. Anal process (Fig. 764) broad and rounded, with marginal serration. Forceps typical of genus. Genitalia (Fig. 765) of holotype elongated; central parameral plate long, median incision of anterior margin very deep; genital lobes normal, virga typical, basal vesicle comparatively small. External parameres medium, almost one-quarter length of paramere, and slightly longer than greatest width of paramere; apically rounded, median section slightly widened. – **Female** similar to male, but ultimate tergite broad, anal process (Fig. 766) triangular, with strongly marginal serration. – **Length** of body with forceps in both sexes: 18–23 mm.

Distribution: West-Central Africa, extending Southwest into Angola.

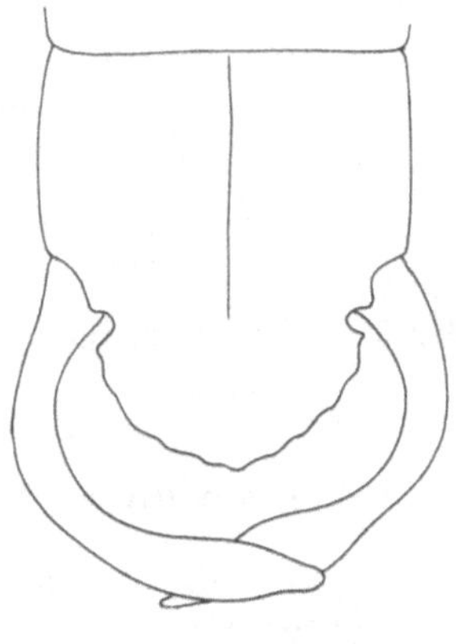

Fig. 764 Holotype ultimate tergite and forceps of *Apachyus murrayi* DOHRN. After STEINMANN.

Fig. 765 Holotype genital armature of *Apachyus murrayi* DOHRN. After STEINMANN.

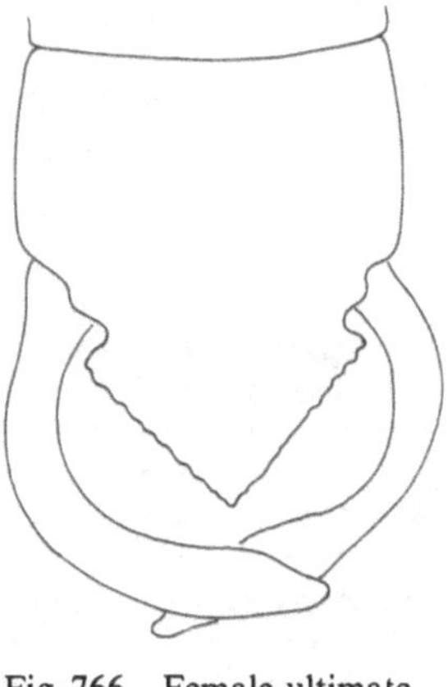

Fig. 766 Female ultimate tergite and forceps of *Apachyus murrayi* DOHRN. After STEINMANN.

4. Apachyus baloghi Steinmann

1977 *Apachyus baloghi* Steinmann, Opusc. zool., Budapest, **13**: 79; fig. 1a (head), 1b (distal part of antenna), 1c (end of abdomen dorsally), 1d (pronotum). – Terra typica: Brazzaville-Congo (Type female: Természettudományi Múz., Budapest).

Description: Female colour pale brown; ultimate tergite and forceps yellow. Head flattened, broad; postfrontal sutures and coronal suture clearly visible, they run along a nearly regular circle. Eyes relatively small, about as long as first antennal joint. Antennae brown, 28-jointed; first joint relatively short, perceptibly shorter than third joint, which is very long, longer than the joints 4–6 together. Pronotum longer than broad; lateral margins highly arcuate, chocolate-brown in colour. Tegmina and wings well developed, brown, the former do not reach the base of hind legs. Abdomen considerably flattened; median longitudinal sulcus present, well marked. Anal process (Fig. 767) long, distinctly longer than wide, its elongated part is acute-angled, its apex pointed; lateral margins slightly arcuate. Forceps longer by a third than the anal process, moderately curved. Penultimate sternite obliquely cut off. – **Male** unknown. – **Length** of body with forceps: 11,5 mm.

Distribution: Congo Republic.

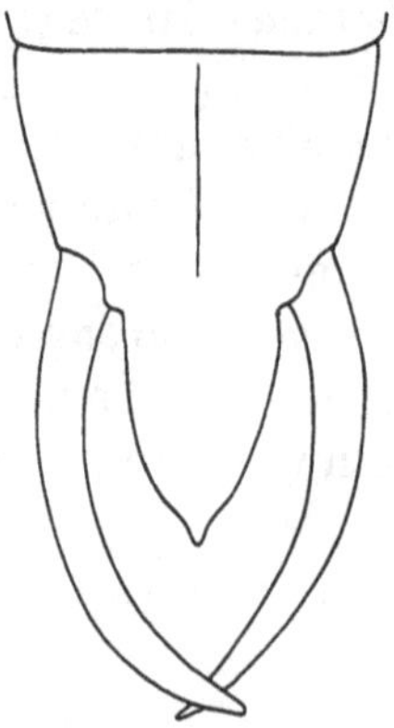

Fig. 767 Holotype ultimate tergite and forceps of *Apachyus baloghi* Steinmann. After Steinmann.

5. Apachyus feae Bormans

1894 *Apachya feae* Bormans, Ann. Mus. civ. Stor. nat., Genova, **34**: 373. – Terra typica: Burma (Type male: Mus. civ. Stor. nat., Genova).

1896 *Apachys pascoei* Kirby, J. Linn. Soc. London, **25**: 521; pl. XX, fig. 1 (male). – Terra typica: Sylhet (Type male: Brit. Mus. nat. Hist., London).

1900 *Apachyus feae* – Bormans, Das Tierreich, Berlin, **11**: 13; fig. 9a (male), 9b (ultimate tergite with forceps of female), Ost-Asien.

1910 *Apachyus feae* – Burr, Fauna of Brit. India, Dermapt. London, Bombay, 33; fig. 1 (anal process and forceps of male), 2 (anal process and forceps of female), 93 (male). Bhutan, Sikkim, Assam, Burma, Tonkin.

1965 *Apachys feae* – Brindle, Ann. Mag. nat. Hist., London, (13) **8**: 442; fig. 17 (male genitalia), 3 (male anal process), 4 (anal process of female). India, Burma, Vietnam.

Literature: Bey-Bienko 1959; Burr 1911; Kirby 1904; Popham & Brindle 1968; Sakai 1970.

Description: Male colour dark chestnut, abdomen reddish brown, shaded with black; legs yellow, wings straw-coloured. Head broad, depressed; posterior margin straight; postfrontal sutures and coronal suture well marked; antennae 40-jointed; first joint relatively short, shorter than distance between antennal bases; second joint transverse, third about twice as long as wide. Pronotum narrow, and rounded anteriorly; longer than wide, converging before the posterior margin, which is truncate; median longitudinal furrow distinct. Tegmina well developed, more or less twice as long as pronotum; posterior margin straight, ample, smooth. Wings ample, coriaceous, and membranous parts scarcely differentiated. Abdomen broad, depressed, parallell-sided; rugulose, with a pair of smooth spots in the middle of each segment. Ultimate tergite ample, longer than wide, with a shallow median depression. Anal process (Fig. 768) pentagonal, that is, the sides are at first convex, slightly emarginate at the base; outer angles marked by a short point, then converging to an apical point. Forceps of *Apachyus*-type, regularly arcuate in the basal third, the inner margin straight and outer margin gently arcuate. Penultimate sternite hidden by the 8th, which has a faint median longitudinal sulcus and is produced posteriorly into a rather short obtuse triangle, not reaching the anal process. Genitalia (Fig. 769) broad; paramere with median incision on anterior margin; deep; genital lobes well developed, with virgae long. External parameres broad and short. – **Female** similar to male, but anal process (Fig. 770) lanceolate, that is, more strongly emarginate at the base so that the sides form a sharp angle near the base. – **Length** of body with forceps: male: 45–47 mm, female: 31–46 mm.

Distribution: Bhutan, India, Sikkim, Assam, Burma, and Vietnam.

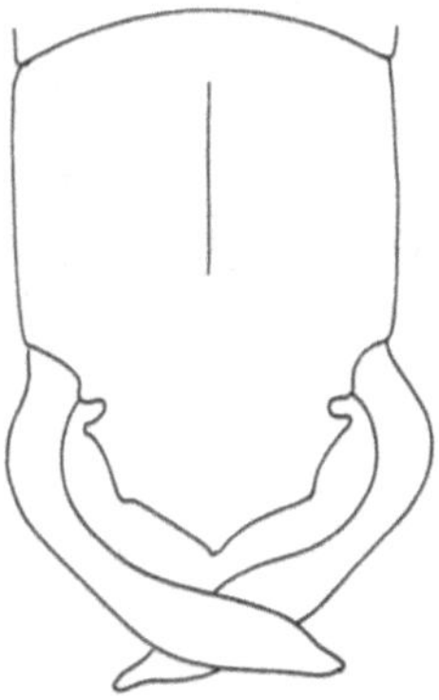

Fig. 768 Male ultimate tergite and forceps of *Apachyus feae* BORMANS. After BRINDLE.

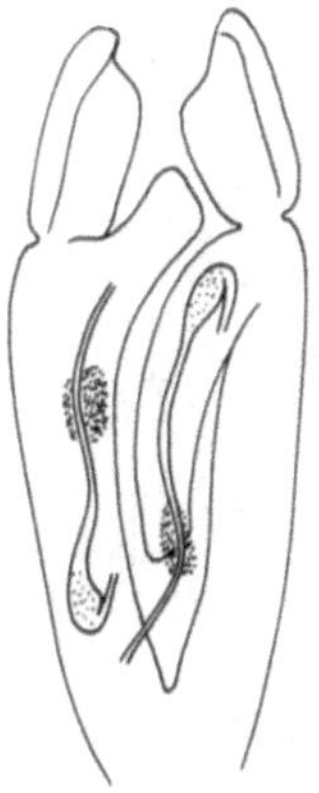

Fig. 769 Male genitalia of *Apachyus feae* BORMANS. After BRINDLE.

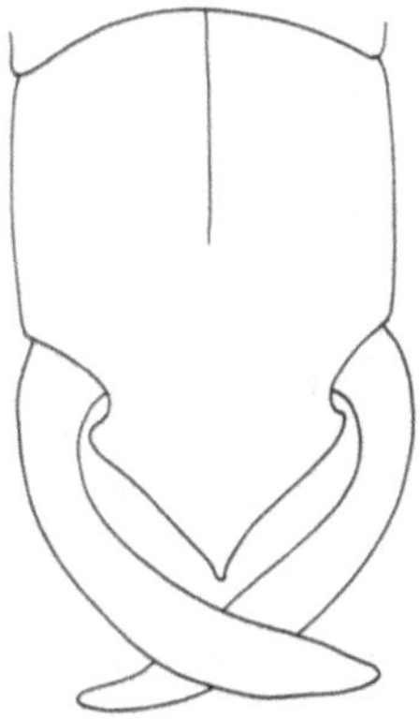

Fig. 770 Female ultimate tergite and forceps of *Apachyus feae* BORMANS. After BRINDLE.

6. Apachyus chartaceus (DE HAAN)

1842 *Forficula chartaceus* DE HAAN, Verh. nederlandse Overs. Bezitt., Orth.: 239; pl. 23, fig. 7 (male). – Terra typica: Sumatra (Type male: Rijksmus. nat., Hist., Leiden; gen. prep. No. 730, det. Dr. H. STEINMANN).

1863 *Apachya charcateus* – DOHRN, Stettiner ent. Ztg., **24**: 43. Borneo.

1911 *Apachyus chartaceus* – BURR, Genera Insectorum, Bruxelles, **122**: 45.

1965 *Apachys chartaceus* – BRINDLE, Ann. Mag. nat. Hist., London, (13) **8**: 443; fig. 11 (male anal process), 12 (anal process of female), 19 (male genitalia). Java, Borneo, Sumatra.

1980 *Apachyus chartaceus* – STEINMANN, Fol. ent. hung., Budapest, (33) **41**: 341, fig. 39 (holotype anal process with forceps), 40 (holotype genital armature).

Literature: BOESEMAN 1954; BORELLI 1926; BORMANS 1900; BURR 1908; DUBRONY 1879; KIRBY 1904; POPHAM & BRINDLE 1968; SAKAI 1970; STEINMANN 1975, 1981.

Description: Male colour brownish yellow; head, pronotum and legs yellow; antennae brown, tegmina, wings and abdomen dark brown; forceps reddish brown. Head flat, broad, visible broader than pronotum. Postfrontal sutures and coronal suture present; antennae 40-jointed; first antennal joint long, but a little shorter than length of head behind eyes or distance between antennal bases; second transverse, third joint long, but shorter than scape. Eyes relatively small, typical of genus. Pronotum long, longer than wide, of *Apachyus*-type; posterior margin straight, with semicircular anterior half; median longitudinal furrow well marked. Tegmina long, coriaceous; somewhat longer than wings. Abdomen depressed, almost parallel-sided, or a little expanded to last tergite. Ultimate tergite broad, a little depressed ventrally, median longitudinal sulcus well marked medially. Anal process (Fig. 771) almost rounded, with marginal serration

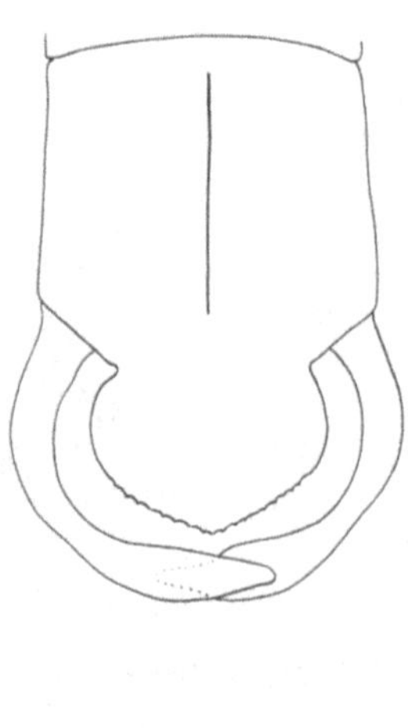

Fig. 771 Male ultimate tergite and forceps of *Apachyus chartaceus* (DE HAAN). After BRINDLE.

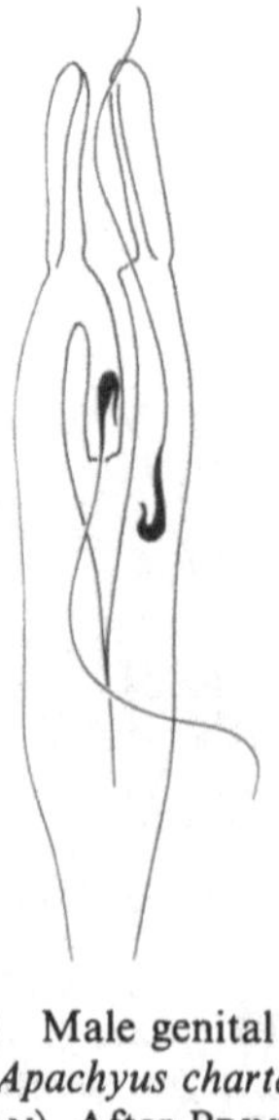

Fig. 772 Male genital armature of *Apachyus chartaceus* (DE HAAN). After BRINDLE.

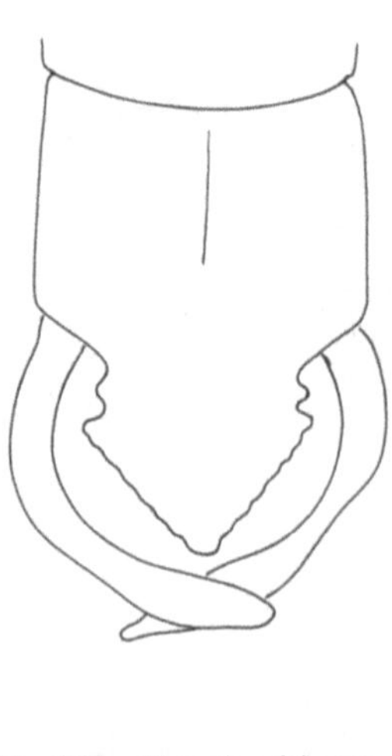

Fig. 773 Female ultimate tergite and forceps of *Apachyus chartaceus* (DE HAAN). After BRINDLE.

prominent. Penultimate sternite typical, triangular. Genitalia (Fig. 772) specific; paramare narrow, V-shaped, median incision of anterior margin deep and narrow; genital lobes small, virga long with specific basal vesicle. External parameres finger-shaped, as in fig. 772. – **Female** similar to male, but anal process (Fig. 773) triangular. – **Length** of body with forceps: male: 21–26 mm, female: 20–24 mm.

Distribution: Java, Borneo, Sumatra, Malay Archipelago.

7. Apachyus javanus VERHOEFF

1902 *Apachyus javanus* VERHOEFF, Zool. Anz., Leipzig, **665**: 200. – Terra typica: Java (Type female: Mus. Naturkunde, Berlin).

1954 *Apachyus sumatranus* BOESEMAN, Zool. Verh., Leiden, **21**: 59; fig. 16 (male), 16b (apical part of abdomen and forceps of male in ventral view). – Terra typica: Sumatra (Type male: Rijksmus. natuurl. Hist., Leiden; gen. prep. No. 728, det. Dr. H. STEINMANN).

1965 *Apachys javanus* – BRINDLE, Ann. Mag. nat. Hist., London, (13) **8**: 443. Only females are known.

1965 *Apachys sumatranus* – BRINDLE, Ann. Mag. nat. Hist., London, **8**: 443.

1979 *Apachyus javanus* – STEINMANN, Fol. ent. hung., Budapest, **32**: 158; fig. 18 (head and pronotum of male), 19 (ultimate tergite with forceps of male), 20 (male genitalia; gen. prep. No. 490, det. Dr. H. STEINMANN, male new).

1980 *Apachyus javanus* – STEINMANN Fol. ent. hung., Budapest, (33) **41**: 341; fig. 37 (male ultimate tergite, and anal process with forceps), 38 (male genitalia).

Literature: BURR 1908; KIRBY 1904; POPHAM & BRINDLE 1968; SAKAI 1970; ZACHER 1911.

Description: Male colour pale, yellowish brown, the dry preparation whitish, with discoloured yellowish brown patches on vertex and pronotum. Head rather wide, conspicuously wider than anterior widened section of pronotum. Postfrontal sutures and coronal suture distinct. Eyes small, but more or less equal to the length of head behind eyes. First antennal joint medium, only slightly shorter

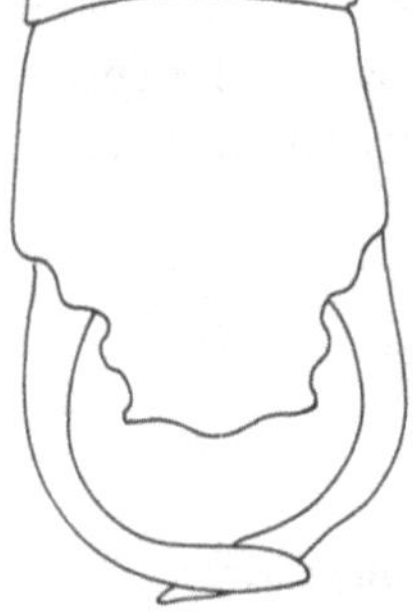

Fig. 774 Male ultimate tergite and forceps of *Apachyus javanus* VERHOEFF. After STEINMANN.

Fig. 775 Male genital armature of *Apachyus javanus* VERHOEFF. Original.

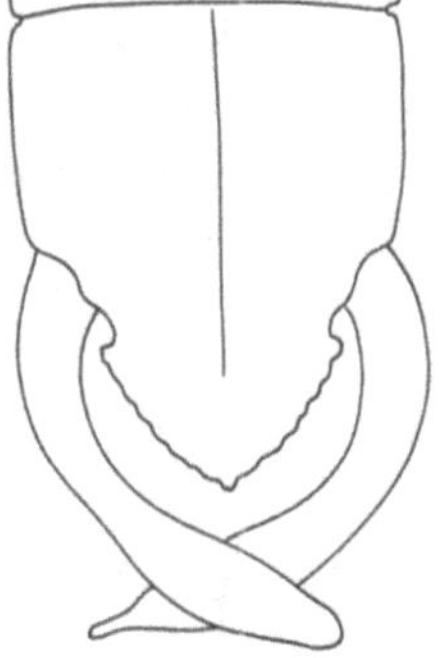

Fig. 776 Female ultimate tergite and forceps of *Apachyus javanus* VERHOEFF. After STEINMANN.

than frontal section between antennal bases. Vertex towards occiput rounded in an arc, posterior section of occiput nearly straight, truncate. Pronotum longer than wide; anterior section slightly widened and rounded, lateral margins of posterior section nearly parallel. Tegmina and wings well developed. Abdomen parallel-sided, ultimate tergite and forceps of *Apachyus*-type, very similar to species typica. Anal process (Fig. 774) transversely truncate posteriorly. Genitalia (Fig. 775) very characteristic; paramere V-shaped, median incision of anterior margin deep and wide; paramere expanded to median portion; genital lobes small, with little chitinous virgae. External parameres relatively small, obtuse. – **Female** similar to male, but anal process (Fig. 776) triangular, with marginal serrations. – **Length** of body with forceps: male: 16–18 mm, female: 17–18 mm.

Distribution: Java, Sumatra, and Luzon.

8. Apachyus peterseni Borelli

1925 *Apachyus peterseni* Borelli, Boll. Mus. Zool. Anat. comp. Univ. Torino, **40**: 36; fig. 1 (male). – Terra typica: Australia: New South Wales (Type male: Mus. Ist. Zool. sist., Torino, holotype without penis).

1926 *Apachyus australiae* Tillyard, Insects of Australia and New Zealand: 112. – Terra typica: Australia (Type male: unknown locality).

1965 *Apachys peterseni* – Brindle, Ann. Mag. nat. Hist., London (13) **8**: 444. New South Wales.

Literature: Popham & Brindle 1968; Sakai 1970.

Description: Very large species. **Male** distinctive by the black and white coloration of the anterior part of the body, the abdomen reddish-brown. Head flat, postfrontal sutures and coronal suture distinct. Eyes normal, shorter than length of head behind eyes. First antennal joint comparatively long, but shorter than distance between antennal bases. Pronotum longer than wide, with median longitudinal furrow. Tegmina and wings well developed, tegmina coriaceous. Pronotum black, with a large triangular area of the pronotum white. Wings greyish with posterior part brownish black, femora black with large basal oval whitish patch on each one dorsally. Abdomen normal, parallel-sided, depressed; anal process (Fig. 777) specific, posterior margin truncate, but with a small blunt

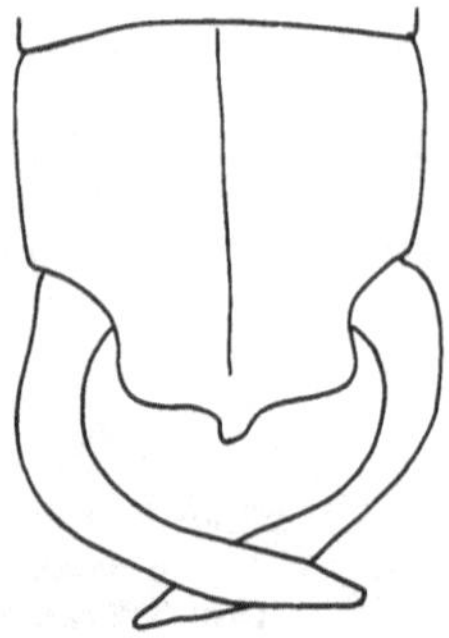

Fig. 777 Male ultimate tergite and forceps of *Apachyus peterseni* Borelli. Original.

tubercle medially. Forceps of *Apachyus*-type. Genitalia unknown. – **Female** unknown. – **Length** of body with forceps: 40–42 mm.

Distribution: Australia: New South Wales.

9. Apachyus queenslandicus MJÖBERG

1924 *Apachyus queenslandicus* MJÖBERG, Ark. Zool., Uppsala, **16**: 5; pl. 1, fig. 4 (ultimate tergite with forceps of male). – Terra typica: Australia: Queensland (Type male: Naturhist. Riksmus., Stockholm).

1965 *Apachys queenslandicus* – BRINDLE, Ann. Mag. nat. Hist., London, (13) **8**: **444**; fig. 7 (anal process of male), 8 (anal process of female).

Literature: HEBARD 1933; POPHAM & BRINDLE 1968; SAKAI 1970; SJÖSTEDT 1933.

Description: Male colour brownish, shiny; cuticle coriaceous; body strongly depressed, only the inside of the alae yellowish, underside slightly lighter brown. Head flat, slightly transversally impressed; eyes rounded. Antennae filiform, 32-jointed; first joint long, but shorter than distance between antennal bases; second very short, third more than three times as long, 4th–7th of about the same length, 8th–13th gradually longer, more than twice as long as wide, 14th–32th still longer, all very finely covered with grey pubescence. Pronotum longer than wide, rounded anteriorly, with median longitudinal furrow. Tegmina and wings well developed. Abdomen more or less parallel-sided; ultimate tergite broad, median longitudinal sulcus present. Anal process (Fig. 778) broad, forceps typical of genus. Genitalia unknown. – **Female** similar to male, but anal process (Fig. 779) longer, the extreme margin with several indistinct, obtuse teeth. – **Length** of body with forceps: in both sexes: 24–27 mm.

Distribution: Australia: Queensland.

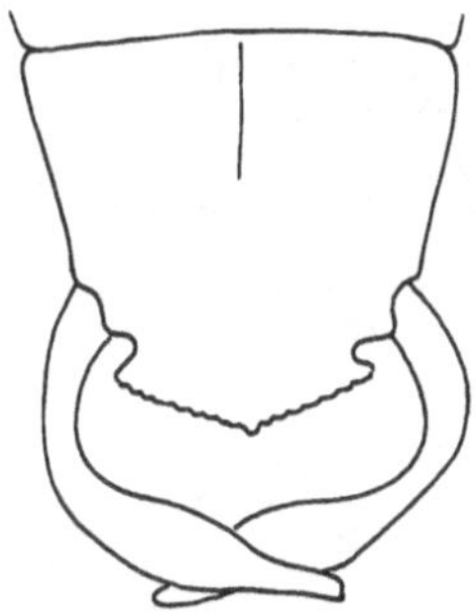

Fig. 778 Male ultimate tergite and forceps of *Apachyus queenslandicus* MJÖBERG. After BRINDLE.

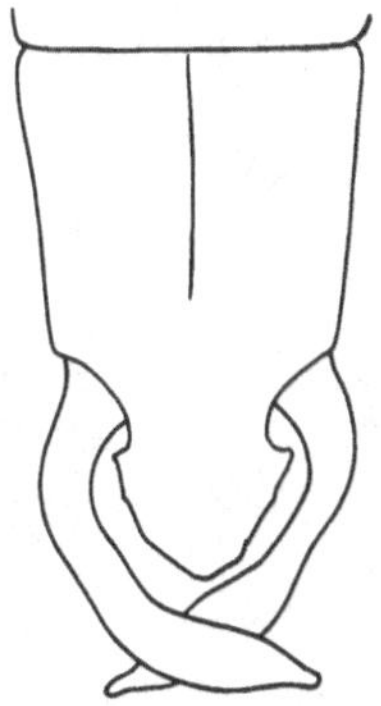

Fig. 779 Female ultimate tergite and forceps of *Apachyus queenslandicus* MJÖBERG. After BRINDLE.

10. Apachyus athertonensis MJÖBERG

1924 *Apachyus athertonensis* MJÖBERG, Ark. Zool., Uppsala, **16**: 6; pl. 1, fig. 5 (anal process of male). – Terra typica: Australia: Atherton (Type male: Naturhist. Riksmus., Stockholm).

1965 *Apachys athertonensis* – BRINDLE, Ann. Mag. nat. Hist., London (13) **8**: 444; fig. 5 (anal process of male).

Literature: HEBARD 1933; POPHAM & BRINDLE 1968; SAKAI 1970.

Description: Male colour dark brown, recognizable by the reddish posterior part of head, the two elongated yellow patches on the tegmina, and the two rows of reddish spots on the abdominal tergites. Head very flat and broad; eyes prominent, coarsely faceted; back of head rufous, distinctly set off by a deep curved line, sides parallel; posterior angles rounded. Postfrontal and coronal sutures distinct. Eyes normal, typical of genus. Antennae 33-jointed, with fine greyish pubescence; first joint very strongly developed, more than twice as long as wide, second very short, third about three times longer than second. Pronotum very characteristic in shape, slightly longer than wide, disc deeply impressed, with deeply marked median longitudinal furrow. Tegmina shiny, with two large, yellowish-white, oblong spots, external parts of alae yellowish at the anterior and posterior margin. Abdomen more or less parallel-sided, shiny, distinctly punctate, with two rows of fulvous spots near the posterior angles. Anal process (Fig. 780) specific, with straight median section; triangular in dorsal view. Forceps not of *Apachyus*-type, not much bent, with small tubercles near the base. Genitalia unknown. – **Female** unknown. – **Length** of body with forceps: 20 mm.

Distribution: Australia: Queensland.

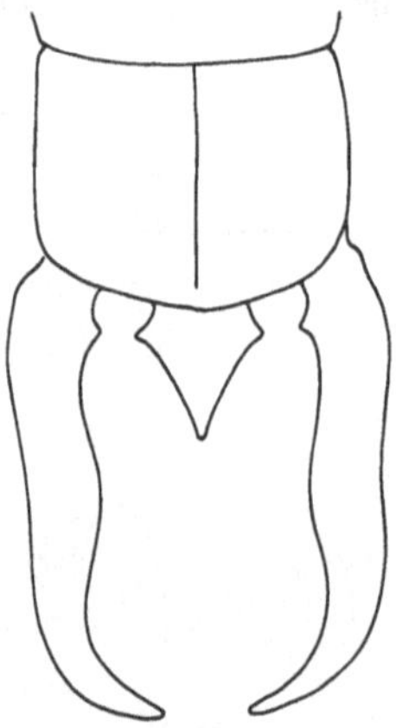

Fig. 780 Holotype ultimate tergite and forceps of *Apachyus athertonensis* MJÖBERG. After MJÖBERG.

11. Apachyus beccari DUBRONY

1879 *Apachys beccari* DUBRONY, Ann. Mus. civ. Stor. nat., Genova, **14**: 349. – Terra typica: New Guinea (Type male: Mus. civ. Stor. nat., Genova).

1900 *Apachyus beccari* – BORMANS, Das Tierreich, Berlin, **11**: 13.

1965 *Apachys beccari* – BRINDLE, Ann. Mag. nat. Hist., London, (13) **8**: 443; fig. 18 (male genitalia), 9 (anal process of male), 10 (anal process of female).

Literature: BOESEMAN 1954; BURR 1911, 1915; GÜNTHER 1930; HEBARD 1933; POPHAM & BRINDLE 1968; SAKAI 1970; STEINMANN 1975, 1981.

Description: **Male** colour various, abdomen and legs yellow, the wings darkened laterally; pronotum yellowish or yellowish white on the disc and brown laterally; tegmina pale yellow medially, lateral margins broadly brown. Wings greyish yellow. Head flat, broad; postfrontal sutures and coronal suture distinct. Antennae multisegmented; first joint long but shorter than distance between antennal bases; second joint transverse, third long, but a little shorter than first. Pronotum about as long as wide, of *Apachyus*-type; lateral margins rounded, posterior margin straight; median longitudinal furrow distinct. Tegmina well developed, shiny; wings fully developed, very long. Abdomen parallel-sided; abdominal tergites punctate; ultimate tergite very broad, anal process (Fig. 781) triangular, but very short, with marginal serration. Forceps of *Apachyus*-type, slightly expanded apically. Genitalia (Fig. 782) normal; paramere with median incision at anterior margin; genital lobes well developed, virgae long. External parameres large and obtuse. – **Female** similar to male, but anal process rounded on posterior margin as in fig. 783. – **Length** of body with forceps in both sexes: 18–22 mm.

Distribution: New Guinea, New Britain.

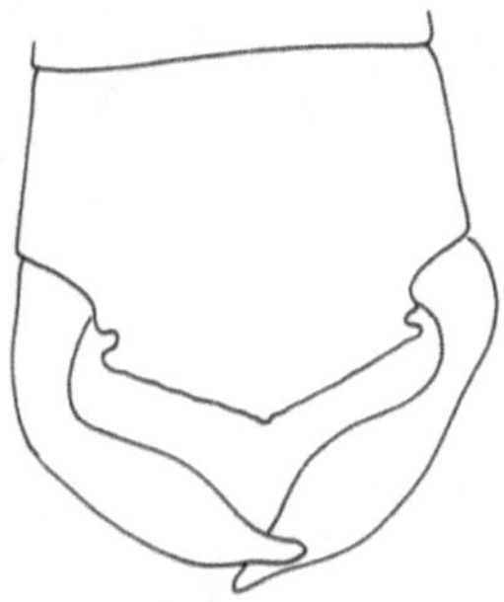

Fig. 781 Male ultimate tergite and forceps of *Apachyus beccari* DUBRONY. After BRINDLE.

Fig. 782 Male genitalia of *Apachyus beccari* DUBRONY. After BRINDLE.

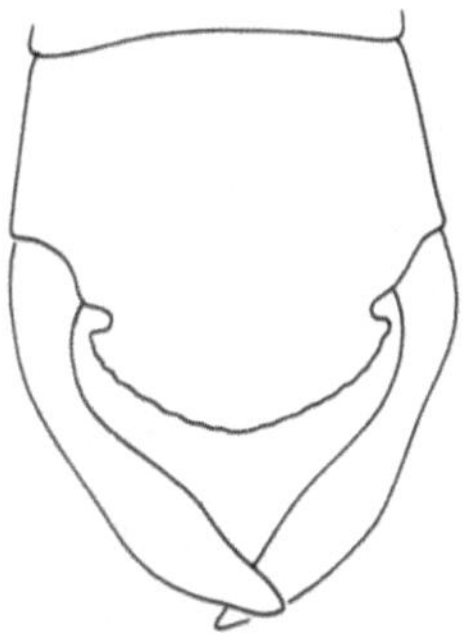

Fig. 783 Female ultimate tergite and forceps of *Apachyus beccari* DUBRONY. After BRINDLE.

Index

Das Tierreich erscheint ab Teilband 102 mit festem Einband. Seit 1958 sind erschienen

Lfg.	Bearbeiter	Titel	Ersch.-jahr	Umfang
72	H. Wermuth	REPTILIA: *Helodermatidae*	1958	16 Seiten, 9 Figuren
73	M. Beier	ORTHOPTERA: *Tettigoniidae* (Pseudophyllinae I)	1962	XII, 468 Seiten, 245 Abbildungen
74	M. Beier	ORTHOPTERA: *Tettignoniidae* (Pseudophyllinae II)	1960	396 Seiten, 241 Figuren
75	K. Odening	TREMATODA (Digenea): *Plagiorchiidae* III (Haematoloechinae) und *Omphalometridae*	1960	IV, 77 Seiten, 60 Figuren
76	K. Sanft	AVES (Upupae): *Bucerotidae*	1960	IV, 176 Seiten, 106 Figuren
77	R. Kilias	GASTROPODA PROSOBRANCHIA: *Tonnidae*	1962	63 Seiten, 47 Figuren
78	St. W. Gorham	GYMNOPHIONA, *Liste der rezenten Amphibien und Reptilien*	1962	X, 25 Seiten
79	R. Mertens	HELODERMATIDAE, VARANIDAE, LANTHANOTIDAE. *Liste der rezenten Amphibien und Reptilien*	1963	X, 26 Seiten
80	H. Wermuth	GEKKONIDAE, PYGOPODIAE, XANTUSIIDAE. *Liste der rezenten Amphibien und Reptilien*	1965	XXII, 246 Seiten
81	J. A. Peters	COLUBRIDAE (Dipsadinae). *Liste der rezenten Amphibien und Reptilien*	1965	VIII, 18 Seiten
82	J. Illies	PLECOPTERA, *Katalog*	1966	XXX, 632 Seiten, 20 Abbildungen
83	R. Mertens	CHAMAELEONIDAE. *Liste der rezenten Amphibien und Reptilien*	1966	X, 37 Seiten
84	C. Gans	UROPELTIDAE. *Liste der rezenten Amphibien und Reptilien*	1966	29 Seiten
85	St. W. Gorham	ASCAPHIDAE, LEIOPELMATIDAE, PIPIDAE, DISCOGLOSSIDAE, PELOBATIDAE, LEPTODACTYLIDAE, RHINOPHRYNIDAE. *Liste der rezenten Amphibien und Reptilien*	1966	XVI, 222 Seiten
86	H. Wermuth	AGAMIDAE. *Liste der rezenten Amphibien und Reptilien*	1967	XIV, 127 Seiten
87	H. Wermuth	CORDYLIDAE (Cordylinae + Gerrhosaurinae). *Liste der rezenten Amphibien und Reptilien*	1968	X, 30 Seiten
88	F. Haas	BIVALVIA: *Unionacea*	1969	X, 663 Seiten, 5 Abb.

Lfg.	Bearbeiter	Titel	Ersch.-jahr	Umfang
89	A. F. Stimson	BOIDAE, (Boinae + Bolyeriinae + Loxoceminae + Pythoninae). *Liste der rezenten Amphibien und Reptilien*	1969	XI, 49 Seiten
90	H. Wermuth	ANGUIDAE, ANNIELLIDAE, XENOSAURIDAE. *Liste der rezenten Amphibien und Reptilien*	1969	XII, 41 Seiten
91	M. Fischer	HYMENOPTERA: *Braconidae* (Opiinae I)	1972	XII, 620 Seiten, 463 Figuren
92	R. Kilias	GASTROPODA PROSOBRANCHIA: *Cymatiidae*	1973	VIII, 235 Seiten, 149 Figuren
93	P. Banarescu & T. T. Nalbant	PISCES, TELEOSTEI: *Cyprinidae (Gobioninae)*	1973	VII, 304 Seiten, 154 Figuren, 19 Karten
94	P. Zwick	PLECOPTERA, *Phylogenetisches System und Katalog*	1973	XXXII, 465 Seiten, 75 Abbildungen
95	E. Duellmann	HYLIDAE, CENTRO-LENIDAE, PSEUDIDAE. *Liste der rezenten Amphibien und Reptilien*	1976	XIX, 225 Seiten
96	M. Fischer	HYMENOPTERA: *Braconidae* (Opiinae II/Amerika)	1976	etwa 1038 Seiten, 890 Figuren
97	S. G. Kiriakoff	LEPIDOPTERA NOCTUI-FORMES: *Agaristidae I*	1976	IX, 180 Seiten, 39 Figuren
98	S. G. Kiriakoff	LEPIDOPTERA NOCTUI-FORMES: *Agaristidae II*	1976	VIII, 165 Seiten, 52 Figuren
99	S. G. Kiriakoff	LEPIDOPTERA NOCTUI-FORMES: *Agaristidae III*	1976	VI, 86 Seiten, 26 Figuren
100	H. Wermuth & † R. Mertens	TESTUDINES, CROCO-DYLIA, RHYNCHO-CEPHALIA. *Liste der rezenten Amphibien und Reptilien*	1976	XXVII, 174 Seiten
101	D. E. Hahn	ANOMALEPIDIDAE, LEPTOTYPHLOPIDAE, TYPHLOPIDAE. *Liste der rezenten Amphibien und Reptilien*	1980	XII, 93 Seiten
102	H. Steimann	DERMAPTERA *Catadermaptera I*	1986	XIV, 343 Seiten
103	A. Kaltenbach	SAGINAE *Saltatoria-Tettigoniidae*	1986	VIII, 92 Seiten
104	M. Fischer	HYMENOPTERA *Opiinae III*	1987	XV, 734 Seiten
105	H. Steinmann	DERMAPTERA *Catadermaptera II*	1989	XIX, 504 Seiten

Kolibri 360 GmbH
Ferdinand-Jühlke-Straße 7
99095 Erfurt, Deutschland
produktsicherheit@kolibri360.de